# The Timber-Frame Home

# The Timber-Frame Home

## Design · Construction · Finishing

Tedd Benson

The Taunton Press

*Cover photo: Brian E. Gulick*

All photos by the author unless otherwise noted.
Illustrations by Joel McCarty, using a computer-aided system from AutoDesk, Inc.,
Sausalito, California. Gratitude is due them, in general, and to Patricia Harris, in
particular, for technical support and encouragement.

©1988 by The Taunton Press, Inc.
All rights reserved

First printing: March 1988
Second printing: February 1989

International Standard Book Number: 0-918804-81-7

Library of Congress Catalog Card Number: 87-72009

Printed in the United States of America

A FINE HOMEBUILDING Book

FINE HOMEBUILDING® is a trademark of The Taunton Press, Inc.,
registered in the U.S. Patent and Trademark Office.

The Taunton Press, Inc.
63 South Main Street
Newtown, Connecticut 06470

*To the ladies who make me rich in love:*
*Christine, my wife, and Corona and Emily, my daughters.*

**Acknowledgments**

Many, many people—family, friends, colleagues, clients and strangers—have contributed to this project, most willingly, some unwillingly, some unwittingly, but all, I trust, to good effect. So, I have many thanks to give.

. . .to my mom and dad, Mr. T.M. and Mrs. Mary Lou Benson, who taught me all I know about high standards. Through their lives they showed me that there is no other kind of quality until there is quality in life. They set a standard of living that remains for me a clear, but elusive, goal.

. . .to the nameless ancient masters of timber-frame craft and architecture, for making the task of attempting to describe techniques and standards for modern timber-frame buildings daunting, but worthwhile.

. . .to the members of the Timber Framer's Guild of North America for their vitality, skill and commitment, and for making the timber-frame revival a reality.

. . .to the men and women of Benson Woodworking Company for doing without me at my best and for putting up with me at my worst. It is to their credit, and theirs alone, that our company has thrived during the creation of this book. They are remarkable human beings, generous and forgiving, every one.

. . .to clients past and present who have welcomed me into their homes for discussion and photography. In my quest for clear illustrations for this book, I have rearranged their furniture, interrupted their lives and irritated their pets. Only they will know whether it has been worth it.

. . .to Brian Smeltz for his help with photo editing.

. . .to the good Doctor Robert L. (Ben) Brungraber, for boundless energy, enthusiasm and insight. His review of and advice on the engineering and joinery sections were invaluable.

. . .to Kathy Nerrie for minding the company in my absence and in my presence.

. . .to Joel McCarty, who did much more than make the drawings for this book. He was always there. He gave me encouragement, counsel and even words in my times of need.

. . .to Laura Tringali, my genial but demanding editor, who was with this book from the whisper of an idea to its conclusion five years later. She scrutinized every line on every page, asking hard questions, reformulating ideas and discovering order and light where I saw only chaos and mist. Throughout, she was steady and consistent when I was foundering and mercurial. Laura is much more than an editor; she is a friend.

I am deeply indebted to all of the above and many more.

# Contents

# Introduction

This book is about a modern timber-frame homebuilding system. It developed because our company decided it wanted to build homes, not just frames. As one of our staff designers wrote in a letter to a client, "No amount of breathtaking timber framing will compensate for a building that is difficult to move around in, hard to heat or cool, expensive to maintain or displeasing to the eye."

Knowledge of this building system did not come easily. I began timber framing in 1974, seeing it as an opportunity to bring more craftsmanship into homebuilding. Through the first years I was fully absorbed in attempting to master sharp framing chisels and tight joints, trying to tame timbers. But eventually, when I looked up from my blissful work, I began to notice that a fine timber frame alone did not a happy client make. Although our clients certainly appreciated our work, they also had other concerns. Would there be enough bedrooms and storage? Would the frame make sense with the floor plan? What about insulation? Would the electrician know how to run the wires? Should there be a solarium? What about plumbing? What would all this cost? It was disturbing to my clients and myself to find that we could consistently offer neither specific answers to these important questions nor proven construction strategies. We were winging it far too often. I soon realized that it was time to shift my emphasis from the timber frame to the timber-frame house.

So I began to spend less time working wood and more time at the drawing board—through the years, I've been fortunate to have had a group of dedicated associates whose fine timber-framing skills have made it easier for me to put my tools aside. I embarked on marathon strategy sessions with clients, contractors and manufacturers of products we wanted to use. It soon became clear that every design and construction precept needed to be reconsidered, for most conventional construction methods and details are inappropriate or inadequate for timber-frame buildings. In time I came to understand we would have to invent many new solutions for our houses rather than adapt existing ones. I came to understand that modern timber-frame house design needed a revival to parallel the revival of timber joinery.

I undertook my first book about timber framing, *Building the Timber Frame House,* because I feared the craft I had grown to love might fail to find a secure place in modern homebuilding. I was afraid timber framing would not capture the imagination of others as it had mine, that my framing chisels and slicks would lie unused in a drawer, that I would become a dancer without a beat. I feared these things might happen if the traditional values, standards and techniques used to create some of the finest buildings of the world fell prey to modern impatience and misguided technology. That first book was my rallying cry, but I didn't know if anyone was listening.

I could have spared myself the anxiety. Not only were people listening, but I was not, as I had thought, a voice crying in the wilderness. The unique merits of timber-frame structures needed no herald. In our company's first few years, I knew of only one other enterprise that shared our goals and enthusiasm. That company was run by my good friend and colleague Ed Levin, who is still a partner in the company, now called Paradigm. Now there are well over 200 companies specializing in timber framing in all parts of the country, and the number of timber-frame homes built each year is in the thousands.

My fear now is that in the joy of reviving the craft of timber framing in its purest form, it would be all too easy to lose sight of the perspective that the frame is, after all, just one component of a building. I am concerned that potential timber-frame homeowners, drawn to the inherent beauty and durability of the frame, will be so intimidated by the prospect of finishing it off that they settle for a lesser type of housing. And that clients who do persevere will wind up with a white elephant that fails to fulfill its true design potential. I worry that the standards of workmanship of the frame will not be carried throughout the rest of the house, due to the inexperience of tradespeople with this type of building system. If these things happen with enough frequency, there is a good chance that timber framing will inadvertently be lumped together with dome-making as just another interesting but out-of-the-mainstream building form.

A properly designed and built timber-frame home is an intriguing, exciting mix of ancient building method and modern technology. It's a model of energy efficiency and comfort that is sensible as well as beautiful. To help ease the transition between the requirements of housing today and the craft as it was in the days when it reigned as the predominant building method, I have written this book. It is my attempt to integrate timber framing with the best attitudes, ideas and technology of modern times, to make timber-frame housing a viable and accessible alternative in the 20th century. The book describes the procedures created by my staff and me (with much help from many colleagues and professionals in other fields) to design, enclose and make livable the frames we build. But in addition to presenting a compilation of typical design and construction solutions for timber-frame homes, I have also given the principles inherent in each solution. There are hundreds of factors to be considered in building a house, and no two houses are ever exactly alike. Local climate and geology influence foundation design. The budget might demand a choice between a dream kitchen and a dream roof. As homebuilders and homeowners, our only hope is in a solid knowledge of existing choices and in the ability to plan for hard and careful compromises. So, as well as being about how to do, this book is about how to think, how to plan and how to balance the scales.

Inevitably, you will find chunks of personal prejudice and opinion between these pages, for to sail through the many decisions required for good homebuilding, it's necessary to be propelled by the strong winds of committed values and firm beliefs. Why such philosophical talk about something as worldly as homebuilding? Because what is worldly about homebuilding is that it happens on this earth, it uses natural and man-made materials and it requires money. The rest of homebuilding has to do with beliefs, feelings, spirit and passion. Certainly shelters can be made without these, but they probably will not be the kind of structures that speak positively and warmly to future generations of occupants. For most people, homebuilding is full of the most sensitive emotions. It separates them from their life's earnings, either to a good or ill end. It can bring families together or tear them apart, for homebuilding can be a dark and dangerous sea full of shoals and turbulent currents. I believe a well-designed and well-constructed timber-frame house is worth the voyage, and I offer this book to help chart the course.

Tedd Benson
Alstead Center, N.H.
January, 1988

# I. Timber-Frame Home Development

**A properly designed and built timber-frame house is an intriguing mix of an ancient building system with modern technology. Simple, classic architecture yields exciting living spaces.**

There are two important challenges for those who build timber-frame homes today. Both concern all those associated with the project, for the timber-frame housebuilding process is more like a chorus line than a solo dance, requiring all parties to play an active role. The first challenge is to cherish and nurture the values and standards set centuries ago, to understand in both spirit and substance the legacy of durable, classic timber-frame building that we have inherited from our forefathers. What is it that makes a structure so fine that it survives numerous centuries? Is it the construction, the architecture or both? What makes people love a building enough to repair and restore it? To give direction to our modern efforts, it's necessary to first thoroughly investigate the historical precedents. To know where we should go, we need to know where the craft has been.

The second challenge is to bring the timber-frame house into the 20th century. This requires a careful analysis of the limitations and opportunities inherent in the building system. Then we need to take a fresh look at modern homebuilding requirements, picking carefully from current standards and materials, and casting off conventions when necessary to create the highest quality work.

In this chapter, we'll look at both challenges, beginning with an exploration of the development of the timber-frame house through history and the attributes that give it timelessness. You won't find a chronicle of each event in

the evolutionary path here, just some of the high points. I wish only to capture a sense of the fine thread that united timber buildings through so many centuries and in so many parts of the world, for it is just this thread we should weave into the fabric of our contemporary houses. Then we'll conclude the chapter with a discussion of the goals, problems and opportunities in timber framing today.

## The legacy

What is apparent through even a cursory look at the history of building is that dwellings always clearly reflect the relationship of people to their world. Instead of following a single course, timber-frame buildings therefore evolved simultaneously along with rising and falling civilizations on several continents. They responded to technological advances, the availability of resources and to social prejudices (always a mirror of the people who built them). Despite this, out of the history of timber framing emerges a continuity of architectural patterns and styles, and a consistency in human values. There are, for instance, many more similarities than differences between an oriental structure like the Kondo Temple in Japan and a European structure like Westminster Hall in England. Both buildings were constructed as a monument to a religious ideal, and the timbers in each were both structure and decoration. Each building demonstrates mastery of a complex craft in a simple rectilinear form with a steeply pitched roof. The focal point in each building is the cantilevering of forces from the roof to the walls; the finely crafted roof timbers not only draw the eye upward to the heavens, they create a sense of balance and mystery, reflecting the forces of God and the universe. One more thing they both have in common: Though many centuries old, they both still stand.

The first timber-frame buildings appeared at about the time of the birth of Christ. They probably evolved from skeletons of tied-together poles around which skins or mattings were wrapped. Such dwellings were usually the homes of nomadic peoples. When permanent settlements became desirable, pole skeletons slowly gave way to more stable and durable timber structures. If we define a timber frame as a self-supporting network fastened with wooden joinery, then there are two significant events in its evolution. The first was the creation of the mortise-and-tenon joint between 500 B.C. and 200 B.C., which meant the tools and technological sophistication existed to work the wood. The second occurred sometime in the 10th century, when a framework was developed that was rigid enough to support itself on top of the ground. Prior to this, the posts of a house were stuck directly in holes and stabilized by compacted earth, a technique that allowed a minimum of interconnection between timbers, but also caused the posts to rot. Seemingly a small achievement, building a self-supporting structure was actually quite revolutionary, as it required that timbers be organized so posts would remain rigid while supporting the beams. Joinery had to become more sophisticated to serve both binding and supporting functions. Constructed in this way, buildings became more durable and expressive, and demanded greater skill to create.

Westminster Hall, built in England around 1395, is a simple but finely crafted form. The focal point is the cantilevering of roof forces. (Illustration by Marianne Markey.)

The network of posts and beams in a timber-frame house forms a structure that is rigid and self-supporting. (Design Associates, Inc., Architects.)

The development of timber-frame buildings differed from region to region, depending on many factors, such as the availability of material. In Egypt, Persia and Greece, where timber was limited, early temples were built with wooden columns and lintels; their forms were the basis for later stone temples. The scholar Vitruvius, writing at the end of the first century B.C., apparently saw the last of the wooden temples. He described the Temple of Ceres as a timber building with brick and terra-cotta infilling and decoration. Because of the scarcity of material and fear of fire, later temples and public buildings were built of masonry when possible, but timber framing had evolved into an accepted building method and was used for dwellings and less significant public buildings. According to Hansen, editor of *Architecture in Wood*, Vitruvius described the construction, calling it *opus craticiom,* or "timber framing."

Vitruvius also wrote that the best building timbers in his day were oak, elm, poplar, cypress and fir—surprisingly similar to our contemporary opinion. Some of these were occasionally imported from Africa, Syria and Crete, but importation of lumber was expensive and reserved for only the most important buildings. In addition, knowledge of new woods tended to make local builders less satisfied with what they had at hand. Palm trunks did not make good timbers after all, and other readily available woods were worse, as Martin Briggs describes in his book, *A Short History of the Building Crafts* (see bibliography, p. 229): "Sycamore is neither very strong nor supple; and of acacia and tamarisk it can only be said that they are the least unsuitable of native trees for building purposes" (p. 116). Not surprisingly, since a growing understanding was combined with a growing scarcity of wood, timbers began to be replaced by masonry in the Middle East.

By contrast, in parts of northern Europe that were heavily wooded, a building style emerged around the ninth century in which the timber framework was vertically infilled with thick wooden planks or more timbers, making a veritable fortress of wood. Also in heavily wooded areas, timber framing was occasionally used along with log building—for example, some Swiss chalets have horizontal logs laid between the joined timbers. Where log construction was dominant, timber framing was still used for the roof system.

The skills and tools required for log building were certainly similar to those for timber framing, but in other ways the differences were vast. Early log building was rudimentary, using simple corner joints and the natural length of the logs to determine building dimensions. Stacking logs is a much more basic means of creating a structure than joining vertical and horizontal timbers into a rigid skeleton, and the aesthetics and feel are significantly different. Log walls look massive and earthbound; as the weight of the wood and gravity keep the structure together, so the

As in ancient times, modern buildings sometimes combine timbers and logs, as does this Japanese-style house designed and built by Len Brackett.

heavy horizontal lines seem to hold it snugly to the ground. The feeling of early log buildings, usually built over pits and with few penetrations in the walls, was probably somewhat like a cave of wood, a decided disadvantage in the minds of people who were working their way out of the pits and caves and into the light. By contrast, timber framing allowed early builders to use fewer pieces of wood, and required the use of structural engineering to keep buildings erect. Timber frames also have a much lighter appearance than log buildings, and rise out of the ground with a more vertical statement. Demanding the use of some complex joinery and geometry, timber framing forced the mind of man to expand.

In the less forested areas of Europe and for buildings of grand purpose, timber framing dominated log building by the Middle Ages. The ancient use of skins to cover the frame gave way to infilling with wattle-and-daub (a mud and plaster system) or later, with brick. In both cases, the structure was visible to both the interior and exterior, instead of just to the inside. The architects and inhabitants of these buildings obviously found the structure fascinating and, in an effort to keep up with the Joneses, made the frames ever fancier and denser with timbers. Through the

centuries wood became scarcer (especially in the British Isles), but still frames displayed more and more timbers, not as a requirement of the structure, but to trumpet the wealth of the occupant. Over the years frames have become more conservative in their use of timbers, but only because vanity has found other forms of indulgence.

It is easy to understand why, at any time previous to modern time, local resources had significant impact on construction techniques. People simply did not settle where there weren't sufficient building materials, and when they built they presumably used the handiest material. Structures thus naturally reflected the landscape— bricks were used to a greater degree when there was clay in the soil, stone when the shape and quantity of rocks allowed, and timbers when there were sufficient trees. In the limestone areas of Great Britain, timber framing was never fully practiced because there just wasn't enough wood to make it practical. Here, as in many other parts of the world, masonry and timbers were used together, the massive stone walls supporting soaring timber-frame roofs. Stone protected the timbers from the deteriorating effects of the ground, and timbers made the great spans. Most of the great halls and cathedrals were a mix of the two materials.

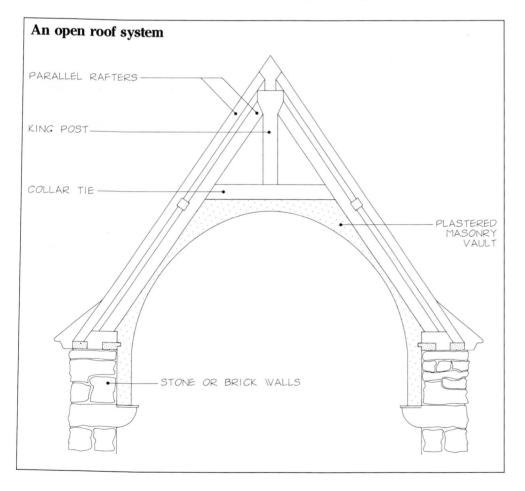

**An open roof system**

PARALLEL RAFTERS

KING POST

COLLAR TIE

PLASTERED MASONRY VAULT

STONE OR BRICK WALLS

By the Middle Ages, the craft of timber framing was fully developed and, by the standards of the time, timber-frame homes required the least amount of labor. Timber was therefore preferred even when it was relatively scarce, and the condition and variety of available trees often determined the shape and style of the frame. According to Martin Briggs, "The feature of Lycian carpentry was the short lengths of timber employed, for the trees that grow on the steep Lycian slopes have gnarled boles and twisted trunks." William the Conquerer's Domesday Book of 1086, an inventory prepared for taxes, indicates that England, although once almost completely forested, was by then only 15% forested. (Three centuries later, the figure was closer to 10%.) Medieval buildings often displayed crooked, twisted timbers, which are now practically a definition of that period's architectural style. But it wasn't a fad propagated by some early rendition of *House Beautiful*—it was all they had.

During this period, English carpenters developed ingenious systems for spanning great widths without long timbers, and created an almost astonishing array of scarf joints (p. 47) to splice timbers longitudinally. Scarce resources sparked creativity and, partly because of this, some of the greatest carpentry of all time is evident in the timber-frame roofs of the Middle Ages. Still it should be noted that the open-trussed roofs would be stronger with a timber locking the walls together; a scarf joint is not as strong as a timber that doesn't need one.

**The gnarled and twisted timbers common in medieval buildings indicate the quality of resources, not the architectural fashion. (Photo by Rob Tarule.)**

In France, long timbers were unavailable by the second half of the 14th century, and so multi-storied, jettied building became common. Here successive stories were stacked upon short posts rather than being joined to posts that passed through all the floor levels. (While horizontal beams can be scarfed, this is a problem in vertical posts because of the extreme compression loads they bear.) To keep the structure rigid, these buildings required bracketing as well as bracing to secure posts to beams. For joinery considerations, horizontal beams would project beyond vertical posts at the outer walls; a timber bracket (much like a shelf bracket) was used in the corner between the post and the projecting beam to reinforce the connection. In a later outgrowth of this type of construction, the jetties were cantilevered. Especially popular in the cities, this architectural style accentuated the building's lines and created more space in the floors above the narrow street. But because it resulted in unsanitary conditions in the darkened streets below, jettied construction was ultimately outlawed.

**When timbers became scarce in Europe, tall buildings were constructed by stacking successive stories using short posts and overhanging beams. The timber brackets in this 15th-century German building, which resemble shelf brackets, reinforce the structure. (Photo by Chris Madigan.)**

As surely as timber limitations influenced building design, so did timber abundances. When the colonists came to this country, Europe had been denuded of timber resources. The settlers suddenly found themselves in the midst of virgin forests with trees long enough to run the length of almost any building. Although they did not immediately devise new techniques to take advantage of this wealth, colonial Americans eventually developed a unique building style that featured long timbers for posts, plates, rafters and tie beams. These buildings were strong, simpler to construct than their old-world counterparts and, ironically, used fewer timbers.

The kind of building that evolved on the free soil of this country represents one of the most important developments in timber-frame housing. Even though framing techniques were refined over several thousand years and in all

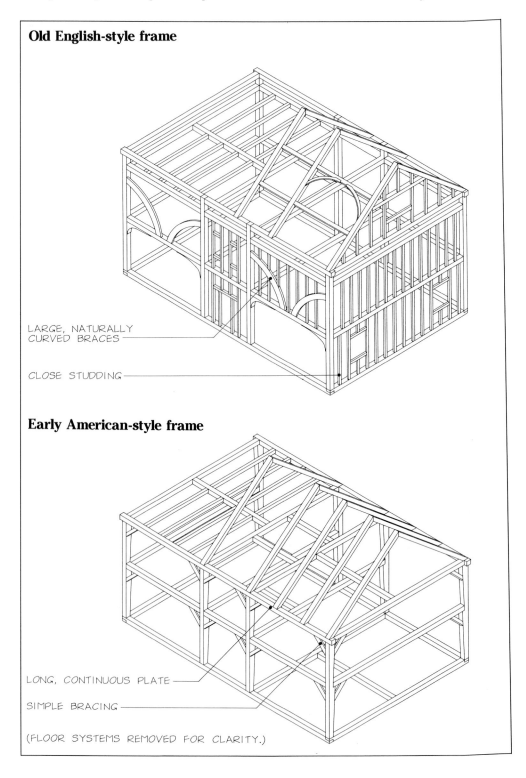

**Old English-style frame**

LARGE, NATURALLY
CURVED BRACES

CLOSE STUDDING

**Early American-style frame**

LONG, CONTINUOUS PLATE

SIMPLE BRACING

(FLOOR SYSTEMS REMOVED FOR CLARITY.)

parts of the world, it appears to be here, in America, that a high-quality timber-frame house style developed specifically to accommodate the common man. In societies with a clearly defined class hierarchy, the finest efforts were invested in the religious institutions and copied at the next level down; similarities between cathedrals and castles are not coincidental, nor between castles and manor houses. How many and how fast building improvements filtered down depended on the distribution of wealth and the organization of the society, but timber-frame house construction of any consequence did not generally make it to the common levels. This is not to say that some form of timber framing was not practiced at these levels, only that since so few structures survived, clearly they were greatly inferior. The Swiss chalets (some of which are completely timber-framed) and the 15th-century timber-frame townhouses common to Belgium, Germany and the Netherlands, probably come closest to being true residential styles.

The sad fact is that many of the wonderfully picturesque manor houses in Europe were surrounded by the hovels of the peasantry. Decent housing was to be enjoyed only by the wealthy, and only because others suffered for it. The feudal society in Japan was even more divisive. Commoners might have wanted to build houses like those built for nobility or the warrior class, but were forbidden to. The restriction was class, not money, although if a commoner had enough money it was sometimes possible through bribery to obtain a desired architectural element, such as a decorative gate or sliding panels. Feudal building restrictions were not removed until 1867, and a consistent residential Japanese building style has been developed only in the last 120 years. Although Japanese timber-frame houses are quite different from Western styles, they evolved along a similar path, including the eventual casting off of the oppressive notions that brought the craft of timber framing (but not timber-frame housing) to its highest moments. It is coincidental that the barriers over which the common people of two completely different societies peered in envy and misery were overcome within a short time of each other. When the colonists stepped onto the shores of America,

The contemporary-American style of timber-frame housing tends toward open spaces and, even in small rooms, a feeling of spaciousness.

and when the Tokugawa shogunate collapsed in Japan in 1867, it became possible for the common people of these countries to attempt to live in dignity and comfort.

This newfound freedom did not immediately lead to a revolution in architecture or frame construction, for there was no real dissatisfaction with the buildings, just with their occupants. So the first timber-frame buildings in America looked an awful lot like the ones left behind in England, and the homes the free Japanese built greatly resembled the Minka buildings they had been forbidden to own. Indulging in forbidden fruit was probably one reason for this, but it's also natural to embrace familiar archetypes. It took a long time for more independent styles to develop.

In America, the colonists learned the hard way about the effects of the climate on foundation and enclosure systems. Extreme changes in temperature heaved the soil, and it was soon obvious that an exterior cladding was necessary to protect the building from the weather. The first timber frames were oak, which was the wood used in England, but later frames used a great variety of species. Eventually, not only were frames redesigned to take advantage of the abundance of virgin trees, they were also redesigned to accommodate new raising techniques. In England, small groups of professionals made the frames for most houses, using joinery and assembly methods that would allow individual pieces to be placed into a frame. But in America, a great spirit of cooperation and neighborliness brought whole communities together for raisings, and timber frames were designed in large units that could be preassembled, then raised in a single day.

Frames also became more utilitarian. Being on free soil meant that class distinctions were dissolving; even the leaders and the wealthy would have been reluctant to act like the oppressive nobility of the old country. So in spite of the rich supply of timber, builders used their knowledge to make strong frames without wasting material or labor on embellishments and showy styles. Simplicity was the hallmark of the Early American timber frame.

Early American timber-frame houses also shared a similar design, derived from generally accepted notions about function that had evolved through many centuries. In his book *Home*, Witold Rybczynski points out that these notions were vastly different from our own current requirements (see bibliography, p. 229). The issues of comfort and privacy, for example, barely existed at certain times in history, and played little part in house design. Subsistence was usually much more to the point, and the first homes often housed animals as well as people because they were critical to survival. The plan of St. Gall, on the facing page, reveals the way homes looked around 850 A.D. The basic ingredient was a central living area, or hall, in which all activities took place. At its center was a firepit—no chimney, just a hole above the pit for smoke to escape. On the outside of the living area were rooms for sleeping and for animals. That's it. None of the house plans show a second-floor living area, probably because it would have been too smokey for habitation. These early plans are rather ingenious, for the central room is insulated by the animal and bedroom areas. The outer wing is terrific insulation and probably the first air-lock entry. Since structural requirements were given priority, it was also a nicely integrated plan. The living area is really the entire main timber structure and the outer wing simply leans against it as a shed-like substructure. Every room is defined by posts and beams. Still, this was not a home that was warm or comfortable in the way we think of those concepts today: It was a place for cooking and working and sleeping—in short, surviving.

Eventually, the development of the chimney made the living space more tolerable and allowed a second floor to become common. But the central hall stayed and simply became grander as architecture began to be used to demonstrate social position. The hall of the late Middle Ages was a soaring and much more highly embellished version of the smoke-filled hall that preceded it. The technology of the chimney and the improving craftsmanship of the carpenters refined the hall, but there was still no attempt to organize for privacy or for specific activities. When other rooms were added for living, there wasn't a hallway for passage between; you had to go through one room to get to the other. In the castles and manors, workrooms were separated from living areas, but this was done more to segregate servant from master than for convenience.

The first American homes also began as a hall and a parlor. Between them was a chimney with a fireplace facing each room. The second floor was a copy of the first with bedrooms on either side of the chimney. Framing was simple. The house was narrow and tall, usually four posts long and two posts wide. Because people had to have shelter quickly, and usually had only minimum means, the home was designed to expand as needs grew and finances allowed. The frame was easy to make and raise, and provided all the living space that was immediately necessary. Later, a lean-to was added to the eave side, which often became the kitchen. This house shape is now referred to as the saltbox, and it was similar in concept to the bedrooms and animal areas in the St. Gall plans. Later houses commonly included the kitchen as a part of the original construction. The house either maintained the saltbox shape or had more room on the second floor, in which case it became a colonial style. This basic plan is so common that any other is considered an exception.

The pattern of the Early American house was still based on survival—the family gathered around the chimney for warmth and to cook; upstairs was for sleeping. Most houses used this plan because it was an easy way to fit the necessary living spaces into the frame. It was also entirely funtional. Instead of grand expanses, there were low ceilings on two floors, allowing as many spaces as possible to fit within the frame and keeping those spaces warmer. It was a style that reflected the distinctly different idea of life in the new country.

Until the mid-1800s and the advent of stud framing, the timber-frame house continued to evolve and reflect the changing lifestyles of many different kinds of people from New England to the South and out to Ohio. Soon the houses had many different kinds of living areas with hallways between major rooms for privacy. There were timber-frame farmhouses and timber-frame townhouses. There were elegant homes and simple homes, and many styles flourished in the variety and abundance of the expanding new world.

## A building from the plan of St. Gall

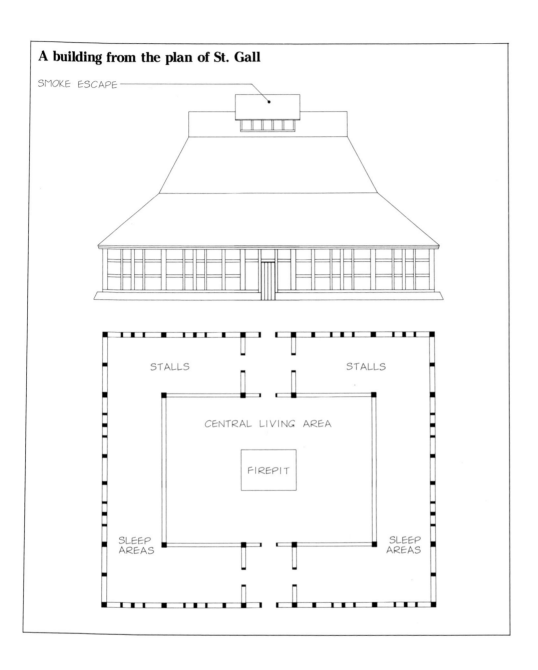

SMOKE ESCAPE

STALLS

STALLS

CENTRAL LIVING AREA

FIREPIT

SLEEP AREAS

SLEEP AREAS

**Timber-frame houses adapt well to modern times. The building method can be used to produce graceful homes that are as energy-efficient as they are beautiful. (Photo by Brian E. Gulick.)**

## The modern timber-frame house

As with most things, the modern timber-frame building system and the home it supports has had to respond to our place and time. If timber framing had not lain fallow these past 150 years, it would probably have been altered radically by the great social and scientific forces of the 20th century. It is just as well, perhaps, that it was put aside before being compromised by the age of machines. What we're left with is an opportunity to bring timber framing back in its very purest form. But the revival of interest in the joiner's craft has not been enough to make the timber-frame house thrive. For modern appeal, the whole concept of the house had to be considered. And the timber-frame house has done well, adapting nicely to the paraphernalia of our era while still providing security, pleasure and comfort—ingredients too often lacking in the modern home.

In this section, we'll look at some of the goals in building a modern timber-frame house. As well as including practical construction objectives, such as good energy performance, these goals include the elusive ideals that allow a house to provide an extraordinary living experience, to enhance our physical, intellectual and emotional lives. This is where we'll begin. While these intangibles are rarely on the list of requirements for a new house, they should be, for when a dwelling exudes them, it becomes a place of growth and comfort, and it becomes a home.

Homes today contain a host of modern conveniences and have become veritable cocoons of controlled weather systems. There was a time when winter was a life-threatening experience and getting clean was too hard a task to undertake more than once a month. Today's home has freed us from worry about mere survival, and a vast majority of the population has more time than ever before to explore other aspects of life. Unfortunately, designers and builders spend so much time making sure a house contains all the right stuff that they expend practically no energy on its soul. Too often, houses that liberate people from harsh weather and the chores of daily living do little else. They do nothing for the senses, nothing for the spirit and nothing to enhance or ennoble the time and energy set free.

Even something as basic as comfort is rarely considered by modern architecture, points out Witold Rybczynski in *Home.* "During the six years of my architectural education, the subject of comfort was mentioned only once. It was by a mechanical engineer whose job it was to initiate my classmates and me into the mysteries of air conditioning and heating. He described something called the 'comfort zone,' which, as far as I can remember, was a kidney-shaped, crosshatched area on a graph that showed the relationship between temperature and humidity. Comfort was inside the kidney, discomfort was everywhere else. This, apparently, was all that we needed to know about the subject. It was a curious omission from an otherwise rigorous curriculum; one would have thought that comfort was a crucial issue in preparing for the architectural profession, like justice in law, or health in medicine" (p. vii).

Creating a comfortable space requires more than a soft chair, and making comfortable houses is not automatic to any form of construction. Comfort comes from qualities in the building that offer a sense of security, balance and order. Perhaps today more than ever there is a need for

houses to offer these amenities, because although our lives have changed in many ways, under the surface not much is different. The anxieties today come from different sources, but they create the same needs. Our forefathers fought off freezing winters and wild animals; the future, however uncertain, was clearly in their hands. In today's wilderness, the adversaries are not so well defined. We struggle for security, but we are almost always at the mercy of others, and this causes instability as surely as a beast howling at the door. Much about life is disconcertingly temporal, and there is comfort in buildings that transcend the ephemeral qualitites of life and conventional measures of strength.

This the timber-frame building system does quite naturally. As John Burroughs, the naturalist, writes in his essay "Roof-tree": "If the eye could see more fully the necessities of the case,—how the thing stands up and is held together, that it is not pasteboard, that it does not need to be anchored against the wind,—it would be a relief. Hence, the lively pleasure in what are called 'timber houses', and in every architectural device by which the anatomy, the real framework, of the structure, inside or out, is allowed to show, or made to serve as ornament. The eye craves lines of strength, evidence of weight and stability" (*The Writings of John Burroughs.* Vol. 7, *Signs and Seasons* [Boston and New York: Houghton Mifflin Co., Riverside Press, 1886], p. 271).

**Exposing the framework of the timber structure to the interior lends a feeling of security and balance. The timber-frame structure also becomes the mounting surface for the solarium glass in this house.**

As Burroughs suggests, the beauty and art of architecture often lie in the form and function of the structure itself. Viewers are captivated not by the road that a bridge supports but by the splendor of the interaction of cables, trusses, beams and ties. The wholeness of the parts performs the magic. Architects have long understood that a pleasing sense of structure can be an important ingredient in design. If the framework itself is not attractive, the architect might try to imitate structural features with finish details, for instance, by building up the exterior corner boards so they look like supporting columns or decorating the building with imitation timbers. Designers who specialize in restaurant decor have the latter trick down to a science. By my own estimation, probably 75% of the restaurants in this country feature exposed fake timbers. Adding nonsupporting frame members after a building is complete is ridiculous, and some of the results are nothing short of ludicrous. Why would anyone feel that a plastic 4-in. by 8-in. "beam" spanning 80 ft. would lend a feeling of stability to the atmosphere? Such design error is simply not possible in a timber-frame building because the sturdiness and beauty of the structure are integral to the system.

Our forefathers reinforced this inherent strength by building, even overbuilding, structures that would have a life beyond their own generation. Relying on hand tools and hard labor, they did things right the first time so that work would not have to be repeated. There is a vigorous sense of integrity in these unyielding buildings, which survive despite their age, and to me overbuilding seems to pose only one threat: the product might last too long. Small sin, for as a consequence the world is blessed with enduring pyramids, temples, cathedrals, bridges, houses and barns.

As our forefathers built for the future, so must we today. But beyond it being ecologically important that today's houses serve later generations, it is also psychologically uplifting to know that in the worst of times a house will stand and that it will have value and purpose beyond our lifetimes. It's a very uncertain feeling to have the building you live in shake in every wind. I lived in a house that seemed to visibly deteriorate with each passing day, and I can categorically say that it caused insecurity and frustration. I began to wonder if it wasn't I who was leaning and decomposing. On the other hand, I grew up in a house that had been built at the turn of the century by a man who had made his wealth mining gold; he used the highest standards of the time, lovely woodwork and detail. When the house passed eventually to my parents, who made their wealth in children (or was it their poverty?; there were eleven of us), it was given some tough tests. But when the snows and winds came, we always felt safe. When there were low spirits and little money, the house itself offered encouragement of better days to come. My feelings about good homebuilding were born in that great old house and nothing less has been satisfactory since.

In design, there are many goals we may strive for to enhance and refine the essential strength and durability offered by the timber frame. Our ancestors, in building for the future, also designed for the future, and here, too, we can follow their lead. As we have seen in the first section of this chapter, floor plans were not so much personally appropriate as they were reflections of typical living patterns and commonly held views about how houses should be arranged—houses were not designed on the basis of architectural whim or passing fancy, as are many today. The shape of buildings also followed principles that were felt to have universal appeal. For their classic, enduring architecture, colonial Americans used simple, graceful lines that were historically familiar to them and that promised to hold strong into the future.

**Old timber-frame structures were built for the future, not to satisfy an architectural whim or passing fancy. They were based on design principles felt to have universal and enduring appeal.**

Modern timber-frame builders should also strive for classic lines, respecting the role the medium plays in design. Without a doubt, connecting large timbers with mortise-and-tenon joinery imposes limitations on the form and style of a building. You will not soon see a timber-frame dome, and if you do, stay out of it, for structurally it would be like a fish out of water. In timber-frame buildings, the walls generally should not be much higher than the posts (whose height depends on the available length of timbers). The joints are best as orthogonal intersections. The frame is rectilinear, and when it is assembled the timbers form a cohesive alliance, gaining strength through balance and symmetry. Timber frames are disinclined to perform feats for the purpose of architectural curiosity, but this is not to suggest that building with timbers means limited design opportunity. A study of Japanese architecture or a look at any of Cecil Hewett's books on English building construction (see bibliography, p. 229) will demonstrate that, within the constraints imposed by the timbers, there is challenge enough to design graceful and beautiful houses that are the essence of domestic architecture.

Using structure as an aesthetic consideration presents some interesting design opportunities, as we will explore in later chapters. Posts and beams can be used to define areas. Frame members can be deliberately placed to create a desired feeling. For instance, a large summer beam might be located over the center of a living room to create a feeling of unity, symmetry, strength and stability. This would be a formal place in which people would group together. Another space might be left open to the roof to create a sense of expansiveness and drama. Here timbers could be arranged for a lighter, more playful effect. This might be a place for fun and informal entertaining. When these and other techniques are used effectively, when the frame and the floor plan intertwine, then each area of the house is naturally well decorated and feels complete, even during construction, when the house is still just a frame silhouetted against the sky.

**In a timber-frame building, structure is also an aesthetic consideration. The summer beam in this living room lends an air of strength, as well as providing visual interest in the ceiling.**

Timber-frame houses are being enthusiastically accepted because they come with good credentials from the past. The task now is to design and build modern homes that are simple but flexible, classic but not out of touch. Perhaps it is obvious by now that I think we should take our cue from the American tradition, adapting and inventing a new style while still respecting the archetypal forms. For example, to be affordable for the average person, Early American houses were arranged simply, usually containing four bents and three bays. Our timber-frame houses tend to follow the same pattern, yet within this basic form we have created hundreds of different floor plans. We still rely

on a central chimney, attaching a small wood stove or furnace to it to heat almost entirely any of our well-insulated houses. And we still often build with the saltbox shape (or variations of it), orienting the low wall to the wind and the high wall to the sun for good energy performance.

Although we don't have the great virgin forests colonial American builders had, our techniques and framing styles are also similar. We use the same kinds of wood, albeit in smaller and shorter pieces. Frames through history changed to suit available materials, and modern builders should be ready to adapt. For instance, a frame designed for oak timbers under 16 ft. in length should not be the

**Timber-frame buildings, even when modern in design, display classic lines.**

same as for one using fir timbers available in 40-ft. lengths. Since we have cranes available to lift timbers, assembly and raising techniques can be similar to those used in colonial America even though we can't (or for safety don't choose to) muster a community of people for a raising. The use of a crane can and should affect the engineering and layout of a frame in the same way that the helping hands of neighbors altered frame design for the colonists. With the crane, we have the opportunity to make frames stronger by assembling more pieces together in a single unit.

There are some situations, however, for which there is no historical precedent, and modern builders must invent viable 20th-century construction alternatives. A realistic route must be provided for each plumbing pipe and electrical wire. Allowance must be made for bathroom fixtures, kitchen appliances and a heating system. It is necessary to consider the requirements of smoke and burglar alarms, stereo systems and telephones. Rooms must accommodate king-size beds, microwave ovens, hot tubs and computers. Some of our clients have the security system of Fort Knox; others don't even bother to lock their doors. One house has an exercise room up on the roof, another has a bomb shelter down in the basement.

Fortunately, the timber-frame building system solves more problems than it creates. Timber framing is naturally compatible with passive solar heating, for example, and with its goal of keeping areas open to one another so that heat can move through the house without fans and ducts. Since the loads of the building are directed to a few large posts instead of to many small framing members, timber-frame houses lend themselves to open floor plans. The necessary expanses of south-facing glass can be installed between posts without difficulty. Glass and timbers complement each other, and from the very beginning we have tried to capitalize on this relationship in our houses. There is no prototype in history for making timber solariums, so we have invented this technology using new ideas but attempting to adhere to old standards.

The most important asset of the timber frame to modern building technology is the basis it provides for an extremely effective insulating system. With the frame bearing all the building loads, the insulation can be a completely separate system. This has the benefit of minimizing thermal breaks and simplifying construction. After some experimentation, we discovered a product called a stress-skin panel, which is a sandwich of interior and exterior sheathing with a core of rigid foam insulation. Using this bit of high technology, we are now able to wrap the timbers in unbroken insulation, achieving the same energy performance as a super-insulated platform-framed house, but without going to nearly the same extremes. In fact, many people express an interest in our homes because of the buildings' remarkable energy efficiency, not because of a primary attraction to the timber frame itself. To us, developing this insulating system felt like a major breakthrough, vastly different from what we knew of European or traditional American enclosure techniques. But ironically, it was not really a new concept at all, as it resembles the primitive method of enclosing a frame with animal skins.

As mechanical and design innovations for the modern timber-frame house have evolved, it has become apparent that in many areas the procedures that are used in standard construction are not necessarily transferable. To keep construction efficient while maintaining traditional standards, it has been necessary to adapt materials and invent new methods. Some of these developments take timber-frame house builders outside the information available in architecture and construction guides, and there are risks involved when going against the grain of conventional practice.

National and local building codes often pose problems for the construction of timber-frame homes because they were written to govern other building techniques. When the building official is presented with an unusual situation, the tendency is to lean on the letter of the law instead of the spirit. The result is that innovative ideas face a rocky

The stress-skin panel is basically a sandwich of interior and exterior sheathing on either side of a core of rigid foam insulation. The panels are installed outside the frame, wrapping it in an unbroken layer of insulation.

road even though they may be better than code requirements. For example, some building codes demand that all structural members be stamped by a lumber grader to indicate that they meet the stress grade. This is well and good until you want to use oak for the frame. To date, the only grading rules for oak are for appearance—allowable stress grades have not been developed. Therefore, although you might reasonably demonstrate that the timber frame you intend to build exceeds the code requirements by a factor of three, a building inspector has the right to prevent you from building it. Wiring is another example. Code requires wires to be in the center of the insulation cavity to prevent them from being punctured by nails entering from either side. This seems like a right-minded rule, but not for every circumstance. We have a situation where the safest and easiest place to put the wire is on the inside of the stress-skin panel surface and the outside of the timber. Yet the strictest interpretation of the code suggests that we have to put the wire in a more difficult and vulnerable position. In such cases, most reasonable building inspectors will use their own judgment or require an engineer's stamp on the building plans.

A similar kind of problem is often caused by people in the building trades who fear the unknown. To compensate for their insecurity, they raise their prices, even though the task may not be any more difficult or complicated—just different. To prevent this sort of costly ignorance, consumers must be thoroughly educated about construction procedure specific to timber-frame houses. They need to be prepared to educate potential contractors and subcontractors, and know when the price does not accurately reflect the difficulty of the job.

Dealing with codes and construction trades can make building a timber-frame house more vexing than it need be. But as this type of construction becomes more and more popular, the problems will begin to disappear. Codes will adapt regulations to govern these new situations and tradespeople will become familiar with the new materials and techniques. In the meantime, it is necessary for those who wish to construct a timber-frame house to become familiar with building details and the standard solutions to various situations that may arise. In part, that is the very reason for this book.

In my experience, building a timber-frame house takes more time than building a conventional house. First of all, the plans need to be more detailed because nothing can be taken for granted. Building inspectors have a tendency to want to see more on paper so that they can satisfy themselves that code is being met. Tradespeople will need to have more explained by way of drawings so they can understand procedure and the interface of materials. Owners who are overseeing their own projects usually keep the job under control by limiting the number of subcontractors on the site at any one time, and so time is traded for quality control.

Another factor that inhibits a quick pace is that owners tend to choose labor-intensive details. Once people see the craftsmanship of a fine timber frame, they usually prefer finish work more in keeping with the quality of the frame. High-quality materials are harder to get than cheap materials (even if they are not more expensive) and good work takes more time (and is definitely more expensive) than bad work. The moral of the story is that nothing good comes easily. A timber-frame home is built to unusually high standards and costs more time and usually more money, too.

Since I'm ending this chapter with a dose of realism, I'll mention another downside fact. A timber frame is made of wood, and wood—especially in timber dimensions—cracks, shrinks and twists. Any timber above 4 in. thick cannot be reliably or economically dried in kilns because it takes too long. Air-drying is also impractical for the same reason. Even when it is possible to make the frame of dry timbers, seasonal shrinking and swelling results in dimensional change or surface checking. We have made frames of timbers that were dried in a kiln and a few from timbers that were resawn from larger timbers over 100 years old. In both cases, there certainly was not as much movement as in frames built from green wood, but there was definitely visible change as the frame acclimated to its new environment. It is just the nature of wood. If you want a door that will never warp or shrink, get one of the new metal or fiberglass doors. Although a shirt made of polyester can be made never to wrinkle, there is something about cotton that is unlike any man-made fiber; if you like the feel and texture and the individual character of natural materials, it is difficult to accept substitutes. Buying quality sometimes means accepting the wrinkles, and the timbers in a frame definitely "wrinkle." This can be seen as a drawback, but it is also proof that they are the real thing.

**The shrinking and checking in timbers are like wrinkles in cotton—evidence of the unique character of a natural material.**

# II. The Structure

**Uncompromising standards of workmanship and engineering allow timber-frame buildings to stand for centuries. This connection consists of a post, two summer beams (to the right and left of the post) and a bent girt (at top).**

Sound timber framing requires standards of workmanship and engineering that are uncompromising. Weak examples of the craft have either not survived at all or survive marginally, ravaged by forces they were not equipped to withstand. But properly designed and well-constructed timber-frame buildings stand sturdily, almost defiantly, despite centuries of use. Better than words, far better than textbooks, these many thousands of surviving timber-frame buildings silently instruct.

Critical to building timber-frame structures to the highest standards is a broad understanding of theory and good practice; next come some creativity and flexibility. For each frame starts out as not much more than a blank canvas, made up of the hopes and dreams of the future occupants (brought down to earth by budget limitations and site requirements). During design, there is nothing about the shape or flavor of the building that dictates more than the most general arrangement of timbers. When the timbers are arranged, the wood species still has to be selected and the timbers individually sized. Information about the arrangement, species and size of the timbers does not automatically lead to an understanding of the joints that will be used. And when the joinery is determined, decisions still must be made about embellishments and finishes. Of course, all of these considerations must be juggled along with the architectural concerns of the building. Compare this decision-making process with the methodology of con-

ventional stud framing, where once you have decided where the walls go, you put them all together in the usual way—16 in. on center, nailed to top and bottom plates.

To me, timber framing has more in common with the design and making of wooden furniture than it does with conventional building procedure. In the same way that a handcrafted chair is more than a place to sit, a timber frame is more than its structure. In both cases it is virtually impossible to separate design from construction, and there are few pat answers to any given question. Be it chairmaking or timber framing, to make successful decisions about woods, finish, style, function and construction, you must first understand the structural requirements of the craft. To achieve the necessary overview, we'll look here at engineering and the physical characteristics of the timber frame. In the next chapter we'll experiment with frame design, and in Chapter IV we'll explore home design. Although we'll be discussing frame design separately from house design, it is important to realize that both must evolve together when plans are being drawn (see p. 67).

## The anatomy of a timber frame

A typical timber frame can be divided into four major systems: walls, floor, roof and cross sections called bents. Walls are the linear compositions of timbers that run parallel to the ridgepole and bents are the major structural elements perpendicular to the walls. A simple frame might

**Typical parts in a timber frame
(showing two roof systems)**

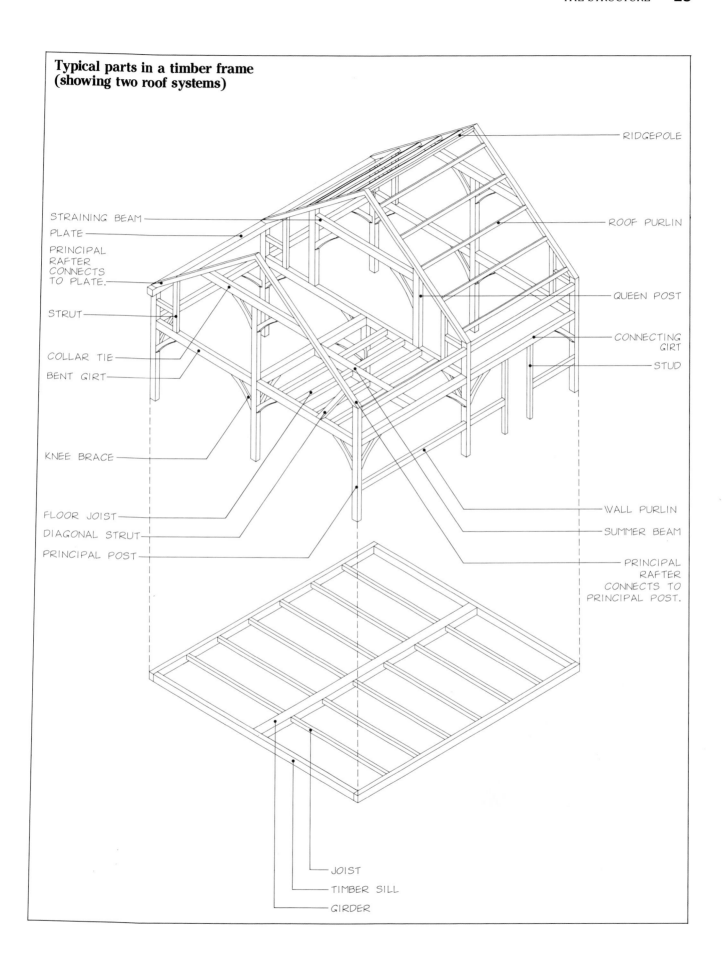

RIDGEPOLE

STRAINING BEAM

PLATE

PRINCIPAL
RAFTER
CONNECTS
TO PLATE.

ROOF PURLIN

STRUT

QUEEN POST

COLLAR TIE

CONNECTING
GIRT

BENT GIRT

STUD

KNEE BRACE

FLOOR JOIST

WALL PURLIN

DIAGONAL STRUT

SUMMER BEAM

PRINCIPAL POST

PRINCIPAL
RAFTER
CONNECTS TO
PRINCIPAL POST.

JOIST

TIMBER SILL

GIRDER

contain three or four bents, two walls, a floor system on one or two levels, and a roof system inclined toward the center equally from each wall. It could be this simple, but there is also no limit to how complex each system might be. The systems can be further subdivided into the vertical, horizontal or inclined timbers that make up each unit.

Historically and geographically there is some inconsistency in nomenclature, but basically the vertical members are posts, struts and studs. Struts can also be inclined, and are used in the roof system as a secondary support for rafters. Studs are secondary posts, primarily used as nailers or for decoration. The horizontal members, or beams, are girts, sills, plates, girders, joists, collar ties, tie beams, ridgepoles and purlins. Girts are the horizontal members in bents or walls that span between major posts. Sills support the post bottoms at ground level, while plates support rafters; girders are major beams that run between sills. Inclined members are generally either rafters or braces. The identification of most pieces in the frame is further refined by function and location. Therefore there are bent girts and connecting girts, roof purlins and wall purlins, principal rafters and common rafters. A few other timbers have interesting names: queen posts, king posts, prick posts, crown posts and even samson posts, as well as summer beams, hammer beams, anchor beams and dragon beams.

The timbers within the walls, floor, roof and bents interact to form a self-supporting structural unit. This unit is often a complex web consisting of many hundreds of pieces, each absorbing and transferring a share of the building load. The accumulating loads are collected and passed on through the frame from the minor members (which tend to fall in the center of the building) to ever larger timbers, until eventually they are received by the principal posts. From here, the loads are passed to the foundation and ultimately distributed to the earth below. To design a frame properly, each timber must be positioned to respond to specific loads and sized to resist the forces that will act upon it. Loads must be directed to the posts without putting undue stress on any timber. Every joint must be planned to transfer the load effectively from one timber to another, and to keep the entire frame rigid enough to stand erect.

When function comes together with form in a well-designed timber-frame building, the frame itself can become a work of art. It takes on the natural balance, symmetry and beauty that are the essence of timber framing. The intelligent placement and careful connection of each timber component coupled with the athletic way loads are relayed through a frame result in a mighty combination of muscle and intellect that is bound to impress.

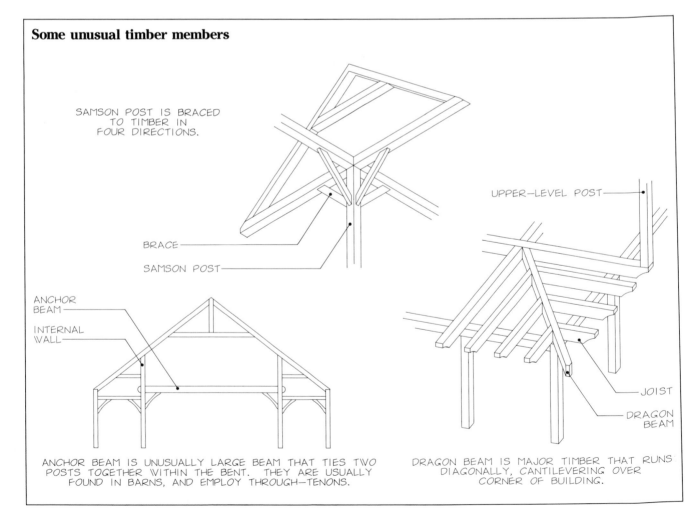

**Some unusual timber members**

SAMSON POST IS BRACED TO TIMBER IN FOUR DIRECTIONS.

BRACE

SAMSON POST

UPPER—LEVEL POST

ANCHOR BEAM

INTERNAL WALL

JOIST

DRAGON BEAM

ANCHOR BEAM IS UNUSUALLY LARGE BEAM THAT TIES TWO POSTS TOGETHER WITHIN THE BENT. THEY ARE USUALLY FOUND IN BARNS, AND EMPLOY THROUGH—TENONS.

DRAGON BEAM IS MAJOR TIMBER THAT RUNS DIAGONALLY, CANTILEVERING OVER CORNER OF BUILDING.

# The forces

Reading this chapter will not turn you into a structural engineer, but if designing timber frames required the full knowledge of one, I would not be writing this now. On the other hand, it is essential that timber-frame designers have a good understanding of the forces that will bear on a frame and their effects. It's also important to know how to calculate loads and to size timbers. These calculations require the use of lengthy formulas or tables, which, although admittedly complicated, are not beyond the reach of the determined person. They are beyond the scope of this book, however, and so I refer interested readers to my first book, listed in the bibliography (p. 229), for that information. I also must stress that it is critical to recognize that point at which the services of a licensed engineer are required. Any frame worth cutting, joining and assembling is worth an investment in proper engineering.

Still, don't let science override common sense. I learned this lesson the hard way, after designing and building an office for my parents' mail-order business. It was not a timber-frame building, but we used timbers for internal posts and carrying beams. Wanting to impress my parents with my knowledge, I carefully calculated each timber size and demonstrated to them that though the beams would have to be very large to support the supplies on the second floor, the posts to support the beams could be quite slender. We could save money on the posts, I boasted. My mother told me several times that the posts looked too small. When I started to argue the math, she stopped me and said, "Those posts are just too small." She didn't need formulas to tell her that. The skinny posts stayed (because my mother has the patience of a saint), and the building was completed and piled high with the paper of their business.

One day, when my mother happened to be away from her usual place in the office, the elderly lady across the alley backed out of her garage, had an anxiety attack with her foot on the gas pedal, and crashed into the side of the office. The car came far enough through the wall to start quite a nice domino effect with the shelves of papers and boxes. The slender posts broke like matchsticks, and the forces that saved my mother (and the driver) had nothing to do with my timber sizing. While I could not have anticipated a car coming through the wall, I have learned to pay attention to gut-level intuition.

Now we'll look at the four primary types of forces acting upon the frame members: compression, tension, shear and bending moment. Although just about any exception is possible, each of these forces tends to be associated with particular timbers or framing circumstances, as you will see in the following discussion.

**Compression** Loads bearing on posts are resisted by an equal reaction on the opposite end, resulting in a compression force in the timber. Posts that support a girt, plate or rafter end receive the load from the top, but the load could be received from a horizontal beam at any point along the length of the post. Although heavy loads are often directed into the posts, seldom is there a compression problem. Wood is strongest along the grain, and posts are usually relatively large because they must accommodate the joinery of intersecting members. When there is a failure in compression, it usually comes in the form of buckling—lean on a wooden yardstick with the weight of your body and it will buckle sideways and break. So even though compressive failure of a timber post is unlikely, avoid using long, slender timbers or long unsupported posts.

Where many members intersect and wood has been removed for the joints, a post could buckle under an extreme compression load. The calculation for the maximum load in compression should be done using the smallest timber section left after the joints are cut. Most formulas assume that the post is acting alone, but the beams that join to the post can add structural value back to the post by serving as buttresses, helping to prevent buckling even of slender posts.

**Tension** A force pulling a timber from either end causes tensile stress, which amounts to two forces playing tug-of-war with a timber. It's usually not the timber that fails in this circumstance, but the joint. Joints always include the end of at least one timber, where it is difficult to generate substantial resistance to withdrawal. (Tensile strength is achieved through the use of pegs or a locking geometry, as discussed on p. 42.) Therefore, if a timber is subjected only to tension, the cross-sectional area needed for the joint would determine the width or depth of the timber. But there is usually more than this one force at work, which will affect the choice of joinery and timber size.

## Forces acting on timbers

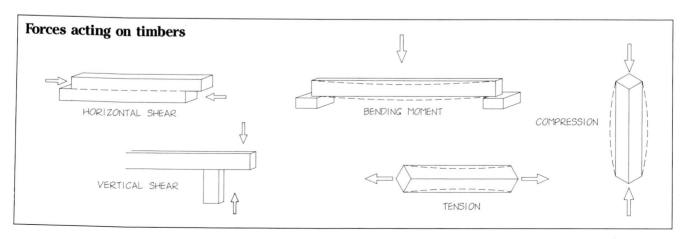

HORIZONTAL SHEAR

VERTICAL SHEAR

BENDING MOMENT

TENSION

COMPRESSION

A king post is a good example of a timber under tensile stress from the weight of the girt at its base, while also receiving a load from the rafters through the king-post braces, as shown below. (Collar ties between rafters must also resist the potential for opposing rafters to spread at their bases.) Knee braces, which play a major role in keeping the frame rigid, are often subjected to significant tensile stress. Though not always the case, the loads on the braces usually come from a force pushing against the side of the building; this puts the first brace to encounter the load in tension and the opposite brace in compression. Failure in brace connections under tension is fairly common, and frames can be made a great deal stronger if they are designed with a compression member (instead of a tension member) to react to the load. (See p. 37 for more discussion about braces.)

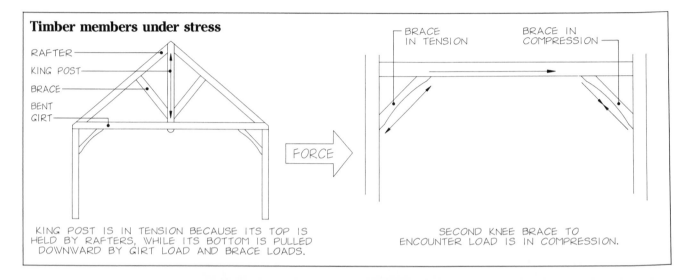

**Timber members under stress**

RAFTER
KING POST
BRACE
BENT
GIRT

KING POST IS IN TENSION BECAUSE ITS TOP IS HELD BY RAFTERS, WHILE ITS BOTTOM IS PULLED DOWNWARD BY GIRT LOAD AND BRACE LOADS.

FORCE

BRACE IN TENSION
BRACE IN COMPRESSION

SECOND KNEE BRACE TO ENCOUNTER LOAD IS IN COMPRESSION.

**The king post is under tensile stress and also receives loads from its braces. The wood of the frame is stained oak; the ceiling boards are red cedar.**

**Shear and bending moment** A pair of scissors perfectly demonstrates shearing action: Two forces slide by each other and destroy the material in between. Shear stress in horizontal timbers can be vertical (across the grain) or horizontal (along the grain). Unlike masonry materials, which are comparatively weak in resistance to shear forces, wood is so resilient that vertical shear failure almost never happens unless the timber is defective or eventually rots. It would occur if the load were so massive that the timber actually broke, with the wood fibers torn apart at the point of connection. Since the strength of the timbers at their joints is critical to frame strength, the joints must also be designed to resist vertical shearing. An example of a situation in which vertical shear might be a problem would be in a garrison frame, in which the beam cantilevers beyond the first-story wall and carries the weight of the upper post and roof beyond the support of the lower post (see drawing below at left). If the overhang were very short—short enough to eliminate bending problems—cross-grain shear would be a significant concern. Especially because of joints in the beam at the intersection of the lower wall and the overhang, this situation requires careful evaluation.

Horizontal shear (along the grain) is a much more common problem and results from adjacent wood fibers being pulled and pushed in parallel but opposite directions. This kind of shear failure in a timber would probably be the result of an extreme load having produced an excessive bend (or bending moment). Bending subjects the upper and lower portions of the timber to conflicting forces of tension and compression; horizontal shear, or a slipping of fibers along the grain, usually occurs in the neutral zone between the two forces. But the great elasticity of some species allows even a timber with a long span and an excessive load to bend quite a bit, like a bow, before failing. Actually, shear failure along the grain is more likely to happen when beams are relatively stiff and are forced to bend. Support a yardstick on chairs 30 in. apart and it will bend a great deal under a load; put a large load on the yardstick with a 12-in. span and it is much more likely to break. So horizontal-shear analysis takes into consideration relative stiffness and structural properties of the species as well as the load.

When establishing the size of horizontal timbers, three possibilities must be considered: shear stress (most probably causing horizontal splitting at the beam ends); deflection, a value of elasticity set by building codes to ensure that structures will be stiff enough to be serviceable; and bending stress, which could cause fiber failure. Fiber failure occurs in a timber when the tensile stresses caused by bending become too great, so the fibers simply break across the grain. Pure fiber failure is not very common and would probably be accompanied by some shearing along the grain as well.

---

**Vertical shear situation**

POST CARRYING
SIGNIFICANT LOAD

BEAM CANTILEVERS PAST
FIRST-STORY WALL.

VERTICAL SHEAR FAILURE
COULD OCCUR HERE.

---

**Horizontal shear and bending moment**

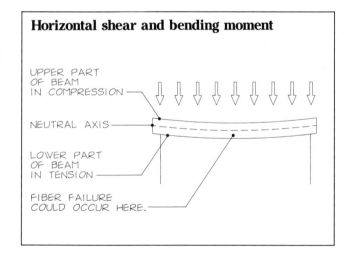

UPPER PART
OF BEAM
IN COMPRESSION

NEUTRAL AXIS

LOWER PART
OF BEAM
IN TENSION

FIBER FAILURE
COULD OCCUR HERE.

**Loads and loading**   The transfer of loads through frame members is an essential engineering consideration in timber framing. Understanding how loads are absorbed and the way in which the tribulations of one timber are distributed to others is critical to a well-designed frame. Because the load is transferred at joint intersections, much will need to be explained about the functions of various kinds of joinery, which we'll do later. The point to note here is that all the components of the frame must work individually and collectively to gather the loads and direct them to the posts without staggering under the effort.

All the horizontal timbers carry a load, if only the weight of the timber itself. In most cases, they also carry the weight of other materials, such as flooring, roof planking and shingles. These materials make up the dead load on the frame, which is almost always evenly distributed over the surface. In addition to dead load there is live load, which consists of all those things other than the building materials—snow, people, furniture and wind. The number to plug into the equation for live load is determined by the local building code. Beams that bear simple live and dead loads would be floor joists, summer beams, purlins and connecting girts.

Because loads pass through the frame like streams converging on a river, many timbers must accept more than one type of load. For example, when a post bears on a beam, it could be transferring a significant load from another area of the frame to that one point. In other cases, there might be two or three points along a beam where loads are applied through vertical or diagonal members. Concentrating the load in this manner further complicates beam sizing because the load type (evenly distributed, one-point, two-point or three-point) must be considered as well. Beams within a frame might support posts that could support other beams that perhaps support a floor, so that when the load is finally accumulated, the beam that must bear it all is asked to perform a formidable task. To do this, it will probably need to be a formidable beam. If, when designing a frame, you try to avoid such extremes with simpler, more direct load routes, the frame will be stronger and easier to build.

You can't begin to consider timber sizing without a decision about the wood for the frame, because there are dramatic differences in the load-resistance values of different species. A relatively weak species (red cedar, for example) can certainly be used, but it should be designed with short spans—it would be silly to design a frame with 16-ft. spans and then decide to use pine or cedar. Aesthetics of the different species also have a bearing on frame design. Timbers made from light-colored wood and clear grain might be more closely spaced than those made from a more dominant wood, such as oak or fir. For these reasons, design the frame to fit the wood and not the other way around.

Never be afraid to oversize a timber-frame member, and never underestimate what a frame might have to withstand. If a frame could be isolated within a protective bubble, it would only have to be strong enough to hold itself together, but real life demands that frames be designed with the assumption that snow will be on the roof, wind will blow against the walls and people and things will weigh heavily upon the floors. We are also asked by code and by common sense to plan for the worst: Someday the snow may be hip deep and very wet; someday the wind might be a hurricane; on any day, the inhabitants might hold a raucous party, and certainly the kids will jump from beds. When sizing timbers, err on the side of endurance, and you probably will not have erred at all.

**Light-colored woods are less likely to appear overbearing than dark woods, even when spaced closely together. The timbers in this bedroom, here still under construction, are white cedar treated with a clear penetrating oil. (Brock Simini, Architects.)**

The darkly stained oak in this house dominates the decor. The beam spanning the kitchen is 16 ft. long. (Photo by Bob Gere.)

# The roof system

More than any other factor, the arrangement of timbers in the roof determines whether a frame will be built with walls or bents. Indeed, frames are often defined by the type of roof they support. The roof is usually the most difficult aspect of the frame, and figuring the angles and the lengths of the pieces is almost always the most demanding mathematical calculation. In addition, the roof generally contains the most challenging joinery and can require sophisticated engineering. Therefore, we'll consider the structure of the frame from the top down.

There are four classic timber-frame roof types: principal and common rafters, principal and common rafters with purlins, common rafters, and principal rafters with purlins.

Deciding which to use might only be a matter of personal preference, but more commonly the choice involves structural factors, including the shape and pitch of the roof, loading, wood species, available timber length and the floor plan.

**The principal and common rafter roof** In this system, the principal rafters are directly in line with the principal posts so that their burden is directly transferred. Smaller common rafters are spaced evenly between. The rafters either connect to plates that ride on top of the outside wall posts or are interrupted by a principal purlin (or sometimes two) on each roof slope. Of course the rafters

The roof structure is usually the most challenging aspect of the frame, requiring difficult joinery and complex mathematical calculations. The turned post and collar ties form the center of a cross-shaped building.

are sized according to loads and spans, but it would not be unusual to use principal rafters that were 8x10 (or even larger) and common rafters as small as 4x6 spaced about 4 ft. apart.

This system makes the most sense to me when used with purlins. First, using purlins makes it possible to break up the span of the common rafters, so shorter (and more readily available) timbers can be used. Second, by spanning from bent to bent, the purlins absorb even increments of the roof load and transfer it toward the principal posts. Some of the burden carried by the common rafters is also transferred by the purlins to the principal posts, with the remainder bearing on the plates. In this type of roof, the purlins might well be the largest timbers because they are typically unsupported in their span between rafters.

The problem with a principal and common rafter roof with purlins is that it demands some tricky assembly, since the purlins have to be joined to the principal rafters before the common rafters can be installed. To position the purlins thus requires that a couple of brave (or foolhardy) individuals shinny up rafters that are not stiffened by any other members. Also, the lower common rafters can be hard to install because the joints to the purlin and the plate have to be engaged simultaneously. Still, the interplay of timbers is beautiful, and the system is without structural compromise.

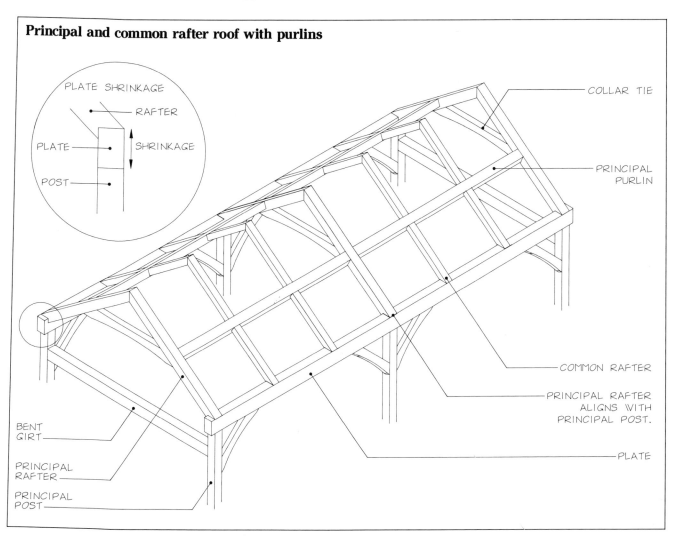

**Principal and common rafter roof with purlins**

**The common rafter roof**   This type of roof is easy to install, but has a different problem. Here the rafters are joined to the plates at the bottom and at the top either to a ridgepole running the length of the building, or simply to each other in pairs. Each rafter needs to be long enough and strong enough for the entire span. The rafters are usually spaced 3 ft. to 4 ft. apart, with no particular regard to the location of principal posts. The problem is that there is no interaction between the roof timbers and the timbers in the rest of the frame. With only one possible timber running perpendicular to the rafters (the ridgepole), it isn't easy to stiffen the roof in that direction, and the timber frame begins to act as two units rather than one. Beyond the collar tie, common rafter trusses don't usually contain additional timbers for reinforcement against spreading; therefore, there is a tendency for increased outward thrust at the plate. Because of these considerations, common rafter roofs are most suitable for small buildings.

**Common rafter roof**

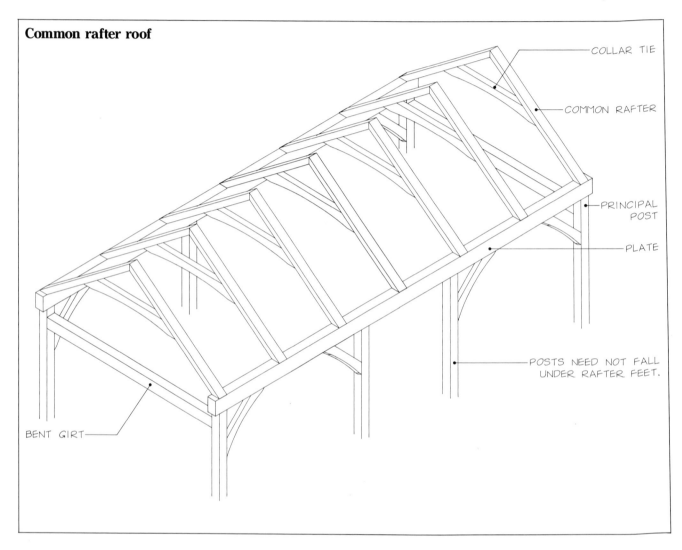

COLLAR TIE

COMMON RAFTER

PRINCIPAL POST

PLATE

POSTS NEED NOT FALL UNDER RAFTER FEET.

BENT GIRT

**The principal rafter and purlin roof** This last roof option is the most popular with our company. Here rafters join directly to the tops of principal posts and are connected to each other with a series of purlins spaced about 4 ft. apart. The system has many benefits, not the least of which is that it eliminates the need for top plates. It's difficult today to get timbers that are adequately long or stable for these plates, so they often have to be scarfed together. When the forests were filled with tall, virgin trees and timbers were hewn by hand, it was almost always possible to make plates to span the length and width of any building, and it would not be at all difficult to get any number of timbers of rafter length. With this kind of material available, and with so much handwork involved, the joiner's concern 200 years ago was not timber length but the number of timbers required. Because a common rafter roof uses many fewer pieces than a purlin roof, I suspect this was the main reason it was favored at that time. (A 36-ft.-long frame with 20-ft.-long rafters would require 20 common rafters on 4-ft. centers, but with purlins and principal rafters the frame would need about 33 members.)

We stay in business because we have a measure of practicality: Our timbers are purchased from a sawmill rather than hand-hewn. Our sawyer can easily handle purlins, which are usually no longer than 16 ft., but if we asked him for two 42-ft.-long plates, four 30-ft.-long bent girts and thirty-two 26-ft.-long rafters, he'd probably try not to laugh, but he would know for certain that we had given over completely to fantasy. If he could get timbers like that he wouldn't be able to saw them, and if he could get them and saw them, it's likely they would be of questionable quality.

Integrating the rafters into the bents allows the raising of large, complete units as well as greater possibilities in bent design. These bents have been assembled in preparation for a raising by crane.

So we use principal rafters and purlins, and benefit from the other advantages of the system. Because the rafters are part of the bents, we can raise large, complete units. (If there were a plate between posts and rafters, the bents, plates and roof would have to be assembled separately.) By extending interior posts to the roof, we can support splices in the rafters if they are needed to make up the complete length. Also, because the rafter ends join directly to the tops of the principal posts, the frame does not shrink in height. If you remember that wood shrinks considerably across the grain but very little along the grain,

you can see that the height of the building would change as the top plates shrank, as shown in the drawing on p. 31.

**Other roof timbers** Regardless of the roof system used, rafters generally require the assistance of other timbers both to bear loads and to resist spreading at the eaves. Roof load is calculated as the hypothetical weight of the materials plus the potential load of the snow or wind. The amount of outward thrust at the eave is a function of the load and the angle of the rafters. As the angle gets shallower, thrust increases. In a simple situation, with relative-

The principal rafter and purlin roof eliminates the need for top plates by joining the rafters directly to the posts.

ly short rafters and a small load, a collar tie is often sufficient to limit both sagging and spreading, although it is unlikely to do both simultaneously. The collar tie works in compression to reduce rafter sag and works in tension to bind the pair of rafters together and reduce their tendency to push outward on the walls. The burden of the load remains with the rafters, but the collar tie helps them support it. When principal rafters are used, a collar tie (or other system) is almost always necessary because the loads are so large on these few timbers. With common rafters, collar ties are sometimes used only with every other rafter pair. A horizontal timber joined between queen posts, called a straining beam, performs like a collar tie (see drawing, p. 23). Diagonal struts joined from a bent girt to the rafters limit outward thrust while allowing the central area between rafters to be kept clear of timbers for use as storage or living space.

As the width of a building increases and the rafter spans lengthen, it becomes more difficult for a collar tie to do either of its tasks effectively. To hold the rafter feet together, the collar tie must be low enough in the roofline to prevent outward thrust, but this requires that its length increase along with the width of the building. And the longer the collar tie, the more difficulty it will have supporting compression loads without buckling. The solution is to bring in additional timbers to support the rafters.

There are some relatively standard combinations of posts, struts, braces, girts and ties used to make the roof structurally stable. A king-post truss uses a central single post, as shown in the drawing and photo on p. 26; if it joined to a collar tie instead of to the rafter peak, it would be a crown post. The king post is usually connected to a bent girt, but might only span between the peak and a collar tie instead. The classic king post has diagonal braces joining to the rafters, creating a triangle that works to make the assembly function as a single roof truss. Since the lower horizontal beam of the truss (known as the bottom chord) is hung from the king post, there is no need for an intermediate support post.

Queen posts are similar to king posts, but they work in pairs. Typically, they rest on the bent girt and join either to a collar tie or directly to the rafters. Queen posts are commonly used when it is necessary to create living space under the central area between rafters. Because queen posts transfer the roof load directly to the bent girt, they must be placed carefully so the point loads they impose do not overburden the girt. It's best to position them as closely as possible to the supports beneath the bent girt. If the queen posts do not join to a collar tie, a straining beam between posts would serve the same purpose. Knee braces are used between the queen posts and the straining beam to support loads and stiffen the assembly.

A collar tie, a horizontal member used to bind rafters together, helps the rafters resist spreading, but can also serve to prevent them from sagging. These collar ties are in turn supported by pairs of queen posts.

In another type of roof assembly, elaborate bracing is used to create trusses that span the width of the building without interior supports or a horizontal beam connecting the walls. The hammer-beam truss and the scissors truss are two examples. The finest examples of these techniques were executed in the great cathedrals of Europe, and demonstrate how complex structure and intricate decoration can combine to turn a common craft into art of the highest order. The most notable example of a hammer beam, in Westminster Hall in England, spans 65 ft. and the mechanics of it have yet to be fully explained. Basically, tension and compression forces are passed through a series of timbers and resolved at the base of the rafters and in the outer wall without the need for a bottom chord or collar tie. The scissors truss was used as a way to frame a vaulted masonry ceiling between timber rafters.

Each of the arrangements we've just discussed is designed to keep the roof from sagging or spreading, to get the load from the rafters to the principal posts efficiently. It's that simple. But when the needs of the structure are put in the pot with aesthetic considerations, an endless variety of savory combinations can be served, and often the arrangement of timbers in the roof becomes the most pleasing and dramatic part of the frame. There's something exciting about the way the forces are pulled in by one timber and passed off to another—the interplay can't help but become a focal point of the frame. In other types of construction, the roof is an unattractive assembly of gussets, steel plates and nails that has to be hidden. Because of this, the space under the rafters is often either totally unused or used only as an attic. When this happens, you not only lose space that was figured into construction costs, but you also lose an area of high visual appeal. The interest created by the roof framing and the extra space under the eaves are two distinct attributes of a timber-frame house.

The hammer-beam truss was developed in the Middle Ages for use in cathedrals. In modern timber framing, the concept is revived for large, open living areas.

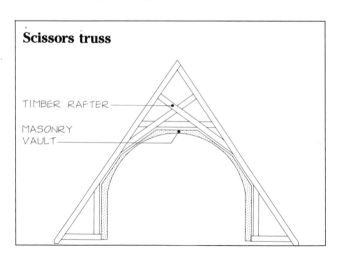

**Scissors truss**

TIMBER RAFTER

MASONRY VAULT

Very little space in a timber-frame house is wasted. Rooms under the roof often have interesting shapes and offer many decorating possibilities. (Photo by Brian E. Gulick.)

## The bent and wall systems

Stretching up to shoulder the load from the roof are the principal outside posts. These are tied to the beams of the wall and bent structures. A principle of timber-frame design is that the burden should get to these posts with the least possible effort. When the force is passed from the rafters to an intermediate post and then through a beam in order to get to its destination, it could be necessary to use a beam of considerable size to prevent bending at the point of transfer.

A better way to avoid overloading horizontal timbers is to design intermediate posts within the bent to reach from the deck to the rafters, as shown in the photo on p. 33. (You see this type of bent design a lot in barn construction, because it's an easy and practical way to gain great strength.) Intermediate posts also keep the bent from hinging at the rafters, something we've learned a great deal about in the raising of our frames. Lifting a bent from a horizontal to a vertical position applies stresses the bent may never see again, but it's a fair indication of the strength of the structure. There's a tendency to design a frame in layers simply because the floor plan is usually designed that way, yet when you look at the frame as a complete unit, it's obviously much stiffer if vertical members pass through the levels. In addition, the natural triangulation between the sloping rafters and the long interior posts enhances the frame's resistance to loads exerted against the gable ends and the walls. If you push against a side of a triangle, the adjacent side resists your force by acting in compression, making it nearly impossible to distort the shape. As a matter of fact, a triangular structure can't distort dramatically unless one of the connections fails. Push against a rectangle and the result is much different. The only inherent strength in a rectangle is at the corners, which cannot possibly bear up under a large force.

Structural triangulation is also the reason knee braces are so important in timber framing. Normally located at the upper ends of the posts, the braces connect diagonally to the joining beam and form small triangles at timber intersections. Under load, they work to keep the angles rigid and prevent the building from racking out of square. To ensure that the frame is braced in every direction, braces are always used in pairs on opposite ends of a beam. When one of the pair is in compression, the opposite brace is obliged to be in tension—the brace in compression is attempting to close and the opposite one to open. (See the drawing on p. 26.) The brace in compression has to crush wood to fail, but the brace in tension only needs to pull the joint apart, a relatively easy task. Therefore, while all the braces are necessary, in any given racking-load situation it's probable that the braces in compression are doing the bulk of the work.

Corner posts are the most in need of bracing and the rest of the outside posts are more important to brace than interior posts. Braces are usually made from 3x5s or 3x6s and are generally between 30 in. and 50 in. long. It's hard to get very scientific about brace sizing because there have been no studies on the subject relative to timber framing. The general rules we follow are these: The long side of the brace should be a little less than half the length of the post it braces; braces are always used in pairs; a pair of braces should be used for approximately every 10 ft. in height and no more than every 16 ft. in length around the perimeter of the building. Be much more wary of under-bracing than over-bracing. These are not hard and fast rules, but we try to maintain a conservative stance. After the rigidity of a frame is ensured, the location and size of the braces can become an aesthetic decision.

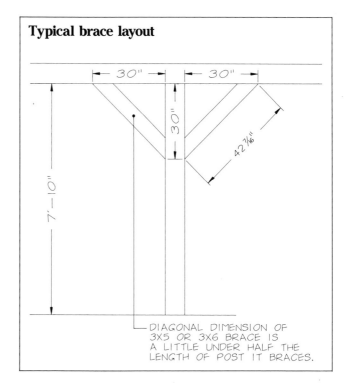

**Typical brace layout**

DIAGONAL DIMENSION OF 3X5 OR 3X6 BRACE IS A LITTLE UNDER HALF THE LENGTH OF POST IT BRACES.

If long, reversed braces can be used without interfering with the other needs of the building, a strong case can be made for joining the brace bottoms to a horizontal timber and their tops to a post. The first brace to encounter the racking load therefore responds immediately with that very effective compressive resistance. (The typical situation, shown in the drawing on p. 26, requires that the first brace to receive the force should work in tension, which is not by any means its forte.) Still, you can't just brace the bottom ends of the posts, leaving the upper 4 ft. or 5 ft. unsupported, and normally it isn't practical to have a 10-ft.-long brace consuming the wall surface. But it makes sense to look for opportunities to use reversed braces, since their efficiency and strength can make it possible to reduce the total number of braces required.

Though diagonal braces primarily oppose lateral forces on the frame, they also help the frame bear vertical loads. The normal floor loads in a frame are first carried by the horizontal members—the joists, girts and summer beams. We determine the dimension of these timbers by first calculating the pounds per square foot to be borne by the timber using figures given to us by the local building codes. By calculating sizing formulas and evaluating the span of the timbers and the way in which load is distributed, we can assign a structural contribution to every timber in the frame once the wood species has been determined. If a beam were to fail, it would probably sag greatly first, then break. We dislike breaking, so we exercise careful control over deflection. But as the span of a timber increases and/or the imposed load becomes heavier or more centralized, to control bending we must generally either reduce the span or use a bigger timber. If knee braces help support the span, however, it should be possible to choose a timber that would otherwise be borderline. Braces channel the vertical load from beam to post, changing the reaction to the forces from bending to compression. And since wood is much more resistant to compression than bending, that's the good news.

The bad news is that although performance and common sense lead us to believe that we understand how diagonal braces work, when engineering the frame we are still not able to assign quantitative values to their load-bearing function. There just isn't enough data yet to put the effectiveness of a brace into the formula. And so, until more testing and analysis can be done, we calculate the size of the timbers as though the braces weren't there. The result is that frames have another built-in guarantee of durability.

**Reversed braces give a compressive resistance to racking loads. Because of the placement of the intermediate connecting girt, these braces can be relatively short; if the girt weren't there, the braces would need to stretch from the top of the posts to the floor, making them impossibly long.**

## The floor system

The timbers in the floor support the loads applied in the spaces between the bent and wall beams. You can therefore look at each space as needing an individual floor system, with the following considerations bearing upon the arrangement and size of the timbers: the distance to be spanned, the loads to be carried, the structural capacity of the joining timbers, and the design goal of laying out the floor plan with a positive relationship to the timbers.

A major requirement of a floor is that it be stiff. Each timber should support its intended load with minimum acceptable deflection. A bouncy floor is an insult to the rest of the frame, and comparable to putting a Chevette engine in a BMW. To avoid bounciness, it is important to make sure that floors are not designed with extreme spans. A study of old barns in England shows that 16 ft. is the most common distance between bents; the study conjectures that the reason for this is to allow two yoked oxen to walk side by side in the bay (the space between bents). After a number of years in the framing business, I am convinced the real reason is that 16 ft. is the maximum practical span for a timber carrying a floor load. Beyond this, the timbers would have to become very large to limit deflection.

There are several common ways to arrange the timbers in a floor, as shown in the drawing below. Using parallel joists spanning between bents or walls is probably the simplest approach. The joists split up the floor load evenly and, at the same time, provide an understated look. But when the loads and spans demand that the joists be excessively large (anything above 6x8, in my view), timbers arranged this way can also seem overbearing. It is then usually more desirable to focus the loads onto a few large timbers. Large spanning timbers such as summer beams are especially useful if they join to posts, thus directly transferring the floor loads. Like other major timbers, these beams can also be used to define living areas. For most situations, we like the interplay between large and small timbers and actually work to divide the floor with large timbers into visually manageable subdivisions. The large spanning timbers often are embellished to give the frame some formality and refinement.

In addition to their load-bearing function, the timbers of the floor actually tie the frame together. If the distance from outside wall to outside wall is more than the length of a single timber (which, of course, it almost always is), there needs to be additional connection to keep the bents from spreading. When good joinery is used, floor joists or summer beams serve this purpose adequately, providing continuous support.

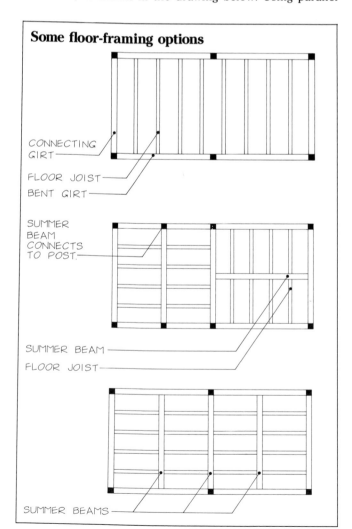

**Some floor-framing options**

CONNECTING GIRT

FLOOR JOIST

BENT GIRT

SUMMER BEAM CONNECTS TO POST.

SUMMER BEAM

FLOOR JOIST

SUMMER BEAMS

**Depending on the size of the timbers, parallel floor joists can be either subtle or overbearing. These, made from fir, affect the feeling of the room without being dominant.**

Good joinery is the basis of
fine timber framing. Here the
continuous post is flared on
two levels to receive beams for
the second and third floors.

# The Collected Writings of Wallace Thurman

## A HARLEM RENAISSANCE READER

Edited by Amritjit Singh
and Daniel M. Scott III

RUTGERS UNIVERSITY PRESS
New Brunswick, New Jersey, and London

Library of Congress Cataloging-in-Publication Data

Thurman, Wallace, 1902–1934.
   [Selections. 2003]
   The collected writings of Wallace Thurman : a Harlem Renaissance reader /
edited by Amritjit Singh and Daniel M. Scott III.
     p.   cm.
Includes bibliographical references and index.
   ISBN 0-8135-3300-7 (cloth : alk. paper)  —  ISBN 0-8135-3301-5 (pbk. : alk. paper)
   1. African Americans—Literary collections.  2. Harlem Renaissance.   I. Singh, Amritjit.
II. Scott, Daniel M., 1960–     III.  Title.
   PS3539.H957  A6  2003
   813'.52—dc21
                                         2002152302

British Cataloging-in-Publication information is available from the British Library.

This collection copyright © 2003 by Rutgers, The State University

Introduction copyright © 2003 by Amritjit Singh

For all other permissions, see acknowledgments.

All rights reserved

No part of this book may be reproduced or utilized in any form or by any means, electronic or mechanical, or by any information storage and retrieval system, without written permission from the publisher. Please contact Rutgers University Press, 100 Joyce Kilmer Avenue, Piscataway, NJ 08854–8099. The only exception to this prohibition is "fair use" as defined by U.S. copyright law.

The publication program of Rutgers University is supported by the Board of Governors of Rutgers, The State University of New Jersey.

Manufactured in the United States of America

His ruminations helped to clarify his own mind. His future method of procedure became more clearly outlined. It would be his religious duty to ferret deeply into himself—deeply into his race, isolating the elements of universality, probing, peering, stripping all in the interests of garnering literary material to be presented truthfully, fearlessly, uncompromisingly.

—Wallace Thurman, "Notes on a Stepchild," c. 1929

Thurman is distinctly a has been—so many people have already buried him. Woe betide 'em when I am resurrected!

—Wallace Thurman to Langston Hughes, c. November 1934

# Contents

# Preface

The need for *The Collected Writings of Wallace Thurman: A Harlem Renaissance Reader* has been felt and expressed for years—and at least one of us has been involved in conversations about putting such a volume together for over a decade. We are happy finally to have brought to fruition a project that we expect will fill a vacuum in Harlem Renaissance literature and scholarship. In preparing the volume, we have been aware of the needs of several audiences it might attract. By providing the complete texts of previously published but uncollected writings as well as the unpublished manuscripts and letters of Wallace Thurman, we intend this *Reader* to serve as a classroom text. The wide range of Thurman's genres and perspectives makes him a logical choice for inclusion in seminar courses on the Harlem Renaissance, or the American 1920s, as well as in courses in American literature, African American studies, and cultural studies. This *Reader* could also provide the basis for an "individual author" seminar on Thurman at either the undergraduate or the graduate level. The general reader interested in culture, literature, theater, and biography will find the volume attractive because of Thurman's centrality to the Harlem Renaissance and the energetic role he played as a social and cultural critic from 1925 to 1934. With a substantive biographical and critical introduction, which contextualizes Thurman in relation to the ever changing scholarship on the Harlem Renaissance, and our carefully researched introductory notes for each of the book's sections, the *Reader* can also make a modest claim as new scholarship.

The *Reader* has been divided into eight parts to highlight the many genres and styles Thurman practiced as he courageously pursued controversial subjects throughout his short and brilliant career. "Part One—Essays on Harlem" comprises the insider's Harlem, a fascinating sociocultural view of Harlem from one of the major players in the Harlem Renaissance. The essays also emphasize important recurring themes in Thurman's writings: the diversity of black America, the consideration of its place in relation to white America, and the complications of African American identity.

"Part Two—Social Essays and Journalism" shows Thurman in his characteristic role of social commentator: sometimes satirical, sometimes funny, sometimes educational, but always insightful. He comments, among other things, on his home state of Utah, U.S. immigration policy, Christmas, and health care in urban hospitals.

Thurman was a delightful correspondent. "Part Three—Correspondence" includes more than one hundred letters that show Thurman in a variety of moods and stances—

incisive, tenacious, ambitious, humorous, sarcastic, despairing, and scathing—as he corresponds with friends such as Dorothy West, Harold Jackman, and Claude McKay, as well as with three other contemporaries—W.E.B. Du Bois, Alain Locke, and Granville Hicks. But this section is most valuable for the light it throws on Thurman's long-standing friendships with fellow iconoclast Langston Hughes and with his white collaborator, William Jourdan Rapp.

"Part Four—Literary Essays and Reviews" brings together literary essays and book reviews originally published in the black press or in periodicals such as *The Bookman, The Independent,* and *The New Republic,* as well as in *Fire!!* and *Harlem: A Forum of Negro Life*—two short-lived magazines Thurman edited. This section also includes an unpublished review by Thurman of his own novel, *Infants of the Spring.*

"Part Five—*Aunt Hagar's Children*" is the full extant text of Thurman's unpublished collection of essays on black American life and some of its most interesting and important personalities. In this collection, modeled on H. L. Mencken's style and approach, Thurman offers an unconventional view not only of historical figures such as Frederick Douglass, Booker T. Washington, and Marcus Garvey but also of himself in an intriguing autobiographical piece, "Notes on a Stepchild."

"Part Six—Poems and Short Stories" includes Thurman's two short stories, "Grist in the Mill" and "Cordelia the Crude," as well as seven of his poems, most of them published here for the first time. While the two stories have been reprinted a couple of times since the 1970s, Thurman's poems are virtually unknown.

In "Part Seven—Plays," we publish for the first time the complete texts of *Harlem: A Melodrama of Negro Life in Harlem* and *Jeremiah the Magnificent,* two full-length plays Thurman wrote with his white friend William Jourdan Rapp. *Harlem* ran on Broadway for an impressive ninety-three nights in 1929, the first play by an African American to have had such a huge success. The section also includes a selection of articles Rapp and Thurman wrote as publicity pieces for *Harlem.*

"Part Eight—Excerpts from the Novels" comprises selections from Thurman's well-known novels, *The Blacker the Berry* and *Infants of the Spring,* as well as from his third novel, *The Interne,* written with a white associate, Abraham L. Furman. The three novels, stylistically quite different from one another, share their exploration of controversial sociocultural issues.

At the time of his death on December 22, 1934, many people described Wallace Thurman as the heart and soul of the Harlem Renaissance. Years later, Dorothy West would proclaim Thurman "the most symbolic figure" of the New Negro Renaissance. In bringing together Thurman's writings for the first time in one volume, this *Reader* opens up the possibility of developing new perspectives on Thurman's multifaceted career and on the Harlem Renaissance in general.

# Acknowledgments

No project of this scope can be completed without the generous support of friends, colleagues, and institutions. We have incurred many debts during the nearly four years of our research and writing.

We record with gratitude the support of many at Rhode Island College, where we both work: the Department of English, especially Joan Dagle, our chair, and the friendly secretaries, Natalie DiRissio and Tessa DaLuz; Richard R. Weiner, dean of Arts and Sciences, for his encouragement; John Nazarian and John Salesses, president and vice president respectively, for approving study leave for Singh to accept a Fulbright Professorship in Berlin during the spring and summer of 2002; and the Faculty Research Committee for liberal grants in two successive years that paid for our research travel to New Haven, New York, New Orleans, Los Angeles, and Washington, D.C. At the Adams Library, we must acknowledge Myra Blank of the Interlibrary Loan Office for her ever ready assistance in tracing and procuring obscure items, as well as Marlene Lopes, Rachel Carpenter, Patricia Brennan, and Carla Weiss for sundry help and advice. Kristen Gagne, Jennifer Wickert, Sonia Chaudhry, and Nicole Wood helped as research assistants in the early stages of the project. Our friend Susan Soltys—who, surrounded by anthropologists, knows well the importance of cultural preservation and textual recovery—helped most generously with the scanning of manuscripts. But it was our graduate assistant Heather Hewitt's eagle-eyed proofreading of those scanned materials and many other kinds of unstinting help that saw us to the finish line. The students in two graduate seminars on twentieth-century U.S. literature taught by Singh during the summer and fall of 2001 helped shape some of our thinking through their debates and writing on Thurman's career.

Our experience at all the research collections we visited or contacted confirms our faith in libraries and librarians. We give special thanks for the perseverance and helpfulness of all the staff at the Beinecke Library at Yale University in allowing us full and friendly access to the James Weldon Johnson Collection and assisting us in every way possible. We are particularly grateful to Patricia C. Willis, curator of American Literature, for answering our queries on a number of occasions. At the Schomburg Center for Research in Black Culture, New York Public Libraries, we must acknowledge the many courtesies extended by Director Howard Dodson, Diana Lachatanere, and Genette McLaurin. We also wish to thank Sylvia McDowell and the staff at the Schlesinger

Library, the Radcliffe Institute for Advanced Study, Harvard University, as well as Sean D. Noel and Nathaniel Parks at Special Collections, Boston University Libraries. At UCLA Libraries, we received kind assistance from Joshua Hirsch, Itibari Zulu, Miki Goral, Octavio Olvera, and many others. Still others who were prompt with help and permissions are: Linda Long at Special Collections, University of Oregon Library; Tara Wenger at the Harry Ransom Humanities Research Center, the University of Texas at Austin; Diane Cooter and Carolyn A. Davis at Department of Special Collections, Syracuse University Library; Joellen ElBashir, curator, Moorland-Spingarn Research Center, Howard University; and Stephen Robinson, Special Collections and Archives, W.E.B. Du Bois Library, University of Massachusetts at Amherst. Other librarians who were quick with suggestions and advice when we needed them most include: David A. Hales, reference librarian, Westminster College, Salt Lake City; Donald E. DeVore, executive director, Amistad Research Center, Tulane University; Kathleen DiPerno at the Tisch Library, Tufts University; Ned Comstock at the Cinema and Media Library, University of Southern California; Frank Orser at the University of Florida Libraries; and Randall Burkett at the Emory University Libraries.

Thanks to Reiner Rohr of the Fulbright Commission in Berlin and Richard Pettit of CIES (Council for International Exchange of Scholars) in Washington, D.C., for facilitating Singh's affiliation with the University of Leipzig and the JFK Institute of North American Studies at Freie University–Berlin. The six-month stay in Berlin allowed time for reflection, writing, and rewriting of the *Reader*'s extensive critical and editorial apparatus. For hospitality and/or assistance of various kinds, Singh would like to thank Winfried Fluck, Heinz Ickstadt, Anne Koenen, Helmut Keil, Rüdiger Kunow, Thomas Claviez, Silke Munsky, Regina Wenzel, Christine Bach, Fabian Hentschel, and Johannes Voelz. The staff members at the fine JFK Institute Library were unfailingly helpful and courteous.

Many friends read all or parts of the introduction and offered useful comments: Anupama Arora, Jay Birjepatil, William Boelhower, Robert E. Hogan, Wolfgang Karrer, Sieglinde Lemke, Günter Lenz, David and Judy Ray, and Samir Singh. Thomas Wirth and Jon Woodson answered questions regarding Thurman's relationship to Bruce Nugent and Gurdjieff's teachings respectively. Zubeda Jalalzai, Modhumita Roy, and Maryemma Graham allowed opportunities to share our ongoing research with audiences at Rhode Island College, Tufts University, and at the Langston Hughes Birth Centennial Conference in Lawrence, Kansas, respectively.

So many friends, family members, and colleagues have been supportive in small and big ways that it is likely this list is less than complete: Gurpreet S. Ahuja and Jasjit Singh, Elizabeth Ammons, Crystal Lynn Bartolovich, John H. Bracey, C. Lok Chua, Sarah Cooper, Talia Danesi, Esha Nyogi De, Gaurav Desai, Gerald Gill, Jitender Gill, Bruce G. Johnson, Ketu Katrak, Gurpreet Kaur, Ulrike Kistner, Judith Kroll, Karen Leonard, Joseph

Litvak, Nellie McKay, William Mulder, Melvyn New, Lawrence Potter, Wesley Quan, Arnold Rampersad, Maureen Reddy, Wilfred Samuels, Barbara Schapiro, Christina Sharpe, Barbara Silliman, Reshma Singh and Jugraj Parmar, Steven Tracy, Cheryl Wall, and Richard Yarborough.

At Rutgers University Press, Leslie Mitchner, Marilyn Campbell, and their colleagues have shown remarkable patience as this project has taken its own time through various stages to discover its shape and character. We owe them thanks for their abiding faith in the project and in our ability to deliver. We must also thank them for their choice of Robin DuBlanc as our copyeditor. We could not have asked for a more careful reader and editor.

The book is dedicated with love to our partners, Prem Singh and James E. Schlageter.

For permission to publish the bulk of the material that makes up this *Reader*—letters to Hughes, Rapp, and Jackman; poems; essays and reviews; and the unpublished manuscripts of *Aunt Hagar's Children, Harlem: A Melodrama of Negro Life in Harlem,* and *Jeremiah the Magnificent*—we thank the James Weldon Johnson Collection, Beinecke Rare Book and Manuscript Library, Yale University.

For permission to publish the letters from Thurman to W.E.B. Du Bois, Alain Locke, and Granville Hicks, respectively, we thank the following libraries and their staff: Special Collections and Archives at the W.E.B. Du Bois Library, University of Massachusetts–Amherst; Moorland-Spingarn Research Center, Howard University; and Granville Hicks Papers, Department of Special Collections, Syracuse University Library.

For permission to publish the three letters from Thurman to Dorothy West written in 1929, we thank the Dorothy West Collection at the Schlesinger Library, Radcliffe Institute for Advanced Study, Harvard University. For permission to publish the September 1934 letter from Thurman to Dorothy West, we thank the Dorothy West Collection, Special Collections, Boston University Libraries.

For permission to publish Thurman's own unpublished review of his novel *Infants of the Spring,* we thank the Harry Ransom Humanities Research Center, the University of Texas at Austin.

# Chronology

| | |
|---|---|
| August 16, 1902 | Henry Wallace Thurman is born to Beulah and Oscar in Salt Lake City, Utah. |
| 1910–1914 | Lives with mother, moving from Salt Lake City to Boise, Idaho (1908), Chicago (1910–1914), and Omaha, Nebraska (1914), where he finishes grammar school. |
| 1914–1918 | Returns to Salt Lake City; because of heart problems, relocates in winter 1918 to Pasadena, California, where he falls victim to the flu epidemic. |
| 1919 | Returns to Utah; graduates from West Salt Lake High School. |
| January–June 1920 | Takes courses as a premedical student at the University of Utah. |
| 1920–1922 | Works at Hotel Utah. |
| Summer 1922 | Moves to Los Angeles. |
| September 1922 | Takes courses for one semester in the School of Journalism, University of Southern California; works at U.S. Post Office; meets Arna Bontemps and Fay Jackson; is appointed associate editor of *The Pacific Defender,* for which he writes a regular column, "Inklings." |
| September 1924 | Founds and edits *The Outlet,* a literary magazine; writes to W.E.B. Du Bois notifying him of the literary venture. |
| March 1925 | *The Outlet* ceases publication. |
| September 1925 | Arrives in New York City on Labor Day; works without pay at Theophilus Lewis's *The Looking Glass.* |
| December 1925 | Begins work at *The Messenger* as short-term replacement for George Schuyler as managing editor; publishes numerous reviews in the periodical. |
| January–May 1926 | Attends Jean Toomer's Harlem study group on Gurdjieff's teachings. |
| Summer 1926 | Lives with Langston Hughes at 314 West 138th Street. |
| August 1926 | Carl Van Vechten's immensely controversial *Nigger Heaven* is published; Thurman is among those defending the author's artistic freedom. |
| November 1926 | Starts working as circulation manager for *The World Tomorrow*; moves to 267 West 136th Street ("Niggeratti Manor" in *Infants of the Spring*); edits and publishes *Fire!!* in collaboration with Langston Hughes, Zora Neale Hurston, Aaron Douglas, John P. Davis, Richard Bruce Nugent, and Gwendolyn Bennett. |

| | |
|---|---|
| 1926 | Writes reviews for *The Messenger* and *Fire!!* |
| Early 1927 | Meets William Jourdan Rapp, his collaborator on the plays *Harlem* and *Jeremiah the Magnificent.* |
| August 1927 | "Negro Artists and the Negro" is published in *The New Republic.* |
| September 1927 | "Nephews of Uncle Remus" is published in *The Independent.* |
| October 1927 | Gets job as an extra—along with Dorothy West and Richard Bruce Nugent—in DuBose and Dorothy Heyward's *Porgy.* |
| Spring 1928 | Thurman's "Negro Life in New York's Harlem: A Lively Picture of a Popular and Interesting Section" is issued as Number 494 of the Haldeman-Julius Little Blue Book series. |
| July 1928 | "Negro Poets and Their Poetry" is published in *The Bookman.* |
| August 1928 | Marries Louise Thompson; they live at 90 Edgecombe Avenue. |
| November 1928 | Edits the first (and only) issue of *Harlem: A Forum of Negro Life.* |
| Late 1928 | Joins editorial staff at MacFadden Publishers; writes stories for *True Story* magazine under the pseudonyms Ethel Belle Mandrake and Patrick Casey. |
| February 1929 | *The Blacker the Berry* is published; *Harlem: A Melodrama of Negro Life in Harlem* opens at the Apollo Theatre On Forty-second Street, New York. |
| April–September 1929 | Lives in Salt Lake City and Los Angeles; writes many letters to Rapp and Hughes; begins the never-completed process of divorcing Thompson; finishes work on his collection of essays, *Aunt Hagar's Children.* |
| | Receives financial assistance from Elisabeth Marbury to work on *Infants of the Spring.* |
| July 1930 | "Langston Hughes Turns from Poetry to the Novel to Give a Sincere Picture of Negro Life in America" is published in *The New York Evening Post.* |
| | Spends much of his time at the Jamaica, Long Island, home of Theophilus Lewis with only occasional trips to Harlem; rewrites *Infants of the Spring.* |
| 1932 | *Infants of the Spring* is published. |
| September 1932 | Begins work at the Macaulay Publishing Company as reader (later serves as editor in chief). |
| Late 1932 | *The Interne* is published. |
| December 1933 | *Jeremiah the Magnificent* is staged for one performance. |
| February 1934 | Travels to Hollywood to write screenplays for Bryan Foy Productions: *Tomorrow's Children* and *High School Girl.* |
| July 1934 | Returns from California and is hospitalized for acute tuberculosis at City Hospital on Welfare Island. |
| December 22, 1934 | Dies; his death is seen by many as the end of an era. |

# A Note on the Text

In 1929, Thurman wrote to his friend Langston Hughes: "Lapses in grammar are certainly to be deplored, but like you, school grammar had no effect upon me whatsoever. Consequently I admit marring what might be good by frequent and really inexcusable gaucheries. . . . I'm hell on bluepenciling the work of others, but much too soft with myself. Auto-intoxication I suppose."

In transcribing manuscripts and letters from both holograph and typescript, all care has been taken to achieve a faithfulness to the original in every possible way. We have silently corrected all typographical errors as well as expanded ampersands and other abbreviations. We have also standardized all spelling to American practice—Thurman's eccentricities as a writer included his occasional fondness for British spelling—and quietly changed "negro" to "Negro" whenever we encountered that word with a lowercase "n." We have chosen, however, to respect Thurman's fondness for odd coinages such as "inspirative," "urgence," and "analyzation" without cluttering the text with "sic." The ellipses appearing throughout Thurman's works are stylistic ones included in the originals; they do not indicate editorial deletions.

The individual works within each part are arranged in chronological order. Where the date is the same, they are alphabetized by title. For the previously published pieces included in Parts One, Two, and Four, we have reproduced the original published texts, noting overlap and recycled material wherever possible. Most of the unpublished works—for example, *Aunt Hagar's Children* in Part Five and the two plays in Part Seven—were available in typescript. The following unpublished pieces have been transcribed from handwritten manuscripts: "The Bump (A Dance They Do in Harlem)" and "Harlem House Rent Parties" in Part One; "Autobiographical Statement" and "Description of a Male Tuberculosis Ward" in Part Two; the articles "My Collaborator," "Casting and Directing *Harlem*," and "The Writing of *Harlem*" in Part Seven.

For *Harlem: A Melodrama of Negro Life in Harlem*, we have chosen to follow the original typescript as prepared by Thurman and Rapp, ignoring the changes made in the director's hand for the Broadway production.

# THE COLLECTED WRITINGS OF WALLACE THURMAN

# Wallace Thurman
# and the Harlem Renaissance

AMRITJIT SINGH

But, say I, gladly would I urge the Negro masses to take an active part in the revolution, just to see them, for one moment emerge from their innate sluggishness, massacre their ministers, and perhaps, in the interim, give birth to a few exceptional individuals capable of arising above the mob, Communism, Christianity, and all other such doctrines to become master intellects and creative giants.

—Wallace Thurman, "The Coming Revolution," in *Aunt Hagar's Children*

Perhaps it is too early to attempt an appraisal of [Thurman's] work in various fields of art. I am ready to give out only one unqualified opinion. He wrote lousy poetry. On second thought, I hazard another opinion without reservation. As editor, novelist, and playwright, he was the most versatile of contemporary Aframerican literary men.

—Theophilus Lewis, "Harlem Sketchbook," January 1935

Acclaimed by both his Harlem Renaissance peers and later historians as the most brilliant and radical of the younger writers of the period, Wallace Thurman is known
today largely through a few magazine pieces and two of his novels, *The Blacker the
Berry* (1929) and *Infants of the Spring* (1932). But he wrote with savvy and success in
almost all genres throughout his short life, often with a fury that was possibly ignited
by a premonition of early death. Beyond the three or four journal articles frequently
mentioned in the endnotes to well-known studies of the Harlem Renaissance, Thurman wrote poems and short stories as well as many more essays and book reviews—
including an unpublished essay collection, *Aunt Hagar's Children*. Thurman was a
delightful correspondent who wrote regularly to close friends and collaborators such
as Langston Hughes and William Jourdan Rapp, among many others. With Rapp, he
coauthored two full-length plays, one of which, *Harlem,* had a successful run on Broadway in 1929. Students of American literature have yet to discover the centrality of his
role in the Harlem Renaissance as a shaper of aesthetic ideas and opinions on controversial subjects.

Thurman, still a neglected figure of the U.S. 1920s, was in many ways very much a
product of the Jazz Age and in other ways well ahead of his time. As a critic, he chal-
lenged the provincialism of all Americans, the double standards and patronizing attitudes with which most whites approached African American art, and the black
bourgeoisie's obsessions with respectability and uplift. It is hard to imagine that Thurman was not aware that—more than anyone else among his peers, including Hughes—
he was claiming for himself a voice and a role not only as a literary critic and theorist
of black writing but also as what we would today call a public intellectual.[1] In *Infants,*
in his essays and reviews, in his work as an editor and journalist, and often in his correspondence, Thurman expected both his black and white contemporaries to confront
the contradictions of "race" as they corrupted art and its evaluation.

Thurman was a provocateur par excellence, inspiring and shaping an assertive
artistic independence among the younger Negro artists that found its clearest expression in Hughes's 1926 essay, "The Negro Artist and the Racial Mountain." Thurman
was an active leader among such peers as Hughes, Zora Neale Hurston, Gwendolyn
Bennett, John P. Davis, Helene Johnson, Richard Bruce Nugent, Eric Walrond, and
Aaron Douglas in their resistance to the prescriptive and overbearing demands on
their art from both whites and blacks, including powerful players like W.E.B. Du Bois
and Alain Locke. The younger writers championed their cause through short-lived
but ambitious small magazines such as *Fire!!* and *Harlem. Fire!!* was especially intended
to provide an unfettered forum for experimental writing by the young writers as a
way of reducing their dependence upon the black press and its narrow aesthetic
preferences.[2]

## Wallace Thurman, Child and Man:
### "Seeking, Observing, Suffering"

A brief recitation of the known biographical facts about Thurman does little to explain the man's complex personality, many elements of which appear to have been shaped by his experiences and family circumstances during childhood and formative years.[3] Thurman was born in Salt Lake City, Utah, on August 16, 1902—the same year as Hughes and Arna Bontemps.[4] Details about how his ancestors ended up in Utah remain murky, but it is known that his maternal grandmother, Emma Jackson ("Ma Jack"), had been in Salt Lake City long enough to have become an active and important member of the Calvary Baptist Church from its inception in 1896. Ma Jack, to whom Thurman dedicated his first novel, was in many ways the most important person in his life. Thurman's father, Oscar, abandoned his family when Wallace was less than a month old.[5] His mother, Beulah—"the goose who laid the not-so-golden-egg" to whom Thurman later dedicated *Infants*—subsequently married and divorced a series of men, changing residences frequently. Ma Jack kept the doors of her home open to both Thurman and Beulah during and between the latter's successive marriages. Even though it occasionally doubled as a "saloon" where alcohol was served without a license, Ma Jack's home must have provided continuity and stability for the young "Wally." Indeed, Thurman remained close all his life to his grandmother; it was she who sent out the notices of his death in early 1935.

Thurman also experienced intermittent periods of poor health throughout his childhood, compounding the problems caused by family instability. According to Thurman's account in "Autobiographical Statement" (Part Two), he started his schooling in Boise, Idaho, at the age of six. Within two months, he was taken ill and became a "pampered invalid" for the following two years, during which he returned to Salt Lake City. He was in Chicago from 1910 to 1914 but finished grammar school in Omaha, Nebraska. "Persistent heart attacks" made it necessary for him to spend the winter of 1918 in the "lower altitude" of Pasadena, California, where, Thurman tells us, he became a victim of the "flu" epidemic. Upon convalescing, he returned to Salt Lake City, somehow managing to finish high school. He matriculated as a premed student at the University of Utah; during his two years there, he had a nervous breakdown, spending one summer in Omaha. Moving to Los Angeles in 1922, he took journalism courses for a while at the University of Southern California and also worked as a postal clerk for three years. His efforts to organize a literary community in Los Angeles were unsuccessful, but he wrote a regular column, "Inklings," for the black newspaper *The Pacific Defender* and for about six months during 1924–1925 edited a little magazine of his own, *The Outlet,* in which he published his own work and that of others.[6]

Lonely as a child and teen, Thurman learned to lean heavily on books and films to entertain himself. In "Notes on a Stepchild" (Part Five), he claims that he wrote his first novel at the age of ten. Having become a "rabid movie fan" by age twelve, he rewrote "the contemporary serial thrillers," and as a college student he spent "many hours composing poems about gypsies, hell, heaven, love and suicide." But his reading was even more intense and erratic than his writing. He read voraciously in Plato, Aristotle, da Vinci, Goethe, Shakespeare, Schopenhauer, Keats, Spinoza, Shelley, Flaubert, Baudelaire, Stendhal, Spencer, Ibsen, Havelock Ellis, Anatole France, Freud, Dostoyevsky, and Mencken, among others. After a "fruitless effort to forge that which he had learned into a harmonious whole, . . . he remained a bewildered eclectic. . . . Stability of thought eluded him. His poetry was tortured and verbose. Mental chaos promised insanity. It was then that he . . . flung himself into the maelstrom of the living present, exploring all provinces of life, communing with every type of person, seeking, observing, suffering."

Thurman made his "hectic hegira to Harlem" in 1925 at the age of twenty-three, arriving there on Labor Day. In Harlem, he formed a circle of friends very quickly, moving almost seamlessly through many jobs and professional roles. Between 1925 and 1929, in addition to writing for the black press, he published articles and book reviews in *The New Republic, The Bookman, The Independent, American Monthly, The Greenwich Village Quill, The New York Evening Post, The Dance Magazine, The New York Times,* and *The New York World*. Serving successively between 1925 and 1927 on the staff of *The Looking Glass, The Messenger,* and *The World Tomorrow,* he landed in 1928 an editorial job at McFadden, publishers of over twenty magazines, including *True Story,* with a combined circulation of 16 million. In September 1932, he joined the Macaulay Company, serving for some time as editor in chief—the only black person during the period to hold such a position.

Thurman's successive quarters in Harlem—one of which was made famous as the "Niggeratti Manor" in *Infants*—were centers of activity and debate for most black young artists and for their friends of diverse backgrounds and sexual orientations. In her affectionate 1970 memoir, "Elephant's Dance," Dorothy West recalled Thurman as "a slight man, nearly Black boy, with the most agreeable smile in Harlem and a rich infectious laugh. His voice was without accent, deep and resonant; it was the most memorable thing about him welling out of his frail body and wasting its richness in unprintable recountings" (77).[7] Thurman supported his young male and female colleagues by publishing their work in the magazines he worked for and by being collegial and helpful in myriad other ways—including getting them, as West tells us, a free meal every now and then from an "unsuspecting downtowner, . . . [who] would expand with simple kindness while Thurman and his coterie went into peals of laughter which the benefactor imagined were peculiar to Negroes after a hearty meal" (79).

Throughout the peak years of the Harlem Renaissance, Thurman lived a self-described "erotic, bohemian" lifestyle. And while he denied his homosexuality in his letters to Rapp, his close friends and roommates knew him to be bisexual. In August 1928, Thurman surprised his friends by marrying a woman friend, Louise Thompson, in what was known in gay circles as a "companionate" relationship, a marriage of convenience. Within a year divorce proceedings were begun.

The year 1929 was perhaps the height of Thurman's professional success in his lifetime: not only did *Blacker*, his first novel, appear, the play *Harlem*, coauthored by Thurman with Rapp, was a major Broadway success. In early 1934, he moved to Hollywood and wrote two screenplays for Bryan Foy's films *High School Girl* and *Tomorrow's Children*. These two films—whose radical treatment of unusual and tabooed subjects resonates with Thurman's earlier writings—allowed him to explore class conflicts in U.S. society. These scripts dealt with white characters, as had his 1932 novel, *The Interne,* a fictional exposé of the appalling conditions in the City Hospital on Welfare Island. Ironically, diagnosed with incurable tuberculosis, Thurman spent the last six months of his life in that very hospital before his death on December 22, 1934. He was thirty-two.

Thurman's short life was characterized by a desperation—amounting at times to despair—to achieve the high artistic standards he had set for himself. In the "maelstrom" of his life and work, "seeking, observing, suffering" in an unmistakably Whitmanic tenor, Thurman attempted to shape the chaos into order in his essays, fiction, and drama. Fed by a desire for literary fame that at times expressed itself as a desperate sense of alienation, Thurman's life was, on the page and in fact, a continual struggle.

In pursuing his literary ideals, he spared no one—not even himself or his Harlem friends—as is evident in this rather exuberant and sweeping judgment from a 1929 letter to Hughes:

Found [McKay's] *Banjo* turgid and tiresome. *Passing* [by Nella Larsen] possessed of the same faults as [her] *Quicksand. Rope and Faggot* [by Walter White] is good for library reference. Nella Larsen can write, but oh my god she knows so little how to invest her characters with any life like possibilities. They always outrage the reader, not naturally as people have a way of doing in real life, but artificially like ill-managed puppets. Claude [McKay] I believe has shot his bolt. Jessie Fauset should be taken to Philadelphia and cremated. You should write a book. Countee [Cullen] should be castrated and taken to Persia as the Shah's eunuch. Jean Toomer should be enshrined as a genius and immortal and he should also publish his new book about which gossip is raving. Bud [Rudolph] Fisher should stick to short stories. Zora [Hurston] should learn craftsmanship and surprise the world and outstrip her contemporaries as well. Bruce [Nugent] should be spanked, put in a monastery and made to concentrate on writing. Gwennie [Gwendolyn Bennett] should stick to what she is doing. Aaron [Douglas] needs a change of scenery and a psychic shock. Eric [Walrond] ought to finish *The Big Ditch* [later *Tropic Death*] or destroy it. I should commit suicide.

While some of the writers Thurman castigates above have now become canonical in U.S. literary studies, many others still remain shrouded in mystery and obscurity. Like these writers, the period or movement known as the Harlem Renaissance is still an unfinished story—even though significant new scholarship during the past three decades, in many disciplines and from various points of view, has helped to gain it a recognition that was until the 1970s largely denied to this rich and complex phenomenon experienced by both blacks and whites in the 1920s and 1930s.

## The Many Harlem Renaissances

When we reconfigure it within the turbulent historical context of African American citizenship, the brief period known as the Harlem Renaissance—for some, only five to six years (1924–1929) and for others, roughly twenty to thirty years, from 1910 (or 1914 or 1919) to 1937 (or 1940)—might appear to be much ado about nothing: a kind of euphoria, a momentary high, in African American cultural history.[8] Many scholars—most notably Harold Cruse—have argued that the Harlem Renaissance failed precisely at the project it set out to accomplish: to extend the civil and human rights of African Americans by demonstrating the quality and quantity of their artistic contributions.[9] Despite such reservations, we must acknowledge the growing prominence of the Harlem Renaissance in our sense of social, cultural, and literary history as well as in the college and university curriculum since the early 1970s.[10]

In embracing new critical approaches and pedagogies, we have become more sharply aware of the reading and academic practices that had once supported the erasure in history and theory of a movement or period experienced most intensely by both black and white Americans in the 1920s.[11] In the 1970s, when scholars such as Nathan Huggins, James O. Young, Margaret Perry, and Amritjit Singh extended and revised the earlier survey treatments of the period by Sterling A. Brown, Hugh M. Gloster, and Robert Bone, they were concerned mostly with remedying that absence—thus inscribing the Harlem Renaissance onto what Van Wyck Brooks views as America's coming-of-age and linking their examination of this historical phenomenon to their new concern with shaping a multiethnic consciousness of American literature.[12] Huggins, in fact, views the black American's confusions over identity as uniquely American and describes them in the post–World War I context as symbiotic with those of white Americans: "White Americans and white American culture have no more claim to self-confidence than black. . . . Blacks have been essential to white identity (and whites to black). . . . they cannot be understood independently." And while for Huggins, "whenever Americans do come of age, they will have gained true insight into themselves by the claiming" of that interdependence, he also recognizes that blacks' willingness to be used by whites to fulfill the latter's "psychic dependency" defines the "deep moral tensions" in U.S. race relations (9, 11, 12, 84).

Today, as Mark Helbling has argued in *The Harlem Renaissance* (1999), there is more than one Harlem Renaissance, as reflected in the readings and methodologies developed since the 1970s. I refer, of course, to several successive and overlapping trajectories. Since the 1980s, feminist scholarship has been the most influential in revising and expanding our sense of the Harlem Renaissance through its interrogation of the peculiar treatment that issues of genre, periodization, class, and gender had received in 1970s scholarship as well as by bringing to the fore many women writers from the 1920s and 1930s previously neglected. The vantage point of the new feminist consciousness of the late 1970s and 1980s allowed a serious and sometimes empathetic examination of the challenges faced by women artists of the 1920s and 1930s as they lived and created in the web of race, gender, and class. This scholarship is embodied in Hull, Scott, and Smith's pioneering anthology, *All the Women Are White, All the Blacks Are Men, but Some of Us Are Brave* (1982) and the cumulative work of writers and scholars such as Alice Walker, Audre Lorde, Sonia Sanchez, Mari Evans, Mary Helen Washington, Barbara Christian, Henry Louis Gates Jr., Nellie McKay, Gloria Naylor, Hortense Spillers, Robert Hemenway, Claudia Tate, Trudier Harris, Houston A. Baker Jr., Calvin Hernton, Gloria Wade-Gayles, Cheryl Wall, Hazel Carby, Michael Awkward, Deborah McDowell, Gloria Hull, Madhu Dubey, Thadious Davis, Maureen Honey, Karla Holloway, Marcy Knopf, and others.

Further, we have benefited immensely not only from the social and cultural histories of the period by Jervis Anderson, Cary Wintz, and David Levering Lewis, but also from the revisionist readings of the "modern" by Houston Baker Jr., Arnold Rampersad, James de Jongh, Michael North, Sieglinde Lemke, and others.[13] And then there is the extensive, still growing new scholarship on music, theater, and the fine arts and on literary and cultural activity away from New York City, as well as several essays on gay and lesbian issues. Just as an essay such as Barbara Smith's "Toward a Black Feminist Criticism" (1977) presaged much of the work by black feminists, Charles Nero's "Toward a Black Gay Aesthetic" (1991) declared yet another new direction in African American literary study. Extending the arguments of Smith and Nero to the Harlem Renaissance, work by Deborah McDowell, Robert Reid-Pharr, Michael L. Cobb, Gloria Hull, and the two essays by Gregory Woods and Alden Reimonenq in Emmanuel Nelson's *Critical Essays: Gay and Lesbian Writers of Color* (1993) offer important rereadings of Harlem Renaissance texts and figures.

With the publication of Thomas H. Wirth's *Gay Rebel of the Harlem Renaissance* (2002), this process of recovery and rereadings is likely to accelerate.[14] Books such as *The Harlem Renaissance in Black and White* (1995) by George Hutchinson and *Terrible Honesty* (1995) by Ann Douglas document the intricate black-white collaboration during the period. While Hutchinson tends to celebrate the Harlem Renaissance as a chapter in the broader U.S. developments in modernism, pragmatism, and cultural nationalism, Douglas views in the 1920s the origins of a new American popular culture that fused "high and low" art and repudiated, she claims, an "outward derivative gentility" in favor of "a boldly masculinized outspokenness." Paul Gilroy's Black Atlantic, transnational conceptions of black culture, and the emerging postcolonialist readings of U.S. imperialism and exceptionalism point also to new possibilities for Harlem Renaissance studies.[15]

And now, we hope, this *Reader* will provide a map of Wallace Thurman's Harlem Renaissance and allow us to consider how it does (or does not) cohere with the other Harlem Renaissances just mentioned—as well as with the conceptions of the period or movement held by his contemporaries such as W.E.B. Du Bois, Carl Van Vechten, Langston Hughes, Victor F. Calverton, Charles S. Johnson, or Alain Locke.

Buried in these possibilities of evolving fresh interpretations is the complicated and often painful history of racial constructions in the United States of which at least some members of the Harlem Renaissance—including Wallace Thurman—showed a nuanced understanding. We believe this *Reader* will invite even further attention to how "race" complicated the lives and careers of several major figures of the Harlem Renaissance—including those who were neither light-skinned nor eagerly assimilationist—and the heavy price that some of them paid for their "Negroness." And while it is important to document—as Hutchinson and Douglas have done in their books— the many levels of black-white cooperation, exploring the intricate texture of the 1920s United States without careful attention to the paternalism, appropriation, and condescension that marked much of white-black interaction during the period would be to run the risk, once again, of allowing a narrative about the souls of black folks to be framed and co-opted by a story about white folks' ways. Some recent commentaries on the Harlem Renaissance by white American scholars show a tendency to concentrate on the fascinating liminality of light-skinned characters or writers—treating it, mistakenly, as the locus of the multifaceted Afro-American identity. And yet all the scholarship on the Harlem Renaissance since the 1970s—including its recovery of lost voices or texts—has definitely prepared us to celebrate the achievements of the Harlem Renaissance in many areas of cultural and public spheres, most especially its early critical role in calibrating and confronting the debates on art and "race" that still continue to rage.

## Wallace Thurman's Harlem Renaissance

Wallace Thurman's Harlem Renaissance is one of the most radical and uncompromising responses from a black writer in the 1920s to the denials of African American personhood and individuality through racialization. The ringing conclusion of Langston Hughes's famous essay, "The Negro Artist and the Racial Mountain" (*The Nation,* June 23, 1926), is quite familiar to most students of the Harlem Renaissance—"We younger Negro artists who create . . ." And since the late 1980s, many of us are equally familiar with Marita Bonner's unraveling of the axis of race, class, and gender in her poignantly lyrical essay, "On Being Young—a Woman—and Colored" (*The Crisis,* December 1925), which had, oddly, not received even a cursory mention in the 1970s scholarship. In many ways, Bonner's essay continues to serve as a corrective epigraph even to the elisions and omissions of some feminist scholarship, whose exclusive focus on sexuality and gender might underplay "race" and/or class. As an intellectual, literary critic, and theorist who, like his nemesis Du Bois, shaped discussions in the public sphere of various issues surrounding racial expression in art, Thurman was more independent, more iconoclastic, more on the edge, than any of his younger contemporaries. And yet, if he had positioned himself so actively and vociferously against the grain, what kind of world did Thurman want to build? What did Thurman's Harlem Renaissance look like or look for?

In his novels, letters, essays, plays, screenplays, poetry, and criticism, Thurman offers a complex portrait of Afro-America in search of its artistic self. More fiercely than any of his peers, Thurman committed himself to individuality and racial transcendence, which is not the same thing as the desire to become white.[16] According to Thurman, the black writer could no longer cater to taste, trend, or tradition, nor was it enough for him or her to rail against the burdens of race in creating art. The artist must start with radical self-examination as a step toward achieving the expressive powers necessary to affect the conditions around him or her. In this light, much of the African American literary tradition might be viewed as an affirmation of full personhood and individuality and at the same time, as a response to the denial of that agency in daily human interaction.

The NAACP—under the leadership of Du Bois, James Weldon Johnson, Walter White, and others—was fighting the abrogation of civil and human rights in socioeconomic and political spheres, including its extreme expression in lynchings. Du Bois's voice rang strong and clear when he wrote in the March 1920 issue of *The Crisis*: "There were 65 persons lynched without trial in the United States during the year 1920. No other civilized country has such a record." Although Du Bois, Johnson, and White all had literary ambitions—and they had significant differences among themselves over style and strategy—they were all primarily focused during the late 1910s and 1920s on gaining ground for the Negro's citizenship rights. Like German Jews in the nineteenth century, American Negroes in the early years of the twentieth century were constantly under pressure—as was evident in the debates that led to black participation in World War I—to prove their credentials for full accreditation as citizens. The results of their efforts were always less than hopeful and often worse, as was demonstrated by the riots that erupted in twenty-five cities during the "bloody summer" of 1919, making blacks more embittered and defiant than ever before.

Thurman believed, however, that racial bigotry would continue in its various manifestations as long as color functioned as a social marker, with the devices for the border patrolling of racialized identities powerfully in place. According to Eleonore van Notten, in his autobiographical "Notes on a Stepchild" and through his voice as heard through Truman Walter in *Blacker,* Thurman rejected the old guard's jeremiad for justice as a waste of time and of the individual's creative potential and suggested a variety of alternative responses. Thurman believed that a rigorous cultivation of individuality would enable black intellectuals to distance themselves from the humiliating consequences of racism and possibly transform a debilitating low self-esteem into "a potentially more empowering sense of social superiority" (238–239). (One should not be surprised to hear some Gurdjieffian notes in some of Thurman's formulations because in early 1926, Thurman attended Toomer's Harlem workshops on Gurdjieff's thought. Thurman—interested as he was in the internal processes of achieving racial transcendence—took Gurdjieff quite seriously for a while.)[17]

This Gurdjieffian influence is noticeable in the strong and clear statement Thurman makes on this subject in the preface to *Aunt Hagar's Children*: It is "the duty of those who have the will to power in artistic and intellectual fields to shake off psychological shackles, deliberately formulate an egoistic philosophy, develop a cosmopolitan perspective, and soar where they may, blaming only themselves if they fail to reach their goal. Individual salvation may prove a more efficacious emancipating agent for his [the Negro's] generation and for those following than self sacrifice and morbid resentment." Rejecting what he called the "herd" or the "horde" in his poems, Thurman's faith in the efficacy of individual cultivation and self-consciousness was so strong that he expressed his admiration for the "not quite achieved, . . . not very wise, a little mad, 'possessed'" individuals invoked in the Maxim Gorki quotation that serves as an epigraph to *Infants*.[18] Even in one of his last letters to Hughes from the City Hospital, Thurman expresses his dread of the "stupidity" and "vulgarity" of the undifferentiated mass. For all we know, Thurman was reincarnated as Woodridge, the English teacher in *Invisible Man* whose words ring in the ears of Ralph Ellison's young protagonist throughout his northern encounters. "Stephen's problems, like ours," exclaims Woodridge with reference to Joyce's young protagonist, "was not actually of creating the uncreated conscience of his race, but of creating the *uncreated features of his face*. Our task is that of making ourselves individuals. The conscience of a race is the gift of its individuals who see, evaluate, record." ([New York: Knopf, 1981], 345–346). In the process, asserts Woodridge, we would have created both a "race" and a "culture."

In Thurman's view, thus, the diminished humanity of African Americans and other marginalized groups is best confronted by an unflinching commitment to the full development of the artistic and intellectual potential within what Du Bois viewed as the "dogged strength" of each black individual. It is possible, for instance, in reading *Infants*, to treat the light-skinned Paul Arbian as "the particular vehicle for Thurman's fears . . . and idealizations" and conclude, as Ross Posnock has indeed done, that Thurman represents "terminal individualism," which "stands in stark opposition to group membership, in which the individual's capacity to choose or devise affiliations is discouraged" (82–83). However, it is equally possible—with Nathan Huggins—to recognize that "Thurman's message was delivered more by Raymond than by Paul," whose decadence and bitter self-contempt foreshadow his bizarre suicide (242). Maybe the distinction that John Edgar Tidwell has developed between Sterling Brown's "rugged individualism" and Robert Frost's would clarify the link in Thurman's thought process between group affiliation and individuality. As Tidwell puts it, "Brown did not enjoy Frost's racial privileges, . . . and Brown's view [of government and New Deal, for example] declared itself against a world view of racial stereotyping. What rescued the dignity of blacks from warped imaginations and projections of difference was sheer will or indomitable spirit" (404).[19]

Through his many outpourings in various genres gathered in this *Reader* on the subject of racial expression in art and literature, Thurman anticipates—as Du Bois, Hughes, and Richard Wright do in their own distinctive ways—the growing recognition today of "race" as a social construction. Much of Critical Race Theory in recent years has demonstrated how "race" is also a legal construction. The work of A. Leon Higginbotham Jr., Derrick Bell, and Cheryl Harris, among others, demonstrates how the operation of law does much more than just legalize "race"—it defines and supports an extensive pattern of domination and subordination within race relations. Patricia Williams has argued, for instance, that legal understandings of "race" should in fact be grounded in the complexity of social experience and not in the false simplicity of formal abstractions. Again, in *White by Law* (1996), Ian F. Haney-Lopez has juxtaposed the legal history of African American and Asian American citizenship to examine how the Supreme Court, in dealing with racial prerequisite cases between 1878 and 1952, chose to ignore the scientific evidence then available and surrendered to popular, common-knowledge versions of "whiteness," preventing millions of people of color from participating fully in the formation of a color-blind community in the United States.

Long before the current debates on essentialism and constructivism, Thurman's essays and book reviews demonstrated how he recognized "race" as neither an "essence" nor an "illusion." In their well-known text, *Racial Formation in the United States,* Michael Omi and Howard Winant attempt to negotiate between the opposing ideas of race as "an *essence*—something fixed, concrete and objective"—and race as "a mere *illusion,* a purely ideological construct which some ideal non-racist social order would eliminate. Challenging both these positions, Omi and Winant define "race" as a concept *"which signifies and symbolizes social conflicts and interests by referring to different types of human bodies."* As Omi and Winant note, in contrast to gender, "although the concept of race invokes biologically based characteristics (so-called 'phenotypes'), selection of these particular features for purposes of racial signification is always and necessarily a social and historical process" (54–55). Thurman, I believe, would have concurred most fervently.

In a similar vein, as part of a new genre of writing called the "philosophy of race," Charles W. Mills has argued in *Blackness Visible* (1998) that "room has to be made for race as both real and unreal: that race can be ontological without being biological, metaphysical without being physical, existential without being essential, shaping one's being without being in one's shape" (xiv). Mills notes that philosophy as a discipline has been most resistant to "multicultural" and feminist revisions because of the assumption that by virtue of its abstractions, philosophy "already generally encompasses the human condition." In philosophy, then, "the writings of the classic Euro-American authors are treated simultaneously as canonical and as raceless and universal"(xii). Examples culled from texts by Mark Twain, William Faulkner, and Richard Wright (which Mills had used in a course on African American philosophy), illuminate "the actual moral code of a white-supremacist system and its ramifications for dominant and subordinated groups." For Mills, these and other U.S. literary texts—including several that Toni Morrison discusses in her *Playing in the Dark* (1992)—"give us moral knowledge of a universe of persons and subpersons, where disrespect for the latter from the former was not only permissible but mandated in order to maintain the social hierarchy, and duties and rights were appropriated accordingly. . . . They describe the necessity for border patrolling, the intricate racial etiquette devised to police this hierarchy" (15–16).

In relation to Thurman and the Harlem Renaissance, Mills's point about the value of literary expression is perhaps the most pertinent. Mills views literature as a valuable resource to help us all understand "the characteristic themes in the black experience: African Americans' need to insist on their personhood, sometimes at the cost of death; the strategies for subverting the order when direct confrontation was judged futile; the importance of black self-respect; . . . and the insistence that one could be culturally black and a person. . . . Yet if whites spontaneously generalize and universalize from their uncontested moral personhood, these moral realities will remain largely invisible to them" (15–16). Thurman—more than any other among his younger contemporaries—appears to have persisted in bringing these challenges to the attention of both whites and blacks during his short decade in and around Harlem.

The two novels by which most readers know Thurman today—*The Blacker the Berry* and *Infants of the Spring*—deal with the challenges of achieving full black personhood against the mind-boggling array of racial specters that haunt both daily experience and artistic creativity. Beyond that, in almost all his essays and book reviews, Thurman appears to be reaching out for a language to assert not only the claims of black personhood but also his own individuality. Sometimes adopting a caustic tone, at other times framed in confessional terms, Thurman's essays on race, writing, art, and social phenomena address the historical denial and negations of black American experience and argue simultaneously for the autonomy of each black person. Thurman's Harlem Renaissance is apparently focused on a persistent search for the individual Negro's personhood and agency, which had been systematically erased by a series of court decisions and legislation that followed the notorious political compromise of 1877 that ended Reconstruction, compelling blacks both in the South and the North to live segregated lives, barred from catching "the full spirit of Western Civilization, . . . [living] somehow in it but not of it," as Wright observes in *Black Boy*.

As with Wright, Thurman's perceptions and insights into the African American experience were shaped by a strong sense of his permanent outsider status. Even the minute details of Thurman's life underscore this outsider status. Starting his life in Salt Lake City, Utah—a most unlikely place for a dark-skinned black boy to be born and raised—Thurman was a bohemian bisexual and a literary jack-of-all-trades. An unflinching secularist, his early and strong rejection of church parallels that of the young Richard Wright. Generally underemployed most of his life, he cultivated a thoroughly contrarian and freethinking position during his Harlem years. As an inveterate outsider, Thurman took a stand in favor of appraising African American literature by purely aesthetic standards and against the "current faddistic interest in things Negroid." And always, Thurman writes against the grain of black bourgeois respectability, which would shackle literary creation by requiring it to present a whitewashed and preapproved idea of African American life.

Stylistically, his critique is never timid. Perhaps because he needed to be heard above the hue and cry of the New Negro fad and the "so-called" Harlem Renaissance ballyhoo, he dresses his opinions in sharp words, scathing tones, and his famously acerbic wit. Thurman arrived in Harlem at a time when the black middle class was resisting primitivistic and realistic representations of black people and life. It was a time when migration and poverty were transforming the social and moral fabric of blacks in the northern cities. Suspended between the impulses to tell the truth and to advocate change, Thurman's writings are caught in the web of questions about the nature of blackness, of black personhood, and of the artist's role in the black community. More than a racial gadfly and literary satirist, Thurman brings both passion and insight to his relentless attempts to raise public debate on the racial specters that haunt and thwart all forms of artistic expression.

Many scholars of the Harlem Renaissance do not share such a view of Thurman as this critically engaged thinker; rather, they see him as the very embodiment of the Harlem Renaissance or as a "strangely brilliant black boy" who remained perpetually misunderstood and unappreciated. Mae Gwendolyn Henderson, for instance, remarks that "in many ways the course of [Thurman's] life parallels the brief, but colorful and intensely creative, period of the Renaissance itself" (147). Beginning with Huggins in the 1970s, critics have found it useful to treat Thurman and his *Infants* as mediating between the Harlem Renaissance and its scholars. Historians and scholars of the period have sometimes taken his insightful and cutting criticism as reason to read Thurman as a misunderstood prophet. As Dorothy West recalled, "he was Wally to his friends, and his sycophants were legion. He could 'dish it out,' and there was no tongue that could return it. Perhaps if Harlem had produced a dozen contemporary minds as keen as his, he would not have drowned his disillusion in drink, and under his leadership something better might have come out of that period than the hysterical hosannas that faded on the subsequent stilly night" (77). Thus, in most accounts of the Harlem Renaissance, Thurman is more often treated as a lens through which to view the movement than as an artist and public intellectual in his own right. Ironically, the Harlem Renaissance's greatest individualist is very seldom appreciated as an individual—for his uncompromising artistic goals or for his sharp critical mind.

To appreciate fully Thurman's Harlem Renaissance, we need also to confront a certain view of his personality that has been part of much discussion and speculation among students of the Harlem Renaissance and among some of the writers from the period who lived beyond the 1930s. Based mostly on Hughes's influential description of Thurman in *The Big Sea* and some anecdotal accounts of the bold and flamboyant gay rebel, Richard Bruce Nugent, this widely prevalent view attributes Thurman's deep inner conflicts as an individual to his unusually dark skin. This conception of Thurman's "self-hatred" has been reinforced by biocritical readings of his bold treatment of intraracial color prejudice in *The Blacker the Berry*. The novel's dark-skinned protagonist, Emma Lou Morgan, has often been identified with the real-life Wallace Thurman, her distaste for the color of her skin assumed to be an echo of her author's similar feelings. But such a reading ignores the very significant differences between the sharp-witted, independent-minded, articulate, bohemian intellectual that Wally was to most people around him and the self-loathing, often-hapless Emma Lou who keeps shunting from one exploitative situation to another throughout the novel.[20] Such a reading also ignores the most mean-ingful difference of all: Thurman made his protagonist a woman. In doing so, he shows an awareness of the sexual and gender issues affecting dark-skinned black women, who faced many more difficulties than their male counterparts. Educated dark-skinned males often married light-skinned women and had successful professional careers.

It is not clear why Nugent, even more than Hughes, continued to circulate well into the 1980s an image of a "Wally" whose "self-hatred" stemmed mainly from his dark skin. In any case, the profiles of Thurman's personality provided by Hughes and Nugent do not chime with several other portraits of Thurman that are available to us—including those by William Jourdan Rapp, Dorothy West, Theophilus Lewis, Granville Hicks, and Thurman himself. Perhaps Nugent's frequently told story of how he left the scene of his first meeting with Thurman (Nugent could not imagine how someone so "brilliant" could be so "black") is more revealing of the young and light-skinned Nugent's own color obsessions than of Thurman's self-contempt.[21]

This is not to suggest that Thurman's psyche and creativity were not affected—even distorted—by how he was sometimes treated by white clerks at railroad ticket windows or theater box offices. Nor could Thurman have been unmindful of the skin-tone obsessions of the black bourgeoisie whose aesthetic values he fervently opposed. Despite his very dark skin and his numerous health problems, Thurman was able in his short and frenetic career to achieve a level of distinction that eluded his many lighter-skinned black male friends—witness his position as chief editor at Macaulay's and Hollywood's invitation (the first such extended to an African American) to write two screenplays. His ability to collaborate with white individuals in warm collegial relationships is well demonstrated in his work with both William Jourdan Rapp and Abraham L. Furman, as well as his friendship with a rich white woman, Elisabeth Marbury, who assisted Thurman financially so he could finish his second novel, *Infants*. Apparently, Thurman sought and achieved warm and equal relationships with several whites. In short, Thurman appears to have been as creative with his color fixations as any black human being in the United States of the 1920s could have been.

We suspect that there are more persuasive explanations for the perplexing schism in Thurman's personality—for schism there was. The man whom Dorothy West remembered as having "the most agreeable smile in Harlem and a rich infectious laugh" was also frequently despondent and prone to excessive drinking. His health was delicate throughout his life, and he must have had an early awareness of his mortality, growing up as he did a sickly, solitary child in an unusually dysfunctional family.

Personal disappointment was compounded by the impossibly high standards for literary achievement he set for himself (and others). During his formative years, his reading of European classics was voracious—but unguided and undigested. In "Notes on a

Stepchild," Thurman admits that most of his precocious reading left him thoroughly confused, "believing this one day, and something else the next, finding verification for all beliefs in literature and in life." He tells us how his attempts "to forge that which [I] had learned into a harmonious whole" drove him close to madness. In *The Big Sea*, Hughes notes Thurman's intermittent despondency about his chances of realizing his ambition of becoming "a *very* great writer, like Gorki or Thomas Mann." Comparing his own work to "the thousands of other pages he had read, by Proust, Melville, Tolstoy, Galsworthy, Dostoyevski, Henry James, . . . [he] found his own pages vastly wanting" (235). Again, he could not resolve the tension he felt between urban excitement and his urgent need for the peace and quiet of the beach or the country. While he was stimulated by the city and its abundant supply of freedom to experience a bohemian lifestyle, he realized that he could complete his writing projects much better in the relative calm of his grandmother's home in Salt Lake City or the suburban environs of his friend Theophilus Lewis's Jamaica, Long Island, home.

Another unresolved element in this schism was the love-hate relationship Thurman had with Harlem throughout his adult life. It is evident from the shifting tone of his essays on Harlem in the 1930s as well as from some possible readings of the ending of *Infants,* that Thurman came to view not only the New Negro Renaissance but also Harlem with increasing skepticism. The narrative movement in the *Infants* captures some of Thurman's initial ecstasy about the place—as expressed in his early Harlem essays—as well as the later ambivalence that is reflected in his letters, his writing of the now-lost play, *Goodbye New York,* and in his extended stays away from Manhattan.

Also at issue is Thurman's deep-seated ambivalence toward his own sexuality. In a 2002 issue of *Transition* magazine, Mason Stokes has suggested that we need a more supple treatment of sexuality in relation to the Harlem Renaissance, many of whose major figures, including Thurman and Countee Cullen, spoke of their homosexuality "in the language of abnormality rather than identity—of pathology rather than newly found liberated homosexuality" (71).[22] The time may indeed to be ripe for a more thorough investigation into Thurman's sexual identity.[23] All in all, one is struck by Thurman's frenetic search, his race against time, throughout his short and brilliant life to find ways to resolve his conflicts and to fulfill the various facets of his multitalented personality.

From our work of collecting Thurman's writings, including almost all of his extant letters, we are persuaded that Thurman was at his best as a literary critic, editor, and public intellectual, as well as a competent writer of novels, plays, and short stories.[24] Critical evaluations of his work might continue to suffer from, among other things, the seeming jack-of-all-trades quality of what he has left behind. In an early 1929 letter to Claude McKay, Thurman describes his career:

[T]hree years have seen me become a New Negro (for no reason at all and without my consent), a poet (having had two poems published by generous editors), an editor (with a penchant for financially unsound publications), a critic (see articles on Negro life and literature in *The Bookman, New Republic, Independent, World Tomorrow,* etc.), an actor (I was denizen of Cat Fish Row in *Porgy*), a husband (having been married all of six months), a novelist (viz: *The Blacker the Berry.* Macaulay's, Feb. 1, 1929: $2.50), a playwright (being co-author of *Black Belt* [later *Harlem]*). Now—what more could one do?

It appears Thurman did not find time during his short Harlem sojourn to develop his preferences among the various genres or fully to hone his experimental techniques. He also had not yet learned to strike a balance in his creative life between the stimulation he needed from the city and the relative quiet of the suburb or small town, which he found he needed desperately to get his writing done. Such calibration on the part of many an artist or intellectual (and he was both in strong measure) requires resources, support systems within family and friends, and the emotional maturity one might gain with extended inner struggle—of which there is definitive evidence in Thurman's autobiographical sketches and personal correspondence. Unfortunately, in the relative absence of resources and support systems, Thurman succumbed to self-destructive drinking binges as he grappled with chronic health problems and finally with incurable tuberculosis.[25]

And yet Thurman strikes us today as a transgressive artist in almost all forms of writing that he attempted—dealing courageously with radical, even forbidden themes such as intraracial color prejudice in *The Blacker the Berry* and the *Staatsgewalt* of forced sterilizations in the film script of *Tomorrow's Children*.[26] But nowhere is his risk-taking more evident than in his remarkable achievement as a critic, opinion maker, and a pioneering theorist on "race" and writing.[27] Thurman published about two dozen essays and reviews in his lifetime—many written for the black press but most of them for well-established white journals such as *The New Republic, Bookman,* and *The Independent.* Since Thurman's own autobiographical sketches as well as the pen-portraits of him by close friends like Theophilus Lewis, Harold Jackman, and William Jourdan Rapp still remain unpublished or uncollected, the full implications of his work and thought for the complex issues surrounding black identity and individuality in the United States have not yet been adequately explored.

Thurman's 1927 essay "Negro Artists and the Negro" (see Part Four, "Literary Essays and Reviews"), as Hutchinson has noted, "deserves rereading on a number of counts" (241). Noting how the provincialism of *all* Americans affects black and white artists alike, Thurman deftly interlinks the cultural complexes of blacks and whites: "Those American Negroes who would not appreciate the spirituals until white critics sang their praises have their counterparts in the American whites who would not appreciate Poe and Whitman until European critics classed them as immortals." As discussed in our introduction to Part Four, Thurman attempts to articulate his theory of a black aesthetic in many of his literary essays and reviews—in especially positive terms in the "Nephews of Uncle Remus"—anticipating in some ways the more celebrated but controversial writings on the subject in the 1970s.

Thurman also worked continually for years on a collection of essays on black leaders and on Harlem life. This collection, entitled *Aunt Hagar's Children,* is being published in this *Reader* for the first time (see Part Five). It is clear from all his essays and reviews that Thurman was greatly at odds with the emerging black middle class, which in the 1920s insisted upon ennobling and dignified representations of blacks in art and literature. As is evident in Langston Hughes's "The Negro Artist and the Racial Mountain," this stance placed the artists of the Harlem Renaissance in an awkward position vis-à-vis a sizable segment of their reading public. Understandably, as Posnock has noted, the black bourgeoisie represented the "major boundary in relation to which authors and texts were situated, . . . perhaps because historically it has been the arena where racial anxieties about authenticity . . . are played out" (80).[28] Thurman participated most eagerly and energetically in the representational debate—"realistic" or "idealized," "sensationalistic" or "uplifting"—that cast a momentous shadow over the intellectual and cultural life of Afro-America in the 1920s. In one way or another, Thurman addresses this debate in all of his essays, because it—as Hughes too makes clear in "The Negro Artist and the Racial Mountain"—impinges on his freedom as an artist. But Thurman's concern goes further than Hughes's, since he wants to expose the more subtle forms of censorship at work. In a way, Thurman's rejection of representational dictates from the black bourgeoisie relates to a similar point Toni Morrison has made in some of her interviews, with reference to attacks from African Americans on *The Bluest Eye*: the insistence that black writers must create "positive images" is based on a dangerous premise that all blacks must think, act, and write the same way.[29]

Thurman, of course, was not the only one among the younger black artists of the period committed to engaging in debate with the black bourgeoisie.[30] All of the younger artists associated with *Fire!!* as well as a few others opposed the programmatic and promotional ideologies of the older generation of black writers, leaders, and intellectuals such as W.E.B. Du Bois, Allison Davis, Aubrey Bowser, and Benjamin Brawley. But for Thurman, arguing against the older generation's insistence on representational didacticism and idealism—for him, indistinguishable from the bourgeoisie's obsessions with uplift and respectability—was the consuming passion of his life. He not only wrote more forcefully and persistently than others on these issues, he also tried to organize the opposition of the younger generation through the publication of both *Fire!!* and *Harlem*.

While the two magazines shocked the older generation as intended, neither lasted beyond the first issue for financial reasons.

For some reason, for Thurman and others, the towering presence of Du Bois made the latter their most cherished target. Today, when we have a much fuller sense of Du Bois's complex perspectives on issues of art, propaganda, and politics, we can note this development with a bit of irony. But as Jack B. Moore has noted, based partly on the discussion of Du Bois's criticism in Arnold Rampersad's *The Art and Imagination of W.E.B. Du Bois* (1976), "Du Bois at once declares that the notion that 'our Art and Propaganda . . . are one' is wrong and in the end harmful, and then elsewhere states that 'All art is propaganda.' . . . While the dual emphases he placed upon the demands of both art and propaganda in the life of the artist may have helped produce in himself a writer-ideologue unusually committed to both art and politics, [Du Bois's] divided focus as a critic could only baffle many young writers" (109).[31]

In "High, Low, Past, and Present," published in November 1928 in the only issue of the magazine *Harlem*, Thurman is not only baffled but "hopping mad" at Du Bois's negative review of Rudolph Fisher's *The Walls of Jericho* for Fisher's alleged failure to provide in his novel "glimpses of better class of Negroes"—"like his mother, his sister and his wife." In the barbed and clear language of which Du Bois himself was capable when writing about issues of politics and citizenship, Thurman articulates his passionate opposition to Du Bois's aesthetic criteria and the risk they pose for the future of African American writing:

> Were [Du Bois] a denizen of "Striver's Row," scuttling hard up the social ladder, with nothing more important to think about than making money and keeping a high yellow wife bleached out and marcelled, one would laugh at such nonsense and dismiss it from one's mind. But Dr. Du Bois is not this. He is one of the outstanding Negroes of this or any other generation. He has served his race well; so well, in fact, that the artist in him has been stifled in order that the propagandist may thrive. No one will object to this being called noble and necessary sacrifice, but the days for such sacrifices are gone. The time has come now when the Negro artist can be his true self and pander to the stupidities of no one, either white or black.

In this passage, as in so many other places in his novels, essays, and letters, one can still hear the strong and resonant voice of Thurman, the critic and public intellectual, directing us even today to insist without concession or compromise on the highest artistic values.

Thurman's Harlem Renaissance is, thus, staunch and revolutionary in its commitment to individuality and critical objectivity: the black writer need not pander to the aesthetic preferences of the black middle class, nor should he or she write for an easy and patronizing white approval. As already noted, Thurman bristles at the thought that he has become a New Negro "for no reason at all and without my consent." Thurman's Harlem Renaissance may have had youthful enthusiasm and unachieved potential, but it is also marked by an evocative sense of its younger writers' desire to leave the past behind them and to step out boldly into the future. In their struggle to reject the intellectual and psychological constraints of the past, Thurman's career and writings point to a future when black writers and black critics might stand in a detached judgment of their people and the sociocultural and economic realities that impinge on their lives. Thurman, it seems, is not so much disappointed in the New Negro Renaissance as he is hopeful of something more. Possibly, his critique is directed more to what was yet to come than to what had already been created. He wants to make sure that African American writing is intellectually honest, carefully crafted, and artistically adventurous. By wrestling courageously in the 1920s and 1930s with the intricacies of racial constructions in the United States, Thurman and some of his younger Harlem Renaissance colleagues have facilitated our ongoing journey toward a much more complex understanding of the issues we still face today as Americans and as global citizens. Thurman's project of racial transcendence was impossible in his time. Even today, not only in the United States but also in most other parts of the world, it remains a daunting challenge.

### Notes

1. I am grateful for helpful conversations about Thurman as a public intellectual and as an early theorist of race and writing with Ulrike Kistner of Johannesburg, who comments wryly that the new emerging realities of "race" in South Africa today—uncannily similar but historically distinctive from the racial formations in the United States—are becoming much harder to explore in a society where the humanities are vanishing fast. In *Color and Culture*, Posnock argues that "[r]ather than latecomers, black writers circa 1900 were arguably the first modern American intellectuals" and suggests that the inescapable tension in their lives between being "race champion and intellectual . . . neither reduces to nor expresses a tension between the political (the race figure) and the aesthetic (the intellectual)" (2, 9). Posnock discusses Thurman as a public intellectual in relation to his "terminal individualism" (82–83). The black writer-intellectual's involvement with race is unavoidable, as suggested by the question Toni Morrison poses in her preface to *Playing in the Dark*: "What happens to the writerly imagination of a black author who is at some level *always* conscious of representing one's own race to, or in spite of, a race of readers that understands itself to be 'universal' and race-free?" (xii).

2. In his editorial for *Harlem*, Thurman suggests that he was attempting to produce magazines such as *Fire!!* and *Harlem* to create forums that would allow black writers to reach a black audience; for black journalists, writing regularly for white magazines was not a feasible option because few blacks would buy a white magazine.

3. In the absence of a book-length Thurman biography, we have relied—cautiously—for information on Klotman, "Wallace Henry Thurman"; Hughes, 233–238; Theophilus Lewis; West; Anderson, 208–210; David Levering Lewis, 193–197, 277–281; Henderson; Wirth, 163–210; and van Notten, who has extensively researched Thurman's early years. During 1990–1991, van Notten interviewed many friends and acquaintances of the fam-

ily in Utah and Idaho to create a somewhat more reliable source in chapter 2 of her book, "Wallace Thurman: Formative Years." One can also glean—with necessary caution—a few facts of Thurman's early life from Bruce Nugent's roman à clef, *Gentleman Jigger*, apparently written around 1930 but not published until 2002. In addition, we have the two autobiographical sketches by Thurman himself, included in this volume: "Autobiographical Statement" and "Notes on a Stepchild." Wherever possible, we have quietly corrected a few errors in the record. In 1926, Thurman was signing his name as "H. Wallace Thurman"; based on this fact, we can hazard a guess that he was baptized as Henry Wallace Thurman and not as Wallace Henry Thurman. There was only one issue each of both *Fire!!* and *Harlem*; there was no second issue of *Harlem*, as has sometimes been stated. It should also be noted that two plays—*Singin' the Blues* (by John McGowan) and *Savage Rhythm* (by Harry Hamilton and Norman Foster)—have been mistakenly attributed to Thurman; the confusion in Klotman's *DLB* essay and elsewhere appears to have stemmed from a typographical error (possibly a missing sentence) on page 128 of Sterling Brown's *Negro Poetry and Drama* (1937).

4. Before she moved to Salt Lake City, Ma Jack had lived for several years in Lawrence, Kansas, the town in which Langston Hughes had spent his early boyhood, as Melvin Tolson has noted in his Columbia University master's thesis, "The Harlem Group of Writers" (cited in Mullen, 121). In a 1929 letter to Claude McKay, Thurman described his ancestors as "pioneer Westerners."

5. Years later, in 1929, after Wallace had been received with "open arms" at the Santa Monica beach hotel run by his father's family, he had an accidental but dramatic meeting with his father. Thurman's brief description of this meeting in his correspondence with Rapp parallels in some ways the better-known and more detailed description of Richard Wright's meeting with his own father at the end of chapter 1 in *Black Boy*. Oscar—although "paralyzed" and sick with " tuberculosis of the throat" himself—had come to visit his own father. Wallace opened the door; the father and son recognized each other immediately. Refusing at first to acknowledge Wallace, Oscar tried later in the day to make up and proposed a follow-up meeting, but Wallace had no interest in meeting his father again.

6. We are still looking for any extant copies of the several pieces he wrote in Los Angeles for *The Pacific Defender* and *The Outlet*.

7. The title of West's memoir alludes to the novelty of Afro-American writing for both whites and blacks, "mingling socially to discuss that elephant's dance, Negro writing, remarkable not so much because it was writing, but because it was Negro writing" (77). The idea of Negro writing as the "elephant's dance"—evocative in some ways of Samuel Johnson's infamous comment on women preachers—definitely fits in with Thurman's sharp critique of both white and black critics frequently praising young black writers for the wrong reasons.

8. Stepto relates genre and periodization in the Harlem Renaissance in "Sterling Brown: Outsider in the Harlem Renaissance?" in Singh, Shiver, and Brodwin, 73–81. In *Color*, Hull observes: "[W]omen writers are tyrannized by periodization, the hierarchy of canonical forms, critical rankings of major and minor, and generalizations about literary periods" (30). Lauter, concerned primarily with the institutional, historiographic, and theoretical factors in the 1920s "that eliminated black, white female, and working-class writers from the canon," offers a helpful insight on periodicity and chronology in relation to the canon: "Dividing experience chronologically tends to accentuate the discontinuities rather than the continuities of life. . . . In some measure, women's lives in patriarchal society have been more fully identified with continuities—birthing, rearing, civilizing children; maintaining family and cultural stability. Indeed, . . . the rituals of female experience—regular, periodic, sustained—are culturally distinguishable from those of males. In this regard, emphasizing distinct chronological or literary periods may be one-dimensional, obscuring what is ongoing, continued. The color line persists, although its conventions and the forms of its literary expression are different in colonial and in modern, urbanized society" (40).

9. I am grateful for my conversations with Sieglinde Lemke on this point. See Lemke, 17. Alain Locke and James Weldon Johnson both argued this point during the 1920s, Johnson being the first one to establish a link between Negroes' civil rights and their artistic output in the 1922 preface to *The Anthology of American Negro Poetry*. Although both developed a modified perspective on these issues in the 1930s, neither of them reversed his position or forsook the importance of racial expression in African American writing. In 1936, Locke stated, "[T]here is no cure or saving magic in poetry and art, an emerging generation of talent or international prestige and interracial recognition, for unemployment or precarious marginal employment" (Locke, "Harlem: Dark Weather-Wane," *Survey Graphic*, August 25, 1936, 467). As late as 1939, Locke defended his views against attacks from John P. Davis and others, claiming his emphasis had always been on the "psychology of the masses" and not on offering any one solution to the "Negro problem." He asserts angrily, "It was the bright young talents of the 20's who themselves went cosmopolite when they were advised to go racial, who went exhibitionist instead of going documentarian, who got jazz-mad and cabaret-crazy instead of getting folkwise and sociologically sober" ("The New Negro: 'New' or Newer: A Retrospective View of the Literature of the Negro for 1938, " *Opportunity*, January–February 1939, 4–10, 36–42; cited in *The Critical Temper of Alain Locke*, ed. Jeffrey C. Stewart [New York: Garland, 1983], 272).

10. For two useful synopses of the debates among critics on the failure or success of the Harlem Renaissance, see Lemke, 17–20, 152; Helbling, 1–18. Cruse, easily the harshest among the critics, blames the Renaissance's failure to define its own nationalist aesthetic on the "ludicrous" integrationist goals of its major protagonists as well as its heavy dependence on white institutions. Huggins's simple conclusion (309) that "the true Negro renaissance awaits Afro-Americans' claiming their *patria,* their nativity" militates against his more complex analyses throughout his book of the interrelatedness of black and white identities in the post–World War I United States. In contrast to Cruse, who views the powerlessness and imitativeness of the Harlem Renaissance writers and intellectuals as a direct consequence of their individualism, Huggins sees the obsession with group identity as a source of a debilitating provincialism. Huggins places a lot of faith in what he extrapolates as Thurman's point in the *Infants*: "[A]rtistic production was an extremely personal, individualistic thing, not to be turned on or off by nationalism of any kind" (241). Describing Cruse's harsh charges as having been "made from the vantage point of the sixties and the seventies," Singh argues that "the failure or success of any literary movement cannot be directly related to the acceptance or rejection of a well-defined ideology." He finds that the patronage of the black middle class was "as mixed a blessing as the white patronage" and suggests that "the majority of Harlem Renaissance novels did not deal with issues central to the black masses" (131–132). As Houston A. Baker Jr. has noted, in focusing on the "tragically wide, ambitious, and delusional striving" (10) of black intellectuals, David Levering Lewis takes a view of the Renaissance's failures that is contrary to Huggins's. For Lewis, "the architects of the Renaissance believed in ultimate victory through the maximizing of the exceptional. They deceived themselves into thinking that race relations in the United States were amenable to the assimilationist patterns of a Latin country" (305–306).

11. For a discussion of this erasure, see Abraham Chapman and "A Brief Review of Previous Research and Criticism," in Singh, 135–138. See also Lauter in note 8 above.

12. I refer here to the lasting impact organizations such as the CLA (College Language Association) and MELUS (The Society for the Study of Multi-Ethnic Literature of the United States) have had on our curriculum and pedagogy. One reflection of the changes since the 1970s is the appeal of an anthology such as the two-volume *Heath Anthology of American Literature* (1990), edited by Lauter et al. In "Race and Gender," Lauter points out that with very few exceptions, black writing was quite dismally represented in "the twenty-one major classroom anthologies (and their revised editions) produced between 1917 and 1950" (24). He notes Bernard Smith's *The Democratic Spirit* (1941) as a major exception. Another major anthology of American literature that attempted substantial

representation of folklore as well as the writings of African Americans and Native Americans is *American Literature: The Makers and the Making* (1973), edited by Cleanth Brooks, R.W.B. Lewis, and Robert Penn Warren.

13. Baker—who has rightly been credited by Helbling and others with moving the discussions of the Harlem Renaissance's failure or success (see note 10 above for more traditional views of these matters) to an altogether new plane—sees as inadequate to the (discursive) history of Afro-American modernism any "*histories* that are assumed in the chronologies of British, Anglo-American, and Irish modernisms" (xvi). Detecting a distinctively "Afro-American *sounding*" in the texts and figures of the Harlem Renaissance—and his uncommon choices include Booker T. Washington, Charles W. Chesnutt, and Du Bois—Baker offers "what is perhaps *sui generic* definition of modern Afro-American sound as a function of a specifically Afro-American discursive practice" (xiv). In his parallel attempt to establish the Harlem Renaissance as a successful project, Hutchinson focuses unrelentingly on "interracialism," and ends up underplaying the often colonial nature of white-black relations in the 1920s United States. Hutchinson chastises Baker for his lack of historical complexity and suggests that Baker, like Huggins, fails to notice that the Harlem Renaissance writers looked more to the "low modernism" of Walt Whitman, Sinclair Lewis, Carl Sandburg, Shaw, and Synge, and not to the "high modernism" of Eliot, Joyce, Yeats, Stein, and Pound (24–25, 12–14, 118–119). In "Langston Hughes and Approaches to Modernism"— originally presented at a 1985 conference and included in Singh, Shiver, and Brodwin—Rampersad had already argued that Toomer, Hughes, McKay, Thurman, Nugent, and Hurston "were as aware as anyone else about the pressure of the modern on their lives and their art" (50) and that in shaping a blues-centered "black modernism," Hughes recognized his affinity with "Whitman, Vachel Lindsay, and above all, Sandburg," sharing "little or nothing of J. Alfred Prufrock's sense of an incurably diseased world" (61).

14. Some of the gay and lesbian scholarship in the 1990s has most probably been shaped in reaction to statements such as the following: "As for the fashionable tendency to assert, without convincing evidence, that Hughes was a homosexual, I will say at this point only that such a conclusion seems unfounded, and that evidence suggests a more complicated sexual nature"; "Amiable, fun-loving, Hughes was yet a sexual blank; his libido, under stimulation or pressure, seemed to vanish into a void"; Hughes was regarded by many of his friends as "asexual, without noticeable erotic feeling for either men or women" (Rampersad, 1:439, 1:289, 1:20). Gerald Early writes in a similar vein about Countee Cullen. After indicating that three scholars—David Levering Lewis, Jean Wagner, and Rampersad—"have asserted that Cullen was homosexual," Early comments: "There is, however, no evidence that Cullen and Jackman were lovers. There is no evidence that Cullen was engaged in any homosexual relations with any figures of the Renaissance. Some scholars have read letters and poems that seem suggestive in this regard but have offered nothing conclusive" (19 n. 21).

15. A few examples of books that explore issues of U.S. imperialism and/or employ postcolonialist perspectives to read U.S. culture and literature include Amy Kaplan and Donald Pease, eds. *Cultures of United States Imperialism* (Durham: Duke University Press, 1993); Amritjit Singh and Peter Schmidt, eds., *Postcolonial Theory and the United States: Race, Ethnicity, and Literature* (Jackson: University Press of Mississippi, 2000); and John Carlos Rowe, *Literary Culture and U.S. Imperialism* (New York: Oxford University Press, 2000).

16. Thurman's forceful grappling with a wide range of issues relating to race and art—as reflected throughout his writings in this *Reader*—makes David Levering Lewis's reading of some Thurman actions and statements quite arguable. "Taken to its logical extreme (which he did)," Lewis avers, "failure to become a great artist meant failure to transcend being a Negro" (278).

17. In an e-mail communication of August 18, 2002, Jon Woodson offers the following view of the Gurdjieffian undercurrents in Thurman's thinking: "Thurman's own individualism is a dyad and he is not necessarily being literal in his treatment of the group/individual

dyad. The rugged individual is as much subject to sleep and unconsciousness as are the members of the herd-horde—though the rugged or possessed individual gets there by a different route. He is perhaps more dangerous because he thinks himself undeluded, . . . Thurman seems to satirize this type of individual in Paul Arbian, but the question remains as to whether Raymond is capable of (awakened) individuality. Since the text does not divulge everything on the surface, there is a case to be made for Raymond as an allegory of awakened individuality—in contrast to Paul's failure." For more on Thurman and Gurdjieff, see Woodson, 10–11, 43–73, 176–177. At the same time, here is Thurman himself on Gurdjieff, in a letter of August 1929 to Hughes: "Ego, my son, is that which intellectuals call what Christians call the soul. It is what Jean Toomer calls the 'immortal I.' . . . that is what he and Gurdjieff had us developing. Ego generally is as much hooey as soul. Thus I could never make a good Gurdjieff disciple."

18. Thurman was quite aware that his own thinking on race and identity, subjectivity and agency, as well as the role of art and representation, was always in flux, still evolving. Thanking Hughes for useful feedback on *Aunt Hagar's Children* in a 1929 letter, Thurman says: "All of my ideas may change overnight. I claim no permanence for them nor infallibility. They are purely expressions of a current mood, guided by certain mental predilections."

19. For a useful discussion of various shades of meaning ascribed to "individual," as distinguished from "individualistic," "self," and "subject," see Winfried Fluck, "The Modernity of America and the Practice of Scholarship," in *Rethinking American History in a Global Age*, ed. Thomas Bender (Berkeley: University of California Press, 2002), 343–366, especially 363–364 n. 22.

20. In her essay "A Female Face," Thadious M. Davis examines "the sociocultural reasons for an emphasis on the representations of the female or the feminine in African American literature" during the decade before Wright's "masculinization of African American fiction" in the late 1930s. Davis stresses the parallels between Emma Lou and Thurman, as she views *Blacker* as an example of a text by a male African American writer who uses "the textual strategy of responding to racial separation and oppression by assuming a female face" (98, 114). In my view, Davis's point about the male writer using "a female face" to explore the problematics of his own identity would have validity even if the fictional details of the heroine's life did not closely match those of the author's own biography. A good example of such a novel is McKay's *Banana Bottom* (1933).

21. Apparently, when Hughes introduced Nugent to Thurman, Nugent decided that someone so dark skinned as Thurman "was not to be trusted," and "so, after a polite ten minutes of torture," Nugent left (Wirth, 170). To his credit, Nugent returned later in the day to apologize to Thurman for his conduct and soon after they became roommates. In an e-mail communication of August 17, 2002, Thomas Wirth indicates that Nugent's insights into Thurman's personality and internal conflicts were founded in Nugent's close observation as Thurman's friend and roommate and not just in Nugent's own color complexes. For Wirth, given all the personal factors affecting his personality, "it is a tribute to Thurman's genius and determination that he achieved what he did and was able to hide his inner turmoil from those who did not know him as well as Nugent did." Nugent narrated his first encounter with Thurman many times: van Notten (171) cites Nugent's 1983 interview with Wirth; see also an interview (286) included in Jeff Kisseloff, *You Must Remember This: An Oral History of Manhattan From the 1890's to World War II* (San Diego: Harcourt Brace Jovanovich, 1989); and the chapter entitled "Meeting Raymond" in Nugent's unfinished novel, *Gentleman Jigger*, which contains a more extended account (Wirth, 168–172). In trying to explain why Nugent persisted in repeating this incident until his death in 1987, Wirth says the following in his introductory note to *Gentleman Jigger*: "The most likely explanation is that Nugent's encounter with Thurman was an epiphany of sorts. It was the occasion for confrontation with the skin-color prejudice which infected 'blue-veined' Washington society in general and Nugent in particular. This prejudice contradicted Nugent's self-concept as a rebel against convention and Washington society; it was also totally alien to his persona in later years" (168).

22. Stokes narrates the fascinating story of Cullen's "pomp-filled" wedding on April 9, 1928, to Yolande, Du Bois's daughter, with his "special friend," Harold Jackman, as the best man, and the reasons for the young couple's quick separation and divorce. He cites at length from Yolande's May 30, 1929 letter to her father: "Shortly after our attempt at reconciliation, Countee told me something about himself that just finished things. . . . When he confessed that he's always known that he was abnormal sexually—as far as other men are concerned then many things became clear. . . . If he was born that way I can't help it. I'm sorry—but I cannot understand it. I think I prefer my own more natural inclinations" (68–69).

23. Apparently, gender identities—like race, color, and class identities—were patrolled by the male middle-class gatekeepers of the Harlem Renaissance. As Henry Louis Gates Jr. has noted, women such as Bessie Smith who attempted to shape more fluid sexual identities "were not often invited to New Negro salons" ("The Trope of a New Negro," 148). With reference to some of the Harlem Renaissance figures, these discussions of gender will surely also include the constructions of black masculinity that have been engaged by books such as Phillip Brian Harper, *Are We Not Men? Masculine Anxiety and the Problem of African American Identity* (New York: Oxford University Press, 1996); and Hazel Carby, *Race Men* (Cambridge: Harvard University Press, 1998).

24. We are still looking for Thurman's review of Locke's *The New Negro*, published in *The Looking Glass* and cited in West; an essay on Du Bois and "The Negro in the Theatre"—two pieces that are missing from the Beinecke manuscript of *Aunt Hagar's Children*; and possibly a few more letters.

25. According to Ann Douglas, "although literary Harlem had its alcoholics [and she singles out Toomer and Thurman among them], it had fewer of them than white Manhattan did. . . . Alcoholism is, among other things, a disease of despair, and it hit the white literary moderns more than it did the black" (90). It's likely, however, that alcoholism among black artists has not been as well documented as among the white.

26. There were almost no roles for black actors in either of the Hollywood films for which Thurman wrote the scripts. As Klotman has noted, Thurman, Hughes, and Zora Neale Hurston all learned that "the Negro may have been in vogue in Hollywood in the thirties, but it was still the cardboard Negro, the Imitation-Judge Priest-Green Pastures-GWTW Negro" ("Black Writer," 91). Regarding Thurman's treatment of intraracial skin-tone prejudice, Donald Petesch notes that "not until [*Blacker*] will readers, black and white, be exposed to a novel whose central theme is the hurt caused by the absorption by blacks of white ideals of color" (157) and the identification of color with beauty. Petesch views *Blacker* as anticipating the significance of this theme in Gwendolyn Brooks's *Maud Martha* (1951) and Morrison's *The Bluest Eye* (1970). Singh (72, 150 n. 2) mentions Jessie Fauset's *Comedy: American Style* (1933) and Chester Himes's *The Third Generation* (1954) as two other novels that extend the bold and pioneering treatment of the tabooed subject in *Blacker*. Petesch's strong discussion of the black bourgeoisie's skin-tone complex and its linkage to status and mobility should be read against the somewhat defensive discussions of the black middle class as a "conceptual problematic" in both Wilson J. Moses and Deborah E. McDowell. As McDowell views it, there might indeed be room for rethinking the ways in which the black middle class has been treated as "a category of disrepute" in some "entrenched critical positions" (xvii). But as Petesch (157–158, 259–260, 265–266) demonstrates through his extended references to James A. Farabee, Gerri Majors, Abraham Kardiner and Lionel Ovesay, Robert Coles, Toni Morrison, Charles W. Chesnutt, and David Levering Lewis, E. Franklin Frazier's representation of the black middle class in *The Black Bourgeoisie* (1957) continues to have a broad relevance for African American studies—considerably beyond the "limited purview" of one specific community of a particular moment.

27. Besides Theophilus Lewis and Langston Hughes, McKay too admired Thurman's acumen as a literary critic. In a letter to Hughes of March 30, 1928, McKay wrote: "I value your opinion above any of the Negro intellectuals—next Wallace Thurman in whom I find some real independent thinking and strength" (cited in Helbling, 122).

28. In a 1927 piece, "These Bad New Negroes: A Critique on Critics," written as a defense against his critics at the invitation of *The Pittsburgh Courier,* Hughes identifies four reasons for attacks from black middle-class readers and critics: the low self-esteem of the "best" blacks, their obsession with white opinion, their nouveau riche snobbery, and their lack of training for reviewing books or art by black and white authors (cited in Rampersad, 1:145).

29. In her 1993 interview with Eleanor Wachtel on CBC Radio in Toronto, Morrison condemns as censorship the actions of African Americans in Atlanta who had taken copies of *The Bluest Eye* off the shelves, proclaiming the book did not include positive images of black life. Morrison retorts with an edge in her voice, "Positive images for whom?"

30. It is worth noting that while both Countee Cullen and Arna Bontemps had friends among the "Niggeratti Manor" group of writers, neither appears to have joined the young Turks' attacks on race leaders and the black bourgeoisie. In the 1940s, Cullen and Bontemps collaborated on the musical *St. Louis Woman,* which was condemned by Walter White and others for perpetuating the demeaning stereotype of the peasant Negro. In contrast to Cullen and Bontemps, the older McKay—more absent than present in New York during the peak years of the Harlem Renaissance—chose to question Du Bois's ability to judge literature: "Nowhere in your writings do you reveal any comprehension of aesthetics and therefore are not competent or qualified to pass judgment upon any work of art" (letter from McKay to Du Bois, June 18, 1928, in *The Passion of Claude McKay: Selected Prose and Poetry,* ed. Wayne F. Cooper [New York: Schocken, 1973], 150).

31. Rampersad's study is based on the premise that Du Bois's "greatest gift was poetic in nature" and all his writings drew "their ultimate power from his essentially poetic vision of human experience and from his equally poetic reverence for the word" (vii).

## Works Cited

Anderson, Jervis. *This Was Harlem: A Cultural Portrait, 1900–1950.* New York: Farrar, Straus, Giroux, 1982.

Baker, Houston A., Jr. *Modernism and the Harlem Renaissance.* Chicago: University of Chicago Press, 1987.

Bone, Robert A. *The Negro Novel in America.* New Haven: Yale University Press, 1965.

Brooks, Van Wyck. *America's Coming-of-Age.* Garden City, N.Y.: Doubleday, 1958.

Brown, Sterling A. *Negro Poetry and Drama.* Washington, D.C.: Associates in Negro Folk Education, 1937.

Cobb, Michael L. "Insolent Racing, Rough Narrative: The Harlem Renaissance's Impolite Queers." *Callaloo* 23.1 (winter 2000): 328–351.

Cruse, Harold. *The Crisis of the Negro Intellectual.* New York: Morrow, 1967.

Davis, Thadious M. "A Female Face: Or, Masking the Masculine in African American Fiction before Richard Wright." In *Teaching African American Literature: Theory and Practice,* ed. Maryemma Graham, Sharon Pineault-Burke, and Marianna White Davis, 98–131. New York: Routledge, 1998.

de Jongh, James. *Vicious Modernism.* New York: Cambridge University Press, 1990.

Douglas, Ann. *Terrible Honesty: Mongrel Manhattan in the 1920s.* New York: Farrar, Straus, Giroux, 1995.

Early, Gerald. Introduction to *My Soul's High Song: The Collected Writings of Countee Cullen, Voice of the Harlem Renaissance.* New York: Doubleday, 1991.

Gates, Henry Louis Jr. "The Trope of a New Negro and the Reconstruction of the Image of the Black." *Representations* 24 (fall 1988): 129–155.

Gloster, Hugh M. *Negro Voices in American Fiction.* Chapel Hill: University of North Carolina Press, 1948.

Haney-Lopez, Ian F. *White By Law: The Legal Construction of Race*. New York: New York University Press, 1996.

Helbling, Mark. *The Harlem Renaissance: The One and the Many*. Westport, Conn.: Greenwood Press, 1999.

Henderson, Mae G. "Portrait of Wallace Thurman." In *The Harlem Renaissance Remembered*, ed. Arna Bontemps, 146–170. New York: Dodd, Mead, 1972.

Huggins, Nathan. *Harlem Renaissance*. New York: Oxford University Press, 1971.

Hughes, Langston. *The Big Sea: An Autobiography*. 1940. Reprint, New York: Hill and Wang, 1963.

Hull, Gloria. *Color, Sex, and Poetry: Three Women Writers of the Harlem Renaissance*. Bloomington: Indiana University Press, 1987.

Hull, Gloria, Patricia Bell Scott, and Barbara Smith, eds. *All the Women Are White, All the Blacks Are Men, but Some of Us Are Brave*. Old Westbury, N.Y.: Feminist Press, 1982.

Hutchinson, George. *The Harlem Renaissance in Black and White*. Cambridge: Harvard University Press, 1995.

Klotman, Phyllis. "The Black Writer in Hollywood, circa 1930: The Case of Wallace Thurman." In *Black American Cinema*, ed. Manthia Diawara, 80–92. London: Routledge, 1993.

———. "Wallace Henry Thurman." In *Afro-American Writers from the Harlem Renaissance to 1940*. Vol. 51 of *Dictionary of Literary Biography*, ed. Trudier Harris and Thadious M. Davis, 260–273. Detroit, Gale: 1987.

Lauter, Paul. "Race and Gender in the Shaping of the American Literary Canon: A Case Study from the Twenties." In *Canons and Contexts*, 22–47. New York: Oxford University Press, 1991.

Lemke, Sieglinde. *Primitivist Modernism: Black Culture and the Origins of Transatlantic Modernism*. New York: Oxford University Press, 1998.

Lewis, David Levering. *When Harlem Was in Vogue*. New York: Alfred A. Knopf, 1981.

Lewis, Theophilus. "Harlem Sketchbook: Wallace Thurman." *New York Amsterdam News*, January 5, 1935.

McDowell, Deborah E. *"The Changing Same": Black Women's Literature, Criticism, and Theory*. Bloomington: Indiana University Press, 1995.

Mills, Charles W. *Blackness Visible: Essays on Philosophy and Race*. Ithaca: Cornell University Press, 1998.

Moore, Jack B. *W.E.B. Du Bois*. Boston: Twayne, 1981.

Morrison, Toni. *Playing in the Dark: Whiteness and the Literary Imagination*. Cambridge: Harvard University Press, 1992.

Moses, Wilson J. "The Lost World of the New Negro, 1885–1919." *Black American Literature Forum* 21 (spring–summer 1987): 65–72.

Mullen, Edward J. *Critical Essays on Langston Hughes*. Boston: G. K. Hall, 1986.

Nelson, Emmanuel. *Critical Essays: Gay and Lesbian Writers of Color*. New York: Haworth Press, 1993.

Nero, Charles I. "Toward a Black Gay Aesthetic: Signifying in Contemporary Black Gay Literature." In *Brother to Brother: New Writings by Black Gay Men*, ed. Essex Hemphill, 229–252. Boston: Alyson, 1991.

North, Michael. *The Dialect of Modernism: Race, Language, and Twentieth-Century Literature*. New York: Oxford University Press, 1994.

Omi, Michael, and Howard Winant. *Racial Formation in the United States: From the 1960s to the 1980s*. 2d ed. New York: Routledge, 1994.

Perry, Margaret. *Silence to the Drums: A Survey of the Literature of the Harlem Renaissance*. Westport, Conn.: Greenwood Press, 1976.

Petesch, Donald A. "Wallace Thurman." In *A Spy in the Enemy's Country: The Emergence of Modern Black Literature,* 157–158, 258–266. Iowa City: University of Iowa Press, 1989.

Posnock, Ross. *Color and Culture: Black Writers and the Making of the Modern Intellectual.* Cambridge: Harvard University Press, 1998.

Rampersad, Arnold. *The Life of Langston Hughes.* 2 vols. New York: Oxford University Press, 1986–1988.

Reid-Pharr, Robert. "Tearing the Goat's Flesh: Homosexuality, Abjection, and the Production of a Late Twentieth-Century Black Masculinity." *Studies in the Novel* 28 (fall 1996): 372–394.

Singh, Amritjit. *The Novels of the Harlem Renaissance: Twelve Black Writers, 1923– 33.* University Park: Pennsylvania State University Press, 1976.

Singh, Amritjit, William Shiver, and Stanley Brodwin, eds. *The Harlem Renaissance: Revaluations.* New York: Garland, 1989.

Smith, Barbara. "Toward a Black Feminist Criticism." *Conditions: Two* (1977): 25–44.

Tidwell, John Edgar. "Two Writers Sharing: Sterling A. Brown, Robert Frost, and 'In Dives' Dive.'" *African American Review* 31.3 (1997): 399–408.

*Transition* 92 (2002). Special issue: "The Gay Harlem Renaissance."

van Notten, Eleonore. *Wallace Thurman's Harlem Renaissance.* Atlanta: Rodopi, 1994.

West, Dorothy. "Elephant's Dance: A Memoir of Wallace Thurman." *Black World* 20.1 (1970): 77–85.

Williams, Patricia. "The Obliging Shelf: An Informal Essay on Formal Equal Opportunity." *Michigan Law Review* 87 (1989): 2128–2151.

Wintz, Cary D., ed. *Black Culture and the Harlem Renaissance.* Houston: Rice University Press, 1988.

Wirth, Thomas H., ed. *Gay Rebel of the Harlem Renaissance: Selections from the Work of Richard Bruce Nugent.* Durham: Duke University Press, 2002.

Woodson, Jon. *To Make a New Race: Gurdjieff, Toomer, and the Harlem Renaissance.* Jackson: University Press of Mississippi, 1999.

Young, James O. *Black Writers of the Thirties.* Baton Rouge: Louisiana State University Press, 1973.

# Part One

# Essays on Harlem

It is evident that Harlem had fascinated Thurman long before he moved there. After his failure to create an artistic community of young black writers in Los Angeles, he arrived in Harlem on Labor Day, 1925. Thurman quickly made it his project to interpret and translate Harlem for others. In the first few years he lived there, Thurman published several articles about Harlem. In profiling this city of surprises, Thurman celebrates Harlem as a teeming, pulsating, complex amalgam of cultures, classes, peoples, and beliefs. In his many writings about Harlem, he strives to show that being black in America is a complex phenomenon. Thurman's awareness of the complexity of Harlem and multifaceted black identity parallels his complicated relationship to the New Negro Renaissance. As both artist and critic, insider and outsider, Thurman constructed a skeptical, ironic view of the figures and phenomena of Harlem in the mid-1920s and early 1930s.

Many of Thurman's published articles on Harlem revel in the color and energy of the place. At times, Thurman sounds like a decadent tour guide, introducing his readers to numbers-running, house rent parties, and "hot men." Often Thurman combines these sensational details with an insistence on the dignity of Harlem, on the inaccuracy of stereotypes about the place, and on the need for white America to know Harlem better. Through a range of subtle gestures, Thurman calls for the realization that Harlem, too, is New York. Sometimes through his lyricism, at other times in his arguments, Thurman highlights differences and similarities between Harlem and New York at large in order to increase the accessibility of Harlem for all of his readers. An example is "Harlem: A Vivid Word Picture of the World's Greatest Negro City" (1927)—first published in *American Monthly* with illustrations by Aaron Douglas—one of Thurman's most lyrical essays on Harlem. He uses dramatic situations and colorful language to paint a sketch of Harlem's movement, creativity, persistence, and hope. He invents words and phrases to re-create the cadences that characterize Harlem. By luxuriating in the sounds and images of Harlem life, the essay achieves Thurman's goal—to show Harlem in all its diversity, contradiction, and excitement. "Harlem Facets," another essay published in 1927, makes a forceful argument for revising the country's view of African Americans: "It is a matter of record that few white people ever see the whole of Harlem. . . . This is partially due to the fact that very few white people really know how like them the American Negro has become. They still cannot comprehend that the Aframerican is assimilating much more quickly than he is being assimilated." Thurman's awareness of the ways in which blacks and whites influence each other is reflected in his insistence on measuring black America with the same yardstick he uses for America in general.

Thurman's "Negro Life in New York's Harlem: A Lively Picture of a Popular and Interesting Section" combines his talent for personal anecdote, factual information, and colorful language to present a three-dimensional view of Harlem. This essay was first published in the fall 1927 issue of *The Haldeman-Julius Quarterly,* a self-described "debunking magazine" and "enemy of sham and hypocrisy." In 1928, the article was reissued as number 494 of the Haldeman-Julius Little Blue Book series, which included several canonical American writers. With sections such as "200,000 Negroes in Harlem," "The Amusement Life of Harlem," and "House Rent Parties, Numbers, and Hot Men," the essay details Harlem's diversity. Much more than a sketch, this work is so committed to Thurman's belief in the richness of Harlem life that it itemizes the community in street-by-street detail. It catalogs objective facts, retells personal stories, and insists upon Harlem's possibilities: "There is no typical Harlem Negro as there is no typical American Negro. There are too many different types and classes. White, yellow, brown, and black and all the intervening shades. North American, South American, African and Asian; Northern and Southern; high and low; seer and fool—Harlem holds them all, and strives to become a homogeneous community despite its motley, hodge-podge of incompatible elements and its self nurtured or outwardly imposed limitations."

In general, the Harlem essays left unpublished at Thurman's death exhibit a somewhat more skeptical view of life in Harlem than his earlier published essays. In "Harlem House Rent Parties," for example, the usually exuberant evaluation of the party's success is qualified by a suggestion of how temporary and fleeting the joy can be: "Despite the freedom and frenzy of these parties they are seldom joyous affairs. On the contrary they are rather sad and depressing. A tragic undercurrent runs through the music and is reflected in the eyes and faces of the dancers." While probably colored by his continuing debt and illness, these later unpublished essays nonetheless show Thurman still at work to make Harlem accessible to the world.

In most of his Harlem pieces, whether lyrical or argumentative, one is impressed by Thurman's sense of humor—sometimes self-deprecating, at other times painfully ironic. In many instances, Thurman uses humor to deflect his own ambivalence toward Harlem. Thurman uses the exuberant exclamations of "fy-ah lawd" in "Harlem: A Vivid Word Picture" or his discussion of gin mills and nightclubs in "Harlem Directory" to keep this ambivalence about Harlem active and in play. His affection for the place nevertheless comes through in this gathering of Harlem writings.

Like many other writers, Thurman had a tendency to recycle his own work, and his Harlem writings show how much he sometimes reworked his essays. Although he wrote about Harlem for a variety of audiences and in a variety of modes, Thurman consistently displays a strong belief in Harlem's promise: "It is a city in which anything might happen and everything does." He stresses the diversity of Harlem again and again. By evoking such a range of types, classes, backgrounds, styles, and outlooks, Thurman finds an exuberant way to resist white America's stereotypes of Harlem and of black life. For Thurman, Harlem is "[a] community cut out for speed and splendor, squalor and wealth, penury and prosperity" ("Harlem Facets"). Like James Weldon Johnson's *Black Manhattan* (1930) and Claude McKay's *Harlem: Negro Metropolis* (1940), Thurman's Harlem writings reveal the place in its many guises, glories, and excesses.

In these pieces, written over some nine years of engagement with the Negro community in New York, Thurman turns Harlem into a metaphor for the black American's complex fate as a human being. In the process, these essays serve collectively as a plea for a multidimensional understanding of black Americans—even as they demonstrate Thurman's talent for insightful and charming prose. Interestingly, these essays appear to stop short of the rejection of Harlem that one can read in the ending of Thurman's novel *Infants of the Spring*.

# Harlem: A Vivid Word Picture
# of the World's Greatest Negro City

*American Monthly,* May 1927

I

Harlem!

Spring on Seventh Avenue. Darkies from the levee moaning, "I got de blues an' I can't be satisfied." Oh, fy-ah lawd. Niggers from the cotton fields of Alabama singing of a white God seated in the sky. An ex-convict from Georgia with a high yaller wife making money by selling goods stolen from a warehouse. He is a "hot" man. A cockney coon from Barbados braying for rights. A former well fed bus boy making "folk songs out of soul sounds." A religious Cuban girl married to a child loving American Negro three quarters Indian. Oh, fy-ah lawd, Spring on Seventh Avenue, lending strength to color and adding fuel to the flame.

I'm goin' to heaben when I die. Cabaret in the cellar, church on the second floor. Deep in the ground sweating forms of men and women, haloed by colored lights and cigarette smoke, laugh, drink and dance. A peroxide blonde with a yellow face and brown arms concocts a new Charleston step. She wears pink bloomers. In her hands are dollar bills. In her left stocking top the addresses of a Chinese merchant and a leering round faced Babbitt from below the line. I likes nigger gals when they're yellow. Fy-ah's gonna burn ma soul.

Cabaret in the cellar church on the second floor. My sons in college done turned Episcopalian. My wife plays bridge with her high yaller friends and makes me eat in the kitchen since we moved on "strivers" row. I'm black. Heaben's goin' to be ma home.

Spring on Seventh Avenue. Harlem on parade. Seventh Avenue big with life and bulging with activity. Masterpieces of flesh, form and color irrigate the avenue. Human lava streams spewed forth from the daylight prejudiced depths of mephitic tenements, seeking air, light and contact. The janitor has been lax. The house has been cold. The landlord is about to raise the rent. Food, heat and electricity, fine clothes and church dues. Men, women and children. The stalwart and the weak. The sedate and the blind. Oh, fy-ah lawd. Spring on Seventh Avenue, lending strength to color and adding fuel to the flame.

School boys loitering by a drug store on the corner of 135th Street. A colored policeman escorts a colored taxi driver to the precinct station house and makes his complaint to a white sergeant. School boys alive with masculinity waiting on the corner for nine o'clock and news of parties. Prostitutes not soliciting this evening walking unnoticed and unmolested side by side with chic college girls home for an Easter vacation. An Abyssinian Jew, black of face and black of beard, blowing his nose on a handkerchief from the five and ten. A teakwood tan barber in his white coat walking to the corner with a coffee brown manicurist dressed in mourning. Two chattering Porto Ricans choking on hot dogs. Spring on Seventh Avenue. Fy-ah's gonna burn ma soul.

Pool hall johnnies in front of the Renaissance Casino watching eagerly for unescorted women to follow through the door. A pregnant woman from Jamaica runs the gauntlet. She wants to see the basket ball game and watch the dancers. Harlem fire has hardened her. A West Indian with a wart on his nose, a mis-fit top coat obscuring his little form, and an ill used cane on his arm, flaunts his sartorial inelegance. People laugh and move on. Monkey chaser. Harlem is the city of refuge and Seventh Avenue is freedman's lane.


## II

There is an urgence in the air, an urgence mad and rhythmic, an urgence inspiring folk to laugh and to walk, to smile and to loiter. Seventh Avenue is a hodge podge of color and forms, flowing along to the tune of jazz rhythms. Lenox Avenue is a defeated dung heap flung out to cover the subway underneath. Fifth Avenue is filthy and stark. Eight Avenue is dominated by the "L." St. Nicholas and Edgecombe are respectable and cold. Black America has a capital. Black America has a cosmopolitan center. Harlem is the capital of black America, and Harlem is rooted deeply in the granite cliffs of upper Manhattan.

Harlem is not to be seen. Or heard. It must be felt. Life there is deeper than laughing externalities, bold fronts, and grim exteriors. Behind a brownstone front may be a clay brick rear. In a sordid tenement may be found a well appointed drawing room. Poet and bootlegger live side by side. Musician and pickpocket eat in the same radio-entertained dining rooms. Preacher and physician, undertaker and dentist, "number" banker and postal clerk, Pullman porter and real estate shark are all aristocrats. Society seldom knows competition. It occasionally knows notoriety or family or achievement or color.

There is filth in Harlem. There is filth all over New York. There is starvation and wealth and food and sickness and death. People are dispossessed, their belongings piled on the pavement. People buy property and raise the rents in order to have period furniture. Baby perambulators choke narrow hallways. Svelte motor hearses mingle with the traffic. Ambulances crash by. Harlem is a ghetto struggling for more room and for more air. Harlem is a ghetto possible only in New York. Harlem is the capital of black America, the greatest Negro center in the world.

Everybody comes to New York. New York puts everybody in his place. Harlem is the Negroes' place. Negroes from the south whose ancestors were African slaves for generation on generation. Negroes whose ancestors were African, and Indian, and pale-face. Negroes from the west, the east, the north. Negroes from Africa, the West Indies, the Bahamas, the Central and South Americas, Cuba, Jamaica, everywhere. Ethiopia has stretched forth her wings. The curse of Ham has seared many folk. The eloquence of missionaries, the prosperity of traders, the urge of relatives, lynching, peonage and the restlessness of man have all helped to make Harlem Harlem and bring to its narrow immensity over 175,000 colored folk.

### III

Spring on Seventh Avenue. It will soon be Easter. Must play the "numbers." If I wins I'll see the "hot" man and get a new suit. If I don't I'll miss a payment on my furniture and get the suit anyway. Watch the clearing house reports. Play fifty cents a day on number 267. My wife dreamt about a white veil on a black child. The dream book says that both a white veil and a black child means number 267. Oh, fy-ah lawd, fifty cents a day.

Spring on Seventh Avenue. Children everywhere. Children dodging taxis, and policemen, and parents, and pedestrians. Children well dressed and subdued. Children well dressed and hoydenish. Children dirty and sad. Children dirty and riotous. Little boys playing leap frog with a water plug. Little girls jumping rope in the middle of the sidewalk. Adolescents writing dirty words on walls. Adolescents with city faces and crowded tenement minds. Babies fed at milk stations. Babies overcherished in pink coverlets and chintz nurseries. Harlem houses are incubators for new Negroes. Harlem sidewalks are breeding boards lined with garbage cans and pregnant with colorful seed.

Spring on Seventh Avenue. Newsboys distributing the latest edition of the World's Greatest Negro Weekly. A fraternal order holding a parade on their way to dedicate a new combination temple, theater, dance hall and hotel. Real estate values soar. Ten Jewish landlords turn their colored tenants over to ten Negro parvenus. A Negro blues singer buys a new Locomobile. There is an urgence in the air. Niggers getting more like white folks every day. Cabaret in the cellar, church on the second floor. Heabens goin' to be ma' home. Y.M.C.A. on the left side of the street. Six gin mills on the other. A church on one corner. An assignation house catering only to white men on the other. An ignorant longshoreman in Apartment 3B, and an educated Pullman porter with a lady physician for a wife next door. Oh, fy-ah lawd. Harlem is the city of refuge. Harlem is the promised land.

# Harlem Facets

*The World Tomorrow*, November 1927.

Harlem is the city of constant surprises, a city of ecstatic moments and diverting phenomena. It is a city in which anything might happen and everything does. It is a multifaceted ensemble, offering many surprise packets of persons, places, amusements, and vocations. It is a cosmos within itself. Life there is not stable and monotonous. Rather it is moving, colorful, and richly studded with contrasting elements and contradictory types.

Harlem is a boundary bursting coop with a population somewhere between 175 and 200 thousand persons, Negroes of all types and classes, a struggling mass of people with varied racial backgrounds, varied capacities for adaptation to a strange and sometimes sinister environment. It is a matter of record that few white people ever see the whole of Harlem. Despite the recent wave of public interest in the place, Harlem is still seen by the white world as a city of coons, cabarets, and black face comedians.

This is partially due to the fact that very few white people really know how like them the American Negro has become. They still cannot comprehend that the Aframerican is assimilating much more quickly than he is being assimilated. If his skin coloring is becoming more white, his mind has already become white in that he thinks, acts, dresses and makes progress in the same way and along the same lines as does the dominating white element in the American environment.

Thus in Harlem we find a community as American as Gopher Prairie or Zenith. A community keenly alert to the cosmopolitan currents swirling around it and through it. A community cut out for speed and splendor, squalor and wealth, penury and prosperity. It permits of everything possessed by that stupendous ensemble—New York City— of which it is a part. Like New York City Harlem is a cosmopolitan city. Its people are as varied and polyglot as could be found anywhere. To the laymen they are all indiscriminately lumped together as "Negroes" or "niggers." To themselves or to a scientific observer they are unclassifiable under any existent ethnic term. The racial complexity of the American Negro is already known. In his veins flows the mixed bloods of the Africans from whom he originally stemmed, the American Indians with whom he intermarried in pre and post slavery days, and of every white race under the sun. And in Harlem this home-grown ethnic amalgam is associated and intermixing with Negroes from the Caribbean, from Africa, Asia, South America, and any other place dark-skinned people hail from.

This makes an interesting and unusual collection. About 40 percent of the Negroes in Harlem are foreign born. The majority of these, 35 or 40 thousand, come from the

British West Indies. Then there are about ten thousand from the Virgin Islands, people forced to seek financial salvation in America because our national prohibition act blasted their rum trade. And there are about 8,000 from Spanish-speaking localities and islands in South and Central America. The remainder are recruited from French possessions in the Caribbean, from Cuba, Africa, Asia and what have you.

These people upon their arrival in New York find themselves segregated in a community the likes of which they have never seen before, and find themselves forced to mingle with other people with distinct cultural and lingual differences. Petty prejudices and race friction arise, the same as in any community where foreign born compete with home born for economic or social supremacy.

The American Negro takes pride in the fact that he is a citizen of the "world's greatest country" and is inclined to be cocky because he has had the advantages of a supposedly superior civilization, with modern plumbing, a universal educational system, and high wages. The foreign born Negro, who will often work for less wages because he is used to a lower standard of living, will quibble with the American Negro because the latter has not been free from slavery as long as he and because he feels that the Aframerican takes such matters as peonage, lynching, and segregation far too casually.

Naturally the various racial groups clash, but fortunately the struggle to live and the amount of mass energy needed to fight the white man's prejudice and discrimination leaves little time for actual intra-racial combat. They express their impatience and disgust with one another in a social or verbal way. The American Negro calls the West Indian Negro a "monkey chaser"; the retaliatory epithets cannot be reproduced here.

This is just one of the many sides of Negro life in Harlem that white people are practically unaware of. It is almost incomprehensible to them that the American Negro should share the American white man's prejudice against foreigners, and that he should vigorously resent their intrusion into his community.

Another aspect of Harlem little known or publicized is the wealth and social security of the upper strata of Negro society. It is taken for granted by most whites that all Negroes with the possible exception of those constantly in the spotlight like Roland Hayes, Robeson, Du Bois, or Weldon Johnson are in a class with their chauffeurs and washerwomen. They do not take into consideration that a large number of Negroes have long been emancipated from meniality, and that many have established fortunes or achieved enviable incomes.

Although the Negroes in Harlem have not been as energetic in commercial fields as have the Negroes of Chicago, Durham, North Carolina, and many other places, there are nevertheless any number of commercially prominent and wealthy Negroes there. There are no Negro bankers in Harlem, but there are a great number of eminently successful real estate operators. There are no longer capitalistic combines like the Overton enterprises in Chicago or the Malone concern in St. Louis, but there are many moneyed entrepreneurs operating and owning minor businesses. Negroes in Harlem do not own a theater or a dance hall of their own, but they do own over $60,000,000 worth of real estate consisting of many luxurious homes, ostentatious apartment hotels, restaurants, drug stores, beauty parlors and haberdasheries. They can be found operating many speakeasies even if they do not own or operate any of the corner saloons, and they own many barber shops even if no grocery stores or meat markets of importance.

However they get their money many Harlem Negroes have much of it. They live in expensively appointed homes and apartments, have maids and chauffeurs, entertain lavishly, send their sons to Columbia, Harvard, Yale and their daughters to Barnard, Vassar, and Wellesley. They attend auctions and invest in antiques and rare objects of art. Their clothes come from Fifth Avenue, and there are a great number of comings and goings in season to and from Europe, Atlantic City, the Maine woods and southern California. A great Negro middle class has been evolved, mercantile persons, forerunners of a future Negro aristocracy, and the founders of fortunes which they are building around nest eggs salted away by the preceding generation of washer women and Pullman porters.

To the white person who views Harlem from the raucous interior of a smoke filled, jazz drunken cabaret this side of Negro life is unknown. It is actually amazing what number of white people will assure you that they have seen and are authorities on Harlem and things Harlemese. When pressed for amplification they go into ecstasies over the husky-voiced blues singers, the dancing waiters, and Negro frequenters of cabarets who might well have stepped out of a caricature by Covarrubias. They can talk for hours about the abandon and physical impressiveness of a Harlem cabaret, the body contortions and hip-wrigglings of Negro dancers and the ecstatic freedom manifested by Negroes out for a gay time.

There seems to be something in these places that the cabarets "downtown" cannot approximate, something that at once thrills and tantalizes the white spectator, leaving him as disturbed as it does amused. But he hardly realizes that the reason he prefers going to a Negro cabaret for a good time is because there is more chance to let himself go in this Negro environment, more chance to lay aside his inhibitions. He is "above the line," and like a country boy in the city, he contributes a looseness to an already lax environment and revels in this hitherto inexperienced physical freedom.

Harlem cabarets were interesting once, and are interesting now to a novice, but their complexion has changed. The frequenters are almost 95 percent white. Negroes have been forced out of their own places of amusement, their jazz appropriated, their entertainers borrowed. There are over a dozen cabarets in Harlem. Should the white patronage be suddenly discontinued hardly three of them could remain open. Negroes spend much money for pleasure, almost as much as they spend for fine clothes, aids to the complexion, and hair pomades, but their cabaret expenditures are neither consistent enough or large enough to warrant the upkeep of such an oversupply as Harlem now has.

The house rent party piano player seems to have a most romantic and colorful career. House rent parties are a Harlem institution. True, they have their precedent in the "Chitlin' Switches" found in the middle western and southern Negro communities, but in Harlem they evolved a technique of their own which renders them indigenous and individual. They owe their origin to the fact that rental fees in Harlem are the last word in exorbitance, and, although tenants sublet every available bit of bed space, another source of income is still necessary in huge lumps to keep off the dispossess notice. Some folk give them weekly, some bi-weekly, some monthly, and others only in time of stress, but regardless of when or how often they are given, music is necessary, and there is not always a musician in the family or among the family's friends. Hence, the genesis of a new division of labor.

Professional givers of these house rent parties generally have more than one instrument to furnish their music, but the rank and file confine themselves to the hiring of only one person to play the piano. This individual, if he is personable, and capable, can play at some such party almost every night in the week. This, of course, stipulates the development of a type, so that all house rent party piano players are easily identified.

They are seldom good-looking, that is handsome, for handsome men of this type are too much in demand as pimps and paramours. They dress flashily in extreme styles. They must have a fair singing voice, a choice repertoire of "wise cracks," parodies, shouts and other such tricks of the trade. They unconsciously become able to regulate the scale of people's emotions, and pick their music accordingly, becoming more and more primitive, more and more vulgar, as the evening advances, and as the effects of corn liquor or synthetic gin become more palpable.

Negroes are still dutiful in their religious worship, still dutiful if less fervent. The old frame structures in which the sisters and brothers would moan and groan and shout with the spirit while ministerial emotionalists would shake the house with sermons on Heaven, Hell, salvation and eternal damnation have given way to stately ecclesiastical edifices in which pentacostalism is frowned upon, and fiery sermons leveled at sinners have given way to polite religious talks. The choirs now appareled in robes awe the audience with their classical hymnology. Sister Susan Brown from the Shiloh Baptist Church in Birmingham is now admonished by swallow-coated ushers to keep quiet during the services, her "amens" and "preach it, brothers" disturb those around her and so punctuate the minister's text that those in the rear of the balcony have difficulty following it. Simple services have given way to elaborate ceremonies. Ministers are college bred and have high-salaried assistants. This has happened in the white churches and it has also happened in the Negro churches.

The Harlem Negro has invested more money in church property than in any other one institution. Not only has he bought buildings from white congregations who picked up their Bibles and fled when they found their church property in the midst of a Negro neighborhood, but he has also built many modern superstructures of his own. And although the church is not as important a social center as it was five or ten years ago, it still furnishes a hub for much of the Negroes' social activity.

## Negro Life in New York's Harlem: A Lively Picture of a Popular and Interesting Section

*The Haldeman-Julius Quarterly*, fall 1927.

### I. A Lively Picture of a Popular and Interesting Section

Harlem has been called the Mecca of the New Negro, the center of black America's cultural renaissance, Nigger Heaven, Pickaninny Paradise, Capital of Black America, and various other things. It has been surveyed and interpreted, explored and exploited. It has had its day in literature, in the drama, even in the tabloid press. It is considered the most popular and interesting section of contemporary New York. Its fame is international; its personality individual and inimitable. There is no Negro settlement anywhere comparable to Harlem, just as there is no other metropolis comparable to New York. As the great south side black belt of Chicago spreads and smells with the same industrial clumsiness and stockyardish vigor of Chicago, so does the black belt of New York teem and rhyme with the cosmopolitan cross-currents of the world's greatest city. Harlem is Harlem because it is part and parcel of greater New York. Its rhythms are the lackadaisical rhythms of a transplanted minority group caught up and rendered half mad by the more speedy rhythms of the subway, Fifth Avenue and the Great White Way.

Negro Harlem is located on one of the choice sites of Manhattan Island. It covers the greater portion of the northwestern end, and is more free from grime, smoke and oceanic dampness than the lower eastside where most of the hyphenated American groups live. Harlem is a great black city. There are no shanty-filled, mean streets. No antiquated cobble-stoned pavement, no flimsy frame fire-traps. Little Africa has fortressed itself behind brick and stone on wide important streets where the air is plentiful and sunshine can be appreciated.

There are six main north and south thoroughfares streaming through Negro Harlem—Fifth Avenue, Lenox Avenue, Seventh Avenue, Eighth Avenue, Edgecombe and

St. Nicholas. Fifth Avenue begins prosperously at 125th Street, becomes a slum district about 131st Street, and finally slithers off into a warehouse-lined, dingy alleyway above 139th Street. The people seen on Fifth Avenue are either sad or nasty looking. The women seem to be drudges or drunkards, the men pugnacious and loud—petty thieves and vicious parasites. The children are pitiful specimens of ugliness and dirt.

The tenement houses in this vicinity are darkened dungheaps, festering with poverty-stricken and crime-ridden step-children of nature. This is the edge of Harlem's slum district; Fifth Avenue is its board-walk. Push carts line the curbstone, dirty push carts manned by dirtier hucksters, selling fly-specked vegetables and other cheap commodities. Evil faces leer at you from doorways and windows. Brutish men elbow you out of their way, dreary looking women scowl at and curse children playing on the sidewalk. That is Harlem's Fifth Avenue.

Lenox Avenue knows the rumble of the subway and the rattle of the cross-town street car. It is always crowded, crowded with pedestrians seeking the subway or the street car, crowded with idlers from the many pool halls and dives along its line of march, crowded with men and women from the slum district which it borders on the west and Fifth Avenue borders on the east. Lenox Avenue is Harlem's Bowery. It is dirty and noisy, its buildings ill-used, and made shaky by the subway underneath. At 140th Street it makes its one bid for respectability. On one corner there is Tabb's Restaurant and Grill, one of Harlem's most delightful and respectable eating houses; across the street is the Savoy building, housing a first-class dance hall, a motion picture theater and many small business establishments behind its stucco front. But above 141st Street Lenox Avenue gets mean and squalid, deprived of even its crowds of people, and finally peters out into a dirt pile, before leading to a carbarn at 147th St.

Seventh Avenue—Black Broadway—Harlem's main street, a place to promenade, a place to loiter, and avenue spacious and sleek with wide pavement, modern well-kept buildings, theaters, drug stores and other businesses. Seventh Avenue, down which no Negro dared walk twenty years ago unless he was prepared to fight belligerent Irishmen. Seventh Avenue, teeming with life and ablaze with color, the most interesting and important street in one of the most interesting and important city sections of greater New York.

Negro Harlem is best represented by Seventh Avenue. It is not like Fifth Avenue, filthy and stark, nor like Lenox, squalid and dirty. It is a grand thoroughfare into which every element of Harlem population ventures either for reasons of pleasure or of business. From 125th Street to 145th Street, Seventh Avenue is a stream of dark people going to churches, theaters, restaurants, billiard halls, business offices, food markets, barber shops and apartment houses. Seventh Avenue is majestic yet warm, and it reflects both the sordid chaos and the rhythmic splendor of Harlem.

From five o'clock in the evening until way past midnight, Seventh Avenue is one electric-lit line of brilliance and activity, especially during the spring, summer and early fall months. Dwelling houses are close, overcrowded and dark. Seventh Avenue is the place to seek relief. People everywhere. Lines of people in front of the box offices of the Lafayette Theater at 132nd Street, the Renaissance motion picture theaters at 138th Street and the Roosevelt Theater at 145th Street. Knots of people in front of the Metropolitan Baptist Church at 129th Street and Salem M.E. Church, which dominates the corner at 129th Street.

People going into the cabarets. People going into speakeasies and saloons. Groups of boisterous men and boys, congregated on corners and in the middle of the blocks, making remarks about individuals in the passing parade. Adolescent boys and girls flaunting their youth. Street speakers on every corner. A Hindoo fakir here, a loud-voiced Socialist there, a medicine doctor ballyhooing, a corn doctor, a blind musician, serious people, gay people, philanderers and preachers. Seventh Avenue is filled with deep rhythmic laughter. It is a civilized lane with primitive traits, Harlem's most representative street.

Eighth Avenue supports the elevated lines. It is noticeably Negroid only from 135th Street to 145th Street. It is packed with dingy, cheap shops owned by Jews. Above 139th Street, the curbstone is lined with push-cart merchants selling everything from underwear to foodstuffs. Eighth Avenue is dark and noisy. The elevated trestle and its shadows dominate the street. Few people linger along its sidewalks. Eighth Avenue is a street for business, a street for people who live west of it to cross hurriedly in order to reach places located east of it.

Edgecombe, Bradhurst and St. Nicholas Avenues are strictly residential thoroughfares of the better variety. Expensive modern apartment houses line these streets. They were once occupied by well-to-do white people who now live on Riverside Drive, West End Avenue, and in Washington Heights. They are luxuriously appointed with imposing entrances, elevator service, disappearing garbage cans, and all the other appurtenances that make a modern apartment house convenient. The Negroes who live in these places are either high-salaried workingmen or professional folk.

Most of the cross streets in Harlem, lying between the main north and south thoroughfares, are monotonous and overcrowded. There is little difference between any of them save that some are more dirty and more squalid than others. They are lined with ordinary, undistinguished tenement and apartment houses. Some are well kept, others are run down. There are only four streets that are noticeably different, 136th Street, 137th Street, 138th Street, and 139th Street west of Seventh Avenue and these are the only blocks in Harlem that can boast of having shade trees. An improvement association organized by people living in these streets, strives to keep them looking respectable.

Between Seventh and Eighth Avenues, is 139th Street, known among Harlemites as "strivers' row." It is the most aristocratic street in Harlem. Stanford White designed the houses for a wealthy white clientele. Moneyed Negroes now own and inhabit them. When one lives on "strivers' row" one has supposedly arrived. Harry Wills resides there, as do a number of the leading Babbitts and professional folk of Harlem.

## II. 200,000 Negroes in Harlem

There are approximately 200,000 Negroes in Harlem. Two hundred thousand Negroes drawn from all sections of America, from Europe, the West Indies, Africa, Asia, or where you will. Two hundred thousand Negroes living, loving, laughing, crying, procreating and dying in the segregated city section of Greater New York, about twenty-five blocks long and seven blocks wide. Like all of New York, Harlem is overcrowded. There are as many as 5,000 persons living in some single blocks; living in dark, mephitic tenements, jammed together, brownstone fronts, dingy elevator flats and modern apartment houses.

Living conditions are ribald and ridiculous. Rents are high and sleeping quarters at a premium. Landlords profiteer and accept bribes, putting out one tenant in order to house another willing to pay more rent. Tenants, in turn, sublet and profiteer on roomers. People rent a five-room apartment, originally planned for a small family, and crowd two over-sized families into it. Others lease or buy a private house and partition off spacious front and back rooms into two or three parts. Hallways are curtained off and lined with cots. Living rooms become triplex apartments. Clothes closets and wash-rooms became kitchenettes. Dining rooms, parlors, libraries, drawing rooms are all pro-faned by cots, day beds and snoring sleepers.

There is still little privacy, little unused space. The man in the frontroom of a rail-road flat, so called because each room opens into the other like coaches on a train, must pass through three other bedrooms in order to reach the bathroom stuck on the end of the kitchen. He who works nights will sleep by day in the bed of one who works days, and vice versa. Mother and father sleep in a three-quarter bed. Two adolescent children sleep on a portable cot set up in the parents' bedroom. Other cots are dragged by night from closets and corners to be set up in the dining room, in the parlor, or even in the kitchen to accommodate the remaining members of the family. It is all disconcerting, mad. There must be expansion. There is expansion, but it is not rapid enough or con-tinuous enough to keep pace with the ever-growing population of Negro Harlem.

The first place in New York where Negroes had a segregated community was in Greenwich Village, but as the years passed and their numbers increased they soon moved northward into the twenties and lower thirties west of Sixth Avenue until they finally made one big jump and centered around west Fifty-third Street. About 1900, looking for better housing conditions, a few Negroes moved to Harlem. The Lenox Avenue subway had not yet been built and white landlords were having difficulty in keeping white tenants east of Seventh Avenue because of the poor transportation facilities. Being good businessmen they eagerly accepted the suggestion of a Negro real estate agent that these properties be opened to colored tenants. Then it was discovered that the few houses available would not be sufficient to accommodate the sudden influx. Negroes began to creep west of Lenox Avenue. White property owners and residents began to protest and tried to find means of checking or evicting unwelcome black neighbors. Negroes kept pouring in. Negro capital, belligerently organized, began to buy all available properties.

Then, to quote James Johnson,

> the whole movement, in the eyes of the whites, took on the aspect of an "invasion"; they became panic stricken and began fleeing as from a plague. The presence of one colored family in a block no matter how well-bred and orderly, was sufficient to precipitate a flight. House after house and block after block was actually deserted. It was a great demonstration of human beings running amuck. None of them stopped to reason why they were doing it or what would happen if they didn't. The banks and the lending companies holding mortgages on these deserted houses were compelled to take them over. For some time they held these houses vacant, preferring to do that and carry the charges than to rent or sell them to colored people. But values dropped and continued to drop until at the outbreak of the war in Europe property in the northern part of Harlem had reached the nadir.

With the war came a critical shortage of common labor and the introducing of thousands of southern Negroes into northern industrial and civic centers. A great migration took place. Negroes were in search of a holy grail. Southern Negroes, tired of moral and financial blue days, struck out for the promised land, to seek adventure among factories, subways, and skyscrapers. New York, of course, has always been a magnet for ambitious and adventurous Americans and foreigners. New York to the Negro meant Harlem, and the great influx included not only thousands of Negroes from every state in the Union, but also over thirty thousand immigrants from the West Indian Islands and the Caribbean regions. Harlem was the promised land.

Thanks to New York's many and varied industries, Harlem Negroes have been able to demand and find much work. There is a welcome and profitable diversity of employment. Unlike Negroes in Chicago, or in Pittsburgh, or in Detroit, no one industry is called upon to employ the greater part of their population. Negroes have made money in New York; Negroes have brought money to New York with them, and with this money they have bought property, built certain civic institutions and increased their business activities until their real estate holdings are now valued at more than sixty million dollars.

### III. The Social Life of Harlem

The social life of Harlem is both complex and diversified. Here you have two hundred thousand people collectively known as Negroes. You have pure-blooded Africans, British Negroes, Spanish Negroes, Portuguese Negroes, Dutch Negroes, Danish Negroes, Cubans, Porto Ricans, Arabians, East Indians and black Abyssinian Jews in addition to the racially well-mixed American Negro. You have persons of every conceivable shade and color. Persons speaking all languages, persons representative of many cultures and civilizations. Harlem is a magic melting pot, a modern Babel mocking the gods with its cosmopolitan uniqueness.

The American Negro predominates and, having adopted all of white America's prejudices and manners, is inclined to look askance at his little dark-skinned brothers from across the sea. The Spanish Negro, i.e., those Negroes hailing from Spanish possessions, stays to himself and has little traffic with the other racial groups in his environment. The other foreigners, with the exception of the British West Indians are not large enough to form a separate social group and generally become quickly identified with the regulation social life of the community.

It is the Negro from the British West Indies who creates and has to face a disagreeable problem. Being the second largest Negro Group in Harlem, and being less susceptible to American manners and customs than others, he is frowned upon and berated by the American Negro. This intra-racial prejudice is an amazing though natural thing. Imagine a community made up of people universally known as oppressed, wasting time and energy trying to oppress others of their kind, more recently transplanted from a foreign clime. It is easy to explain. All people seem subject to prejudice, even those who suffer from it most and all people seem inherently to dislike other folk who are characterized by cultural and lingual differences. It is a failing of man, a curse of humanity, and if these differences are accompanied, as they usually are, by quarrels concerning economic matters, there is bound to be an intensifying of the bitter antagonism existent between the two groups. Such has been that case with the British West Indian in Harlem. Because of his numerical strength, because of his cockney English inflections and accent, because of his unwillingness to submit to certain American do's and don'ts, and because he, like most foreigners, has seemed willing to work for low wages, he has been hated and abused by his fellow-Harlemites. And, as a matter of protection, he has learned to fight back.

It has been said that West Indians are comparable to Jews in that they are "both ambitious, eager for education, willing to engage in business, argumentative, aggressive, and possess a great proselytizing zeal for any cause they espouse." Most of the retail business in Harlem is owned and controlled by West Indians. They are also well represented and often officiate as provocative agents and leaders in radical movements among Harlem Negroes. And it is obvious that the average American Negro, in manifesting a dislike for the West Indian Negro, is being victimized by that same delusion which he claims blinds the American white man; namely, that all Negroes are alike. There are some West Indians who are distasteful; there are some of all people about whom one could easily say the same thing.

It is to be seen then that all this widely diversified population would erect an elaborate social structure. For instance, there are thousands of Negroes in New York from Georgia. These have organized themselves into many clubs, such as the Georgia Circle or the Sons of Georgia. People from Virginia, South Carolina, Florida and other states do likewise. The foreign contingents also seem to have a mania for social organization. Social clubs and secret lodges are legion. And all of them vie with one another in giving dances, parties, entertainments and benefits in addition to public turnouts and parades.

Speaking of parades, one must mention Marcus Garvey. Garvey, a Jamaican, is one of the most widely known Negroes in contemporary life. He became notorious because of his Back-to-Africa campaign. With the West Indian population of Harlem as a nucleus, he enlisted the aid of thousands of Negroes all over America in launching the Black Star Line, the purpose of which was to establish a trade and travel route between America and Africa by and for Negroes. He also planned to establish a black empire in Africa of which he was to be emperor. The man's imagination and influence were colossal; his manifestations of these qualities often ridiculous and adolescent, though they seldom lacked color and interest.

Garvey added much to the gaiety and life of Harlem with his parades. Garmented in a royal purple robe with crimson trimmings and an elaborate headdress, he would ride in state down Seventh Avenue in an open limousine, surrounded and followed by his personal cabinet of high chieftains, ladies in waiting and protective legion. Since his incarceration in Atlanta Federal prison on a charge of having used the mails to defraud, Harlem knows no more such spectacles. The street parades held now are uninteresting and pallid when compared to the Garvey turnouts, brilliantly primitive as they were.

In addition to the racial and territorial divisions of the social structure there are also minor divisions determined by color and wealth. First there are the "dictys," that class of Negroes who constitute themselves as the upper strata and have lately done much wailing in the public places because white and black writers have seemingly overlooked them in their delineations of Negro life in Harlem. This upper strata is composed of the more successful and more socially inclined professional folk—lawyers, doctors, dentists, druggists, politicians, beauty parlor proprietors and real estate dealers. They are for the most part mulattos of light brown skin and have succeeded in absorbing all the social mannerisms of the white American middle class. They live in the stately rows of houses on 138th and 139th Streets between Seventh and Eighth Avenues or else in the "high-tone" apartment houses on Edgecombe and St. Nicholas. They are both stupid and snobbish as is their class in any race. Their most compelling if sometimes unconscious ambition is to be as near white as possible, and their greatest expenditure of energy is concentrated on eradicating any trait or characteristic commonly known as Negroid.

Their homes are expensively appointed, and comfortable. Most of them are furnished in good taste, thanks to the interior decorator who was hired to do the job. Their existence is one of smug complacence. They are well satisfied with themselves and with their class. They are without a doubt the basic element from which the Negro aristocracy of the future will evolve. They are also good illustrations, mentally, sartorially, and socially, of what the American standardizing machine can do to susceptible material.

These people have a social life of their own. They attend formal dinners and dances, resplendent in chic expensive replicas of Fifth Avenue finery. They arrange suitable intercoterie weddings, preside luxuriously at announcement dinners, pre-nuptial showers, wedding breakfasts and the like. They attend church socials, fraternity dances and sorority gatherings. They frequent the downtown theaters, and occasionally, quite occasionally, drop into one of the Harlem night clubs which certain of their lower caste brethren frequent and white downtown excursionists make wealthy.

Despite this upper strata which is quite small, social barriers among Negroes are not as strict and well regulated in Harlem as they are in other Negro communities. Like all cosmopolitan centers Harlem is democratic. People associate with all types should chance happen to throw them together. There are a few aristocrats, a plethora of striving bourgeoisie, a few artistic spirits and a great proletarian mass, which constitutes the most interesting and important element in Harlem, for it is this latter class and their institutions that gives the community its color and fascination.

## IV. Night Life in Harlem

Much has been written and said about night life in Harlem. It has become the *leit motif* of sophisticated conversation and shop girl intimacies. To call yourself a New Yorker you must have been to Harlem at least once. Every up-to-date person knows Harlem, and knowing Harlem generally means that one has visited a night club or two. These night clubs are now enjoying much publicity along with the New Negro and Negro art. They are the shrines to which white sophisticates, Greenwich Village artists, Broadway revelers and provincial commuters make eager pilgrimage. In fact, the white patronage is so profitable and so abundant that Negroes find themselves crowded out and even segregated in their own places of jazz.

There are, at the present time, about one dozen of these night clubs in Harlem—Bamville, Connie's Inn, Baron Wilkins, the Nest, Small's Paradise, the Capitol, the Cotton Club, the Green Cat, the Sugar Cane Club, Happy Rhones, the Hoofers Club and the Little Savoy. Most of these generally have from two to ten white person for every black one. Only the Hoofers, the Little Savoy, and the Sugar Cane Club seem to cater almost exclusively to Negro trade.

At Bamville and at Small's Paradise, one finds smart white patrons, the type that reads the ultrasophisticated *New Yorker*. Indeed, that journal says in its catalogue of places to go—"Small's and Bamville are the show places of Harlem for downtowners on their first excursion. Go late. Better not to dress." And so the younger generation of Broadway, Park Avenue, Riverside Drive, Third Avenue and the Bronx go late, take their own gin, applaud the raucous vulgarity of the entertainers, dance with abandon and go home with a headache. They have seen Harlem.

The Cotton Club and Connie's Inn make a bid for theatrical performers and well-to-do folk around town. The Nest and Happy Rhones attract traveling salesmen, store clerks and commuters from Jersey and Yonkers. The Green Cat has a large Latin clientele. Baron Wilkins draws glittering ladies from Broadway with their sleek gentlemen friends. Because of these conditions of invasion, Harlem's far-famed night clubs have become merely side shows staged for sensation-seeking whites. Nevertheless, they are still an egregious something to experience. Their smoking cavernous depths are eerie and ecstatic. Patrons enter, shiver involuntarily, then settle down to be shoved about and scared by the intangible rhythms that surge all around them. White night clubs are noisy. White night clubs affect weird music, soft light, Negro entertainers and dancing waiters, but, even with all these contributing elements, they cannot approximate the infectious rhythm and joy always found in a Negro cabaret.

Take the Sugar Cane Club on Fifth Avenue near 135th Street, located on the border of the most "low-down" section of Harlem. This place is visited by few whites or few "dicty" Negroes. Its customers are the rough-and-ready, happy-go-lucky more primitive type—street walkers, petty gamblers and pimps, with an occasional adventurer from other strata of society.

The Sugar Cane Club is a narrow subterranean passageway about twenty-five feet wide and 125 feet long. Rough wooden tables, surrounded by rough wooden chairs, and the orchestra stands, jammed into the right wall center, use up about three-quarters of the space. The remaining rectangular area is bared for dancing. With a capacity for seating about one hundred people, it usually finds room on gala nights for twice that many.

The orchestra weeps and moans and groans as only an unsophisticated Negro jazz orchestra can. A blues singer croons vulgar ditties over the tables to individual parties or else wah-wahs husky syncopated blues songs from the center of the floor. Her act over, the white lights are extinguished, red and blue spot lights are centered on the diminutive dancing space, couples push back their chairs, squeeze out from behind the tables and from against the wall, then finding one another's bodies, sweat gloriously together, with shoulders hunched, limbs obscenely intertwined and hips wiggling; animal beings urged on by liquor and music and physical contact.

Small's Paradise, on Seventh Avenue near 135th Street, is just the opposite of the Sugar Cane Club. It caters almost exclusively to white trade with just enough Negroes present to give the necessary atmosphere and "difference." Yet even in Small's with its symphonic orchestra, full-dress appearance and dignified onlookers, there is a great deal of that unexplainable, intangible rhythmic presence so characteristic of a Negro cabaret.

In addition to the well-known cabarets, which are largely show places to curious whites, there are innumerable places—really speakeasies—which are open only to the initiate. These places are far more colorful and more full of spontaneous joy than the larger places to which one has ready access. They also furnish more thrills to the spectator. This is possible because the crowd is more select, the liquor more fiery, the atmosphere more intimate and the activities of the patrons not subject to be watched by open-mouthed white people from downtown and the Bronx.

One particular place known as the Glory Hole is hidden in a musty, damp basement behind an express and trucking office. It is a single room about ten feet square and remains an unembellished basement except for a planed down plank floor, a piano, three chairs and a library table. The Glory Hole is typical of its class. It is a social club, commonly called a dive, convenient for the high times of a certain group. The men are unskilled laborers during the day, and in the evenings they round up their girls or else meet them at the rendezvous in order to have what they consider and enjoy as a good time. The women, like the men, swear, drink and dance as much and as vulgarly as they please. Yet they do not strike the observer as being vulgar. They are merely being and doing what their environment and their desire for pleasure suggest.

Such places as the Glory Hole can be found all over the so-called "bad lands" of Harlem. They are not always confined to basement rooms. They can be found in apartment flats, in the rear of barber shops, lunch counters, pool halls, and other such conveniently blind places. Each one has its regular quota of customers with just enough new patrons introduced from time to time to keep the place alive and prosperous. These intimate, lowdown civic centers are occasionally misjudged. Social service reports damn them with the phrase "breeding places of vice and crime." They may be. They are also good training grounds for prospective pugilists. Fights are staged with regularity and with vigor. And most of the regular customers have some mark on their faces or bodies that can be displayed as having been received during a battle in one of the glory holes.

The other extreme of amusement places in Harlem is exemplified by the Bamboo Inn, a Chinese-American restaurant that features Oriental cuisine, a jazz band and dancing. It is the place for select Negro Harlem's night life, the place where debutantes have their coming out parties, where college lads take their co-eds and society sweethearts and where dignified matrons entertain. It is a beautifully decorated establishment, glorified by a balcony with booths, and a large gyroflector, suspending from the center of the ceiling, on which colored spot lights play, flecking the room with triangular bits of vari-colored light. The Bamboo Inn is the place to see "high Harlem" just like the Glory Hole is *the* place to see "low Harlem." Well-dressed men escorting expensively garbed women and girls; models from Vanity Fair with brown, yellow and black skins. Doctors and lawyers, Babbitts and their ladies with fine manners (not necessarily learned through Emily Post), fine clothes and fine homes to return to when the night's fun has ended.

The music plays. The gyroflector revolves. The well-bred, polite dancers mingle on the dance floor. There are a few silver hip flasks. There is an occasional burst of too-spontaneous-for-the-environment laughter. The Chinese waiters slip around, quiet and bored. A big black-face bouncer, arrayed in tuxedo, watches eagerly for some too boisterous, too unconventional person to put out. The Bamboo Inn has only one blemishing feature. It is also the rendezvous for a set of oriental men who favor white women, and who, with their pale face partners, mingle with Harlem's four hundred.

When Harlem people wish to dance, without attending a cabaret, they go to the Renaissance Casino or to the Savoy, Harlem's two most famous public dance halls. The Savoy is the pioneer in the field of giving dance-loving Harlemites some place to gather nightly. It is an elaborate ensemble with a Chinese garden (Negroes seem to have a penchant for Chinese food—there are innumerable Chinese restaurants all over Harlem), two orchestras that work in relays, and hostesses provided at twenty-five cents per dance for partnerless young men. The Savoy opens at three in the afternoon and closes at three in the morning. One can spend twelve hours in this jazz palace for sixty-five cents, and the price of a dinner or an occasional sustaining sandwich and drink. The music is good, the dancers are gay, and the setting is conducive to joy.

The Renaissance Casino was formerly a dance hall, rented out only for social affairs, but when the Savoy began to flourish, the Renaissance, after closing a while for re-decorations, changed its policy and reopened as a public dance hall. It has no lounging room, or Chinese garden, but it stages a basket ball game every Sunday night that is one of the most popular amusement institutions in Harlem, and it has an exceptionally good orchestra, comfortable sitting-out places and a packed dance floor nightly.

Then, when any social club wishes to give a dance at the Renaissance, the name of the organization is flashed from the electric signboard that hangs above the entrance and in return for the additional and assured crowd, some division of the door receipts is made. The Renaissance is, I believe, in good Harlemese, considered more "dicty" than the Savoy. It has a more regulated and more dignified clientele, and almost every night in the week the dances are sponsored by some well-known social group.

In addition to the above two places, the Manhattan Casino, an elaborate dance palace, is always available for the more de luxe gatherings. It is at the Manhattan Casino that the National Association for the Advancement of Colored People has its yearly whist tournament and dance, that Harlem society folk have their charity balls, and select formals, and that the notorious Hamilton Lodge holds its spectacular masquerade each year.

All of the dances held in this Casino are occasions never to be forgotten. Hundreds of well-dressed couples dancing on the floor. Hundreds of Negroes of all types and colors, mingling together on the dance floor, gathering in the boxes, meeting and conversing on the promenade. And here and there an occasional white person, or is it a Negro who can "pass"?

Negroes love to dance, and in Harlem where the struggle to live is so intensely complex, the dance serves as a welcome and feverish outlet. Yet it is strange that none of these dance palaces are owned or operated by Negroes. The Renaissance Casino was formerly owned by a syndicate of West Indians, but has now fallen into the hands of a Jewish group. And despite the thousands of dollars Negroes spend in order to dance, the only monetary returns in their own community are the salaries paid to the Negro musicians, ushers, janitors and door-men. The rest of the profits are spent and exploited outside of Harlem.

This is true of most Harlem establishments. The Negro in Harlem is not, like the Negro in Chicago and other metropolitan centers, in charge of the commercial enterprises located in his community. South State Street in Chicago's great Black belt, is studded with Negro banks, Negro office buildings, housing Negro insurance companies, manufacturing concerns, and other major enterprises. There are no Negro controlled banks in Harlem. There are only branches of downtown Manhattan's financial institutions, manned solely by whites and patronized almost exclusively by Negroes. Harlem has no outstanding manufacturing concern like the Overton enterprises in Chicago, the Poro school and factory in St. Louis, or the Madame Walker combine in Indianapolis. Harlem Negroes own over sixty million dollars worth of real estate, but they neither own nor operate one first-class grocery store, butcher shop, dance hall, theater, clothing store or saloon. They do invest their money in barber shops, beauty parlors, pool halls, tailor shops, restaurants, and lunch counters.

## V. The Amusement Life of Harlem

Like most good American communities the movies hold a primary position in the amusement life of Harlem. There are several neighborhood motion picture houses in Negro Harlem proper, and about six big time cinema palaces on 125th Street that have more white patronage than black, yet whose audiences are swelled by movie fans from downtown.

The picture emporiums of Harlem are comparable to those in any residential neighborhood. They present second and third run features with supporting bills of comedies, novelties, and an occasional special performance when the management presents a bathing beauties contest, a plantation jubilee, an amateur ensemble and other vaudeville stunts. The Renaissance Theater, in the same building with the Renaissance Casino, is the cream of Harlem motion picture houses. It, too, was formerly owned and operated by Negroes, the only one of its kind in Harlem. Now Negroes only operate it. The Renaissance attracts the more select movie audiences; it has a reputable symphony orchestra, a Wurlitzer organ, and presents straight movies without vaudeville flapdoodle. It is spacious and clean and free from disagreeable odors.

The Roosevelt Theater, the New Douglas, and the Savoy are less aristocratic competitors. They show the same pictures as the Renaissance, but seem to be patronized by an entirely different set of people, and, although their interiors are more spacious, they are not as well decorated or as clean as the Renaissance. They attract a set of fresh youngsters, smart aleck youths and lecherous adult males who attend, not so much to see the picture as to pick up a susceptible female or to spoon with some girl they have picked up elsewhere. The places are also frequented by family groups, poor but honest folk, who cannot afford other forms or places of amusement.

The Franklin and the Gem are the social outcasts of the group. Their audiences are composed almost entirely of loafers from the low-grade pool rooms and dives in their vicinity, and tenement-trained drudges from the slums. The stench in these two places is nauseating. The Board of Health rules are posted conspicuously, admonishing patrons not to spit on the floor or to smoke in the auditorium, but the aisle is slippery with tobacco spew and cigarette smoke adds to the density of the foul air. The movies flicker on the screen, some wild west pictures three or four years old, dirty babies cry in time with the electric piano that furnishes the music, men talk out loud, smoke, spit, and drop empty gin or whiskey bottles on the floor when emptied.

All of these places from the Renaissance to the Gem are open daily from two in the afternoon until eleven at night, and save for a lean audience during the supper hour are usually filled to capacity. Saturdays, Sundays and holidays are harvest times, and the Jewish representatives of the chain to which a theater belongs walk around excitedly and are exceedingly gracious, thinking no doubt of the quarters that are being deposited at the box office.

The Lafayette and Lincoln theaters are three-a-day combination movie and musical comedy revue houses. The Lafayette used to house a local stock company composed of all Negro players, but it has now fallen into less dignified hands. Each week it presents a new revue. These revues are generally weak-kneed, watery variations on downtown productions. If Earl Carroll is presenting Artists and Models on Broadway, the Lafayette presents Brown Skin Models in Harlem soon afterwards. Week after week one sees [the] same type of "high yaller" chorus, hears the same blues song, and applauds different dancers doing the same dance steps. There is little originality on the part of the performers, and seldom any change of fare. Cheap imitations of Broadway successes, nudity, vulgar dances and vulgar jokes are the box office attractions.

On Friday nights there is a midnight show, which is one of the most interesting spectacles in Harlem. The performance begins some time after midnight and lasts until four or four-thirty the next morning. The audience is as much if not more interesting and amusing than the performers on the stage. Gin bottles are carried and passed among groups of friends. Cat calls and hisses attend any dull bit. Outspoken comments punctuate the lines, songs, and dances of the performers. Impromptu acts are staged in the orchestra and in the gallery. The performers themselves are at their best and leave the stage to make the audience a part of their act. There are no conventions considered, no reserve is manifested. Everyone has a jolly good time, and after the theater there are parties, or work according to the wealth and inclinations of the individual.

The Lincoln theater is smaller and more smelly than the Lafayette, and most people who attend the latter will turn up their noses at the Lincoln. It too has revues and movies, and its only distinguishing feature is that its shows are even worse than those staged at the Lafayette. They are so bad that they are ludicrously funny. The audience is comparable to that found in the Lafayette on Friday nights at the midnight jamboree. Performers are razzed. Chorus girls are openly courted or damned, and the spontaneous utterances of the patrons are far more funny than any joke the comedians ever tell. If one can stand the stench, one can have a good time for three hours or more just by watching the unpredictable and surprising reactions of the audience to what is being presented on the stage.

## VI. House Rent Parties, Numbers, and Hot Men

The Harlem institutions that intrigue the imagination and stimulate the most interest on the part of an investigator are House Rent Parties, Numbers, and Hot Men. House Rent parties are the result of high rents. Private houses containing nine or ten or twelve rooms rent from $185 up to $250 per month. Apartments are rated at $20 per room or more, according to the newness of the building and the convenience therein. Five-room flats, located in walk-up tenements, with inside rooms, dark hallways and dirty stairs rent for $10 per room or more. It can be seen then that when the average Negro workingman's salary is considered (he is often paid less for his labors than a white man engaged in the same sort of work), and when it is also considered that he and his family must eat, dress and have some amusements and petty luxuries, these rents assume a criminal enormity. And even though every available bit of unused space is sub-let at exorbitant rates to roomers, some other sources of revenue is needed when the time comes to meet the landlord.

Hence we have hundreds of people opening their apartments and houses to the public, their only stipulation being that the public pay twenty-five cents admission fee and buy plentifully of the food and drinks offered for sale. Although one of these parties can be found any time during the week, Saturday night is favored. The reasons are obvious; folk don't have to get up early on Sunday morning and most of them have had a pay day.

Of course, this commercialization of spontaneous pleasure in order to pay the landlord has been abused, and now there are folk who make their living altogether by giving alleged House Rent Parties. This is possible because there are in Harlem thousands of people with no place to go, thousands of people lonesome, unattached and cramped, who stroll the streets eager for a chance to form momentary contacts, to dance, to drink, and make merry. They willingly part with more of the week's pay than they should just to enjoy an oasis in the desert of their existence and a joyful intimate party, open to the public yet held in a private home, is, as they say, "their meat."

So elaborate has the technique of these parties and their promotion become that great competition has sprung up between prospective party givers. Private advertising stunts are resorted to, and done quietly so as not to attract too much attention from the police, who might want to collect a license fee or else drop in and search for liquor. Cards are passed out in pool halls, subway stations, cigar stores, and on the street. This is an example:

Hey! Hey!
Come on boys and girls let's shake
that thing.
Where?
At
Hot Poppa Sam's
West 134th Street, three flights up.
Jelly Roll Smith at the piano
Saturday Night, May 7, 1927
Hey! Hey!

Saturday night comes. There may be only piano music, there may be a piano and a drum, or a three or four-piece ensemble. Red lights, dim and suggestive, are in order. The parlor and the dining room are cleared for the dance, and one bedroom is utilized for hats and coats. In the kitchen will be found boiled pigs' feet, ham hock and cabbage, hopping John (a combination of peas and rice), and other proletarian dishes.

The music will be barbarous and slow. The dancers will use their bodies and the bodies of their partners without regard to the conventions. There will be little restraint. Happy individuals will do solo specialties, will sing, dance—have Charleston and Black Bottom contests and breakdowns. Hard little tenement girls will flirt and make dates with Pool Hall Johnnies and drug store cowboys. Prostitutes will drop in and slink out. And in addition to the liquor sold by the house, flasks of gin, and corn and rye will be passed around and emptied. Here "low" Harlem is in its glory, primitive and unashamed.

I have counted as many as twelve such parties in one block, five in one apartment house containing forty flats. They are held all over Harlem with the possible exception of 137th, 138th, and 139th Streets between Seventh and Eighth Avenues where the bulk of Harlem's upper class lives. Yet the house rent party is not on the whole a vicious institution. It serves a real and vital purpose, and is as essential to "low Harlem" as the cultured receptions and soirees held on "strivers' row" are to "high Harlem."

House rent parties have their evils; it is an economic evil and a social evil that makes them necessary, but they also have their virtues. Like all other institutions of man it depends upon what perspective you view them from. But regardless of abstract matters, house rent parties do provide a source of revenue to those in difficult financial straits, and they also give lonesome Harlemites, caged in by intangible bars, some place to have their fun and forget problems of color, civilization, and economics.

Numbers, unlike house rent parties, is not an institution confined to any one class of Harlem folk. Almost everybody plays the numbers, a universal and illegal gambling pastime, which has become Harlem's favorite indoor sport.

Numbers is one of the most elaborate, big-scale lottery games in America. It is based on the digits listed in the daily reports of the New York stock exchange. A person wishing to play the game places a certain sum of money, from one penny up, on a number composed of three digits. This number must be placed in the hands of a runner before ten o'clock in the morning as the reports are printed in the early editions of the afternoon papers. The clearing house reports are like this:

| Exchanges | $1,023,000,000 |
| Balances | 128,000,000 |
| Credit Bal. | 98,000,000 |

The winning number is composed from the second and third digits in the millionth figures opposite exchanges and from the third figure in the millionth place opposite the balances. Thus if the report is like the example above, the winning number for that day will be 238.

An elaborate system of placement and paying off has grown around this game. Hundreds of persons known as runners make their rounds daily, collecting number slips and cash placements from their clients. These runners are the middle men between the public and the banker, who pays the runner a commission on all collections, reimburses winners, if there are any, and also gives the runner a percentage of his client's winnings.

These bankers and runners can well afford to be and often are rogues. Since numbers is an illegal pastime, they can easily disappear when the receipts are heavy or a number of people have chosen the correct three digits and wish their winnings. The police are supposed to make some effort to enforce the law and check the game. Occasionally, a runner or a banker is arrested, but this generally occurs only when some irate player notifies the police that he "ain't been done right by." Numbers can be placed in innumerable ways, the grocer, the butcher, the confectioner, the waitress at the lunch counter, the soda clerk, and the choir leader all collect slips for the number bankers.

People look everywhere for a number to play. The postman passes, some addict notes the number on his cap and puts ten cents on it for that day. A hymn is announced by the pastor in church and all the members in the congregation will note the number for future reference. People dream, each dream is a symbol for a number that can be ascertained by looking in a dream book for sale at all Harlem newsstands. Street car numbers, house numbers, street numbers, chance calculations—anything that has figures on it or connected with it will give some player a good number; and inspire him to place much money on it.

There is a slight chance to win, it is a thousand to one shot, and yet this game and its possible awards have such a hold on the community that is often the cause for divorce, murder, scanty meals, dispossess notices and other misfortunes. Some player makes a "hit" for one dollar, and receives five hundred and forty dollars. Immediately his acquaintances and neighbors are in a frenzy and begin staking large sums on any number their winning friend happens to suggest.

It is all a game of chance. There is no way to figure out scientifically or otherwise what digits will be listed in the clearing house reports. Few people placing fifty cents on No. 238 stop to realize how many other combinations of three digits are liable to win. One can become familiar with the market's slump days and fat days, but even then the digits which determine the winning number could be almost anything.

People who are moral in every other respect, church going folk, who damn drinking, dancing or gambling in any other form, will play the numbers. For some vague reason this game is not considered as gambling, and its illegality gives little concern to any one—even to the Harlem police, who can be seen slipping into a corner cigar store to place their numbers for the day with an obliging and secretive clerk.

As I write a friend of mine comes in with a big roll of money, $540. He has made a "hit." I guess I will play fifty cents on the number I found stamped inside the band of my last year's straw hat.

Stroll down Seventh Avenue on a spring Sunday afternoon. Everybody seems to be well dressed. The latest fashions prevail, and though there are the usual numbers of folk attired in outlandish color combinations and queer styles, the majority of promenaders are dressed in good taste. In the winter, expensive fur coats swathe the women of Harlem's Seventh Avenue as they swathe the pale face fashion plates on Fifth Avenue down town, while the men escorting them are usually sartorially perfect.

How is all this well-ordered finery possible? Most of these people are employed as menials—dish washers, elevator operators, porters, waiters, red caps, longshoremen, and factory hands. Their salaries are notoriously low, not many men picked at random on Seventh Avenue can truthfully say that they regularly earn more than $100 per month, and from this salary must come room rent, food, and other of life's necessities and luxuries. How can they dress so well?

There are, of course, the installment houses, considered by many authorities one of the main economic curses of our present day civilization, and there are numerous people who run accounts at such places just to keep up a front, but these folk have little money to jingle in their pockets. All of it must be dribbled out to the installment collectors. There was even on chap I knew, who had to pawn a suit he had bought on the installment plan in order to make the final ten dollar payment and prevent the credit house collector from garnisheeing his wages. And it will be found that the majority of the Harlemites, who must dress well on a small salary, shun the installment house lechers and patronize the "hot men."

"Hot men" sell "hot stuff," which when translated from Harlemese into English, means merchandise supposedly obtained illegally and sold on the q.t. far below par. "Hot men" do a big business in Harlem. Some have apartments fitted out as showrooms, but the majority peddle their goods piece by piece from person to person.

"Hot stuff" is supposedly stolen by shoplifters or by store employees or by organized gangs who raid warehouses and freight yards. Actually, most of the "hot stuff" sold in Harlem originally comes from bankrupt stores. Some ingenious group of people make a practice of attending bankruptcy sales and by buying blocks of merchandise get a great deal for a small sum of money. This merchandise is then given in small lots to various agents in Harlem, who secretly dispose of it.

There is a certain glamour about buying stolen goods aside from their cheapness. Realizing this, "hot men" and their agents maintain that their goods are stolen whether they are or not. People like to feel that they are breaking the law and when they are getting undeniable bargains at the same time, the temptation becomes twofold. Of course, one never really knows whether what they are buying has been stolen from a neighbor next door or bought from a defunct merchant. There have been many instances when a gentleman, strolling down the avenue in a newly acquired overcoat, has had it recognized by a former owner, and found himself either beat up or behind bars. However, such happenings are rare, for the experienced Harlemite will buy only that "hot stuff" which is obviously not second hand.

One evening I happened to be sitting in one of the private reception rooms of the Harlem Y.W.C.A. There was a great commotion in the adjoining room, a great coming in and going out. It seemed as if every girl in the Y.W.C.A. was trying to crowd into that little room. Finally the young lady I was visiting went to investigate. She was gone for about fifteen minutes. When she returned she had on a new hat, which she informed me, between laughs at the bewildered expressions on my face, she had obtained from a "hot man" for two dollars. This same hat, according to her, would have cost $10 downtown and $12 on 125th Street.

I placed my chair near the door and watched the procession of young women entering the room bareheaded and leaving with new head gear. Finally the supply was exhausted and a perspiring little Jew emerged, his pockets filled with dollar bills. I discovered later that this man was a store keeper in Harlem, who had picked up a large supply of spring hats at a bankruptcy sale and stating that it was "hot stuff" had proceeded to sell it not openly in his store, but sub rosa in private places.

There is no limit to the "hot man's" supply or the variety of goods he offers. One can, if one knows the ropes, buy any article of wearing apparel from him. And in addition to the professional "hot man" there are always the shoplifters and thieving store clerks, who accost you secretly and eagerly place at your disposal what they have stolen.

Hence low salaried folk in Harlem dress well, and Seventh Avenue is a fashionable street crowded with expensively dressed people, parading around in all their "hot" finery. A cartoonist in a recent issue of one of the Negro monthlies depicted the following scene: A number of people at a fashionable dance are informed that the police have come to search for some individual known to be wearing stolen goods. Immediately there is a confused and hurried exodus from the room because all of the dancers present were arrayed in "hot stuff."

This of course, is exaggerated. There are thousands of well dressed people in Harlem able to be well dressed not because they patronize a "hot man," but because their incomes make it possible. But there are a mass of people working for small wages, who make good use of the "hot man," for not only can they buy their much wanted finery cheaply, but, thanks to the obliging "hot man," can buy it on the installment plan. Under the circumstances, who cares about breaking the law?

## VII. The Negro and the Church

The Negro in America has always supported his religious institutions even though he would not support his schools or business enterprises. Migrating to the city has not lessened his devotion to religious institutions even if it has lessened his religious fervor. He still donates a portion of his income of the church, and the church is still a major social center in all Negro communities.

Harlem is no exception to this rule, and its finest buildings are the churches. Their attendance is large, their prosperity amazing. Baptist, Methodist, Episcopal, Catholic, Presbyterian, Seventh Day Adventist, Spiritualist, Holy Roller and Abyssinian Jew—every sect and every creed with all their innumerable subdivisions can be found in Harlem.

The Baptist and the Methodist churches have the largest membership. There are more than a score of each. St. Phillips Episcopal Church is the most wealthy as well as one of the oldest Negro churches in New York. It owns a great deal of Harlem real estate and was one of the leading factors in urging Negroes to buy property in Harlem.

There are few new church buildings, most of them having been bought from white congregations when the Negro invaded Harlem and claimed it for his own. The most notable of the second hand churches are the Metropolitan Baptist Church at 128th Street and Seventh Avenue, Salem M.E. Church at 129th Street and Seventh Avenue, and Mt. Olive Baptist Church at 120th Street and Lenox Avenue. This latter church has had a varied career. It was first a synagogue, then it was sold to white Seventh Day Adventists and finally fell into its present hands.

The most notable new churches are the Abyssinian Baptist Church on 138th Street, Mother Zion on 137th Street, and St. Marks. The latter church has just recently been finished. It is a dignified and colossal structure occupying a triangular block on Edgecombe and St. Nicholas Avenues between 137th and 138th Streets. It is the latest thing in churches, with many modern attachments—gymnasium, swimming pool, club rooms, Sunday school quarters, and other sub-auditoriums. When it was formally opened there was a gala dedication week to celebrate the occasion. Each night services were held by the various secret societies, the Elks, the Masons, the Knights of Pythias, the Odd Fellows, and others. The members of every local chapter of the various orders turned out to do homage to the new edifice. The collection proceeds were donated to the church.

St. Marks goes in for elaborate ceremony quite reminiscent of the Episcopal or Roman Catholic service. The choir is regaled in flowing robes and chants hymns by Handel. The pulpit is a triumph of carving and wood decoration. There is more ceremony than sermon.

The better class of Harlemites attends the larger churches. Most of the so called "dictys" are registered as "Episcopalians" at St Phillips, which is the religious sanctum of the socially elect and wealthy Negroes of Harlem. The congregation at St. Phillips is largely mulatto. This church has a Parish House that serves as one of the most ambitious and important social centers in Harlem. It supports a gymnasium that produces annually a first class basket ball team, an art sketch class that is both large and promising, and other activities of interest and benefit to the community.

Every Sunday all of the churches are packed, and were they run entirely on the theatrical plan they would hang out the S.R.O. sign. No matter how large they are they do not seem to be large enough. And in addition to these large denominational churches there are many smaller ones also crowded, and a plethora of outlaw sects, ranging from Holy Rollers to Black Jews and Moslems.

The Holy Rollers collect in small groups of from twenty-five to one hundred and call themselves various things. Some are known as the Saints of God in Christ, others call themselves members of the Church of God and still others call themselves Sanctified Children of the Holy Ghost. Their meetings are primitive performances. Their songs and chants are lashing to the emotions. They also practice healing, and, during the course of their services, shout and dance as erotically and sincerely as savages around a jungle fire.

The Black Jews are a sect migrated from Abyssinia. Their services are similar to those in a Jewish Synagogue only they are of a lower order, for these people still believe in alchemy and practice polygamy when they can get away with it. Just recently a group of them were apprehended by agents from the Department of Justice for establishing a free love farm in the State of New Jersey. They were all citizens of Harlem and had induced many young Negro girls to join them.

The Mohammedans are beginning to send missionaries to work among Negroes in America. Already they have succeeded in getting enough converts in Harlem, Chicago, St. Louis, and Detroit to establish mosques in these cities. There are about one hundred and twenty-five active members of the Mohammedan church in Harlem, practicing the precepts of the Koran under the leadership of an Islamic missionary.

The Spiritualist churches also thrive in Harlem. There are about twenty-five or more of their little chapels scattered about. They enjoy an enormous patronage from the more superstitious, ignorant classes. The leaders of the larger ones make most of their money from white clients, who drop in regularly for private sessions.

## VIII. Negro Journalism in Harlem

The Harlem Negro owns, publishes, and supports five local weekly newspapers. These papers are just beginning to influence Harlem thought and opinion. For a long time they were merely purveyors of local gossip and scandal. Now some of them actually have begun to support certain issues for the benefit of the community and to cry out for reforms in the regulation journalistic manner.

For instance, *The New York Age,* which is the oldest Negro weekly in New York, has been conducting a publicity campaign against numbers and saloons. These saloons are to this paper as unwelcome a Harlem institution as the numbers. Each block along the main streets has at least one saloon, maybe two or three. They are open affairs, save instead of calling themselves saloons, they call themselves cafes. To get in is an easy matter. One has only to approach the door and look at a man seated on a box behind the front window, who acknowledges your look by pulling a chain which releases a bolt on the door. Once in you order what you wish from an old fashioned bartender and stand before an old fashioned bar with a brass rail, mirrors, pictures, spittoons, and everything. What is more, they even have ladies' rooms in the rear.

The editor of *The New York Age,* in the process of conducting his crusade, published the address of all these saloons and urged that they be closed. The result of his campaign was that they are still open and doing more business than ever, thanks to his having informed people where they were located.

At first glance any of the Harlem newspapers give one the impression that Harlem is a hot-bed of vice and crime. They smack of the tabloid in this respect and should be considered accordingly. True, there is vice and crime in Harlem as there is in any community where living conditions are chaotic and crowded.

For instance, there are 110 Negro women in Harlem for every 100 Negro men. Sixty and six-tenths percent of them are regularly employed. This, according to social service reports, makes women cheap, and conversely I suppose makes men expensive. Anyway there are a great number of youths and men who are either wholly or partially supported by single or married women. These male parasites, known as sweetbacks, dress well and spend their days standing on street corners, playing pool, gambling and looking for some other "fish" to aid in their support. This is considered by some an alarming condition inasmuch as many immigrant youths from foreign countries and rural southern American districts naturally inclined to be lazy, think that it is smart and citified to be a parasite and do almost anything in order to live without working.

The newspapers of Harlem seldom speak of this condition, but their headlines give eloquent testimony to the results, with their reports of gun play, divorce actions (and in New York State there is only one ground for divorce) and brick throwing parties. These conditions are magnified, of course, by proximity, and really are not important at all when the whole vice and crime situation in greater New York is taken under consideration.

To return to the newspapers, *The Negro World* is the official organ of the Garvey Movement. At one time it was one of the most forceful weeklies among Negroes. Now it has little life or power; its life-giving mentor, Marcus Garvey, being in Atlanta Federal Prison. Its only interesting feature is the weekly manifesto Garvey issues from his prison sanctum, urging his followers to remain faithful to the cause and not fight among themselves while he is kept away from them.

*The Amsterdam News* is the largest and most progressive Negro weekly published in Harlem. It, like all of its contemporaries, is conservative in politics and policy, but it does feature the work of many of the leading Negro journalists and has the most forceful editorial page of the group, even if it does believe that most of the younger Negro artists are "bad New Negroes."

*The New York News* is a political sheet, affecting the tabloid form. *The Tattler* is a scandal sheet. It specializes in personalities and theatrical sport news.

## IX. The New Negro

Harlem has been called the center of the American Negroes' cultural renaissance and the mecca of the New Negro. If this is so, it is so only because Harlem is a part of New York, the cultural and literary capital of America. And Harlem becomes the mecca of the so-called New Negro only because he imagines that once there he can enjoy the cultural contact and intellectual stimulation necessary for his growth.

This includes the young Negro writer who comes to Harlem in order to be near both patrons and publishers of literature, and the young Negro artist and musician who comes to Harlem in order to be near the most reputable artistic and musical institutions in the country.

These folk, along with the librarians employed at the Harlem Branch of the New York Public Library, a few of the younger, more cultured professional men and women and the school teachers, who can be found in the grammar and high schools all over the city, constitute the Negro intelligentsia. This group is sophisticated and small and more a part of New York's life than of Harlem's. Its members are accepted as social and intellectual equals among whites downtown, and can be found at informal and formal gatherings in any of the five boroughs that compose greater New York. Harlem to most of them is just a place of residence; they are not "fixed" there as are the majority of Harlem's inhabitants.

Then there are the college youngsters and local intellectuals, whose prototypes can be found in any community. These people plan to attend lectures and concerts, given under the auspices of the Y.M.C.A., Y.W.C.A., churches, and public school civic centers. They are the people who form intercollegiate societies, who stage fraternity go-to-school campaigns, who attend the course of lectures presented by the Harlem Branch of the New York Public Library, during the winter months, and who frequent the many musical and literary entertainments given by local talent in Harlem auditoriums.

Harlem is crowded with such folk. The three great major educational institutions of New York, Columbia, New York University and the College of the City of New York, have a large Negro student attendance. Then there are many never-will-be-top-notch literary, artistic, and intellectual strivers in Harlem as there are all over New York. Since the well advertised "literary renaissance," it is almost a Negro Greenwich Village in this respect. Every other person one meets is writing a novel, a poem, or a drama. And there is seemingly no end to artists who do oils, pianists who pound out Rachmaninoff's Prelude in C Sharp Minor, and singers, with long faces and rolling eyes, who sing spirituals.

## X. Harlem—Mecca of the New Negro

Harlem, the so-called citadel of Negro achievement in the New World, the alleged mecca of the New Negro and the advertised center of colored America's cultural renaissance. Harlem, a thriving black city, pulsing with vivid passions, alive with colorful personalities, and packed with many types of classes of people.

Harlem is a dream city pregnant with wide awake realities. It is a masterpiece of contradictory elements and surprising types. There is no end to its versatile presentation of people, personalities, and institutions. It is a mad medley.

There seems to be no end to its numerical and geographical growth. It is spreading north, east, south, and west. It is slowly pushing beyond the barriers imposed by white people. It is slowly uprooting them from their present homes in the near vicinity of Negro Harlem as it has uprooted them before. There must be expansion and Negro Harlem is too much a part of New York to remain sluggish and still while all around is activity and expansion. As New York grows, so will Harlem grow. As Negro America progresses, so will Negro Harlem progress.

New York is now most liberal. There is little racial conflict, and there have been no inter-racial riots since the San Juan Hill days. The question is will the relations between New York Negro and New York white man always remain as tranquil as they are today? No one knows, and once in Harlem one seldom cares, for the sight of Harlem gives any Negro a feeling of great security. It is too large and too complex to seem to be affected in any way by such a futile thing as race prejudice.

There is no typical Harlem Negro as there is no typical American Negro. There are too many different types and classes. White, yellow, brown, and black and all the intervening shades. North American, South American, African, and Asian; Northern and Southern; high and low; seer and fool—Harlem holds them all and strives to become a homogeneous community despite its motley, hodge-podge of incompatible elements and its self nurtured or outwardly imposed limitations.

# Harlem Directory: Where to Go and What to Do When in Harlem

*Harlem: A Forum of Negro Life*, November 1928.

There are four main attractions in Harlem: the churches, the gin mills, the restaurants, and the night clubs. It is not necessary here to define what churches are so we will proceed to give a list of those which attract the largest congregations:

St. Mark's A. M. E., 138th Street and St. Nicholas Avenue.
St. Philip's Episcopal, 133rd Street, between 7th and 8th Avenue.
Abyssinian Baptist, 138th Street, between Lenox and 7th.
Mother Zion, 136th Street, between Lenox and 7th.
Salem M. E., 129th Street and 7th Avenue.
Metropolitan Baptist, 128th Street and 7th Avenue.
St. Mark's Catholic, 138th Street and Lenox Avenue.
Mt. Olivet Baptist, 120th Street and Lenox Avenue.
Grace Congregational, 139th Street, between 8th Avenue and
    Edgecombe Avenue.

And there are innumerable smaller churches and missions, countless spiritualists' rooms, a synagogue, a mosque, and a great number of Holy Roller refuges, the most interesting of which is at 1 West 137th Street.

Gin mills are establishments which have bars, family entrances, and other pre-Volstead luxuries. For reasons best known to ourselves and the owners of these places we will not give the addresses and even were these reasons not personal, there are far too many gin mills to list here. As a clue to those of our readers who might be interested we will tell them to notice what stands on every corner on 7th, Lenox, and 8th Avenues. There are also many such comfort stations in the middle of the blocks.

The best restaurants to go to in Harlem are Tabb's, located at 140th Street and Lenox Avenue, where you can get a good chicken dinner in the Grill Room and have ragtime music while you eat. The Marguerite, on 132nd Street between Lenox and Seventh Avenues, guarantees you a full stomach. Johnny Jackson's at 135th Street and Seventh Avenue; St. Luke's on 130th Street, between Lenox and Seventh, the Venetian Tea Room on 135th Street, between Seventh and Eighth Avenues, and the Blue Grass at 130th Street and Seventh Avenue, are also good bets. If you are broke and want only coffee and rolls or a piece of pie, there are Coffee Pots next to every gin mill or if you should wish *vino* with your dinner there is the La Rosa on Seventh Avenue near 139th Street.

Among the best known Harlem night clubs are the Cotton Club at 142nd Street and Lenox Avenue; the Lenox Avenue Club on Lenox Avenue, between 142nd ad 143rd Streets; Cairo's on 125th Street, between Lenox and Fifth Avenues; the Sugar Cane at

135th Street and Fifth Avenue; Small's at 135th Street and 7th Avenue; Barron's at 134th Street and 7th Avenue; Connie's Inn at 131st Street and 7th Avenue; Club Harlem at 129th Street and Lenox Avenue, and the Bamboo Inn at 139th Street and 7th Avenue. Most of these places with the exception of the Cotton Club and Connie's Inn are fairly reasonable and are generally packed, but if you really desire a good time, make friends with some member on the staff of *Harlem* and have him take you to Mexico's or to Pod and Jerry's or to the Paper Mill. We warn you that only the elect and the pure in heart are admitted to these places.

## Harlemese

Written in 1929 with collaborator William Jourdan Rapp for the playbill distributed at performances of their play, *Harlem*.

Harlem, New York City's Black Belt, has a language all its own. The quarter million colored folks crowded together north of 125th Street hail from many parts of the world and innumerable colloquial expressions heretofore peculiar to Martinique, the Virgin Islands, Cuba, Trinidad, Jamaica, our own Southland, Liberia, Abyssinia, South Africa and Senegal, are being assimilated into Harlemese, and so corrupted in the process that their origins are lost and only confirmed Harlemites can now understand them.

"Ofay" is the term generally used in referring to a white person and it belongs among the legion of expressions whose origins are hopelessly lost.

It is easier to understand why Negroes are known to Negroes as "spades" or "eightballs." But just try to fathom "zigaboo" and "jigwalk."

"Astorperious" is supposed to have originated in Florida. It means "high hat" and is a tribute to the socially prominent Astors. "Hincty" is a companion term connoting superciliousness. And "dickty" is another close relation applied to those people who belong to any select social group: i.e., "Dickty college niggers," "Dickty church niggers," "dickty white niggers," etc.

Probably the most apt term in this group expressing snootiness is "striver." It is the nearest equivalent in Harlemese to *nouveau riche*. And a certain street in Harlem where the swells reside is locally known as "strivers' row."

A most fertile field for the collection of Harlem *argot* is the house rent parties or parlor socials where all the world mixes for a quarter admission. Here, the dancers, while indulging in a mad orgy of gyrations, give vent to ecstatic exclamations.

"Rock, church, rock!" "Oh, do it, you dirty no gooder!" "Shake it and break it!" "Walk that broad!" "What old broad!" "Oh, play it Mr. Man!" and "Oh, sock it!" are but a few of the expressions used to urge on the piano player to renewed frenzy at his keys.

And then there are such priceless expressions as "Get off that dime!" thrown at a couple who dance in a stationary position; and "Mama, come get your little blue-eyed baby" shouted by a very black, six-foot longshoreman to a little "yaller" girl.

West Indian Negroes are known as "monkey-chasers" and "ringtails," because it is bruited about that the favorite native dish in the Caribbean is "monkey hips and dumplings."

A man who commercializes his sex appeal is known as a "sweetback."

One who sells stolen goods is a "hot man" or a "hot stuff man."

Those Negroes who retain their slave psychology and kow-tow to whites are "Uncle Toms" and "handkerchief heads."

"Lawd to-day!" is an exclamation of wonder. And "Out of this world!" means just what it says—past human belief.

"Down to the bricks!" connotes to the nth degree.

"Oh, no, now!" is an exclamation of admiration, and "No lie!" means "It ain't nothin' else but the truth!"

"Jive!" is a synonym for flatter. And to be "high" is to be in the best of spirits.

"Hiney" signified, "Well, we must go into that!"

"Goopher dust," is a blanket term referring to all the various powders used by the voodooists to bring good luck on oneself and bad luck on one's enemies.

"Forty!" "Thirty-eight and two!" and "Righteous!" all mean Okay.

"Two-time" is to double cross, and "picked" means robbed or gypped.

And if one's sincerity is doubted in love, friendship, or business, the doubter will ask the doubted to "make me know it!" and how the doubted responds is "nobody's business."

### Glossary of Harlemisms

| | |
|---|---|
| Rent party | A Saturday night orgy staged to raise money to pay the landlord. |
| Sweetback | A colored gigolo, or man who lived off women. |
| Hincty | Imperious; snooty. |
| Dicty | Highbrow. |
| Monkey-chaser | A West Indian Negro. |
| Love-charm | A good-luck piece designed to attract a particular person of the opposite sex. |
| Goofer-dust | A supposedly magic powder which, if sprinkled judiciously, will rout bad luck, win a recalcitrant sweetheart, or chase a discarded but tenacious one. |
| Monkey-hip eater | A derisive name applied to a Barbados Negro; supposed to have originated with the myth that Barbados Negroes are |

|  | passionately fond of monkey meat, particularly "monkey hips with dumplings." |
| --- | --- |
| Numbers | A gambling game peculiar to Harlem; a sort of lottery based on three figures of the daily Clearing House Statement. |
| Chippy | A tart; a fly, undiscriminating young wench. |
| Mess-around | A whirling dance; part of the Charleston. |
| Hot-stuff man | A seller of stolen goods. |
| 38 and 2 | That's fine. |
| Forty | Okay |
| Righteous | Right-o; correct. |
| Sweet man | A great lover. |
| Big sugar, small sugar | Terms of endearment. |
| Jive or two-time | To Double-cross. |
| Lily-liver | A coward; yellow. |
| Down-to-the-bricks | To the limit. |
| "Mon" | A West Indian Negro's pronunciation of "man," therefore a West Indian Negro. |
| Picked | Gypped, robbed |

## Few Know Real Harlem, the City of Surprises: Quarter Million Negroes Form a Moving, Colorful Pageant of Life

*The New York World*, March 3, 1929; written with William Jourdan Rapp.

Few people ever see the whole of Harlem, New York's Black Belt. Even after the recent wave of public interest, it is still to most whites a city of coons, cabarets, and black-face comedians.

Harlem in reality is a boundary-bursting coop with a population of a quarter million Negroes of all types and classes; a struggling mass of people with varied capacities for adaptation to a strange and sometimes sinister environment.

It is a city of constant surprise; a city of ecstatic moments and diverting phenomena. It is a multi-faceted ensemble, offering many surprise packets of persons, places, amusements and vocations. It is a cosmos within itself. Life there is not stable and monotonous. Rather, it is moving, colorful and richly studded with contrasting elements and contradictory types.

In Harlem we find a community as American as Gopher Prairie or Zenith; a community keenly alert to the cosmopolitan currents swirling around it and through it; a community cut out for speed and splendor, squalor and wealth, penury and prosperity. It permits of everything possessed by that stupendous ensemble—New York City—of which it is a part.

## About 40 Per Cent Foreign-Born

Like New York, Harlem is a cosmopolitan city. Its people are as varied and polyglot as could be found anywhere. The whites indiscriminately lump them together as "Negroes" or "niggers." But they are really unclassifiable under any existent ethnic term, for the racial complexity of the American Negro is astounding. In his veins flows the mixed bloods of the Africans from whom he originally stemmed, the American Indians with whom he intermarried in pre- and post-slavery days, and of every white race under the sun. And then in Harlem this home-grown ethnic amalgam is associating and inter-mixing with Negroes from the Antipodes and Caribees, from Africa and Asia, South America and every other place that dark-skinned people hail from.

About 40 per cent of the Negroes in Harlem are foreign-born. The majority of these foreigners come from the British West Indies. Another large group comes from the Virgin Islands, having been forced to seek financial salvation in America because our national Prohibition Act blasted their rum trade. And the next largest quota comes from the French and Spanish speaking localities and islands in South and Central America.

These foreign Negroes, upon their arrival in New York, find themselves segregated in a community the likes of which they have never seen before, and are forced to mingle with other Negroes having distinct cultural and lingual differences. Petty prejudices and frictions naturally arise. It is the same as in any community where the foreign-born compete with the natives for economic or social supremacy.

The American Negro looks down upon these foreigners just as the white American looks down upon the white immigrants from Europe. The native black man takes pride in the fact that he is a citizen of the "world's greatest country" and is proud that he has had the advantages of a supposedly superior civilization, with modern plumbing, a system of education and high wages.

The foreign-born Negro, although he will often work for less pay because he is used to a lower standard of living, chides the American Negro because the latter has not been free from slavery as long as he and because he feels that the Afro-American takes such matters as lynching and segregation far too casually.

Naturally these various groups within the race clash, but fortunately the struggle to live and the amount of mass energy needed to fight the white man's prejudice and discrimination leaves little time for actual intra-racial combat. The foreigners and natives express their impatience and disgust with one another in a social or verbal way. The

American Negro calls the West Indian Negro a "monkey chaser." The West Indian's retaliatory epithets cannot be printed.

This seething melting-pot of conflicting nationalities and languages is just one of the many sides of Negro life in Harlem of which white people are practically unaware. It is almost incomprehensible to them that the American Negro should share the American white man's prejudice against foreigners and that he should vigorously resent their intrusion into his country.

## Many on Top Economically

Another aspect of Harlem life little known or publicized is the social and economic life of the upper strata of Negro society. It is taken for granted by most whites that all Negroes with the possible exception of those constantly in the spotlight—like Roland Hayes, Paul Robeson, William Du Bois or James Weldon Johnson—are in a class with chauffeurs and washerwomen. They do not take into consideration that a large number of Negroes have long been emancipated from meniality, and that many have established fortunes or achieved enviable incomes.

Although the Negro in Harlem has not been as energetic in commercial fields as have the Negroes of Chicago, or Durham in North Carolina and some other cities, there are, nevertheless, any number of commercially prominent and wealthy Negroes north of 125th Street.

There are no Negro bankers in Harlem, but there are a great number of eminently successful real estate operators. There are no capitalistic combines like the Overton enterprises in Chicago or the Malone concern in St. Louis, but there are many moneyed entrepreneurs operating and owning large and prosperous businesses.

Negroes in Harlem do not even own a theater or dance hall of their own, but they do own over $60,000,000 worth of real estate consisting of many luxurious homes, ostentatious apartment hotels, restaurants, drug stores, beauty parlors and haberdasheries.

Innumerable Harlem Negroes live in expensively appointed homes and apartments, have maids and chauffeurs, entertain lavishly and send their sons to Columbia, Harvard, Yale, while their daughters go to Barnard, Vassar and Wellesley. They attend auctions and invest in antiques and rare objects or art. Their clothes come from Fifth Avenue. They go for vacations in Europe, Atlantic City, the Maine woods and Southern California.

A great Negro middle class is being evolved. They are the mercantile forerunners of a future Negro aristocracy. Fortunes are being built by them around nest eggs salted away by a preceding generation of washerwomen and Pullman porters.

It is quite natural that to the white person who views Harlem from the raucous interior of a smoke-filled, jazz-drunken cabaret, this side of Negro life is unknown.

## No More Shouting

Then there is the religious side of life in Harlem which often completely escapes the white observer. The old frame structures in which the sisters and brothers moaned and shouted with the spirit while ministerial emotionalists shook the house with sermons on Heaven, Hell, salvation and eternal damnation have given way to stately ecclesiastical edifices in which pentecostalism is frowned upon and where fiery sermons leveled at sinners have given way to polite religious talks.

The choir is now appareled in robes and sing from a classical hymnal, Sister Brown from Shiloh Baptist Church in Birmingham is admonished by swallow-coated ushers to keep quiet during the services, for her constant "Amen!" and "Preach it, Brother!" disturb those around her and so punctuate the minister's text that those in the rear of the balcony have difficulty in following it.

Simple services have given way to elaborate ceremonies. Ministers are college-bred and have high-salaried assistants.

## The Great Game of Numbers

The foregoing would indicate that the Negro in Harlem is thoroughly Americanized and that life there is comparable to life in any similar community of whites. This is growingly true, but there are some phenomena peculiar to Harlem alone, phenomena which are inherently expressions of the Negro character before it was conditioned by the white world that now surrounds him, although even here that world influences the manner of expression to some extent.

For instance, there is the game of Numbers, Harlem's most popular indoor sport and the outlet for the Negro's craving for gambling. It is based upon the digits listed in the daily Clearing House Statement. A prospective player chooses three digits (231, for instance) and places them with the runner who acts as intermediary between the players and the bankers.

The banker holds the money bags, pays winners, if any, and allows his runners a commission on all sums they bring in.

The winning number is compounded as follows: Say the daily Clearing House Statement reads thusly for any one day:

| | |
|---|---|
| Exchanges | 2,432,000,000 |
| Balances | 250,130,000 |
| Credits | 1,456,000 |

The second and third digit of the exchanges with the second digit of the balances make up the winning number, which, if the report is as above, would be 435.

A person can play any amount from 1 cent up. The odds paid are 540 to 1, while the chance of picking the right number is 1,000 to 1.

Sometimes days pass without the banker having to pay, as no one picks the winning number. Often, when a number of people pick the correct number, a banker disappears. Also, runners sometimes pocket the bets, not turning them over to the banker. The whole business is illegal, so the number entrepreneurs, like the bootleggers, are open to hijacking, as they can hardly appeal to the police if their runners and collectors are held up.

## House Rent Parties

Since the greater portion of Harlem's quarter million inhabitants play the numbers, and in many cases play heavily—from $1 to $5 daily—the number-runner, considering he works on a commission, is able to earn quite a sum.

Number-running, being both profitable and dangerous, offers sufficient attraction to lead many to take it up as a career. And almost any person one meets walking the streets of Harlem, or entering the hallways of one of Harlem's mephitic tenements, might be suspect.

There is no possible way of finding out how many runners operate, but when one considers the ease with which a number may be placed, and the number of people who are addicted to this form of gambling, there must be in Harlem over a thousand number-runners daily, collecting slips from more than a hundred thousand clients.

Then there are the House Rent Parties, which, like numbers, are an institution peculiar to the Black Belt. These parties owe their origin to the fact that rental fees in Harlem are exorbitant, and, although tenants sublet every available bit of bed space, another source of income is still necessary to keep off the dispossess notice.

Some folk give rent parties weekly, some bi-weekly, some monthly and others only in time of stress. The rent party brings the public into the house at 25 cents a head, and to bring the public in requires a good, "hot" piano player.

Some people have found rent parties so profitable that they have become professional givers of house rent parties, getting their whole income from them. These professional parties usually have more than one instrument to furnish their music, but the rank and file confine themselves to the hiring of only one man to play the piano.

If the piano player is personable, and capable, he can play at some party almost every night in the week. In fact, house rent piano playing in Harlem has become a profession, and the house rent piano player is an easily recognizable type.

They are seldom good looking, that is, handsome, for handsome men of this type are too much in demand as "sweetbacks" and paramours. They dress flashily in extreme styles. They must have a fair singing voice, a choice repertoire of "wise cracks," parodies, shouts, and other such tricks of their trade. They unconsciously become able to regulate the scale of people's emotions, and pick their music accordingly, becoming more and more primitive, more and more vulgar, as the evening advances, and the corn liquor and synthetic gin, which is sold at every rent party, begins to take effect on the dancers.

When the party reaches a climax it is the piano player alone who controls its emotional and physical destiny. Some inkling of this seemingly permeates his being, and a strange barbaric ecstasy emanates from his perspiring body. The dim-lighted rooms will surge with strange rhythms and barbaric quarter-tone beats. The sound of the panting, passionate, hip-wiggling animals will merge with the blue harmonic orgy of the moaning piano strings, interspersed by inspired utterances from the gin-filled genie at the piano to "Shake that thing, Mr. Charlie!" and "Do it, you dirty no-gooder!" Then, when the party has ended, burned out by its own intensity, the musical magician will creep out into the dawn of a Harlem morning, $5 in his jeans, a woman on his arm, and a stomach full of gin.

Truly, New York's Black Belt is a city in which anything might happen and everything does!

## The Bump
## (A Dance They Do in Harlem)

Undated handwritten essay.

There is a dance they do in Harlem called the Bump, a dance which when reproduced in a current Broadway show, offended the aesthetic taste of the police censors and was hurriedly cut out in order to prevent a raid. This idea that the Bump is an indecent, vulgar dance gives those in the know a laugh, for in Harlem and other Negro centers it is a dance to which the dictys are addicted rather than the low brows.

In fact it is bruited about that the dance in question was originated by a group of Negro college students in Washington, D.C. and for a long time was known as the Omega Bump, because those students responsible for it were members of the Omega Psi Phi fraternity at Howard University.

Now the Bump is a favorite dance among all the better class Negroes. They who scorn the frank "scrounching" and "mess-around," they who turn up their noses at the unrestrained hip wriggling of couples in low down Harlem dives, will march out onto a public dance floor, attired in evening clothes and do the Bump energetically and joyously.

The Bump is a dance the name of which is all explanatory. It consists of rhythmic body movements done in unison backwards and forwards. The feet are dragged along the floor. The mid-section of the body perform[s] the essential movements. When done properly it presents a good study in the poetry of motion. When exaggerated it proves that Judy O'Grady and The Colonel's Lady are sisters under the skin. The Bump is best and most beautifully performed to slow music. When done to fast music it not only loses its beauty, it also makes the dancers lose their equilibrium and their breath.

Since the recent influx of whites to Harlem parties, many pale faces have essayed to master this seemingly simple and attractive dance. A few have succeeded. The rest are sights for a peep show.

There are few other typically Harlem dances which are not generally known to downtown whites. Ann Pennington and Tom Patricola made a bit with the Black Bottom in *Geo. White's Scandals,* two years ago. The Charleston has merited international fame. And with Billy Pierce and "Bojangles" Bill Robinson in such demand among musical comedy producers every chorus ensemble on Broadway will soon be doing special tap routines which hitherto could be seen only in Negro shows and cabarets.

It is seldom though that white dancers achieve the spontaneity and ease that characterizes Negro dancers. To the former it is usually work-work for which the preparation has by no means been easy. On the other hand most Negroes who can dance get so much pleasure out of it that they impress people as being exponents of some primitive form of self-expression. There is nothing studied or hard about what they do. They are merely having a good time.

Most Negro dances originate in southern cane brakes and cotton field settlements. They are introduced into the north by black migrationists and find their way into the theatrical world after they have been seen in some gin dive or cabaret. It is thus indeed hard to give credit where credit is due. Any number of people claim the honor of having originated this or that dance. All may have some ground on which to base their claim, for it is very possible that each one, having seen the raw material, has refined it for stage purposes.

Negro performers on the whole are not very original. The steps they do on the stage can be seen on almost any street corner in Harlem, or in any other Black Belt. For it seems characteristic of that class of youth and men who frequent pool halls and loiter on corners, suddenly to break into a dance step the moment there is a lull in the conversation, which considering how long some of them "hold-down" corners is by no means infrequent.

White vaudeville and musical comedy performers are frequent visitors in Harlem cabarets and theaters. The Negro performer initiates his unprofessional brother. The white performer initiates the professional blackamoor. And before you know it such dances as the Shimmy, the Charleston and the Black Bottom find themselves being performed in the most elite dance salons and private parties. One wonders how long it will be before the Bump too will make jaded Park Avenue evenings more hilarious.

# Harlem House Rent Parties

Unpublished.

Harlem house rent parties are more an institution of necessity than one of pleasure. The people who give them need money. The people who attend need some place to go.

Rents in Harlem are high. Negro working men and women rate low salaries. An apartment is acquired, all available space is sublet, all, who are able, work. This is not enough. Their income needs supplementing.

There are also thousands of Negroes among Harlem's quarter million who are footloose, restless and unattached with little money to spend for pleasure and few inexpensive places to go. The bare walls of a two by four room, furnished simply and rudely, and the utter lack of privacy in a rooming house drive them into the streets, the movies, the poolrooms and the house rent parties.

In the beginning house rent parties were spasmodically given by whoever might feel the urge during a poverty-stricken interlude. Then the ease with which the money came in inspired certain astute people to make the giving of house rent parties a regular business. Now both types flourish and these parties are most numerous, especially on a Saturday night.

Cards are printed and passed out in public places, cards of a particular genre, worded only as Harlem house rent party entrepreneurs can word them, frank invitations, spiced with Harlem argot, setting forth the date of the party and the special attractions which make it imperative that one should attend.

Piano players of renown, for the most part unable to read music, improvise slow, sensuous, primitive blues tunes. Corn liquor is either made or bought, diluted with water and served for twenty-five cents a drink. Pigs' feet and slaw, black-eyed peas and rice are also served in the kitchen. A crap game is generally run in a side room—or it may be a game of cards or both.

The party guest pays a quarter at the door, throws his hat and coat where he may, either brings his own girl or acquires one at the party, drinks, dances, and eats until the wee hours of the morning.

It is the duty of the piano player and whatever musicians he may bring with him to control the tempo of the party, to whip the festive crowds into an insensate frenzy and keep things going at a high pitch during the entire evening.

Fights are frequent and expected, but they seldom stop the party for any appreciable length of time and unless there is gun or knife play the dance goes on.

Despite the freedom and frenzy of these parties they are seldom joyous affairs. On the contrary they are rather sad and depressing. A tragic undercurrent runs through the music and is reflected in the eyes and faces of the dancers. Their frankly sensual movements become writhings of despair. Their hoarse shouts of pleasure become cries of pain. The environment in which they live is a steel vise, restricting their natural freedom, depriving them of their spontaneity. Their efforts to fight the restrictions of this environment are most apparent at these rent parties, and are most depressing to an onlooker because they are so futile.

Harlem house rent parties derive from what are known as "chiddlin' switches," held throughout the south, occasions when chitterlings are served while the guests dance. These affairs are still carried on in southern metropolises like Memphis, Birmingham and Atlanta and also in such stockyard towns as Omaha, Kansas City and Chicago where chitterlings (hog guts) are most plentiful.

## Odd Jobs in Harlem

Typewritten manuscript; unpublished.

Harlem is a city of constant surprise, a city of ecstatic moments and diverting phenomena. It is an egregious ensemble, offering many surprise packets of persons, places, amusements, and occupations. It is a city in which anything might happen and everything does. Life there is not stable and monotonous. It is moving and colorful, and richly studded by contrasting elements and contradictory types.

Harlem, a boundary bursting coop, has 175,000 inhabitants—a struggling mass of people with varied racial backgrounds, and varied capacities for adaptation. One sees and marvels alike at the ignorance and ingenuity of those of God's chosen animals gathered there. And marveling at the ingenuity one begins to wonder how all this mass of people earns its bread and board.

Harlem has a vast number of occupations which the environment alone sustains and generates. For instance, there is the "number runner." This person collects number slips from number players, and Numbers, a modernized and abbreviated version of the ancient game of Policy, is Harlem's favorite indoor sport. The number runner has a personal clientele from whom he collects slips bearing the numbers which the players have picked to win for that day. The winning number is compounded from digits appearing in the daily clearing house reports. The runner, after making his rounds turns the slips he has collected over to a person known as the number banker. This person bears all responsibility for paying off winners, and allows the runner a percentage on all the money he collects as well as on a portion of the sums won by the runner's clients. It being illegal to play Numbers, both runners and bankers have occasionally been known to disappear without paying winners, after making a large collection.

There is the respectable housewife who earns as much money running for Numbers as her husband earns as a clerk in the post office. She has a large clientele—personal friends, lodge sisters, fellow choir members, etc. She deals with Harlem's most respectable number banker, and is considered a reliable person with whom to place one's number. One day she reports to a person who has given her a slip bearing number 356 that the slip has been lost. 356 happens to be the winning number for that day. The person who has played five cents on this number is due to receive $27.50. When paying off time comes the expectant winner receives no money. He remembers that this same thing has happened to other of his friends dealing with the same lady. The police thereupon are mysteriously notified that the respectable housewife is in reality a number runner. That good lady is soon being shadowed, and before she can rid herself of the day's collection of slips which she carries in her purse she is arrested and marched off to jail. Her banker bails her out, there is a trial, she receives a suspended sentence. The next day she is back to her clients.

Then there was a West Indian lad earning his way through college as a number runner. One day he was left with a batch of unpaid winning slips by a decamping banker. A crowd of fellow students who should have been winners proceeded to manhandle him. When they had finished with him, he was chased up stairways, across roofs, through a skylight, and down more stairs by a pair of energetic plain clothes men from headquarters.

Truly, number running is an intriguing occupation. It is both profitable and dangerous. It has sufficient attractions to lead many to take it up as a career. There is no possible way of checking up on the number of runners operating in Harlem. Almost anyone is suspect, from your grocery clerk to your best friend, and when one considers the ease with which a number may be placed and the number of people who are addicted to this form of gambling there must be in Harlem over a thousand number runners daily collecting slips from 100,000 clients.

Next to the number runner the house rent party piano player seems to have the most romantic and colorful career. House rent parties, like numbers, are a Harlem institution. They owe their origin to the fact that rents in Harlem are so high that the average working man must resort to all sorts of new schemes in order to maintain a roof over his head. Some folks give house rent parties weekly, some bi-monthly, some monthly, and others only in time of stress. But regardless of when or how often they are given, music is necessary, and there is not always a musician in the family or among the families' friends to volunteer for duty. Hence the genesis of this new division of labor.

The house rent party piano player receives five dollars per night. If he is good, that is full of pep and able to play what is known as low down music, he can play at some such party almost every night in the week. He must have personality, a good singing voice, and a choice repertoire of wise cracks, parodies, shouts and other such tricks of the trade. He must also be able to regulate the tempo of the party, and become more and more primitive, more and more vulgar as the evening advances and as the corn liquor and synthetic gin begin to make the crowd tipsy.

The house rent party usually starts off slowly and gains momentum as the night passes. By the time the party reaches a climax it is the piano player alone who controls its emotional and physical destiny. Liquor has lit the fire, music must fan it into a flame. The piano player knows this and a strange barbaric ecstasy will illumine his smile, invigorate his perspiring body and master his music. The dim lighted room will literally surge with strange rhythms, insinuating and slow, while the dancers, [in]spired by the blue harmonic orgy of the piano keys, will revel in the movements of their bodies.

Street corner orators are not peculiar to Harlem, but probably nowhere else are they as numerous. Every evening during the spring and summer months they mount their soap boxes, or abbreviated stair platforms, and without waiting for an audience begin to talk. They don't have to wait long for a crowd for Seventh Avenue is lined with promenaders and idlers anxious to accept any invitation to loiter.

Harlem street speakers know their audiences in the same way that a house rent party piano player knows his dancers. Speaking about anything which happens to pop into their minds, they are able to stir up their listeners to laughter or tears almost at will. The crowd does not react to what they say so much as how they say it.

The most ubiquitous of these orators is a West Indian from the Barbardoes. Short and wizened he has a Napoleonic complex, and affects a broad brimmed hat which he turns up front and back and wears sidewise upon his head. His corner is at 137th Street and Seventh Avenue. His time is spent in telling the crowd how dumb they are, then daring them to ask him a question which he cannot answer.

This dark Demosthenes spends his days in the reading room of the library looking through encyclopedias for strange facts and universal fallacies so that he can spring them on the crowd at night. If there is an election he speaks in favor of the party that pays him the most. If there is some sensational development in American Negro life he takes the side of the minority. If there is a lynching or race riot he hurls invective at white America. But he never gets so enthused over his subject that he cannot stop at regular intervals in order to take up a collection.

Prohibition has created numerous strange jobs in Harlem. The chain pulling Buddha of the Harlem saloon is one of many who owe their livelihood to the Volstead Act. Hour after hour he sits in the saloon window with a long chain in his hand, which is connected to a bolt on the entrance door. A customer approaches, the chain is pulled, the door pushes open, and the customer enters. When the customer's thirst is gratified, and he is ready to leave, the chain is pulled once more and the door swings outward. And all this time the man in the window remains seated on his box, imperturbable and dumb.

One wonders why he is there. The police pass and wave greetings. Should revenue officers decide to raid the place the bolted door and the man with the chain would only verify their suspicions. The chain could not be held long enough to give the bartenders time to destroy all evidence. And since the door is opened to anyone who approaches whether they are frequenters of the place or not he certainly cannot be on the lookout for spotters.

Vainly have I tried to engage them in conversation in order to find out the history and purpose of their job. Not only did they ignore me, but they also refused the cigar I proffered and the drink I promised. And so they sit, hour after hour, day after day, silent and pensive, moved to activity only when they pull their chain.

Harlem's appetite for unusual food has also developed its share of odd jobs. There was Pig Foot Mary, who had a stand near the subway entrance at 135th Street and Lenox Avenue. She made a small fortune serving pigs' feet, and pigs' ears over a greasy counter. Before she retired to take care of her real estate investments, she had many competitors all over Harlem. None of these, however, possessed the personality and color of Pig Foot Mary and now they are being rapidly pushed to the wall by the more elaborate Coffee Pots and Rotisseries who also handle pigs' feet, a delicacy for which Harlem stomachs crave and clamor.

There is, however, still room for the yam man, and he can be seen pushing his yam filled, glass covered go-cart through the streets, his apron greasy, his derby at a right angle, his left hand busy ringing a bell, his mouth wide open singing in a minor key, "Yams, oh, oh, oh, here comes the yam man." He, too, like Pig Foot Mary, will soon be able to retire.

This story would hardly be complete without some mention of the traveling janitor who is an asset and a necessity. In a place like Harlem where there are so many petty business establishments not prosperous enough to hire a full time person to do their cleaning, yet too prosperous to do it themselves, he is a godsend. Early in the morning he can be seen going from place to place, his bucket on his arm, his mop and broom slung over his shoulder, a determined ally of cleanliness. He will probably have five or six offices to clean, and maybe one drug store or a beauty parlor. The offices will pay him on the average of about $40.00 per month, the larger stores about $60.00.

The best known traveling janitor in Harlem has a system all his own. He plans his work so that it seems as if he is never employed, and yet he cleans more offices than any other janitor, in Harlem. He cashes in on his speed, boasting that there is no one alive who can beat him washing windows or mopping floors. Afternoons he can be seen standing on the corner of 135th Street and Seventh Avenue talking to his cronies. The early evening hours and the early morning hours he attends to his duties.

This janitor has a large family none of whom help him. All the children are being prepared to enter *some* profession, and *some* profession to their father means either a doctor or a lawyer. Dentists he cannot abide, because the only people for whom he works that do not pay him regularly are a pair of young dentists just out of school.

# Part Two

# Social Essays and Journalism

In the following essays, Thurman positions himself as a guide to the world around him, even as he reveals his own uncertainties, biases, and weaknesses. Unlike the essays on Harlem or on literature, these essays bear testimony to the varied interests and particular circumstances that shaped Thurman's development as a writer. Thurman's "Christmas: Its Origin and Significance" probably dates from his years as founder and editor of *The Outlet* in Los Angeles. Most likely, this essay appeared first in *The Outlet* in December 1924 and then in *The Messenger* in December 1925. This historical overview of Christmas is a pleasant change from the caustic tone of much of Thur-

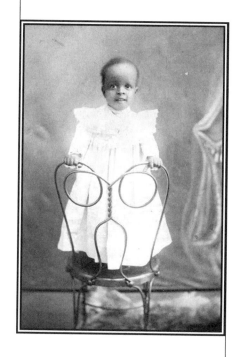

man's other writing. With "In the Name of Purity," Thurman consolidates the hard-hitting, scathing style that became his hallmark. He balances his analysis of American "foolishness" with a darkly articulate sense of humor. The lengths to which officials would go to protect the United States from "corrupting" foreign influence outraged Thurman, who hungered for new ideas and experiences. This lampooning of America's cultural backwardness resonates with H. L. Mencken–like moments of disdain.

Indeed, Thurman's feelings about his own background were complex and ambivalent, as is apparent in "Quoth Brigham Young—This Is the Place," a report on the social and cultural life of his home state. While this piece is critical of Utah, as usual Thurman finds a way to disarm some of the critique with humor. Despite his dissatisfaction with Utah, Thurman uses his origins to highlight his distinctiveness from his Harlem contemporaries, noting that

> when the conversation begins to lag I can always casually introduce the fact that I was born in Utah, and immediately become the center of attraction nonchalantly answering the resultant barrage of questions. I find that I can even play this trick on the same group of persons more than once, for it seems as if

they never tire asking—Do Mormons still have more than one wife?—Do they look differently from other people?—How many wives did Brigham Young have?—Are there any Negro Mormons?—Can one really stay afloat in the Great Salt Lake without sinking?—and thus they continue ad infinitum, and I might also add—ad nauseam.

The two unpublished essays we include here give some insight into Thurman's life. Both pieces demonstrate his ongoing commitment to recording and critiquing his personal experiences. "Autobiographical Statement" probably dates from late 1928 when Thurman, encouraged by William Jourdan Rapp—his collaborator on two plays—began to mine his life for possible material for longer works. Along with "Notes on a Stepchild" (in Part Five—*Aunt Hagar's Children*), this piece provides Thurman's perspective on his development as a writer. While "Notes on a Stepchild" focuses on aesthetic and philosophical issues, "Autobiographical Statement" emphasizes family background and education. From these two pieces together, one gets a clear sense of the ambitions, dreams, and contexts that made Thurman, in Langston Hughes's phrase, that "strangely brilliant black boy, who had read everything, and whose critical mind could find something wrong with everything he read" (235).

"Description of a Male Tuberculosis Ward" dates from sometime after July 1934, when Thurman was hospitalized at City Hospital on Welfare Island, New York, with little hope of recovery. It is clear from letters written to friends in late 1934 that Thurman wrote this piece against the orders of his physicians, who asked him to rest as much as possible. But the writing served as a diversion, as he told Hughes, from being "not only bored but full of holes and pains." Thurman remarks on the irony of being hospitalized in the very facility he had criticized in his last novel, *The Interne,* in a letter to Hughes:

> Whowouldathought I would soon end up in the tubercular ward of the very hospital I damned and god-damned when I wrote *The Interne.* Ironic, I calls it. Or is nature finally avenging art? The hospital is still terrible. Certain conditions have changed for the better—there is more help and less dirt. But the patients are the worst feature. Being bedded among the proletariat is enough to make me or anybody become a rabid lover of the aristocrats. Such dumbness, such utter stupidity, such innate vulgarity, such congenital affinity for dirt and filth.

Despite these harsh words, Thurman struggles in his essay "Male X" to overcome personal irritation in favor of a writer's curiosity about the lives and characters that surround him. He finds in "Male X"—ward number ten for men—an accumulation of men from a variety of backgrounds who form a community. In capturing the tuberculosis ward in his prose, Thurman not only exercises his powers of description and narration but also desires to view each patient as an individual—apart from racial identifications. The Roman numeral "I" at the beginning of the piece suggests that Thurman planned a more extended study of the ward, but the piece is obviously unfinished. It is being printed here for the first time.

These essays address a variety of human topics; they are not centered on the core issues of race and literature as is much of Thurman's other work. Their more personal atmosphere—in which Thurman shows himself to be just as aware of the failings of his own life as, for example, he is of those of Utah or the U.S. immigration officials—gives his readers the opportunity to appreciate the rigor and scope of his intellectual and spiritual makeup.

# Christmas:
## Its Origin and Significance

*The Messenger*, December 1925.

Once more it is the Christmas season. Once more it is the season of good cheer and good hope. Once more it is the time when the spirit of gladness and good will invades the souls of all men and makes them temporarily joyful, charitable, and bountiful.

It is really a pity that the Christmas spirit has to be confined to such a small portion of the year. It is really a pity that the Christmas spirit cannot be intrigued to prolong its visit. What a more pleasant world this might be could we all retain that magnanimous feeling of good fellowship and peace with all men that is characteristic of the Christmas time. And it is also a pity that the charitable organizations, and their donors who make Christmas a time of joy for the little children of the poor, do not realize that these same little children need nourishing food and delicacies every day in the year, and that they would also welcome gay festivities and gifts more than once a year. Some people believe that because they give to the poor at Christmas that they have completed their task, and turn a deaf ear to the pleas that come to them during other times.

When one watches people during the Christmas season one realizes more than ever that man is a creature of whimsical paradoxes, a creature of fleeting, ever changing moods. And one wonders what great force or what great influence is at work within the majority of the earth's people that causes such a sudden, albeit short, change in human nature as we see during the Christmas tide.

Christmas comes on the twenty-fifth day of December, and is supposed to commemorate the birth of Jesus Christ. Actually, like so many other ecclesiastical holidays, Christmas has been celebrated on many different dates and in many different ways. No one knows the exact date of the Christ child's birth. It was celebrated, during the earlier periods, by different people on the first and the sixth of January, the twenty-ninth of March, and even on the twenty-ninth of September.

In the fourth century the season of the winter solstice or the season of the New Year was adopted as the period of celebration, but the western church began their New Year on the twenty-fifth of December, while the eastern church began theirs on the sixth of January. Toward the end of the fourth century, Julius I, a Roman Pope, had his staff of scholars and savants do extensive research, but they could not find out anything new concerning the birth of Christ, and in the maze of tradition and myth connected with it nothing logical could be ascertained, so they decided upon the twenty-fifth of December as the most probable date.

Thus ever since the fourth century Christmas, or the birth of Christ, has been celebrated on that date although historians and astronomers have attempted to prove that this is not the true date of the nativity.

Before the Christmas era there were similar celebrations at a similar time in all the pagan countries. In Rome it was called the Saturnalia, a festival in honor of the god Saturn. During the course of this festival all moral barriers were thrown down. The god of Joy, Bacchus Dionysius, reigned supreme. Even the slaves were given their temporary freedom, and in some cases were treated as equals by their masters. Work and business were suspended, houses were decorated with laurels and evergreen, presents were exchanged, and public games were held in which all the citizens participated. Also every sort of moral license was practiced and countenanced.

In the cold bleak northland where the early Saxons, a fatalistic pessimistic people, lived, there was also held an annual festival in honor of their God Thor that corresponded with our Christmas. Indeed the custom of the burning of the Yule Log comes from them, and it is also from them that we inherited the use of the mistletoe. They had priests called Druids, who worshipped the mistletoe. These priests would go into the forest followed by the people singing and dancing, and when the mistletoe was found (they believed that it was dropped from heaven), they would build huge fires in which they burned cattle and men for sacrifices to the gods. This practice was also carried into England when the Saxons crossed over into that country, and remained an annual affair until replaced by the Roman Saturnalia when the Romans conquered the Anglo-Saxons, and substituted Roman customs for old.

In the Persian Empires of the east, the priests, who strangely were so like the old Druids of England, conducted a similar festival during the winter solstice. And even in the more remote countries, where the earlier history is almost a closed book, can be found traces of a celebration, pagan and horrible to us today, yet which was most certainly the forerunner of our own Christmas.

Most authorities are inclined to believe that the early Christian ministers transferred these pagan celebrations to their own religion as a compromise, a sort of concession to attract converts. Thus traditions became identified with the church after they had been bereft of their uncivilized, revolting features.

As to be expected this caused no end of trouble. Certain religious sects have continually denounced Christmas as being heathen, sacrilegious, and blasphemous. During the Puritan domination of England many laws were passed to abolish this old custom of commemorating Christmas. Anyone who was found not working or who happened to indulge in any form of festivity would be liable to a prison term, and in some cases where they persisted the culprits were beheaded in a nice Christian manner. The pilgrim fathers of our own country also punished people who were so indecorous as to celebrate the nativity of Christ, and even to this day the Calvinistic Baptists do not believe in or celebrate Christmas.

Despite all of this, despite the fact that most of our ideas of celebration have descended from pagan people, despite the fact that we do not know the exact date of the Christ Child's birth nor in fact are we certain that he was ever born, no one will hardly deny that the fleeting Christmas season is a veritable blessing to man. It is the one time in the year that Christians come the nearest to living up to their professed doctrines. Neighbors, heretofore inimical to one another, exchange gifts and delicacies. Wayward children and stubborn parents forget their difficulties. The rich remember that there are poor people deserving of their charity. The laborer and the middle class business man "stick together." It is the only time in the year when everyone has a smile for everyone else. It is the time when everyone seems to be inspired to help their fellow man even if it only lasts a week. It is the greatest time to watch for the manifestations of the teachings of Jesus Christ.

The Christmas spirit that enters the hearts and souls of all men at this time coincides with my idea of religion. I like to think of it as not something tangible from which I shall ultimately receive material gain, but as something intangible, yet beautiful and forceful, capable of inspiring in me a greater love and regard for my fellow man.

Oh if only the Christmas spirit would and could inhabit the mind, soul, and body of all mankind throughout the three hundred and sixty-five days of the year.

## In the Name of Purity

*The Messenger*, April 1926.

I started to entitle this article "The Damn Fool Americans," but decided that a mere damn was too mild, and realizing that if I should happen to make it more emphatic I would be liable to detention in the nearest hoosegow, I discarded the idea. Then, too, I reasoned that all Americans are not to be placed in this category; at least five out of every million are exempt.

The present mood of irritation has been induced in many others besides myself, in fact anyone with an I.Q. registering above that of a moron would react belligerently toward the constant efforts being made to fundamentalize the United States of America. We have had conscientious fundamentalists like Bryan, we have had unscrupulous fundamentalists like Simmons, and now we have deliberate fundamentalists like our Secretary of State, our Secretary of Labor, and their underlings.

The latest gesture on the part of these gentlemen to keep America free from moral blemishes, and to convert it into a sublimated Valhalla is the exclusion of the Countess

Cathcart. No more asinine gesture has been made. This surpasses even the Dayton, Tennessee, circus, and the exclusions of the Karolyis and Saklatavla. As someone has so aptly phrased it: "If the present regime continues much longer anyone whose countenance does not reflect the sublime contentedness and Christian piety of a Texas yokel will be excluded, to say nothing of people with red hair and tilted noses."

The Countess Cathcart was once upon a time impelled by the call of youth to leave her husband and flee to the wilds of Africa with the Earl of Craven. Eventually, they tired of their adventure, the husband of the countess divorced her and named the Earl as co-respondent, while the offending Earl was welcomed back to his family hearth. Thus the incident closed, and thus it would have remained closed had it not been for the official piddling of our addle-headed ninnies in Washington and Ellis Island.

Just why the immigration authorities thought that the Countess could in any way lower the moral standards of America is not quite clear. In the first place the law has been sadly misinterpreted, for neither the Countess nor the Earl were ever convicted of a crime, nor were they even indicted for one; and the law specifically states that the person must be guilty of moral turpitude, meaning this not in the light of one's personal sex behavior, nor in the light of one being supposedly guilty of committing a crime against some European dictator, but in the light of one's being an actual criminal. If being sued for divorce is a crime then 90 per cent of our "pure, white, Nordic, American stock" should be incarcerated immediately. Then, too, the Earl was admitted without question not only once, but twice, and would have remained here undisturbed had not certain persons become inquisitive anent this double standard. Can it be that our governmental authorities are advocates of the "woman must always pay" doctrine, or is it that they feared the Countess might choose to elope with a fatal number of our American husbands?

Seriously, though, such tommyrot should be thoroughly investigated. This sudden alertness is indeed unnatural, and was certainly provoked by other agencies than the mere concern of our immigration authorities about the morals of the Homo Americanus. "Touch my palm with silver, and . . ."

Perhaps the present case may be the means of generating an organized reaction against this drive for a nationwide purity. In the words of a pool hall Johnny: "Who in the hell wants to be pure anyway"; especially, I might add, when purity is measured in terms of hypocrisy. As long as one hides one's supposed impurity, one is according to the standards of our governmental officials, pure. At least that is what I infer from the presented evidence in the Cathcart-Craven case. Had the Countess lied she would have been admitted without further question.

It seems to me that our authorities would at least have something upon the statute books to back them up. In the instance of the Countess Karolyi, a war-time measure prohibiting the admittance of anyone who might preach a doctrine subversive to the best interests of the government of the United States was invoked. This bill was specifically aimed at Bolshevists. The Countess Karolyi is no preacher of Bolshevism nor is she a Communist, and in no way liable to be included under the invoked statute. It so happened that Premier Horthy of Hungary considered Countess Karolyi his own personal enemy so he informed his ambassador to get busy here in the United States since she might influence certain persons against him who were preparing to lend his government some money. The ambassador did get busy, and "Nervous Nellie" Kellogg flew to the rescue, striving to save we susceptible Americans from being inoculated with the deadly germs of Communism.

Thus we are kept pure from ideas and infidelity. Americans must not become cognizant of any doctrines save the existent one, nor must they be contaminated by an adulteress. In this nation of prohibition and prostitution one lone woman might incite us all to Bolshevism, and another might incite or excite us all to indulge in promiscuous sex relations, which is the ultimate reduction ad absurdum, the last word in asinine ridiculousness, and should be hahahed into limbo.

This fanatic fervor to reform is also due to the war, I suppose, as is every other deplorable condition now extant since 1914. We Americans sallied forth to make the world safe for democracy, and returned home still seeing phantoms. Consequently we have been pursuing these phantoms ever since, and it has reached such a stage that it seems that the whole nation will evolve into Don Quixotes attacking moral wind-mills.

First the Anti-Saloon League begged enough money to effect prohibition; then the K.K.K. culled enough coin from the pockets of our illiterati to effect racial purity, female chastity, and America for Americans; next the leading fundamentalists of the country began to plead for funds to establish a memorial to William Jennings Bryan in order to carry on that worthy's campaign against science and evolution; now our government authorities are evidently receiving their salaries to give vent to personal grudges or to cater to the prejudices of our financial giants. And all in the name of purity.

## Quoth Brigham Young—
## This Is the Place

*The Messenger*, August 1926.

I am fully aware what Brigham Young had in mind when he uttered the above enthusiastic statement, yet try though I may the most enthusiastic thing that I can find to say about my home state and its capital city is that it invariably furnishes me with material for conversation. It does not matter to whom I am talking, whether it be Jew or Gentile, black or white, Baptist or Episcopalian, thief or minister, when the conversation begins to lag I can always casually introduce the fact that I was born in Utah, and immediately become the center of attraction nonchalantly answering the resultant barrage of questions. I find that I can even play this trick on the same group of persons more than once, for it seems as if they never tire asking—Do Mormons still have more than one wife?—Do they look differently from other people?—How many wives did Brigham Young have?—Are there any Negro Mormons?—Can one really stay afloat in the Great Salt Lake without sinking?—and thus they continue ad infinitum, and I might also add—ad nauseam. Nevertheless it is amusing at times, and, as I say it is for this reason alone that Utah has one warm spot in my rather chilled heart, for whenever I stop to remember the many dull hours I spent there, and the many dull people I spent them with, even that aforementioned warm spot automatically begins to grow cold.

Utah was a wilderness composed of ore laden mountains, fertile valleys, and desert wastes frequented only by trappers and Indians when the Mormons, an outlaw religious sect believing in and practicing polygamy, settled there. These Mormons had trekked over half of the continent in search of a spot where they could found a settlement, earn their livelihood from the soil, and indulge in their religious peculiarities unmolested by their pernicious brethren in God who insisted that they practice other religious peculiarities. They had been run out of Illinois, they had been run out of Missouri and Kansas and they had forged their way over miles of Nebraska prairie land, miles of Wyoming sage brush hills, and miles of mountain trails before they finally stood on a peak overlooking the beautiful Salt Lake valley, surrounded by the Wasatch range of the Rocky Mountains, and cheered when their intrepid leader, Brigham Young, shouted: This is the place!

Once they had found this suitable site the Mormons, under expert leadership, founded their mundane Zion, named their townsite Salt Lake City after the great inland sea nearby, christened the crooked river that ran around the city's outskirts—the Jordan, cultivated the rich farm lands, carried on a profitable trade with the Indians, began to raise stock, and started the construction of their sacrosanct religious temple and tabernacle, which stand today as monuments to their super-achievement.

Things hummed in the new town. Cattle carts lumbered down rocky mountain trails carrying the big stones that were being used in building the temple. Gold rushers, bound for the coast, stopped and sometimes stayed if they felt like braving the arrogant hostility of the Mormon fathers. Square blocks of land were apportioned off to the various churchmen, who energetically erected primitive homes for themselves and their wives. The great tithing square on the site where the renowned Hotel Utah now stands, teemed with people pouring in from the surrounding countryside to pay their tithes, while the public watering ground, where the Salt Lake City and county building was later built, was crowded with overland wagon trains, and Mormon visitors from nearby settlements, for Zion had soon overflowed and mushroom towns appeared overnight in the immediate vicinity. Zion flourished, Zion grew wealthy, and Zion grew more holy per se.

However, all mundane paradises seem subject to an invasion by the devil's forces, and the Mormon Zion was no exception, for the devil's forces soon came in the persons of non-Mormons, derisively called Gentiles. Like most gold seekers of their day (and are gold seekers ever any different?) they wanted only the chance to garner gold—damn how they got it or how they suffered meanwhile. So Zion was invaded, and Zion soon succumbed to a wave of prosperity, progress and prostitution, and the transcontinental railroad, which had its east-west junction near Ogden, was the most telling blow dealt by the Gentiles.

The result was pitiful. Thousands of easterners came pouring in to see whether or not these Mormons really had horns, and finding that they were not so endowed by nature decided to stay and break down the Mormon wall around the natural wealth of the state. The Mormons put up a brave battle while Brigham Young lived, but after his death there was a complete debacle. Utah was finally forced to come into the Union, and for coming in she had to abolish polygamy, and lose her individuality, for from that day on Utah was just another state, peopled by a horde of typical American booboisie with their bourgeoisie overlords, and today Utah is a good example of what Americanization and its attendant spores can accomplish.

I have as yet made no mention of the Negro, and this article is supposed to fit into the series called–"These Colored United States." For the moment I wish to quibble, and assert that there are no Colored United States, *id est,* no state in the Union where the Negro has been an individual or vital factor. As George S. Schuyler is so fond of saying all Aframericans are merely lampblack whites steeped in American culture (?) and standardization. When it comes to such localities as Harlem, the south-side black belt of Chicago, the Central Avenue district of Los Angeles, the Seventh Street district in Oakland, the North 24th Street district in Omaha, the Vine Street district in Kansas City, the Beale Street district in Memphis, and similar districts in Atlanta, Charleston, New Orleans, Houston, El Paso, Richmond, Birmingham, et cetera, one might write of these as colored cities, for it is there that the Aframerican spirit manifests itself achieving a certain individuality that is distinguishable from that achieved in similar white districts despite all the fervent protests of Brother Schuyler to the contrary. What I am leading to is this, that to write of "These Colored United States" is to be trying to visualize a phantom, for in state lots the Negro, save in such southern localities where the population is greater than the white and even in these one can only pick out certain communities to dissect, is a negative factor contributing nothing politically, historically, or economically. He only contributes sociological problems.

The above paragraph is rather rash, and perhaps I should temper it somewhat, and confine myself to the northeastern and northwestern states, for I am not so sure that the Negro has not made some contribution at least economically, in the southern states, but neither am I so sure that this has not been swallowed up beyond the point of recognition by the whites who most certainly hold the power. And now I find justification for having such a series of articles even if they are rather far-fetched, for Negroes need to be told of past achievements and present strivings. They need this trite reminder to stir them, and to urge them on to greater achievements. They must develop a race pride, and so they must be told of what they and their foreparents have achieved. I am sorry then that I have to write of the Utah Negro, for there has been and is certainly nothing about him to inspire anyone to do anything save perhaps drink gin with gusto, and develop new techniques for the contravention of virginity.

There is little difference between the few Negroes in Utah and their middle class white brethren. The only difference is one of color, and those Aframericans who have been in the state longest have done everything in their power to abolish even this difference. Miscegenation was the common thing for years, and until a state law was passed prohibiting intermarriage the clerks at the county court house were kept busy signing up fair ladies with dusky men. Then when the prohibitive law was passed the roads to Wyoming and Montana were crowded until those commonwealths also passed anti-miscegenation laws. What is more it reached such proportions that even as late as 1915 there was in Salt Lake a club catering only to Negro men and white women, and, when I was last there, which was a year ago, there were three super-bawdy houses that I knew of, where white ladies of joy with itching palms cavorted for the pleasure of black men only.

This situation was of course not peculiar to Utah alone. It was also true of most western states, and the "Manassa" group of the middle-west was far more notorious than any like group Utah has produced. However, this happened only because the population of Utah was considerably less than that of some of her sister states. Statistics will readily prove, I believe, that comparatively speaking the intermixing of races was as great or greater in Utah than in any other western state.

But to get to another point—There were two Negroes in the first overland Mormon train, a man and his wife (he had only one, for Mormons did not believe that a Negro could ever enter into Heaven as an angel, and that since because of Ham's sin he was to be deprived of full privileges in Heaven, he was not entitled to enjoy the full privileges of a good Mormon on earth), who were servitors to Brigham Young. A little later other vagabond souls, eager to escape the terrors of both the pre and post civil war South, drifted in and remained if they found employment. Then still others were caught in the contemporary westward drift of American population, and entered into the "Bee" state as gamblers, gold-seekers, prostitutes, and home servants. And later, during the ascendancy of the Gentile regime there was quite an influx of Pullman car porters, dining car men, hotel waiters plus more pimps and prostitutes. This population was for the most part transient, but a few of them accidentally during drunken moments or temporary physical ecstasy settled there and commenced the raising of families, which families are now members of the Utah Negroes' *haute monde*.

Until the war had inspired the northern migration of southern colored people, there were few of what is known as respectable Negroes in the whole state. These strived hard to cling together, and they generally did except upon the matter of religion, which I might boldly add herein, has done more to keep the American Negro at variance with himself than any other agency. Some folk were Methodists, some were Baptists. Then some Methodists would turn Baptists, and some Baptists would turn Methodists. Moreover, some Methodists and some Baptists would grow discontented and there would be rumors of a split, and most times these rumors would develop into actualities. At the present time there are three Negro churches in Salt Lake City, which has a population of about 1,800 colored people. Only about 500 of these are of the church-going variety, and imagine their strength divided as they are between two Baptist and one Methodist Churches.

Salt Lake City and Ogden have the largest Negro communities, and of these two Salt Lake has the greater population, but one would never believe this after walking through the streets of the two cities, for one can walk for hours in Salt Lake without meeting a colored person, while in Ogden one will meet any number in the downtown district. This is due to the fact that the Negro population of Salt Lake has not become centralized, and there is no Negro ghetto, while in Ogden almost the entire Negro population is centered around the railroad yards and depot, because almost the entire Negro population of Ogden is engaged in fleecing the transient railroad porters and dining car waiters out of as much money as possible while these men are in the town. The only other place in Utah where there is an appreciable colored settlement is at Sunnyside, in the southern part of the state, where some two or three hundred men are employed in the coke ovens.

In the glorious state of Utah, there are no representative Negro institutions of note save the deluxe gambling clubs and whore houses in Salt Lake and Ogden. The churches are pitiful and impotent. There are no Negro professional men. There are no Negro publications, not even a church bulletin. There are no Negro business houses. There are no Negro stores. There are no Negro policemen, no Negro firemen, no Negro politicians, save some petty bondsmen. There are a few Negro mail carriers, and the only Negro mail clerk in the state passes for Spanish or something else that he isn't in order to keep his position and not be forced to become a pack laden carrier. Most of the Negroes in the state are employed on the railroad as porters and dining car waiters, or else in the local railroad shops, or else earn their livelihood as janitors, hotel waiters, and red caps, thereby enabling themselves to buy property and become representative bourgeoisie.

Negroes are rigorously segregated in theaters, public amusement parks, soda fountain, and eating places. This too seems to be a result of the post world war migration of southern Negroes to the North which was accompanied by a post world war wave of Kluxism and bigotry. The earlier Negro settlers experienced little of these things. They were welcome in any of the public places, but as the Negro population grew, and as the Gentile population grew, so did prejudice and racial discrimination. Until now the only thing that distinguishes Utah from Georgia is that it does not have jim-crow cars. Last year there was even a lynching—the second in the history of the state.

Add to this the general dullness and asininity of the place and the people, and you will understand why a writer (who was also born in Utah) in a recent issue of the *American Mercury* declared that there was not an artist in the entire state, and that if one was to stay there he would soon be liable to incarceration in the insane asylum at Provo, or else buried in one of the numerous Latter Day Saint cemeteries. I was there for a short time last summer, and sought to buy my regular quota of reading matter. I asked for a *New Republic* at every downtown newsstand in Salt Lake City, and out of ten stands only one had ever heard of it. I made equally vain searches for *The Nation, The Living Age, The Bookman, The Mercury,* and *The Saturday Review of Literature.* At the only stand that had ever heard of these publications the proprietor advised me to pay him in advance and he would order them for me as he did for a few other of his customers who were crazy enough to read such junk. He capped it all by inquiring whether or not I was a Bolshevist.

Thus is Utah burdened with dull and unprogressive Mormons, with more dull and speciously progressive Gentiles, with still more dull and not even speciously progressive Negroes. Everyone in the state seems to be more or less of a vegetable, self-satisfied and complacent. Yet I suppose that Utah is no worse than some of its nearby neighboring states, which being the case the fates were not so unkind after all—I might have been born in Texas, or Georgia, or Tennessee, or Nevada, or Idaho.

## Autobiographical Statement

Unpublished; probably written in late 1928.

My parents and grandparents were pioneer Westerners who settled finally in Salt Lake City, Utah, thus enabling me to be born, twenty-six years ago, within the protective shadows of the Mormon Temple and the Wasatch Mountains.

I first entered school at the age of six in the city of Boise, Idaho. Within two months I was taken ill and for the next two years was a pampered invalid. Meanwhile I had returned to the city of my nativity only to leave there after another two years to move to Chicago where I remained from 1910 until 1914. Omaha, Nebraska was my next stopping off place. It was there that I finished grammar school and was a high school freshman. This done I once more went back to Salt Lake. Persistent heart attacks made a lower altitude necessary so off I went to spend a winter in Pasadena, California. Came the "flu" epidemic of 1918, I succumbed and on convalescing returned to my home town. Somehow or other by this time I had finished high school and had matriculated at the University of Utah. Two years there, a pre med student, then a nervous breakdown, a summer trip to Omaha, a "hobo" trip back to Salt Lake. Then Los Angeles again, three years a postal clerk, two simultaneous years a student at the University of Southern California, sudden inspiration, decision to be a writer, and in 1925 a hectic hegira to Harlem.

Thus is my checkerboard past. Three years in Harlem have seen me become a New Negro (for no reason at all and without my consent), a poet (having had two poems published by generous editors), an editor (with a penchant for financially unsound publications), a critic (see articles on Negro life and literature in *The Bookman, New Republic, Independent, World Tomorrow,* etc.), an actor (I was a denizen of Cat Fish Row in *Porgy*), a husband (having been married all of six months), a novelist (e.g., *The Blacker the Berry*, Macaulay's, February 1, 1929: $2.50), a playwright (being co author of *Black Belt*). Now, what more could one do?

## Description of a Male Tuberculosis Ward

Handwritten notes in the Thurman folder at the Beinecke Library at Yale University; written sometime after July 1934.

I
—

Male X possessed one rather amazing peculiarity. Despite its sunlit spaciousness, it bore more than a slight resemblance to a tenement house railroad flat.

Male X knew no privacy. It was at once a rectangular hospital ward and an elongated corridor, through which there continually passed a variegated procession of septic people and anti-septic paraphernalia.

Physically, there remained little else to criticize. True, 25 beds did tax the available space. But there was not too much congestion. And the sunlight and air drawn from the six windows was sufficient to please even the most captious, which was all to the good. For Male X was dedicated to the care and cure of pulmonary tuberculosis.

It was seldom that any of the 25 beds were empty. Neither transfers to other sanatoria or to the morgue or outright releases served to prevent an overflow to other wards. Which, of course, is not surprising for a rather small ward in just one of a huge municipality's many hospitals.

The patients were worthy of a novel. Without exception they were drawn from the most indigent of the proletariat. Poor Irish and ghetto Italians were in the majority. Bewildered central European peasants, a few Negroes and Porto Ricans, an occasional Oriental completed the racial amalgam. Surprisingly, or was it, only the Jew was not tolerated, spoken of in harsh terms if mentioned.

There was a Scotchman, too. He was 67 years old. He was a self-confessed dipsomaniac and a braggart teetotaler. There was more than a streak of the Baron Munchhausen in his mental makeup. By his own admission he had never been to Scotland, having been born in Canada. Yet, as a child, he had played around that very Loch Ness contemporaneously in the news.

Butte, Montana, had so been named by him. For this contribution to American municipal nomenclature, our Scotchman had received $2,500. Naming cities seemed to have been a particularly profitable hobby during his youth. He had also named several townsites in Canada and never was his remuneration less than $1,500. This hitherto unknown and pecunious industry caused quite a furor in Male X. But it was nothing compared to the near riot which followed the Scotchman's amazing statement that in 1914 he was transported from the U.S. overseas to fight the Germans as a member of the American Legion!

It can be seen that things were not dull in Male X. There was seldom a general outburst of hilarity and little horseplay. Neither was there much of the "male talk" characteristic of a number of men isolated in one room.

Nor was there an aura of deep depression. The atmosphere was one difficult to analyze. There was definitely an emasculation. Most of the patients were over 40. Two or three were deep in senility. The few younger men were wan, lacking in vitality or color. This illness had completed a process which the factors of their lives had initiated. Beaten by poverty, worn out physically from labor which rewarded them only with occupational diseases, they all had the bewildered mien of Jesus Christ on the cross. It was as if they all cried out in unison: "My God—why hast thou forsaken me?"

They were no longer men. And even if the disease was allayed and some semblance of health returned: What could they do? Where could they go in this depression ridden world?

There was Foley, a jolly Irishman with a conventional jolly brogue. For 19 years he had followed the sea. Then employment scouts for the steel mills had lured him from his rather pleasant berth on an English tramp steamer to the frigid pleasantness of a Pennsylvania steel mill.

The money was all the employment scouts had promised. But the company store was an insatiate delver into the pay envelope. And the lust for beer after long hours being baked in front of a steel furnace further depleted the money earned.

Ill health followed. He lost his job. He became a seasonal worker, laboring in rail and lumber yards or on section gangs or in construction camps when his health so allowed.

He tried to return to sea. There were no berths. He longed to return to his native land there to pass his last days. But the union had urged him to become an American citizen.

There is little more to add to his story. Tuberculosis germs provided a period. A free ward in a municipal hospital was almost a welcome asylum.

# Part Three

# Correspondence

In nearly all the letters from Thurman that we have gathered in this section, one can clearly see Thurman's love for letter writing. Whether asking for a loan, complaining about his pending divorce from Louise Thompson, praising the rural life, or critiquing a recently read novel, Thurman is consistently readable and often controversial. Thurman's correspondence reveals his aspirations, his flaws, and his pretensions more clearly than anything else he wrote. He is always willing to be both critical and self-critical—and he usually does both with a sense of humor. In his letters Thurman adopts a number of tones and manners—sometimes bashful or haughty, at other times intense or frivolous, but always engaging and entertaining.

Dating from right before their 1926 collaboration on *Fire!!*, the correspondence between Thurman and Langston Hughes provides a glimpse of these two key figures of the Harlem Renaissance at the beginnings of their careers. In the early letters, Thurman is already a fan and advocate of Hughes's work, and this advocacy apparently never wavered. In his letters to William Jourdan Rapp—his coauthor of the plays *Harlem: A Melodrama of Negro Life in Harlem* and *Jeremiah the Magnificent,* Thurman demonstrates a relationship marked by ease and trust, a remarkable feat for a friendship between a dark-skinned black man and a white man whose early years do not indicate any special depth of interest in African American community or culture. Thurman's communications with Rapp and Rapp's responses to Thurman are amazingly free of the condescension and patronage that often marked the many otherwise well-meaning relationships between blacks and whites during the Harlem Renaissance. In contrast to the friendly but professional tenor of the correspondence with Rapp, Thurman's letters to Harold Jackman and Dorothy West convey a strong sense of the shared experience of race and of their common goals as young artists. Thurman is open with them at a level of felt emotion and raw nerves that gives his letters a "funkiness" entirely missing from his communications with Rapp. The few letters we have from Thurman to W.E.B. Du Bois, Claude McKay, Alain Locke, and Granville Hicks demonstrate the ambition and industriousness that characterized Thurman's quest for a literary career.

## Langston Hughes

Thurman corresponded regularly and as an equal with Langston Hughes. Here are two young writers—born in the same year and having come to Harlem from the West—who had a clearer understanding than most others among their Harlem Renaissance peers of their artistic ambitions and preferences. Thurman rarely holds back in these letters, talking freely and without hesitation about his chronic health problems, his financial woes, and his feeling that Louise Thompson was extorting him for alimony and expenses in their divorce proceedings. He narrates to Hughes details about his works in progress, his collaborations with Rapp, his negotiations with greedy producers regarding the Broadway production of the play *Harlem.*

Hughes's letters to Thurman are open and friendly, too, but Thurman must have known early on in their relationship that Hughes did not have the same desire to be as completely open. One only has to recall Thurman's portrayal of Hughes as Tony Crews in both *The Blacker the Berry* and *Infants of the Spring* to realize Hughes's reticence. In *Infants,* for example, Crews is depicted as a "smiling and self-effacing, a mischievous boy. . . . Tony was the most close-mouthed and cagey individual Raymond had ever known when it came to personal matters. He fended off every attempt to probe into his inner self and did this with such an unconscious and naïve air that the prober soon came to one of two conclusions: Either Tony had no depth whatsoever, or else he was too deep for plumbing by ordinary mortals."

Indeed, in some of the letters Thurman seems to be tantalized by Hughes's reticence. There are mystifying personal, possibly sexual, allusions in the correspondence, as, for instance, in Thurman's letter of February 6, 1926, "I have not yet had what I told you I needed to make me happy. . . . Perhaps I never will. I don't often weep on people's breasts, and I know that you sincerely hope that I don't choose you again." In the postscript to the same letter, Thurman adds, "I have not yet forgiven you for tearing up my picture. I feel that there was more reason than the one you gave. Hence I have sealed the pieces up in an envelope and labeled it 'Ashes of Vengeance.' It shall be preserved." Another indication in this correspondence of their personal closeness from 1926 to 1929 is Thurman's use of personal salutations and nicknames to address Hughes: "Hello there," "Mon Ami," "Dear Lank," "Cherie Ami," "Lang," "Dear Lank-y-yank-yank," and so on.

But quite apart from the importance one might or might not attach to such personal elements in their correspondence, it is obvious that Hughes admires this brilliant dark-skinned man and listens with respect to Thurman's freely and generously offered critique in letters and elsewhere of his poetry and short fiction. At the same time, Hughes must have quickly realized that while Thurman was an accomplished critic of poetry, he was not a good poet. Apparently, Hughes answers Thurman's repeated requests for Hughes's comments on his own poems with polite silence. But Thurman continues to trust Hughes as a reader of his essays and seems to have benefited from his critiques. For example, Hughes read *Aunt Hagar's Children* and returned it to Thurman with his comments on this collection of revisionist biographical essays about well-known African American figures. In an August 1929 letter, Thurman thanks Hughes for his "pointed notes," stating, "I do agree with you about using Negro in so sweeping a sense. Don't worry. I shall make the necessary qualifications. . . . I am glad you like the book. It should stir up some interesting comment if nothing more, and it is valuable to me in that having done this primary work[,] . . . I can now clarify certain personal problems and proceed on my novel with new vigor."

As a friend, Thurman supported Hughes's goal of getting a degree from Lincoln University and encouraged him to finish his autobiography while he was still at this "backwoods college." When, in 1925–1926, Thurman worked for a few months as an editor at *The Messenger*—replacing George Schuyler, who was on assignment in the South for *The Pittsburgh Courier*—he published Hughes's poems along with work by many others, including Edward Silvera, Arna Bontemps, Helene Johnson, Dorothy West, Claude McKay, and Georgia Douglas Johnson. A bit later, he told Hughes that if he had "any prose short-stories on hand lying dormant . . . I can get you ten dollars for one. *The Messenger* is quite flush, and I would like to see you benefit before we go broke again." Hughes published four stories at this time in *The Messenger*—"Luani of the Jungles," "The Young Glory of Him," "The Little Virgin," and "Bodies in the Midnight." In *The Big Sea*, Hughes quotes Thurman, who had characteristically described Hughes's stories as "very bad stories" but "better than any others they [Thurman and Schuyler] could find" (234). Interestingly, Hughes did not later reprint any of these stories except for "Luani."

Hughes had very likely helped Thurman adapt to Harlem life soon after the latter's arrival in New York on Labor Day, 1925. Hughes had introduced Thurman to Bruce Nugent, who—after his initial disbelief that anyone as dark-skinned as Thurman could be so brilliant—soon became Thurman's close friend and roommate. In the early years of the Hughes-Thurman friendship, Thurman's studios at 314 West 138th Street and then at 267 West 136th Street were the regular haunts of Hughes and other young writers and artists. While there has been much scholarly speculation about the actual whereabouts of "Niggeratti Manor," Thurman's letters to Hughes from March 1928 and July 1929—as well as one of his letters to Claude McKay—establish that the address was 267 W. 136th Street.

Discussion about the production of the first and only issue of *Fire!!* comprises a sizable portion of Thurman's 1926 correspondence with Hughes. Thurman was undoubtedly the leader of the group of younger writers who wanted to assert their artistic independence against both W.E.B. Du Bois's didacticism and Alain Locke's prescriptive cultural pluralism by publishing this literary magazine. In the end, however, there was no readership for such a magazine—*Fire!!* turned out to be a commercial disaster. And while Hughes seemed to have helped more than many others with expenses, the correspondence establishes that it was Thurman who ended up with most of the bills. He continued to pay the bills for three or four years from his meager income—long after the *Fire!!* experiment had been reduced to ashes.

Beginning in early 1930, there is a tapering off of Hughes-Thurman correspondence, probably signaling a shift in the warm and engaging relationship the two had maintained for nearly four years. Scholars have suggested that perhaps Hughes and Thurman had had a falling out over the play *Mule Bone*, co-authored by Hughes and Zora Neale Hurston, which apparently Thurman had agreed to rewrite without Hughes's consent. While this may have been an element (the letters are silent on the subject), we suspect that the reasons for the change in their relationship are much more complicated. By 1930, Hughes had had more success in his career than Thurman and may have felt that Thurman was not as important to him, although there is evidence that he continued to admire Thurman's brilliance long after the latter's untimely death in 1934. Also, Hughes may have been frustrated as a friend by his inability to help Thurman stop his self-destructive drinking. There is no evidence that Hughes responded to the last few letters Thurman wrote in 1934 to him from his deathbed in the City Hospital. Arnold Rampersad suggests that while Hughes did ask Carl Van Vechten to send some magazines out to Thurman, Hughes himself "seldom replied to the pathetic letters" Thurman wrote (1:302). While there is certainly a pleading tone in these last letters, they are not "pathetic." There is always a spark in Thurman's letters—even at this point in his life. In his last letter to Hughes, Thurman manages to be surprisingly defiant, witty, and funny; his postscript is almost an epitaph: "Thurman is distinctly a has been—so many people have already buried him. Woe betide 'em when I am resurrected!"

## William Jourdan Rapp

Thurman's letters to William Jourdan Rapp mostly date from 1929, when Thurman traveled to the West to visit family in Utah and explore creative possibilities in Hollywood. Perhaps he also wanted some distance from Harlem and from his disintegrating marriage to Louise Thompson. After the successful run of the New York production of *Harlem: A Melodrama of Negro Life in Harlem*, Thurman probably felt financially secure and artistically adventurous enough to undertake such a journey. He left New York on April 6, 1929, traveling to Detroit, Chicago, Salt Lake City, Los Angeles, and then back to Salt Lake City. It was a time marked by the anxiety of coming home to Salt Lake City; the optimism and disappointment of his Hollywood stay, during which he met his long-estranged father, Oscar, in Santa Monica; the negotiation of his divorce from Thompson; worries over his grandmother's health; and continuing work on a number of projects. He maintained a busy schedule in spite of several periods of illness and convalescence. The sheer number of letters Thurman wrote to Hughes, Rapp, West, and others marks this 1929 trip west as a particularly productive period in his life.

Thurman's letters to Rapp are filled with details about their projects and Thurman's attempts to interest Hollywood in a film version of their play. In Los Angeles from late April to early July 1929, Thurman met with a number of studio representatives. When they showed little interest in his play, he attempted to sell them some of his other projects and finally worked on film ideas they gave him; Hollywood producer James Cruze wanted Thurman to write screenplays that would feature the black actress Evelyn Preer. While in Hollywood, of course, Thurman saw some films. He was especially impressed with the all-black musical *Hallelujah,* which suggested to him the success black material could have on film. Although this 1929 trip to Los Angeles didn't bear fruit, Thurman returned to Los Angeles in 1934—shortly before his final hospitalization—and cowrote two screenplays for producer Bryan Foy.

It is clear in the letters from 1929 that Thurman held Rapp in high regard. In the unpublished essay "My Collaborator" (see Part Seven), Thurman writes: "He is the most energetic man I know being happiest when there is much work to do. He is generous of his time and money, and I have never known him to be bored or pessimistic even when *Black Belt* was in the hands of producers who, as St. John Ervine would say, ought to be in the cloak and suit business. His enthusiasm is most infectious, affecting even blasé Broadwayites."

Rapp clearly brought a lot of playwriting experience to their collaboration, having written several plays. He had connections in the New York theater and literary worlds. Thurman brought Rapp a wealth of fresh, exciting material about Harlem that catered to the growing curiosity about the place and its denizens.

They not only collaborated on plays, they also influenced each other's careers significantly. In 1928 and 1929, Thurman and Rapp worked together on *Harlem* itself and also on some dozen articles about *Harlem*. These articles—meant to drum up interest in the play—reflect the depth of the two men's friendship as well as their commitment to demonstrating the relevance of their work. They would again collaborate on *Jeremiah the Magnificent,* and doubtless Rapp helped Thurman publish in *True Story* magazine under pen names such as Patrick Casey and Ethel Belle Mandrake. It is just as clear in these letters, however, that Thurman is struggling with the nature of his collaboration with Rapp. He vacillates between rejecting some of Rapp's ideas and longing for his approval. For instance, Thurman refuses to write the novelization of *Harlem*—leaving it for Rapp's wife, Virginia, to do—until early July, when he takes it up enthusiastically.

Thurman's need for money may have conflicted with his more profound need to establish aesthetic boundaries apart from Rapp. While his greatest popular and financial success as a writer came with the help of Rapp, Thurman had a strong sense of individuality. It is clear in the letters that he wrestled with the nature of the collaboration throughout the summer of 1929, even as he worked with Rapp on the novelization of *Harlem* and on *Black Cinderella*—the third in the *Color Parade* trilogy of plays he and Rapp planned—and on a new project about the Mormon prophet Joseph Smith. At the same time, he finished work on a collection of essays, *Aunt Hagar's Children* (Part Five) and worked on the first draft of what would become *Infants of the Spring* (Part Eight).

Thurman and Rapp's relationship was personal as well as professional. When he left New York in April 1929, Thurman entrusted his financial and divorce-related business to Rapp's care. A particularly interesting letter is the one dated May 7, 1929, in which Thurman relates the circumstances surrounding his arrest for indecency in New York City—an incident Thompson threatened to use against him in divorce proceedings. A number of other letters deal with the details surrounding the divorce and the arrangements Thurman and Rapp were making for Thompson to travel to Reno to file for divorce.

The later letters between Thurman and Rapp reveal an increasing strain in their relationship—perhaps connected to Thurman's habit of asking to borrow money. Thurman hesitates to make more requests of Rapp, but he must. His last letter—written from City Hospital in 1934—suggests how lonesome and in need of intellectual stimulation he was in his final months.

W.E.B. Du Bois, Claude McKay, Alain Locke,
and Granville Hicks

Thurman's letters to Du Bois, McKay, Locke, and Hicks hint at the professional focus and personal ambition with which he approached his role as the editor of *Fire!!* and *Harlem: A Forum of Negro Life*. In them, Thurman lists his literary achievements but also examines his failures. Written some six months after his arrival in Harlem, the letter to Du Bois inquires about an opening on the editorial staff of the NAACP forum, *The Crisis,* to which Du Bois sent a polite reply indicating no positions were available. Thurman's is a respectful letter that testifies to the influence Du Bois held in 1920s Harlem. The letter reveals Thurman's goal to work in literary criticism, a wish he would fulfill with the publication of *Fire!!* in November 1926. As *Fire!!* would show, though, Thurman and Du Bois viewed the role of art in black life quite differently. Thurman would go on to be a leader of the younger generation of Harlem Renaissance writers who tried to reject Du Bois's call for socially responsible and uplifting art.

In his two letters to McKay, Thurman addresses the better-known McKay as a fellow artist who had rejected the staid didacticism advocated by Du Bois and others. Although McKay lived in Europe, his poetry and his novel *Home to Harlem* (1928) had made him a major literary presence for Harlem Renaissance writers. Appealing to McKay to submit something for his magazines, Thurman catalogs his own adventures since arriving in New York in 1925 and expresses his awareness of how much his literary criticism has ruffled the feathers of the respectability-obsessed black middle class. From these letters, it is clear that Thurman saw McKay as a comrade in the aesthetic wars of the period.

The letter to Locke—one of the deans of the Harlem Renaissance along with Du Bois and Charles S. Johnson—is revealing on several scores: by the time of this letter, late 1928, Thurman seems to be comfortable enough with Locke to joke about the surprising news of his marriage and to invite Locke to visit him and his new wife next time Locke visits New York. More importantly, however, the letter shows how much Thurman's conception of his second magazine, *Harlem: A Forum of Negro Life,* differed from that of *Fire!!,* published two years earlier. Perhaps in an effort to avoid the commercial disaster of *Fire!!,* Thurman makes every effort to assure Locke that his second magazine would appeal to a broad audience: "We are not confining ourselves to any group either of age or race. I think that is best. *The Crisis* and *The Messenger* are dead. *Opportunity* is dying. Voilà here comes *Harlem,* independent, fearless and general, trying to appeal to all." Locke responded to Thurman's letter by sending him a piece for *Harlem.* Entitled "Art or Propaganda?" the short essay defends Locke's aestheticism against the call for social relevance from Du Bois and others.

Thurman's letter to Hicks is the expression of a mature writer who is clearly unhappy with his latest effort. Thurman offers Hicks—editor of *The New Masses* and a well-regarded critic—an unsparing critique of *Infants of the Spring* from the perspective of the author himself. Blaming financial pressures on him to produce the novel, Thurman assures Hicks that he has an idea for another novel, on which he will take his time: "But I have what I think is a tremendous idea for another Negro novel. And that is the one, I'm going to mull over and nurture, for if it is properly done it will redeem my self-esteem, and give me faith enough in myself to continue my scribblings." This Negro novel, however, was never written.

## Harold Jackman

The two letters to Harold Jackman record something of Thurman's reading habits and musical tastes. He itemizes his summer reading list, reviews a classical concert he heard on the radio, and offers his view of Carl Van Vechten's *Parties: Scenes from Contemporary New York Life* (1930). These letters also reveal something about Thurman's friendships during the 1930s. He counts Jackman among his closest friends—along with "Bunnie," who is Harald Jan Stefansson. A white Canadian, Stefansson was Thurman's lover and the model for Jorgenson, a minor character in *The Blacker the Berry* and a major one in *Infants of the Spring.*

Written during a summer getaway, these letters relate how Thurman spent his time during his occasional stays at the Jamaica, Long Island, home of his friend Theophilus Lewis. A well-known theater and cultural critic, Lewis had given Thurman a job at his magazine, *The Looking Glass,* in 1925 when he arrived in New York and had recommended him in 1926 for the managing editor's job at *The Messenger.* One of Thurman's most steadfast friends, Lewis offered his home to Thurman as a retreat from the intensity of Harlem. (In fact, Thurman writes to Jackman of a play he is writing called *Goodbye New York.*) Thurman maintained a close relationship with Lewis's son, Theodore, teaching him piano and about literature and art.

## Dorothy West

Thurman probably met writer and essayist Dorothy West and her cousin Helene Johnson (whose work Thurman praises in his essay "Negro Poets and Their Poetry") during the time he lived at 267 West 136th Street in "Niggeratti Manor." While there are only a few letters from Thurman to West, two of those included here demonstrate the respect and warm admiration the two shared. In his letter from around August 1929, Thurman advises West on the best way to live her artistic and erotic life. Thurman comes across as a true cosmopolitan—a man of the world who has seen much and experienced more, a slightly older male who would like his two young, innocent women friends—Dorothy West and her cousin, Helene Johnson—to break out of their sheltered existence and open themselves to new sexual and cultural possibilities.

The letter dated September 12, 1934 is a poignant document of the life he led while hospitalized at City Hospital. This letter—addressed from "Rake's End"—testifies to the great friendship Thurman felt for West. Even as he tried to recuperate in the tuberculosis ward of City Hospital, Thurman couldn't help but give West advice about her new role as literary director of *Challenge*. Later, West would reflect her own high regard for Thurman in her well-known reminiscence, "Elephant's Dance." In this tribute to Thurman and to his importance to the Harlem Renaissance, West concludes: "He was our leader, and when he died, it all died with him."

## Letters to Langston Hughes

Don't forget to send me that promised poem for *The Messenger.*
Feb. 6, 1926

Dear Langston:

I have experienced another great tragedy since you left—I failed to win *The Nation* poetry prize or to even gain an honorable mention. I hope to recover.

I trust that you suffered no ill effects from your New York spree, and that you were in time for your dinner engagement. I also trust that you were really amused THAT morning and not bored. I love to give pleasure. Perhaps I am an incarnation of the cosmic clown. Do you agree?

I have not yet had what I told you I needed to make me happy, but on second thought I remember that I did not tell you specifically what it was. Perhaps I never will. I don't often weep on people's breasts, and I know that you sincerely hope that I don't choose you again.

No news, Bruce and I had luncheon together the day you left. I am taking him to the Gurdjieff tonight. And by the way I shall begin to take Gurdjieff seriously from now on.

Madame Walker was tickled pink (?) over your inscription. She just must meet you, and will try to arrange to do so as soon as you return.

You don't like long letters, and I have already bored you enough.

> Wallie
> 666 St. Nicholas, Apt. 4

Send mail to house, I have not yet forgiven you for tearing up my picture. I feel that there was more reason than the one you gave. Hence I have sealed the pieces up in an envelope and labeled it " Ashes of Vengeance." It shall be preserved. W. T.

--------

> Friday, 11 P.M.
> Langston—

I really feel guilty in answering this letter so soon, and in not being able to enclose even a portion of the $20 that is of such momentous necessity à vous. However I am afraid that if I had $20 xtra I should never be at home concerning myself over other people's misfortune. Thus it is that adversity provokes a virtuous and philanthropic reaction in even the most Nietzschean person. I know now why the poor proletariat are always the most altruistic of people. They hope that their desire to help others will prove boomerangial, and result in their own aggrandizement.

Considering all things why not finish your autobiography while at Lincoln. Perhaps you will be able to finish it here. 314 is at your disposal, and the environment hereof is quite conducive to creative writing—so private—so quiet—so comfortably accoutered. Even my novel is beginning to assume an interesting pre-natal appearance.

By the way—if you happen to have any prose short-stories on hand lying dormant but able to be facilely transposed from your fertile brain I can get you ten dollars for one. *The Messenger* is quite flush, and I would like to see you benefit before we go broke again.

Yes I saw Raquel, and I am still undecided why I was so pleased at her performance. Analyzing all she is not a great actress nor a compelling songster. The continental glamour probably contributes to the general hypnotism, but even devoid of that I am afraid one would find oneself succumbing to her peculiar brand of cataleptic anesthesia induced by a highly vibrant personality. (Whee—I sound like Dr. Frank Crane on a Volstead spree.) Yes—I liked her.

Saw Bruce this noon for the first time in over two weeks. He is on a hunger strike—and he looks it.

I am planning to see *The Bride of the Lamb* mañana. I have a sympathetic white friend who treats me to dinner and theaters occasionally, all in the interest of lessening interracial conflict. He is quite serious about it all—while I—I, heartless wretch, am interested only in the meals we have at Craig's with Paul and Essie Robeson. He is quite anxious to know you so I can assure you of one good dinner when you come anyhow. His name is Leland Pettit. He is organist at Grace Chapel—quite a friend of Countee's too I believe or *was*.

I was quite interested in your spiritual expression at the end of your latest letter. Now just where is the key. Oh yes I am in favor with these modern soul expressions but yours told me nothing, which might after all connote that it is great art, but being a nebulous critic I shall say it is rotten. Come on over soon—I wish to see you—To hell with money and care—

Wallie

### Frustration

Oh the music I would sing
Were I some more vital thing
Than a phantom formed from dust
Saturate with mortal lust.
I would sing gargantuuam songs
Reverberating as the gongs
Vulcan forged for Jove to smite
With his thunderbolts of might;
Oh my voice would always be
Clear without cacophony,
Culling from a thousand throats
Silver threads and golden notes
As certain as some finite loom

Culling threads from Juno's womb;
But the strictures on my soul
Will not let my songs unroll,
Will not let my lyric need
Violate a cosmic creed,
And the maiden life therein,
Vanquished by the world's vile din
Strikes one note, then sings no more,
Death has seized another whore.

## God's Edict

Let the wind-rolled waves tell the tale of the sea,
And the talkative pines tell the tale of the tree;
Let the motored hum of an automobile
Tell the modern tale of power and steel.

Let the blithesome chirp tell the tale of the bird,
And the moo and the baa tell the tale of the herd;
Let the purring cat and the barking dog
Contribute true tales to the cosmic log.

Let the grunting pig tell the tale of the stye,
And the lumbering yak tell of mountains high;
Let all these things abide by rule,
But let man's tale tell the tale of a fool.

I consider these the best I have done, which does not say much. Let me know frankly what you think of them. I know you don't like my style, but fire.

I am prepared.

W.T.

––––––––

Mon Ami—

"No Images" is indeed a fine poem—undoubtably [*sic*] one of your best, and potent proof of my contention that your potential is unpredictable.

Where in the hell did you get the name? And did you hope to fool me?

However I won't expose you—Just accept my congratulations—

Wallace

––––––––

Fire—Devoted to Younger Negro Artists
314 W. 138
New York City

Dear Lank—

The *Fire!!* meeting was great. *I* read your poetry and made a grand hit. From now on I substitute pour vous.

Hard luck has hit us all. I paid Kevin his $175. I have paid $75 on the other debt. We still owe $250.

I am a week behind in room rent. Eat when I get a free meal. Ho hum. When will I see you. Anxious to talk.

Wallie

———

Langston—

My idea of a foreword. What think you? Change or rewrite and special to me immediately. Must be in press by Wed. night or early Thurs. morning.

All other material ready. We could use another story. We will have 44 pages instead of 48 and perhaps 500 copies. That seems more prudent.

Special this to *W. T.* office

52 Vanderbilt R. 410, immediately.

*Flaming,*
Wallie

*Fire!!*—flames and burns and sears penetrating far beneath the superficial items of the flesh to boil the sluggish blood.

*Fire!!*—a cry of conquest, savage and sincere, invigorating those who sleep and enlivening they who linger in the quiet places dozing.

*Fire!!*—thrusting sprightly light between steel bars, challenging the court, and burning frame-wood opposition with a cackling chuckle of contempt—

"Fiyah, Lawd, Fiyah
Fiyah gonna' burn ma' soul"

W. T.

———

Fire—Devoted to Younger Negro Artists

Hello again—

It is 2:30 A.M. Just after I mailed the first letter the printer called saying they could have me all proofs by 4:30 P.M. Sunday. I went after it, came back, and corrected the whole damn thing—then dummied the magazine—alone.

Zora had a date. Jeanette was in South Norwalk. Bunnie could not be found. Neither could Bruce. Aaron eluded me. Hence I toiled until I am about ready to scream all sorts of *Fire!!* calls.

It would look fine. Will not use Bruce's drawings. Would cost $60 per drawing.

Am using his blue headed cabaret woman off my wall. No room for more. Neither will Johnnie's story be included. No room and it is not so good.

Must have $100 by Wednesday. How much can you help me out *(??)*

God damn *Fire!!* and all the editors—

> Near unto death,
> Wallie

Used "Sweat"—it *is* great. More when I am not hysterically tired—*Fire.*

———————

Fire—Devoted to Younger Negro Artists
*Fire letters sent O.K.*

Dear Langston:

I am still here though weak. We will have an "ad" in *The New Masses* and in *Opportunity*. I'm waiting to hear from a few others. Also may get book ads from certain publishers—trying Knopf, Boni, Boni-Liveright, Viking.

No word from Gwenny. No further word from John. Have telegraphed sir John again and again. Proofs all read. Making dummy Monday. You must get in to see page proofs. Will let you know when ready. Then we will need our money too—can send exact figures in a day or two.

Will you be able to make any personal contribution to the final sum?

*Fire* must burn this time.

Hooray for the potential backer, can he not be touched for an immediate loan of about $100?

I know how you hate to borrow money, but I fear between the two of us we must borrow $200 between now and the 15th . . .—$100 of it is needed by the 10th . . .—Ye Gods! My hair is already gray.

If I don't hear from Gwenny and John I shall—well—want to cast them from our presence. Gwenny has not even obtained a sub or sent in a patron's name, yet she will strut most grandly when pointed out as one of *Fire!!*'s editors. Malice boils in me.

Enclosed is a copy of the completed foreword. I hope you like it: I have done two editorial essays. One on the reaction of Harlem Negroes to Carl's *Nigger Heaven,* another one on the question of whether there is or is not a Negro art. Quite proud of results. Hope they please you too.

I am running "Sweat" under Zora's name and the play under a pen-name she chose for herself. She is not—never has been—keen about the play. Johnnie's story is the weakest link in the chain. I almost, since reading the proof, regret that it was set up. Bruce's drawings will be done in color on a different paper and inserted in the binding. Added cost negligible and the contrast will heighten the effect. Bruce refuses to aid even in the dummying and proof-reading. Bunny and Jeanette helped me. I am to dummy-up at Zora's Monday. Bunny and Jeanette will be on hand again. It seems as if *Fire!!* won't burn without Nordic fuel.

Saw *Broadway*—was enthralled. *Black Boy* closed Sat. *Deep River* a week ago. Negromania on Broadway checked. Can you rush—"O! Blues"?

*New Yorker* music critic talks about Mimi Heager, John Alden Carpenter *y usted*. Says your "verse was notably self-conscious."—now what in the hell does that mean. I am sending it to you.

Have had no excitement since that Thursday a week ago. Had a quiet, rainy weekend with Jeanette in South Norwalk last week. That is why I missed you. Have spent this week on *Fire!!*—and all the rest will be so spent until the *Fire!!* is blazing hot.

Let newspaper releases out. I still have cut. Johnnie must have kept the pictures we rushed to him.

Return these letters immediately.

Also work tout-de-suite—or muy pronto—

> Your kindled friend,
> Wallie

### Foreword

> Fire—flaming, burning, searing and penetrating far beneath the superficial items of the flesh to boil the blood

> Fire—a cry of conquest in the night warming those who sleep and revitalizing those who linger in the quiet places dozing

> Fire—melting steel and iron bars, poking *livid* tongues between stone apertures and burning wooden opposition with a cackling chuckle of contempt

> Fire—weaving vivid hot designs upon an ebon bordered loom and satisfying pagan thirst for Beauty unadorned . . . the flesh is sweet and real . . . the soul an inward flush of fire . . . Beauty? Flesh on fire . . . flesh on fire in the furnace of life—blazing.

> Fy-ah,
> Fy-ah, Lawd.
> Fy-ah gonna burn ma soul.

———

The World Tomorrow
October 30, 1926

Hello Langston,

Well at last Langston *Fire!!* is in press. The galley proofs will be ready on Thursday. I will need some more material so I am writing another sketch or rather re-writing that Harlem one about the man and woman. Perhaps I will get another poem or two from someone who is not already represented, and I am searching for another short story. Do you think it would look too Zoraish to run her play and story too? John sent his, also the checks. He still owes us $16.00 however. I have specialed him for it.

We won't need any more money until about Monday week, or rather about the tenth. Altogether we will need about $200 more. Where will we get it? I shall send the things to Harry Block, and I am still trying to get subscriptions. I hope you can get everyone at Lincoln and environs to subscribe.

We won't be able to use Bruce's drawings. It will add at least $75 to the cost. It will mean 25% added to the press work because the whole magazine will have to be run through the press twice. I just had an idea when I wrote that. Why should we have to run them through the press twice when we merely want them pasted in? Yes we will be able to do it after all. I must have been dumb when the printer was telling me that. As things are now we have only 32 pages of material. Instead of 48 pages we will have 44. You see I am trying my damnedest to cut the price down. Will tell you definitely what it costs in about three days. I am leaving town for the weekend, may be gone to Wed. as Tuesday is a holiday, and I do not have to work Monday unless I wish to. If I don't get my desk cleared a trifle, however, I shall come back to N.Y. on Monday.

Nothing exciting is happening. I went on a wild party with Peggy, Otto, Jeanette, George and Lillian last night. Peggy stayed at the house. We forgot to set the alarm and went to sleep about four. At ten I woke up with Mrs. Peterson sweeping the hall out. You should have seen my efforts to slip Peggy by her. It was very funny. Peggy wishes to become my common law wife. Figure that, would you?

But enough. Best regards. Write soon. Your secretary is a wow.

Sincerely,
Wallie

This letter is typed as badly as the one you sent me. Overlook it. I am not yet sober.

————

The World Tomorrow

Hello Langston—

Well I have the page proofs. And I have a new overcoat, thanks to *The World Tomorrow* who handed me a check for that purpose so sorry did they feel for my freezing carcass. And I have a check for $15 from Gwendolyn. Now—I am waiting to hear from Alfred Knopf, Harpers, Bonis, etc. for an advertisement. I should know by five today.

I got together $75.00. Don't ask me how. $15 from Gwenny, some subscriptions and some borrowing. If you can get $25.00 anyplace, anywhere between now and Monday please send it to me muy pronto, and please try to get it. *Fire!!* should be ready by Tuesday or Wednesday. How can you get in to autograph copies? That will have to be done.

The hold-up? It was 2:30 Sunday morning. I had gone to Seventh Avenue for coffee after dummying *Fire!!* all day and night. On my way back just about half way the block a fellow stepped up and asked me to help him get his friend to a taxi, the friend was laying in a dark corner drunk. Being a good samaritan I went. The friend stuck a gun in my belly and the other friend went through my pockets relieving me of all I had then they ordered me out of my overcoat, and hit me on my valuable head when I protested. Next they got in a waiting taxi and drove off. That is all.

Harvey Carson Grumbine is a white writer who has written books, and is a professor. I told you that we should have investigated at first. Now it will cost extra money to take his poem out and put in another for the pages are all made up. Just one more complication and I will be ready to blow up.

Have had two bad checks which I had to make good to *The World Tomorrow.*

*Fire!!* is certainly burning me.

    Wallie

————

    The World Tomorrow

    Dear Langston:

I am in one hell of a mess again, or should I say yet. I want to begin work on the next issue but dare not while I have so damn much worrying me now. I obtained a thirty day extension on Jeanette's note, and hope to have the entire amount by the 25th of this month. On top of that the last $300 I borrowed has to be repaid in monthly payments of $25.00 each. So far you know all of this has been coming out of my salary, leaving me in some pickle. I am just now growing resentful, and wondering why I should have shouldered the full responsibility of this finance business. I am so tired of people hounding me for money and holding me responsible that I would like to slink off into some far corner and commit hari-kari. I am writing a tearful letter to Gwenny too, and I do hope some of you will realize my plight and do what you can to get me out of it. Because of *Fire!!* I am also in danger of losing my job. Kirby is certain that I am a lewd young thing given to lewd thoughts. He pumped his secretary who told him I was carrying all of the girls in the office out to Harlem cabaret parties and up to my room. Of course he bristled, and electric sparks are in the air. Even if I do not get fired because of *Fire!!, The World Tomorrow* is about to go broke and I will be jobless anyhow.

If I do proceed with another issue of *Fire!!,* as I am aching to do, I must insist that the price be 50 cents. And $2.00 per year. If you will promise to aid me in these monthly $25.00 payments and if the rest will cooperate just a little bit to help the two of us I don't mind even losing my job. I can prostitute myself long enough to get out another issue of *Fire!!*

Can I expect anything at all from you before the 25th! Either on the $25.00 payment or on the large debt? Write to me, will you. I am certainly in the dumps.

    Wallie

Dec. 8, 1927

Cherie Ami—

I feel quite ashamed of myself, not having written before, but as usual life has been so damn complicated. First—Boils—five of them—right where I sit down—all had to be lanced—$5 per—much pain no work for ten days—there—starvation—and now a swollen thyroid gland and infected tonsils—expect to go to hospital for double operation next week. Truly I am akin to Lazarus.

The play is being passed around at last. Several enthusiasts interested. Agent promises to have producer within 10 days.

And I forgot to add that my protégé from Calif. who now has a scholarship at the Julliard Foundation is on my hands. No job, no money from home. And the League for Mutual Aid is about to drive me nuts; all save Agnes.

However I can still find life amusing and laugh.

Bruce has a *new* suit.

Service is back.

Know nothing else.

Write me a long, cheerful bit—

Wallace

———

[c. early 1928]

Langston:

I refuse to salute you more intimately. Here lay I, in bed with the "flu"; and you spent your vacation here without seeing me. Blame the bell. I also feel that you too are in fault. Where is your door-key?

I am up again—weak in the knees, but happy. Crosby Gaige has taken my play. 'Twill be on Broadway soon. Rehearsals may start within ten days. Also I have a job—reader for Boni-Liveright. And I am signed up for three more plays, number one of the three I already outlined, synopsed (if there is such a word) and begun. The novel is about re-written. I am to re-submit it for prize within 15 days, but I don't think I will. A and C Boni are a bunch of crooks. My novel when ready for publication goes to MacMillans, just to be different. Will explain it all to you when and if I ever see you again. Maybe next time you come to N.Y. the Omegas won't be having a convention and aspiring lady sorts won't need so much escorting around. Then mayhap you will see some of your not so dicty niggeratto friends. (Isn't that a nasty trio of sentences.)

It is a novelty to be eating regularly once more—I hope it continues.

*Fire* debts—!!!! I am still being harassed. I paid off one completely. But the Mutual Aid is still waiting and fretting.

No news or scandal. I am practically in seclusion. Not a bit well. Tonsils swollen—neck puffed out like a balloon on one side. Dizzy in the head, etc., etc. Soon as my inflammation goes from tonsils out they come. Dastards.

Thanks for the New Year's card. Excuse the blot. Send me a book of some kind as a peace offering and write me damn you, write me—

> Wallie

———

[267 W. 136th St., N.Y.C.]
Tuesday [c. early 1928]

Hello there!—

Just finished Claude's book—'Tis the best novel any Negro has published. Just like that. And I too await the holocaust. Good thing he's in Marseilles!!!! Decadent is no name for our Renaissance. Coin another word.

You know I'm not good enough to be in Countee's ball. I mean marriage. What I had in mind to tell you was that Otto cut some fellow up at the Hamilton Lodge Ball! Ehh Eh!— Whee—. Unfortunately I forgot the date and missed it—the ball I mean.

Rehearsals start tomorrow. Play all cast—and what a cast! Laura Bowman, Lawrence Crever, Leo Kirkpatrick, Bessie Allisson (Cordelia), Eddie Thompson, Ernest Whitman, Carroll Boyd, Peekabo Jimmie—and more!!

My protégé is ill. I'm having his mother send for him.

Bruce is elated over prospects of tour with *Porgy*. Paul Robeson is great as Crown. Jack Carter got temperamental. Walked out, now on way to Hollywood with Dolores de Rio's husband.

Scandal!!

More scandal—

I'm in love!

Don't ask?

I really meant to write a sensible letter for once. Our correspondence is so fit-sy and starty. Bromo Seltzer with a manure base.

You're a dear about—what's his name—Kerlin. Soon as royalties come in I'll clean up the rest. Carl, Jean, James Weldon.

But Claude's book—oh how I want to review it. And nobody asks me to.

Working on another play and on an auto-biography *not to be published*. Then *267* is my next opus.

What does you do these days? Literarily speaking?

Coming over soon—Wolff and I am to collaborate on a novel about Lincoln

> Beware—
> Wallie

267 W. 136
Friday [c. May 1928]

Langston—

I haven't a friend in the world.

I have been laid (lain) up with the "flu" for three days.

No Money. Room rent unpaid. No food. No nothing.

Why haven't I heard from you?

*The Bookman* has accepted my article. Will be paid by the 15th or so—if possible (and since your debts are akin to mine, it probably isn't) lend me 5 or 10 until that check comes in.

Am to do a bit for *The New Yorker*—must pass—Portrait of A'Lelia on order after first "bit." Also are assignments from *The Nation* and a chance to do "The Negro and the Theatre" for *Theatre Arts*.

Now if I can only keep from starving.

Let us hear from you pronto—

Wallie

————

[c. summer 1928]

Dear Langston –

I have disappeared! Yes I have. Once more. Had to do it. Harlem was too hectic for me. You really don't know. So here I am in the wilds of somewhere no one knows. And if you tell I'll—write unfavorable criticisms of all your future pieces. See the June *Bookman* if it ever comes out. Also procure the May issue of *The Dance*. Bruce's drawings are priceless. So is my article.

Promise you won't reveal my whereabouts to anyone. You—Bunnie—Jeanette and Rapp are the only ones who know. You don't know yet, do you? Well I am in Mamaroneck, N.Y.!!! You're a dog with fleas.

Before I left Harlem someone entered my room and departed with all my clothes save the ones I had on.

The play is off that is till fall. Open Labor Day—on Broadway—unless they change their minds again. Disgusting business.

The second show is coming fine. For an experiment I submitted my novel to Doubleday Denon. The reader's report was priceless. "This is a very articulate novel for a Negro." And on "It has none of the décor or sensationalism of *Nigger Heaven* and *Home to Harlem*." And—"It is a good novel but would advise against publishing. Would not be a commercial success. However we may consider it."

Screw 'em.

Meanwhile I am making it longer for Boni-Liveright. And it is *not* a good novel but will be before published.

Have you read *Dark Princess*?

Has Fred Bair been to see you yet? I log-rolled you to him—quite sincerely tho, I assure you.

Lost your last letter. Will send you a *Fire!!* Send it on where you asked me to.

Raymond says hello. He wants to write to you. I say—why not? Now after reading this he says he's gonna' write to you. Fy-ah gonna' burn ma soul.

Yes I am at his home—

The song is ended. Laugh, Clown, Laugh—

And write me soon—

> Wallie
> 51 Madison St.
> Marmaroneck N.Y.

You are sworn to secrecy!!

Please mail the enclosed. It isn't a plot.

————

Tuesday

Dear Langston:

*Harlem* is going fine. Here is a clipping which may interest you. I am still waiting for an answer to my last presumptuous and imperious letter. The sales are stunning and the letters of praise are overwhelming. If *Fire!!* burned my soul will *Harlem* release it from the eternal incinerator? Or will the present boom only sink me lower in the depths of perdition.

> Sincerely,
> Wallie

————

> 308 E. Ninth South
> Salt Lake City, Utah
> Tuesday [December 5, 1928]

Dear Lank-y-yank-yank:

Forgive my informal salutation. 'Tis merely the result of having spent two weeks out here in the wild west. I am under the impression that I wrote to you en route, but maybe I didn't since I just received a letter from you which had recently been sent to Harlem, and since I have not yet received one from you at my present abode.

About *Harlem*? I know nothing. I became thoro-ly disgusted with the whole business, sent the material for the second issue to press with a note of resignation near the first page, but I have not yet been able to ascertain whether or not there is to be the second issue. It's a long story I'll tell you sometime.

Can I let you in on a secret, one which you will not even divulge to your new and as yet unidentified inamorata? Ostensibly I am in Salt Lake because my mother is ill. I received two hectic telegrams advising me to leave at once. I dashed to my publishers. They gave me the necessary $150, and I left. Tehee. On my arriving in Salt Lake my mother was of course quite surprised to see me as was everyone else save Mignon, my first love, whom I had wired to send me those telegrams. It's almost too good to keep, but I swear I'll murder you if you ever breathe it to a soul.

I just had to flee. I was fed up on New York, on the magazine, and on married life (yes it's all over, but I dare you to mention that either). There was no way I could escape everything unless some crisis presented itself, so there being none in the offing, I lied like a gentlemen and with Mignon's aid (you should and perhaps will know Mignon some day—it won't be long now) precipitated a crisis of my own. Nevertheless on arriving home I found my mother not ill but married again for the sixth time. Now I ask you????????????

However to all intents and purposes she is now convalescent, and has received notes of sympathy from almost every-one I knew in New York. My wife, her mother, my publishers, my collaborator and my producers. Am I not a shameless wretch? I really should develop a conscience just to experience a moment of contrition. But . . . self preservation is the first law of nature, and I either had to flee or else be lost. Do you see how much depends on your treating this information with the strictest (is there such a word?) confidence. As it stands I am saved and no one else is hurt. But should my plot be exposed . . . ye Gods. Father Confessor Hughes hold thy tongue. I told you alone because I knew of no one who would appreciate it more.

And for God's sake don't fall too deeply in love or rather fall as deeply as you wish but keep your senses and flee marriage as you would flee the seven plagues of Israel. All the time I have been saying that artists should never marry I was talking from theory only. Now I can preach the truth with a vengeance. Another long story I will tell you when we meet again.

Have been receiving telegrams urging my immediate return to New York. *Black Belt,* nomad that it is, has been bought by another producer, the third. Says he, rehearsals start next week. Will expect you to be here. Says I, when the play opens I shall return, and not before. However I expect I will be back shortly after the holiday season. I plan to spend a week in Chicago with Bruce and Service both of whom are now playing there in their respective shows. I'll leave here shortly after Xmas according to my present plans. (. . . and the end is not yet)

However if I get the money I may end up in California. There really is no telling what I will do. Meanwhile I live in exile. Have not even had a drink since my arrival. I have done quite a bit of writing and gobs of reading. Good meals, plenty of sleep, cod liver oil to put flesh on my bones and build up my nervous system, meditation and quiet. Dost envy me? The only things of interest in Salt Lake are an epidemic of influenza, and the attitude of the older folk here toward me who have snubbed me with a vengeance because of an article I wrote three years ago in *The Messenger* about my home town and state. It seems as if I said that most of Utah's early Negro settlers were whores, gamblers and washerwomen, which is true of any of these western states. Somehow they got the idea that I said all of Utah's Negroes from the beginning until now were whores and gamblers. They didn't object to washerwomen. The younger folk have all been enjoying shot gun weddings since I left with the result that those whom I used to push around in a baby carriage are now pushing baby carriages of their own.

This is enough for one sitting. Write me soon and lengthily. I send this air mail. Take a hint. Tell me about this new and all consuming love and what on earth did you mean about lippish poetry? My memory fails me. Hooray for Yolande. What kind of house was that you found in Baltimore? Your writing is not at all times legible, ahem. Why Boston society and then again why not?

Adios,
Wallie

————

Lang—

Back. Busy as hell. *Black Belt* in rehearsal opening in Bronx, Feb. 4. Come up. Broadway opening Feb. 18. *Blacker the Berry* out Feb. 1—your copy should reach you next week.

How is the love affair? I am quite interested.

Just tried to read Eugene Gendon's novel. It is *vile.* (shush) No news about *Harlem.* Quite anxious to see you.

Yes I'm at 90 Edgecombe.

Woe is me,
Wallie

————

1538 Fifth Street
Santa Monica, Calif.
[c. May 1929]

My own dear Langston:

Such a salutation seems essential, considering the way I have neglected to write to you. Please believe that for a while it was only because I was under such a mental and emotional strain that I could write no one, and that recently I have been too busy. Those are the only excuses I have to offer. I'm sure you can appreciate them.

Life has been most full since last I saw you. Divorce in Reno soon. Alimony bleeding me of my excess royalties which I had banked on to keep the wolf away from the door. A trip to Detroit, then to Salt Lake, and then to Los Angeles, where I spent one hectic week, attending movie parties, swilling Hollywood gin and falling madly in love with a most worthless but adorable person. I also received a smack in the eye, necessitating having a doctor take three stitches in the lower lid. I went to a formal dance, met a lovely lady, asked her to go cabareting. She had her Romeo with her. I was to pick her up after he had bid her farewell. I got there too soon. Gin had rendered me myopic. I did not see him lurking in the shade. A most wild adventure in true western style.

Then I had enough. I retired to this picturesque little beach town and began to write. Have done oodles. Take a dip each day, lie on the beach for hours reading, eat, sleep, and motor. All for the purpose of recovering my mental and physical health. I am still in love however, although, thank God, the object of my affections was called back to New York and is due to be gone for three weeks. By that time they will either have other interests or I will be recovered. Let us pray.

Harlem is not in it compared to Hollywood. Why not run out for the summer after the strenuous work of graduating is over. It is only $151 a round trip. I would return with you. You could rest here on the beach, living quite cheaply, and then make occasional sorties to Los Angeles and Hollywood, only 18 miles away. Everyone here has cars and they are most generous in putting them at your disposal. You would enjoy seeing the movie studios, and I don't know what else. There is so much.

New Yorkers, white and colored, are here in droves. Even I may get a chance to write a talking movie. And such fabulous salaries. Come West, young man.

And bye the bye what wish ye for a graduation present?

Have you read *Black Magic*? I thought Aaron's pictures (illustrations rather) far superior to the text. Am awaiting copies of *Banjo, Passing,* and *Rope and Faggot.* What are you doing? I have started another play, and also my next book, which, sotto voce, will be a third person narrative of mine own life and times.

Now please write me. I promise to answer. The storm is o'er. Peace is here. And consider seriously a trip out here. On the way back we could see the glories of the west, Salt Lake, the Rockies and on your way here you could see the Grand Canyon. And me at the other end.

> Can I offer more,
> Yours,
> Wallie

———

Friday [c. May–June 1929]

Dear Langston:

It was good to hear from you again. I am most glad all is over and that you are now an accredited A.B. What plans have you for next winter?

Money my dear boy is something I have not. My prosperous period is over, and even now I languish in Santa Monica trunk packed awaiting ticket money from Bill so I can go to Salt Lake. While friend wife takes a trip to Reno and has a nice little pile of hubbie's royalties. The show flopped terribly in Chicago before closing. Thanks to Negro propaganda. Don't ever marry. Had I remained single I would be financially secure for the next year. As it is I am broke, attorneys are over a thousand dollars richer, and Louise has a couple of thousand, with two more in sight should I ever make any more money. In fact I am almost glad to be broke again. Every person I know save you, Jeanette, and Harold Jackman have borrowed money from me. And all are now pouting because I cannot lend additional sums. I lent out almost eight hundred dollars. Paid $2,200 worth of debts. I return to New York (maybe) to starve again next winter unless I get another break. Such is life.

Have been having glorious time here. Met lots of nice people. Been to Mexico, Catalina and in the mountains. Had a party which has set the town on ears. Made whoopee in Hollywood. William Still is here. Ditto Abbie who has hatted me. I can't find where she lives although I have sent her three telegrams to theater. Paul Robeson is due here this next week. But I hope to leave no later than Tuesday. Am working on my novel, have finished book of essays, and am also working on a novelized version of *Harlem,* a new play, and a story for James Cruze to do in the movies. And meanwhile have found time to immerse myself in Marcel Proust and do a lot of other reading.

Found *Banjo* turgid and tiresome. *Passing* possessed of the same faults as *Quicksand. Rope and Faggot* good for library reference. Nella Larsen can write, but oh my god she knows so little how to invest her characters with any life like possibilities. They always outrage the reader, not naturally as people have a way of doing in real life, but artificially like ill managed puppets. Claude I believe has shot his bolt. Jessie Fauset should be taken to Philadelphia and cremated. You should write a book. Countee should be castrated and taken to Persia as the Shah's eunuch. Jean Toomer should be enshrined as a genius and immortal and he should also publish his new book about which gossip is raving. Bud Fisher should stick to short stories. Zora should learn craftsmanship and surprise the world and outstrip her contemporaries as well. Bruce should be spanked, put in a monastery and made to concentrate on writing. Gwennie should stick to what she is doing. Aaron needs a change of scenery and a psychic shock. Eric ought to finish *The Big Ditch* or destroy it. I should commit suicide.

Don't mind such raving. I am distraught today, emotionally, and for no reason. When I get calm I'll write a long letter about adventures of a niggeratti in Southern Cal. They are immense. Meanwhile write me at 308 E. 9th South St., Salt Lake City, Utah.

As always,
Wallie

————

308 E. 9th South
Salt Lake City
Friday [c. July 1929]

My dear Langston:

Write me more say you, send me a measly postal card, me who has been quite ill for the past month and unable to do anything but think about what I would be writing would the doctor let me.

There is nothing to say. I am here to benefit from the high altitude and my grandmother's care. I live on milk, eggs and cod liver oil. It is damnably hot. There is no news. I am in no mental mood to wax eloquent over trifles.

What are you doing? You are such a very mysterious person, so discreet in your reports of intimate matters. By intimate I mean literary and personal plans of course. Will you teach this fall? Are you working on a book? Or is it none of my damned business. Remember I am to be the literary historian of the niggeratti, and if you remain so damned impersonal and non confidential how can I keep track of your growth. Remember I still have faith in that unpredictable and illimitable potential. God what a child I have been.

When do you return to Harlem? My own return is indefinitely postponed. I must get well. I must finish a talkie. And thus it goes. Write me please in the old vein. I object to shorthand notes without the shadow of reality behind them.

Affectionately if peevish,
Wallie

Shoot this to N.Y. as soon as possible.

Thanks.

————

Monday [c. July 1929]

Dear Langston:

Three cheers. But I am so dumb. I sent you the carbon copy of "Du Bois" which I had not corrected *en toto,* and kept the one I meant to send you in a portfolio. Some of the items you mentioned had already been changed, especially the paragraph about the wedding which was most certainly in bad taste and written in heat as was most of the article, the only one of the group in which I believe I was more personal than objective. For that reason I am re-writing the whole piece, or should say have rewritten it. Lapses in grammar are certainly to be deplored, but like you, school grammar had no effect upon me whatsoever. Consequently I admit marring what might be good by frequent and really inexcusable gaucheries. I have taken great pains with this present work, for the very reason which you mention, for I realize that I cannot shout J'Accuse and at the same time expose myself as a nincompoop. There were some awful and really unforgivable grammatical errors in *B. the B.* I found them too late. So this time in addition to careful revision, I intend to have several people look over the manuscript and make suggestions about things which may have managed to escape me. Among them, and primarily, is yourself. I am sending you the entire volume in manuscript, not to make notations as in the present case, but to actually correct anything which offends your eye or ear as far as words and sentence structure go. If this be imposition . . . I'm sorry. Perhaps you are too busy. I hope not. And when you finish, there will be an addressed envelope to send same on to the publishers. You realize of course that I am going over it again before it goes to the printer. And that I shall have a trained editor read the proof. I'm hell on bluepenciling the work of others, but much too soft with myself. Auto-intoxication I suppose. Anyway I hope to improve with age and experience. That's all any of us can look forward to. The only reason I keep trying is that I firmly believe that I am aware of my own limitations and thus have some possibility of overcoming them. When I lose that illusion I join the suicidal maniacs.

All of the book is here except the *Bookman* article of which I have no copy, and which the publication itself has been promising me for the last three months. I hope one will soon be available. And I am still working on the theater article having recently found some new and interesting material.

Don't hesitate to let me know frankly your opinion of this measly effort. I shall anxiously await your reaction.

That's all for this time. I am terribly tired and have been overdoing things the past week against doctor's orders. He has threatened to send me back to bed unless I desist. But who are doctors that I should heed them?

East Ninth South is a fine street, very wide, and well proportioned, being one of the town's main boulevards. You should see it. No more would you scoff. I shall, in my way, let you pay the postage on this to New York, not having any idea what it would cost. Add it to the other sums I owe you. Mayhap I'll get another break, *Harlem* is going on the road soon, and send you a lollypop of some sort in appreciation for the many things you have done for me, directly and indirectly. But I grow maudlin.

Affectionately,
Wallie

(Return the enclosed immediately air mail special, Please! With criticism and comment!)

————

Thursday [c. July 1929]

My dear Langston:

The picture of you, alone and languishing in that deserted, decrepit and no doubt mephitic dormitory, wondering why you cannot become ill, and available to any stray prowler who might wish to investigate, or steal or rape is more than I can bear. Even so astute and dry a fellow as a literary historian must be provoked to tears. Like Jeremiah I fill the Salt Lake woods with my lamentations, and like John on the Isle of Patmos I have revelations.

One would need them to get anything definite out of your letter. You will teach when you become old, poor, and indigent. That is to be in the remote future. And fall is remote. You practice prose technique waiting for an inspiration. And whether that inspiration is recalcitrant or constant is left to my own imagination. You are in the final analysis the most consarned and diabolical creature, to say nothing of being either the most egregiously simple or excessively complex, person I know. Pee on you!!

The invalid is improving. He can now get to his literary labors. He does write. Damn the doctor, although the M.D. may be the best literary critic after all. He probably was thinking of the public when he advised me to stop writing. That just dawned upon me, egoist that I am. Anyhow . . . did I tell you that I have completed a volume and that it is already in the hands of my publishers? Its title is *Aunt Hagar's Children.* In it are the articles I had in *The Bookman, The Independent, The New Republic, The Dance Magazine, Haldeman-Julius Quarterly,* plus much new material, including interpretive biographical sketches of the Messrs. Douglass, Washington, Garvey and Du Bois. There is no doubt but that I will be blacklisted forever now by the N.A.A.C.P. Wait till all read what I have said of the honorable Du Bois.

My new novel is proving almost too much of a task. I intend taking about a year to finish it and that year will soon be gone, and still I lurk far this side of the halfway mark. It is my final salute to 267 and in it I have incorporated all the inhabitants and stray visitors of that remarkable melange.

And the talkie goes not at all. I can't get into it. I don't want to do it. I am damn sorry I ever promised. I have done the synopsis and sent it on to Bill to finish or at least partially collaborate putting in the essentially moviesque touches. All I want is the check. And if this is materialism . . . o.k. . . . for I do so want return fare to Harlem . . . or at least to Gotham. I doubt if I will live in Harlem when I return. Reasons best known to myself. No questions asked.

And of all things I have to rewrite a novelized version of *Harlem,* which I at first refused to write and which Bill sold then could not write himself and which his wife wrote and sent to me and which is so terrible I shall now have to do the whole thing in order to fulfill the contract. Zounds.

Thus are the lamentations of a young literary historian, young in age but hoary with experience. You get checks for book reviewing and German translations!!!!!!!! How I envy you. But what are the book reviews and where did they appear? Send me copies or at least information. I would like to see them.

If you do not believe I have been ill read the following which came like Minerva out of the head of Zeus while he lay abed, ice bag on head, hot water bottle on feet, and a copy of *Cane* on his breast:

### On Meeting a Genius
(To Jean Toomer)

I
I was a wilderness;
Prickly heat points
Hurtled through my body,
Brain cells like thirsting sand-dunes

Knew no inner incandescence;
Emotion was a mirage well,
Instinct entombed, ego diffuse and sere.

II

You came.
I flung myself upon you,
The lemon of your body
Merged into my blackened skin.
The lips "too thin for kissing"
Searched my own, transmitted flame.
I burned. I was content.
Knew Jesus was a leper,
Knew Christ would spring from knotted roots
Beneath the troubled earth;
Knew roots were chains
Suffused with sap . . . informed with teats
Natal wombs of super-men.
Your navel cord had pierced my umbulus.
My blood was red, then pale,
My mind a lambent milky way
Intangible and vague;
The soul of me was amber dust
My body knew no life.
There was nothing but the ego
To confound eternity.

You don't know what it means? Neither do I, but please refrain from showing it to any psychoanalyst. Is there not something in that sect's lore known as "wish fulfillment"? Alors.

Anent Hollywood. Nina Mae McKinney and Daniel Haynes, one from the chorus of *Blackbirds,* the other Jules Bledsoe's understudy in *Show Boat* are the co-stars of King Vidor's *Halleluliah,* a marvelous picture when I saw the 18 reel preview. However it is still being cut and reassembled. Doubtless it will be terrible when released, so artistically astute are those who cut and supervise. Nina Mae (pronounced Neenah if you please) is a marvelous little dumb kid, who fell prey to all of Hollywood's male and female satyrs. She actually bragged in public about having been to bed with Pepper, a lady who is cousin to Marion Davies, and grinned exultantly when said Pepper smacked her face at a party for having playfully kissed one of the males (a real one) present. She and Dan sing some songs that are bound to be great hits. Irving Berlin. At the End of the Road. The Flat Foot Waiter's Ball. And there is a revival scene and baptizing that will thrill you through and through.

Step-in Fetchit is a marvel. You should know him. He is exactly the same off as on. A slow, lazy southern type with a drawling voice. He has three Cadillacs, two chauffeurs, a butler, valet, cook, and two maids. He earns five hundred dollars per week. All his cars are equipped with banners, announcing to the world at large that they belong to Step-in Fetchit Fox movie star. He wears an aviator's helmet and leather jacket as do his wife and chauffeurs. He drinks wine tonic by the case. He is now being sued for $100,000 for breach of promise by a young lady who claims that he failed to marry her after having obtained the license and "sampled it." The sample evidently did not meet Step's requirements. He is a devout Catholic, gives a tenth of his earnings to the church and makes early mass. He has no money two days after pay day, and cannot understand why movie stars like himself with money and everything should have to die the same as poor folk.

In addition to these headliners there are countless senegambians in Hollywood who sing and talk for the white stars, remaining hidden behind a screen while the movie morons move their lips. But it took James Cruze to put on the final touch. Needing voices for a chorus scene in his new picture *The Great Gabbo* he tried out hundreds of whites. All voices were flat. In desperation he hired a colored chorus, white washed them, placed them in the background and gave the sign for them to sing.

And of course Eva Jessay and her chorus from N.Y. including George McClean and Eddie Coleman are in *Halleluliah,* and another priceless little lady known as Hot-Shot. You should migrate to Hollywood.

Now have I done my duty, flippant Puck? However much I cavil I did enjoy your ra-ta-tat-tat persiflage. Like shrapnel it enlivened me. Send me more.

Affectionately of course,
Wallie

————

Wednesday [c. August 1929]

My dear Langston:

Fideles Achates.

You don't know how much I appreciate your taking time to read *Aunt Hagar's Children* and your pointed notes. I do agree with you about using Negro in so sweeping a sense. Don't worry. I shall make the necessary qualifications. I did not necessarily mean in the Washington one that those young bucks had not heeded that which I know they have not heard. I meant that because those who did hear caviled instead of aiding, these young bucks heard nothing and now pound the pavements and pimp. I am glad you like the book. It should stir up some interesting comment if nothing more, and it is valuable to me in that having done this primary work (crude as it necessarily is in part) I can now clarify certain personal problems and proceed on my novel with new vigor.

Mephitic: Poisonous, foul, odiferous, deadly.

Umbulus is your belly button, through which the umbilical cord transmits nourishment from the mother to child and which is severed at birth by the physician.

Ego, my son, is that which intellectuals call what Christians call the soul. It is what Jean Toomer calls the "immortal I." I used it in that sense in the poem, because that is what he and Gurdjieff had us developing. Ego generally is as much hooey as soul. Thus I could never make a good Gurdjieff disciple.

See where Sidney Peterson, Dorothy's brother, has married a white frail?

And if you have a copy of that *Bookman* article please send it to me. I will see that it is returned. *The Bookman* claims they have recently moved and can't find any until they get straightened out.

As for "Tribute" I intend to re-write it *en toto*. I too sensed its weakness, but was unable to remedy it at the moment. I certainly shall mention Chas. S. [Johnson], having completely forgotten him just then.

"Terpsichore" is not going into the book. It really says nothing at all, I having been requested to write it to order about something of which I really know nothing. Having received the fifty bucks some years ago, there is no reason for perpetuating its progenitor.

The cabarets are of course as they were when the article was written some three years ago, I shall add a footnote.

Why not: vibrant, rich and steady?

I write this early in the morning so that I can mail it while on my way downtown to the pawnshop. Unlike you I distribute suits among the town's pawnbrokers, having bought several during my period of prosperity. And there they shall stay until the Shuberts are finally forced by The Dramatist Guild to pay me my back royalties on the second company of *Harlem*. They owe us some four weeks. Why they hold it up I do not know, unless, I remember your retelling of their iniquities some years ago (Kaj Gynt was their author, was she not), they are awaiting my return to N.Y. in order to exact tribute. Then they should send my railroad fare. I'll assure them I have a willing disposition when over a thousand dollars is at stake.

And why remember Vonnie and Minnie?

And why not invite a fellow hobo to go to the east shores of Maryland? I'd love to see the moon dripping with nigger babies.

My expedition to the pawn broker is necessary because I went out raising whoopee Sunday with a crowd and in my cups, or should I say lager for we drink nothing out here but home brew, the whiskey being *mephitic* (References to Emily Post will be censored, they being relics of different articles), I, to continue, ordered three rounds of beers, then had to ask for credit. My grandmother of course would never give me money to pay liquor debts, and I cannot think up a good lie, she having bought me a street car pass, a supply of stamps and paper, ink, and vaseline for my hair. She feeds me and sleeps me. Why should I need three dollars? Hence I slip my tuxedo to the pawn shop.

Such is the life of Aunt Hagar's stepchild.

The novel is going fine, taking on some semblance of what I have in mind.

And I guess there is no more to say. Only I beg of you remember me out here in exile, drinking home brew, with no interesting companions, no love affairs, no excitement, plenty of cod-liver oil, milk and eggs, and a few pawn tickets . . . then write me of your adventures . . . or just drop an occasional note to cheer me up. God knows I need it.

My grandfather has just lost his "life-time job" without notice and he is far from dead. My grandmother goes to the hospital in a few days to have a cataract removed from her one good eye. The other being covered long ago. My mother has left her sixth husband for the fourth time in two weeks. The atmosphere is depressing. If I only had train fare to N.Y.? But I will have it, damn it, day by day it comes closer and closer. Page Mr. Coué. Or should I become a disciple of Krishnamurti? I also left California owing a board bill. Ah me. Don't forget the *Bookman* article. I have some new poetry which would surprise you.

Affectionately, Wallie

———

308–E 9th S.
Wednesday

Dear Langston:

Severest critic and friend to give you the title Merton of the movies gave his wife. (Meaning no offense, apologies to John Riddell.)

Yesterday I sent you the entire manuscript of *Aunt Hagar's Children* to read over then send on to Macaulay's. An enclosed letter explains a great deal not mentioned herein. One thing: in the piece entitled "Tribute," there is a line near the bottom of the page one which reads: He loves youth and trusts to them and their intellects. . . . Please change that to: He loves youth and trusts to youth's courage and intellect. . . . I had made a note of it but don't believe I made the actual change. No doubt when it comes back for final revision I will find much more. I hope so.

With that gone I have began work on a perfectly intriguing play, again in collaboration with Wm. Jourdan Rapp. It is my idea. *Sultan Smith* is the title, it being woven around the experiences of Joseph F. Smith, founder of the Mormon religion. The scene of action is Nauvoo, Ill., where for one brief moment the Mormons lingered before being driven out and forced to migrate to Salt Lake. It was at Nauvoo that Smith received his vision authorizing polygamy. That is the kernel of the play. I have some gorgeous ideas for its development. Bill too is enthusiastic. We ought to do something unusual and mayhap good with the extant material I have delved into the history of Smith and the Mormon church like a laborious titmouse. I have even read the entire Book of Mormon. Keep this under your hat. I'll let you know what progress is made. The Marcus Garvey play is scheduled for late fall production. *Harlem* takes to the road again in September. Perhaps the sun will shine again.

And the movie story remains unfinished. I'll be damned if I can do it. And Bill has not been much help so far. Mayhap he will have a brilliant idea. I hope so. We need the shekels. Mr. James Cruze had the temerity to say that my first version read like a *Dial* story, which is what I had already said. And of course no *Dial* story could be talkied. Ah, me. By the way *The Dial* is no more. And did . . . what did . . . that awful poem on genius . . . did I say poem? . . . pardon me . . . mean to you?

I hope you understand my idea in doing such a book as *Aunt Hagar's Children*. I really do not enjoy carping and caviling, but I do believe a certain amount of it has to be done to sift out the chaff from the wheat, to become platitudinous. I have no illusions about my being widely read by the people to whom primarily it should be of interest, the young intellectuals. All I want to do is get some things off my own chest, and perhaps beget discussion, which will be perhaps of some aid to those of us who wish to make the most of our talents. Mayhap I am somewhat of a crusader. It's a thankless task, but if one feels the urge why not? There is so little to do in life that is interesting. And certain Negroes as well as certain whites, have to be pricked to develop. All of my ideas may change overnight. I claim no permanence for them nor infallibility. They are purely expressions of a current mood, guided by certain mental predilections. Were they not to change somewhat with the coming of maturity and the growth of mind I would soon give up the fight. Needless to say I am as skeptical about my own harshly phrased opinions as I am of other people's. Love.

Wallie

————

24 West 129th Street
August 13, 1931

My dear Langston:

This is to certify that because of conditions over which we have no control and which were entirely unforeseen, I hereby release you from our proposed collaboration, and also certify that the idea is yours to which you alone have a right. I shall make no attempt to appropriate the same, which is a dramatization of the Scottsboro, Alabama, Trial of the nine Negro boys held and sentenced to death on a charge of rape.

Very truly yours,
Wallace Thurman

————

316 West 138 St., NYC
[c. January 1932]

Dear Langston:

Thanks for the Xmas card.

But where in the hell are you now? I want to send you a copy of my new novel, which is being released February 12. It is lousy, but as a matter of record, and as a monument to my inherent adolescence, I think you ought to have one.

Getting prolific. Have another book scheduled for March. An experiment. Novelization of a play I recently finished, entitled *The Interne*. Wherein Mr. Thurman gives city operated hospitals particular hell. If you ever hear of me being ill, don't let them send me to any city hospital. I am sure to be given the black bottle.

The play incidentally was practically sold to Wm. A. Brady. The contracts were drawn up. The sets were designed. The leading role had been cast. And Mr. Thurman had spent one more hectic week, getting six hours' sleep in five days, revising said play for Mr. Brady's benefit. Rehearsals were supposed to start immediately. When what happened? Mr. Brady decided he couldn't give an immediate production. That he should wait until fall. And offered the poor author $500 for a six months' option. Poor author just collapsed. But his agent politely told Mr. Brady to go to hell. I understand two other producers are acutely interested, but I'm not until I get a signed contract.

Have been working like hell. Holding down two jobs. Working from nine until five at Macaulay's. And working from six until ten for Wm. Jourdan Rapp, editing mind you, actually editing, *True Story* magazine. It's been a great experience. I have learned something about magazines, about public tastes, about many technical problems, and about American moronity at its best(?). Of course Mr. MacFadden had no idea I was doing this work. But Mr. Rapp couldn't do it. He having taken on more jobs than one human could possibly handle. So in stepped little Wallie . . . the ghost editor. And I've got material for a satire which when I grow up ought to get me a first page review in *The Times Literary Supplement.*

In addition to working 11 hours a day, I have been turning out this hospital novel, and revising the hospital play. There has been little time for play. I haven't been to but one party since September . . . and that was on New Year's Eve when I got royally drunk, and had the honor of being forcibly ejected from Mike's by the bouncer who happened to be Bob Hewlett, Harlem's baddest old man. Of course I was in the company of Lincolnites, Ray Meddows and Edgar Tompkins. Never will I forgive you for going to Lincoln. All the tragedies of my life in Harlem have been precipitated or in some way involved with people I met because you would go to Lincoln University. Bah!

New York is dull. Harlem is duller. I lived in style on Riverside Drive for 3 months, then came back, to of all places, Mrs. Peterson's. She welcomed me with open arms. We get along magnificently. She thinks I have reformed. Since I have little company, and no wild revels. The poor woman doesn't realize I'm merely getting old.

Speaking of age, you have a birthday soon. Greetings on your tercentenary, or should it be tertiary heliogalabus?

Write me. Send me your address or where I can send you an *Infants of the Spring.*

Bruce had a party last night. I haven't seen him for months. Don't know what kind of party it was. But guess all had a good time.

My Marcus Garvey play is finally to be produced in the fall. Or so Chester Erskin says. He is trying to get Paul for the lead. And Rose. And Cora La Redd. What a cast! Which may also include Jack Carter and Wallace Thurman!

And so I ramble.

And so to bed.

    Wallie

I'll like to have a copy of *Scottsboro.*

And a beautiful lady from the I.L.D. just dropped in and said she must find you.

And Broadway Brevities reports that you have finally heeded the call to go to Holly-wood. How come?

———

Tuesday [February 1934]
—Aboard train

Hello Langston:

Are you still there?

Have been sold down the river to Hollywood. On my way to be grist in the mill.

Must see you!

Louise, Mildred, Mollie (she sublet my apt.) all say "Hello, Bum" or maybe it was "Com-rade." Will be at Hotel Dunbar—41st Street and Central Ave.

Wire or write immediately. We must get together—

Arrive there Wed. at 8 A.M.

Some fun, eh, kid?

Wallie

———

WESTERN UNION TELEGRAM
LOS ANGELES CALIF APR 24 1934

LANGSTON HUGHES
CARMEL CALIF

FAY AND I MARRIED TODAY WILL SEE YOU NEXT WEEKEND SEND
CONGRATULATIONS TELL NOEL

WALLY
947P

———

WESTERN UNION TELEGRAM
LOS ANGELES CALIF APR 29 1934 900A

LANGSTON HUGHES
BOX 1582 CARMEL CAL

DONT YOU REALIZE DIFFERENCE BETWEEN GIN AND MARRIAGE
ARRIVING NEXT WEEKEND STOP SEND OK

WALLACE
944A

———

Tuesday [July 10, 1934]

Dear Langston,

I know you said he's a dirty dog and I'll never get my money, etc, etc.

I'm not a dirty dog but a damn sick one. Shortly after I wired you I collapsed, went to bed, called the Doctor. He rushed me to the hospital and here I am. In fact, I understand, I'll be here at least six months (if I don't die) then six more months in a sanatorium. Yes, at last I'm a genius. I have T.B.

After wiring you other things began to happen. Some lawyers sued me for divorce fees and has my income from trust fund all tied up. And Macaulay's had a strike the day I was to start to work. I refused to be a strike breaker and the boss, my lawyer and trustee, also fell out with me.

No wonder I'm in the hospital. The workers also got sore because I didn't picket and me with a temperature of 102.

Where are you? What's doing? Do write. It's damnable being buried alive like this. If you're in L.A. ask Loren to drop me a line. Also Fay—Also Arna. Anybody.

Nurse says I have written long enough. I must rest and rest and rest so that old hussy of a left lung will get well.

Another example of the wages of sin. Can't you hear the darkies preach—

Well—

So long

    Wallie

Send Louise my address please.

    City Hospital
    Welfare Island, N.Y.C., N.Y.

Your money will be a liquid asset soon. W. T.

Claude McKay wants to start a magazine. (Teehee)

————

    City Hospital, Male—Four
    New York City, N.Y.
    Thursday [September 1934]

Well:

And so:

Dear Langston—

I lay here in this damn bed for over three months wondering why you had not written to me. I was certain I had dropped you a line telling you of my incarceration. But from you there was nothing but a great silence.

*The Daily Worker* and *The New Masses* gave me the shudders. I had visions of you being ridden out of Carmel on a rail. (There's a title for a song: "Ridden out of Carmel on a Rail.") More, I could envision you spluttering around in a vat of tar and feathers. I was even composing your epitaph, to say nothing of coining a title for your biography. But alas the California vigilantes neither jailed, lynched or framed you.

Now why the hell didn't you write before?

Well, sir, whodathought when I was in Carmel that I would soon end up in the tubercular ward of the very hospital I damned and god-damned when I wrote *The Interne*. Ironic, I calls it. Or is nature finally avenging art?

The hospital is still terrible. Certain conditions have changed for the better—there is more help and less dirt. But the patients are the worst feature. Being bedded among the proletariat is enough to make me or anybody become a rabid lover of the aristocrats. Such dumbness, such utter stupidity, such innate vulgarity, such congenital affinity for dirt and filth. If this be capital's fault, capital must be some villain.

So you've been a'flying? Ain't it grand? You'd better fly back and get your watch.

So you like to rest. So do I. But not rest complicated by such little items as:

    1 pneumo-thorax a week (a mere surgical operation)

    4 doses of cough syrup daily

    3 doses of iron-quinine-strychnine tonic daily

    3    "    of some unidentified poisonous liquid

Hypodermic needles of various lengths stuck in various parts of your anatomy upon the slightest provocation.

Rest did you say? Come and try it. I'm not only bored but full of holes and pains. Literally. I feel like a cross between a guinea pig and a pincushion. And I always thought tuberculosis was a romantic disease!!

Hey—hurry up and finish that book. I like it. And your potential needs bolstering.

I have never received a copy of your latest opus, Mr. Hughes. As an invalid and debtor may I demand an autographed copy? I'm talking about *The Ways of White Folks* which I decided not to like long before it was published.

I have not seen a *Crisis* for years. Evidently, I am no longer considered a race man. I can't even get anyone to bring me an *Amsterdam News*.

I have very little company and most of that is dull. But anyone from the outside is an improvement over those already inside. Come on over. I receive every afternoon from 1–3 except Saturday.

Well, I'm all in. I had much more to say but the nurse says halt and my strength is gone.

Write me soon—Hurry back to N.Y.

Have you any good books you don't want?

    Always,
    Wallie

I think you ought to be a second cook!

Why not get T.B. and join me for the winter. Rest (?), peace, solitude, writing . . . Food (free).

———

[c. November 1934]

Dear Langston:

Your letter of ages ago was most amusing. I must be forgiven for my delay. I had a damn relapse and when I have a relapse boy I have a good one. It was worse than my original come-down (nice word).

Just a note. Too weak and jittery to write more. Have read not quite everything. Use your imagination. Is there any *Death in Venice* around?

Had a letter from Arna. Poured cold water on Fay's project.

Have very little company. Drop in sometime.

And write me lengthily???—How's the book and literature á la Hughes in general?

Always,
Wallie

Thurman is distinctly a has been—so many people have already buried him. Woe betide 'em when I am resurrected!

## Letters to William Jourdan Rapp

[c. April 1929]

Dear Bill:

At the last moment before leaving I decided to bring the actors' story along with me and work on it today. In rereading it I find I have nothing that has not already been printed in the papers. We have already exploited the various careers and native backgrounds. Perhaps you have something else in mind for me to do with that material. All I have are birth places, education, and past careers, and those of interest have already been given space in several clippings which I found last night. Cordelia—Kid Vamp—Basil—the extras. If you have another angle for me to approach this, let me know immediately.

This seems doomed to be a letter of impotence. I put in four hours this morning on the novel and it just won't come off. Instinctively, and I hope you will understand, I rebel against doing it and with certain mental attitudes of course find it impossible. I have not talked frankly with you about this before, primarily because I wished to do what I could and secondly because I felt I owed it to you to do it whether I wished to or not. I hope you will understand that this is no mere whim. Even discarding my temperamental objections to novelizing the play as quickly and as haphazardly as it would have to be done, there is also the fact that I am completely stale on Harlem and all its more native characteristics. I could not, I know now, write a straight narrative rehashing all that has been done before. Nor do I feel certain that there would be a public for such a work. All we have said in the play has been said by others in novels—not from our angle of course, but the varied manifestations of the decadent Harlem milieu—sweet backs, numbers, rent party, etc.—have been done to death by both low and highbrows. Unfortunately, I have not your gift of being able to dash things off whether they appeal to you or not. And I also hesitate to strain myself doing something utterly distasteful when I am itching to do something more to my liking.

I thus go into detail assured that you can understand my point of view. I don't defend it as being sensible and I am heartily sorry that I did not mention it to you before, yet I feel that I did right in trying to do something before shrugging my shoulders. The first was easy—Jasper's letter—his marriage and the excitement in Carolina on the receipt of the letter—easy if not the narrative you wanted. But the moment I attempt to write or visualize what to write once the family arrives as I did this entire morning—I become lost and heartily sick of the attempt.

If you feel that you could do the job alone—just letting me provide you with atmospheric bits and suggestions, I would be most willing to relinquish any rights I might have in the novel to you. But as for my doing it—I don't think I can, despite my sense of duty to you.

Under separate cover—I enclose the list of names and copy of letter. Some of the addresses will have to be found in the telephone book.

Please write frankly Bill about my defection. I have, and would appreciate your doing the same.

Am also anxious to hear how things are. What about the Chicago company?

Feeling much better. Had 12 hours' sleep and two long walks on the boardwalk—one last night, another this noon. And what an appetite I have developed. Am now preparing to start rewriting the Douglass sketch for Mencken and will also get to work on *Harlem Cinderella*. More anon when I hear from you—

> Sincerely, and a trifle ashamed,
> Wallie

(I discover I haven't Swerling's address. Will you see that he gets this???)

————

> Dear Bill:

Thanks for the letter and statement.

I sent you an air mail letter after leaving Chicago enclosing a clipping from *The Detroit Free Press* and full of details concerning alimony to Louise and also asking you to see Swerling about my payments to him.

His fee is $350. He said I could pay him something on account and the balance in easy payments. He mentioned no figures so I wish you would either telephone or see him and get definite figures, making things as easy for me as possible. I am also to pay Ernst, Louise's attorney, $150.

On Swerling's advice in order to keep down litigation and protect any future income I promised to pay Louise $50 per week for a year—totaling about $2,500. It galls me to do it, but it is probably cheap for my experience and mistake. Also Swerling suggested that payments not begin until after divorce was granted. I ignored his suggestion because Louise claimed to be very broke. On reflection I can see no reason for being so soft-hearted considering her present tactics.

I also wish you would discuss this matter with him and let me know what the two of you think best.

Meanwhile I plan to leave for L.A. the last of this week or first of next, so please wire me at my expense deposits for this week. I am taking for granted you will get both checks same time this week. I am praying for both numbers one and two. After this week I have an arrangement I wish you to make concerning check from number two company, but more of that anon.

For less prosaic news—*Cinderella* is coming to life. I hope Act One will surprise you and it will if I can keep up my present stride—and I intend to. Also I have an idea for a show I am trying to do alone. Will let you know of progress.

I appreciate your having to be troubled with sundry matters of mine and hope I won't have to trouble you any further. I have told Swerling that I will be responsible for Louise's checks. There is no need for you being drawn into this—

> Sincerely—Wallie

———

Saturday

Dear Bill:

I had, as I told you, about ten more ideas for skits, none of which have developed into anything interesting or unusual. Our field is limited, very limited, and there is a great danger of repetition. That has been my difficulty. And in the back of my mind is the doubt that we can sustain the interest of an audience in the type of thing we plan to do. Why not discuss what has been done already with the composers or anyone else who might offer suggestions and see just what our chances are. Meanwhile, I will re-work the ideas I have.

I was thinking that our present outline might be just a trifle too highbrow. It would have to be diluted with so much of the usual stuff that I doubt it would be worthwhile. No one to date has fully exploited the possibility of Negro voices in an operetta or a musical comedy with a sustained plot while revues have been done to death.

As for *Cinderella* I have attacked her with renewed interest and hope. The sledding is hard but I hope to have something (Act I) finished before I leave for Los Angeles sometime after the 20th.

I am anxious to know just what kind of business we are doing both in N.Y. and Detroit. Also please let me know just how much money you deposit for me so that I can keep my accounts straight.

I have had a week of peace and rest. You can hardly imagine how differently I feel. The gods willing I can now settle down and do some of the work I have so long wanted to do.

I will appreciate too your sending me any clippings which might appear.

How is Virginia coming along with the novel. Give her my regards and best wishes.

And please keep a sharp eye on young Abe. God how I distrust him—

Sincerely,
Wallie

P.S.— Do you know of anyone I might see in L. A. in reference to doing some dialogue for colored talking pictures. Both Fox and Lasky are specializing in Negro two-reel films.

————

[c. April 1929]

Dear Bill:

I am afraid that I am losing my sense of humor, for I find myself less able to laugh at things and more inclined to let them depress me. Right now I am at a very low ebb. I am indeed sorry that I did not flee to Europe—but perhaps even then things would not be well.

This should be one of the more consoling periods in my life. Instead, I find myself more harassed than ever before.

I am fighting hard to refrain from regarding myself as a martyr and an outcast. I wish you could take my place in Negro society for about a week. Even on the train I was beset by a Pullman porter for my dastardly propaganda against the race. And here at home a delegation of church members (at my grandmother's request) flocked in on me and prayed over me for almost an hour, beseeching the Almighty to turn my talents into the path of righteousness. All of which is amusing until the point of saturation is reached.

It has always been my theory that a Negro who achieved personal success could fight his way past racial barriers. I still believe in that despite my own disillusionment on many occasions. A number of things happened around the theater and New York after the show opened which gave me quite a shock. Because I make no protest people seem to think I am oblivious to certain things. For an example (about which I have said noth-ing before because nothing is more despicable to me than a Negro who continually has an over-sensitive chip on his shoulders), five different times I have bought seats for myself to see *Harlem*—including opening night—and though I asked for center aisle seats (as much as a week in advance) not yet have I succeeded in not being put on the side in a little section where any other Negro who happened to buy an orchestra seat was also placed.

I appreciate of course the customs and phobias which make such things seem advisable. And it is only because I have been overwrought recently that the thing has stuck in my mind. Then this morning I went down to make my reservations to go to Los Angeles— and though said reservations had been made over the telephone I was refused upon making an appearance because I wished to ride on a "crack train." That floored me, for never before in all my traveling—and I have traveled quite a bit—have I ever been refused reservations on any train no matter how "crack" it happened to be. Not to be outdone I sent a messenger boy down and got them anyhow. I now await my boarding the train. I am sure it was just the particular ticket agent, for I have been on the same train over and over again. N.Y. is heaven compared to the rest of the country, despite certain unpleasant experiences one has. At least people don't stare at you or jump away as if you were a leper. Coming west three people left the observation car in protest to my being there.

Can it be that prejudice is increasing out in the provinces? Here in Salt Lake just 10 years ago there was no segregation whatsoever and now Negroes are segregated à la Georgia everywhere except on street cars. A taxi man even refused to drive me home from the depot!!! Now I ask you!

To continue my jeremiad:—when *Harlem* opened I was $2,200 in debt and after the first week at the Apollo every debtor almost assumed I could pay him in full immediately. You can imagine how I have been harassed. I have been paying out at least $150 a week—my only luxury has been my trip west. I have of course saved nothing. Now *Harlem* seems to be petering out. I am still considerable in debt. I have a $500 counsel fee and am expected to pay $50 per week alimony and for a divorce. Is it any wonder I quote:

> Darkling I listen; and for many
>     a time
> I have been half in love with
>     easeful death.

With a much less poetic meaning than Keats had in mind.

However I am still able to write, have a fund of fresh ideas, and hope to have the leisure and peace for a while at least to get some of them on paper.

Did Swerling perchance tell you about the scandal friend-wife unearthed and used as a bludgeon to make me pay off. If not I'll tell you the dirty details in my next. You probably do not realize what an infamous person I am.

Now, I intend to protect myself debts or no debts, alimony or not. For the week of April 22—mail me the N.Y. royalty check to L.A. I will send address. Deposit the Detroit check to my checking account. The next week mail me N.Y. check as before but deposit 2nd company check to a savings account in the Title Guarantee. I am going to send you signature specimens and you can hold pass book for me depositing road company royalties every week. Maybe it had better not be done in my name. I don't know, I would like it known that I have sold out all royalty rights for second company and that I derive no income from it what-so-ever. Maybe you had better start the account in your name making only such provision as is necessary to protect me should anything happen to you. Then again to make things less complicated it could still be in my name and no one save you and I need know of it. That seems the best plan.

Act One of *Cinderella* est finis. I will revise and type during my first week in L.A. and shoot it to you. It looks much better than other versions.

Pardon this lengthy ramble. Will wire you address in L.A. Leave here Friday unless I'm thrown from the train.

As ever,
Wallie

———

WESTERN UNION TELEGRAM
LOS ANGELES CALIF 21
1929 APR 21 PM 11 22

WM JORDAN RAPP
322 WEST 72 ST NEW YORK NY

FEELING FINE PAY LOUISE MAIL ME NUMBER ONE CHECK DEPOSIT NUMBER TWO IN CHECKING ACCOUNT WILL BEGIN WORK SOON AS ACCUMULATED GOING TO WATCH THE SCREENING OF *HALLELULIAH* TOMORROW GLAD BUSINESS IS PICKING UP LETTER FOLLOWING

WALLIE

———

Tuesday, May 7 [1929]

Dear Bill:

Here you are, for better or for worse. I am not as enthusiastic as I was when the first burst of creation was upon me. Rereading and revising left me dubious. There is only one item which grants satisfaction. As bad as this is it is far superior to other versions. At last I have made the people a little human. I hope you can work with this. If there are any scenes you want me to build, just say the word and I'll get to [it].

After one hectic week in Los Angeles and Hollywood I retired to the sheltering and quiet roof of my paternal grandfather. And for the first time in my life was accepted by my father's family. They run a beach hotel and I wired for reservations without announcing who I was. They greeted me with open arms, excoriated my father who I find is not dead but paralyzed and still living in Los Angeles, and have made me most welcome. It seems they had read the *Literary Digest* and being assured that I was not after any of my grandfather's money decided it would be quite all right to let me retain the name of Thurman. Anyway it's nice down here. I take two dips a day in the ocean, sun baths on the beach and spend the rest of the time reading and writing. I have started my autobiography and am mailing a couple of my autobiographical sketches away.

The vagaries of the show business intrigue me. Hollywood is much more mad than Broadway. I was introduced around the studios. M.G.M. officials immediately decided that perhaps I could do a story suitable for Nina Mae McKinney, star of *Halleluliah*. Then they got cold feet and decided to do nothing until the box office returns on *Halleluliah* were known. Pathé has a couple of Negro comedians under contract now starring in some of Hugh Wiley's stories. They asked me could I do something on the same order making it more authentic. I told them that there was nothing authentic whatsoever in

Hugh Wiley's work, that there was good authentic Negro comedy available without resorting to caricature and that the Harlem scene was more intriguing than any plantation. Thereby, I think, talking myself out of a nice contract. I was even told that I was too intellectual. They had expected me to be like their Negro actors and were a little surprised at my line of talk. However there is still a slender chance that I may do something for Pathé, and now Fox wants another picture for Clarence Muse, who was in *Hearts of Dixie*. All however depends on the success of *Halleluliah*. Moran and Mack are making a Harlem picture. Another company is filming *Rent Party* also laid in Harlem. And I can't get inside dope on it to save me. I was told all around that all were eager to see a script of *Harlem*. They had had good reports but no approach from the producer or agent.

Speaking of producers. I hope you are keeping a close watch on friend Blatt. Either he is cheating on us or observers are easily fooled. I had many reports from Detroit, one from a boy whose brother worked as porter in the Shubert Lafayette. His report of the full houses certainly belied the reported grosses. I know of course how easily a layman can be fooled or exaggerate. But it pays to keep one's eyes open.

Hope *Harlem* in New York can last through May. And that number two lasts forever. What about the road chances of number one in the east? What are plans if it closes for the summer?

Thanks for keeping me in touch with things and for taking hold of my alimony et al. As long as New York company continues send me that check and deposit number two, letting me know what you have done about the matter. I do wish you would send Adelaide Schulkind, League for Mutual Aid, $100 out of number two check this week, thereby closing out a long indebtedness to them, and leaving me owing no one but my grandmother whose $1,100 is about paid. Then use your discretion about payment of lawyer's fees. You know how to handle that much better than I do. Just manage if possible to let me have a little stake in the bank. I am, I understand, to pay Louise's fare to Reno and a $300 lawyer fee there. I will not under any circumstances stand for more. I couldn't. She'll pay her own living expenses while there out of the $50 per week and she will get back as best she can. One moment. The summer rates are on until September. If she wishes to buy a round trip ticket *that way,* o.k., because it will be only a few dollars more.

Now for the scandal. In 1925 a young colored lad anxious to enter a literary career came to New York. He had a little stake which was soon gone. He found no job. He owed room rent and was hungry (not offered in extenuation of what is to follow but merely a statement of facts). One night he got a job as relief elevator operator, just for one night. He worked. The next night he returned hoping to work again. Failing, he returned homeward. At 135th St., he got off the Subway, and feeling nature's call went into the toilet. There was a man loitering in there. The man spoke. He did more than speak, making me know what his game was. I laughed. He offered me two dollars. I accepted. Two plain clothes men, hidden in the porter's mop closet, rushed out and took the two of us to jail. Night court. I was fined twenty-five dollars or three days. The man got six months. He was a Fifth Avenue hair dresser. He had been picked up before, and always of course as the aggressor. I gave a false name and address, then sent a special delivery letter to the only friend I had in New York. He borrowed the money, gave it to a minister friend, who came down and got me out after I had spent 48 hours in jail. Only two people thus knew it. This minister took a great interest in me. And to my surprise I discovered that he too belonged to the male sisterhood and was demanding his pound of flesh to keep silence. I cursed him out, told him he could print it in the papers if he dared and saw him no more. Meanwhile of course he had told his scandal. By some quirk of fate it reached Louise just at the time she was fighting me for money settlement. She told Ernst. He verified the story, and they threatened to make charges that I was homosexual, and knowing this and that I was incapable of keeping up my marital relationship, had no business marrying. All of which Louise knew was a lie. The incident was true, but there was certainly no evidence therein that I was homosexual and Louise also knew that though there had been sexual incompatibility it had been her fault, not mine. She had to have an operation while I was in Salt Lake . . . remember . . . in order to make an entry possible, and because by that time I had lost all sexual feeling for her and though there was a consummation of the sexual act, I was blackmailed thusly. The alibi being that she had been so upset by this vile disclosure that it had ruined her life. And such is my tale of woe. Doesn't it read like a novel? You can understand now what a mental state I was in during those last few weeks in New York, and why I had to get away. And you can also imagine with what relish a certain group of Negroes in Harlem received and relayed the news that I was a homo. No evidence is needed of course beyond the initial rumor. Such is life.

I am anxious to hear what you have to say about musical comedy. From *Variety* I glean that those now coming into New York are not so hot and are flopping right and left, which makes me eager to do something new and entirely out of the beaten path. What shall it be? And *Jeremiah*? What news of my most beloved work?

How is Virginia's opus coming along? And how is Harl. Give him my regards. And it will take you a whole day to answer all my questions, so for the moment I close.

> So long,
> Wallie
> 1538–5 Street (Santa Monica)

————

WESTERN UNION TELEGRAM
LOS ANGELES CALIF 15 216A
1929 MAY 15 AM 5 33

WILLIAM JORDAN RAPP
322 WEST 72 ST NEW YORK NY

SEE *NEW YORK AGE* MAY ELEVENTH HAVE BLATT AND ERSKIN TAKE ACTION SUCH BALDERDASH SHOULD BE STOPPED AM QUITE DISGUSTED ABOUT WHOLE MATTER COURT ACTION WILL CURE THIS DISEASE MAILED LONG LETTER TODAY ANGER HAS REVIVIFIED CREATIVE URGE WILL WILLINGLY HELP IN ANY LIBEL SUIT

> WALLIE

———

Tuesday

Dear Bill:

I am somewhat despondent today. For the past week I have been unable to write a word, and those things on which I had worked before seem (seem nothing) *are* quite bad, no finish, no gusto. I do not know what can be wrong. I have torn up my 8,000 word Frederick Douglass essay, and will begin it again. I had one finished on Du Bois and another on Garvey, but neither are worth the time and paper on which they were written. I have also written about 4,000 words on my autobiography, but they too are going into the waste paper basket. I lie on the beach all day in the sun waiting for the urge to return. It doesn't, so I come home and spend half the night forcing things out. The result is what might be expected. I am about ready to climb down off the water wagon and go on a spree for about ten days. But I know that dissipation will do no good either. Something is cold within me. I have no enthusiasms, no life. I would like nothing better than to go off into a trance and stay there indefinitely. Physically I am more fit than I have been in many a day. Maybe I am too bloody healthy for my creative powers to function.

Your letter and enclosures received yesterday. They make me reverse some of the things I said to you in my last letter. Continue sending me the checks minus payments to Louise. And do send the League for Mutual Aid that $100. I think that Louise is out of luck as far as Reno and a divorce is concerned. I simply cannot afford it. And unless some unexpected money appears she *is* out of luck. So please make no disbursements for railroad fare or anything else.

I will probably remain here through the summer. I can live quite cheaply. My room costs me five dollars per week. My board is about one dollar per day. Laundry is negligible and my one trip to Los Angeles per week to which I limit myself costs nothing. Someone always drives me back and forth so I save the fifty cents per round trip fare. And I am always invited to dinners or else entertained. And even if I decide to get off the wagon which is unlikely for another month at least, all I have to do is visit certain friends at the studios. Not only do they give you all you can hold to drink, but also insist on giving you two or three bottles to carry home. My daily dips cost nothing either, nor the sunbaths or library. This *is* the promised land. All of which is told you for the purpose of saying that I am soaking away as much money as possible, my only outlay on this end being the weekly checks I send home to my grandmother in order to clear up my debts to her. So far, so good, maybe I won't have to court starvation in New York next winter after all. And I will not risk it under any circumstances to pay for Louise's trip and divorce. I have already been bled enough or will be if the 2,500 is paid, plus N.Y. counsel fees.

About my being western agent. I am willing to try and will begin as soon as copies arrive. I have entry to two studios, Pathé and M.G.M. and to a lesser extent to Fox. The rest are harder to get into than heaven. But I will see the powers that be in the three aforementioned and strive for an audience in others. It is quite likely we will get a break. My only drawback is that I have not the slightest idea how one bargains about such matters. I'll merely get offers and wire them back, if any, awaiting instructions.

*Cinderella* went air mail special about three days after telegram was sent. Its delay was due to the fact that I dropped it in an automobile while out on a drive and missed it not until the people brought it back. Then I hied to postoffice, had it weighed and sent on.

The Lafayette Players are here at the Lincoln Theatre having quite a run. The management, connected with West Coast Theaters, want me to do a play. They have even gone so far as to give me a copy of Paul Laurence Dunbar's *The Sport of the Gods* to consider dramatizing. There is a great demand here for colored shows. Not one has appeared here, save these Lafayette Players who have a very large white and black following. The proposition is that I'll get a third cut of the show. If successful in the neighborhood Lincoln, it is to be taken downtown and presented in other coast cities. I don't like Dunbar's novel. It is quite bad, but there is the germ of a play in it. It seems as if these people own the dramatic and movie rights to Dunbar's works. But I won't do this novel. It is dated for one thing, occurring in the beginning of the century. That of course, since I said it, is no drawback. It could easily be modernized, but these people are soaked in the ordinary stock melodrama, and already have suggested that I make the play as syrupy and happy endingish as possible. I doubt if it would be worthwhile anyhow. And I am not enthused about writing a play for a stock company where the casting would be a matter of expediency rather than expertness. Outside of Bowman there is no one in the cast worth a second look. Their weekly bills are amusing. Tenth rate melodramas written for whites and how these Negro actors do act in them. They did *Jekyll and Hyde* last week. I almost had hysterics, but everyone else seemed to enjoy it. The week before they did *Should a Woman Tell* and before that *Alias the Deacon*. Evelyn Preer, weighing twice as much as she did, does all the leads: pure young girl, etc. It's immense.

No new news on my own possible tie-up with movies. Am to go to M.G.M. again tomorrow. I don't know why. Write soon.

Wallie

————

Dear Bill:

When I received your wire from Cavalier Beach I had my suspicions. Therefore your news did not surprise me. You know that I speak sincerely when I say that I am most glad the grand event occurred, and that I am bubbling over with good wishes. Here's luck and congratulations to you both.

The Reno business is o.k. As I figure things out Mr. Blatt owes us four weeks' royalties which will more than cover the expense. I am glad that you took this matter under consideration and acted in a sensible manner. Nevertheless I would like to have a nice poke at the honorable Louise's jaw. She seems intent on making me as miserable as possible, writing ridiculous letters to my folks and doing all she can to alienate friends of mine from me. She flatters me. I could never be the villain she has tried to make some people believe existed. Her latest is to write to Miss Randolph (whom you will remember) who had asked her in all friendliness to lunch. Louise responded in a hysterical letter, alleging that I had flaunted my friendship with Jeannette to her during our married life, and striven hard to make her (Louise) believe that Miss Randolph and I were committing adultery. Fortunately, Jeannette had enough sense to know that I wasn't such a moronic cad. She dismissed the matter and renewed her invitation. Needless to say it was turned down.

Am on pins and needles awaiting word from Hollywood and Culver City. Things look favorable. I had to send for another copy of *Harlem*. Chet and Ed replied that it was on the way. Have an appointment at Fox tomorrow. Will see James Cruze next week. My only worry now is what to do if one responds and the others seem about to bite. Should first come be first served or should I play ball and wait for further developments? I am a poor salesman and a worse bargainer. I shall however call upon the shades of my Jewish ancestors to stand by.

The novel has not arrived yet, but I shall devote all my time to it once it is here. Is not Hearst a possible purchaser of serial and syndicate rights? To me he seems like the best bet. His papers run such things and *Harlem* would be a novelty as well as sufficiently sensational to intrigue his editors and readers.

My own novel is coming along nicely. I am now working on the last essay for my miscellany. And by the way please rush me a copy of *Dance Life* containing my article. I have them all now but that. It is necessary that I receive it as soon as possible as Furman is waiting for the completed volume. And by the way again may I not borrow *Color Parade* for a title? It fits and if we are not going to use it for our collection of three plays I would like to use it for this collection of articles on Aframericana. If you say no then get ye to work and suggest other possible titles. I'm stumped beyond that. The only other one I have in mind is *Chiaroscuro* which is rather highbrow and obscure.

Mencken has had three of my biographical essays for over three weeks. Maybe he's going to take them. They return things fast at his office unless considered favorably. I have shot another to *Harper's,* and *Plain Talk* and *The Bookman* are also being bombarded.

The sun is shining for the first time in over two weeks. So I think I'll celebrate by taking a day off on the beach.

Best regards to Virginia. Write soon, taking especial notice of various queries and requests.

So long,
Wallie

————

Dear Bill:

I hardly know what to say about this Louise business. There is only one thing about which I am certain and that is: I will succumb to no more sentimental pleas that "we discuss matters between ourselves in an adult way." That's the reason I frankly stated my position to her and told her that I could not see the way clear at present to pay her expenses to Reno. I am tired now of submitting passively to her demands as if I had no side at all. I will not pay her way to Reno and if she will not sign the agreement I will not make her another payment. This may be foolhardy in view of the fact that she may be able to soak me for more in the future, but as I figure it out, Ernst will act immediately, probably advising her to start suit for separate maintenance in New York. Which will of course avail them nothing. I have the trump cards. She cannot get any judge to grant her that which I have not. We have our agreement drawn up to prove that I tried to do my share. If she brings up this homosexual business, she will get nowhere. She has already admitted to Swerling that after her operation all was well sexually between us and neither can she bring up that knowledge of that one incident in my life disgusted her so that she was unable to have further intercourse with me, for she did, on the very day I left New York, April the sixth. If she will sign the agreement, I will do my darnedest to keep up those payments with the proviso that as soon as the shows begin their road tour I will furnish the necessary lucre for a divorce, and if she, at that time, is bound to some certain place because of a job, will see that she has the grounds to get her divorce where she may. If she won't sign the agreement . . . to hell with her. I take this stand because the whole affair has degenerated into a case of pure extortion. She has written me asking that, if I can't send her to Reno, I pay her expenses to Berkeley, California, so that she can have a vacation. I wonder why she thinks I would do what she asks when I am unable to do what had been originally planned. Moreover, why $600 to go to Reno? . . . plus fifty per week? Nothing doing.

I too realize the importance of getting the divorce as soon as possible, but I won't be a softie in every respect. You might have Swerling deliver my ultimatum, then plan our action from their next move.

Met James Cruze, who is quite anxious to see script of *Harlem*. Has long wanted to do a first class colored movie and showed me countless stories he has considered. He wants to star Evelyn Preer, which is alright by me so long as he buys the movie rights. Will wire you immediately any developments. Bowman is also here and he likes her, so I made a grand little sales talk on how he could make use of both of them in *Harlem*. Believe it or not I still have conferences on the average of once a week with various Pathé officials. I have come to the conclusion that they must like to talk to me, or that they are trying to gain some knowledge. I thus talk quite guardedly, for fear should I be led into a discussion of Negro life, a Dictaphone might be around which could relay my talk to a voracious scenario hound. This may sound fanciful, but you don't know Hollywood.

All my biographical sketches are finished and on their way to various magazines, also a couple of other essays. The novel too is coming along fine. You cannot imagine how nice it is to be isolated for a time, reading and writing as the mood hits you, far away from the wrangling mob.

Had a dramatic meeting with my father last Sunday. He did not know I was anywhere around. He came down to visit his father. I opened the door. Mutual recognition. I almost fainted from the sight of him . . . he is paralyzed and has tuberculosis of the throat and I do not know what all. The most pitiful albeit nauseating sight I have seen in many a day. He was more than shocked to see me, in fact he almost dropped in his tracks but recovered sufficiently to deny knowing me when someone asked him the question. The effrontery of this after all the emotion he had displayed and from a man in his condition shocked everyone into silence. He shifted about uncomfortably for a moment, then asked me, was I not from Salt Lake. I answered in the affirmative. He then asked me my mother's name. I got up and started from the room. He then said, Oh yes, he knew me, I continued out. Before he left he came to my room and asked me to visit him as he was anxious to talk to me. I assented to get rid of him, but I hope I never lay eyes on him again, and I won't on my own volition. All in all I would say that my life is one hell of a mess. How can one have any respect for one's fellow human beings? It might also interest you to know that I have lost most of my New York friends. Speculations about my income have urged any number of them to write me for money. Refusal has made enemies out of them, which is what I expected and for which I am glad. Even Bruce wrote me from London asking me to cable him $100!!! As Walter Winchell would say . . . I gave vent to a huge belly laugh.

I am completely out of debt . . . even to my grandmother, and I now have on deposit here almost seven hundred dollars to which I shall hold tenaciously. I will however finance no Reno trip until we know something definite about the Chicago company. If it stops I will need all I have and what is due to see me through the summer, return to N.Y. and live until road tour starts. If movie rights are sold . . . o.k. or if Pathé officials should finally cease talking so much and produce a contract. They are congenial fellows, but oh so dumb, and it is only their money in which I am interested, not their company and gin.

That's about all for this time.

> So long,
> Wallie

———

Saturday

Dear Bill:

No need to certify checks. I deposit them to account anyhow. The one fourth cut is o.k. by me. When does it begin. On the fourth week? Hope things will work out all right with number two. Meanwhile I am heartily sick of the publicity attendant upon closing number one. I am deluged with clippings and letters. It seems as if I "offered no objection to my white partner firing the cast, treating them like children," etc. It is all very annoying.

Have been working like mad for the past week. As I said in a wire to you I got angry over those stories and was thus roused from my apathy. All biographical sketches finished and on way to magazines. Duplicates going to Macaulay's for publication in book form. Will include articles from *New Republic, Bookman* and *Independent* also in this volume. Doing two more essays, one on Negro in theater, another on the Negro problem for same book. It will thus cover a great deal of ground and have a unifying note too. *Dance Life* article will also be included. Am polishing up those which have appeared in magazines. And have managed to get considerable done on novel.

Have been nowhere recently except to spend an evening with Ransome Rideout (who wrote *Going Home,* and is here with M.G.M.). Anxiously awaiting copies of *Harlem.* Do not entirely agree with American Playwrights decision that since *Harlem* was not a smash it would not be good business to do *Jeremiah* first. Consider the differences in the play. And *Jeremiah* should to my idea be done early in the season. My view is probably colored by personal interests. Let me know what show you get hold of and all other details of your progress. Awaiting novel and second act outline of *Cinderella.* Will take especial pains to make it long enough this time. And will work on novel to exclusion of everything else once it arrives.

I sent agreement back to Swerling. Reno trip out of question. I have communicated with Louise to this effect. And considering the dwindling of my income when one fourth cut goes into effect how can fifty a week payments continue? There will undoubtedly have to be a compromise. I can figure on this of course when I know just what average weekly income will be when cut takes effect. I am not going to release the little money I have on hand and thus risk being broke immediately upon my return to New York . . . agreement or no agreement. To all intents and purposes I have not a penny on hand and any payments will be based on what I earn, the total sum of course to be paid as agreed, but perhaps not as quickly.

Are you not going to take a vacation this summer? Where do you plan to go? Why not consider California?

> That's all for this time,
> So long,
> Wallie

————

Friday [c. July 1929]

Dear Bill:

Talk about extremes. Physically I seem to vacillate from one to another. One month ago I was complaining about being too bloody healthy to work, and now, I seem to be shrinking away into skeleton. Gone are all my proudly gained pounds and zest for life. And meanwhile I wonder what it is all about. It seemed to have started when I caught a chill while cavorting in the Pacific. Then this damnable "unusual" weather aggravated the condition. The result is I have a nasty cough, frequent chills and fever and a suspicion of a spot on my left lung. I am certain though that a change of altitude will take care of all that. Hence I go to Salt Lake soon as possible. It's not a matter of rest. I still get almost too much of that. It's the climate and of course mental attitudes.

Have Blatt get copies of *Harlem* to Tiffany-Stahl and R.K.O. and Asher and Green. The latter are about to do *Black Boy* with Paul Robeson. I don't know how to get in touch with any of them. And I will probably commit mayhem if I have to discuss *Harlem* with any more Israelite garment makers. Negro movies are at a complete standstill however. The fate of *Halleluliah* will decide all. And until this Equity business is straightened out, there will be little experimenting or new production plans. Moreover there is actually a deal of resentment among the white movie contingent at the working of Negro actors. They want to black up. God save us. Can you imagine Colleen Moore playing Cordelia, or Douglas Fairbanks leaping around as Jeremiah? Or maybe John Barrymore since Ethel is going to black up and hoof in a dramatized version of Scarlet Sister Mary.

Another item: I was told at M.G.M. that were it possible for me to pass for white, i.e., should I have happened to be a mulatto, I could possibly have become connected with their staff. Such is the case with Jo Trent in the musical department. They did not mind my being a Negro, oh, no, it was just that I was too obviously one. And such are the things which add to the gaiety of Wallace Thurman.

I am also about ready to arm myself with hand grenades and annihilate all magazine editors. I sent "Frederick Douglass" to *Harper's*. They sent it back with the regular rejection slip but with the words "too long for our purposes" underlined. Heigh ho, said I, it must have possibilities, so I sent it to *Hersey's Famous Lives*. They have just returned it without even a printed rejection slip. I told you that Mencken returned "Garvey" and "Du Bois" with a rather nice and encouraging little note. I sent "Du Bois" to *The Century*, "Garvey" to *Plain Talk*. The first was returned this morning and again no rejection slip. The second is still away. And "Booker T. Washington" is at *Scribner's*. I am going to impose upon you (my occupation for the past two, or is it three, years?) and send them on to you for your editorial judgment, giving me your opinion as to what is wrong if anything and how I could best improve them. My own theory is that magazine editors being what they are and eternally accepting the men I have dared to interpret and criticize as the only ones capable of discussing Negro questions, dismiss me as a young upstart speaking without authority. On the other hand, the things may not be as good as I think they are. Hence I call upon you. You understand what I want. No rewriting except as done by myself, but editorial criticism, because they must be good before going into a book. The Douglass one is biographical rather than interpretive like the other three, mainly because so little is known of Douglass in comparison with Du Bois, Washington, and Garvey. But, if you think they are good or will be with a little more revision, to the publisher they go despite magazine rejection slips. Look for them under separate cover. As I aforesaid, two are now out.

I am following the Williamses relentlessly through their adventures in Harlem, and strangely, am interested once more in them as human beings. About half finished. May have present portion typed next week and send it on.

Does *Harlem* take to the road again in the fall or is it definitely consigned to the storehouse?

Apropos of films, present ground work may prove valuable in case *Halleluliah* goes over and there is demand for colored features, despite rumored blacklisting of Harlem by Czar Hayes. And I imagine *Jeremiah* would be a riot in the movies. I am almost willing to bet it could be sold now.

Had a glorious Fourth of July. Remained in bed all day and listened to the fireworks in the street and the detonation of battleships moored in the harbor. Read or rather reread a volume of *Swann's Way* by Proust, wrote a poem which is pretty awful, and added about five hundred words to story I am doing to submit to James Cruze.

Give my best regards to Virginia and Harl. I will be most glad to see all of you once more. And perchance is Mrs. Thurman in Reno?

As ever,
your pestiferous collaborator,
Wallie

———

Thursday

Dear Bill:

Thanks for the fifty. I carelessly had not kept track of my checking account and had overdrawn it. As I said in [my] night letter, I sent my grandmother five hundred dollars of the money I had to hold until fall, which she is doing so well, she refused to send me even fifty to run me over until I had received another check from you. Have been having all sorts of difficulties. Caught a chill while swimming and have been considerably under the weather for the past ten days. The local doctor claims my left lung is infected, and from the cough I have you would imagine I was in the last stages of T.B. I despise this climate. One hot day, six cold ones, all damp. As soon as things can be brought to some conclusion in Hollywood I will leave for Salt Lake, I hope, no later than this time next week.

The syndicate interested in doing *Harlem* out here all seem to have disappeared. Have they perchance communicated with Blatt as I told them to? If not he might get in touch with Harry Wurtzel, 1509 Vine Street, Hollywood. He was the guiding spirit. They ran me to death for about a week, then all at once seem to have lost interest. Claimed price was too high. Then the manager of the Lafayette Players heard of their interest and said he wanted to do it. I have heard no more from him either since I made terms known. Enclosed is letter from Pathé. Block was gone on the play, but evidently sub-officials have dissuaded him. And since this letter came I have been unable to even get into Pathé studios to see anyone. M.G.M. alone has given no answer. Cruze claims he could not possibly gamble on getting past the censors and he would not risk ruining the play to appease them. He then asked me to do him an original story so that he might star Evelyn Preer. This I have undertaken to do, realizing that a skeleton is all any movie director wants. I am trying now to get to Universal and submit the play to them. Sent a copy to Famous Players' Lasky this morning. And thus it goes.

Suppose Louise is in Reno by now. Thank goodness. And again let me tell you how much I appreciate your guiding and interest in this messy business. Received a letter from Swerling along with a Confession of Judgement which I am returning all signed to him today. From his letter I gleaned that he does not think very highly of me. He said: "Mrs. Thurman will leave for Nevada in a few days. I might add that the whole deal went through only because Bill Rapp came to the rescue and advanced $600 to finance the trip to Nevada and legal expenses there. He has been exceedingly nice about the whole business and has afforded the most indisputable proof of genuine friendship."

Therein praising you and spanking me. I chuckled.

Have done quite a deal of the novel, although I have kept so busy dashing about Hollywood, waiting all day in studios to see such and such a person that I have not had much time to spend on it. Two weeks in Salt Lake should see it finished, as I will drop everything else once the opportunity presents itself.

That's all for the nonce. Best regards to Virginia. Will wire you any new developments, and hope Blatt comes through with royalty checks in time for me to have train fare to Salt Lake next week. Sent my essay on Douglass to *Hersey's Famous Lives*. Mencken returned his two finally saying "They have interested me a great deal, but unfortunately, I have so many articles in type dealing with the situation of the Negro in America that I can't take them." Blah, blah. However undaunted I shipped them to other magazines. May yet have to call upon your aid in disposing of them. And may I ask: Where is *Harlem*?

> Gratefully,
> Wallie

––––––––

Dear Bill:

There is no reason for this letter, except that I feel I must unburden myself, and choose you for the victim because there is no one else. I have been in bed ever since the third of July, inert save for occasional spasms of writing which sadly display my hysteria, and attacks of coughing which render me baleful and a hypochondriac. The wind blows, the sun shines fitfully, all about me people move, laugh and sing. The radio in the next room drives me to distraction. There was a slight earthquake yesterday morning. The doctor doses me with cod-liver oil and reluctantly leaves a prescription for a sleeping draught. Meanwhile I send you another hectic telegram and plan to fly away from California forever.

Decidedly I have much to complain of in myself, and I am astonished and alarmed by my condition. Do I need a change of air? I cannot tell. Is it my sick body that weakens my will and mind, or is it a spiritual cowardice wears out my body? I do not know. But what I do feel is an immense discouragement, a sensation of unbearable isolation, a perpetual fear of some remote disaster, an utter disbelief in my capacity, a total absence of desire, an impossibility of finding any kind of interest.

Thus Baudelaire's *cri de coeur* to his mother on the 31 of December 1857. How amazing that I should just discover it, particularly during a period in my life when it exactly expresses that which I feel the urge to stay.

The doctor says: Relax, cease reading, forget you own a fountain pen. He might as well advise me to cease living. To be completely inactive would generate madness or suicidal mania, neither of which I wish to play host to just now.

The diagnosis is of course: too much introspection. Continual fumbling over the past, berating self for innate lack of will and self control, consciously unreeling the past as I do in my new novel what Proust succeeded in doing in his volumes. And too there has been the irritation of not being able to devote all my time to one thing. Too much stray diffusion of energy.

I have you see worked on the essays rather assiduously, only to have them rejected as chronicled in my last. I have completed about half of *Harlem* and done about 30,000 words on my new novel. Then this James Cruze story, which no doubt, I will complete in first draft, then shoot to you, should you care to collaborate and put on the finishing touches. I am certain my final version would make a better *Dial* story than a movie scenario.

And I have been reading far too much: Proust, Joyce, Dostoyevsky, Shakespeare's tragedies, Ibsen, Molière, Hardy and Swift. Is it any wonder I'm depressed and enfeebled both mentally and physically?

Another thing: I was never created to forego companionship. And I have had damn little of it out here. I miss New York. I miss the chats I used to have with certain kindred spirits. I miss the occasional mad nights experienced in Harlem. And yet I know should I be in New York no work would be done. I am determined to remain away until the initial draft of my novel is completed, the book of essays finally fit for publication, and *Harlem* novelized. And if you say so, until *Cinderella* has taken on some semblance of acceptability. Two more months should see this done, should my spiritual intensity prolong itself and my physical disabilities prove mainly mental and momentary. For such I believe them to be.

I am learning to hate. I hate the world in general, Negroes, all my friends save you, Jeannette and Stefansson, all my relatives save my grandmother. A lamentable state of mind, stewing, fermenting, becoming bloated with feeling that finds no outward expression save in lugubrious, silly epistles like the present one—

Wallie

———

Tuesday

Dear Bill:

The trip was hot and tedious. I recklessly treated myself to a compartment, and remained abed the entire 24 hours, practically nude. It was expensive but comfortable. Having escaped from the capricious climate of Southern California, I hope for signs of immediate recovery. Yesterday I felt terrible. It was quite warm as it has been today, and my being in bed has not made it any cooler. However after tomorrow I shall have a hammock on the porch—me, books, paper, fountain pen, cod liver oil and all.

Louise was here Thursday and spent the entire day with my grandmother. From what I gather she sobbed all over the place, asked to be adopted into the family, etc., etc. but the best bit is this. It seems, indirectly stated, that my friends are the cause of it all, and, directly stated, that 'tis no one else but Bill Rapp engineering the divorce. It seems that you have practically driven her to Reno as well as prevented me from acquiescing to a reconciliation. Be not surprised if you are made the defendant in a suit accusing you of alienation of affection!

However she lost what ground she gained by bringing up the money question. Her statements as to my income were so incredible that my grandmother doubted most of her previous lamentations. When blankly asked about the alimony, she alleged I could pay her $200 per week and not miss it. Imagine her surprise when my grandmother showed her a W.U. receipt for my railroad fare home, not telling her of course it was from a meager sum she was keeping for me. She, I am told, immediately ceased lamenting her lost love, and talked incessantly about what difficulties she would have living in Reno. It must have been immense. She even claimed that she would look for work—menial labor—unless something unforeseen happened. It hasn't and it won't.

Meanwhile in Santa Monica I was presented with a bill from the N.Y. Telephone Company for $50.96 for April and May. I politely told the collector to find Mrs. Thurman and resubmit the bill to her. What colossal nerve! I haven't, you know, lived in the apartment since February.

Imagine my joy to discover that little has to be done on the last half of *Harlem*. Save for certain emendations and additions, it seems about K.O. I've done considerable. Three more days and I hope to turn it over to a typist. Not bad, eh! Did you receive the sketches? "Booker T." on way soon as typed. And please get me 2 copies June 1928 *Bookman,* containing my article on Negro poetry. Better verify the month. I've written to them twice for copies—to no avail. A damned inefficient office? And also Bill, if possible, please send me a copy of Leonard's article on the Negro in the theater. I don't know where it appeared. Moreover, I would appreciate a three months' subscription to *The Morning* and *Sunday World, The Times Book Review,* and *Books,* which you can pay for out of my accrued royalties when they are delivered unto thee.

Thanks! I'll ask for nothing else for an age!

Best regards to Harl and Virginia. Tell the latter I hope she won't want to murder me for the liberty I have taken with certain portions of her manuscript.

> As always, Wallie
> 308 E. 9th So. St.

————

Thursday [c. July 1929]

Dear Bill:

There is a lot to be said in answer to your last two letters, but first let me get this off my chest. It's about the biographical sketches. I appreciate your criticism but am loathe to accept it *en toto*. Contrary to your appraisal, I think the Du Bois one the best of the group with the Douglass as the least successful and interesting, thereby no doubt giving proof of my stubbornness and vanity. I purposely composed them as they stand, and do not believe I would be able to change the style. Should I make them replicas of the ordinary popular biographical sketch I would defeat my purpose in trying for the first time in history of Negro writing to be critical, subjectively truthful, and unconventional in my appraisal of these men. Perhaps it is presumptuous for one as little known as myself to speak *ex cathedra,* but someone has to do it, and to date no one else has. Considered as parts of the book I have in mind they seem to me adequate and necessary although I would be the last to suggest that they could not be improved. The entire book, every essay, is more or less an expression of my own personality, and in it is formulated, rather immaturely I admit, premises for a new philosophy for a new generation of younger Negroes. Admitted that perhaps I am aiming too high, admitted that perhaps I am not yet ripe for such a task, but I had to do it, and do it in the way it now stands. Thus while I am willing to rewrite in the interests of literary composition, I am not willing to do a *volte face* and produce interesting but commonplace sketches of Negroes who for the most part are immune to intellectual criticism or appraisement. I am expressing rather poorly perhaps what I have in mind, and of course I am not overlooking the possibility that I may be all wet. Nevertheless I think once you see the entire volume, preface and all, you will understand why those sketches had to be written as they are. Shoot them back to me anyway. I already have notes for revision, and I am anxious to get the volume to Furman and get the reaction of his readers. As you say, they may not do for magazine reproduction although I have had another letter from Mencken in which he again expressed regret for not being able to use them because he was overflooded with articles on Negro life, and wished to preserve a balance, adding that was I unable to find another market to re-submit them to him at a later date. He seemed to sense what I was trying to do and thought it a necessary and good thing. I want you to see the entire volume, then once more send me your reaction. I want you to understand that I do appreciate and understand your criticism even if I do cavil and dissent.

I received your long letter, it following close on my heels from California. Coming as it did when I was striving to do what you did in the letter, it meant a great deal to me. The curse of being the neurotic son of neurotic parents is that one is such a weakling and succumbs so easily to mental ills which soon become manifest in one's body. Had I not snapped out of it there is no doubt but that I would have become in short order an incurable consumptive. As it is I keep my mind entirely out of such personal channels, foregoing the intense and pleasurable suffering of introspection, get plenty of fresh air, drink a quart of milk per day, eat plenty of fresh eggs, fruits and vegetables, and continue my codliver oil although I am certain it adds materially to my discomfort in this hot weather. I had planned a hiking and camping trip into a nearby canyon with a group of former school chums this week but the doctor advised against it. I shall however go next week, and, coincidental with your advice, I have a friend who has a fruit farm on the outskirts of town, and I intend to spend a week with him sometime next month . . . picking fruit. I fear for the trees, the picked fruit and certain portions of my anatomy which might suffer when the earth calls its own and I tumble from rickety stepladders.

Of writing I have done little, almost nothing. And the heaviest reading since my arrival in Salt Lake has been *Alice in Wonderland.*

But I am beginning to feel the itch. I intend to finish up *Harlem* before the week is out, hoping to get it to you by the first of August.

There are days when I feel fine, followed by days when I am too weak hardly to lift a fountain pen. The weather, of course, it is damnably hot here, is in some degree to blame. But on the whole I am improving rapidly, picking up flesh and getting a sparkle in my eye to say nothing of a healthy outlook on life in general and myself in particular.

Glad to hear about *Harlem*. Hope you get a good road company together. And Blatt's movie idea listens well, but I wonder just how much success he will have. The Cruze story gets nowhere at all. I shall type out a synopsis of what I have in mind and shoot it to you sometime this week. And is there any truth in the rumor that Robert Levy, manager of the Lafayette Players, has obtained the coast rights for *Harlem*? Such is the information his star Evelyn Preer writes to me. Anxiously awaiting all the numerous packets I asked for, and repeating my request that you send me the sketches as quickly as possible, and be as ruthless with the manuscript of the entire volume as you were with these excerpts. I shall close. My best regards to la famille. I now retire to my porch hammock.

Affectionately,
Wallie

————

[c. July 1929]

Dear Bill:

And that briefly is what I have had in mind. Is it any wonder progress has been so slow? Cruze wants a part as I have said for Evelyn Preer, where she can wear clothes, dance, sing, and generally occupy the spotlight. This story by the way is simply a slightly camouflaged life story of one Norah Holt Ray, who is one of Negro America's most interesting characters. She however has had one more husband than Nannie, he too was wealthy, being of all things an executive secretary at the Bethlehem Steel Plant. And the lovely Norah got *beaucoup de l'argent* from both wealthy husbands. She is now the rage of Monte Carlo, singing there in an exclusive nightclub. She was the original for Lasca in Van Vechten's *Nigger Heaven,* but he only caught her on the fly as it were. Should one give a faithful depiction of the lady the censors would all die of apoplexy so astounding and a-moral are her escapades. She has figured in innumerable divorce and alienation of affection suits, all with prominent and wealthy Negro men, and she has a whole slew of white suitors. While in New York she resides at the Ritz or Plaza and makes all the first nights with Carl Van Vechten!! But the story is partially ruined because Cruze wants only Negroes in it. However, do you think we can inject novel drama into the skeleton on the other page? I have in mind a woman who seems utterly ruthless once she gets going, but who is shown at the end to be entirely the slave of the most worthless person with whom she has ever come into contact. Thus getting sympathy for a rather hard character. The morons should eat that up especially in the movies, if given a deal of darky dancing, nigger comedy, and coon shouting. And I would also like to give views of three different sectors of Negro society. The upper crust, the lower Pullman porter class, and the theatrical group, centering largely around the latter, exposing the usual jealousies, lack of talent (subtly of course), and the queer types which abound in this milieu. There is as you know nothing quite so interesting as backstage among Negro actors. And our Negro strivers!! The movie of course can only cover so much of this. I intend saving Nannie or Norah rather for a novel. She's a great character.

I realize now, Bill, that I should have explained fully about those essays in the beginning. I see your point of view. It's what I asked for of course . . . an editorial criticism. And I shall watch out for a surplus of invective.

You know I have been anxious to work on *Sultan Smith* for a long time, and think we can make a much better show out of it than we could out of *Cinderella*. (And as an afterthought . . . Wouldn't *Cinderella* make a good movie story? The stage atmosphere could be brought in by having the ugly duckling, played by Miss Preer, find fame and fortune on the stage? . . .) And since I am now enthroned in the citadel of Mormonism, where all available records can be found there seems no better time than now. I shall hie myself to the library and the Mormon library and read all I can find on the subject. Then I shall make out a cast of characters and begin work.

I feel like a million dollars this morning. Spent Sunday, Monday and Tuesday in the canyon, hiking, chopping wood, fishing, roughing it in general. Slept all day Wednesday, and awoke this morning fully determined to finish *Harlem* before the day is done, stopping only to get this letter off to you.

And while in the woods and mountains I wrote scads of poetry none of which however is worth a second look. These poetic fancies of mine are of value only in as much as they make me cultivate rhythms and exercise my vocabulary. I am by no means a poet. Note:

### Stars

Earth turns
Sun sinks westward into sea
And the night brings forth
Your terse banality.

Insensible to mental probing
Lost to day, immune to night
Sparkling infinitudes
Mere architectural curly-cues
Signifying nothing.

Which last line is a good criticism of the entire poem. And in my more eloquent moments I turn out emotional effusions like this:

## On Meeting a Genius

(To Jean Toomer)

I

I was a wilderness
Prickly heat points
Hurtled through my body
Brain cells like thirsting sand-dunes
Knew no inner incandescence;
Emotion was a mirage well,
Instinct entombed, ego diffuse and sere.

II

You came.
I flung myself upon you
The lemon of your body
Merged into my blackened skin
The "lips too thin for kissing"
Searched my own, transmitted flame.
I burned. I was content.
Knew Jesus was a leper,
Knew Christ would spring from knotted roots
Beneath the rumbling earth;
Knew roots were chains
Suffused with sap . . . informed with teats
Natal wombs of super-men.
Your navel cord had pierced my umbulus:
My blood was red, then pale,
My mind a lambent milky way
Intangible and vague;
The soul of me was amber dust
My body knew no life.

There was nothing but the ego
To confound eternity.

Now I ask you! Don't fret, I'll spare you any more. But such is the result of spending three days in a canyon with a copy of *Cane* by Jean Toomer, *Thus Spake Zarathustra,* and *An Anthology of World's Poetry.*

Can it be possible that the Shuberts are living up to their reputation of being the Pirates of Broadway. Did I ever tell you the harrowing tales the little Swedish lady, Kaj Gynt, author of *Rang Tang* told Langston Hughes and me anent the Shuberts? Well it seems that they were to produce her show, but they could never come to terms. She had imported her sweet patooties, a scene designer, to do the sets. The Shuberts looked upon him and said no. She would not let the show be produced unless said S. P. did said sets, and unless Miller and Lyle were the stars. The redoubtable frères had the two black-amoors under contract and would not release them. Came an impasse. The lady grew bitter, and oh the tales she told. It seems as if the Shuberts were lecherous creatures, insatiable satyrs, who kept assignation couches in their office and before closing a deal with any actor, writer, or what not, be they male or female, must subject them to the indignities of this couch. She especially warned Langston and me against them, saying Jake, I believe, was especially fond of colored youths and gave us an example of one poor Negro chap who got into his coils and whose father came all the way from Georgia armed with a shot gun then climbed to the roof garden apartment of said lecher and ran him all the way to his yacht parked somewhere in the East River. You cannot really appreciate this unless you hear it as originally told, Swedish accent, European frankness, and blushing ellipses. The lady repeated her tale of filth to every comer, and gave them examples to fit their own particular case just as she had to us. There seemed to be no end to the Shuberts' iniquities. And thus I wonder not at their holding up our monies. I only hope you don't have to risk rape in order to collect.

Enough of this foolishness. I now turn my typewriter to better uses. I received the *Times* but no *World* as yet. Best regards to votre famille.

As always,
Wallie

Had a letter from Lyon in Paris.

———

Saturday

Dear Bill:

I announced in my letter to you the other day that I intended to finish *Harlem*. I have. The entire thing has gone to my typist. You shall have it soon as she finishes her work.

Having written or rather re-written the first half, i.e. that portion which occurs before the actual opening of the play, and having had it typed, I tore up the thing and rewrote it completely. I had, you see, elaborated on Virginia's version, and as a result had produced a long, windy, pedantic opening which by no means was interesting or compatible with the latter half. This time I condensed both versions, Virginia's and mine, having as little text as possible before the actual opening of the play where Virginia has so woven in the dialogue that it moves quickly and interestingly. My preamble has been made as short as possible. It really is all that is necessary without courting repetition.

Having finished said novel I went downtown in search of a typist. There are no colored stenographers about. Undaunted I found a public stenographer in one of the downtown offices. Smilingly I presented her with my manuscript. The lady took it not. With hostility she regarded me. And icily informed me that she was much too busy to take *any* work. Still smiling I departed, came home, packed up the manuscript, took it to the postoffice and sent it to a girlfriend of mine in Los Angeles. She is a good typist, reasonable and speedy. And so much for *Harlem*. Except . . . don't you think Virginia should get full credit? I really have done very little, and doubt if you will find much more to do. Outside of some condensing in the first quarter and the adding and cutting of individual lines throughout the rest, I have done absolutely nothing. And berate myself for not having had it done a long time ago.

I still feel fine and while awaiting your verdict on the Cruze synopsis [will] do some more work on my novel. I have also almost finished revising the sketches except the Garvey one of which I have no copy, and which I have not yet received from you. And I still spout poetry, but I will spare you any examples. They are all pretty terrible.

And my room is crowded with books on Joseph Smith and the early Mormons. I even have a Book of Mormon, confessions of one of Brigham's wives and much other juicy material, both scandalous and serious. Some emancipated Mormons I know here have aided me in gathering material, and I have gone directly to the Church library for the rest. Give me a week and I will have a cast of characters and ideas enough to begin work or at least to transmit to you so we can develop a continuity.

That is all I suppose. Have you any definite idea when the road company of *Harlem* begins its activities? Or any new news concerning *Jeremiah*? And have you perchance heard of Hemsley Winfield's Negro Art Theatre in Cherry Lane where he is presenting *Salome* with himself in the title role? I saw him essay this once before. It was immense. Just imagine our former Jimmie attired in veils, and doing the famous dance, finally stripping to a g-string. *Variety* had some amusing comment on the proceedings. And I notice that *Bamboola* closed, thirty grand in the red. *Variety* says that Cooper was associated with Blatt in the producing of *Harlem*. Who is Irving Cooper? I do not remember him.

That is all for this time. Best regards of course to the Rapp household.

> Jubilantly, for a change,
> Wallie

———

Tuesday [August 13, 1929]

Dear Bill:

Your letter was delayed somewhat, you having sent it to 309 instead of 308. I am working on the new scene. Hope to have it in the mails by tomorrow at the latest. And I am still awaiting *Harlem* from my L.A. typist.

Things are a trifle depressing here at home right now. My grandfather has just been fired without notice at the end of a vacation from his "life-time job" and he is far from dead. And my grandmother is preparing to go to the hospital to have a cataract removed from her one good eye, having been blind in the other some time. Meanwhile my mother adds to the gaiety of things by leaving her sixth husband for the fourth time in two weeks!!!!!!! Give the little girl a hand.

I feel fine physically, and only wish I was out of Salt Lake. It is damn lonesome here, there being no one here of interest. I spent the weekend on a friend's fruit farm, but did little work. It being much nicer to lounge around in the shade and watch the others. I did enjoy feeding the pigs, chickens, rabbits and ducks. And oh how I devoured freshly picked fruit.

Have some more *Smith* notes. Will send them in my next by which time I hope to have them typed. And I eagerly await the movie story. I have sent the book of essays to Macaulay, that is *Aunt Hagar's Children*. The novel is about three quarters done, with considerable re-writing in order. But I grow weary of it and have put it down for a while. It is damnable hot. And why should you have earthquakes in New York?

Friday I will be twenty seven. The years are passing much too swiftly. Should Macaulay send me an advance, which is doubtful, or should the Shuberts pay us, I shall try to leave here the end of the month. Having ridden in luxury from L.A. here, and having paid doctor bills both here and there, my reserve fund has dwindled to the point where I barely have my return fare. And with things critical here, my grandfather is planning to open up some sort of business in Elko, Nevada, god alone knows what or why, I must do a little bit toward keeping up with expenses. Of course they own their home, and have all of their winter's fuel and staple foods put by, and I happen to know they are far from being poor, but you know how elderly people cry wolf, being afraid that they might go broke in their extreme old age. Thus I am at the mercy of the Shuberts, for while I can stay here contentedly this month so long as I keep busy working, it will be a hard job to remain away from N.Y. much longer. Do you realize it has been five months since I left Grand Central Station?

Am writing to Miss Fishbein's agents, telling them the whole story. Be sure to let me know how *Halleluliah* is received in N.Y. I believe it to be a superb picture, unless it has been ruined in cutting, I saw eighteen reels, and parts of it are far superior to the average movie.

That's all for the present. Look for Cordelia's demise anytime after this arrives. My best regards to Virginia and Harl.

> As always,
> Wallie

The *Bohemian*? And Leonard's article?

———

> Dear Bill:

Herein is my first contribution to the beginnings of *Sultan Smith*. Have immersed myself in Mormon history. And am raring to go. Can you send any information you might want. Just ask questions.

Also enclosed is the final word from M.G.M. The high mogul of the story department has finally said his piece.

Just had a letter from Miss Fishbein, asking me to send her agents Hollywood's information about to whom the play has been submitted, etc. I don't know what to answer her not having heard from you. I have had final "noes" from M.G.M., Pathé, Fox and

James Cruze. They are welcome, if you say so, to go further. But what of Blatt's plan? I'll send them a copy if you say so, but I have little hopes for them unless they can interest some of the independents from whom we would not get much of a price.

*Harlem* has not yet returned from L.A. Will shoot it to you soon as it arrives. Finally finished revising manuscript and sent *Aunt Hagar's Children* to Macaulay. As a consequence of my recent flurry of work I am a little worn out. Intend to rest a few days, then begin again.

Received sketches, *Times* and *World*. Many, many thanks.

Curses on the Shuberts and that is all I suppose until I hear from you,

> Best regards,
> Wallie

————

Sunday midnight [August 1929]

Dear Bill:

I am rather in the dumps tonight so I write to you as an antidote, not that I have anything to say but because in writing you I can well imagine how you would bustle me back into good humor and thus effect a vicarious cure. So be it.

Things are rather sad around here. My grandmother refused to remain in the hospital any longer and came home today. Her eyes are still bandaged and it is still doubtful if she will ever be able to see again. She feels quite pessimistic about the matter, and it is hard for me to see the one person in the family who seemed immune from neuroticism finally giving in. She even talks wildly about committing suicide should her sight really be gone. My mother has moved in to look after her. I am of course of no use whatsoever, having gone to the hospital to visit her once and becoming so nauseated that I almost had to remain as a patient. The doctor is vague but hopeful. If we can keep her spirits up I think she will pull through. I hope so. I could never bear seeing her blind. Thus I stand poised ready to take flight on a moment's notice. I only await to hear from Furman. If he will advance me some money so that I can eat and sleep once in New York, I'm on my way.

The line between tragedy and comedy is both variable and thin n'est-ce pas? A humorous thing occurred today. Naturally having heard that my grandmother was home all her brothers and sisters in God as manifested in the Baptist Church flocked in. Even the minister, a typical oleaginous and pompous illiterate. Of course he had to pray and being an apt inflamer of the emotions reached eloquent heights, so much so that he infected my grandmother especially when he asked mercy for the blind and afflicted. The dolt. Well I exploded and interrupted the dear brother's prayer. Everyone was aghast but the suave man of God. He tried to squelch me. I ordered him out of the house and, while my mother hastened my grandmother out of the room, told him some profane facts. Had it not been for the presence of the good sisters he no doubt would have knocked me down; anyhow, I got him out, and silenced the sisters when they protested.

My ostracization among polite colored circles in Salt Lake will now be complete. But I have dared that praying hypocrite to return to this house until my grandmother is more emotionally and physically stable. She of course is quite peeved at me, but being peeved at me has taken her mind off herself. So be it.

Before my grandfather went away he fixed up the screen porch on the rear of the house for me. I have my bed, my books and all out in the open. Hence I can type without disturbing anyone all night if I wish and I am sleeping and writing in the open. I feel like a million dollar McFadden disciple.

I finished the first draft of my novel about five A.M., yesterday morning. Writing it has been an adventure. I stood as one apart and watched it issuing forth from Wallace Thurman. It is the first thing I have ever let write itself, playing amanuensis to some inner urge. This sounds like hooey but it's true. Parts of it I shall not have to retouch. They are surprising even to the author. Other portions need lots of work. They shall have it. To have done this much gives me more inspiration to continue writing than anything else. I only hope I haven't mesmerized myself.

No letter this week at all from New York. But I have F.P.A. and the rest of *The World*. And some kind anonymous soul mails me bundles of Winchell and *The New Yorker*. And my newsdealer saves me *Variety*. By the way how goes Mr. Blatt's *Murder in the Subway*? Expect a wire announcing my arrival at any time. All I want is protection against starvation and from doing a Bruce by having an address "In the gutter in front of the Provincetown Theatre."

> Regards to all,
> Wallie

————

Dear Bill:

Before I had a chance to mail the foregoing lugubrious notes and lamentations, something did happen to arouse me from my lethargic state. You could never guess what it was. A meeting and a hundred mile train journey with my erstwhile spouse. Catch your breath. Came a telegram addressed to my grandmother announcing that she (Louise) had been called east because her mother was seriously ill. She was to pass through Ogden at noon that day. Odgen is some thirty six miles away. No one could go from here except me. I went. Was greeted by tears and a request to ride up the road aways at her expense. Which of course I did. Nothing of significance happened. I was impersonal and wary. The thought of all my money having been wasted and me here in exile dying to return to New York was sufficient to hold my emotions in check. I don't know what is to be done now. She of course loses the money paid to the Reno lawyer so she said, and she also said that she did not expect to be able to return as she would have to go to work and take care of her mother now down with cancer of the stomach. What is more she suggested that I might lend her some money or get some for her. Which I considered a rank display of crust. Howsomever that's what happened. What does one do now, having staved off a reconciliation which was angled for most subtly? If I had the money

I would put in for the divorce right here in Salt Lake and have no trouble getting it at all. And god knows I wish it could be over.

I went as far as Green River, Wyoming, with her and remained there five hours waiting for a return train. It was some town. A typical wide-open western town, with saloons, gambling halls, dance halls, and bordellos running at full sway. It is a railroad division point town, populated largely by Greeks and Mexicans with a smattering of whites and Negroes, of the latter being the underworld czar of the town. His name is Cat-eye. And his joint is a joy to behold. Whores of all races, white, black, mulatto, Mexican, and what have you. A bar almost a half block long. A dance hall. A hotel (?). Poker, dice and black-jack, roulette, and fan-tan. I spent a glorious five hours exploring and almost missed my train. I arrived back in Salt Lake full of pep and set to work.

I am also mailing you, for better or for worse, *Harlem*. The typist was delayed because in the interim she had a baby. Her bill, for *Harlem* not the baby, is fifteen dollars, which, I don't happen to know, may be reasonable or expensive. Her name is LaVelle Jones, if you wish to make the check out to her. Having received this typed copy I am amazed at its shortness. Mayhap, I expect, to be ordered by Czar Rapp to write about twenty thousand more words. God forbid. But it is too short. In condensing the first half so that it would not drag or suffer by comparison with the last, taken wholly from the play, I have only 82 typewritten sheets. I await your verdict, keeping the original and the carbon, ready to set to immediately upon receiving a suggestion from you as to how it can be lengthened if need be. Of course it could be padded, the book I mean, with pictures. It's too racy now to pad with words, but of course new episodes might be added. Let me know about this immediately so we can get it off our hands. I am heartily sick of *Harlem* and all things Harlemese. Which does not mean however that I am not willing to do my damnedest to make this latest offspring an acceptable reality.

*The Bookman* arrived this morning. Many thanks. You're an angel. And there seems to be nothing else to say. It's raining like hell. I am all alone in the house. My grandfather having gone to Nevada, and my grandmother having gone to the hospital. I, the dog, and three canaries, eke out a lonely existence. It seems strange not to have had a letter from you for so long a time. But I know you are busy, and keep my peace.

 Best regards,
 Wallie

————

Dear Bill:

Thanks for the clippings which came this morning. And the novel was mailed yesterday. I am indeed sorry that I have delayed this so long, only to find when completely typed that it is still inadequate. I can however do any necessary adding at a word from you. It won't take me any time to lengthen any portion you believe can stand it. I too hope we can have it out by the time the road show goes out.

Hurrah for *Halleluliah*. Evidently they did not ruin it entirely in the cutting room. I had certainly expected them to change the ending, for there were many murmurs against it.

Pollyanna as it is, they wanted the girl to reform and become a missionary along with Zeke. Don't you like Nina Mae? Wouldn't she make a good movie Cordelia?

I am resting content, doing a great deal of work on my novel. I hope to have the first draft completed by the end of next week. There will be months of rewriting, but the groundwork will be laid, and revision can go on unhampered by minor worries about story, character development, etc.

By the second week in September I hope to leave here. My grandmother will be out of the hospital about the first. I will stay one week with her and start towards Gotham. I still have railroad fare, and hope to keep it. After that . . . maybe the Shuberts' treasurer will have returned. I still am wondering what can be done about my separation and divorce agreement. Feeling fine. Will be completely up to par when I do get to N.Y. Otherwise I would isolate myself, despite the misery, for another month.

> Love to the Rapp cohorts,
> From,
> Wallie

––––––––

Wednesday [c. 1930]

Bill:

I was in hopes I'd have a chance to talk to you last night privately about the things I now write here.

As you say I do have the "darnedest" luck and continual doses of it have just about broken my spirit of resistance.

I do hate having to keep worrying you with my troubles, for I know you have enough of your own but I guess it cannot be helped.

The reason I wanted you to speak to Phillips was because I am under the impression they refused me merely because it was me. As much as I try to think otherwise I'm certain had not the question of color come up, the advance would have been forthcoming. Lawrence had pestered them to death for money and I believe they thought—and by they I mean the auditor, Coburn—that I too wanted to bleed them just because I thought it possible. I am certain he still thinks I wanted the money to buy gin with and shoot craps. It was he who vetoed the thing and who said there was no use of my speaking to Lewis as he had done so already and Lewis had said no.

Now of course after reflection I am glad you didn't speak to Phillips about it. It wasn't his fault and there is really no sense in my becoming as inveterate a pest as was Lawrence.

But something has to be done. What you can afford to give me won't keep me eating regularly. That is part of my trouble now—too poor and too little food. My system can't keep up. I can't keep up on one meal a day. At present I owe two weeks room rent. Today I had to have an x-ray taken of my foot. Tonight the doctor is coming to lance my throat again. All of which means money—if not for the doctor fee—at least for food, medicine and the x-ray.

I have been to every person I could think of who might lend me money and to no avail—the people I know are for the most part broke too. The folks at home are having too much trouble of their own. So I'm stumped. Something must be done or I'll be done for. So if any solution presents itself to you I don't care what the future obligations will be for me. But I can't go on like this being harassed for money and half hungry. I know you personally are doing your darnedest, I feel like a crook bleeding you as is, and if you can arrange with someone to loan me a small sum even in weekly installments it'll help things immeasurably. I didn't mean to be so long winded and I hope this S.O.S. is not as distressful as it sounds.—

Wallie

————

Saturday [c. August 1934]

Dear Bill:

I was particularly anxious to see you last week. Some business I wanted you to take care of. It turned out all right in case you get a belated telephone call or something relayed from me. One of the doctors was kind enough to give me a lift.

Sorry, I didn't get a chance to talk to you the day you came out. But now that I'm writing, I'll put down the trivia I had in my mind.

Not having anyone in the world to notify if anything went wrong, I took the liberty of giving your name and I wish you would make it official (unless you object) by dropping a note to the office here saying you are to be notified in case of serious developments or death. You needn't mention that I inspired the note. It's merely a precaution. If I should die, I don't care to be rushed off to the autopsy room, then Potter's Field without someone knowing so that the folks at home can be notified. I'm keeping them in the dark about my illness. They have enough of their own. But—well—I think you understand. Please don't think I'm planning to die or something. Far from it. I merely wish to be prepared for a probable emergency.

One other request and I'll leave you alone. I know you're busy. I also know it isn't good policy for you to rush over here too often with a young baby at home. That's what Furman says, anyway. But it so happens you can do me a great favor if you will. All of my Harlem friends have evidently already consigned me to the grave so conspicuous are they by their absence. (Which however is no surprise to me.)

Now the favor: please get me the August issues of *The Mercury, Vanity Fair, Esquire, Story* and mail them over to me. I'll repay you shortly when I finish transferring some money I have, so I can get my hands on it.

I feel pretty good today but was a sick child for awhile. The treatment is worse than the disease as far as I am concerned. But the doctors are swell to me and doing all they can to pull me through. It was lucky for me Al Shay made me come out here and luckier you advised me to stay.

In addition to the new magazines I asked for, I'd appreciate any old periodicals, if any, you have lying around and don't want. All I can do is read and with no visitors it's hard to get reading matter. Best regards to the family and to Doc.

Wallie—

## Letters to W.E.B. Du Bois, Claude McKay, Alain Locke, and Granville Hicks

March 8, 1926
W. E. Burghardt Du Bois
*The Crisis Magazine*

Dr. Du Bois:

I have heard that you have or either will have a vacant position on your editorial staff, and I am writing to you for the purpose of obtaining a personal interview in regards to the same if such be the case.

I received my education at the University of Utah and the University of Southern California. I have worked as associate editor of *The Pacific Defender* in Los Angeles, and edited *The Outlet,* the first western Negro literary magazine. Since coming to New York to continue my development as a writer I have edited *The Looking Glass* during its lifetime, and am now connected with *The Messenger.*

I am very anxious however to connect myself with some organ where my work will be of a more literary nature since that happens to be my primary interest. I would appreciate an interview and consideration if you have an opening.

Very truly yours
H. Wallace Thurman

————

February 3, 1928

Dear Claude McKay:

You will no doubt wonder what prompted me to write to you. Just this . . . Harold Jackman let me read your last two letters to him. In one you warmed my heart by saying you enjoyed my contributions to *Fire!!* Like most of the writer craft I am vain, therefore anyone who admires my work is a potential spirit mate. But more than that I am really anxious to correspond with you. We might have something to tell one another, something to communicate which may prove good and inspirational.

Just now I can only talk about myself, a sensible procedure I believe in this instance since you undoubtedly know so little about me. I am a 25 year old, American born Negro. Salt Lake City, Utah knew me first. I came to New York two and one half years ago from Los Angeles, California to start on my career as a writer. Meanwhile I had lived in Chicago, and Omaha, attended the University of Utah and the University of Southern California, edited and published a magazine in Los Angeles, known as *The Outlet,* which used up all the money I had saved and could save for several years and convinced me that I must make a hasty hegira to Harlem, which at that time was just coming of age. I came to Harlem knowing one individual. Since then I have struggled and starved and had a hell of a good time generally. I have become a poet, because I once wrote a poem which gained honorable mention in an *Opportunity* contest and was used by the honorable Mr. Braithwaite in one of his anthologies. I have also become a critic, because I wrote two articles, one for *The New Republic* and one for *The Independent,* in which I tried sincerely to debunk this Negro literary renaissance . . . you can imagine how popular I am among the Negro literati and since I also attacked the Negro bourgeoisie in *The New Republic* you can also imagine how popular I am among that sect of Aframericans. I have become an actor too, because I was a member of the mob in the Theatre Guild's production of *Porgy,* until just a few weeks ago. I have become a novelist also, because I am about to have one published by Boni-Liveright. But most important of all and probably the most thrilling, at least it is to me, I have become a playwright, really, having just had my first play, done in collaboration with a white friend of mine, accepted for Broadway production by Crosby Gaige, one of the biggest of the producers. Outside of that I have done very little since coming to Harlem two and one half years ago, except learn many variations on old vices, increase my capacity for holding synthetic gin and become more and more a pariah among my own people, although the publicity attendant upon being a successful (?) playwright has tended to make them look upon me with more favor than hitherto. The rest you can read in my autobiography, ready for fall publication, 1930.

By now you must either think me a highly conceited ass or just a plain damn fool. I assure you I am neither. In fact I am really something much worse which you will discover later. I am anxiously awaiting *Home to Harlem,* my critical claws ache to get into it. I have already had a chance at your poetry in an article on Negro poetry due to be published in *The Bookman* soon. Will see that you get a copy. Now I think that I have written about enough for the first time. Let me hear from you if you feel the urge to write. I shall be glad to answer and tell you whatever scandal or news (are not the two synonymous in the U.S.?) comes to hand.

Pour le present, je suis fini, (Oh yes I took French in high school)

Sincerely,
Wallace Thurman

———

October 4, 1928
*Harlem: A Forum of Negro Life*

Dear Claude:

Things do happen. Since I last heard from you several things have happened to me. First, I got married, which action I, who never and still do not believe in marriage for an artist of any type, will not try to explain. It's just one of those unexplainable things that happens even to the best of us. My only point of extenuation is that I happen to have married a very intelligent woman who has her own career and who also does not believe in marriage and who is as anxious as I am to avoid the conventional pitfalls into which most marriages throw one. I assure you ours is a most modern experiment, a reflection of our own rather curious personalities.

Next my new editorial responsibilities. Talk about your answers to a maiden's prayer . . . This is it. Here ever since *Fire!!* I have been praying and wishing for a chance to edit another magazine without having to worry about financial backing. And along it comes at the least expected moment. Good backing and carte blanche to do anything I choose with the editorial end without interference from the backers or the business department.

I am going to try to make *Harlem* a distinctive magazine. It will cater to no particular class or group nor will its contributors be entirely segregated. Its pages are open to anyone who has something to say as long as they say it with some degree of literary merit.

I am going to feature fiction, poetry, essays, literary and economic studies, debates on racial and non-racial issues, articles on current events, and special departments on the theater and current literature. How does that sound to you? And what is so encouraging is that the publishers have promised me that if I put the first three issues over with any success whatsoever that beginning with the January issues I can begin paying my contributors for their material. What could be sweeter?

Of course I want you to send me some material, anything you have on hand or might care to prepare. And the sooner you send it the more I will appreciate it and the more fervent my prayers will be for you at night while I kneel on my bare floor after my midnight gin rickey to have my daily conference with old Jehovah.

My novel, *The Blacker the Berry*, is scheduled to appear this winter. Announcements will be made soon. I saw an announcement of your new novel, *Banjo*. I envy you that title.

Did you happen to see the July *Bookman*. I had an article in there on Negro poetry and poets from Phillis Wheatley down to you, Countee and Langston and all the rest. I thought it was quite a well-balanced and fair piece of work, but I have had nothing but brick bats thrown at me since it appeared. Especially what I said about you has aroused much antagonism. I contended that you had more mental depth and more emotional depth than any of your contemporaries and that you were the only truly revolutionary poet of the lot, which fact is obvious to anyone who can read, and despite some other things which I have said about you which were not quite so complimentary, I have been accused of being your press agent. Can you beat it? Countee was so upset about the whole matter that he wrote a special paragraph to be included in the *Dark Tower* "laying me low." Langston alone, the little angel, took my criticism of him with equanimity and practically agreed with my estimate of him. Such is the life of one who tries to be critically truthful and bring some order into a literary stampede in which some who should survive are about to be exterminated with those who should have never been born.

I have already started the study of Du Bois which you mentioned in your last letter. In fact I had collected the material last summer. I plan to do a series including him, Garvey, Weldon Johnson, Booker T., Moton, Madame Walker, Frederick Douglass, et cetera. It ought to make an interesting collection especially when written by whom Du Bois calls, the young upstart.

There is not much more news. I am head over heels preparing the first issue of *Harlem* which will appear during the first week of November. I wish I had been able to let you know sooner so that you could have had something in the first issue. It hardly seems possible now since I must go to press by the 20th, but if you can get something here before that time I wish you would, or at least send me something, anything, as soon as possible. Don't wait too long to write.

Sincerely,
Wallace

———

October 3, 1928

Dear Dr. Locke:

I was indeed sorry that I did not know you were in town on last Sunday or I certainly would have looked you up or at least have been present at the Douglases when you were there. I hope that by now you have fully recovered from a hectic time in Europe and the startling news of my marriage to Louise on your return. It was not fair to shock you like that, was it?

About *Harlem*. I suppose Aaron explained about the venture. I am mighty glad of the chance to be able to edit a magazine and let someone else worry about the financial end, in fact, after *Fire!!*, that is the only way I would ever venture forth again.

I know you are very busy but that does not keep me from asking you to send me a short article of some kind for our first issue due out during the first week in November. That means that all material must be in our hands no later than the 15th or 20th of October, but surely you can dash off a little something for us by then, on any subject you choose.

*Harlem* is to be a general magazine, containing verse, fiction, essays, articles on current events and debates on racial and non racial issues. We are not confining ourselves to any group either of age or race. I think that is best. *The Crisis* and *The Messenger* are dead. *Opportunity* is dying. Voilà here comes *Harlem,* independent, fearless and general, trying to appeal to all.

Louise and I have an apartment at 90 Edgecombe, Apt. 56. Our phone number is Edgecombe 1979. I hope you call us and visit us when you come to N.Y. Also let me hear from you as immediately as possible about my request and what you think in general of the magazine.

No doubt you heard that my novel is due out in a couple of months, and that I am expecting my play to go into rehearsal almost any day again. And that I have another play going the rounds, and a book of plays scheduled for late winter. Aside from that I have been quite idle.

Awaiting an early reply,
With thanks,
Wallace Thurman

————

January 30, 1932

Dear Granville Hicks:

I was delighted to hear from you again. It's years, actually, since I have had a note from you, and I hesitate to count the number of years since we met. I have been an interested follower of your magazine and literary supplement pieces. And I really did want your frank and honest opinion of *The Infants*.

I agree with you. It is a most disappointing novel. I didn't realize what a job I had undertaken until the first draft had been accepted by the publisher, and I had pocketed the advance. I tried my darnedest to back out of the deal. I wanted another year to work on the book. But I had spent the five hundred dollars. I was broke. The publisher wanted his manuscript. So there you are.

I am not devoid of blame of course. A little more concentrated effort and even in the short time allotted for revision, I could have done a much better job. I was too close to my material. I didn't completely realize its possibilities, nor how little I had contributed to reportorial observation. But now we have it. There is nothing for me to do, but work like hell on my next, and complete it to my satisfaction if it takes fifteen years. No more will I try to meet a publisher's schedule. Which is a lie, as I am doing the same thing right this minute.

I have been commissioned to novelize a play of mine. I don't know what kind of novel it is going to be. It concerns itself with the experiences of a sensitive young medical graduate who enters a city hospital to intern for a year. I don't know how much you know about city operated hospitals. But conditions are appalling. And the interne who leaves college filled with ideals and ethics has a rude awakening. This is my first attempt to do something which does not concern Negroes. It may be my last.

But I have what I think is a tremendous idea for another Negro novel.

Since I saw you last, I have had one play, *Harlem*, produced. I have four others finished. One has been held by every producer in New York at one time or the other in the past four years. It is now scheduled for fall production. Another was bought by Belasco, but he died before he could give it production and no other producer seems impressed. I was commissioned to do a third by Chester Erskin, but he has been unable to produce it, as his backers have played him false. The hospital play which I am novelizing may be bought any minute by Wm. A. Brady. He has been considering it for two weeks. One day he says he will give it an immediate production, but when my agent shows up with the contracts, Mr. Brady is either drunk, or has some magnificent excuse to delay longer. You really should have some experiences with Broadway producers. I can only marvel.

I didn't mean to ramble on so. But I was glad to hear from you, even if you didn't like *The Infants*.

Best regards,
Wallace Thurman

## Letters to Harold Jackman

Thursday [c. summer 1930]

Dear Harold:

Here is the expected clipping. Does this go into the library archives? Or is it for your own private collection? I needn't state my preference. It really is a good review.

I have gone stale. Cannot write a word. And this with the Belasco play already overdue and over three quarters finished. Also with about twenty thousand words done on the "new" (I must stress that newness) novel. More I have some short stories all started and in a first draft, but evidently my brain and emotions are rebelling from too consistent labor. I bide my time, impatiently I admit, for I am rather in the habit of working now, and spend my time reading. Such things as *Point Counter Point, The Counterfeiters, A School for Wives* which I do not like, *Death of Venice, Madame Bovary, The Idiot,* which just about cover my favorite novels of all times. All I need to do now is read *Upstream, The Case of Mr. Crump, Winesburg, Ohio, Jurgen,* your long awaited *Magic Mountain, Of Human Bondage,* and *Penguin Island.* That is Thurman's course in novel reading.

And I have a new occupation. One which intrigues me too much to last. I have taken over the cultural training of Lewis' son. I adore the youngster. Everyone claims he is mine because he resembles me quite closely, and he is the most pleasingly precocious child I have ever known. Yesterday I went down town and picked out a grand piano for Lewis to buy. It is to be delivered tomorrow. Which means that I shall once more go in for piano in a big way and begin xx (damn this typewriter) Theodore on his elementary scales and fingering. Imagine me such a paternal elegant. It's a role I am enjoying heartily. Perhaps, after all, these past four hectic years in Harlem have not left me in a rut. I may still have my former capacity for experiencing sea changes. In which case, there is no telling what might happen next. Liquor I will always like, but my motto now is civilized tippling. Jeanette and I went to Bunnie's for dinner the other night when I was in town. The four of us, his sister made the fourth, had a glorious time, all out of three drinks each, my first in three months. I experienced that glorious mental stimulation, which liquor alone can provide, without going native, and it appealed to me so much, that except for rare occasions I never expect to gorge myself again.

As for other pursuits which used to occupy a great deal of my time, I have more or less outgrown them. And most of the people I know and used to enjoy bore me so that I could be most happy never to see them again in life. And that goes for all my former associates except, of course, Jeannette, Bunnie, and yourself. I visited Helene the night Georgia arrived from Denver and had to leave hurriedly because I became, frankly, sick of my stomach. Thus does Wallace Thurman enter into the vale of middle age. Please don't imagine I am claiming any great reformation or that I am saying that this state of mind is going to be permanent, but I have always gone in for things until I exhausted myself then dropped them. And this may be true of the erotic bohemian life I have been leading in Harlem. As far as I know I shall remain in Jamaica until I can go to Europe or some other fresh place. It may be that I shall be working this winter. In which case I may have to move back to town to escape the rigors of commuting. But, if I happen to have no regular job, in Jamaica I hope to stay. And that's that.

I listened in to the Stadium concert last night. It was seventy five percent Gershwin. "The Rhapsody," "The Concerto," "The American in Paris." George Jean Nathan once said that Gershwin had heard only one Negro song . . . "The Memphis Blues." I agree. "The Concerto" is spotty and echoes in the "Rhapsody." It shows a great lack of imagination. And "An American in Paris" is impossible. At least to me. Only the "Rhapsody" stands out as being something to cheer about, and there are amateurish areas throughout it. However it is a brave adventure which more than compensates for its occasional lapses. I have in mind a magnum opus for the theater. A Negro thing of pantomime, singing, dancing, and sheer music which will use the spirit and form of Gershwin's "Rhapsody." More of that anon, when I have mastered all theatrical technique. Then, and then only, will I cast it out the window. I have learned that much from writing a novel. One cannot very well experiment until one has a conventional foundation and rigid training. I have tried to make the novel elastic without having first learned the boundary lines so that I could steer a clear course. The results you know. Sheer chaos.

And since I sat down to write you a short note, and have rambled this far, I shall try once more to finish Act Three of *Goodbye New York*. Maybe I'm inspired again. I'm sorry you haven't kept your word and been out to visit me. I won't scold, nor say some of the sarcastic things my fingers are aching to spell. I'll say, instead, so long,

    Wallie

Will be in N.Y. Saturday. Luncheon with Chet and some theatrical moguls. Dinner and theater with Bunnie, his girl friend, and his sister. So I doubt if I will get to Harlem. If I do, I shall run in to say hello but the chances are slim. The subway ride is intolerable to me—now that I'm a child of the open spaces—and I really have no magnet to draw me Harlemward unless we could have an hour or so together. And my schedule hardly permits of that—to say nothing of yours—

    Wallie

————

Midnight: Thursday

Dear Harold:

*Parties* arrived and has been read. I searched its pages for a note from you. There was none. Or perhaps it remained invisible. I did not like Carl's latest effusion. Frankly I believe it to be his worst novel to date. All of his limitations are relentlessly exposed. His superficial xxxxxxxxx ity (pardon the hiatus) in this novel does not serve its purpose as it did in *The Blind Boy* or in *Firecrackers*. David was the only person in the novel I liked or whom I could visualize. The others . . . save perhaps for Roy and Donald were much too fantastic or shadowy. The Graf just didn't click and she needed to. There is much more I could say, to make myself more clear but I am tired tonight and words won't come.

Have you perchance read Claire Sheridan's *Naked Truth*? If you haven't do so immediately you can obtain a copy. Claire Sheridan you may know is an English sculptress, a rather sapient, witty, and frankly mad egoist, who has doubled in journalism. It has been said that her sculpture too was journalistic. I can well believe it. But she is a person, or at least in her book she makes herself a person I adore. Her madness has not the ethereal quality of Shelley's nor is it a syphilitic madness like de Maupassants. She is Byronic if anything. Her escapades with Lenin, Trotsky, Charlie Chaplin, Mustapha Kemal, and countless other outstanding individuals of this generation will both amuse and educate you. And what Mussolini tried to do to her, according to her story, is too delicious to miss. Imagine a man like *Il Duce* attempting rape on a body.

I am very busy . . . too busy. Writing, practicing the piano, teaching Theodore, reading. I have read I suppose about ten books in the past week. Hast read *Revelation*? It's a new novel by a Frenchman, and concerns a mother who has adored her son, and who finds on his death that he was a homo-sexual. The thing is so deftly handled that gaucheries of style are forgotten, I mean prose style. It irritated me. Others have seemed to like it, but to me it belonged what I call "female" writing. The sort of thing most women writers are guilty of, and which more and more men adopt. Someday when I am less tired I will explain that too.

And now I must go to bed. I have been up since seven this morning, and have not had an idle moment. I have even been to town twice, that is downtown Jamaica, which is a good mile walk, which means I probably walked four miles. Labor Day I sprained my ankle playing tennis, today I sprained my wrist for the second time playing the piano. Sprain is rather strong used here. Anyway I am uncomfortable. What matter terminology?

Adios, Salutis, salute,
Wallie

Is there anything at all you can do about *Magic Mountain*? This is a *cri de coeur*—

W

## Letters to Dorothy West

1538 Fifth Street
Santa Monica, Cal.
June 5, 1929
Miss Dorothy West
470 Brookline Ave.
Boston, Mass.

Dear Dot:

Both your letter and one from your mother stayed at the Hotel Somerville I don't know how long before they were forwarded here to my beach retreat. You know how prompt Negro hotels are in such matters.

About your stories. The reader at Macaulay's liked the two she read. She agreed with me that there was not enough for a volume, and that the second story was in need of more work. It reads as if written rather hastily and although good needs polishing. Send me number three if you wish. I will shoot the whole thing back to them. Should they accept, all is well. If not, *Harper's* is the next best bet. I will guarantee you an immediate answer from Macaulay's. I have the other stories with me. I await the third. There is nothing more definite that I can say. The polishing can come later. The material is there, for any publisher to judge, and I can vouch for your ability to put on the finishing touches.

Glad you are seeing the world and living so fully. That's what one needs. I have lived too fully. Consequently, I am now living the life of a monk, isolated on the beach, reading, writing, sleeping, and dipping in the Pacific. It's ideal. Tell Bruce his letter too just reached me yesterday and that I must say "Impossible." He'll know what I mean. Will write him fully later. Will be most glad to see you. Hope to be in New York by September. When do you return? Helene must not believe in writing. She didn't answer me. When you answer this write to 308 East 9th South St., Salt Lake City, Utah. Mail can always reach me from there and I know not how long I will be here. Much love,

Wallie

————

Midnight Thursday [c. August 1929]

My dear Dot:

Your letter came this afternoon. I answer during my first free moment. My grandmother is in the hospital having had a cataract removed to save the sight of her one good eye. My grandfather has migrated to Nevada after having lost his life time (?) job here in Salt Lake. Hence I am the official housekeeper. I feed the canaries, superintend their baths, feed the dog, and switch him temperately when he nips the heels of passersby, vacuum the rugs, prepare my own breakfast (dine in a restaurant), and today even cut the lawn as well as watered. It, the last item, was indeed a novel experience. It being my first attempt, the lawn's appearance eloquently incriminates me as an amateur.

My dear Dot I do not want you to become promiscuous nor to sacrifice your virginity purely because I ventured the opinion that it seemed to me your stories lacked passion and that your virginal state might be in some vague way responsible. I may, you know, be all wrong in both my opinions. Were I more certain I would be more rash with my prescriptions for immediate treatment. For if I was really assured that your limitation was the direct cause of being virginal I would see that you lost the status of a maiden even if I had to officiate at the ceremony.

But I grow facetious. What I do want you to do is not be like Helene, and for your intellect's sake get rid of the puritan notion that to have casual sexual intercourse is a sin. It's a biological necessity my dear. More tragedies result from girls clinging to their virginity than you would imagine. Physically, mentally it is bad after a certain age. Celibacy is certainly admirable under certain conditions and at certain times, but sex is after all but an expression of bodily hunger and must be appeased like the hunger of the stomach. Not immoderately of course, for gluttony is always harmful to one's physical and mental organs. But when one is hungry one should eat, and an 18 year fast may bring about chemical disturbances as the 18 day diet brings on acidosis. I don't say just saunter forth and give yourself to the first taker. I only say don't repress yourself, nor violently suppress your sex urge, just because you are Puritan enough to believe that hell fire awaits he who takes a bite of the apple, unless you are profoundly inoculated with the illusion of love. You might be too sensible to entertain that illusion, but you will certainly be subject to chemical affinity, and when the call comes do not wait to decide if this is *the* man. *The* man may never materialize or else be halted in his rush to you. And unplucked fruit soon loses its fragrance and rots. Be discreet but be adventurous is a good motto for the literary tribe.

And about Helene? I wish you would tell me more. This is not idle curiosity. But I still do not know how she hurt her leg, and I do feel partially responsible for all that may have happened to her since I urged her into Harlem and wrote her a rather silly letter after leaving her in Detroit.

About Simone. I cannot guess who she is positively, nor do I believe she is specifically any one person. I saw in her a composite of Cheryl and Georgette Harvey (God forgive me this patent disparity, but she seemed to have la belle Harvey's power and force of will and Cheryl's aesthetic pretensions). Am I by any chance lukewarm? And did I not sense something of Edna Thomas in Bersis. And Crown's Bess in Jessica. And is it possible on recapitulation to say that Simone is all Cheryl?

After Macaulay's let's try *Harper's* and all the rest. Somebody has to be congenial. I have written most eloquently to Mr. Furman of Macaulay's damning the easily published and untalented Fausets and Larsens and pleading for recognition of potential talent from a newer and renovated generation.

Do write me again fulsomely and immediately. It's damn lonesome here and I do enjoy hearing from those whom I isolate as particularly beloved companions. The talkie lies dormant half finished. I'm not imbecilic enough yet for the movies, but the mood may come. And meanwhile my novel progresses apace.

Most affectionately,
Wallie

Love to Helene.

————

308 E. 9th South St.
Salt Lake City, Utah
Friday [August 30, 1929]
Miss Dorothy West
Oak Bluff, Mass.

Dear Dot and Helene:

It is measly writing you a single note for two, but I have been quite ill for the past month, and am now in no mood to write or even think consecutive or coherent thoughts.

I am worried about you Helene. The leg? How on earth did it happen? Is that all Harlem could do for you?

And Dot, so cosmopolitan now. A traveled lady planning to spend the winter in Paris!! The idea of a colony intrigues me although such things generally turn out to be stupid unless colorful personality abounds, and then it grows tedious and unproductive. However should I be able I too shall flee to Europe this fall.

I await the story. I am all eager to see it. Shoot it to me.

And Helene? What new poems are written or conceived? I would enjoy receiving some copies of recent effusions.

Have finished a book of essays entitled: *Aunt Hagar's Children*. It contains essays on Negro writers, Du Bois, Garvey, Washington, Douglass, and divers other subjects. And my new movie is progressing nicely if slowly, and I am writing a talking movie. Please write to me both of you. I promise to write individual letters next time. Much love,

As always,
Wallie

————

Rake's End
Tuesday, September 12, 1934
Miss Dorothy West
243 W. 145th St., NYC

My Dear Dorothy:

Your letter made me a trifle blue at first then amused me when I realized that despite your belated knowledge of life in the raw, you were as naïve as ever.

Do you think I'm taking a rest cure in a country club? Write you an article? The doctor should see your request. I happen to be sick, Dorothy, flat on my back in bed, and ten days ago my chances of living were quite slim. I suddenly took a turn for the better after every T.B. expert in N.Y. (it seemed to me) had been called in for consultation.

Writing is out. In the first place I have hardly enough energy to sustain normal physical functions. Even this letter must be short and written in installments. The nurse is timing me diligently. So after that outburst, I'll rest awhile.

*Two Hours Later.* Nice of you to send me *Challenge*. Frankly, I was not impressed by the contents. I do not mean to be a dog in the manger decrying that which I failed to assist. (But even then I was ill, very ill, and just living on nerve and whiskey until I was ready to flop in a hospital. You see I knew this was coming, knew when it started.) But I don't wonder at your being afraid of competition. To date you've expressed only the dead spirit of seven or eight years ago, ignored the pulsations of a new day. As I have said before—the magazine lacks significance. (Damn that nurse. See you later.)

*Again.* It lacks personality. There is no gusto, no verve. It is too "pink tea" and la-de-da—too high schoolish (this last applies to most of the material).

I hope I'm not being too cruel. But there is a field for a magazine called *Challenge*—if it lives up to its name.

I accept your explanations, but it is mighty dreary lying ill in a gruesome environment and seldom seeing a familiar face. Without Helene and Georgia I should have gone mad. As well as gone wanting for the little things only your friends can do for you.

I must stop. Drop me an occasional note at least—

> Regards to Mildred.
> Wallie

# Part Four

# Literary Essays and Reviews

One of the primary aims of this *Reader* is to establish Thurman as a key figure from the 1920s and the 1930s in the continuing debates about black aesthetics and the role race plays in the work of U.S. writers of all backgrounds. Thurman's essays and reviews, here gathered in one place for the first time, confirm his already-known assertions for artistic independence, his rejection of black bourgeois notions of respectability, and his interrogation of the double standards that often marked the commentary of white reviewers and critics on African American writing.

According to Huggins, more than anyone else among his Harlem Renaissance peers, Thurman struggled in vital ways with many issues that African American artists have been compelled to confront throughout the twentieth century: "the Negro as artist or advocate, the writer as individual or race man, art as self-expression or exposition of ethnic culture" (*Harlem Renaissance,* 193). These essays and reviews—written primarily during the decade of Thurman's active association with Harlem (1925–1934)—represent an important legacy and convey a complex sense of the thinking that he shared with Langston Hughes. Hughes's "The Negro Artist and the Racial Mountain" (1926) remains the best-known articulation of the younger writers' attitudes and beliefs from the period. Unlike Hughes, however, Thurman developed the critical vocabulary and theoretical frameworks that would make him an indispensable part of the African American critical tradition. Some of these essays, if they had been more accessible sooner, would probably have found a place in books such as Addison Gayle's *The Black Aesthetic* (1971) and Winston Napier's *African American*

*Literary Theory* (2000). More relentlessly than anyone else during the period, Thurman employed his theoretical paradigms to critique not only all Harlem Renaissance writing, including his own, but a fair amount of new writing by white American writers that had a bearing on peoples of African descent. If there was a critic-in-residence at the Niggeratti Manor, it was undoubtedly Thurman.

Thurman's primary target in his literary criticism was the expectation that art be "respectable" and "constructive." As Thurman saw it, this view not only limited the artist's choice of subjects but also distorted the quality and range of his or her representation of such subjects. As he argued in "Notes on a Stepchild" (Part Five), he saw no reason why African American artists could not achieve "universality" through choosing to depict the lives of their own people. Like Hughes, Thurman wanted African American art to remain free of both white and black readers' expectations about the "proper" subject and scope of black art. In "Fire Burns," for example, Thurman lashes out at writing that closely corresponds to such expectations, considering it demeaning to writers and readers alike, generating and supporting stereotypes about blacks that ultimately constrain both individuality and creativity. Thurman's "Negro Artists and the Negro" is a lively attack on the impact of provincialism on American art and artists. Thurman worries that the growing canon of the New Negro Renaissance is welcomed by the black bourgeoisie for the wrong reasons. Pointing out how the "experimental" tenor of Hughes's poetry and of *Fire!!* have remained unappreciated among blacks, he notes that it is only when white music critics began praising the "beauty of the spirituals . . . that Negroes joined in the hallelujah chorus." However, he ends this essay on a hopeful note for the development of a black aesthetic with the work of a few "[s]erious and inquisitive individuals" who are "isolating, interpreting, and utilizing those things which seem to have a true esthetic value."

Thurman's call for a black aesthetic involves cutting through what he calls "propaganda"—writing that sets itself the task of "reflecting" social conditions as well as treating the Negro as a "sociological problem rather than as a human being." For Thurman, a black aesthetic would not necessarily involve protagonists recognizable to black readers as black or a readerly identification with "black" heroes. Nor does it involve a simple effort of cross-racializing—sketching figures as whites and then coloring them in with black. For example, in "Nephews of Uncle Remus," Thurman denounces literary achievement that is "racial rather than literary." In "High, Low, Past, and Present," he says of W.E.B. Du Bois that "the artist in him has been stifled in order that the propagandist may thrive" and builds a recurring contrast between literary art and propaganda. So, at least in 1928, Thurman draws the contours of a black aesthetic in negative terms. The way in which he militates against the comfort of his readers is reminiscent, in some ways, of Brecht's contemporaneous conception of epic theater. In doing this, Thurman also anticipates the aesthetic logic of Richard Wright's

*Native Son,* insofar as the latter rejects the sentimentality that might allow "bankers' daughters" to read his book "and weep over and feel good about" it. Like Du Bois before them, both Thurman and Wright also attempted to shape a racial discourse that would not exclude the participation of whites nor offer them an easy escape from acknowledging their responsibility for a pervasive white supremacy. (Du Bois, however, was much more aware of his two audiences—one black, the other white.)

Like Du Bois and Wright again, Thurman views "race" as a social construction, declaring in "Nephews of Uncle Remus" (1927) that "anthropologically the American Negro is really no Negro at all. He retains such an ethnological classification only because there seems to be nothing else to call him. There are probably no pure-blooded Negroes in America." Further, he underscores the hybridity of the African American experience by noting that the "the American Negro has absorbed all of the American white man's culture and cultural appurtenances." Thurman appears to grapple with a variety of racial designations in search of a classification that can provide a mold for the lived experience of a racialized group and a source for artistic productions that strive for "universality." He rejects both particularity and universality if they are wielded as exclusive terms, but seems to embrace the two notions in their conflictual relationship. For example, he asserts that "no Negro has written about his own people as beautifully or as sympathetically as has DuBose Heyward," the author of *Porgy.* Thurman wants to see both particularity and universality continually contending with each other in the context of American literature: "The Negro author can . . . introduce a Negro note into American literature . . . , but in America this can be done by a white writer as well as by a black one." In his striving for artistic universality inflected by "racialized" experience, Thurman's vision is not unlike Hughes's invocation of the "low down folks" ("proletariat," in Thurman's parlance) in "The Negro Artist and the Racial Mountain." Through this process, Thurman seeks liberation both from restrictive racial bonds and from a sense of inferiority. Based on these considerations, it would be interesting to examine the contexts in which Thurman uses the following terms, often differentiated in his texts: Negro; American Negro; negroid; nigger; Nordics, near Nordics, and non-Nordics; whites; the American white man; and Aframerican.

Published in *The Independent* in September 1927, "Nephews of Uncle Remus" is also one of the rare instances in which Thurman outlines the notion of a black aesthetic in positive terms, although he stops short of programmatic elaboration. Thurman wields his black aesthetic, using similes of warfare and weaponry, to deflate his audience's expectations and sense of respectability and to have "the effect of bird shot rather than that of shrapnel." This essay is the most explicit statement of Thurman's own position on racial expression in literature and art. He grapples with the paradoxical and yet complementary parameters of particularist black experience and universalist aesthetic categories.

While Thurman obviously admired the civil rights commitments of W.E.B. Du Bois, Walter White, and others, he rejected their notions of art as largely a weapon for propaganda. On more than one occasion in 1926, Du Bois had asserted the need for black writers to abjure artistic freedom in favor of creating portrayals of black life that would effectively challenge and contradict "the net result of American literature [in picturing] the twelve million black Americans as prostitutes, thieves, and fools" ("The Negro in Art," *The Crisis,* February 1926, 165). For Du Bois, then, "all art is propaganda and ever must be, despite the wailing of the purists. . . . whatever art I have for writing has been used always for propaganda, for gaining the right of black people to love and enjoy. . . . I do care when propaganda is confined to one side and the other side is stripped and silent" ("The Criteria of Negro Art," *The Crisis,* October 1926, 296). In his essays and reviews and in his work as the editor of both *Fire!!* and *Harlem,* Thurman argued with equal passion the opposite point of view.

Thurman's method of complicating his own argument for unbridled artistic freedom even in the wake of massive civil rights challenges can best be seen in his review of Walter White's *Flight* (1926), intriguingly entitled "A Thrush at Eve with an Atavistic Wound." He begins by setting uncompromisingly high standards for all writing, declaring the writer "who has the suitable material and fails to do it justice" as the "more oafish offender" than the writer who "tries to write without first having suitable material." Placing Walter White in the former category, he claims a bit disingenuously that he has no quarrel with the propagandist school to which Mr. White belongs but bemoans the absence of "genuine performance." A bit later, Thurman squarely refutes Du Bois's line of thinking by declaring: "All art no doubt is propaganda, but all propaganda is most certainly not art. And a novel must, to earn the name, be more than a mere social service report, more than a thinly disguised dissertation on racial relationships and racial maladjustment." Pointing out that Mimi Daquin, the protagonist of White's *Flight,* "should have been an intense, vibrant personality instead of the outlined verbal puppet that she is," he still recommends this book of "one-sided propaganda" to one and all: "It will irritate the Nordics, induce thought and provide argumentative material for the near Nordics, and salve the aching stings of the non-Nordics." He wittily ends his short review by adding: "And now I hear that inevitable question: 'What more could *Flight* have done if it had been one of your so-called works of art?' Which question I refuse to answer in so short a space."

In the unpublished review of his own *Infants of the Spring,* Thurman applies his usual exacting critical standards to his fictional satire on the frustrations and challenges black artists face. Although in the tradition of Walt Whitman's self-promoting reviews of *Leaves of Grass,* Thurman's essay is brutally frank about the aesthetic and political pressures that he and his fellow Harlem writers experience: "no Negro novelist can expect intelligent critical guidance or encouraging appreciation. This is not, mind you, an excuse for a bad novel, nor the empty caviling of a disgruntled author." As elsewhere, Thurman demonstrates here a sharp awareness of the factors that continue to weigh on the African American artist's creativity.

Indeed, in all of his literary essays, Thurman manages most engagingly to suggest issues relating to race and art that continue to involve our reading and evaluation of U.S. writing—thus preparing us for the more recent discursive analyses by Richard Wright, Ralph Ellison, Amiri Baraka, and Toni Morrison.

# Review of *Black Harvest,*
# by I.A.R. Wylie

*The Messenger,* May 1926.

I.A.R. Wylie is at once observant and anticipatory. She has seized upon the pregnant spectacle offered by the savage black legions, which the civilized European powers utilized in their late war to end war, and has limned an interesting and ingenious tale.

Every time a white novelist undertakes to write of Negro characters, that novelist is immediately accused of not having written understandingly of his black subjects. At times this criticism seems indeed justified, and at other times it seems like mere petulant quibbling. Any novelist, i.e., any sincere novelist, treats all subjects whether they are white or black, in the light of his experience. It is the insincere novelist that sees his subjects in patterns, and it is he who should be ignored. Then too, no race of people is exactly what it believes itself to be any more than the southern Negro is what either Octavius Roy Cohen or the latter day abolitionists would have you believe him to be. It takes a brave writer indeed in this day and time to attempt to write about any race save the one to which he belongs, and that writer who will not only write contemporaneously about some other race, but will also write speculatively about the future of that race deserves a croix-de-guerre for braving fire, for to fire he will most certainly be subjected.

The present author centers her tale about a "Jung Seigfried," a black product of a rape committed by a Senegalese soldier under the French flag upon a German prostitute. This mulatto Messiah has gained the support of Negroes both in America and Africa, and, at a word from him the entire black population of this world is prepared to march to the tune of "Die Wacht und der Rhine." By this grand gesture I gather that both the now conquered Germans and the now dominated blacks will achieve freedom.

However, "Jung Seigfried" is as we know, a black man, and his white conspirators are the conventional type of whites, which means that they are as fully aware of his color as they are aware of his power and ability. They realize that without him theirs is a lost cause, yet it seems that they prefer to remain subjugated rather than admit this man to be their equal.

All of this wrecks havoc upon the nerves of our slightly hysterical Hans Felde, especially when his desire for a woman—a white woman—becomes pitifully potent.

It is to this that our fellow Negroes will object, and mutter the accusation that these terrible white novelists just won't write truthfully about Negroes. Well, what if they don't, as long as Negroes won't write truthfully about themselves or won't recognize themselves when they are presented truthfully?

Hans Felde is not a far-fetched figure. Born, reared and subtly frustrated in and by the milieu the author creates he could not have acted or reacted any differently towards his environmental stimulus. As a character in *Black Harvest,* Hans Felde is a truthful one, and that is all that is necessary, for why should he be a general concoction of what Negroes believe to be Negroid virtues?

Yet this novel is far from satisfying or complete. As entertainment or as controversial stimuli it is indeed good, but one expects a little more from a volume so ambitious in theme and so bristling with positive character electrons. Thus *Black Harvest* remains one of those books that everyone should read, speculate upon, discuss, and then forget.

## A Thrush at Eve with an Atavistic Wound: Review of *Flight*, by Walter White

*The Messenger*, May 1926.

I do not know which is considered the greater literary criminal, he who writes or rather tries to write without first having suitable material or he who has the suitable material and fails to do it justice. In my opinion it is the latter who is the more oafish offender, for it is surely he who is the most inconsiderate of his expectant audiences, and the most persistent prostitutor of his art.

Now, the Negro is supposedly experiencing a cultural and spiritual renaissance, supposedly emerging from restricted zones to air himself in the more exclusive, the more esoteric ateliers. He is supposed to be developing a new type, which type is in turn destined to be a serious contender in the universal race struggle for supremacy. All of this is very well, very interesting, and very necessary, I suppose, to the scheme of things, and it has all been accomplished by a salient use of a salient weapon—i.e., propaganda.

Mr. White is one of the most salient users of this salient weapon, and it is partly because of his strenuous efforts that the alleged inferior Negro has been pushed from the unnoticed back ranks of the national chorus to a principal position on the polyglot American stage. This, remember, has already been done, and the propagandist school to which Mr. White belongs hopes to accomplish much more with this same efficient weapon. I have no quarrel with them on that score, even if I do doubt the continued potency of their weapon, and even if it does seem to me that it is about time for the next step, about time for the ballyhooing to cease and for the genuine performance to begin. However, my only quarrel with this school is that they are wont to consider their written propaganda as literary art. All art no doubt is propaganda, but all propaganda is most certainly not art. And a novel must, to earn the name, be more than a mere social service report, more than a thinly disguised dissertation on racial relationships and racial maladjustment.

Mimi Daquin, the central figure in Mr. White's latest lucubration, should have been an intense, vibrant personality instead of the outlined verbal puppet that she is. She could have been an individual instead of a general type; as it is Mimi is never more than an alphabetic doll regaled in cliché phrases, too wordy sentences, and paragraphs pregnant with frustrated eloquence. When one thinks of the psychological possibilities of such a theme as the author had to work with, one is almost appalled by the superficial and inadequate treatment that the theme receives. As a novelist Mr. White seems somewhat myopic, and it is this narrow vision that keeps his propaganda just outside the gates of literature-land. Mimi, with her complex racial heritage, her complex and heterogeneous social milieu, and her duo-racial urge, could have been as complete and as great a literary creation as, say, Emma Bovary, Nana, Candida, or the more contemporaneous, if less great, Clara Barron.

But there is no need to continue this tearful jeremiad, for by this time everyone knows that in my opinion Walter White is primarily a propagandist, an earnest one at that, burning up with the desire to show that his people are not inferior merely because their skins are dark, and consumed by the hope that his social service reports will not only assist in the fruition of this desire, but that they will also perchance become staunch literary survivors instead of ephemeral firecrackers. As a premier writer of propaganda, Mr. White deserves a gilt-edged palm, but as a writer of literature he is still the propagandist distinguished only because it is a new departure for an American Negro to brazenly throw lighted Roman candles into the public market places.

However, I recommend that *Flight* be read by all Nordics, all near Nordics, and all non-Nordics, and that it be read frankly as one-sided propaganda. It will irritate the Nordics, induce thought and provide argumentative material for the near Nordics, and salve the aching stings of the non-Nordics. It will also be of inspirative value to ambitious blacks and keep the talented Negro to the forefront, unless the saturation point has already been reached.

And now I hear that inevitable question: "What more could *Flight* have done if it had been one of your so-called works of art?" Which question I refuse to answer in so short a space.

## Editorial

*The Messenger*, July 1926.

It is fitting, proper, and necessary for even a magazine to take stock of itself and become intimate in regards to its hopes and ambitions. This serves to bring the editors of the publication closer to its readers, and also serves to give the readers of a magazine an insight into what they may expect from the future.

With Negro labor shaking off its complacency, and shackling servitude; with Negro artists energetically scaling the heights of Parnassus; with new forces shaping the destiny of the Negro's future, it is indeed necessary that some organ keep the public apprised of current trends, and, perhaps, serve as an *agent provocateur* for new energies and new aspirations.

There is much to inspire any Negro editor to fruitful labor in this present day, and, since the Negroes themselves seem to be growing more and more cognizant of the necessity in their being more appreciative and more loyal to their race publications, the editors are anxious to extend themselves in order to measure up to the demands of the day.

In keeping with this, the editors of *The Messenger* are striving to be constantly on the *qui vive,* and can promise their patrons much enervating food for future delectation. In these pages you will find and be able to follow the romantic rise of organization awareness of various labor groups. You will also find expert dissertations on the economic current of Negro life, which current is manifesting itself more and more as THE current. And, moreover, you will find ample evidence of the contemporary cultural renaissance.

We have tooted our own horn quite eloquently, and now await the crowd which we hope will assemble to watch the announced performance. Walt Whitman once said that to have great poets it was necessary to have great audiences. Did we think it necessary Whitman's implied plea could be paraphrased quite aptly, but we prefer to let the reader do his own paraphrasing.

## Review of *Blues: An Anthology,* edited by W. C. Handy

*The Messenger,* July 1926.

It was inevitable I suppose that during the present epidemic of Negrophilia the blues should finally come in for their share of honor and respectability. Negro poetry, Negro spirituals, and Negro prose (?) have all had their day of analyzation and acclaim, but the blues, which of them all have had the most telling effect upon American life, have been left to the last.

For many long years variations of the blues motif have been seeping into the main American cultural stream. It has been borrowed and berated, damaged and damned. It has caused a distinct change in vocal, orchestra, and dance technique, and to some extent has even influenced certain poetry and prose. Fond parents have shivered when they happened to hear their absorbing offspring moan haunting melodies, have shivered, protested, and then shivered again as they in turn succumbed to this savage musical mesmerism, and in turn became shaking addicts of the shimmy or calisthenic addicts of the Charleston. For all these things are the blues responsible, and once having exterminated the ragtime of the mauve decade, they precipitated the jazz era, and furnished America with her first contribution to musical art. True, there has been an evidently endless argument anent the question of whether jazz is music or not. Gilbert Seldes, whom Abbe Niles quotes in his brilliant introduction to this volume, seems to have the best definition thus far formulated. He says, in essence, that jazz is not music, but rather an American form of playing music. This must be narrowed down to modern music, for jazz strains entered into the music of some of our most revered old masters.

W. C. Handy is the acknowledged father of the blues, even though such acknowledgement has come rather late. Not so very long ago there was quite a furor aroused because some Broadway mogul had, in arranging a blues melees, included everyone except the man who first unearthed this folk music, and who first arranged it for public delectation. Handy was always the musician, and his keen ear both appreciated and remembered the vagrant ditties of his vagrant associates. He eavesdropped upon these dark troubadours while they worked in the cotton fields and on the levees or while they played upon street corners, saloons, and bawdy houses, and the rhythmic harmonies he heard upon these occasions were later sublimated and glorified into such pieces as "The St. Louis Blues," "Mr. Crump," and their many followers and derivatives.

Handy's life and accomplishments are ably set forth by Abbe Niles in his introduction. In addition to this Niles also gives us an analytic survey of the birth and development of the blues, as well as a treatment of their melody motifs, and their harmonic structure. He also supplies us with much invaluable detail concerning their propagators. His introduction is one of those rare scholarly treatises, pregnant with information, yet remarkably free of pedantics.

Mention must be made of the Covarrubias illustrations done to the tempo di blues. To be both modern and brief—they are immense. Some of our more stiff-necked bourgeoisie brethren will no doubt avert their eyes hurriedly from such blasphemous and misrepresentative pictures, but the fact remains that they are terse, excellent and realistic mood interpretations.

All in all *Blues* is a worthy monument to W. C. Handy, and a worthy anthology for all those interested in assembling evidence of Aframerican influence upon American culture, and most important of all, besides being a beautiful and worthy volume, it just fits the piano rack, and will prove an invaluable aid to parlor entertainment.

# Review of *Heloise and Abelard,*
# by George Moore

*The Messenger,* July 1926.

Any book written by George Moore is preordained by his tutelary muse deities to be both audacious and delightful. I make this statement rather boldly as I have not as yet read the entire literary output of the author, but I have read *Esther Waters, Confessions of a Young Man, Conversations in Ebury Street, Hail and Farewell,* and the volumes now being reviewed, which appear for the first time in a popular edition although they were published in a *de luxe,* limited edition some years ago. And even though I do not know the names of any other books written by him I stand prepared to back up my first statement that they are all audacious and delightful.

Now I realize that those two adjectives can be as easily and as dogmatically applied to works of dubious literary value, and that within themselves they note no literary pre-eminence. It is the rare author indeed who can gain the contemporary critical stamp as an indubitable producer of classic literature, and still be termed delightful, even if he be audacious; yet, George Moore is such a *rara avis.*

Heloise and Abelard rank with Troilus and Cressida, Dante and Beatrice, Romeo and Juliet, as tragic lovers, who, *ipso facto,* have gained immortality. They lived in France during the Middle Ages when the Catholic Church dominated and directed the destinies of the civilized world, in the days when the western Crusaders sought to vanquish the eastern Saracens—sought to retrieve the lost Holy Land, in the days when even a slight deviation from the existent religious tenets was punishable, as was the widely renowned witchcraft of the day, by burning at the stake, flagellation or decapitation, in the days when the *haute monde* spoke Latin leaving the national vulgate to be developed by the *hoi polloi.*

Those were interesting (to us) old days, but hardly conducive to a roseate romance such as Heloise and Abelard, perfect lovers as they were, should have experienced, and it is the trend of the times that frustrates the flowering of their devotion, and foredooms them to taste bitter gall instead of soothing ambrosia.

George Moore is a master of the literary presentation of amours whether they be his own subjective experiences (real or imaginary according to Mr. Moore's whims) or whether they be far removed objective phenomena as in the present case. He is also a master at introducing amusing anecdote and erudite digression. Thus we find our main theme being studded with realistic presentations of France during the Middle Ages; studded with realistic presentations of her people, her customs, her pastimes; studded throughout with true literary gems. We also are treated with peeps into the famed

Courts of Love that furnished amusement and carnal satiation to the bored elite during the time, and we get glimpses into the vagabond life of the southern troubadours and the northern *trouvères* who made their way over the bandit-infested highways to sing in the various wayside inns as well as at the various courts of chivalrous knights and flirtatious ladies. And, to our delight, we get a cross-section of the lives of the monks and nuns in the monasteries and convents. These monks were gay old dogs who, in between muttering Te Deums, and indulging in religious persecution, drank much wine and dawdled with lithe dancing maidens, and minstrel boys, who the author tells us were sometimes preferred to the girls "because their love was more ardent."

If the above material seemed more important to me than the tragic course of the title characters' *affair du coeur,* it does not necessarily indicate that the author has written an intellectual treatise rather than an emotional love tale. I must confess, however, that before I had reached the end of the second volume, I was not very interested in Heloise and Abelard (which I might add parenthetically may denote the presence of a despised and for the most part suppressed pedantic strain in the reviewer), and even fate's final gesture of frustration—the luring away of their child to join the children's battalion of Crusaders, the castration of Abelard by his jealous church brethren—failed to excite me. I was far more interested in Moore's unsentimental presentation of the Middle Ages' intellectual and emotional life.

Yet Moore's pair of lovers are by no means of the slushy variety. Their phenomenal devotion to one another is as attributable to the more tangible intellectual and physical forces as it is to the less tangible psychic and supposedly God-given forces that are generally paraded out to explain the passions that exist between men and women. I was much more attached and much more interested in Heloise than I was in Abelard, for she seemed to be the finer of the two. She was at least the finer drawn, and I am almost convinced that Moore was also prejudiced, for the gestures she makes are the gestures of a super-woman while the gestures of Abelard are for the most part purely masculine and purely normal. Women are always I suppose more intense realists than men, and Heloise is a super-realist, for never once does she attempt to mask her true aims and ambitions beneath altruistic or soothing illusions. She submits to Abelard without benefit of clergy because she felt passion's urge, and because she loved him. She at first refused to marry him because she realized that to bring about a fruition successful of his budding career it was necessary for him to remain in the Church, which of course postulated bachelorhood. She consented to marry him only when she was practically assured that the same would save his life and at the same time remain a secret thereby not interfering with his precious (to her) development. Her entrance into the convent, her patient awaiting for their re-union, her heroic acceptance of the loss of her child, and her final gesture of choosing to remain Abelard's emotional and physical slave when he finally comes to her a broken, disabled sterile monk all serve to place stars in her crown. The Heloise limned by George Moore is undoubtedly a true saint devoid of cant and hypocrisy, and set up in my personal shrine as really worth reverence.

## Review of *Teeftallow,*
## by Thomas Sigismund Stribling

*The Messenger,* July 1926.

I was reading *Teeftallow,* when a friend of mine wished to know "What it was all about." I answered that the author was giving a realistic depiction of life among the inhabitants of the Tennessee hill country. "Does he damn them?" my friend asked with interest. "Well," I hesitated, "not exactly" and then I sought to formulate my nebulous thought, but before I could do so my friend had interjected "No, I guess not, for God has done that already."

*Teeftallow* is to Lane County, Tennessee, what *Main Street* is to Gopher Prairie, Minnesota, and what *Barren Ground* is to the decadent backwoods of Virginia. The cover blurb informs us that herein Mr. Stribling "argues no cause, pleads for nothing, suggests no change." He doesn't, but one cannot help feeling that as a creator of characters the author is somewhat akin to the creator of our cosmos, which is to say that once he gets the toe of his boot working regularly on one's posterior regions he forgets to remove it for any appreciable period.

Abner Teeftallow, the title character, is a hill-billy raised in the poorhouse, and "with a brain unspoiled by book learnin'; a judge for a granddaddy, and a crazy woman for a mammy." With this auspicious ancestral heritage and environmental influence Abner is precipitated into a typical Tennessean milieu at the age of seventeen, and left to "sink or swim." That anyone could survive in this morass of moral whoremongers, religious fanatics, and uninspired ignoramuses is more than one can imagine. Yet Abner, wholly in tune with his environment, does survive, and acts as a catalytic guide for the readers through this hot-bed of fervid fundamentalism. Meanwhile he experiences an adolescent sex affair, a whipping at the hands of the "white-caps" whom he himself had helped to organize, and a second love affair which results in his losing a recently acquired and long withheld fortune.

If the people of Tennessee are truthfully delineated in this volume, one need no longer wonder about the phenomenal success of Bryanism and other such contemporary hokum. In a land where the solitary school-teacher aligns himself with a fanatical preacher to save the young lambs of the land from being taught that their grandfathers were monkeys; in a land where the church deacon is also the country usurer, and charges two per cent above the legal rate of interest on loans; in a land where the chief charity worker and Christian beacon light will go out of her way to inspire mob spirit, and then hound a "fallen" girl to the point of desperation; in a land where as a sequential reaction to a religious revival, Kluxism, fueled by the fiery fanaticism of frustrated old maids, and non-satiated married men and women runs riot, rides bootleggers and ladies of joy out of a town on rails, tars and feathers chicken-thieves, hangs a white man as the premier *pièce de resistance* of the celebration, and then, characteristically attempts to vent the rest of their sublimated passion upon the girl Abner has seduced; as I say, in such a land one would expect anything to happen.

Thus are the people with whom Mr. Stribling populates Teeftallow's world. A hard primitive lot, yet nevertheless the sons of our pioneers and the fathers of our future Americans. Even when one grows pessimistic concerning them one cannot help but feel somewhat sympathetic, for one realizes that the fault lies not so much within as it does without, and that is the barrenness and uneventfulness of their environment that renders them so petty, so pitiful to us more sophisticated people. These people gossip because they have no other subjects of conversation save the activities of their neighbors. They indulge in religious fanaticism of the most fervid kind, because they must have some outlet for their repressed urges and their riling inhibitions. They lynch a hapless black or an unlucky white because they have no other form of passionate celebration in which to participate. They disrespect the law because nine-tenths of the law they see functioning is either totally asinine or totally ineffective.

It is to Mr. Stribling's credit that he takes account of these things. Thus he escapes being assailed as a myopic or partial observer; thus he gives the impression of having spied upon these people and of having penetrated into their every native emotion, their every concrete thought. After which he seems to have been able to present his material entirely from an objective viewpoint seldom letting any subjective prejudice or reaction color his vigorous and vital narrative.

## Notes: Short Review of *Flight*, by Walter White

*The New Republic*, September 1, 1926.

When a Negro writes with passion and understanding of the peculiar problems and vicissitudes of his own race, he is likely to produce a book which commands attention and respect. This, be it understood, Mr. White has once achieved. But the present volume does not fall in that class. Mr. White seems to feel that he can handle his people from the same angle that one would use in treating a group of whites, and yet create a special interest by assuring us from time to time that they are colored—and by introducing the dramatic episode of Negro persecution. He fails, inevitably. The main outline of this story is not essentially the problem and struggle of the Negro. It might, with a few omissions including the wholly unconvincing denouement, be hung about a group of whites. And it is neither a very consistent nor a very stirring story. That part of the book which fits the Daquin family into its Louisiana background of radiant sunlight, vivid gardens and Creole indolence is full of charming bits. The rest—the girl's career— is entirely disappointing.

## A Stranger at the Gates: A Review of *Nigger Heaven*, by Carl Van Vechten

*The Messenger*, September 1926.

When I first heard of the author's proposed novel of Negro life in Harlem I immediately conjectured that while the whites would relish it, especially since the current faddistic interest in things Negroid was still a flourishing reality, that Negroes themselves would anathematize both the book and its author. I predicated this conjecture on the fact that most colored people with whom I came into contact bristled belligerently at mention of the title. And some of those who had been most active in showing the author the sights of Harlem crowed about ingratitude, Nordic duplicity, et cetera, as long as the object of their wrath remained downtown, and promptly forgot it when he was uptown, which gave me a chance to speculate upon African duplicity. However, since reading the volume I find myself forced to cancel my first conjecture and come forward with another, which is, that whites may find it a trifle tedious at times, but that Negroes will accept it so warmly that even the detested "Nigger" in the title will be forgotten, and, I would not be surprised should some of our uplift organizations and neighborhood clubs plan to erect a latter-day abolitionist statue to Carl Van Vechten on the corner of 135th Street and Seventh Avenue, for the author has been most fair, and most sympathetic in his treatment of a long mistreated group of subjects. True, some of his individual characters may seem tarnished, but the race as a whole emerges as a group of long suffering martyrs, deservant of a better fate. No one in Harlem could have presented their case more eloquently and with more finesse. Truly Gareth Johns, you deserve an especial Spingarn medal.

Yet once the characters cease discussing the "Negro problem," once they cease spouting racial equality epigrams, and anti-racial discrimination platitudes, the novel begins to move—begins to pulsate with some genuine rhythms peculiar (objections from George S. Schuyler) to Harlem alone, and, thereupon holds the reader rapt until he reaches the dramatic climax.

Mr. Van Vechten seems at his best in the cabaret scenes, the charity ball, and the Black Mass (and let me insert a question here—where, oh where is this Black Mass in Harlem, for it is too good to be merely a figment of the writer's imagination—I mean too good in the sense that it should not be such a selfish, subjective creation, I, for one, would love to see).

An invisible band, silent at the moment they had entered this deserted room, now began to perform wild music, music that moaned and lacerated one's breast with brazen claws of tone, shrieking, tortured music from the depths of hell. And now the hall became peopled, as dancers slipped through the folds of the hangings, men and women with weary faces, faces tired of passion and pleasure. Were these the faces of dead prostitutes and murderers? Pleasure seekers from the cold slabs of the morgue?

"Dance!" cried Lasca. "Dance!" She flung herself in his arms and they joined this witches sabbath.

This all occurs in "a circular hall entirely hung in vermilion velvet" . . . with . . . "a flavor of translucent glass" through which flows "a cloud of light—now orange, now deep purple, now flaming like molten lava, now rolling sea waves of green."

And the two cabaret scenes surpass even the above in their realistic evocation of an individualistic setting, surging with soul shreaving individualistic rhythms.

I shall not dwell upon the chronicle of events that compromise the main motif of the novel proper. I found the trimmings and trappings more genuine and more gracefully done than the thesis. The affair between Mary Love and Byron Kasson struck me as rather puerile. Show me any normal pair of even cultured lovers in Harlem skipping down bridle paths in a park, and show me any girl in Mary's milieu who is as simple as Mary in the matters of an *affair du coeur*. The tragedy of the myopic Byron who wished to write but could find nothing to write about the while a veritable ocean of material was swirling about him, and his passionate plunge with the fiery and exotic Lasca Sartoris into a pool of physical debauch is of sterner stuff and much more calculated to draw the reader back to the volume for a second reading.

*Nigger Heaven* will also provide high Harlem with a new indoor sport, namely, the ascertaining which persons in real life the various characters were drawn from. Speculations are already rampant even before a general circulation of the book, and I have heard from various persons whom each character represents with far more assurance than the author himself could muster.

All in all *Nigger Heaven* should have a wide appeal and gain much favorable notice. As I say, some "ofays" may find arid stretches in it, especially where they are being lampooned in good old N.A.A.C.P. fashion, but the pregnant picturization of certain facets of Negro life in Harlem will serve to make at least partial amends for that. Mr. Van

Vechten has done what Russett Durwood, in the novel, predicts to Byron someone would do—i.e., beat the Negro litterateurs to a vibrant source pot of literary material which they for the most part have glossed over. And he has done it so well in places that despite the allegation of certain Manhattan sophisticates that he wrote the book merely to prove that there are Negroes who read Paul Morand, Jean Cocteau, Edmond Gosse, Louis Broomfield and Cabell, and who can feel at home in sumptuous settings of tiger-skin covered beds, magenta and silver cushions, taffeta canopied dressing tables, Sèvres china, and furniture which is a "Bavarian version of Empire," he has also laid himself liable to being referred to, in the provinces, as another Negro writer.

## Fire Burns:
## A Department of Comment

*Fire!!* November 1926.

Some time ago, while reviewing Carl Van Vechten's lava laned *Nigger Heaven* I made the prophecy that Harlem Negroes, once their aversion to the "nigger" in the title was for-gotten, would erect a statue on the corner of 135th Street and Seventh Avenue, and ded-icate it to this ultra sophisticated Iowa New Yorker.

So far my prophecy has failed to pan out, and superficially it seems as if it never will, for instead of being enshrined for his pseudo-sophisticated, semi-serious, semi-ludicrous effusion about Harlem, Mr. Van Vechten is about to be lynched, at least in effigy.

Yet I am loathe to retract or to temper my first prophecy. Human nature is too per-verse and prophecies do not necessarily have to be fulfilled within a generation. Rather, they can either be fulfilled or else belied with startling two-facedness throughout a series of generations, which, of course, creates the possibility that the fulfillments may out-number the beliements and thus gain credence for the prophecy with posterity. Witness the Bible.

However, in defending my prophecy I do not wish to endow Mr. Van Vechten's novel with immortality, but there is no real reason why *Nigger Heaven* should not eventually be as stupidly damned by the majority of Harlem's dark inhabitants. Thus I defiantly reiter-ate that a few years hence Mr. Van Vechten will be spoken of as a kindly gent rather than as a moral leper exploiting people who had believed him to be a sincere friend.

I for one, and strange as it may sound, there are others, who believe that Carl Van Vechten was rendered sincere during his explorations and observations of Negro life in Harlem, even if he remained characteristically superficial. Superficiality does not neces-sarily denote a lack of sincerity, and even superficiality may occasionally delve into deep pots of raw life. What matter if they be flesh pots?

In writing *Nigger Heaven* the author wavered between sentimentality and sophistication. That the sentimentality won out is his funeral. That the sophistication stung certain Negroes to the quick is their funeral.

The odds are about even. Harlem cabarets have received another public boost and are wearing out cash register keys, and entertainers' throats and orchestra instruments. The so-called intelligentsia of Harlem has exposed its inherent stupidity. And *Nigger Heaven* is a best seller.

Group criticism of current writings, morals, life, politics, or religion is always ridiculous, but what could be more ridiculous than the wholesale condemnation of a book which only one-tenth of the condemnators have or will read. And even if the book was as vile, as degrading, and as defamatory to the character of the Harlem Negro as the Harlem Negro now declares, his criticisms would not be considered valid by an intelligent person as long as the critic had had no reading contact with the book.

The objectors to *Nigger Heaven* claim that the author came to Harlem, ingratiated himself with Harlem folk, and then with a supercilious grin and a salacious smirk, lolled at his desk downtown and dashed off a pornographic document about uptown in which all of the Negro characters are pictured as being debased, lecherous creatures not at all characteristic or true to type, and that, moreover, the author provokes the impression that all of Harlem's inhabitants are cabaret hounds and thirsty neurotics. He did not tell, say his critics, of our well bred, well behaved church-going majorities, nor of our night schools filled with eager elders, nor of our brilliant college youth being trained in the approved contemporary manner, nor of our quiet, home loving thousands who hardly know what the word cabaret connotes. He told only of lurid night life and of uninhibited sybarites. Therefore, since he has done these things and neglected to do these others the white people who read the book will believe that all Harlem Negroes are like the Byrons, the Lascas, the Pettijohns, the Rubys, the Creepers, the Bonifaces, and the other lewd hussies and whoremongers in the book.

It is obvious that these excited folk do not realize that any white person who would believe such poppy-cock probably believes it anyway, without any additional aid from Mr. Van Vechten, and should such a person read a tale anent our non-cabareting, church-going Negroes, presented in all their virtue and glory and with their human traits, their human hypocrisy and their human perversities glossed over, written, say, by Jessie Fauset, said person would laugh derisively and allege that Miss Fauset had not told the truth, the same as Harlem Negroes are alleging that Carl Van Vechten has not told the truth. It really makes no difference to the race's welfare what such ignoramuses think, and it would seem that any author preparing to write about Negroes in Harlem or anywhere else (for I hear that DuBose Heyward had been roundly denounced by Charlestonian Negroes for his beautiful *Porgy*) should take whatever phases of their life that seem the most interesting to him, and develop them as he pleases. Why Negroes imagine that any writer is going to write what Negroes think he ought to write about them is too ridiculous to merit consideration. It would seem that they would shy away from being pigeon-holed, so long have they been the rather lamentable victims of such a typically American practice, yet Negroes would have all Negroes appearing in contemporary literature made as ridiculous and as false to type as the older school of pseudo-humorous, sentimental white writers made their Uncle Toms, their Topsys, and their Mammies, or as the Octavius Roy Cohen school now make their more modern "cullud" folk.

One young lady, prominent in Harlem collegiate circles, spoke forth in a public forum (oh yes, they even have public forums where they spend their time announcing that they have not read the book and that the author is a moral leper who also commits literary sins), that there was only one character in *Nigger Heaven* who was true to type. This character, the unwitting damsel went on, was Mary Love. It seems as if all the younger Negro women in Harlem are prototypes of this Mary Love, and it is pure, poor, virtuous, vapid Mary, to whom they point as a typical life model.

Again there has been no realization that Mary Love is the least life-like character in the book, or that it is she who suffers most from her creator's newly acquired seriousness and sentimentality, she who suffers most of the whole ensemble because her creator discovered, in his talented trippings around Manhattan, drama at which he could not chuckle the while his cavalier pen sped cleverly on in the same old way yet did not-could not spank.

But—had all the other characters in *Nigger Heaven* approximated Mary's standard, the statue to Carl Van Vechten would be an actualized instead of a deferred possibility, and my prophecy would be gloriously fulfilled instead of being ignominiously belied.

## Negro Artists and the Negro

*The New Republic*, August 31, 1927.

When the Negro art fad first came into being and Negro poets, novelists, musicians and painters became good copy, literate and semi-literate Negro America began to strut and to shout. Negro newspapers reprinted every item published anywhere concerning a Negro whose work had found favor with the critics, editors, or publishers. Negro journals conducted contests to encourage embryonic geniuses. Negro ministers preached sermons, Negro lecturers made speeches, and Negro club women read papers—all about the great new Negro art.

Everyone was having a grand time. The millennium was about to dawn. The second emancipation seemed inevitable. Then the excitement began to die down and Negroes as well as whites began to take stock of that in which they had reveled. The whites shrugged their shoulders and began seeking for some new fad. Negroes stood by a little subdued, a little surprised, torn between being proud that certain of their group had achieved distinction, and being angry because a few of these arrived ones had ceased to be what the group considered "constructive" and had in the interim produced works that went against the grain, in that they did not wholly qualify to the adjective "respectable."

Langston Hughes was the major disturbing note in the "renaissance" chorus. His first volume of verse, *The Weary Blues,* introduced him as a poet who was interested in artistic material rather than in sociological problems. He went for inspiration and rhythms to those people who had been the least absorbed by the quagmire of American *Kultur*, and from them he undertook to select and preserve such autonomous racial values as were being rapidly eradicated in order to speed the Negro assimilation.

*The Weary Blues* did not evoke much caustic public comment from Mr. Hughes' people. Negroes were still too thrilled at the novelty of having a poet who could gain the attention of a white publisher to pay much attention to what he wrote. Quietly, and privately, however, certain Negroes began to deplore the author's jazz predilections, his unconventional poetic forms, and his preoccupation with the proletariat. But they were hopeful that he would reform and write in a conventional manner about the "best people."

Mr. Hughes' second volume, *Fine Clothes to the Jew,* a hard, realistic compilation, happened to be published while Negroes were still rankling from Carl Van Vechten's novel of Negro life in Harlem, *Nigger Heaven.* It seemed as if this novel served to unleash publicly a store of suppressed invective that not only lashed Mr. Van Vechten and Mr. Hughes, but also the editors and contributors to *Fire!!*, a new experimental quarterly devoted to and published by younger Negro artists. Under the heading "Writer Brands *Fire!!* as Effeminate Tommyrot," a reviewer in one of the leading Negro weeklies said: "I have just tossed the first issue of *Fire!!*—into the fire, and watched the cackling flames leap and snarl as though they were trying to swallow some repulsive dose."

*Fire!!*, like Mr. Hughes' poetry, was experimental. It was not interested in sociological problems or propaganda. It was purely artistic in intent and conception. Its contributors went to the proletariat rather than to the bourgeoisie for characters and material. They were interested in people who still retained some individual race qualities and who were not totally white American in every respect save color of skin.

There is one more young Negro who will probably be classed with Mr. Hughes when he does commence to write about the American scene. So far this writer, Eric Walrond, has confined his talents to producing realistic prose pictures of the Caribbean regions. If he ever turns on the American Negro as impersonally and as unsentimentally as he turned on West Indian folk in *Tropic Death,* he too will be blacklisted in polite colored circles.

The Negro plastic artists, especially Aaron Douglas and Richard Bruce [Nugent], are also in disfavor, Douglas because of his advanced modernism and raw caricatures of Negro types, Bruce because of his interest in decadent types and the kinks he insists on putting upon the heads of his almost classical figures. Negroes, you know, don't have kinky hair; or, if they have, they use Madame Walker's straightening pomade.

Moreover, when it first became popular to sing spirituals for public delectation, the mass of Negroes objected vigorously. They did not wish to become identified again with what the spirituals connoted, and they certainly did not want to hear them sung in dialect. It was not until white music critics began pointing out the beauty of the spirituals, and identifying the genius that produced them, that Negroes joined in the hallelujah chorus.

Negroes are, of course, no different in this from any other race. The same class of Negroes who protest when Mr. Hughes says:

> Put on yo' red silk stockings
> Black gal.
> Go out an' let de white boys
> Look at yo' legs.
> . . . . . . . . . . . . .
> Put on yo' red silk stockings, gal
> An' tomorrow's chile'll
> Be a high yaller . . .

have their counterpart in those American whites who protest against the literary upheavals of a Dreiser, an Anderson, or a Sandburg. And those American Negroes who would not appreciate the spirituals until white critics sang their praises have their counterpart in the American whites who would not appreciate Poe and Whitman until European critics classed them as immortals.

The mass of American Negroes can no more be expected to emancipate themselves from petty prejudices and myopic fears than can the mass of American whites. They all revere Service, Prosperity and Progress. True, the American Negro may be the more pitiful figure, since he insists on selling every vestige of his birthright for a mess of potage.

The American Negro feels that he has been misinterpreted and caricatured so long by insincere artists that once a Negro gains the ear of the public he should expend his spiritual energy feeding the public honeyed manna on a silver spoon. The mass of Negroes like the mass of whites, seem unable to differentiate between sincere art and insincere art. They seem unable to fathom the innate differences between a dialect farce committed by an Octavius Roy Cohen to increase the gaiety of Babbitts, and a dialect interpretation done by a Negro writer to express some abstract something that burns within his people and sears him. They seem unable to differentiate between the Uncle Remus tales and a darky joke told by Irvin Cobb, or to distinguish the difference, in conception and execution, between a *Lulu Belle,* with its cheap gaudiness and blatant ensemble, and an *All God's Chillun Got Wings* by a sympathetic, groping Eugene O'Neill. Even such fine things as a *Porgy* or a *Green Thursday* are labeled inadequate and unfair. While *Nigger Heaven*—ask Carl Van Vechten!

Negroes in America feel certain that they must always appear in public butter side up, in order to keep from being trampled in the contemporary onward march. They feel as if they must always exhibit specimens from the college rather than from the kindergarten, specimens from the parlor rather than from the pantry. They are in the process of being assimilated, and those elements within the race which are still too potent for easy assimilation must be hidden until they no longer exist.

Thus, when the publishers of Mr. Hughes' second volume of verse say on the cover that "These poems, for the most part, interpret the most primitive types of American Negro, the bell boys, the cabaret girls, the migratory workers, the singers of Blues and Spirituals, and the makers of folk songs," and that they "express the joy and pathos, the beauty and ugliness of their lives," Negroes begin to howl. This is just the part of their life which experience has taught them should be kept in the background if they would exist comfortably in these United States. It makes no difference if this element of their life is of incontestable value to the sincere artist. It is also available and of incontestable value to insincere artists and prejudiced white critics.

The Negro artist is in a no more enviable position than is the emerging, or sometimes, for that matter, even the arrived artist, of other races or countries. He will receive little aid from his own people unless he spends his time spouting sociological jeremiads or exhausts his talent in building rosy castles around Negro society. He will be exploited by white faddists, and sneered at by non-faddists. He will be overrated on the one hand, and under-praised on the other.

Neither is the position of the bourgeois Negro an enviable one. Fearing as he does what his white compatriots think, he feels that he cannot afford to be attacked realistically by Negro artists who do not seem to have the "proper" sense of refinement or race pride. The American scene dictates that the American Negro must be what he ain't! And despite what the minority intellectual and artistic group may say, it really does seem more profitable for him to be what he ain't, than for him to be what he is.

The first literary works that came out of the so-called "Negro renaissance" were not of the riling variety. *Cane,* by Jean Toomer, was really pre-renaissance, as it was published too soon to be lifted into the best-seller class merely because its author was a

Negro. And, as Waldo Frank forewarned in his introduction to *Cane,* Jean Toomer was not a *Negro* artist, but an *artist* who had lost "lesser identities in the great well of life." His book, therefore, was of little interest to sentimental whites or to Negroes with an inferiority complex to camouflage. Both the personality of the author and the style of his book were above the heads of these groups. Although *Cane* reeked with bourgeois-baiting revelations, it caused so little excitement among the bourgeois sector of Negro society, save in Mr. Toomer's hometown, Washington, D.C., where the main criticisms were concerning his treatment of Negro women.

*Fire in the Flint* by Walter White, *Flight* by the same author, and *There Is Confusion* by Jessie Fauset were just the sort of literary works both Negroes and sentimental whites desired Negroes to write. The first, a stirring romantic propaganda tale, recounted all the ills Negroes suffer in the inimical South, and made all Negroes seem magnanimous, mistreated martyrs, all southern whites evil transgressors of human rights. It followed the conventional theme in the conventional manner. It was a direct descendant of *Uncle Tom's Cabin,* and it had the same effect on the public. White latter-day abolitionists shook their heads, moaned and protested. Negroes read, boiled and bellowed.

Less sensational and more ambitious, the second novel from this author's pen sought to chronicle the emotional and physical peregrinations of a female mulatto with such a preponderance of white blood in her veins that she could be either Caucasian or Negro at will. Miss Fauset's work was an ill-starred attempt to popularize the pleasing news that there were cultured Negroes, deserving of attention from artists, and of whose existence white folk should be apprised.

All of these works of fiction, as well as the two outstanding works of non-fiction, *The Gift of Black Folk* by W.E.B. Du Bois, and *The New Negro,* edited by Alain Locke, that appeared during the heated days of the "renaissance" were considerate of the Aframerican's *amour propre,* soothing to his self-esteem and stimulating to his vanity. They all treated the Negro as a sociological problem rather than as a human being. I might add that only in *The New Negro* was there even an echo of a different tune. The rest were treatises rather than works of art.

These works were all designed to prove to the American white man that the American Negro was not inferior *per se* and therefore, were honored and blessed by Negroes.

*Color,* a volume of verses by Countee Cullen, was also conventional in theme and manner. True, Mr. Cullen was possessed by a youthful exuberance that occasionally flamed with sensual passion, but for the most part he was the conventional Negro litterateur in all respects save that he had more talent than most of his predecessors. He could say:

> Yet do I marvel at this curious thing:
> To make a poet black, and bid him sing.

or wish

> To do a naked tribal dance
> Each time he hears the rain

and finally

> Once riding in old Baltimore,
> Heart-filled, head-filled with glee,
> I saw a Baltimorean
> Keep looking straight at me.
>
> Now I was eight and very small,
> And he was no whit bigger,
> And so I smiled, but he poked out
> His tongue, and called me, "Nigger."
>
> I saw the whole of Baltimore
> From May until December;
> Of all the things that happened there
> That's all that I remember.

This last poem was enough to endear Mr. Cullen to every bourgeois black soul in America, as well as to cause white critics to surpass themselves in calling attention to this Negro poet's genus, a thing far more important to them than his genius. And, since Mr. Cullen, unlike his contemporary, Mr. Hughes, has not and perhaps never will seek the so-called lower elements of Negro life for his poetic rhythms and material, and since, he, too, assumes the conventional race attitude toward his people rather than an artistic one, he will probably remain endeared to both bourgeois black America and sentimental white America, more because of this attitude than because of his undisputed talent or his intense spiritual sensitivity.

Fortunately, now, the Negro art "renaissance" has reached a state of near sanity. Serious and inquisitive individuals are endeavoring to evaluate the present and potential significance of this development in Negro life. They are isolating, interpreting, and utilizing those things which seem to have a true esthetic value. If but a few live coals are found in a mountain of ashes, no one should be disappointed. Genius is a rare quality in this world, and there is no reason why it should be more ubiquitous among Blacks than Whites.

## Nephews of Uncle Remus

*The Independent*, September 24, 1927.

It is too bad that Negro literature and literary material have had to be exploited by fad finders and sentimentalists. Too bad that the ballyhoo brigade which fostered the so-called Negro art "renaissance" has chosen to cheer and encourage indiscriminately anything which claims a Negroid ancestry or kinship. For as the overfed child gags when forced to swallow an extra spoonful of half sour milk, so will the gullible American public gag when too much of this fervid fetish known as Negro art is shoveled into its gaping minds and mouths, and the Afro-American artist will find himself as unchampioned and as unimpressive as he was before Carl Van Vechten commenced caroling of his charms in *Vanity Fair* and the *Survey Graphic* discovered the "New Negro" in 1924. Less so; for even in the prerenaissance days any Negro who achieved something in literature was regarded with wonderment even by the emancipated white intellectuals. But now so many Negroes have written a book or a story or a poem that the pale-faced public is no longer astonished by such phenomena.

There has, of course, been some compensation. Negroes have become more articulate and more coherent in their cries for social justice, and they have also begun to appreciate the advantages of racial solidarity and individual achievement. But speaking purely of the arts, the results of the renaissance have been sad rather than satisfactory, in that critical standards have been ignored, and the measure of achievement had been racial rather than literary. This is supposed to be valuable in a social way, for it is now current that the works produced by Negro artists will form the clauses in a second emancipation proclamation, and that because of the Negro artists and their works, the Afro-American will reap new fruits of freedom. But this will be true only inasmuch as the Negro artist produces dignified and worth-while work. Quick, tricky, atmospheric bits will be as ephemeral as they are sensational, and sentimental propaganda, unless presented in a style both vigorous and new, will have the effect of bird shot rather than that of shrapnel. Which is to say that the Negro will not be benefited by mediocre and ephemeral works, even if they are hailed by well-meaning, but for the moment, simple-minded, white critics as works of genius.

There is a constant controversy being waged as to whether or not there can be in America something specifically known as Negro literature. One school says yes, pointing out the fact that there are inherent differences between whites and blacks which will cause the artistic works of the two groups to be distinctly different. The other school contends that there can be no such thing in America as an individual Negro literature; that the Afro-American is different from the white American in only one respect, namely, skin color, and that when he writes he will observe the same stylistic conventions and literary traditions.

This latter school seems to have the better of the argument. It is true that anthropologically the American Negro is really no Negro at all. He retains such an ethnological classification only because there seems to be nothing else to call him. There are probably no pure-blooded Negroes in America. The present-day American colored man is, as the Anthropological Department of Columbia University has advanced, neither Negro, Caucasian, nor Indian, but a combination of all three with the Negroid characteristics tending to dominate. Furthermore, the American Negro has absorbed all of the American white man's culture and cultural appurtenances. He uses the same language, attends the same schools, reads the same newspapers and books, lives in the same kind of houses, wears the same kind of clothes, and is no more keenly attuned to the African jungle or any of its aesthetic traditions than the average American white man.

It is hard to see, then, that the Negro in America can produce an individual literature. What he produces in the field of letters must be listed as American literature, just as the works of the Scotchman Burns or the Irishman Synge are listed as English literature. The Negro author can, by writing of certain race characteristics and institutions, introduce a Negro note into American literature, just as Yeats, Colum, A. E., Lady Gregory, and others introduced a Celtic note in English literature, but in America this can be done by a white writer as well as by a black one, and there is a possibility of its being done much better by the former, in that he may approach his material with a greater degree of objectivity. For example, no Negro has written about his own people as beautifully or as sympathetically as has DuBose Heyward, the author of *Porgy,* purely because the Negro writer, for the most part, has seen fit to view his own people as sociological problems rather than as human beings, and has written of little else save the constant racial struggle between whites and blacks. Consequentially, he has limited himself and left a great deal of fresh vital material untouched.

It was to be hoped that the renaissance would make the Negro writer more aware of the value of the literary material provided by his own people. During this time, DuBose Heyward wrote *Porgy,* Julia Peterkin wrote *Green Thursday* and *Black April,* but as yet no Negro novelist has taken a hint from these white writers, let alone tried to surpass them.

The question arises: Must the Negro author write entirely of his own people? So many of the young Negro writers have asked this and then wailed that they did not wish to be known as *Negro* artists, but as *artists*. They seemed to think that by writing of white people they could produce better work and have a less restricted field of endeavor. In trying to escape from a condition their own mental attitude makes more harrowing, they forget that every facet of life can be found among Negroes, who being human beings, have all the natural emotional and psychological reactions of other human beings. They live, die, hate, love, and procreate. They dance and sing, play and fight. And if art is the universal expressed in terms of the particular, there is, if he has the talent, just as much chance for the Negro author to produce great literature by writing of his own people as if he were to write of Chinese or Laplanders. He will be labeled a *Negro* artist, with the emphasis on the Negro rather than on the artist, only as he fails to rise above the province of petty propaganda, or fails to find a means of escape from both himself and his environment.

The pioneer work done by Jean Toomer in freeing himself from restrictive racial bonds and letting the artist in him take flight where it will, and the determination with which Langston Hughes went about depicting his people, their rhythms and institutions, make these two appear to be the most important literary figures the renaissance has produced. They, of all the young Negro writers, have seemed to be the most objective and the most aware of the aesthetic value of literary material in their race. Their work is distinctive because it contains vestiges of that Negro note which will characterize the literature produced by truly talented, sincere Negro writers. Neither of them is worried by an inferiority complex which makes him wish to escape his race by not writing of it and at the same time binds him more tightly to the whipping post he would escape. The trouble with Mr. Toomer is that he has written too little. The trouble with Mr. Hughes is that he has written too much, in that he seems to lack, where his own work is concerned, that discriminating sense of selection which makes the complete artist as critical as he is creative. Urged on by a faddistic interest in the unusual, Mr. Hughes has been excessively prolific, and has exercised little restraint. The result is that his work is uneven to an alarming degree, and makes one fear that soon he will expend all of his spiritual energy in bubbling over in his adolescence rather than conserving himself until he matures. He needs to learn the use of the blue pencil and the waste-paper basket.

Eric Walrond can be classed with Mr. Toomer in that when writing of his own people he treats them objectively and as human beings. His *Tropic Death* was marred only by the writer's inability to master completely his style. He was attempting to forge a new method of expression, to escape, as Jean Toomer escaped, from the staid conventionality of stereotyped prose; and he succeeded to a degree that almost made one overlook those passages which were incoherent or tiresome. Nevertheless, it must be admitted that had his style been less esoteric or more controlled, the subject matter would have been more effective, a large order when one remembers such stories as "The Vampire Bat" and "The Black Pin."

Rudolph Fisher has written one very good short story of Negro life in Harlem, and a few more which, while vivid with local color, are weak in other respects. "The City of Refuge," which appeared in *The Atlantic Monthly* almost three years ago, and was later incorporated in *The New Negro*, that handbook of the renaissance edited by Alain Locke, was one of the first short stories written by a Negro about Negroes which did not follow the conventional formula. It told the story of a Southern Negro's coming to Harlem, the city of refuge, and of his struggles to adapt himself to a strange and complex environment. Here was drama of a new order, drama and color and life, a story as good as it was unusual. One waits expectantly for Mr. Fisher's first novel.

There are many more Negro writers who could be discussed in detail if the space allowed. Zora Neale Hurston is a good story-teller, but an indifferent craftsman. Walter White and Jessie Fauset have produced nothing out of the ordinary, or very good. James Weldon Johnson belongs to the elder generation of Negro writers, as his *Autobiography of an Ex-Colored Man,* recently reissued by Knopf, will prove. Despite Carl Van Vechten's enthusiastic introduction, there seems to be no real reason for reprinting this volume in the Blue Jade Library. It is well written, yes, and the author predicts the popularity of ragtime, but its content seems uninspired and stereotyped. Fifteen years ago it might have been a little more than just a well-written book; today it merely makes one wonder why the author didn't call it the autobiography of a colored man rather than that of an ex-colored man. Mr. Johnson's *God's Trombones* is of a different species. The majesty and eloquence of the sermon poems included therein, especially "Go Down Death" are searing and unforgettable. In these poems Mr. Johnson has been the artist rather than the propagandist, and *God's Trombones* will be remembered long after the *Autobiography* has been forgotten, even by sentimental white folk and Jeremiah-like Negroes.

Countee Cullen has published three volumes, *Color, Copper Sun,* and *The Ballad of the Brown Girl,* all of which tend to convince the critical investigator that although Mr. Cullen can say things beautifully and impressively, he really has nothing new to say, nor no new way in which to say it. He follows the tradition of a literary poet. His lyrics of love and death are reminiscent of Keats, Housman, and Edna Millay, and one gets the impression that Mr. Cullen writes not from his own experience, but from vicarious literary experiences which are not intense enough to be real and vital. His poetry is not an escape from life in the big sense, but from the narrow world in which he has been caged, and from which he seems to have made no great effort to escape.

Undoubtedly Mr. Cullen has more talent than any of his colored contemporaries. What he lacks is originality of theme and treatment, and the contact with life necessary to have had actual, rather than vicarious, emotional experiences. He is still young, and perhaps when some of the fanfare aroused when it was discovered that a *Negro* could write beautiful conventional verse has died down, he will be able to work out his own destiny. He has far to go if only he is given the chance and stimulation.

In a recent issue of *Opportunity, Journal of Negro Life,* there are reviews of two recent books. Both of the reviewers were white poets, both books were by Negro poets. The first was of Countee Cullen's *Copper Sun,* and in dithyrambic prose the reviewer went to every possible verbal excess in praising the volume. According to him Countee Cullen is *the* thing in American poetry—he equals Housman and surpasses Edna Millay. There is more such lyric nonsense, all done in good faith, no doubt, but of no value whatsoever either to Mr. Cullen or to the race to which he belongs. The second review, while less turgid and more restrained, was excessive in its praise of *God's Trombones,* and in neither piece was there any constructive criticism or intelligent evaluation.

These two reviews are typical of the attitude of a certain class of white critics toward Negro writers. Urged on by their desire to do what in their opinion is a "service," they dispense with all intelligence and let their sentimentality run riot. Fortunately, most of the younger Negro writers realize this, and as most Negroes have always done, are laughing up their sleeves at the antics of their Nordic patronizers. It is almost a certainty that Countee Cullen and Langston Hughes know full well that there are perhaps a half dozen young white poets in these United States doing just as good if not better work than they, and who are not being acclaimed by the public or pursued by publishers purely because a dash of color is the style in literary circles today.

If the Negro writer is to make any appreciable contribution to American literature it is necessary that he be considered as a sincere artist trying to do dignified work rather than as a highly trained dog doing trick dances in a public square. He is, after all, motivated and controlled by the same forces which motivate and control a white writer, and like him he will be mediocre or good, succeed or fail as his ability deserves. A man's complexion has little to do with his talent. He either has it or has it not, and despite the dictates of spiritually starved white sophisticates genius does not automatically descend upon one because one's grandmother happened to be sold down the river "befo' de wah."

## Negro Poets and Their Poetry

*The Bookman,* July 1928.

Jupiter Hammon, the first Negro in this country to write and publish poetry, was a slave owned by a Mr. Joseph Lloyd of Queens Village, Long Island. Hammon had been converted to the religion of Jesus Christ and all of his poems are religious exhortations, incoherent in thought and crudely executed. His first poem was published in 1761, his second, entitled "An Address to Miss Phillis Wheatley, Ethiopian Poetess in Boston, who came from Africa at eight years of age and soon became acquainted with the Gospel of Jesus Christ," in 1768.

This Miss Phillis Wheatley, who had been bought from a slave-ship by a family named Wheatley in Boston Harbor and educated by them, wrote better doggerel than her older contemporary Hammon. She knew Alexander Pope and she knew Ovid—Hammon only knew the Bible—and she knew Pope so well that she could write like a third-rate imitator of him. Phillis in her day was a museum figure who would have caused more of a sensation if some contemporary Barnum had exploited her. As it was she attracted so much attention that many soft hearted (and, in some cases, soft headed) whites and blacks have been led to believe that her poetry deserves to be considered as something more than a mere historical relic. This is an excerpt from her best poem:

> *Imagination!* who can sing thy force?
> Or who describe the swiftness of thy course?
> Soaring through the air to find the bright abode,
> The empyreal palace of the thund'ring God,
> We on thy pinions can surpass the wind,
> And leave the rolling universe behind:
> From star to star the mental optics rove,
> Measure the skies, and range the realms above.
> There in one view we grasp the mighty whole,
> Or with new worlds amaze th' unbounded soul.

She never again equaled the above, far less surpassed it. And most of the time she wrote as in the following excerpt from "On the Capture of Major General Lee." (This poem would warm the heart of "Big Bill" Thompson of Chicago; he really should know about it.) A captured colonial soldier is addressing a British general:

> . . . Oh arrogance of tongue!
> And wild ambition, ever prone to wrong!
> Believ'st thou chief, that armies such as thine
> Can stretch in dust that heaven-defended line?
> In vain allies may swarm from distant lands,
> And demons aid in formidable bands.
> Great as thou art, thou shun'st the field of fame,
> Disgrace to Britain, and the British name!

She continues in this vein, damning the British and enshrining the Americans until she reaches a climax in the following priceless lines:

> Find in your train of boasted heroes, one
> To match the praise of Godlike Washington.
> Thrice happy chief in whom the virtues join,
> And heaven-taught prudence speaks the man divine!

Thomas Jefferson is quoted as saying that "Religion has produced a Phillis Wheatley, but it could not produce a poet. Her poems are beneath contempt." Nevertheless, Phillis had an interesting and exciting career. The Wheatleys carried her to London, where her first volume was published in 1773. She was exhibited at the Court of George III, and in the homes of the nobility much as the Negro poets of today are exhibited in New York drawing rooms. She wrote little about slavery, which is not surprising considering that save for her epic trip across the Atlantic in a slave-ship, she had never known slavery in any form. She often mentioned her homeland and once spoke of herself as

"Afric's muse," but she was more interested in the religion of Jesus Christ and in the spreading of piety than in any more worldly items, save perhaps in her patriotic interest for the cause of the American colonists.

Heretofore every commentator, whether white or black, when speaking of Phillis Wheatley, has sought to make excuses for her bad poetry. They have all pointed out that Phillis lived and wrote during the eighteenth century, when, to quote from the introduction to White and Jackson's *Poetry of American Negro Poets* "the great body of contemporary poetry was turgid in the style of debased Pope." It would be too much, they continue, to expect "a poet of Phillis Wheatley's rather conventional personality to rise above this influence." In his preface to *The Book of American Negro Poetry,* James Weldon Johnson contends that "had she come under the influence of Wordsworth, Byron, Keats, or Shelley, she would have done greater work." Does it smack too much of lese-majesty to suggest that perhaps Phillis wrote the best poetry she could have written under any influence, and that a mediocre imitation of Shelley would have been no less mediocre than a mediocre imitation of Pope? Phillis was also influenced by the Bible, but her paraphrases of the scripture are just as poor as her paraphrases of "debased Pope."

Phillis died in 1784 and until Paul Lawrence Dunbar published his *Oak and Ivy,* in 1892, American Negro poetry stayed at the level at which she had left it, although there must have been over one hundred Negroes who wrote and published poetry during this period. Most of them came into prominence during and after the Civil War, and were encouraged by abolitionists to write of their race and their race's trials. Frances Ellen Harper is probably the best of this period. One volume, *On Miscellaneous Subjects,* was published with an introduction by William Lloyd Garrison. Over ten thousand copies were circulated. Mrs. Harper also wrote and published *Moses, a Story of the Nile,* in verse covering fifty-two closely printed pages. Many of her contemporaries were equally ambitious. Length was a major poetic virtue to them.

It seems highly probable that these people wrote in verse because neither their minds nor their literary tools and backgrounds were adequate for the task of writing readable and intelligent prose. They could be verbose and emotional in verse, and yet attain a degree of coherence not attainable when they wrote in prose. George M. Horton is a good illustration. He was born a slave in Chatham County, North Carolina, in 1797. It is said that he "was not a good farm worker on account of devoting too much time to fishing, hunting, and attending religious meetings." He taught himself to read with the aid of a Methodist hymn book and a red-backed speller. In 1830 he secured work as a janitor at Chapel Hill, the seat of the University of North Carolina. Here he made extra money writing love poems for amorous students. Desiring to obtain his freedom and migrate to Liberia, Horton, aided by some of his white friends, published a volume of verse entitled *The Hope of Liberty,* but the returns from the sale of this volume were not sufficient for his purpose. But he remained more or less a free agent, and was allowed to hire himself out instead of having to remain on his master's plantation. In 1865, a troop of Federal soldiers, who had been quartered in Chapel Hill, was ordered North. Horton left with them and went to Philadelphia where he eventually died.

Here is a sample of his prose: "By close application to my book and at night my visage became considerably emaciated by extreme perspiration, having no lucubratory apparatus, no candle, no lamp, not even lightweed, being chiefly raised in oaky woods." And here is a sample of his verse:

> Come, Liberty, thou cheerful sound,
> Roll through my ravished ears,
> Come, let my grief in joys be drowned
> And drive away my fears.

Further comment would be superfluous.

After the Civil War, the Negro found himself in a dilemma. He was supposed to be free, yet his condition was little changed. He was worse off in some respects than he had been before. It can be understood, then, that the more articulate Negroes of the day spent most of their time speculating upon this thing called freedom, both as it had been imagined and as it was in actuality.

However, none of the poetry written at this time is worthy of serious critical consideration. It was not even a poetry of protest. Although Negro poets objected to the mistreatment of their people, they did not formulate these objections in strong, biting language, but rather sought sympathy and pled for pity. They wept copiously but seldom manifested a fighting spirit. The truth is, only one American Negro poet has been a fighting poet, only one has really written revolutionary-protest poetry, and that is Claude McKay, who will be considered later.

Paul Lawrence Dunbar was the first American Negro poet whose work really merited critical attention. Dunbar was the son of two ex-slaves, both supposedly full-blooded Negroes, a fact flagrantly paraded by race purists, to controvert the prevalent Nordic theory that only Negroes with Caucasian blood in their veins ever accomplish anything. He was born in Dayton, Ohio, June 27th, 1872. His father had escaped from his master and fled to Canada, but later returned to the States and enlisted for military service during the Civil War in a Massachusetts regiment. Dunbar may have inherited his love for letters and writing from his mother, whose master had often read aloud in her presence.

Dunbar attended the public schools in his home town, and was graduated from the local high school, where he had edited the school paper. Then he found employment as an elevator operator. In 1892 he delivered an address in verse to the Western Association of Writers, and shortly afterwards he published his first volume, *Oak and Ivy.* In 1896, through the subscription method, he was able to publish another volume entitled,

*Majors and Minors.* William Dean Howells wrote a most favorable review of this volume and later paved the way for Dodd, Mead and Company to publish *Lyrics of Lowly Life,* for which he wrote an introduction. Meanwhile Dunbar had visited England, and had become a great friend with Coleridge-Taylor, the Negro composer, with whom he collaborated on many songs. On his return to the United States, another friend, Robert G. Ingersoll, helped him to get a position in the Library of Congress. He was only able to keep this job two years, for meanwhile he had developed pulmonary tuberculosis, and despite pilgrimages to such lung-soothing climates as the Adirondacks, the Rockies, and Florida, he finally succumbed to the disease and died in Dayton, Ohio, on February 9, 1908.

From 1892 until the time of his death Dunbar published five volumes of verse, four volumes of collected short stories, and four novels. Not only was he the first Negro to write poetry which had real merit and could be considered as having more than merely sentimental or historical value, but he was also the first Negro poet to be emancipated from Methodism, the first American Negro poet who did not depend on a Wesleyan hymn-book for inspiration and vocabulary. Most of the poets preceding him were paragons of piety. They had all been seized upon by assiduous missionaries and put through the paces of Christianity, and their verses were full of puerile apostrophizing of the Almighty, and leaden allusions to Scriptural passages.

Yet Dunbar was far from being a great poet. First of all, he was a rank sentimentalist, and was content to let surface values hold his interest. He attempted to interpret the soul of his people, but as William Stanley Braithwaite has said, he succeeded "only in interpreting a folk temperament." And although he was, as William Dean Howells affirmed, the first "man of pure African blood of American civilization to feel Negro life aesthetically and express it lyrically," neither his aesthetic feeling nor his expression ever attained enough depth to be of permanent value.

Dunbar is famous chiefly for his dialect poetry. Yet he often regretted that the world turned to praise "a jingle in a broken tongue." He was ambitious to experiment in more classical forms, and to deal with something less concrete than the "smile through your tears" plantation darky of reconstruction times. Here perhaps was his greatest limitation. Being anxious to explore the skies, he merely skimmed over the surface of the earth.

After Dunbar, there was a whole horde of Negro poets, who like him, wrote in dialect. The sum total of their achievement is zero, but happily, in addition to these parasitic tyros there were also two new poets who had more originality and more talent than their contemporaries. And though neither of these men produced anything out of the ordinary, they did go beyond the minstrel humor and peasant pathos of Dunbar, and beyond the religious cant and doggerel jeremiads of Dunbar's predecessors. One of these men,

William Stanley Braithwaite, is best known as a student and friend of poets and poetry rather than as a poet. He has yearly, since 1913, issued an anthology of American magazine verse, and has also published some academic studies of English literature.

The second, James Weldon Johnson, achieved little as a poet until recently, when he published *God's Trombones,* a volume of Negro sermons done in verse. His first volume, *Fifty Years and Other Poems,* contains little of merit. The title poem, which recounts in verse the progress of the race from 1863 to 1913, has, because of its propagandist content, been acclaimed as a great poem. No comment or criticism is necessary of this opinion when part of the poem itself can be quoted:

> Far, far the way that we have trod
> From heathen kraals and jungle dens,
> To freedmen, freemen, sons of God,
> Americans and Citizens.

Mr. Johnson, it seems, has also been fairly intimate with Methodist hymn books.

His sermon poems, while at times awkward and faulty in technique, have an ecstatic eloquence and an individual rhythm which immediately place them among the best things any Negro has ever done in poetry. Although this may not be saying much, and although, as a poet Mr. Johnson may not be adequate to the task of fully realizing the promise of these sermon poems, he has at least laid a foundation upon which a new generation of Negro poets can build. He will have to be remembered as something more than just a historical or sentimental figure. He, like Dunbar, is an important, if a minor bard; and if the Negro poet of the future is to make any individual contribution to American literature, he must derive almost as much from the former's *God's Trombones* as from the latter's *Lyrics of Lowly Life.*

To consider all the Negro poets who since 1913 have lifted up their voices in song would necessitate using an entire issue of any journal. It is not only an impossible task but one not worth the time and space it would require. For our present study we will touch only the high spots, passing over such people as Fenton Johnson, whose early promise has never been fulfilled; Joseph Cotter, Jr., who, it is alleged by most critics in this field, would have been a great poet had he lived but whose extant work belies this judgment; Georgia Douglas Johnson, whose highly sentimental and feminine lyrics have found favor; Arna Bontemps, who specializes in monotonous and wordy mystic evocations which lack fire and conviction, and Helene Johnson, who alone of all the younger group seems to have the "makings" of a poet.

But taking up the contemporary triumvirate—McKay, Cullen, and Hughes—all of whom have had volumes published by reputable houses and are fairly well known to the poetry-reading public, we have poets of another type. Each one of them represents a different trend in Negro literature and life.

Claude McKay was born in Jamaica, British West Indies, where he received his elementary education, served a while in the constabulary, and wrote his first poems. A friend financed his journey to America to finish his scholastic work, but McKay found himself at odds with the second-rate schools he attended here, and finally fled to New York City where he became a member of the old *Masses, Seven Arts, Liberator* group of radicals and artists. During this period he received a legacy which, he tells us, was spent in riotous living. Broke, he attempted to make a living by washing dishes, operating elevators, doing porter work—the usual occupations engaged in by Negro artists and intellectuals.

McKay's first volume was published while he was still in Jamaica, a compilation of folk-verse done in the native dialect. The Institute of Arts and Science of Jamaica gave him a medal in recognition of this first book. It is in many ways remarkable, and in it the poet gives us a more substantial portrait and delves far deeper into the soul of the Jamaican than Dunbar was ever able to in the soul of the southern Negro in America.

McKay's latter poetry is often marred by bombast. He is such an intense person that one can often hear the furnace-like fire within him roaring in his poems. He seems to have more emotional depth and spiritual fire than any of his forerunners or contemporaries. It might be added that he also seems to have considerably more mental depth too. His love poems are not as musical or as haunting as Mr. Cullen's, but neither are they as stereotyped. His sonnet to a Harlem dancer may not be as deft or as free from sentiment as "Midnight Nan" by Langston Hughes, but it is far more mature and moving. All of which leads us to say that a study of Claude McKay's and of the other better Negro poetry convinces us that he, more than the rest, has really had something to say. It is his tragedy that his message was too alive and too big for the form he chose. His poems are for the most part either stilted, choked, or overzealous. He could never shape the flames from the fire that blazed within him. But he is the only Negro poet who ever wrote revolutionary or protest poetry. Hence:

> If we must die, let it be not like hogs
> Hunted and penned in an inglorious spot,
> . . . . . . . . . . . . . . . . . . . . . . . . . . . . .
> Oh, kinsman! we must meet the common foe!
> Though far outnumbered, let us show us brave,
> And for their thousand blows deal one deathblow!
> What though before us lies the open grave?
> Like men we'll face the murderous, cowardly pack,
> Pressed to the wall, dying, but fighting back.

There is not impotent whining here, no mercy-seeking prayer to the white man's God, no mournful jeremiad, no "ain't it hard to be a nigger," no lamenting of or apologizing for the fact that he is a member of a dark-skinned minority group. Rather he boasts:

> Be not deceived, for every deed you do,
> I could match—out match; Am I not Africa's son,
> Black of that black land where black deeds are done?

This is propaganda poetry of the highest order although it is crude and inexpert. Contrast it with these lines from Countee Cullen's sonnet "From the Dark Tower":

> We shall not always plant while others reap
> The golden increment of bursting fruit,
> Nor always countenance, abject and mute
> That lesser men should hold their brothers cheap; . . .

Countee Cullen is the symbol of a fast disappearing generation of Negro writers. In him it reaches its literary apogee. On the other hand, Langston Hughes announces the entrance of a new generation, while Claude McKay, glorious revolutionary that he is, remains uncataloged. For two generations Negro poets have been trying to do what Mr. Cullen has succeeded in doing. First, trying to translate into lyric form the highly poetic urge to escape from the blatant realities of life in America into a vivid past, and, second, fleeing from the stigma of being called a *Negro* poet, by, as Dunbar so desired to do, ignoring folk-material and writing of such abstractions as love and death.

There is hardly anyone writing poetry in America today who can make the banal sound as beautiful as does Mr. Cullen. He has an extraordinary ear for music, a most extensive and dexterous knowledge of words and their values, and an enviable understanding of conventional poetic forms. Technically, he is almost precocious, and never, it may be added, far from the academic; but he is also too steeped in tradition, too influenced mentally by certain conventions and taboos. When he does forget these things as in his greatest poem, "Heritage":

> What is Africa to me:
> Copper sun or scarlet sea,
> Jungle star or jungle track,
> Strong bronzed men, or regal black
> Women from whose loins I sprang
> When the birds of Eden sang?
> *One three centuries removed*
> *From the scenes his fathers loved,*
> *Spicy grove, cinnamon tree,*
> *What is Africa to me?*

and the unforgettable:

*All day long and all night through,*
*One thing only must I do:*
*Quench my pride and cool my blood,*
*Lest I perish in the flood,*
*Lest a hidden ember set*
*Timber that I thought was wet*
*Burning like the dryest flax,*
*Melting like the merest wax,*
*Lest the grave restore its dead,*
*Not yet has my heart or head*
*In the least way realized*
*They and I are civilized.*

or his (to illustrate another tendency):

I climb, but time's
Abreast with me;
I sing, but he climbs
With my highest C.

and in other far too few instances, he reaches heights no other Negro poet has ever reached, placing himself high among his contemporaries, both black or white. But he has not gone far enough. His second volume is not as lush with promise or as spontaneously moving as his first. There has been a marking time or side-stepping rather than a marching forward. If it seems we expect too much from this poet, we can only defend ourselves by saying that we expect no more than the poet's earlier work promises.

Mr. Cullen's love poems are too much made to order. His race poems, when he attempts to paint a moral, are inclined to be sentimental and stereotyped. It is when he gives vent to the pagan spirit and lets it inspire and dominate a poem's form and context that he does his most impressive work. His cleverly turned rebellious poems are also above the ordinary. But there are not enough of these in comparison to those poems which are banal, though beautiful.

Langston Hughes has often been compared to Dunbar. At first this comparison seems far-fetched and foolish, but on closer examination one finds that the two have much in common, only that where Dunbar failed, Langston Hughes succeeds. Both set out to interpret "the soul of his race"; one failed, the other, just at the beginning of his career, has in some measure already succeeded.

The younger man has not been content to assemble a supply of stock types who give expression to stock emotions which may be either slightly amusing or slightly tragic, but which are never either movingly tragic or convincingly comic. When Langston Hughes writes of specific Negro types he manages to make them more than just ordinary Negro types. They are actually dark-skinned symbols of universal characters. One never feels this way about people in Dunbar's poetry. For he never heightens them above their own particular sphere. There is never anything of the universal element in his poems that motivates Mr. Hughes's.

Moreover, Langston Hughes has gone much farther in another direction than any other Negro poet, much farther even than James Weldon Johnson went along the same road in *God's Trombones*. He has appropriated certain dialects and rhythms characteristically Negroid as his poetic properties. He has borrowed the lingo and locutions of migratory workers, chamber-maids, porters, boot-blacks, and others, and woven them into rhythmic schemes borrowed from the blues songs, spirituals and jazz and with them created a poetic diction and a poetic form all his own. There is danger in this of course, for the poet may and often does, consider these things as an end in themselves rather than as a means to an end. A blues poem such as:

> I'm a bad, bad man
> Cause everybody tells me so.
> I'm a bad, bad man,
> Everybody tells me so.
> I takes ma meanness and ma licker
> Everywhere I go.

or:

> Ma sweet good man has
> Packed his trunk and left.
> Ma sweet good man has
> Packed his trunk and left.
> Nobody to love me:
> I'm gonna kill ma self.

may be poignant and colorful but the form is too strait-laced to allow much variety of emotion or context. The poems produced are apt to prove modish and ephemeral. But when this blues form is expanded, as in:

> Drowning a drowsy syncopated tune,
> Rocking back and forth to a mellow croon,
>    I heard a Negro play.
> Down on Lenox Avenue the other night
> By the pale dull pallor of an old gas light
>    He did a lazy sway . . .
>    He did a lazy sway . . .
> To the tune o' those Weary Blues.
> With his ebony hands on each ivory key
> He made that poor piano moan with melody.
>    O Blues!

Swaying to and fro on his rickety stool
He played that sad raggy tune like a musical fool.
    Sweet Blues!
Coming from a black man's soul.

the poet justifies his experiment, and finds at the same time the most felicitous and fruitful outlet for his talent.

Mr. Hughes, where his race is concerned, is perfectly objective. He is one of them so completely that he, more than any other Negro poet, realizes that after all they are human beings; usually the articulate Negro either regards them as sociological problems or debased monstrosities. To Mr. Hughes, certain types of Negroes and their experiences are of permanent value. He is not afraid of, nor does he ignore, them. He can calmly say:

Put on yo' red silk stockings,
Black gal.
Go an' let de white boys
Look at yo' legs.
. . . . . . . . . . . . .
An' tomorrow's chile'll
Be a high yaller.

or:

My old man's a white old man
And my old mother's black
. . . . . . . . . . . . . . . . . . .
My old man died in fine big house.
My ma died in a shack.
I wonder where I'm gonna die,
Being neither white nor black?

and reach the heights of his achievement in "Mulatto," one of the finest and most vivid poems written in the past few years. But Mr. Hughes has also written some of the most banal poetry of the age, which has not, as in the case of Mr. Cullen, even sounded beautiful.

The future of Negro poetry is an unknown quantity, principally because those on whom its future depends are also unknown quantities. There is nothing in the past to crow about, and we are too close to the present to judge it more than tentatively. McKay is exiled in France, an alien and a communist, barred from returning to this country. Once in a while a poem of his appears, but the period of his best work in this field seems to be at an end. Langston Hughes and Countee Cullen are both quite young, as poets and as individuals. Neither can be placed yet, nor can their contributions be any more than just intelligently commented upon. Whether they are going to or will continue to go in the right direction is no more than a matter of individual opinion. All of us do know that as yet the American Negro has not produced a great poet. Whether he will or not is really not at all important. What does matter is that those who are now trying to be great should get intelligent guidance and appreciation. They seem to have everything else except perhaps the necessary genius.

## Editorial Essay

*Harlem: A Forum of Negro Life*, November 1928.

In the past there have been only a few sporadic and inevitably unsuccessful attempts to provide the Negro with an independent magazine of literature and thought. Those magazines which have lived throughout a period of years have been organs of some philanthropic organization whose purpose was to fight the more virulent manifestations of race prejudice. The magazines themselves have been pulpits for alarmed and angry Jeremiahs spouting fire and venom or else weeping and moaning as if they were either predestined or else unable to do anything else. For a while this seemed to be the only feasible course for Negro journalists to take. To the Negro then the most important and most tragic thing in the world was his own problem here in America. He was interested only in making white people realize what dastards they were in denying him equal economic opportunities or in lynching him upon the slightest provocation. This, as has been said, was all right for a certain period, and the journalists of that period are not to be censored for the truly daring and important work they did do. Rather, they are to be blamed for not changing their journalistic methods when time and conditions warranted such a change, and for doing nothing else but preaching and moaning until they completely lost their emotional balance and their sense of true values. Every chord on their publicist instrument had been broken save one, and they continued raucously to twang this, unaware that they were ludicrously out of tune with the other instruments in their environment.

Then came the so-called renaissance and the emergence of the so-called new (in this case meaning widely advertised) Negro. As James Weldon Johnson says in the current issue of *Harper's* magazine: "The Negro has done a great deal through his folk art creations to change the national attitudes toward him; and now the efforts of the race have been reinforced and magnified by the individual Negro artist, the conscious artist. . . . Overnight, as it were, America became aware that there were Negro artists and that they had something worthwhile to say. This awareness first manifested itself in black America, for, strange as it may seem, Negroes themselves, as a mass, had had little or no consciousness of their own individual artists."

Naturally these new voices had to be given a place in Negro magazines and they were given space that hitherto had been devoted only to propaganda. But the artist was not satisfied to be squeezed between jeremiads or have his work thrown haphazardly upon a page where there was no effort to make it look beautiful as well as sound beautiful. He revolted against shoddy and sloppy publication methods, revolted against the patronizing attitudes his elders assumed toward him, revolted against their editorial astigmatism and their intolerance of new points of view. But revolting left him without a journalistic asylum. True, he could, and did, contribute to the white magazines, but in doing this almost exclusively he felt that he was losing touch with his own group, for he knew just how few Negroes would continually buy white magazines in order to read articles and stories by Negro authors, and he also knew that from a sense of race pride, if nothing more, there were many Negroes who would buy a Negro magazine.

The next step then was for the artist himself to produce this new type of journal. With little money but a plethora of ideas and ambition he proceeded to produce independent art magazines of his own. In New York, *Fire!!* was the pioneer of the movement. It flamed for one issue and caused a sensation the like of which had never been known in Negro journalism before. Next came *Black Opals* in Philadelphia, a more conservative yet extremely worthwhile venture. Then came *The Quill* in Boston which was to be published whenever its sponsors felt the urge to bring forth a publication of their own works for the benefit of themselves and their friends. And there were other groups of younger Negroes in Chicago, Kansas City and Los Angeles who formed groups to bring out independent magazines which never became actualities.

This last development should have made someone realize that a new type of publication was in order. The old propagandistic journals had served their day and their generation well, but they were emotionally unprepared to serve a new day and a new generation. The art magazines, unsoundly financed as they were, could not last. It was time for someone with vision to found a wholly new type of magazine, one which would give expression to all groups, one which would take into consideration the fact that this was a new day in the history of the American Negro, that this was a new day in the history of the world and that new points of view and new approaches to old problems were necessary and inescapable.

*Harlem* hopes to fill this new need. It enters the field without any preconceived editorial prejudices, without intolerance, without a reformer's cudgel. It wants merely to be a forum in which all people's opinions may be presented intelligently and from which the Negro can gain some universal idea of what is going on in the world of thought and art. It wants to impress upon the literate members of the thirteen million Negroes in the United States the necessity of becoming "book conscious," the necessity of reading the newer Negro authors, the necessity of realizing that the Negro is not the only, nor the worst mistreated minority group in the world, the necessity of sublimating their inferiority complex and their extreme race sensitiveness and putting the energy, which they have hitherto used in moaning and groaning, into more concrete fields of action.

To this end *Harlem* will solicit articles on current events, essays of the more intimate kind, short stories and poetry from both black and white writers; the only qualification being that they have sufficient literary merit to warrant publication. *Harlem* will also promote debates on both racial and non-racial issues, giving voice to as many sides as there seem to be to the question involved. It will also be a clearing house for the newer Negro literature, striving to aid the younger writers, giving them a medium of expression and intelligent criticism. It also hopes to impress the Negro reading public with the necessity for a more concerted and well-balanced economic and political program. It believes that the commercial and political elements within the race are just as in need of clarification as the literary element and will expend just as much energy and time in the latter fields as in the former.

This is *Harlem*'s program, its excuse for existence. It now remains to be seen whether the Negro public is as ready for such a publication as the editors and publishers of *Harlem* believe it to be.

## High, Low, Past, and Present:
## Review of *The Walls of Jericho, Quicksand,* and *Adventures of an African Slaver*

*Harlem: A Forum of Negro Life*, November 1928.

I had already written a review of *The Walls of Jericho* and was about to send it into the printer when I chanced to open a copy of *The Crisis* for November, and found therein a review of the same work by W.E.B. Du Bois. The following paragraph set my teeth on edge and sent me back to my typewriter hopping mad. Listen to this:

> Mr. Fisher does not yet venture to write of himself and his own people: of Negroes like his mother, his sister, and his wife. His real Harlem friends and his own soul nowhere yet appear in his pages, and nothing that can be mistaken for them. The glimpses of better class Negroes which he gives us are poor, ineffective make-believes. One wonders why? Why does Mr. Fisher fear to use his genius to paint his own kind, as he has painted Shine and Linda? Perhaps he doubts the taste of his white audience although he tries it severely with Miss Cramp. Perhaps he feels too close to his own to trust his artistic detachment in limning them. Perhaps he really laughs at all life, and believes nothing. At any rate, here is a step upward from Van Vechten and McKay—a strong, long, interesting step. We hope for others.

The more I reread the above lines the more angry and incoherent I became. I was not so much worried about the effect such a narrow and patronizing criticism would have on Mr. Fisher or on any other of Dr. Du Bois' audience who might take it seriously, as I was concerned for what it tokened for the reviewer himself. Were he a denizen of "Strivers' Row," scuttling hard up the social ladder, with nothing more important to think about than making money and keeping a high yellow wife bleached out and marcelled, one would laugh at such nonsense and dismiss it from one's mind. But Dr. Du Bois is not this. He is one of the outstanding Negroes of this or any other generation. He has served his race well; so well, in fact, that the artist in him has been stifled in order that the propagandist may thrive. No one will object to this being called noble and necessary sacrifice, but the days for such sacrifices are gone. The time has come now when the Negro artist can be his true self and pander to the stupidities of no one, either white or black.

Anyone with a knowledge of literature and the people who write it should know that when the truly sincere artist begins to write he does not take into consideration what the public might say if his characters happen to be piano movers or his wife and sister. He is drawn toward certain characters and certain situations which interest him and which seem worth writing about. There happens to be no Will Hays or Judge Landis of literature to say: "Nay, nay, dear scribe. Don't you dare write about such and such a situation or such and such a character. They're not nice. What will the best people in the community think?" It happens that most writers have all been able and brave enough to say, "To hell with what the best people in the community think," because they know that generally speaking, the best people in the community do not think at all. The entire universe is the writer's province and so are all the people therein, even lower class Negroes, and if they happen to attract the writer there is no reason why he shouldn't write about them. Nor is it implied here that all Negro writers should write only of "the half world above 125th Street," for such an implication would be just as ridiculous as the one being constantly made by Dr. Du Bois.

*The Walls of Jericho* is a disappointing book from the standpoint that Mr. Fisher's short stories, published off and on for the past three years, have led one to believe that his first novel would be a more unusual piece of work. And it *is* unusual in one respect, being the first novel written by a Negro wherein the author handles his theme and writes with enviable ease; the first novel written by a Negro which does not seem to be struggling for breath because the author insists upon being heavy handed either with propaganda as in *Dark Princess* or with atmosphere as in *Home to Harlem*. Mr. Fisher keeps his proportions well, almost too well, and despite what Dr. Du Bois says, does not give us any "ineffectual make believes" when a "better class Negro" appears on the scene. But after that what have you? Some brilliant bits of authentic dialogue, some biting caricatures, *viz*, Miss Cramp, but no sustaining characterizations or anything vital, truly indicative of the gifts Mr. Fisher so ably displayed in one of the best short stories of Negro life ever written: "The City of Refuge."

Had this novel been written by some of the lesser lights, or greater ones for that matter, among Negro authors, one's applause would be less constrained. But here is the case where the author lays himself open to criticism not because he has not been good, but because he has not been good enough. However, I am glad that Mr. Fisher has this off his chest, glad that he has proven that it is not necessary for a Negro writer to moan and groan and sweat through a book simply because he is a Negro, and I hope he comes across in the near future with something that will not be a let-down from the man who could write the remarkable short story mentioned above. More, I even hope he takes Dr. Du Bois' suggestion and applies his artistic detachment to his own kind and I hope he includes Dr. Du Bois in his gallery of characters. Then the fun will really begin, and I know of no one better than Mr. Fisher to do this as it should be done.

The author of *Quicksand* no doubt pleases Dr. Du Bois for she stays in her own sphere and writes about the sort of people one can invite to one's home without losing one's social prestige. She doesn't give white people the impression that all Negroes are gin drinkers, cabaret hounds and of the half world. Her Negroes are all of the upper class. And how!

Nevertheless, one has to admit that the performance here is a little less impressive than Mr. Fisher's. Not because of her people or because of the milieu in which they move, but purely because the author seems to be wandering around lost, as lost as her leading character who ends up by doing such an unexpected and unexplainable thing that I was forced to reread the book, wondering, if in my eagerness to reach the end, I had perhaps skipped a hundred pages or so. But no, such had not been the case. Helga does get blown into the gutter and Helga does let herself be carried away by a religious frenzy to the point where she marries a Southern minister and spends the rest of her life having babies. This would have been all right for anyone except the Helga to whom Miss Larsen had introduced us, and even then it would have been all right had the author even as much as hinted that someday her character might do either the expected or the unexpected. But for the most part all Helga ever does is run away from certain situations and straddle the fence; so consistently, in fact, that when she does fall on the dark side, the reader has lost all interest and sympathy, nor can he believe that such a thing has really happened.

Captain Canot was a jolly old soul and in this narrative almost convinces one that the slave trader was a much maligned and noble creature. It wasn't his fault that there was a slave trade. Could he help it that Nordic and Latin tradesmen fomented civil strife among African tribes and bought the vanquished from the victors with brilliantly colored cloth, German glass, English rum and American tobacco? Not at all. He was just trying to make a living and having a good time while doing so. You even suspect that Captain Canot had a good time writing this narrative and can almost hear the reverberation of what Walter Winchell would call his belly laughs, as he thought of the gullibility of the human race in general and of the readers of this book in particular.

This is a book that almost anyone could enjoy, even a bitter twentieth-century Negro, for it contains a wealth of information and makes good and interesting reading despite the rather turgid and bombastic style. The drawings by Covarrubias alone are worth the price of the book. If Captain Canot makes the crossing of the slaves in the middle passage seem like a luxurious and interesting event, Covarrubias makes one realize just what a brutal and indescribable experience it really must have been, and what is more interesting, his delineation of African types is no less than masterful.

Captain Canot, were he alive now, would probably be a guiding spirit among rum runners, for he would never be content to participate in any but a lawless pursuit. It would please him to exchange shots with the revenue cutter, please him to drop a boat load of liquor into the ocean rather than have it taken by government officers. He did as much with slaves and, despite his holy protestations that he was always kind and generous to those with whom he was entrusted, the reader can easily sense that a load of contraband slaves were to him of no more human consequence than would be a load of contraband liquor.

Yet there is no doubt much truth in what he says. Students in this field have long known that no one was more assiduous in selling their brethren into slavery than certain African tribes. It was their means of punishment to the criminal, their means of ridding the tribe of its enemies and of the unfit, their means of making away with recalcitrant parents, husbands, wives, and children. And one also can not deny that there is much truth in the author's statement that the better the slaves were cared for on their voyage the more money they would draw at the auction block. At any rate slave trading was a profitable and adventurous business and one can hardly condemn a man of Captain Canot's caliber for entering into it with so much zeal and enthusiasm. He at least made the most of his opportunities and we most certainly thank him for this book. Can one say more?

## Langston Hughes Turns from Poetry to the Novel to Give a Sincere Picture of Negro Life in America

*The New York Evening Post*, July 28, 1930.

Langston Hughes's novel, *Not Without Laughter* (Knopf, $2.50), will confound many soothsayers and prematurely pessimistic critics. For the poet has proven that he can write prose adequate for the story he has to tell, and he has not concerned himself either with Harlem's Midnight Nans or the southland's cottonfield sweet men. He has depicted, rather, as the blurb on the jacket so conveniently states, a "deeply human picture of a simple people, living in a typical Kansas town, meeting, as best they can, the problems of their destructive, complex environment." He has done this, that is, for 249 of his 308-page novel. Remember this reservation for future reference.

Aunt Hagar Williams is an aged Negro woman, a pious, homespun, ample-bosomed individual, beloved by all with whom she comes into frequent contact. She lives her simple life, immersed in the steam and suds of her washtubs, dedicated to the welfare of her children, to the grace of God, and to the service of her neighbor should he be in need of her assistance. And, as a reward for this exemplary conduct, makes such an impression upon the community in which she lives that notice of her death merits, "in small type on its back page," the following paragraph from *The Daily Leader*: "Hagar Williams, aged colored laundress, of 419 Cypress Street, passed away at her home last night. She was known and respected by many white families in the community. Three daughters and a grandson survive."

One of these surviving daughters, Tempy, is a victim of that lamentable American disease from which there seems to be no antitoxin: Keeping up with the Joneses, the Joneses in Tempy's case being bourgeois whites. With a smattering of what she believes to be culture, a few parcels of income property, and a husband who works for the Government, Tempy spends her time trying to convince herself and the world at large that she is neither part nor parcel of the peasant environment in which she happened to be born. While Tempy sounds as if she might be interesting, she is actually the most poorly delineated character in the novel. The reason is obvious. Being constantly surrounded by a legion of Tempy's prototypes, Mr. Hughes has not been as objective in this portrayal as he might have been had Tempy, like most of his other characters, been more indigenous to the novel's milieu.

The youngest daughter, Harriet, is a naïve hedonist, early and easily seduced by flashy clothes, torrid dancing, spontaneous song fests, stimulating beverages and amorous dalliance. Frustrating all efforts to keep her in school or to make her amenable to mental labor, she gleefully, if not always successfully, defies the wrath of Aunt Hagar's God and Charlestons down a gaudy, primrose path.

The other daughter, Annjee, is the mother of Hagar's one grandchild, Sandy. Annjee is distressingly normal. She works hard, complains but seldom, has no desire to be either a Tempy or a Harriet, and loves her husband, the irresponsible Jimboy, intensely, constantly, albeit he is subject to the "travelin' blues," and walks off from Annjee and her child whenever the spirit so moves him.

It is with Sandy, the son of this haphazard union, whom *Not Without Laughter* primarily concerns itself. Around him most of the action is woven, through his eyes much of the story is reflected, and it is the problem of Sandy's adjustment to his environment, a problem more or less peculiar to the maturing Negro child, which occupies many pages of the novel.

It is not Sandy, however, but Aunt Hagar, Harriet, Jimboy, and Annjee who remain alive in the reader's mind once the novel is completed. And of these Aunt Hagar is by far the most indelible, a fact which explains the reference made to page 249 some paragraphs ago. For when Aunt Hagar dies and crosses over Jordan, the reader is hard put to retain his former high pitch of interest. Simultaneously with her death, Harriet, Jimboy and Annjee also fade into the background. It is too great a bereavement. Either the novel should have managed to end itself at this point or Aunt Hagar's demise should have been postponed. Certainly her heav'nly home could have done without her for a few more years. Her presence is vitally necessary to the more mundane *Not Without Laughter.*

Following the death of his grandmother, Sandy is forced to sojourn with the dicty Tempy, his mother having decided to trail the truant Jimboy. Tempy keeps him in school, urges him to abjure bad company, and, being a myopic disciple of Dr. W.E.B. Du Bois, attempts to invest him with ideals diametrically opposed to those disseminated and held dear by his dead grandmother, a loyal, if vague, disciple of Booker T. Washington. Sandy seems destined to pursue a middle course, and we leave him optimistically facing the future, appropriately misunderstood by Annjee, obligingly relieved of Tempy, and, surprisingly enough, encouraged and aided by Harriet, whose aversion to the straight and narrow path has brought her a measure of fame and fortune on the stage.

*Not Without Laughter* is an enviable first performance. Belonging to a more decorous school, it lacks the dramatic intensity, the lush color and tropical gusto of Claude McKay's *Home to Harlem.* It also lacks the pallid insipidity and technical gaucheries of certain other contemporary novels by Negroes which are best left unnamed and forgotten. But it does present a vivid, sincerely faithful and commendable picture of peasant Negro life in a small American town, a town which vacillates uncertainly, spiritually and physically, from one side of the Mason Dixon line to the other.

In a moment of post-college exuberance, while considering Mr. Hughes's first book of poems, *The Weary Blues,* the present reviewer stridently declaimed that its author was possessed of "an unpredictable and immeasurable potential." After five years, the language might be more simple and more choice, but the spirit of the phrase would remain unchanged. For, as in his poetry Mr. Hughes carried a beacon light for Negro poets, he now, with this volume, advances to the vanguard of those who have recourse to the novel in an earnest endeavor to depict the many-faceted ramifications of Negro life in America.

# Books to Read: Review of *Southern Road*, by Sterling Brown

*The National News: The Newsmagazine of Colored America*, June 2, 1932.

Many have wondered who would be the next Negro poet, as yet unpublished, to join the triumvirate of McKay, Cullen and Hughes. Negro novelists far outnumber Negro poets, despite the fact that our Negro publications are heavily laden with the effusions of innumerable minor poets, none of whom however seem important or talented enough to merit book publication. And recently, our three outstanding Negro poets have turned to fiction, a natural phenomena but one which is liable to result in a dearth of good poetry and a flood of bad novels.

Skeptically, I approached this volume. I had read several poems by the author in *The Crisis* and in *Opportunity*. None of them impressed me. I either passed them off as being imitations of Langston Hughes or else adjudged them as being interesting but of no especial significance.

And now, having read and re-read the poems collected in *Southern Road,* I wish to make a public apology. That is, I will if Mr. Brown allows me to punctuate my apologies by throwing my hat in the air. Either the author has been holding out on his prospective admirers or else I deliberately overlooked the more worthy of his efforts which have seen magazine publication. For certainly I have read nothing of his before which prepared me for some of the excellent poems in this present volume.

Turn to "When de Saints Go Ma'ching Home," a poem built on the foundation provided by a familiar hymn, for a genuine example of Negro folk poetry. And if that doesn't convince you that Mr. Brown is a poet to be reckoned with, linger over "Johnny Thomas," or chuckle over the "Slim Greer" sagas:

> Down in Atlanta,
> De white folks got laws
> For to keep all de niggers
> From laughin' outdoors.
>
> Hope to Gawd I may die
> If I ain't speakin' truth
> Make de niggers do deir laughin'
> In a telefoam booth.
>
> Slim Greer hit de town
> An' de rebs got him told,—
> "Dontcha laugh on de street,
> If you want to die old."

They show Slim the telephone booth. Shines are lined up by the hundreds, waiting for their turns to laugh. This so tickles Slim that he could contain himself no longer. He fought his way through the crowd,

> Pulled de other man out,
> An' bust in de box,
> An' laughed four hours
> By de Georgia clocks.

> Den he peeked through de door,
> An' what did he see?
> *Three* hundred niggers there
> In misery.—

If the above quoted lines and the above mentioned poems do not convince you that here is a fresh and original talent, I'll take a most serious vow (and if you don't believe it's serious ask any of my friends or enemies), to wit: I'll never drink another of Louis' whiskey sours.

I'm not being intentionally flippant. It's just that I'm too excited to write prosaically about a book of poems in which there is little that is conventional or stereotyped. My excitement does not lead me to exclaim that Sterling Brown will emerge as the major poet of his generation. I only wish to go on record as saying that here is a poet who has intelligence, who has humor, sensitiveness, and the spirit of adventure. He can mold a poem after a pattern of his own devising, and he can also turn out a neat sonnet. He is not restricted by propaganda. He does not shy away from material which is essentially negroid. Nor does he whine. And he has the courage of a McKay when he writes a poem like "Strong Man":

> The strong men . . . coming on
> The strong men gittin' stronger.
> Strong men . . .
> Stronger . . .

Impatiently I await more poems from Mr. Brown. This first volume promises much, as much if not more than the extensively ballyhooed first volumes of certain of his contemporaries. What that tokens for the future none of us can tell. But something keeps singing to me that *Southern Road* is but the beginning, a highway which will lead its author to an enviable peak of poetic achievement. Or must I give up drinking whiskey sours?

# Review of *Infants of the Spring,*
# by Wallace Thurman

Unpublished; c. 1932

*Infants of the Spring* is a novel which the author was impelled to write. The characters and their problems cried out for release. They intruded themselves into his every alien thought. And assumed an importance which blinded him to their true value.

The faults and virtues of the novel, then, are the direct result of this inescapable compulsion. And now that his book is a printed reality, the author muses: Had I waited five more years before setting this material down, it might have been more refined and seemed less jejune. But it might also have lacked earnestness and spontaneity necessary for vivid presentation. And I might never have been able to produce anything else without these present unworthy characters willfully insinuating themselves, wreaking havoc.

All of which is an acceptable excuse for having written an unsatisfactory novel. But most certainly no excuse for having allowed it to be published.

Bohemianism, in the popular sense of the term, came to the present author too late to be the youthful fourteen-year-old adventure it should have been, and too early to be clarified by twenty-five-year-old maturity. It came during that uncertain middle period of his life (he was eighteen then) when it was alternately fascinating and repellant, more the former than the latter, but sufficiently both to unbalance him. The result is *Infants of the Spring,* a novel which undoubtedly has contributed much to the author's individual growth, but which he fears will do little to impress a critical public.

Critical public? Hardly. It would be more apt to apply the adjectives "prejudiced," "myopic." For this public, with lamentably few exceptions, is so stereotyped in its mental attitudes, and so intent on surface values, that, for the moment, no Negro novelist can expect intelligent critical guidance or encouraging appreciation.

This is not, mind you, an excuse for a bad novel, nor the empty caviling of a disgruntled author. It is rather a barbed arraignment of those who classify as bad any creative work which does not concur with their biased and pre-conceived illusions of the material presented.

Negroes, themselves, resent any novel, no matter how meritorious, which does not deal with what they call "the better class of Negroes." Meaning the semi-literate bourgeoisie to whom keeping up with the Joneses implies doing nothing of which bourgeois whites might disapprove. Someday a harassed Negro author is going to accept the challenge and truthfully depict this sector of Negro society. And then! . . . well . . . I only hope none of my contemporaries deprives me of the pleasure.

Literate whites, too, have peculiar ideas concerning the material Negro authors should utilize. The more emancipated they are supposed to be from mob delusions, the more insistent they are that the Aframerican novelist limit himself to one certain type of character, the earthy, naïve Negro to whom life is just a bowl of cherries, but who is ultimately strangled by the pits.

The reason for all this, of course, is that few whites and blacks can conceive of Negroes being human beings, subject to the same variations—emotional, physical, and mental—as are the rest of God's chosen anthropoids. The poor black brother has been a problem for so long, the world in general seems to forget that the business of living in our present day civilization can humanize even a transplanted and highly synthetic Ethiopian.

# Part Five

# *Aunt Hagar's Children*

Thurman worked on the manuscript of *Aunt Hagar's Children* during his sojourn in Salt Lake City and Los Angeles from April to September 1929. His caustic and uncompromising analyses of middle-class pretensions, always part of his writing, are consolidated and intensified in this compilation of essays on black life in the United States. Comprised of both new essays and ones recycled from *The Bookman, The Independent, The New Republic, The Dance Magazine, The Haldeman-Julius Quarterly,* and *Fire!!, Aunt Hagar's Children* is notable for the extent to which Thurman foregrounds himself and for his unrelenting critique of pretension and self-delusion. Thurman arranged the essays so that they start with the most individual topics and move toward the most collective ones. The early essays present his personal perspective, followed by the biographical/ historical essays that establish an intellectual foundation, and the final essays bring that perspective to bear on the pressing questions of race, identity, and community in contemporary life.

    Written in the tradition of H. L. Mencken, *Aunt Hagar's Children* assembles many ideas and postures that can be traced to Mencken's iconoclasm. As

Eleonore van Notten has noted, Thurman's Menckenian tenets "include Thurman's position as observer-participant among his peers; his eagerness to display the superiority of his intellect; his combined scepticism, opportunism, egoism, and self-reliance; his preoccupation with the self; his loathing of sentimentality and provincialism; and his criticism of traditional middle-class morality and respectability" (111). As Charles Scruggs has discussed, Mencken influenced many black writers and published some of them—especially George Schuyler—in *The American Mercury.* Thurman told Rapp in his letters that he had sent three of his *Aunt Hagar* essays to Mencken, although they were not published in *The American Mercury.* Nonetheless, with Thurman's pervasive Menckenian streak, he can be counted among those black writers who, as Mencken saw it, knew from experience that "beneath the smug surface of American life was a core of rottenness" (Scruggs, 25). Thurman surely had the "urbane persona" that Mencken wanted black writers to cultivate, a persona founded on the following principles: "You are a civilized man; the racist is not; do not fight with him to grant you your humanity, but laugh at the loss of his; do not admit inferiority in any sense, but claim, directly or indirectly, that you are superior" (Scruggs, 66).

There are other influences behind Thurman's *Aunt Hagar's Children.* The name "Hagar" invokes the passages in Genesis 16 and 21 that tell of Abraham's servant, who was presented to him by his barren wife, Sarah, so that he might produce a male heir. Hagar's son, Ishmael, is prophesied to be a wild and embattled outcast—but also predicted to become the father of a great nation. As Steven Tracy has noted in his unpublished essay "'Did You Ever Dream Lucky'": "Hagar is a vulnerable woman, a foreigner and a servant, who nevertheless demonstrates strength and commitment to her son in the face of horrendous mistreatment." "Hagar" apparently also refers to the popular blues song "Aunt Hagar's Children's Blues," written by W. C. Handy and Tim Brymn in 1921 and inspiration not only for Thurman's "Hagar" but for Hughes's character of the same name in his novel *Not Without Laughter* (1930). In all likelihood, both the blues reference and the more well-known biblical story inspired Thurman's title, suggesting Thurman's sardonic view of black America as intellectually impoverished but possessing enormous cultural potential all the same.

In some ways, in fact, *Aunt Hagar's Children* is entirely about the multiplicity and flexibility of perspectives on race. Even as he critiques black America, Thurman displays a remarkable range of tones and perspectives: from the matter-of-factness of a piece like "Frederick Douglass," to the stinging critique in "Booker T. Washington," to the bemused and utterly lighthearted humor of "Marcus Garvey." Surely in *Aunt Hagar's Children* there is more than one Thurman, as he pays homage to Mencken by emulating him and also comments on the tragicomedy of black life in the United States.

Long before the work of Richard Wright or Ralph Ellison, Thurman's *Aunt Hagar's Children* presents the problematic position of a black man who adopts the radical skepticism of a Mencken and directs that skepticism toward developing a critique of his own community. It is a particularly difficult position. As Thurman writes to Hughes in a letter from summer 1929: "I hope you understand my idea in doing such a book as *Aunt Hagar's Children*. I really do not enjoy carping and caviling, but I do believe a certain amount of it has to be done to sift out the chaff from the wheat, to become platitudinous. I have no illusions about my being widely read by the people to whom primarily it should be of interest, the young intellectuals."

In the "Author's Preface" and "Notes on a Stepchild," which offer many insights into Thurman's authorial viewpoint, Thurman encapsulates some ideas that recur throughout the book: individualism, exploration, and the ennobling powers of creativity. He also conveys the powerlessness artists often feel in effecting social change. The author, he claims, "has no panaceas to offer, no sure fire theories for the solution of the race problem." But he is convinced that those solutions are "more and more . . . a personal matter for each and every Aframerican." Self-exploration and the interconnection between texts, cultures, and peoples are the themes of "Notes on a Stepchild." As in his literary essays and reviews (Part Four), so in "Stepchild" Thurman confronts the impact of race upon artists and their creativity: "He tried hard not to let the fact that he had pigmented skin influence his literary or mental development. . . . Negroes live, die, hate, love and procreate. They dance, sing, play and fight. And if art is the universal expressed in terms of the particular he believed there was, if he had the talent, just as much chance for the Negro author to produce great literature by writing of his own people as if he were to write about Chinese or Laplanders."

The biographical essays—"Tribute," "Frederick Douglass," "Booker T. Washington," "Marcus Garvey"—elaborate on the themes of intellectual curiosity and self-exploration that were introduced earlier in the personal essays. These essays comprise portraits of Thurman's personal heroes and villains. "Tribute" is a thank-you note to Alain Locke and James Weldon Johnson, whom Thurman views as godfathers to the younger writers of his generation. In "Frederick Douglass," Thurman introduces Douglass to his readership. As he remarks in letters from the summer of 1929 when he wrote the piece, he is surprised that so many people do not know Douglass. For him, Douglass is a model of individualism, achievement, and self-motivation. He writes a solid biography of the man whom he calls unmatched in black history for his " faculty for standing alone in any crisis, his willingness to change a course of action if one which seemed more suitable presented itself." Although "Booker T. Washington" is a biographical piece like "Frederick Douglass," it has an argumentative tone because Thurman considers Washington's ideas outdated; while they were appropriate for their time, they are unimaginative when set next to the ideals of a Marcus Garvey. Thus, "Marcus Garvey" is a lively portrait that emphasizes both Garvey's strengths and weaknesses. Undoubtedly, this picture of Garvey anticipates Thurman's work on the play *Jeremiah the Magnificent,* which he coauthored with his *Harlem* collaborator, William Jourdan Rapp, probably around the same time. While he admires Garvey's imagination, Thurman highlights how contemporary black leaders disapproved of Garvey. For Thurman, what makes Garvey interesting is the extent to which he was both great and a failure: "He had the emotional power and imagination to conceive grandiose programs . . . , but he lacked the mental depths to regulate" them.

In contrast to the opinionated biographical essays, the essays on contemporary matters demonstrate just how contradictory and complex Thurman could be. In them, he is both an exacting realist and a dreamer, a cynic and an optimist, willing to believe in a mass movement and determined to go his own way as an individual. In "Draw Your Own Conclusions," "The Perpetual Bugaboo," "The Coming Revolution," "So This Is Harlem," and "Terpsichore in Harlem," Thurman offers the kind of cultural analysis and topical writing he had done for years, but not without a trace of Menckenian skepticism. "Draw Your Own Conclusions," for instance, pits the pretensions of the masses against the efficiency of one educated, self-motivated young woman. In comparing group behavior to individual initiative, Thurman puts his hopes in someone doing something sensible about racial disharmony. "The Perpetual Bugaboo" analyzes the anxiety white and black Americans have about racial mixing. For Thurman, this anxiety masks hypocrisy and resists what he considers a social inevitability. "The Coming Revolution" takes the form of an imaginary dialogue with a Communist friend. Like the two essays that precede it, "Coming Revolution" is a celebration of exceptional individuals. In an ironic tone, Thurman calls for the time when the black masses might "emerge from their innate sluggishness, massacre their ministers, and perhaps, in the interim, give birth to a few exceptional individuals capable of arising above the mob, Communism, Christianity, and all other such doctrines to become master intellects and creative giants." After the intense passion of the preceding essays, "Terpsichore in Harlem" is a human-interest essay, examining the dances of Harlem and their origins. It includes a discussion of one of Thurman's favorite Harlem phenomena, the house rent party.

In attempting to emulate Mencken, Thurman walks occasionally into dangerous racial and personal territory. He admits to Hughes that what he is attempting in *Aunt Hagar's Children* is difficult and risky. He acknowledges even the incomplete and transitory nature of his own ideas:

> All I want to do is get some things off my own chest, and perhaps beget discussion, which will be perhaps of some aid to those of us who wish to make the most of our talents. Mayhap I am somewhat of a crusader. It's a thankless task, but if one feels the urge why not? There is so little to do in life that is interesting. And certain Negroes as well as certain whites, have to be pricked to develop. All of my ideas may change overnight. I claim no permanence for them nor infallibility. They are purely expressions of a current mood, guided by certain mental predilections. Were they not to change somewhat with the com-

ing of maturity and the growth of mind I would soon give up the fight. Needless to say I am as skeptical about my own harshly phrased opinions as I am of other people's.

In the anthology *Classic Fiction of the Harlem Renaissance* (1994), editor William Andrews notes that Thurman was an "excellent debunker of the pretension and self-deception of the Harlem Renaissance" (381). Though it did not find publication in his lifetime, *Aunt Hagar's Children* is a remarkable testimony to Thurman's persistent skepticism and his capacity to make momentous demands upon himself and his people.

## A Note on the Text of *Aunt Hagar's Children*

As it appears in the James Weldon Johnson Collection in the Beinecke Library, the table of contents for *Aunt Hagar's Children* is: "Author's Preface," "Notes on a Stepchild," "This Negro Literary Renaissance," "Negro Poets and Their Poetry," "Tribute," "Mr. Van Vechten's Jurors," "Frederick Douglass," "Booker T. Washington," "W.E.B. Du Bois," "Marcus Garvey," "Draw Your Own Conclusions," "The Perpetual Bugaboo," "The Coming Revolution," "So This Is Harlem," "Terpsichore in Harlem," and "The Negro in the Theater." In preparing this edition of the text, however, the editors have chosen to exclude essays that can be found elsewhere in this book. The reader will find "Negro Poets and Their Poetry" and "Mr. Van Vechten's Jurors," which is a very slight revision of "Fire Burns," in Part Four. "So This Is Harlem" is "Negro Life in New York's Harlem" and can be found in Part One. Unfortunately, no extant versions of "W.E.B. Du Bois" and "The Negro in the Theater" have been found. Fortunately, one can sample Thurman's views on Du Bois in "High, Low, Past, and Present" (Part Four).

## Author's Preface

This is essentially a youthful book, being the record of a young Negro, who belongs to a new generation which is just beginning to speak for itself, and who is concerned herein with the task of boring into himself as well as testing the sap of certain trees standing on the edge of the racial forest.

The inner forest itself remains largely unexplored, this particular young Negro believing that too much surface exploring has already been done with no appreciable results.

He has no panaceas to offer, no sure-fire theories for the solution of the race problem. In fact he considers those items to be of secondary importance, for it seems to him that the so-called Negro problem resolves itself more and more into a personal matter for each and every Aframerican. Those who can escape will. Those who cannot must suffer the consequences.

He believes it to be the duty of those who have the will to power in artistic and intellectual fields to shake off psychological shackles, deliberately formulate an egoistic philosophy, develop a cosmopolitan perspective, and soar where they may, blaming only themselves if they fail to reach their goal. Individual salvation may prove a more efficacious emancipating agent for his generation and for those following than self sacrifice or morbid resentment.

There has been in the past far too little of the former and far too much of the latter. The Negro has been so busy bemoaning his fate, so busy placing the entire responsibility for his failures on Marse George that he has not yet stopped to take stock of himself. This volume pretends to be a step in that direction.

Some of the material included herein has previously appeared in *The Bookman, Fire!!, The Haldeman-Julius Quarterly, The New Republic, The Dance Magazine,* and *The Independent.* The author wishes to thank the editors of those various publications for permission to reprint that which first appeared in their columns. There has been some re-writing done in order to bring portions of the material completely up to date. And certain articles have been dissected, then welded to other portions, for the purpose of preventing meaningless repetition.

It will be noted that of the four essays dealing with Negro leaders, three are largely interpretive while one is almost wholly biographical. This was thought necessary because Frederick Douglass is too great a man to remain in obscurity, only because recently no one has taken the pains to write about him.

A faithful friend to whom this book is dedicated has objected that the whole is more illuminating of the author than of those about whom he writes. The only defense the culprit can give is that he has tried to say something which he felt impelled to say, and which he fervently believes should be said. Except for certain stylistic improvements he would allow for no change whatsoever, although he is certain that he will not be so positive about the matter five years hence.

## Notes on a Stepchild

At the age of ten he wrote his first novel. Three sheets of foolscap covered with childish scribbling. The plot centered around a stereopticon movie he had seen of Dante's *Inferno*. It concerned the agony of a certain blonde woman. At the age of twelve, being a rabid movie fan, he began to re-write the contemporary serial thrillers, and was more prodigious with death defying escapades for his heroes and heroines, than the fertile Hollywood scenarists. On entering high school he immediately lost all interest in writing, and did not regain it until his last two years in college, when he spent many hours composing poems about gypsies, hell, heaven, love and suicide.

Being a precocious youth he had read all juvenile books at the age when most children are still fingering grade school primers. In high school his favorite authors were Harold Bell Wright, Zane Grey and Marie Corelli. The Alger period had been passed. For Nick Carter, Opie Read and the Wild West thrillers he had no use whatsoever. All efforts by English teachers to interest him in Dickens (save for *The Tale of Two Cities*), Thackeray, Emerson, Longfellow, Scott and the other high school reliables failed miserably.

Co-incidental with his poetry writing period, he turned to literature and at the same time wandered fitfully "through the palaces of thought," linking the masters of one with the hosts of the other. He toured Shakespeare's two worlds, grinned with the gnomes, flitted about with the sprites, clowned with Launcelot Gobbo, suffered with Hamlet and Lear, and was completely seduced by the sonnets. He stumbled through the dialogues of Plato and decided to emulate Socrates, tried to comprehend the comprehensiveness of Aristotle and DaVinci, returned Bacon, Hegel, Berkeley, Hume and Kant to their shelves, largely unread. Discovered in Goethe a universe of culture as he had found in Shakespeare a universe of human being. Then growing more and more mature, becoming more and more able to assimilate and understand, brooded with Schopenhauer, found beauty in Keats, serenity in Spinoza, and through contact with Shelley and Nietzsche, generated the essential sparks needed to set his own mind and spirit on fire.

Through Huneker he learned of Max Stirner and became an egoist, worshipping at the shrine of the superman (Mencken had caused him to become a confirmed Nietzschean), and in Artzibashev's *Sanine* found a hero he could idolize. His volume of Emma Bovary became a well thumbed bible. Baudelaire, St. Beuve and Stendhal also became major prophets on his literary Patmos. His college professors ridiculed his taste, but, having little respect for their opinions, he continued his miscellaneous and haphazard browsings. From Lafcadio Hearn he learned of Herbert Spencer. Huneker guided him to Ibsen. Brandes enlarged his knowledge of Shakespeare and the giants of the nineteenth century. He learned of nature's cussedness from Hardy, of intellectual lampooning and the value of the classics from Anatole France, of the mysteries of sex from Havelock Ellis and Freud, and, finally, from morbid sessions with Dostoyevsky, that "compassion was the chief and perhaps only law of human existence."

Meditation then, and a fruitless effort to forge that which he had learned into a harmonious whole, dismissing all which seemed redundant or of no value. But he remained a bewildered eclectic, believing this one day, and something else the next, finding verification for all beliefs both in literature and in life. Stability of thought eluded him. His poetry was tortured and verbose. Mental chaos promised insanity. It was then that he left college, ceased living completely in the abstract literary past, and flung himself into the maelstrom of the living present, exploring all provinces of life, communing with every type of person, seeking, observing, suffering.

He was going to be a writer, a realist, refining all of life's experiences within himself, and giving, as Meredith had said he wished to give, a picture of "earth with an atmosphere."

Taking as a motto Huysmans' "I record what I see, what I feel, what I have experienced, writing it as I can, *et voilà tout,*" he began his first novel, spending his nonwriting hours trying to find a master among the contemporary realists, a man in his own generation before whom he could bow down and burn incense. There was Joyce, the Joyce of *Dubliners* and *The Portrait of the Artist,* and of certain portions of *Ulysses* . . . but Joyce was mad. He had had too much genius, too much impenetration, and his drillings into the subconscious had let loose a flood which his conscious mind had been unable to dam back or to filter. The gods had destroyed Joyce because he had divined too much.

There were others of course. Reymont, Undset, Hamsun and Rolland, to none of whom he felt any spiritual kinship. He found Proust illuminating but difficult, Cocteau ephemeral and capricious, Mann too deliberately Teutonic except in *Death in Venice* and *Tonio Kroeger,* and admired Gide primarily as a man who showed continual growth.

There was much to be said for the technical felicity and intellectual acuteness of people like Virginia Woolf, Aldous Huxley, Willa Cather, James Branch Cabell and Ernest Hemingway. And again there was much to be said for the earnestness and power of Dreiser, and for the subterranean meanderings of D. H. Lawrence. But there was not one of these whom he felt to be completely worthy of being designated as *Le Maitre.*

Could it be that the *Zeitgeist* was not conducive to the fullest fructification of literary talent? Could it be that literature was experiencing more than its periodic slump, that the fires of contemporary revolt were only flashes from smoldering faggots? Or was it that the democratization of art and letters was making authors sterile and ridiculous as the democratization of government had turned the world into a blatant comic opera? (Menckenian echoes!)

Not having stopped to analyze himself as he had stopped to analyze others, nor doubting the sanity and correctness of his own judgements, he allowed himself to grow distraught, put his half finished novel aside and decided not to become a writer after all. He saw himself going mad like Joyce, becoming incoherently bewildered like Gertrude Stein or Sherwood Anderson, becoming a dealer in intellectual monsters like Aldous Huxley, or preciously pornographic like the young intellectuals. The current of events in his everyday life did little to make his dilemma less complicated. He was a failure in life as well as in letters. He wondered if it was possible for a man to transcend the *Zeitgeist* and forge out a new path for himself. Perhaps he could, but at the end of that path might stand a brick wall or a consuming oblivion. He thought of destroying himself, remembered Beddoes's apology for self annihilation, and at the same time felt impelled to cry out with John Davidson:

> I dare not, must not die; I am the sight
> And hearing of the infinite; in me
> Matter fulfills itself; before me none
> Beheld or heard, imagined, thought or felt;
> And though I make the mystery known to men,
> It may be none hereafter shall achieve
> the perfect purpose of eternity;
> It may be that the universe attains
> Self knowledge only once; and when I cease
> To see and hear, imagine, think and feel,
> The end may come . . .

No, he was too much the egoist to desire death. He must live. Yet why? What assurance did he have that he was different from any hundred or thousand others? What proof did he have that he would achieve the "perfect purpose of eternity"? What was that purpose? How could he be certain even that for all his self assurance, his reading and attempts to think independently (which he had most certainly not done), he had anything new to say or any new way to say that which had been said before? He remembered Whistler's jibe at Wilde about a sponge remaining a sponge. Perhaps self murder was the easiest way out after all. He could make it a grand gesture and thumb his nose at both man and the Council of the Immortals. The matter certainly merited consideration, but, even on the verge, the act of suicide was better viewed as a dramatic possibility than as a finished certainty. The argument was ended. The egoist bore the victor's palm. The novel neared completion.

## II

He tried hard not to let the fact that he had pigmented skin influence his literary or mental development. He had no desire to be further tormented in life by a bellicose race consciousness. Color prejudice had never inspired within him any blind hatred or anger. Not caring to be a propagandist, he had tried to view the whole problem objectively, tracing things to their roots and distributing the blame to whatever quarter deserved it. His experiences as a Negro in America had not made him bitter. Even as a child reading *Uncle Tom's Cabin,* the death of little Eva had caused him more sorrow than Tom's sessions at the whipping post. He had no particular urge to punish people whose ignorance caused him personally only momentary discomfort. He did not hate all white people, nor did he love all black ones. He found individuals in both races whom he admired, and for whose right to live as they wished he was willing to fight. He was not interested in races or countries or people's skin color. He was interested only in individuals, interested only in achieving his own salvation and becoming if possible a beacon light on Mount Olympus.

This did not mean as so many seemed to think that he had no affection for his own people or desired to break away from them entirely. There was within him none of that typical Negroid yearning to lose his distinguishing characteristics. He had discovered too that the majority of those Negroes who spent the greater part of their time professing a love of race in one breath, and denouncing whites in another, was, for the most part, insincere and ignorant demagogues, no matter how many Phi Beta Kappa keys they strung across their vests, or how many academic degrees they initialed behind their names. They were almost without exception cheap and envious imitations of those whom they professed to despise.

He sympathized with the condition of the peasant Negro in the South and with the dilemma of all Negroes who found themselves caged in and inhibited by color prejudice. He also sympathized with every other minority group in the world, some of whom he knew were more to be pitied than the American Negro. He had consciously detached himself from any local considerations, striven artfully for a cosmopolitan perspective. He knew that there was a certain amount of discomfort, a certain amount of interference, inevitably to be expected from one's fellow men, no matter what happened to be one's color or race or environment. And it was a matter of experience that he had and would suffer from the hands of the black mob as much as if not more than he had and would suffer from the hands of the lily whites.

He had only a passing, pigeon-holed interest in those hapless blackamoors who happened to be lynched or burned in the backwoods of Dixie, yet he had cried and fumed over the fate of Sacco and Vanzetti. This was easily explained, for had Sacco and Vanzetti been Negroes, or had any of the Senegambian victims of Judge Lynch been potential supermen or symbols of supreme values, his reaction would have been just as positive. He was passive only when the characters in a given drama were merely unimportant supers, carrying neither spear nor torch. His interest was not concentrated upon the masses, but upon individuals, individuals in whom could be found germs of truth and beauty. Few, too few, American Negroes of this type had appeared upon the horizon, surely none had been lynched, and those who had had the potentiality of becoming great had thwarted their own destinies by a lack of vision or by weakly succumbing to outside pressure. For every emancipated Jean Toomer or belligerent Claude McKay, there were, as in every other race, several million begging, cringing, moaning nonentities.

Thus he summed up his own people, and thus he had been most surprised to realize that after all his novel had been scorched with propaganda. True, he had made no mention of the difficulties Negroes experience in a white world. On the contrary he had concerned himself only with Negroes among their own kind, trying to interpret some of the internal phenomena of Negro life in America. His book was interesting to read only because he had lain bare conditions scarcely hinted at before, conditions to which Negroes choose to remain blind and about which white people remain in ignorance. But in doing this he realized that he had fixed the blame for these conditions on race prejudice, which manifestation of universal perversity hung like a localized cloud over his whole work.

He was determined not to fall into this trap again, determined to free his art from all traces of inter-racial propaganda. He intended to continue writing about Negroes. He had very little sympathy for those young Negro writers who queried: Must the Negro author write entirely of his own people? . . . then wailed that they did not wish to be known as *Negro* artists but as *artists*. They seemed to think that by writing of white people they could produce better work and have a less constricted field of endeavor. He realized that they were trying to escape from a condition their own mental attitudes made more harrowing, and in so doing were inclined to forget that every facet of life could be found among their own people, and that Negroes, being human beings (a thing both whites and blacks seemed to forget), had all the natural emotional and psychological reactions of other human beings. Negroes live, die, hate, love and procreate. They dance, sing, play and fight. And if art is the universal expressed in terms of the particular he believed there was, if he had the talent, just as much chance for the Negro author to produce great literature by writing of his own people as if he were to write about Chinese or Laplanders. He would be labeled a *Negro* artist, with the emphasis on the Negro rather than on the artist, only as he failed to rise above the province of petty propaganda or failed to find an efficacious means of escape from the stupefying *coups d'états* of certain forces in his environment.

There was this additional difficulty. Both white and black critics had a certain conception of how books about Negroes should be written. The white critics maintained that most Negroes in literature were only blackface whites, unwittingly telling the truth. They continued their preposterous lamentations by alleging that there were inherent differences between whites and blacks which would cause the artistic works of the two groups to be distinctly different and thus give rise to a body of Negro literature.

He dissented from the popular belief on this point, for he realized that the Aframerican was different from the white American in only one respect, namely, skin color, and that when he wrote he would observe the same stylistic conventions and literary traditions. The present day American Negro, except for a fastly disappearing minority, was really no Negro at all, retaining such an ethnological classification only because there seemed to be nothing else to call him. Anthropologists had already pointed out that what was denoted as being a Negro was neither Negro, Caucasian nor Indian, but a combination of all three with the Negroid characteristics tending to dominate. Furthermore, he realized, that the American Negro had absorbed all the American white man's culture and cultural appurtenances, that he used the same language, attended the same schools, read the same newspapers and books, lived in the same kind of houses, wore the same kind of clothes, and was no more keenly attuned to the African jungle or any of its aesthetic traditions than the average American white man.

He could not see then how the Negro in America could produce an individual literature. He reasoned that what was produced in the field of letters by Negroes would be listed as American literature, just as the works of the Scotchman Burns or the Irishman Synge were listed as English literature. He did believe that by writing of certain race characteristics and institutions, the Negro author could introduce a Negroid note into American Literature, as the Irish had introduced a Celtic note into English literature, but he also believed that in America this could be done by a white writer as well as by a black one, and that there was a possibility of it being done much better by the former, in that he might approach his material with a greater degree of objectivity. Outside of Jean Toomer and Eric Walrond no Negro had yet written as beautiful or moving a story of his own people as had Julia Peterkin or DuBose Heyward. And only Claude McKay had been as penetrating.

His ruminations helped to clarify his own mind. His future method of procedure became more clearly outlined. It would be his religious duty to ferret deeply into himself—deeply into his race, isolating the elements of universality, probing, peering, stripping all in the interests of garnering literary material to be presented truthfully, fearlessly, uncompromisingly. A new laboratory experiment had been inaugurated, a fuliginous St. George had set out to kill the dragons which lurked on the road impeding his progress toward the summit of Mount Olympus.

# This Negro Literary Renaissance

I

There has been, we are told, a literary renaissance in Negro America. As proof we have a shelf of books recently published by Negro authors. These authors furthermore have won prizes, fellowships, critical acclamation and public applause. It is not important or congenial to mention that none of these works have been very good, or that most of the authors have no talent whatsoever. Nor is it considered good manners to query: How can there be a rebirth without an original awakening? A score of books by Negro authors have been published, therefore there has been a literary renaissance. Do not dare inquire if this renaissance is the result of mass rumblings or the revolt of a lunatic fringe. It exists. That is enough. Its champions must not be asked embarrassing questions.

What has happened is really no renaissance at all, nor is this movement an autonomous one. It is in reality but a legitimate outgrowth of a national phenomena, a microcosmic circle in a macrocosmic vortex. American literature now having become autochthonous, it is only natural that among other home-grown subjects suitable for literary treatment, native authors should also discover the American Negro. And it is also only natural that the Negro should in some measure become articulate and discover himself.

Precisely the same thing is happening in Negro America as has already happened in the far west (finding expression in Twain, Miller, Ward and Harte), in the middle west (note Norris, Dreiser, Sandburg, Anderson, Masters, Herrick, et al.), and concurrently in the south (vide Peterkin, Stribling, Heyward, Glasgow). It has also happened among the immigrant minority groups (read Rolvaag, Lewisohn and others). And because the same nation wide wind has chanced to stir up some dust in Negro America, whites and blacks deafen us with their surprised and vociferous ejaculations. Renaissance? Hell, it's a backwash!

II

When the Negro art fad first came into being and Negro poets, novelists, musicians and painters became good copy, literate and semi-literate Negro America began to strut and to shout. Negro newspapers reprinted every item published anywhere concerning a Negro whose work found favor with critics, editors, or publishers. Negro journals conducted contests to encourage embryonic genius. Negro ministers preached sermons, Negro lecturers made speeches, and Negro club women read papers—all about the great new Negro art.

Everyone was having a grand time. The millennium was about to dawn. The second emancipation seemed inevitable. Then the excitement began to die down and Negroes as well as whites began to take stock of that in which they had reveled. The whites shrugged their shoulders and began seeking for some new fad. Negroes stood by, a little subdued, a little surprised, torn between being proud that certain of their group had achieved distinction, and being angry because a few of these arrived ones had ceased to be what the group considered "constructive," having in the interim, produced works that went against the grain, in that they did not wholly qualify to the adjective "respectable."

Langston Hughes was the first disturbing note in the "renaissance" chorus, although his initial volume of verse, *The Weary Blues,* did not evoke much caustic public comment from the Negro reading public. They were still too thrilled at the novelty of having a poet who could gain the attention of a white publisher to pay much attention to what he wrote. Quietly and privately, however, certain of them began deploring the author's jazz predilections, his unconventional poetic forms, and his preoccupation with the proletariat. But they were hopeful that he would soon reform and write in a conventional manner about the "best people."

Mr. Hughes' second volume, *Fine Clothes to the Jew,* a hard, realistic compilation, happened to be published while Negroes were still rankling from Carl Van Vechten's novel of Negro life in Harlem, *Nigger Heaven.* It seemed as if this novel served to unleash publicly a store of suppressed invective that not only lashed Mr. Van Vechten and Mr. Hughes, but also the editors and contributors to *Fire!!,* a new experimental quarterly devoted to and published by. younger Negro artists. Under the heading "Writer Brands *Fire!!* as Effeminate Tommyrot," a reviewer in one of the leading Negro weeklies said: "I have just tossed the first issue of *Fire!!*—into the fire, and watched the cackling flames leap and snarl as though they were trying to swallow some repulsive dose."

*Fire!!,* like Mr. Hughes' poetry, was experimental. It was not interested in sociological problems or propaganda. It was purely artistic in intent and conception. Hoping to introduce a truly Negroid note into American literature, its contributors had gone to the proletariat rather than to the bourgeoisie for characters and material, had gone to people who still retained some individual race qualities and who were not totally white American in every respect save color of skin.

There is one more young Negro who will probably be classed with Mr. Hughes when he does commence to write about the American scene. So far this writer, Eric Walrond, has confined himself to producing realistic prose pictures of the Caribbean regions. If he ever turns on the American Negro as impersonally and as unsentimentally as he turned on West Indian folk in *Tropic Death,* he too will be blacklisted in polite colored circles. And it is hardly necessary to add that the authors of *Home to Harlem, Walls of Jericho,* and *The Blacker the Berry* have also been castigated and reviled.

The Negro plastic artists, especially Aaron Douglas and Richard Bruce [Nugent], are also in disfavor; Douglas because of his advanced modernism and raw caricatures of Negro types. Bruce because of the kinks he insists on putting upon the heads of his fantastic figures. Negroes, you know, do not have kinky hair; or, if they have, they use Madame Walker's straightening pomade.

Moreover, when it first became popular to sing spirituals for public delectation, Negroes objected vigorously. They did not wish to become identified again with what the spirituals connoted, and they certainly did not want to hear them sung in dialect. It was not until white music critics began pointing out the beauty and power of the spirituals, that Negroes joined in the hallelujah chorus.

Negroes are, of course, in this respect the same as any other race. The same class of Negroes who protest when Mr. Hughes says:

> Strut and wiggle
> Shameless Nan
> Wouldn't no good fellow
> Be your man

have their counterpart in those American whites who protest against the literary upheavals of a Dreiser, an Anderson, or a Sandburg. And those American Negroes who would not appreciate the spirituals until white critics sang their praises have their counterpart in the American whites who would not appreciate Poe and Whitman until European critics classified them among the immortals.

The mass of American Negroes can no more be expected to emancipate themselves from petty prejudices and myopic fears than can the mass of American whites. They all revere Service, Prosperity and Progress. Furthermore, the American Negro feels that he has been misinterpreted and caricatured so long by insincere artists that once a Negro gains the ear of the public he should expend his spiritual energy feeding the public honeyed manna on a silver spoon. The mass of Negroes, like the mass of whites, are unable to differentiate between sincere art and insincere art. They are unable to fathom the innate differences between a dialect farce committed by an Octavius Roy Cohen to increase the gaiety of Babbitts, and a dialect interpretation done by a Negro writer to express some abstract something that burns within his people and sears him. They are unable to differentiate between the Uncle Remus tales and a darky joke told by Irvin Cobb, or to distinguish the difference, in conception and execution, between a *Lulu Belle*, with its cheap gaudiness and blatant ensemble, and an *All God's Chillun Got Wings* by a sympathetic, groping Eugene O'Neill. Even such fine things as a *Porgy* or a *Scarlet Sister Mary* are labeled inadequate and unfair. While *Nigger Heaven*—ask Carl Van Vechten!

Negroes in America feel certain that they must always appear in public butter side up in order to keep from being trampled in the contemporary onward march. They feel as if they must always exhibit specimens from the college rather than from the kindergarten, specimens from the parlor rather than from the pantry. They are in the process of being assimilated, and those elements within the race which are still too potent for easy assimilation must be hidden until they no longer exist.

Thus, when the publishers of Mr. Hughes' second volume of verse say on the cover that "These poems, for the most part, interpret the more primitive types of American Negro, the bell boys, the cabaret girls, the migratory workers, the singers of Blues and Spirituals, and the makers of folk songs," and that they "express the joy and pathos, the beauty and ugliness of their lives," Negroes begin to howl. This is just the part of their life which experience has taught them should be kept in the background if they would exist comfortably in these United States. It makes no difference if this element of their life is of incontestable value to the sincere artist. It is also available and of incontestable value to insincere artists and prejudiced white critics.

Of course the position of the bourgeois Negro is not an enviable one, for fearing as he does what his white compatriot thinks, he feels that he cannot afford to be attacked realistically by Negro artists who do not seem to have the "proper" sense of refinement or race pride. The American scene dictates that the American Negro must be what he ain't! And despite what the minority intellectual and artistic group may say, it really does seem more profitable for him to be what he ain't, than for him to be what he is.

The first literary works that came out of the so-called "Negro renaissance" were not of the riling variety. *Cane,* by Jean Toomer, was really pre-renaissance, as it was published too soon to be lifted into the best-seller class merely because its author happened to be a Negro. And, as Waldo Frank forewarned in his introduction to *Cane,* Jean Toomer was not a *Negro* artist, but an *artist* who had lost "lesser identities in the great well of life." His book, therefore, was of little interest to sentimental whites or to Negroes with an inferiority complex to camouflage. Both the personality of the author and the style of his book were above the heads of these groups. Although *Cane* reeked with bourgeois-baiting revelations, it caused little excitement among that sector of Negro society, save in Mr. Toomer's home town, Washington, D.C., where the main criticisms were concerning his treatment of Negro women.

*Fire in the Flint* by Walter White, *Flight* by the same author, and *There Is Confusion* by Jessie Fauset were just the sort of literary works both Negroes and sentimental whites desired Negroes to write. The first, a stirring romantic propaganda tale, recounted all the ills Negroes suffer in the inimical South, and made all Negroes seem magnanimous, mistreated martyrs, all southern whites evil transgressors of human rights. It followed the conventional theme in the conventional manner. It was a direct descendant of *Uncle Tom's Cabin,* and it had the same effect on the public. White latter-day abolitionists shook their heads, moaned and protested. Negroes read, boiled and bellowed.

Less sensational and more ambitious, the second novel from this author's pen sought to chronicle the emotional and physical peregrinations of a female mulatto with such a preponderance of white blood in her veins that she could be either Caucasian or Negro at will. Miss Fauset's work was an ill-starred attempt to popularize the pleasing news that there were cultured Negroes, deserving of attention from artists, and of whose existence white folk should be apprised.

All of these works of fiction, as well as the two outstanding works of non-fiction, *The Gift of Black Folk* by W.E.B. Du Bois, and *The New Negro,* edited by Alain Locke, which appeared during the heated days of the "renaissance" were considerate of the Aframerican's *amour propre,* soothing to his self-esteem and stimulating to his vanity. They all treated the Negro as a sociological problem rather than as a human being. I might add that only in *The New Negro* was there even an echo of a different tune.

Countee Cullen's two volumes of verse were also conventional in theme and manner. True, Mr. Cullen was possessed by a youthful exuberance which occasionally flamed with sensual passion, but for the most part he was the conventional Negro litterateur in all respects save that he had more talent than most of his predecessors. But, unfortunately for the Negro middle class, in addition to these "respectable" authors, there also emerged a group of "non-respectables" to join Langston Hughes. Gleefully did they write and full of the devil were their published works.

Recently however, Negroes have been somewhat mollified because in addition to Claude McKay's primitive delineations, they have also had respectable volumes by Jessie Fauset and Nella Larsen, somewhat mollified because they can hide *Home to Harlem, The Walls of Jericho,* and *Banjo* behind *Quicksand, Plum Bun,* and *Passing.* Thus they imagine that all is not yet hopeless. Thus they reason that after all white people may not be made privy to but one side of their lives. Thus they hide their heads in the sand and leave their posteriors conveniently in the air for further fannings.

## III

It is too bad that Negro literature and literary material have had to be exploited by fad finders and sentimentalists. Too bad that the ballyhoo brigade which fostered the so-called Negro art "renaissance" has chosen to cheer and encourage indiscriminately anything which claimed a Negroid ancestry or kinship. For as the overfed child gags when forced to swallow an extra spoonful of half-sour milk, so will the gullible American public gag when too much of this fervid fetish known as Negro art is shoveled into its gaping minds and mouths, and the Aframerican artist will find himself as unchampioned and as unimpressive as he was before Carl Van Vechten commenced caroling of his charms in *Vanity Fair* and the *Survey Graphic* discovered the "New Negro" in 1924. Less so; for even in the prerenaissance days any Negro who managed to write a grammatical sentence was regarded with wonderment even by the emancipated white intellectuals. But now so many Negroes have written a book or a story or a poem that the pale-faced public is no longer astonished by such phenomena.

There has, of course, been some compensation. Certain Negroes have become more articulate and more coherent in their cries for social justice, and a few have also begun to appreciate the advantages of racial solidarity and individual achievement. But speaking purely of the arts, the results of the renaissance have been sad rather than satisfactory, in that critical standards have been ignored, and the measure of achievement had been racial rather than literary. This is supposed to be valuable in a social way, for it is now current that the works produced by Negro artists will form the clauses in a second emancipation proclamation, and that because of Negro artists and their works, the Aframerican will reap new fruits of freedom. But this will be true only inasmuch as the Negro artist produces dignified and worthwhile work. Slip-shod, tricky, atmospheric bits will be as ephemeral as they are sensational, and sentimental propaganda, unless presented in a style both vigorous and new, will have the effect of bird shot rather than of shrapnel. Which is to say that the Negro will not be benefited by mediocre and ephemeral works, even if they are hailed by well-meaning, but for the moment simpleminded, white critics as works of genius.

During the hey-day of the renaissance, there appeared two book reviews in *Opportunity, Journal of Negro Life.* Both of the reviews were written by white poets, both books were by Negro poets. The first was of Countee Cullen's *Copper Sun,* and in dithyrambic prose the reviewer went to every possible verbal excess in praising the volume. According to him Countee Cullen was the Keats of American poetry—equalling Housman and surpassing Edna Millay. There was more such lyric nonsense, all done in good faith, no doubt, but of no value whatsoever either to Mr. Cullen or to the race to which he belongs. The second review, while less turgid and more restrained, was excessive in its praise of *God's Trombones,* and in neither piece was there any constructive criticism or intelligent evaluation.

These two reviews are typical of the attitude of a certain class of white critics toward the Negro writer. Urged on by their desire to do what in their opinion is a "service," they dispense with all intelligence and let their sentimentality run riot. Fortunately, most of the younger Negro writers realize this, and as certain Negroes have always done, are laughing up their sleeves at the antics of their Nordic patronizers. It is almost a certainty that Countee Cullen and Langston Hughes know full well that there are perhaps a half dozen young white poets in these United States doing just as good if not better work than they, and who are not being acclaimed by the public or pursued by publishers, purely because a dash of color is the style in literary circles today.

If the Negro writer is to make any appreciable contribution to American literature, it is necessary that he be considered as a sincere artist trying to do dignified work rather than as a highly trained dog doing trick dances in a public square. He is, after all, motivated and controlled by the same forces which motivate and control a white writer, and like him he will be mediocre or good, succeed or fail as his ability deserves. A man's complexion has little to do with his talent. He either has it or has it not, and, despite the dictates of spiritually starved white sophisticates, genius does not automatically descend upon one because one's grandmother happened to be sold down the river "befo' de wah."

Fortunately, now, the Negro art renaissance has reached a state of near sanity. Serious and inquisitive individuals are endeavoring to evaluate the present and potential significance of this development in Negro life. They are isolating, interpreting, and utilizing those things which seem to have a true esthetic value. If but a few live coals are found in a mountain of ashes, no one should be disappointed. Genius is a rare quality in this world, and there is no reason why it should be more ubiquitous among blacks than whites.

## IV

Charles W. Chesnutt and Paul Lawrence Dunbar were the first Negro novelists. Dunbar's novels are of no value whatsoever, being poorly conceived, poorly executed and poorly written. And although Chesnutt's novels have little literary value, they are interesting as psychological studies, the man having revealed therein his own peculiar complex.

He concerned himself mainly with those folk, who like himself were more white than black, yet nevertheless found themselves inextricably bound to the least desirable race. This, to him, was the most tragic aspect of the many sided race problem, and, so deeply did it goad him, that he spent his writing years, composing distorted, lifeless, fictional tracts, wherein he pled, passionately if indirectly, for his own admittance into the white race.

For that reason and that alone his books are of interest to the student of the Negro psyche.

## V

There are, as is pointed out in another portion of this section, two groups of contemporary Negro novelists, characterized by black America as the respectable and the damned. It is an interesting commentary on the essential humanness of the Negro mob and on its typical mass intelligence to realize that only among the damned is there any show of promise, any kernel of talent.

The respectable novelists are Jessie Fauset, Nella Larsen, and Walter White. Miss Fauset once wrote a novel entitled *There Is Confusion,* which is also an apt criticism of everything the lady has written. And it was probably no idle accident which caused her second novel, *Plum Bun,* to be typographically heralded in one of the Negro newspapers as Plum Bum.

Nella Larsen is superior to her respectable female comrade in that she has interesting ideas and an easy, readable style. She also has a great desire to be a daring lady novelist, which desire, however, seems doubtful of attainment, if her published novels are to be taken as guide posts toward any subsequent work. There is the germ of a good story in both *Quicksand* and *Passing,* but so ineptly does the lady develop her story, so devoid of life are her characters, and so childish are her denouements that the critical reader even begrudges her her original conception.

Walter White is a good journalist. His error is that of so many of his clan. Having a keen reportorial-trained eye and ear, and being able to write a good newspaper story, he is eventually seduced into writing a novel. If this novel should happen to be about things with which, in the pursuance of his craft, he has been intimately concerned, it is more than likely to be a sincere and passionate document, as was Mr. White's *Fire in the Flint*; sincere and passionate, if preachy, given to pointing out obvious morals, devoid of living characters, and not properly proportioned. Then, should he write a novel far afield from his journalistic purlieus, as did Mr. White in his second novel, *Flight,* he makes a public exhibition of all his innate limitations.

Had Mr. White been a novelist rather than a journalist, the heroine of his second novel might have been one of the great characters in American fiction, for her creator, being in color her male counterpart, would have been able to make us privy to what the Negro who passes for white actually feels and experiences. As it is, this type of person still awaits an able interpreter, having been maltreated by Miss Fauset, Miss Larsen, and Mr. White, not to mention the numerous white novelists who have also shot the bolt without locking the door. And while on this question of novels concerning Negroes who cross the line, let us ask: when will some novelist emerge courageous enough to give a truthful delineation? To date, it has become a literary convention to have these fictional passers cross over into the white world, remain discontented, and in the final chapter hasten back from whence they came.

There are several thousand Negroes who each year lose their racial identity, and of this number less than one per cent return to their native haunts. There is in real life none of that ubiquitous and magnetic primitive urge which in fiction draws them back to their own kind. This romantic reaction is purely an invention of the fictioneers, and like sheep, they all make use of it, blindly following the leader over the cliff of improbability.

Taking up the so-called damned Negro novelists, we find Claude McKay and Rudolph Fisher (I leave to others the author of *The Blacker the Berry*), who to date are the only Negro novelists who seem to have both talent and an artistic perspective, the only ones whose characters are, to quote Dostoyevsky, not "puppets and walking dictionaries."

McKay is volcanic, emitting colorful streams of lava in such quantities that he smothers his story and almost suffocates his reader. No scalpel suffices, he must use an ax, chopping off huge chunks of novel and fervid atmospheric material, which he never thins out or trims. But he is courageous, sincere, and penetrating. He comprehends his own people and does not hesitate to set down that which he knows. His most glaring fault is a lack of discipline and concentrated effort. There is little evidence of pruning or revision. And he has a regrettable O'Neillish tendency to be the philosopher rather than the poet. This latter tendency was the cause of certain arid sections in the lush *Home to Harlem,* and it also was responsible for causing certain portions of the latter half of *Banjo* to sound suspiciously like a tract issued by the National Association for the Advancement of Colored People, although it must be admitted that in McKay's polemic there was more breadth than in the thing to which it has been compared.

Of all the Negro novelists, Rudolph Fisher wields the lightest pen, flooding the pages neither with the red ink of propaganda not with the purple ink of passionate novelties. And he is, moreover, the most able storyteller in the group. Where McKay is diffuse and incoherent, Fisher is concise and clear. But where McKay has managed to blurt out something, Fisher has said nothing whatsoever.

In *The Walls of Jericho* he told a good story. He presented us with some amusing and lifelike characters. But he never stirred us, never provoked more than a mild smile. For that reason his first novel was an interesting but disappointing performance. It happened that Mr. Fisher had previously written some of the best short stories yet written about Negroes, and it may be pertinent to suggest that perhaps the short story is his medium rather than the novel. But at any rate here is a man to be watched, one of the few potential literary headliners Negro America has yet produced.

## VI

Eric Walrond is the unknown quantity among Negro authors. None is more ambitious than he, none more possessed of keener observation, poetic insight or intelligence. There is no place in his consciousness for sentimentality, hypocrisy or clichés. His prose demonstrates his struggles to escape from conventionalities and become an individual talent. But so far this struggle has not been crowned with any appreciable success. The will to power is there, etched in shadows beneath every word he writes, but it has not yet become completely tangible, visibly effective.

*Tropic Death,* his only published volume, leaned too heavily on folk lore and folk dialect. There was no adequate refining, although in every story the author latched on to some essential characteristic of the peasant West Indian, and every once in a while would blind the reader with painfully illuminating passages. The element of universality was also consistently suggested.

But the author, it is concluded, had the same struggle with his material that he had with his prose style. He was not yet completely able to master them. They both eluded him at the very moments when he most needed to hold them under control. As a consequence his work was abortive and obscure. As he could not transcribe West Indian dialect so that it would be completely comprehensible to the reader, so he could not successfully open up the conduits of his consciousness and let his reader receive the full flow of expression. He knows what he wants to say, and how he wants to say it, but the thing remains partially articulated. Somewhere there is an obstruction and though the umbilical cord makes frequent contacts, it never achieves a complete connection.

Thus he remains an unknown quantity, with his power and beauty being sensed rather than experienced. It is for this reason that his next volume is eagerly awaited. Will he or will he not cross the Rubicon? It is to be hoped that he will, for he is too truly talented, too sincere an individual and artist to die aborning.

## VII

*Cane* by Jean Toomer is the most unheralded and artistic book yet written by an American Negro, the only one so far which can sincerely be considered as a contribution to the high places of the nation's literature. It has had little vogue. Its author belongs to the left wing of American art and letters, and to date only the left wing critics have seemed to appreciate him.

"He is imitative," some cry, "on every page of *Cane* are echoes of Waldo Frank." These critics should reread both Frank and Toomer, then they should also study the other leaders of the impressionistic, expressionistic schools. They might then become apprised of other influences in Toomer's development.

Compare any page of Frank and Toomer. Frank's staccato prose is hard; it glistens like stalactites, and is not always pleasing to the ear. Toomer's prose is warm, mellow, pulsing with fire and passion. It is:

> Redolent of fermenting syrup,
> Purple of the dusk

For example:

Nigger woman driving a Georgia chariot down an old dust road. Dixie Pike is what they call it. . . . The sun, which has been slanting over her shoulder, shoots primitive rockets into her mangrove-gloomed, yellow flower face. . . .

(The sun is hammered to a band of gold. Pine needles, like mazda, are brilliantly aglow. No rain has come to take the rustle from the falling sweet-gum leaves. Over in the forest, across the swamp, a sawmill blows its closing whistle. Smoke curls up. Marvelous web spun by the spider sawdust pile. Curls up and spreads itself pine-high above the branch, a single silver band along the eastern valley. . . . A girl in the yard of a whitewashed shack not much larger than the stack of worn ties piled before it, sings. Her voice is loud. Echoes, like rain, sweep the valley. Dust takes the polish from the rails. Lights twinkle in scattered houses. From far away, a sad strong song. Pungent and composite, the smell of farmyards is the fragrance of the woman. She does not sing; her body is a song. She is in the forest, dancing. Torches flare . . . juju men, greegree, witchdoctors . . . torches go out. . . . The Dixie Pike has grown from a goat path in Africa.

*Night.*

Foxie, the bitch, slicks back her ears and barks at the rising moon.)

Now compare that with any passage from Waldo Frank's *Holiday*!

Toomer's growth is apparent throughout *Cane*, a volume by the way which defies classification, it being a miscellany containing prose sketches, poems, short stories, and bastard experiments. In the beginning we feel him grasping for a firmer hold on his medium. His step is slow. The ebb and flow of words is not well timed, and sometimes they stumble over their own subtlety. But by the time one finishes "Kabnis," the last piece in the book, one knows the author has emerged victorious, and one cries: More! More!

Toomer is a mystic and a true artist, groping, searching, delving deep within, soaring to the constellations. There is music in his prose, subtle harmonies, plaintive melodies, insinuating rhythms. And above all he is the most emancipated and intelligent Negro yet to appear. Not for him the dark morass of races. He stands on a mount . . . eyes cast alternately earthward and alternately heavenward, relentlessly searching for some meaning in the meaningless universe.

# Tribute

There are two men for whom the younger generation of Negro intellectuals and artists should be thankful, two Negroes of an older generation who are remarkable because they alone profess to understand and to aid in anyway possible those in revolt. The usual Negro of prominence has no time to waste on young upstarts. He either dismisses them with a laugh or else criticizes them severely and unintelligently because they do not happen to look upon him as the god-head of all that is right and good.

The two exceptions are Alain Locke and James Weldon Johnson, both of whom, in different ways, have been invaluable to the emerging talents in the younger generation. Whatever one may think of Dr. Locke's ballyhoo anent the New Negro, he must admit that it was through Dr. Locke that many of these "New Negroes" gained public hearing, and that it was because of his encouragement that many of them have achieved even the few things now listed on the credit side of the ledger.

Dr. Locke does not always understand, but he always sympathizes. He may regret the Beardslyan tendencies of the younger group, but he never attacks them as being moral lepers. He knows that extremes are necessary before a medium can be reached. They may be regrettable and not exactly in good taste, but considering what has gone before, reflecting upon the usual restraining inhibitions and complexes of Negro authors and artists, he is willing to let things take their course, not blindly though, for quietly and subtly he goes around, making a suggestion here, and plotting out a course there, clarifying, instructing and advising. He is the guardian saint of the talented young Negro, a man who loves youth and entrusts to youth's courage and intellect the solving of future problems in a much more happy way, perhaps, than the men of his own generation have been able to solve the problems of the past.

James Weldon Johnson, on the other hand, concerns himself with the dilemma of the younger Negro intellectual, and though he does not necessarily defend he does try to understand and explain. The Negro author has been attacked from all sides by people of his own race, and he has also had to combat the prejudices and forgone conclusions of white editors, white publishers, and white readers. Mr. Johnson has been most energetic in being an elder spokesman for the younger generation. He has looked at their dilemma from all sides, investigated the whys and wherefores, and set his findings down, dispassionately, objectively. When Negroes claimed that white publishers would only publish books by Negro authors who concerned themselves with the lower classes, he quietly wrote an article in *The Crisis,* listing all the recent books published in this field, disproving the myopic allegation. And in recent issues of the various quality magazines he has taken up other aspects of the problem which faces the Negro author, and which he is sagacious enough to realize are the problems to be expected by any truly talented, sincere artist.

The activities of these two men have done a great deal toward preventing the younger Negro from becoming blindly bitter toward the entire older generation of living Negro leaders. For it is from these two alone that they have to any extent received the sympathy, the encouragement, and sane criticism, which they most certainly need if they are to mature and survive.

# Frederick Douglass:
# The Black Emancipator

I
___

Frederick Douglass, born forty-four years before the beginning of the Civil War, was the greatest Negro of his time and has not yet been surpassed. Before him there were only desperate slave leaders like Nat Turner of Virginia and Denmark Vesey of Carolina, who, inciting their fellow slaves to rebellion, made futile fights for freedom. Since Douglass there have been Booker T. Washington, W.E.B. Du Bois, and Marcus Garvey, none of whom, however, have had his capacity for achievement, his faculty for standing alone in any crisis, his willingness to change a course of action if one which seemed more suitable presented itself, his tolerance, or his sense of humor.

Douglass was born on a Maryland plantation. His mother was a Negro woman whom he tells us had features "resembling those of King Ramses the Great," and who was the only slave in that vicinity who could read and write. His father's identity has never been clearly established, but he was thought to have been white. Douglass spent the first seven years of his life in a little cabin on the edge of his future master's plantation. This cabin was the property of his grandmother who it seems "was held in high esteem," allowed unusual privileges, and noted throughout the countryside as fisher-woman, gardener, and nurse.

At the age of seven, young Frederick was delivered to his master's plantation, there to begin his actual life as a slave. The first five years proved easier than those which were to follow. He was still too young to work in the fields, and his greatest hardships resulted from insufficient food, inadequate clothing, and the cruelty of his black supervisor Aunt Kate, who "like everybody in the South seemed to want the privilege of whipping some-body else." What schooling he received was biblical in nature and was administered by a notable old slave character known as Uncle Isaac. This aged gentleman was known as both Doctor of Divinity and Doctor of Medicine; "His remedial prescriptions embraced four articles: for diseases of the body, epsom salts and castor oil; for those of the soul, the Lord's prayer and a few stout hickory switches."

Douglass remained on the plantation for five years, at the end of which time he was loaned to one of his master's relatives, who lived in Baltimore. His duties were inconse-quential, and consisted mainly of acting as companion and protector to his new mas-ter's infant son. He did this job so well and made himself so agreeable around the house that his white mistress took a great liking to him. It was her custom nightly to read pas-sages out of the Bible to her husband. Young Frederick was completely fascinated by this ritual. A new world was opened to him. He too wanted to be able to hold a book in his hand and make sense out of the jumble of letters printed therein. Counting on the kindliness and affection which his mistress manifested toward him, he asked her to teach him how to read. This idea intrigued her and she set about doing it immediately. Her protégé's progress was remarkable, so remarkable in fact that the lady felt impelled to tell her husband of her handiwork. On being appraised of Frederick's learning to read, the master became enraged, and, warning his wife that "if you give a nigger an inch he will take an ell," enlisted her aid in forbidding young Frederick ever to read again. But the "nigger" having already been given the inch proceeded to "take the ell."

His problem now was to find something to read. By shining shoes in the streets of Baltimore, he was able to earn money enough to buy himself a *Webster's Dictionary* and a *Columbian Orator.* This latter volume contained speeches "redolent of the principles of liberation." And it was from this source that he gleaned his first ideas concerning the right of all men to enjoy personal freedom and to pursue, unimpeded by shackles, what happiness life might hold for them. He never ventured forth into the streets without his two textbooks, and soon, with the aid of his playmates, had mastered their contents.

He furthered his education by listening in to his master's conversations. On several occasions he heard the word "abolitionists," always mentioned in a manner which made him realize that it connoted something inimical to his master and his master's friends. Seeking its meaning, he consulted his dictionary, but this enlightened him but little. "It taught me that abolition was the act of abolishing, but left me in ignorance at the very point where I most wanted information, and that was, as to the thing to be abolished." Undaunted, however, he continued his quest, and in the news columns of *The Baltimore American* found "the incendiary information denied him in the dictionary," learning for the first time that there were white people in the world who looked upon slavery as an inhumane menace, and that there was a possibility that through them he and all his people might one day be emancipated.

His old master died, and Young Frederick was sent back to the plantation to be apportioned off with the beasts of the fields to the various heirs. He was, however, returned to Baltimore where he remained until there was a family quarrel which resulted in his being taken back to the plantation. He tells us:

> By a law which I can comprehend, but cannot evade or resist, I am ruthlessly snatched away from the home of a fond grandmother and hurried away to the home of a mysterious old master; thence I am snatched away to the eastern shores to be valued with the beasts of the field, and with them divided and set apart for a possessor; then I am sent back to Baltimore, and by the time I have formed new attachments and have begun to hope that no more rude shocks shall touch me, a difference arises between brothers and I am again broken up and sent to St. Michael's; and now from the latter place I am footing my way to

another master, where I am given to understand that like a wild young working animal I am to be broken to the yoke of a bitter and lifelong bondage.

After young Frederick has been returned to St. Michael's for the second time, following the dispute of the two brothers, one of whom had lent him to the other, it was thought necessary to teach him a lesson. It was known he could read and write and it was known that he was conducting classes among the slaves. Fearing that he might have learned too much to be docile and obedient, he was hired out for a period of one year to Master Covey, who had an enviable reputation of being a "nigger breaker."

But young Frederick was not be broken, although for a time it seemed as if the desired end had been accomplished. Forced to labor both day and night in the fields, insufficiently clad against the cold or heat, fed on food hardly fit for pigs, and being beaten so regularly that the bruises inflicted by one were never allowed to heal, Frederick soon lost all ambition to further his education or to teach those around him. He became so despondent that for the moment he thought of liberty as something never to be attained. He also began to lose what little faith he had had in God and in the goodness of religion, for Master Covey, like most of the other slave overseers, was Douglass' brother in the Methodist church.

The first six months of young Frederick's year under Master Covey proved to be the hardest, for at the end of that time, Frederick having failed to make his master intercede for better treatment, took matters into his own hands. Master Covey attempted to administer his daily lashing. To his surprise the whip was snatched from his hands and he was thrown to the ground. From that day on, there were no more fresh whip wounds on Frederick's back.

At the end of the year Frederick was returned to his master's plantation, and immediately began to make plans for his escape. He took several of the younger slaves into his confidence, taught them to read and to write, and with them planned to make a dash for the nearest free state. But on the eve of their departure some one of the band turned traitor, and, instead of having their freedom, Frederick and his disciples found themselves in jail. Not being able to make any of the conspirators confess, it was decided that Frederick alone was guilty and dangerous. His friends were released; he was left in jail under the impression that he was to be sold to the slave traders and shipped south.

In a few days though, he too was set free and sent back to Baltimore, this step being taken by his master in order to protect his life against the whites in the vicinity of St. Michael's who had made up their minds "to get that fly nigger, Fred." Back in Baltimore, he worked in the shipyards until the white laborers, who objected to working with him, gave him many severe beatings. He then persuaded his master to allow him to become a free agent, hiring himself out wherever he could find work, and depositing the money he earned in his master's hand every Saturday night. But at the same time he managed to earn enough extra money to finance his carefully planned northward flight.

Obtaining a seaman's pass, and disguising himself in a sailor's uniform, Frederick began this second bolt for freedom. It must have been a most thrilling experience. In the first place he in no way resembled the man described in the pass, and, had the train officials read it closely, Douglass would surely have been found out and handed over to the authorities. Twice during the course of his journey he found himself face to face with white men who were friends of his master, and, at another time, while on a ferry boat, met a Negro who knew him and who insisted upon talking to him about his old home and friends. Moreover, at this time, the territory adjacent to free and slave states was infested with kidnappers, always on the alert to pounce upon fugitive slaves and return them to their masters. Frederick, however, managed to reach New York in safety, and was joined almost immediately by a free young colored woman from Baltimore whom he married. Then, being warned that of all places New York was the most unsafe and unsympathetic for a fugitive slave, the newlyweds fled further north to New Bedford, Massachusetts.

II
____

Having taken refuge in this latter city, and having been assured that there was no danger of his being turned over to his former master by the citizens of Massachusetts, Frederick set about adapting himself to his new environment. His first action was to find for himself another name. He had been christened Frederick Augustus Washington Bailey, but felt that he had better shorten as well as change this to one which would not easily be recognized by someone from the vicinity of his master's plantation. A colored man in New Bedford, who befriended him throughout this period, suggested the name of Douglass, he having been favorably impressed by a character of that name in Scott's *Lady of the Lake.* Acting upon this suggestion, Frederick Augustus Washington Bailey became Frederick Douglass, which name he retained for the rest of his life.

There was so much in this new port of freedom which was strange and incomprehensible to Douglass. He had always been taught that wealth was possible only to those people who owned slaves, and he had noticed that the only poor white people in the South were those who did not specialize in human chattel. His only textbooks had been the dictionary, *The Columbian Orator,* and the Bible, and they had in no way given him any information concerning the economic conditions of people in other parts of the country. He was most surprised to find some people of wealth and others with sound and regular incomes. He had expected to find only a mass of poor whites.

With a comparatively safe place in which to live, a wife and a new name, Douglass next began to look for a job. At first he was discouraged, because he found prejudice operating against him whenever he endeavored to find work at his old trade of ship building. But he went about doing odd jobs, reveling in the fact that what he earned was his and not his master's, until he finally obtained a steady job in an oil refinery where he worked for three years.

He had not been in New Bedford very long before he became a subscriber to the *Liberator,* edited by William Lloyd Garrison. In its columns he found that for which he had long been seeking—propaganda for the abolition of slavery. For three years he read this paper, learning by rote the aims and ideals of the abolitionist movement. During this time he also attended all the anti-slavery meetings held in New Bedford and in the near vicinity. Then too, he continued his studies, read widely, and passed on to his fellow black men the propaganda and learning which he himself had acquired.

At the end of this time (1841), he decided to attend an anti-slavery convention in Nantucket. This was his first vacation from menial labor since his arrival in New Bedford three years before, and it proved to be a lasting one. Sitting unobtrusively through the various sessions of the convention, believing himself to be unheard of and unknown, he was most surprised when he was called upon to speak about his slave experiences by one of the leaders of the convention, who had heard him lecture before a group of colored people in New Bedford. Complying, although a little frightened and bewildered, Douglass soon found himself to be of value to the abolitionists, for at the close of the convention he was asked to become one of their traveling lecturers.

At first Douglass was used only as exhibit A, the fugitive slave. It was his job to mount the platform and relate his slave experiences. Then the speaker of the evening would deliver a scathing denunciation of the slave system, point out its vices and plead for support of the abolitionist movement. But the more Douglass traveled, spoke, read, and came into close contact with intelligent beings, the less willing he became to appear as a mere exhibit with no message of his own to deliver. He soon began to make elaborate speeches, telling what he thought of slavery, and drawing conclusions from that which he had seen and experienced. His natural talent as an orator became evident. His audiences sat spellbound before the eloquent speeches of this ex-slave. His coworkers grew alarmed and repeatedly importuned him to play his appointed role and not to usurp theirs. People began to doubt his story. His erudition and oratorical power were not in keeping with the popular conception of an illiterate slave. But Douglass knew that he was growing, that he was destined to be more than exhibit A, and, regardless of the consequences, refused to dam the newly released oratorical flood.

However, to disprove the cries of fraud which were plaguing him and the white abolitionists, he wrote *Slave Narrative,* telling frankly who he was, who his master was, and all the circumstances of his former bondage. Then he plunged more deeply into the abolitionist movement, going into Rhode Island at the time of the Dorr Rebellion, combating the principles or rather the compromises which it sought to establish, and doing a great deal to bring about the abolitionization of that state. In 1843, the New England Abolitionist Society fostered one hundred anti-slavery conventions throughout the states of New Hampshire, Vermont, New York, Ohio, Indiana, and Pennsylvania, in most of which Douglass took an active part.

Places of meeting were denied him and his comrades in many of the cities in which they were scheduled to speak. In Grafton, Massachusetts, where Douglass went alone, "there was neither house, hall, church, nor market-place" in which he could speak. Determined to make the town's unwilling citizens hear his message he conceived a novel plan. "I went to the hotel and borrowed a dinner bell with which I passed through the principal streets, ringing the bell and crying out: !Notice! Frederick Douglass, recently a slave, will lecture on American Slavery, on Grafton Common at seven o'clock. Those who would like to hear of the workings of slavery by one of the slaves are respectfully invited to attend." A crowd gathered to hear what he had to say, and afterwards the largest church in the town was opened to him.

In traveling from place to place, Douglass encountered many other difficulties. Hotels and restaurants barred him with "We don't want niggers here." When he traveled by steamship he was not allowed to sleep in a cabin but must make himself as comfortable as possible on the deck or atop the cargo and baggage. But Douglass was not one to submit docilely to segregation. On the Eastern Railroad from Boston to Portland there was attached to each train a "mean, dirty and uncomfortable car" for colored people. Douglass always bought a first class ticket and always took his seat among the first class passengers, refusing to submit peaceably to that which he believed to be "the fruit of slaveholding prejudice." On one occasion, the conductor of the train, having failed to make Douglass ride in the Jim Crow car, called upon six men to put him where he belonged. "They however," Douglass tells us, "found me much attached to my seat, and in removing me I tore away two or three of the surrounding ones, on which I held with a firm grasp, and did the car no service in some other respects."

This escapade so aroused the executives of the company that they would no longer allow passenger trains to stop in Lynn, where Douglass lived, despite the fact that it inconvenienced many of his white fellow townsmen. When attacked for its stubborn attitude by the abolitionists and others, the officials of the railroad company answered, "that until the churches abolished the 'Negro pew' the railroad company could not be expected to abolish the Negro car." Douglass, of course, agreed with them, and it was not long before both the churches of Massachusetts and the railroad company did away with Jim Crowism on their respective properties.

When Douglass and his co-workers ventured out of New England, they ofttimes met with armed mobs formed purposefully to break up anti-slavery meetings. But Douglass remembered the resolution he had made while "being broken" by Master Covey, and, whenever he encountered physical opposition, fought back as long as he was able. A mob would break into his meeting and order him to stop speaking; Douglass would attempt to win them over by his eloquence. This failing, he would resist them until superior numbers could overpower him and drag him away. He was possessed of unusual physical strength; his labors on the plantation fields and in the shipyards of Baltimore had hardened his body and steeled his muscles. He was equal to resisting any mob and seldom suffered a more severe beating than he was able to inflict upon others. Fists failing, he would not hesitate to avail himself of a club or some other bludgeon and proceed to fight his way to safety, leaving behind him many cracked skulls as mementos of the encounter.

All of which, of course, made him most unpopular with the general population. For, no matter how well intentioned people might be towards the thing for which he stood, they did not like the idea of a black man so belligerently defending himself against white men or their institutions. Color prejudice knew no Mason-Dixon line, and Douglass, always in the limelight, always fighting either with his fists or with words for what he believed to be right, became singled out as a "dangerous nigger" who should be gotten rid of.

Then too, the publishing of his slave narrative which had revealed his true identity, made it very possible for him to be kidnapped and returned to his master. His position became so precarious indeed that his friends hastened to get him out of the country. Thus, in 1844, we find him seeking "refuge in monarchial England from the dangers of republican slavery."

III
———

He stayed in England for about two years, traveling throughout the British Isles, visiting Ireland, Scotland, and Wales. At this time there were two great questions agitating the British public; the repeal of the corn laws, and the repeal of the union between England and Ireland. The entire country was in a turmoil. Parliament was the scene of many bitter and brilliant debates. Richard Cobden and John Bright led the anti-corn law contingent. Sir Robert Peel, prime minister, turned his back on the landed aristocracy and openly admitted his conversion to the doctrines of Cobden and Bright. Daniel O'Connell, the Irish orator, pled the cause of his people and fought out the battle of Catholic Emancipation. While Disraeli, riding in on the "tide which led to fortune," became the hope of those whom Peel had deserted.

Douglass sat on the sidelines and observed everything. He knew both Cobden and Bright and was their guest at the Free Trade Club in London. He listened to speeches by Peel, Disraeli, Lord John Russell, Lord Brougham and others, made notes on their various styles of oratory and obtained ideas to be used in the perfection of his own speech-making. Daniel O'Connell, the Irish Liberator, welcomed him to Ireland and introduced him as the "Black O'Connell of the United States." William and Mary Howitt invited him to be their house guest and it was at their home that he met Hans Christian Anderson, whose writings his hosts had translated into English, and Sir John Bowering, poet and diplomat, who, in telling Douglass about the Orient, enabled him to get a more cosmopolitan idea of the general condition of suppressed minorities. In Edinburgh he met and conversed with George Combe, author of *Combe's Constitution of Man,* which Douglass had read years before. He also met Thomas Clarkson, who, along with Granville Sharp, Thomas Fowell Buxton, and William Wilberforce, had "inaugurated the anti-slavery movement for England and the entire civilized world."

Douglass was much impressed and benefited by his reception in England. The contacts he made broadened his intellectual outlook, gave him courage, and inspired him on to greater activity. The seeming absence of race prejudice made him more certain than ever that such conditions could also be obtained in the United States once the slaves were set free and allowed to take their place in American civilization as human beings.

While he was in England both the World's Evangelical Alliance and the World's Temperance Union held international conventions, at which many prominent American divines were present, solely, it seemed, to receive endorsement for that American institution which Douglass had determined to fight. These eminent men of God were much disturbed and embarrassed by the presence of this former slave, especially when he was called upon to speak before both conventions, and, with his logical eloquence, made their attempts to establish the Christian character of the slaveholders seem far fetched and ludicrous.

Many people urged Douglass to remain in England, but he felt it his "duty to labor and suffer with my oppressed people in my native land." Steps were taken to procure his freedom. It was established that his master would release him for 150 pounds sterling. This amount was collected and sent to America. Douglass' manumission papers were obtained and he was enabled to return home a free man.

## IV

Numerous people in England had signified their willingness to do something to aid both Douglass and his work. Thinking over their offer, he decided that the best thing they could do would be to give him sufficient money to obtain a printing press and establish a weekly paper. He believed that "perhaps the greatest hindrance to the adoption of abolition principles in the United States was the low estimate . . . placed upon the Negro as a man; that because of his assumed natural inferiority, people reconciled themselves to his enslavement and oppression as being inevitable, if not desirable. The grand thing to be done, therefore, was to change this estimation, and demonstrate his capacity for a more exalted civilization than slavery and prejudice had assigned him." He also believed that such a paper would serve to awaken Negroes to their own latent talents and possibilities for achievement, "enkindle their hope of a future, and develop their moral force."

Upon his return to the United States, he enthusiastically communicated his idea to his white associates and friends, and was surprised when they insisted that such a paper could not succeed, that others far more experienced in journalism than he had attempted to do the same thing and failed, and that his talent was for speaking rather than for writing. But Douglass was not to be discouraged. Leaving New England, where he would not seem to be competing with the two other abolitionists' journals published there, and where everyone seemed opposed to his plans, he settled in Rochester, New York, and there began the publishing of his proposed paper, *The North Star.*

Douglass' publication was four years old when he created a sensation by dissenting from certain principles to which William Lloyd Garrison and most of the abolitionists were committed. Heretofore he had accepted all of their tenets without question. It had never occurred to him to question the validity of any of their principles. But now, standing alone and having attained intellectual maturity, he began to think things out for himself.

The abolitionists had set up a slogan saying: "No union with slaveholders," but Douglass soon came to the conclusion that it would not be necessary to dissolve the union in order to abolish slavery, that the mere dissolving of the union, the mere separation of the North from the South, would in no way touch the core of the problem. He also could not see any virtue in people not exercising their constitutional right to vote, because he believed that they could win with the ballot that which they most sought. Neither could he agree with the majority opinion that slavery was authorized by the Constitution:

> My new circumstances compelled me to re-think the whole subject, and study with some care not only the just and proper rules of legal interpretation, but the origin, design, nature, rights, powers, and duties of civil government, and also the relations which human beings sustain to it. By such a course of thought and reading I was conducted to the conclusion that the Constitution of the United States—inaugurated "to form a more perfect union, establish justice, insure

domestic tranquility, provide for the common defense, promote the general welfare, and secure the blessings of liberty"—could not well have been designed to maintain and perpetuate a system of rapine and murder like slavery, especially as not one word can be found in the Constitution to authorize such a belief. Then, again, if the declared purposes of an instrument are to govern the meaning of all its parts and details, as they clearly should, the Constitution of our country is our warrant for the abolition of slavery in every state of the Union.

Douglass lost many of his most influential friends because of this uncompromising deviation from their principles. But he stuck to his guns and managed to keep his newspaper going "from the autumn of 1847 until the union of the States was assured and emancipation was a fact accomplished."

This was no easy task. Ofttimes he was hard pressed for money. Again he encountered opposition on all sides. *The New York Herald* even advocated that the people of Rochester banish him to Canada. "There were times," he says,

> when I almost thought my Boston friends were right in dissuading me from my newspaper project. But looking back on those nights and days of toil and thought, compelled to do work for which I had no educational preparation, I have come to think that, under the circumstances, it was the best school possible for me. It obliged me think and read, it taught me to express my thoughts clearly, and was perhaps better than any other I could have adopted. Besides it made it necessary for me to lean upon myself, and not upon the heads of our Anti-Slavery church. To be a principal and not an agent.

In addition to the editing and distribution of his newspaper, Douglass also continued his speechmaking, traveling throughout the North creating favorable sentiment toward the abolition of slavery. This period just before the beginning of the Civil War was "pregnant with great events." It was the time of the Free Soil Convention at Buffalo, which marked the organization of a polyglot anti-slavery party, the leaders of which, however, did not realize how far reaching that which they had begun would prove to be. It was also during this time that the fugitive slave law was passed, "a bill undoubtedly more designed to involve the North in complicity with slavery and deaden its moral sentiment than to procure the return of fugitives to their so-called owners." There was much more pro-slave legislation inspired by southern members of Congress to protect what they called the God-given constitutional property rights of their constituents. But the battle was to continue, marked by such climactic events as the Dred Scott decision, the repeal of the Missouri Compromise, the Kansas-Nebraska bill, the border war in Kansas, the John Brown raid at Harper's Ferry, and finally culminating in the Civil War.

This was the greatest period in the career of Frederick Douglass. With speech and pen he kept his ideas before the public, denouncing those legislators who would protect that which he believed should be abolished, delivering scathing rebuttals to their debates, logically destroying their spurious arguments with documented facts. He was indefatigable, relentless, and without fear. Nothing could stop him. He had set out to help set his people free, and in the chaotic interlude which followed his return from England to the beginning of the Civil War, he did everything that was humanly possible for him to do in order to realize his ambition.

He not only took an active part in all events of national importance, but also fought race prejudice and segregation throughout the North, and made his home in Rochester a northern terminus for the underground railroad, harboring countless fugitive slaves until he could arrange to smuggle them into Canada. And he was proud of the record that not one slave who came into his hands was ever apprehended. When his daughter was sent to a private school in Rochester, she returned to tell him that she was kept in a room to herself and not allowed to mingle with the other students. Douglass immediately investigated. He was told that the parents of the girls objected rather than the pupils themselves. He personally interviewed all of these parents and found only one who opposed his child going to school with a "nigger." This was enough, however, to convince the principal of the school that no colored pupil should be admitted.

*Uncle Tom's Cabin* was published, and Douglass became acquainted with Harriet Beecher Stowe. He was much impressed by her book. He called it "a work of marvelous depth and power. Nothing could have better suited the moral and humane requirements of the hour." Before she sailed for England, where she had been invited to lecture, she called on Douglass and asked him to submit some plan to her whereby she could further aid the cause. Douglass wrote her a long letter in which he suggested that she get sufficient funds to endow an industrial school for Negroes, saying:

> Colored men must learn trades; must find new employments; new modes of usefulness to society, or . . . they must decay under the pressing wants to which their condition is rapidly bringing them. We must become mechanics; we must build as well as live in houses; we must make as well as use furniture; we must make bridges as well as pass over them, before we can properly live or be respected by our fellow men. We need mechanics as well as ministers. We need workers in iron, clay, and leather. We have orators, authors, and other professional men, but these reach only a certain class, and get respect for our race in certain select circles. To live here as we ought we must fasten ourselves to our countrymen through their every day cardinal wants.

Thus he anticipated Colonel Armstrong, who founded Hampton Institute, and provided Booker T. Washington with an answer to give to Dr. Du Bois, when the latter attacked him for his educational theories. Although Mrs. Stowe did not carry out her original plan, Douglass nevertheless kept preaching this doctrine and lived to see his ideas given practical application, though not as completely as he had wished for and knew necessary.

Douglass was also friendly with John Brown, who unfolded to him his plan for establishing a cordon of guerilla troops throughout the Alleghenies to set fire to plantations, destroy crops, and incite the slaves to rebellion, and thus by force drive slavery out of county after county, until eventually one, then two, then all states should know it no more. Douglass, as first, did not agree with Brown that "slavery was a state of war," and thought that Brown's plan was doomed to failure. But later he changed his mind, and although he would not actively enter into Brown's belligerent plots, he did, nevertheless, become more and more convinced that slavery could not be abolished peacefully, and from then on reiterated again and again that "slavery could only be destroyed by bloodshed."

Just before the John Brown raid on Harper's Ferry, Douglass had a rendezvous with the old man at Chambersburg, Pennsylvania. Brown pled with him to take an active part in the insurrection. This Douglass refused to do because he believed that the attempt to capture the arsenal "would be an attack upon the federal government," and would be fatal to the slaves Brown planned to help run away. Douglass knew that Harper's Ferry was a veritable "steel trap," and warned Brown that he would surely be surrounded and captured. But the old man could not be dissuaded. The two friends separated. Douglass started back to Rochester. Brown to Harper's Ferry, his death, and martyrdom.

Immediately following the historic John Brown raid, a cry was set up throughout the entire country to get Douglass. It was alleged that he had been a member of John Brown's raiding party, and it was known that he and Brown were friends and had been in constant communication. Although it was impossible to prove his presence at Harper's Ferry, it was not impossible to prove that he had been a co-conspirator. Douglass eluded capture, fled to Canada, and sailed to England from Quebec.

He was very despondent and discouraged: "Slavery seemed to be at the very top of its power; the national government with all its powers and appliances were in its hands, and it bade fair to wield them for many years to come. Nobody could see then that in the short space of four years its power would be broken and the slave system destroyed."

In this frame of mind he made his second flight from his home country, feeling certain that his was indeed a lost cause, that the blood necessary to wash out slavery would never be shed, that the work he and others had done to make the North abolition conscious had all been futile and that he would be forced to remain away from the United States for the rest of his life, a despised "nigger exile." He was more discouraged than he had ever been before except during those first six months he had spent with Master Covey, discouraged despite Emerson's prediction "John Brown's gallows would become like the cross," and his own past certainty that the trend of current events prophesied the end of slavery once and for all.

## V

After a sudden flurry of activity, the Congressional committee, which had been formed to investigate the insurrection at Harper's Ferry, abandoned its efforts and discontinued its search for conspirators. Douglass was called home soon afterwards by the death of his eldest daughter.

Upon returning, he found himself in a maelstrom of political activity. He took the stump for Abraham Lincoln, and was especially interested in calling the threat which had been made by southern politicians that if their candidate should be defeated the southern states would immediately secede. Abraham Lincoln was elected, the South was inconsolable and the North, even to President Lincoln, made efforts to compromise and bring about national amity once more. The southern states remained adamant. They wanted the whole hog or none; the result was the Civil War.

Douglass says: "From the first, I, for one, saw in this war the end of slavery; and truth requires me to say that my interest in the success of the North was largely due to this belief. True that this faith was many times shaken by passing events, but never destroyed." His faith was shaken when Secretary Seward made the statement that "terminate however it might, the status of no class of people of the United States would be changed by the rebellion—that the slaves would be slaves still, and the masters would be masters still." And again when General Butler and General McClellan gave warning that any attempt to free the slaves would be "suppressed with an iron hand." And still again when President Lincoln blamed the Negro for the war, and supported those generals who stationed Union soldiers around Virginia plantations to prevent slaves from escaping, paying no attention to the fact that certain troops of Union soldiers were more interested in kicking fugitive slaves out of their camp than they were in shooting rebels.

Douglass fought all this with his usual frankness and fervor. He wrote letters and sent them all over the United States and to England. He made many speeches and wrote in his paper, all, more or less on this one subject: "that the Union cause would never prosper till the war assumed an anti-slavery attitude, and the Negro was enlisted on the loyal side."

He stood practically alone. Northern soldiers did not care to fight side by side with Negroes, in fact they threatened to lay down their arms and go home if such a thing should come to pass. Furthermore, public opinion in the North was strictly organized against turning the rebellion into a war of abolition. It was feared that the border states, too, would join the secessionists and that the Union would inevitably be destroyed.

But if the North would not make use of the Negro, the South did, using him to build forts and trenches and to keep its commissaries supplied. Douglass declaimed: "It is the cotton and corn of the Negro that makes the rebellion sack stand on end, and causes a continuance of the war. Destroy these and you cripple and destroy the rebellion." But his cry was ignored, nor could he or his friends make the powers of the government listen to their advice. It was not until after the disasters at Bull Run, Ball's Bluff, Big Bethel, Fredericksburg, and the Peninsular that the Union leaders decided to change their tactics and began to swerve in the direction Douglass had been pointing out to them since the beginning of the war.

Finally Negro troops were used by the Union, ostensibly as laborers, but once on the battlefield they were handed firearms and admitted into the ranks. Douglass immediately began a recruiting campaign, sending out a proclamation in which he told his fellow blacks that it was "better even to die free than to live slaves." His own sons were among the first to enlist, and Douglass proved to be an indefatigable worker for the Union cause. He stopped only once, and that was when he learned that the Negro troops were not being put on equal footing with the white ones, that they were not being paid the same salaries, nor being promoted as they deserved, nor being exchanged as prisoners of war as had been promised.

He immediately petitioned for an audience with President Lincoln and laid before him the fact that the government had not kept its promise. Lincoln could tell Douglass nothing definite and hedged on all the questions put to him. Douglass says: "Although I was not entirely satisfied with his views, I was so well satisfied with the man and the educating tendency of the conflict, I determined to go on recruiting."

From his audience with President Lincoln, Douglass next went to see Secretary of War Stanton, who also assured him that "justice would ultimately be done." He then offered Douglass a post as assistant adjutant, and though accepted, the commission never came. When Douglass inquired about its delay, he was told to report to General Thomas in the Mississippi Valley. Douglass replied that he would report when he received the commission and, since it was not forthcoming, he remained at home.

It was not until January 1, 1863, the date of Lincoln's Emancipation Proclamation, that the Civil War actually became a war of abolition. Douglass and his friends were jubilant, although later, on scrutinizing it more closely, he discovered that "it was not a proclamation of liberty throughout all the land, unto all the inhabitants there of" but that "its operation was confined within certain geographical and military limits. It only abolished slavery where it did not exist, and left it intact where it did exist. It was a measure apparently inspired by the low motive of military necessity." Douglass also realized that "the proclamation itself was like Mr. Lincoln throughout. It was framed with a view to the least harm and the most good possible in the circumstances, and with especial consideration of the latter. It was thoughtful, cautious, and well guarded at all points. While he (Lincoln) hated slavery, and really desired its destruction, he always proceeded against it in a manner the least likely to shock or drive from him any who were truly in sympathy with the preservation of the union, but were not friendly to Emancipation."

Shortly after the issuance of this Emancipation Proclamation, Lincoln sent for Douglass and discussed with him the possibilities of peace, and the complete emancipation of the slaves. He remarked: "That the slaves were not coming so rapidly and so numerously as I had hoped." And Douglass told him that this was probably due to the fact that the masters had not told their slaves about the Proclamation. Thereupon, he suggested that he be allowed to organize a band of colored scouts who were to spread the news of the Emancipation throughout the South. Lincoln was enthusiastic and gave Douglass the necessary authority, but as the Union victory soon became apparent the plan was never carried out.

When the war was over and the slaves fully emancipated, Douglass at first declared that "Othello's occupation was gone." That for which he had been fighting was accomplished and for the moment he did not realize how much more work there was to be done, and that although the Negro "was free from the individual master, he was still a slave of society." He spent the rest of his life much as he had spent the earlier part, speaking, writing, holding conferences with political leaders, trying to better the condition of his people. He pled for full political equality saying: "If a Negro knows enough to fight for his country, he knows enough to vote; if he knows enough to pay taxes for the support of the government, he knows enough to vote; if he knows as much when sober as an Irishman knows when drunk, he knows enough to vote." He warned against the carpetbaggers and their disastrous regime. He fought to get the Negro admitted to full baptismal rights in the Baptist church. He fought to get them admitted into the various institutions of learning and to have schools erected to meet their especial needs. He was elected as a republican committeeman from Rochester to attend the Loyalists Convention in 1866. He later began another newspaper, and moved to Washington, D.C., so that he could be in close touch with the political moves of the nation. He was appointed president of the Freedman's Bank, and did not realize until too late that his name was being used as a smokescreen to obscure the bank's inevitable failure.

Douglass designates the decade from 1871 to 1881 as a rounding out period, "crowded . . . with incidents and events which may well enough be accounted remarkable. . . . My early life not only gave no visible promise, but no hint of such experience." During this time he was sent to Santo Domingo "as a member of the council for the government of the District of Columbia," and chosen "as elector at large for the State of

New York." He was also appointed to carry the electoral vote from New York to the National Capitol, and was appointed by President Rutherford B. Hayes to the office of Marshal of the District of Columbia, and later by President Garfield to the office of Recorder of Deeds. Moreover, he was asked to make speeches at many important national events, becoming recognized as the leading orator of his time.

Many other political and diplomatic plums were offered to him, but Douglass usually declined and sent another whom he thought more competent in his place. He did accept the post of Minister to Haiti, which was tendered him despite almost universal disapproval, but did not hesitate to resign when he felt that the officials of the government in Washington had not showed him the proper respect and confidence.

While he was Marshal of the District of Columbia, a number of Negroes began exhorting their people to leave the South and come North. Douglass fought this wholesale exodus with all the power he had at his command even though he was opposed by most of his own people. He knew that the Negro would be better off sticking to the farm lands of the South than he would be crowded into the tenements and factories of the North, that he would have less competition, and that, given the proper instruction, would have greater opportunity for amassing property and wealth, as well as greater political advantages. But they did not heed him and the exodus continues.

After the death of his first wife he married again, this time a white woman. Immediately he was set upon by both whites and blacks, called a traitor to his race and a "white hope." His answer publicly was as follows: "My first wife was the color of my mother and my second wife the color of my father; you see, I wanted to be perfectly fair to both races."

In 1886 Douglass once more went to Europe, toured the Continent and also visited Egypt. On coming back to America he took part in the women's suffrage movement, saying: "If the whole is greater than a part; if the sense and sum of human goodness in man and woman are greater than that of either alone or separate, then this government that excludes women from all participation in its creation, administration, and perpetuation demeans itself."

He was appointed Commissioner for the Haitian Republic to represent that country at the World's Fair in Chicago in 1893, and while there made two of the most notable addresses of his career. After the exhibition he went on a speaking tour throughout the West, then returned to Washington where his home, in the suburb of Anacostia, became a place of pilgrimage for countless people, both white and black. Douglass received them all and was generous with both his money and his time, especially where the young men and women of his race were concerned.

He had always said that he wanted "to fall as the leaf in the autumn of life." His wish was granted. On February 2, 1895, he had attended a women's suffrage meeting early in the day, and that night was planning to speak at a Negro church. While on his way out of the house to his carriage, he suddenly dropped to his knees. Looking surprised he exclaimed: "What does this mean?"—then fell upon the floor dead.

## Booker T. Washington

Booker T. Washington was your true story book American, a poor boy, self-educated, never admitting defeat, getting his name in the history books, and remaining through it all a celebrated common man, noted for his horse sense and ability to keep his head out of the clouds. Being severely practical he was suspicious of all highbrows. He put more faith in human nature than he did in books, read people instead of philosophical tomes. Democracy connoted to him what it connotes to the average American. He actually believed it existed and that eventually the black man would share in its blessed usufructs. He believed there was good in all men, and that this goodness, being a gift of God, would eventually prevail and put the devil in rout. A typical American, but unlike most typical Americans he was sincere and singularly free from buncombe.

There was nothing complex about the man. To the contrary he was the personification of simplicity. On superficial examination he seems to have been somewhat shallow. Mentally he was, being too much of the people to plumb depths far beneath their level. Within him lived the true spirit of Jesus Christ. He felt kindly toward all people, picked the mote from his own eye, cast no stones, dismissed all manifestations of human perversity as being the work of the devil, and looked upon the difficulties of himself and his race as being rungs in Jacob's ladder leading to a heavenly home.

He had an almost unbelievable faith in himself and in the rest of God's none too perfect human animals. To those who believed they were aiming more highly than he, it seemed as if he was possessed of narrow vision. They did not stop to consider that his particular narrow beam of light pierced further into the baffling murkiness of the future than their landscape flooding searchlights, or, to change the metaphor, that the undernourished maw of the America Negro needed staple foods before it could digest Savarin delicacies.

Washington began his career during a critical period in the history of the American Negro. The slaves had been freed only to experience a more bitter bondage. Where they had subsisted on bacon rinds and corn pone, they now had difficulty in obtaining stray crusts. The entire responsibility of taking care of themselves was invested no longer in some outside agency. All they had was their God and the Emancipation Proclamation, neither of which provided for their physical needs or made life less arduous than it had before Lincoln's hesitating *beau geste*. Then came the reign of the carpetbaggers.

Negroes, by dint of superior numbers and clever jockeying by unscrupulous whites, knew for one brief moment a power they could neither exercise nor appreciate. It had come generations too soon. Chaos begat chaos. Poor whites and aristocratic whites, one ignorant, the other shortsighted, grew alarmed at the prospects of black supremacy. Punitive measures were the results. The Ku Klux Klan instituted a reign of terror. The blacks saw themselves either completely annihilated or else once more fettered to the auction block, once more knowing the lash and cruel overseer.

The situation called for common sense, and as is usually the case in such crises, it was the last thing anyone considered utilizing. Negro leaders resumed their pre-emancipatory thundering, taking their cue from Frederick Douglass, whose dramatic career was drawing to a close, and whose former methods of combat were of no constructive avail in the present tournament. Douglass, however, had also sensed that the Negro needed practical adjustment to the current state of affairs more than he needed rebellious yeomen, but Washington was the only Negro of potential influence and position who seemed to heed that particular portion of the great emancipator's message.

Without dramatics, seemingly unconcerned about more pressing matters, Washington went to Tuskegee and began laying the foundation for his industrial school. At Hampton Institute, where he had been both student and teacher, certain principles had been absorbed by him. They were, to wit, that the Negro needed to know how to work for himself, that he needed to appreciate the value of producing as well as consuming, the value of fashioning necessities from nature's products.

This did not listen well to Negroes. The higher orders thought he should give attention to the more sensational facets of the problem, that he should proclaim with them for an immediate granting of full political and social rights to the American Negro. The lower orders had had enough of menial labor. They wanted then, as so many do now, to live as their former masters had lived, without going through the tortuous process of natural development. Not for their children did they desire careers as hewers of wood and toilers of soil. Rather they saw visions of innumerable Aframerican professional men, attired in swallow tail coats and high hats, living in mansions, waited upon by servants, and presumably earning the necessary money by practicing on one another.

Washington listened to neither group, nor did he swerve an inch out of the path he had chosen. His initial efforts were local in scope. He gained the good will of the southern whites, who were so pleased to be informed that someone was going to furnish them with trained farm hands and house servants that they did not stop to realize what his program meant to future generations. Thus proving themselves to be as blind as Washington's Negro contemporaries. Many of Washington's first students slipped away horrified upon discovering that they were about to be taught how to work with their hands. They had entered the school under the impression that they would be taught how to conjugate Latin verbs and be informed with other cultural fauna, which they were neither prepared to utilize or assimilate. Nevertheless there were those who remained. Tuskegee grew. Its founder and principal merited sufficient encouragement and material assistance to keep his own morale at a high pitch. There were days of discouragement and days of joyous light. There were false steps and mistakes, but Tuskegee continued to grow.

Then came the Atlanta Exposition address in the fall of 1895. Never before had such an opportunity come to an American Negro, an opportunity to speak before an audience of southern whites in the very bowels of the southland. The world cupped its ears to listen. Negro America was on tenterhooks. Would this man, their spokesman, scourge the Pharisees in the temple or would he turneth away wrath with a soft answer? He did the latter, and although his honeyed words soothed the ears of his white listeners, they also carried a sting. Reading this address, it is difficult to comprehend why the more intelligent Negroes of the time became so outraged as much as why the southern whites accepted it so complacently. All must have been baffled by his sincere naiveté.

He said in part: "Ignorant and inexperienced it is not strange that in the first years of our new life we began at the top instead of at the bottom, that a seat in congress was more sought than real estate or industrial skill, that the political convention or stump speaking had more attraction that starting a dairy farm or truck garden." His advice to both whites and blacks was: "Cast down your bucket where you are. . . . in making friends, friends in every manly way of the people of all races by whom we are surrounded." He reminded both races of their debt to one another, then said: "In all things that are purely social we can be as separate as the five toes yet one on the hand of all things essential to mutual progress." And a few minutes later slyly quotes:

> The laws of changeless justice bind
> Oppressor with oppressed,
> And close as sin and suffering joined
> We march to fate abreast.

Negroes in the south "shall contribute one third of its intelligence and progress . . . one third to the business or prosperity . . . , or . . . shall prove a veritable body of death, stagnating, depressing, retarding every effort to advance the body politic."

> No race that has anything to contribute to the markets of the world is long in any degree ostracized.
> It is right that all privileges of law be ours, but it is vastly more important that we be prepared for the exercise of these privileges.

This was his message, but unfortunately not enough Negroes heeded to make his accomplishments as great as they should have been. His was the type of genius which needed a large public following completely to establish its power, but those Negroes who could have helped him consoled themselves by throwing stones and seemed surprised when there was no answering barrage.

Tuskegee is still growing, but its founder is dead. His experiments are being faithfully carried out, but there is need for ten Tuskegees instead of one. As Washington foresaw, Negroes prefer the stinking slums of the north to the sunny fields of the south.

They prefer to branch out blindly rather than "cast their buckets" in the south and begin fighting their problems at their source. A doctor or lawyer, no matter how inefficient or poor, is honored above a truck farmer. A bootlegger or sweetback is the idol of countless young bucks who would fare better picking cotton than pounding the pavements of Harlem, Philadelphia, St. Louis, Detroit or Chicago. Frederick Douglass advised them to stay in the south. Washington tried to show them how this could be done to the material advantage of all concerned. And that the majority of their people heeded them not is more a reflection upon the race than upon the two greatest men the race has yet produced. Many of Washington's theories are untenable today, they being formulated to meet conditions, which existed contemporaneously with their sponsor. The man was singularly well equipped to lead his own generation. It is indeed a pity that he was a prophet almost entirely without honor among his own people. However, it is interesting to note that he has become the patron saint of Negro Babbitts, who, as Mencken recently pointed out, may be discovered to be sheathing victorious sabres in their golf sticks and malacca canes.

## Marcus Garvey

Marcus Garvey has been variously designated as charlatan, mountebank, savior and fool. It probably never occurred to the individual authors of these labels that the man, being somewhat of a genius, was a combination of all four.

Garvey was the first Negro leader to capture the imagination of the masses, the first whose program was designed to appeal more to the black rabble than to the intelligent Negro or the philanthropic white. Ignoring and outraging both these agencies, he went directly to the underdog. His success was phenomenal; his failure dramatic and inevitable.

The scope of his original program was stupendous. Africa for the Africans. All the black men of the earth spring from common African stock. Therefore let them all join together in order once more to regain their native land. There they can establish a government of, for and by the blacks. Thus the Negro problem will be solved. Thus the Negro can assert his right to be a man among men.

How this was to be accomplished was never clearly formulated. Africa might be the property of the Africans, but it was certainly being held in trust by non-Africans who had no intentions of transferring their guardianship to Garvey. Of the two independent bits of territory, Abyssinia and Liberia, the first adhered to a closed-door policy which excluded foreign blacks as well as whites, and the second was too much in debt to America and American financiers to risk incurring their displeasure by becoming a colonization center for empire building Garveyites.

It was more than Garvey's colonization plans however which won him his phenomenal American following. The idea of an empire was somewhat hazy and the empire itself intangible. Nevertheless, it was a remarkable something to look forward to, and while plans were being made for its formation, Garvey gave his followers tangible evidence of its future glory by dressing them in colorful court regalia, bestowing high sounding titles on all his aides and officials, and by beginning a campaign to make his people "race conscious," proud of being black.

Thus he provided a satisfactory substitute for the various secret organizations which Negroes revel in, and at the same time sounded a new note, which was to prove his one great contribution to the American Negro. Garvey insisted that his people should cultivate their natural characteristics and stop imitating the surrounding whites. Negroes, he declaimed, should be proud of their black skins, thick features, and kinky hair. They should strive for race purity, and instead of becoming whiter and whiter every generation as was their wont, do a *volte face* and cultivate blackness.

His followers were to be known as *black* men and *black* women. The word "colored" was to be thrown into the dung heap along with skin bleaching ointments and hair straightening formulas. There were to be no calendars or pictures in Negro homes on which appeared white faces. Moreover, he reasoned, since most theological authorities are agreed that God is spirit, and since it is the way of most men to visualize their deity as an anthropomorphic Being in their own image, black men should do likewise. God was a Negro who bore a striking resemblance to Marcus Garvey. Jesus too was a Negro, and many pamphlets were published which proved, with the aid of the scriptures, that the crucified Nazarene and Simon the Cyrenian were more than brothers under the skin. Flaxen haired angels and blue eyed cherubs were discarded for kinky haired blackamoors. Heaven became a glorified Harlem, in which black was the prevailing color, and where white men, no doubt, would undergo a transformation once they arrived.

This on the surface seems ridiculous and of no consequence, and it was on such grounds that Dr. Du Bois and other Negro leaders based their attacks, not realizing that Garvey's idea of heaven was as logically developed as that of any Christian, Mohammedan, or Buddhist, and that its development was of inestimable value to the suppressed black minority. Hitherto, the American black man, surrounded by a white world, and the victim of a deep rooted inferiority complex, was able to see himself only in terms of the American white man. He and his leaders depended upon "good white people" for all they attained and all they hoped to attain. Few were independently ambitious. Few had any hopes of earthly salvation. Their only solace was that when they died, their white souls would be liberated from their black bodies and thus experience no difficulty in being admitted to the white man's heaven.

Garvey, unlike most Negro leaders, was proud of his color. He spent no time bemoaning his black skin or proudly prating of any possible Indian or Caucasian ancestry, which he hoped would make him less the despised outcast. Nor did he sit on his haunches and cry out against the white man for denying him equal economic, political or social rights. Rather he declared: America is a white man's country. Naturally the white man is going to give his kind the best available. Being in an alien land, what else can the black man expect but crumbs and bones. Retrieve your own country. Set up your own government. Serve your own flag. Quit being a suffered beggar and become an independent man.

A brave program. What a pity there was no chance for its ultimate success.

## II

Marcus Garvey was born in Jamaica, [British West Indies], August the seventeenth, 1887. His parents were both black Negroes, a point he insisted upon stressing. At one time, it is reported, his father was a man of wealth, but having dissipated his fortune, finally died in the poor house. Garvey was reared and educated by his mother. Tiring of his native home where the blacks were ostracized by both whites and mulattos, he traveled throughout the South and Central America, then sailed for Europe where he visited England and France, in hopes of finding a people not subject to color prejudice. To his dismay, he was forced to come to the hitherto unvoiced conclusion that a black man was a black man in any white man's country, and that he could hope for no more in one than in another. It is interesting to note here that Claude McKay, another Jamaican, expresses a similar disillusionment with European countries in his recent novel, *Banjo.*

Garvey's theory was given further impetus by reading Booker T. Washington's *Up from Slavery,* and by coming into contact with Africans who told him of the horrors of native life under the contemporary European regime. Disheartened, but ambitious, he returned to Jamaica "doomed," he says, "to be a race leader." He asked himself: "Where is the black man's government?" "Where is his king and kingdom?" "Where is his president, his country, and his ambassador, his army, his navy, his men of affairs?" Unable to find them, he declared: "I will help make them." And he began by organizing the Universal Negro Improvement Association which was to skyrocket its founder to dizzying heights of power and affluence only as suddenly to drop him into Atlanta's Federal Penitentiary.

III

He began his campaign in Jamaica just before the war. He met with so little success that he soon decided to transfer himself and his ideas to America. Before sailing he had some correspondence with Booker T. Washington, who seems to have been the only outstanding American Negro who placed any credence in Garvey's program. However, Washington died before Garvey arrived, so it is impossible to predicate how far the founder of Tuskegee would have cooperated or dissented from the Jamaican's colonization dreams.

Once in the United States, Garvey held conference with the various Negro leaders. They all dismissed him with a laugh and suggested that he return to Jamaica. Ignoring their discouraging counsel, he took as a nucleus the West Indian population of New York City, and formed the first American unit of his U.N.I.A. This was soon split into political factions and disrupted. Undaunted, he began again, only once more to lose half his members, and have them set up a rival organization. The third start was to have more success. With *The Negro World* as his official organ, Garvey for the first time began conducting his organization campaign on a grand scale, touring the country, establishing branch offices in every large city, and laying plans for future activities.

Money rolled in. The black rabble who heretofore had supported only their men of God dug into their jeans, and gave bountifully of their earnings. For, according to Garvey, they were soon going back to Africa, an Africa pregnant with sufficient natural resources to make them all independently wealthy. They were soon to shake American dust from their feet, and, from the recesses of their future kingdom, be able to form a government which would demand retribution from any country whose people should be so careless as to use rope or faggot on a hapless Senegambian. And in the meantime they could parade around in brilliantly colored robes, and address one another in terms consonant with their noble rank.

With the millions Garvey had amassed, he established a number of enterprises for the purpose of raising more capital, giving employment to his disciples, and at the same time making them more able to shift for themselves once the shores of Africa were reached. There was the Universal Steam Laundry, the Universal Tailoring and Dressmaking Department, the U.N.I.A.'s chain of groceries, *The Daily Negro Times,* and last, but by no means least, since it was his greatest and most unfortunate venture, the Black Star Line.

Of the aforementioned enterprises few ever progressed beyond an initial stage of development. An inadequately trained personnel, and lack of intelligent management, doomed most of them to an early failure. There was for instance the doll factory, dedicated to the making of black dolls for black babies. This factory was located in the U.N.I.A. building a floor below the one on which Garvey had installed his printing presses. The building was old, the floors weak, the presses heavy, and it was not long before both the doll factory and the presses were a mass of wreckage.

The Black Star Line was Garvey's *pièce de resistance*. His plan was to procure a fleet of ships which were to ply between America, the West Indies and Africa, carrying colonists, and establishing an intercontinental trade route. First he purchased the S.S. *Yarmouth* at a cost of $165,000. This was to be the merchant flagship of the Black Star Line. Its history has a Gilbertian flavor which makes it one of the major maritime comedies.

To begin with the officials of the U.N.I.A. who made the purchase, bargained with the agent, added a sum to the original cost, and split the proceeds among themselves. The maiden voyage was to Cuba. On the way it struck a reef, which rendered it more unseaworthy than it had been when first purchased. On its return, the captain pronounced it unsafe for a second voyage but either changed his mind or had it changed by others when it was discovered that its cargo was to be a shipment of whiskey.

The Eighteenth Amendment was soon to go into force. Local distilleries were making every effort to ship the major portion of the stock on hand out of the country in order to prevent confiscation. One of these hard pressed companies chartered the *Yarmouth,* and, although known to be unseaworthy, it set sail with its valuable cargo. Soon after its departure Garvey was the recipient of one of the most classic cables on record. It read: We are drunk; we are sinking.

The cargo had evidently been too much of a temptation for a crew destined to return to a dry America, especially since it was known that the ship could not safely reach its goal heavily laden as it was. When they had drunk their fill, the remaining excess cargo was tossed overboard. This was done in a place conveniently crowded with empty tugs, whose owners seem to have known in advance just when and where their services would be needed. Who received the profits from this transaction is not a matter of record, but it was the members of the U.N.I.A. who had to make good the loss.

Garvey made a trip to Central America in the interests of his organization. While he was gone his officials were supposed to have paid a twenty five thousand dollar deposit to the U.S. Shipping Board on the S.S. *Orion*. Meanwhile, Garvey's enemies were influential enough to hold up his visa and keep him out of the country for a longer time than he had planned. When he was finally admitted and returned to headquarters, he found that the S.S. *Orion* was still undelivered, and, on investigating, discovered that only $5,000 had been deposited, although the books showed a deficit for the entire amount. While, it is alleged, he was pursuing his investigation, an additional amount was mysteriously paid in, bringing the total up to $22,500. It later became known that the

$11,000 had been borrowed from the Massachusetts Bonding Company of New York, also to be applied to the original deposit fund. This was repaid, making a total of $36,500 drawn from the treasury of the Black Star Line to pay a $23,500 deposit on a ship which was never delivered. The partial deposit is still in the coffers of the U.S. Shipping Board.

The next step was to buy a private Yacht, the S.S. *Kanawks,* from the multimillionaire, Colonel H. H. Rogers. This addition to the fleet cost $69,000, plus a refitting expenditure of $25,000. Shortly after its purchase it too was wrecked and left at anchor in a Cuban harbor. Later a riverboat, the S.S. *Shadyside,* was purchased at a cost of $35,000. What part this latter vessel was to play in trans-Atlantic shipping will perhaps remain one of the eternal mysteries. But this was not all. The S.S. *General G. W. Goethals* was also purchased from the Panama Railroad Company at a price of $100,000, and re-equipped at a cost of over $60,000. At least these were the figures which appeared on the ledgers of the organization. It is no secret that those officials who had figured in the purchase of the first ship also had their share of future booty. Considering all this, it is no wonder the Black Star Line became insolvent, no wonder its president found it necessary to continue soliciting funds in its name when the company itself was no longer existent. And it was because of this fiasco that Garvey ran afoul the Federal authorities, was found guilty of an attempt to defraud through the mails, and received a five year sentence in the big house at Atlanta.

During Garvey's career he collected over five million dollars from the members of his race. Save for the generous salaries and expense accounts allotted to himself, it is doubtful whether Garvey did any further grafting, for, by the time his henchmen had pocketed their cleverly begotten rakeoffs, there was not much left for the Emperor. It is no wonder they were all so willing to join in with the prosecution during their chief's trial. Had they done otherwise, Garvey might have had a number of assistant janitors in Atlanta to make his own duties less arduous.

## IV

It has been noted that Garvey's rival Negro contemporaries did not approve of his organization. Garvey even claimed that they were behind an unsuccessful attempt to have him assassinated. Such men as W.E.B. Du Bois of the N.A.A.C.P. and A. Phillip Randolph, editor of the supposedly socialistic *The Messenger,* joined hands for the first time to down a common enemy. Their zeal in fighting him does not at all times seem to have been purely for the benefit of the race. It is not entirely a fabrication to postulate that there was also no little jealousy and envy of Garvey's success. How dare this "little, fat, black man; ugly, but with intelligent eyes and a big head," to quote Mr. Du Bois, come along and get more followers and money in one year than they had been able to garner in a decade? They appreciated the more ridiculous aspects of his dream and righteously denounced them. But they did not and would not appreciate its more cogent elements, which they might well have incorporated into their own programs. Could it be that there was some truth in Garvey's allegation that: "These fellows represent a small group led by Du Bois, who believe that the race problem is to be solved by assimilation, and that the best program for the Negro is to make himself the best imitation of the white man and approach him as near as possible with the hope of jumping over the fence into the white race and be completely lost in another one hundred years"?

Everything possible was done to eliminate Garvey from the American scene, even to the drafting of a letter to Attorney General Daugherty in which the outraged signers, all Negroes of prominence, accused their arch enemy of every conceivable crime. It was alleged that he was stirring up ill feeling between the races through the medium of his "race consciousness" campaign. This from men, who, by exploiting the Negroes' wrongs in their various journals, books and newspapers, did more to keep racial feeling at a high pitch than Garvey could ever have done by telling Negroes to worship a black God rather than a white one.

They also alleged that Garvey was working hand in hand with the Ku Klux Klan. Positive evidence of Garvey's dealings with the Klan cannot be found. It is known by his own admission that he once held a two hour conference with Grand Imperial Wizard Simmons. What transpired has never come to light. However, it is reasonable to assume that Simmons and Garvey had much in common. Both were agreed that this was a white man's country and that the black man was here only through sufferance. Garvey had said that no white man was good enough to rule a black one and that no black was good enough to rule a white one. The Klansmen were undoubtedly delighted to hear of Garvey's plan to rid the country of Negroes by shipping them all back to Africa. Neither Simmons nor Garvey was intelligent enough to realize the non-feasibility of this solution. Fundamentally, their ideas were the same, and the Imperial Wizard no doubt expressed satisfaction with the wisdom shown by the Imperial African Potentate. They had more in common than either probably realized at the time. The fundamentals of their "inspired" theories were based on the same premises.

Regardless of Garvey's shortcomings, and they were legion, his opponents rarely showed themselves his intellectual superior when on the field of battle. They constantly made use of slanderous personal items when they had a plentitude of factual evidence on which to base their arguments. They constantly made slurring allusions to his blackness of skin, his alien birth, and lack of formal education, thereby playing directly into Garvey's hands when he accused them of being ashamed of their Negro origin and proud of the fact that they had been regimented in a white man's school.

Garvey not only had trouble with these outside forces but he was also constantly at war with his own chieftains. Considering the man, it is not surprising that his suddenly attained power, his self-conferred robes of honor and noble titles would make him arrogant and egotistical. What man, about to become Emperor of all Africa, and ruler of four hundred million blacks by his own count, could be expected to keep an humble mien? He brusquely alienated himself from those whose advice might have delayed his tobogganing to ruin, and perversely placed himself in the hand of a few sycophants, who eventually played Judas to his Messiah.

Like many other Monarchs he also had his matrimonial troubles. He divorced his first wife after six weeks of conjugal strife. He claims it was because of her "crookedness." She claims it was because of his secretary, whom he subsequently married.

## V

Garvey did much to awaken "race consciousness" among Negroes, much to make the black masses aware of their latent possibilities for depending more upon themselves and less upon their "white folks." The alleged Negro renaissance, which has been responsible for the growing number of suddenly articulate Negro poets, novelists, et cetera, owes much to Garvey and to his movement. He laid its foundation and aroused the need for its inception. How much of a scoundrel the man actually was is hard to decide from the garbled testimony of either himself or his accusers. The conclusion is, that he was sinned against quite as much as he sinned. He had the emotional power and imagination to conceive grandiose programs for which fundamentally there was a real need, but he lacked the mental depths to regulate or properly proportion these programs once they were in operation.

## VI

Garvey is missed in Harlem. Life there has not been so colorful since the days of his glory. One misses his Majesty's royal parades when he, surrounded by his courtiers, and followed by his black legionnaires, his black cross nurses, and other imperial units, triumphantly appropriated Seventh Avenue. Once in a while the remaining handful of Garveyites stage a parade, or a rally, or a fight (for they are hopelessly divided into hostile factions), but the old glamour is gone, their gaudy uniforms are dilapidated and worn, their flag and banners of red, black and green have become symbols of impotence rather than of power. And Liberty Hall, the ramshackle temple of freedom, which was to go down in history as another cradle of independence, has been sold at auction, available to the Garveyites only through the sufferance of the man who now owns the property. It is now the council chamber of wrangling groups, for the most part West Indians, whose disparate interpretations of Garvey's dreams sometimes result in such noisome combats that people living in the neighborhood are compelled to send in a riot call.

Garvey is gone, deported to Jamaica, barred from England and Canada, but his spirit still survives and his will to power is as indomitable as ever. If you believe otherwise obtain any copy of *The Negro World,* still published in Harlem, and read his weekly manifestoes. He still believes that Africa should be returned to the Africans, and that he is destined to be its emperor. It is indeed remarkable that even a primitive child in this modern age can so long keep faith in Santa Claus.

## Draw Your Own Conclusions

Recently in a southern California town, Negro students at the local high school were amazed and flabbergasted, on entering the gymnasium in pursuit of their scholastic duties, to find a sign informing them that no longer would they be allowed to use the swimming pool.

Home they went with the astounding news. Enraged parents spread the evil tidings. A mass meeting was called. Indignant ministers, professional men, and local satellites of the National Association for the Advancement of Colored People gathered to discuss this latest flaunting of their civil rights as taxpayers and citizens of the state of California.

There was much violent talk, very little order, few constructive suggestions. Finally a chairman was nominated, but the meeting broke up in a wrangle, without any plan of action having been agreed upon. Reason? It could not be decided who would fill the most honorable position on the protesting committee and thus receive the credit for something yet to be accomplished.

Meanwhile, a lone young colored woman, recently graduated from college, and engaged in editing a local news magazine, quietly interviewed the principal of the offending institution, then sought out the superintendent of schools. Both gentlemen disclaimed any knowledge of this new development. There had been no order for such an insulting sign. Such procedure was contrary to their policy and illegal. She mentioned having a lawyer look into the matter. They informed her that the sign would be removed on the next day. It was.

Not knowing this, the eminent and aggressive Negroes called another mass meeting, drew up many resolutions, drafted letters of protest. They even decided who should get the honor when the matter was finally settled satisfactorily. Imagine their surprise and chagrin to learn that an upstart young woman had done quietly in an hour what it probably would have taken them six months to do, and where she had not spent a penny outside of carfare, they were already soliciting public contributions!

## The Perpetual Bugaboo

Any discussion of social equality between the white and black citizens of the United States always amuses me, it being the most ludicrously considered facet of the so-called Negro problem. Your average white man, quailing, it is alleged, lest he suffer a loss of sexual prestige should Negro men be allowed indiscriminately to mingle in pale face society, immediately gives voice to that old war cry: Would you want your daughter to marry a Negro? Your average Negro shouts: We don't want social equality, we want economic equality. Which is a redundant phrase, because, in the land of the stars and stripes, money talks, and any Negro having it will experience no difficulty in surrounding himself with white associates.

It has been my privilege to know any number of white men who believe in the principle of social equality so long as the newly created milieu is not augmented by white women. They, themselves, adore Negro women, but tolerate Negro men only for appearances' sake. It has also been my privilege to know any number of Negro men afflicted with the same complex, as I have known white women who consort with Negro men for economic aggrandizement, and Negro women who consort with white men for the same potent reason. I have also known people of both races who intermingled and intermarried simply because they happened to respect or love one another.

Then there are Negroes who loudly disclaim any belief in social equality, but who never lose an opportunity to be seen in the company of whites, and there are whites who among their own kind damn the practice which they indulge in secretly. It is, as I have said, a most amusing and complicated situation.

The common Negro and the common white man probably never think of social equality as such. This type of Negro is satisfied to remain among his own kind, and feels uncomfortable if he does otherwise. And the idea never occurs to the common white man. He may know Negroes on the job, and might live adjacent to them. But what of it? They meet, speak to one another, and go on about their business. Neither one has to put extra locks on his doors at night to keep their respective women folk from being raped. Nor do surreptitious mulatto children appear because their wives happen to exchange recipes over the back fence.

It is of course impossible for two races, however dissimilar, to live in close proximity, even briefly, without intermingling, and eventually becoming intermingled. Wholesale social equality may never become *de rigeur,* but there are constant encroachments of white on black and black on white, with the aggressiveness about evenly distributed. And it is these encroachments, minute and undercover, which will eventually solve the color problem in America, despite the race purists.

Whiter and whiter every generation is the slogan of black America, and they achieve this, not so much by marrying white women and white men, as by marrying the mulatto results of previous black white unions. And as lighter Negroes become more and more the rule rather than the exception, as chemists keep producing efficacious bleaching agents, and as Sun Tan powder darkens the lily whites, so will the population of black America decrease to increase that of white America.

This might of course be the text for a moral sermon. But why? Assimilation will eventually be a *fait accompli.* White people began it, and are really the most active in continuing that which they started, despite the suspicious ravings of the Ku Klux Klan. And strangely enough it will be the paradoxical doings of the very people in both races who are most vociferous in their condemnation of such a possibility, aided and abetted of course by the quiet humans who go about, following their natural impulses, uninhibited by sociological bugaboos. The purity of the white race is not endangered by those Negroes Mrs. Hoover might entertain as a routine matter in the White House, nor by those Negroes who are welcomed in the intellectual and artistic salons of New York, nor by those Negroes who are allowed to occupy orchestra seats in the theaters or Pullman cars on railroads. Rather it is endangered by the people who hysterically build high fences, then when darkness descends dig holes underneath, through which they can crawl on their bellies. And this goes for both races.

## The Coming Revolution

I have a Communist friend. He orates for hours on the coming revolution, trying to convert me to the gospel of Leninism. For, says he, the revolution is inevitable, the culture of Communism is predestined to spread and become as world dominating as the culture of Christianity, and Lenin is to become another mythical messiah redeeming once more the lowly and the meek. Thus, he continues, the Negro must align himself with the Communists, for with them and them alone will he ultimately become a free human being, with equal political and social rights.

I listen intently, for the idea of a revolution intrigues me, being one of those paradoxical pacifists, seduced by Carlyle, who has always regretted not having been alive during the French Revolution. But I cannot profess a passionate interest in the benefits my friend promises those who join hands with these new religious crusaders. Nor can I wax optimistic over the heavenly future promised to the Negro should he too ally himself with the working classes and help destroy the present regime. Yet I feel certain that should the revolution come during my lifetime, which I believe to be unlikely, I would most certainly abjure Negroes to become one hundred per cent communistic.

Still I have no faith in revolutions as such, for human beings are human beings, regardless of the philosophical *obiter dicta* of their leaders, and my readings in history have practically assured me that the revolutionist of today inevitably becomes the reactionary of tomorrow. Communism listens well in theory, and as an experiment is something to demand respectful attention, but, as I tell my protesting friend, the substitution of some new fanaticism for one which seems to be dying from inner mortification does not necessarily augur the coming of a millennium. I cite him examples, and ask him to be more explicit in his elucidations concerning the benefits mankind will receive should so-called communism be substituted for so-called democracy. Why, I ask him, should we expect any more from such a possibility than we have already received from the substitution of so-called democracy for aristocracy, or from the ascendancy of so-called Christianity over paganism? He hedges and once more returns to the race problem.

Under the communistic regime, he perorates, the doctrine of equal social and political rights for Negroes would be so thoroughly inculcated into the masses that they would soon accept it as a natural course of events. Inter-marriage would be urged to bring about assimilation and closer relationship between the races. Whites and blacks would soon be brought to comprehend that race prejudice had been created by the capitalists in order to keep the lower orders continually at odds with one another. White, black, yellow and red workingmen would guide the ship of state. The brotherhood of man would at last have been achieved.

I listen to the above, and, when he stops for breath, allow myself the pleasure of a yawn and a smile. And, before he can begin again, I remind him of the voices of the past whose words he is echoing. I outrage him by suggesting that his faith in the masses is a rather naïve illusion, and sarcastically ask him to tell me when the masses have ever been anything but sheep following the leader who happened to ring the loudest bell. But, say I, gladly would I urge the Negro masses to take an active part in the revolution, just to see them, for one moment emerge from their innate sluggishness, massacre their ministers, and perhaps, in the interim, give birth to a few exceptional individuals capable of arising above the mob, Communism, Christianity, and all other such doctrines to become master intellects and creative giants.

If, I continue, your communistic revolution can, by its upheaval give birth and freedom to a handful of such superior beings in America, black and white, bring it on. And although I do not promise to become cannon fodder for any purpose, I will most certainly help ring the bell in order to attract those sheep who might come into the fold.

## Terpsichore in Harlem

Ten years ago, a dim, noisy subterranean cellar deep in dark Harlem; raucous laughter; moans and shouts; dancers—grotesque shadows silhouetted against a dark background by red and blue electric bulbs; music, slow stirring music—loud, discordant music coming from a piano accompanied only by a drum. The pianist is the dominating figure in the room. He has no sheets of music before him—he wouldn't know what to do with them if they were there. He plays by ear alone, improvising if he happens to forget the proper sequence of some melody he has heard another pianist play.

His music becomes more mad and more intense as the evening progresses, and he depends more and more upon his own creative powers. The dancers seem to be in a frenzy. Their movements are frantic and inspired. There is rhythm and masterful coordination of limb and body movement. There are no conventions considered. Couples do as they please, moving whatever portions of their bodies they choose in any way they choose, inventing new steps and introducing complicated rhythms into their dance movements.

The evening wears on. There is much noise and joy and drinking. Men who have worked on the docks or in the ditch all day perspire as freely during the night from a different sort of exertion. Women who have been cooking or washing or cleaning in some white woman's kitchen or house now lord it over their associates in their midnight rendezvous. Prostitutes and pimps maintain a superior attitude. Here is low Harlem at its best, unspied on, unfettered and alive.

It was in such places, found not only in Harlem, but in Chicago, St. Louis, Baltimore, Philadelphia, Kansas City and Omaha, that many of our new popular dance steps and rhythm motifs were first introduced. They were seldom born in such places; rather, these Negro cabarets in northern centers were a sort of clearing house where the Negroes from southern plantations and Mississippi levee camps merge their unsophisticated, semi-primitive dance music and dance tunes with the current northern output.

Fresh from the South, fleeing to the North to find the economic and spiritual freedom promised them there, and finding only new bonds, new barriers and new problems, they sought relief in dance, song, drink, and night life just as they had done in their former homes. Eager not to be considered back numbers, they would attempt to contribute something new to the evening's fun, some tricky foot shuffle or rhythmic body movement which had long been in vogue in their part of the country but which was as yet unknown in their new environment. This step would be picked up and broadcast by other inmates of the cabaret until finally it would reach the theaters in the Negro districts. In a short time some white performer would master its intricacies and present it in big time vaudeville or in a revue. And if it proved fascinating enough there would soon be ballroom adaptations.

It was in this way that we received the Shimmy, Black Bottom, Charleston, Camel Walk and many other popular dances which are universally known as American, but are really miscegenated progenies with the Negro characteristics tending to dominate. Vernon Castle admitted that all of his innovations in the field of the dance had been inspired and built upon Negro dances he had watched and imitated, and he always used a Negro orchestra.

All of this, however, is well-known. What one wishes to know now is whether or not the Negro will continue to make salient contributions to the modern American dance. There are those who fear that his continual Nordicization will preclude any such possibility, for so rapidly is back America assimilating the manners and customs of white America that most of the particularly Negroid characteristics, such as an almost divine feel for rhythm and a most enviable spirit of irresponsible spontaneity, are now found only among a fast disappearing lower class. Northward migration has increased the store of sophisticated Negroes and aided standardization.

One, however, has a faint hope that there are still elements in black American life which retain some of their native vigor and traits, and that there are still some places in northern cities where this element will continue to gather and to dance in its own inimitable fashion. With this in mind one begins a round of the cabarets which in the olden days were festive mines for one searching for material. There are over a dozen of these places in Harlem, the greatest Negro center in the world, of all sizes, shapes and degrees of respectability. Yet the visitor looking for new dance rhythms or new anything will be disappointed, for it seems that now the Negro cabaret, especially in Harlem, and the same can be said for other northern cities, is maintained for whites rather than for blacks. Instead of a mob of Negro dancers cutting new steps and inventing new dance rhythms as was the case ten years ago, one now finds a mob of ambitious whites energetically trying to do the old Negro dances. For every Negro dancer seen on the floor there will be from ten to twelve white ones, and those Negroes seen are what the lower class of Harlemites call "dictys," which means that they are different from whites only in the matter of skin color. They have money, culture, fine clothes, and a copy of Emily Post on their library tables.

In at least half of the Harlem cabarets the Negro patronage is so negligible that when Negroes do attend they are liable to find themselves segregated or else advised that they are not welcome. The only Negroes around are the entertainers, the jazz band

and the waiters. But the question is asked: "Don't these entertainers and waiters burst forth into spontaneous dance steps and thus perhaps create something new which can be copied and utilized?" Unfortunately, the answer is "No." The steps they do are seldom, if ever, original. They are the steps one can see in any third-rate vaudeville house or in any white night club. These people are being paid to give their white patrons what they are attuned to, the only difference being that they probably do what dances they do with more abandon than white performers are able to achieve. The creation of new dances requires the creation of a new music with new rhythmic swings. It will be remembered that in the olden days before Negroes became a fad and Carl Van Vechten discovered Harlem, Harlem cabarets did not sport symphonic bands, but depended upon a single piano player aided, perhaps, by a drummer, who played unwritable concoctions of his own rather than stereotyped melodies from Tin Pan Alley. Them was the days—but now—the orchestra is a dozen strong, attired in tuxedoes, and either plays jazz variations of the classics or else specializes in high-toned symphonic arrangements. The dancers are for the most part pale faces from "below the line," i.e., downtown New York.

Obviously then one will not find here the nuclei for new popular dances so one begins to look around for other hunting grounds. Next there are the dives and speakeasies where it is said that the lower element of black Harlem has now retreated. It is in these places that most of the Negroes who frequented the cabarets before white folk in evening dress, ginger ale at one dollar a bottle and two dollar couvert drove them away. It is in these places that the pianist of olden days who "plays by ear" is found; he who starts out with some popular melody he has picked up and ends by playing something of his own, usually something which cannot be transcribed to the conventional musical staff. Here are all the elements then for the thing we wish to find, but there is very little dancing in these dives and speakeasies; the patrons are too busy defying the Prohibition edict. They drink and talk and sing occasionally, but there is little dancing. For one thing, the places are too small, and for another dancing generally leads to riotous abandon, which gives birth to noise, which in turn might attract some extra-alert Prohibition agent and cause a complete cessation of all activities.

Where then is one to find a clue? One has been to a cabaret and found that it is "passing" for white. One has visited the various dives and speakeasies and discovered that drinking rather than dancing is the most popular pastime. There are a few more leads, one of the most important of which is the many groups of little yellow, brown and black boys who occasionally gather on the sidewalks and amuse themselves and passersby with their dance antics.

Here at least seems to be something one can observe profitably. These youngsters seem to be suffused with the spirit of rhythm. The movements of their bodies, legs and arms are well coordinated and their entire activity seems free from any obvious effort. What is more, they achieve all of their effects without the aid of music save as is provided by their whistling, humming and clapping of hands. Right now they seem much more interested in doing the Charleston and Black Bottom than anything else, but such Charlestoning one has never seen before or ever thought possible.

Probably the most profitable lead of all is the "house rent parties." These owe their origin to the fact that when Negroes first moved into what has now become Negro Harlem, white landlords raised rents so high that Negroes were unable to pay them and their other expenses without supplementing in some way their regular incomes. This was done, first, by renting rooms, and second, by giving these Saturday night parties, charging twenty-five cents admission and serving food and drink at stipulated prices.

This practice spread and continued until now these parties are one of the most colorful of Harlem institutions. The guests begin to gather about ten o'clock. Since about nine-thirty the musicians, usually a pianist and a drummer occasionally aided by a saxophonist, have been tinkering with their instruments, trying out new blues harmonies, and improvising new "slow drag" melodies. Meanwhile, they have also been seeking inspiration and spiritual warmth in the gin bottle which their host has given them or which perhaps, taking no chances, they have brought with them.

The guests arrive, deposit twenty-five cents each with the doorman, divest themselves of their hats and coats, then swarm into the rooms which have been cleared for dancing. The musicians begin to play in earnest, the host sees to it that his guests make frequent trips to the kitchen oasis where they are served gin or corn whiskey in coffee cups in between platefuls of "hopping John," a West Indian dish composed of black-eyed peas, rice and pigs' feet.

In one corner of the room a couple is doing the "bump," the name being a literal description of the dance. They glide along slowly, there is little foot movement. The "bump" is a body dance as shocking to a conventional neophyte now as was the "shimmy" ten years ago. Another couple is doing the "mess around," their bodies reveling in the ecstasy derived from the rhythmic circular movement. The "mess around" is also a body dance, and the couple is standing transfixed beneath the solitary red globe which provides the light, bouncing on the balls of their feet, while the mid section of their bodies go round and round. Still another couple is doing the "fish tail," dipping to the floor and slowly shimmying into an upright position then madly whirling a moment before settling into a methodical slow drag one-step.

As the evening progresses there are more variations from the stereotyped ballroom steps. Then enters the Negro vaudeville performer, who makes it his business to patronize the most colorful of these parties whenever he can, and once there become a part of the crowd, observing their every action and following as best he can the most original and most striking of their dance steps. The next days he spends his time reproducing and refining them. Then he calls in his partner and teaches the finished product to him. The next week he thrills an audience in some theater in the Negro district with his new steps. Other performers imitate him, until finally some white performer on a big time vaudeville circuit appropriates what he has seen a less well-known performer do, labels it with a catchy name and presents it as his own. In a few more months scandalized society matrons object to dashing debutantes disturbing the decorum of their fashionable dances by reproducing refined versions of the mad, stark, dance rhythms first seen in a Harlem "house rent party." Thus the cycle, and thus can we account for the various contemporary dance crazes, each one a little more mad and daring than its predecessor. Decidedly the Negro influence on the American dance is not waning, nor is the creative instinct in Negroes as standardized as a superficial observation of them seems to indicate.

# Part Six

# Poems and Short Stories

Very likely, Thurman wrote more poems than the seven we have included below. There are references in his correspondence to rejection slips for poems he had sent to various magazines as well as to his submissions for poetry contests at *The Crisis* and *Opportunity* that he did not win. Only three of the seven poems below were published during Thurman's lifetime. The most curious of all his poems is the unpublished "On Meeting a Genius (To Jean Toomer)," which Thurman copied, with slight variations, into letters to both Langston Hughes and William Jourdan Rapp.

Thurman's confidence as a critic allowed him to judge himself and his colleagues harshly regardless of the genre they practiced. Raymond in *Infants of the Spring* echoes Thurman when he describes himself and other Harlem Renaissance writers—with the single exception of Jean Toomer—as "mere journeymen." But Thurman's self-doubt casts its longest shadow on his poetry. Even though he appears to have written poems throughout his life, he often admitted that he was not much of a poet. In this opinion he was not alone; in his obituary of Thurman in the New York *Amsterdam News* (January 5, 1935), Theophilus Lewis declared: "Perhaps it is too early to attempt an appraisal of [Thurman's] work in various fields of art. I am ready to give out only one unqualified opinion. He wrote lousy poetry."

As a poet, Thurman could barely get an acknowledgment from either Rapp or Hughes—fellow writers with whom he had warm and open friendships for years. For example, after copying "Frustration" and "God's Edict" (the latter in a very different version from the one published) in a 1926 letter to Hughes, Thurman adds, "I consider these the best I have done, which does not say much. Let me know frankly what you think of them. I know you don't like my style, but fire. I am prepared." Hughes did not react at all, it seems. However, as a critic of Hughes's poetry, Thurman commanded his friend's attention and even respect.

"The Last Citadel," "God's Edict," and "Confession"—all published in 1926—are reminiscent in style of short poems by Stephen Crane, a U.S. writer of an earlier generation who, like Thurman, died young of tuberculosis. The modernist imagism of "The Last Citadel" contrasts with both the cumulative rhetoric of "God's Edict" and the attempt to define a paradox in "Confession." As Eleonore van Notten notes, both "The Last Citadel" and "God's Edict" can be related to "the Menckenite polarity between the isolated individual and the inferior mob" (122). "Stars," "Frustration," and "Untitled"—all sent in letters to friends—document Thurman's search for a poetic voice of his own, experimenting with mood and language. In its striking display of psychosexual tension and orgiastic imagery, "On Meeting a Genius" might reveal more of Thurman's own personality than that of the poet whose creativity the "I" in the poem hopes to invite into himself.

The two short stories included here have a more familiar history than do any of Thurman's poems. "Cordelia the Crude"—first published in the only issue of *Fire!!*—supplied the basis for the plot of the play *Harlem,* which Thurman later wrote with Rapp and which had a successful run on Broadway. The story has been often reprinted and praised for its realistic, even sensational, treatment of difficulties faced by a black southern family soon after they arrive in the "city of refuge." While the story is of a piece with much of the Harlem Renaissance fiction that explores issues of internal migration and generational conflict, its amoral treatment of sexuality—not unlike Theodore Dreiser's in *Sister Carrie*—is both new and shocking in African American writing. Accompanied in the magazine by Richard Bruce Nugent's overtly homoerotic "Smoke, Lillies and Jade!," "Cordelia the Crude" testifies to Thurman's desire to shock his readers and to interrogate bourgeois notions of "good taste."

"Grist in the Mill"—a sardonic treatment of race in the tradition of Kate Chopin and Charles W. Chesnutt—has been seldom reprinted since its original publication in the June 1926 issue of *The Messenger* and has received scant critical attention. The treatment of Colonel Summers, a "relic" from the pre–Civil War days, and his "encounters" with Zacharia, a black migrant worker, is an ironic, at times hilarious, exposé of "pure blood" ideologies and their human consequences. From the "protective cloak of meekness" that Zacharia has adopted, one would think that this outsider has assimilated the ways of southern blacks, but he "was continually laughing at all those about him, both white and black" and, unlike southern Negroes, he "suffered little" and "laughed much." As van Notten suggests, "here, behind the mask, is one of Thurman's favorite characters, the urban self-reliant black who lives by his wits and feels no racial affiliation" with other blacks (113). Once better known, the story is likely to be counted among the best of those early texts in U.S. literature that interrogate the claims of narrow identity politics.

# POEMS

## The Last Citadel

*Opportunity*, April 1926.

There is an old brick house in Harlem
Way up Fifth Avenue
With a long green yard and windows barred
It stands silent, salient,
Unconquered by the surrounding black horde.

## Confession

*Opportunity*, July 1926.

I called you human tumbleweed
And chided you for sowing seed
Of misanthropic malcontent;
Yet I suspect my savage breast
Would never nurture seeds of rest,
Even if you sowed them there.

## God's Edict

*Opportunity*, July 1926.

Let the wind-rolled waves tell the tale of the sea,
And the talkative pines tell the tale of the tree;
Let the motored purr of the automobile
Tell the hum-drum tale of power and steel.
Let the blithesome chirp tell the tale of the bird,
And sad, low sounds tell the tale of the herd;
Then enthrone man on the dunce's stool
And let his tale be the tale of a fool.

## Untitled

Letter to Langston Hughes, circa 1928.

. . . Your eyes are fixed and blind
And there is deafness in your ears
Your lips are closed and mute
They know no words, bespeak no fears
You are immune to tears
You hear no sound you feel no jar
You quietly submit
To every touch . . .

## Frustration

Letter to Langston Hughes, circa June 1929.

Oh the music I would sing
Were I some more vital thing
Than a phantom formed from dust
Saturate with mortal lust.
I would sing gargantuan songs
Reverberating as the gongs
Vulcan forged for Jove to smite
With his thunderbolts of might;
Oh my voice would always be
Clear without cacophony,
Culling from a thousand throats
Silver threads and golden notes
As certain as some finite loom
Culling threads from Juno's womb;
But the strictures on my soul
Will not let my songs unroll,
Will not let my lyric need
Violate a cosmic creed,
And the maiden life therein,
Vanquished by the world's vile din
Strikes one note, then sings no more,
Death has seized another whore.

## On Meeting a Genius

(To Jean Toomer)

Letter to William Jourdan Rapp, circa July 1929.

I
___

I was a wilderness;
Prickly heat points
Hurtled through my body,
Brain cells like thirsting sand-dunes
Knew no inner incandescence;
Emotion was a mirage well,
Instinct entombed, ego diffuse and sere.


II
___

You came.
I flung myself upon you,
The lemon of your body
Merged into my blackened skin.
The lips "too thin for kissing"
Searched my own, transmitted flame.
I burned. I was content.
Knew Jesus was a leper,
Knew Christ would spring from knotted roots
Beneath the troubled earth;
Knew roots were chains
Suffused with sap . . . informed with teats
Natal wombs of super-men.
Your navel cord had pierced my umbulus.
My blood was red, then pale,
My mind a lambent milky way
Intangible and vague;
The soul of me was amber dust
My body knew no life.

There was nothing but the ego
To confound eternity.

### Stars

Letter to William Jourdan Rapp, circa July 1929.

Earth turns
Sun sinks westward into sea
And the night brings forth
Your terse banality.

Insensible to mental probing
Lost to day, immune to night
Sparkling infinitudes
Mere architectural curly-cues
Signifying nothing.

# SHORT STORIES

## Grist in the Mill

*The Messenger,* June 1926.

This is indeed an accidental cosmos, so much so, that even the most divine mechanism takes an occasional opportunity to slip a cog and intensify the reigning chaos. And to make matters more intriguing, more terrifying, there seems to be a universal accompaniment of mocking laughter, coming from the ethereal regions as well as from the more mundane spheres, to each mishap whether that mishap be experienced by a dislodged meteor, a moon-bound planet, a sun-shrunken comet, or a determined man. All of which serves to make this universe of ours a sometimes comic spectacle, serves to push all unexpected cosmic experience just over the deviating border line that divides the comic from the tragic, for there is always something delightfully humorous in an accident even if that accident be as earthly, as insignificant (cosmically speaking) and as fatal as was the accident of Colonel Charles Summers, the second, of Louisiana.

Colonel Charles Summers, the second, was a relic; an anachronistic relic from pre–civil war days, being one of those rare sons of a dyed-in-the-wool southern father who had retained all the traditionary characteristics of his patrician papa. Even his aristocratic blood had escaped being diluted by poor white corpuscles making him indeed a phenomenal person among the decadent first families of the decadent south. Colonel Charles Summers, the second, was your true reborn Confederate, your true transplanted devotee of the doctrines of Jeff Davis, your true contemporary Colonel Charles Summers, the first, even to the petty affectation of an unearned military title, and a chronic case of pernicious anemia.

It was on one of those placid days when a wary human is always expecting the gods to play a scurvy trick upon him, one of those days when smiling nature might be expected to smirk at any moment, one of those days when all seems to be too well with the world that the first act of Colonel Summers' accident occurred. He should have sensed that all was not to be well with him on that day, for long, lonely isolated years of living with and nursing a dead ideal, had made him peculiarly attune to the ever variant vibrations of his environment. He had little companionship, for there were few kindred souls in the near vicinity. His wife, to him, was practically non-existent, being considered a once useful commodity now useless. He had no children, had not wanted any, for fear that they would become too seared with the customs and mannerisms of the moment to complacently follow in his footsteps. He could not abide the poor white or mongrel aristocrats who were his neighbors. He shrank from contact with the modern world, and preserved his feudal kingdom religiously, passionately, safeguarding it from the unsympathetic outside. Hence he communed with himself and with nature, and became intensely aware of his own mental and physical reactions and premonitions, so aware, in fact, that he privately boasted that no accident could befall him without his first receiving a sensory warning, but of course, he forgot to be aware at the proper time, even though the day was rampant with danger signals.

Mrs. Summers was an unemotional ninny, being one of those backwoods belles whom the fates failed to attend properly at birth. Her only basis of recognition in this world at all was that she was a direct descendant of an old southern family. She was one of those irritating persons who never think about a thing nor yet feel about it. Rather she met all phenomena dispassionately, practically, and seemed to be more mechanical than most other humans. When an interne from the hospital brought her the news that only a blood transfusion would save her husband's life, she accepted that without the slightest suggestion of having received a shock; and she had accepted the news of his sunstroke, induced by walking beneath a torrid, noonday Louisiana sun, and which had resulted in an acute aggravation of his chronic illness in the same "well, that's no news" manner.

"Blood transfusion," she stated rather than queried, "well, why not?"

"We thought, madame," the interne was polite, "that you might have been able to suggest—"

She gave a little shrug, the nearest approach to the expression of an emotion that she ever allowed herself. "I am no physician," and the door was closed deliberately, yet normally.

The hospital staff was thus placed in an embarrassing dilemma, for there was no professional blood donor available, and no volunteers forthcoming either from the village center or the outlying plantations. One must suffer from not having friends as well as suffer from having them—so the Colonel's life line continued to fray, his wife made perfunctory visits, and continued to appear disinterested, while the hospital staff pondered and felt criminal, not too criminal, you know, for they remembered that the Colonel had most insultingly refrained from ever donating to their building or upkeep fund, yet they could not let him die while there was a possible chance of saving him, so being both human and humanitarian they played a joke on the doughty old Colonel and at the same time saved his life.

Zacharia Davis had a suppressed desire, and the suppression of that desire was necessarily more potent than the desire itself, for Zacharia wished to make a happy hegira to the northland, and being in the south decided that it was best to keep this desire under cover until such time came that he would have what he called the "necessary mazuma."

Zacharia had been born and schooled in Illinois, and had been perfectly willing to remain there until a certain war-time conscription measure had made Mexico seem more desirable. Once in Mexico he had remained until the ink of the Armistice signatures had been dry five years, and then he had recrossed the border into the cattle lands of western Texas. There he had parked, and attempted to amass sufficient coin to enable him to return to and dazzle Chicago's south side black belt, but an untimely discovery of a pair of loaded dice on his person during an exciting crap game had made it necessary for him to journey by night and by freight to the cane-brake country of Southern Louisiana.

Here he had occupied himself by doing odd jobs about the various plantations and village shops, and by gambling down by the river bottom at night with the rice field and cane-brake laborers. He avoided trouble either with his fellow black men who were somewhat suspicious of this smooth talking "furriner," and with that white portion of his environment that demanded his quiescent respect. Consciously he adopted a protective cloak of meekness, and at first glance could not be distinguished from the native southern blacks; in fact, only a keen analyst could have discerned that Zacharia was continually laughing at all those about him, both white and black, and that he, unlike the southern native Negro suffered little, even unconsciously, and laughed much. Then, too, Zacharia washed the hospital windows every Saturday morning, and was thus drawn into the little comedy in which Colonel Summers was to play the star role.

Meanwhile there were other factors working to deter Zacharia from ever realizing the fruitation of his desire. The sheriff of the parish had finally decided to clamp down on the river bottom gambling activities where there had lately been a siege of serious cuttings and fatal shootings. Of course, normally it did not matter if all the "coons" insisted upon killing one another, but it did matter when northern migration was at its highest peak, and labor was both scarce and valuable to the plantation owners and it was at the instigation of these persons that the sheriff was moved to act. He planned his raid secretly and carefully, seeking the aid of the local K.K.K., and the more adventurous villagers. They had no intention of using firearms or of jailing any of the game participants. Neither did they have any intention of stopping the games completely. They merely hoped to lessen the attendance, and to inspire caution in those who would attend about the advisability of carrying firearms and knives.

Of course, it was in line with Zacharia's general luck that he should be in the midst of a winning streak on the night when the eager vigilantes swooped down upon the river bottom rendezvous like revengeful phantoms in the moonlight, and proceeded to do their chosen duty. And, of course, it was in line with Zacharia's general procedure to forgo immediate flight in order to gather up forsaken cash piles.

There was so much confusion there in the damp darkness. The fiery, white demons reveled in the raucous riot they had created, while the scuttling blacks cursed and cried out against the lash sting and the club beat. The rendezvous was surrounded, there was only one means of escape, the river, and the cornered ones shrank back from its cold, slimy, swift currents. Hysteria descended upon the more terrified. Knives were drawn, and temporarily reflected the gleaming moonlight as they were hurled recklessly into the mad white-black crowd. Periodic pistol shots punctuated the hoarse shouts of the conquerors and the pained moans of the vanquished. Torchlight flares carried by the invaders gave the scene the color and passion of a Walpurgis night. Marsh grass was trampled, its dew turned red by dripping blood. And the river—the muddy river— became riotous with struggling men, and chuckled to itself as an occasional body was unable to withstand the current, or unable to reach the other shore.

There was much more confusion there in the damp darkness. Bleeding heads emitted free blood streams, more groans, more blood, and then grew still, grew horribly inanimate. Wounded bodies squirmed and moaned. The flares were all extinguished. The river was once more quietly rippling undisturbed by super-imposed freight. The round-up had commenced, the injured whites were being carefully carried into the village hospital, while the wounded blacks were being dragged to jail. Thus the night wore on, and seemed a little weary of having witnessed such a carnal spectacle, such elemental chaos.

Among those hapless blacks who regained consciousness in the crowded jail was Zacharia who was nonchalantly nursing a cracked head, and a sock full of coin. Being in jail was no novelty, nor was having a cracked head an entirely new experience, but the sock of money, the sock that contained his pecuniary emancipation, the sock that contained the "necessary mazuma," ah, that was new, saliently new, and comforting.

The town was in an uproar. A deputy sheriff had died from a knife wound inflicted by some infuriated black during the conflict. No one had expected any of the invaders to come back wounded. No one had considered that the cornered colored man might stand at bay like a wild, jungle animal, and fight back. Everyone had considered the whole episode as an unusual chance to sock a few niggers upon the head, and to flay a few black hides with a long unused lash, but instead most of the blacks had fought their way to freedom, only a mere handful of the more seriously wounded were in custody, and they were being claimed by their plantation employers. Moreover, the hospital was overcrowded with wounded whites, and now, this death, this death of a white man at the hands of a nigger. Of course, some one had to pay. The plantation owners were not willing to part with any of their hired help, considering the cultivation of the rice and sugar cane crop of more importance now than the punishment of some unknown assailant. Oh, yes, catch some one and punish him, but don't take this nigger of mine, who is one of my best workmen, seemed to be the general attitude.

No one came to claim Zacharia, and he remained in his cell, awaiting to be released, and amusing himself meanwhile by trying to compute in his mind just how much money his beloved sock, so carefully hidden away, contained. No one came to claim him and finally he was accused of having murdered the deputy sheriff.

The trial was conducted rather leisurely. There was no hurry to cash in on the mob's vengeance. Their call for blood had been satiated by that river bottom battle. It was enough that they had a victim in custody whom they could torture at will, and whom they could put to death legally. Thus Zacharia found himself a participant in a mock trial, found himself being legally railroaded to the gallows, found himself being kept away from freedom—from Chicago—when he had the cash, the long desired cash. He was too amazed at first to realize just how completely he had been enveloped by a decidedly hostile environment. Realization came slowly, and noticeably. His bronze colored face grew wan and sickly. His beady eyes became more and more screwed up until it seemed as if they would completely retreat into the protective folds of their wrinkled sockets. Even the firm lower lip, his one sign of forceful character, drooped, and mutely asked for pity.

He was found guilty, and made ready to take his journey to the state penitentiary where he would be held until the date set for his hanging. The date of his departure drew near, and Zacharia became pitifully panic stricken. The four walls of his lousy cell seemed to be gyrating mirrors sordidly reflecting his certain doom. The bars running diagonally across the cell door and standing upright in the cell window all seemed to assume the personality of ballet dancers attired in hemp, and forming twirling circular figures, lunging at him with menacing loops. Everything choked him, his food—the air—even thought. Incipient nausea tortured him. And then one thought flashed across his mind, lingered there, shimmering with the glorified heat of potential hope. A spasm of grotesque smiles distorted the uneven, thick features, and the quivering lips called to the guards, and begged them to send for Colonel Summers.

Had Zacharia asked for anyone else besides Colonel Summers his request would have been either roughly refused or rudely ignored, but to have a condemned "nigger" ask for old stuck up Colonel Summers, well, well, well, what a chance for some fun at the Colonel's expense. The question was would the Colonel come. In all probability he wouldn't. Since he had recovered from that last illness of his, he had drawn more and more into himself. His wife had imported a sister for company, but the Colonel continued to tramp about his plantation, continued to commune with himself.

It was sheer accident that Colonel Summers happened to be in town on the same day that Zacharia had asked to see him. His wife's sister had had an attack of indigestion. In fact, it seemed to Colonel Summers that she was always having an attack of something. And she was always having prescriptions filled, always dispatching a servant to the drug store. Damned frump, the Colonel called her. Worrisum bitch, was what the black servants called her. However, on this day she had sent for medicine twice, and each time the little black boys had come back with the wrong brand, so impatient at both his sister-in-law, the stupid black boys, and the crafty druggist, the Colonel went into town himself.

Of course, once there the Colonel did the usual thing, *id est* wandered aimlessly about the streets and enjoyed himself by cursing the activities of these ambitious, pettily so, of course, poor trash. And in his wanderings he walked past the jail, was hailed, stopped to see what the insolent fellow wanted, gaped slightly when he heard, and without a word, or without an idea why he did so except that his pride would not let him appear to be placed at a disadvantage, strode into jail, and asked to see Zacharia.

Fifteen minutes later the amused eavesdropping guards and jail loiterers rushed into the cell passageway to see the Colonel striking through the bars with his cane, perspiring dreadfully, his face inordinately infused with blood, and to find Zacharia cowed against the further wall, his face a study in perplexity and pleading, his lips whimpering, "I didn't lie, I didn't lie, it was me, it was me," on and on in ceaseless reiteration.

The surprised and amused men plied the old Colonel with questions in a vain effort to find out what was wrong, but the old southern gentleman was incoherent with rage, and sick, both in body and in mind. He seemed on the verge of collapse and the more solicitous men in the group attempted to lead him into the warden's quarters where he could lie down. Someone even suggested a doctor, but all were overruled by Colonel

Summers, who had meanwhile regained some of his strength and cried out, "The hospital, the hospital," and to the hospital the men carried him, not knowing that he did not wish to go there for treatment, or that he was seeking for verification—verification of what the doomed Zacharia had told him.

Twenty-four hours later he was taken home, babbling, unconscious, and pitiful. The hospital authorities had verified Zacharia's statement, and Colonel Summers now knew that it was the black man's blood that saved his life.

It commenced to rain about twilight time. Colonel Summers suddenly sat up in his bed, the most ambitious move that he had made in a week. He was alone in the room, alone with himself, and his fear, alone in the defeated twilight.

The rain drops increased in volume and velocity. Colonel Summers threw the covers back, struggled out of the bed, and staggered laboriously to the panel mirror set in his closet door. Eagerly, insanely he peered into it, and what he saw there evidently pleased him, for the drawn features relaxed a trifle, and only the eyes, the weak, pitiful eyes, remained intensely animate as they peered and peered into the mirror. Then his strength gave out, and he sank with a groan to the floor.

The rain drops began to come down in torrents, urged on by a rising wind. Colonel Summers once more drew himself up with the aid of the door knob, and once more peered and peered into the mirror. By this time he had ripped his night shirt from him, and stood there naked, his wasted body perspiring from the effort. Soon his strength gave out again, and as he sank to the floor there was a peaceful half smile striving for expression on his pained and fear-racked face.

"Still white, still white," he muttered, and then more loudly, "still white, still white, still white," the voice became hoarse again, "still white, thank God, A'hm still white."

Night came, greeted by the whistle of the frolicsome wind and the ceaseless chorus of the scampering rain drops. The bedroom became dark, and once more gaining consciousness the naked Colonel crawled across the carpeted floor to the nearest window. The darkness frightened him, he was seeking for light, and since the interior offered none, he sought for it or a reflection of it through the window panes. But on the outside was also the black night plus the cachinnating rain drops, and the playful wind. He shrank back in abject terror only to be confronted with the same terrifying darkness behind him.

He looked out of the window once more. A flash of lightning provided the wanted light, but it brought no release, brought only additional terror, for the tree tops, glistening wet and swaying with the wind, assumed the shapes of savage men, rhythmically moving to the tune of a tom-tom, rhythmically tossing to the intermittent thud of the reverberating thunder.

"Darkies," he murmured, and tried to draw from the window, "My God— darkies." Then the scene changed. His insane eyes set in a bearded skull conjured up strange figures when the lightning flashed. Each tree assumed a definite personality. That broken limb dangling from the tree just beyond the fence was Zacharia, and as it gyrated wildly in the mad night, it seemed to whisper to the wind. "He is my brother, my brother, my brother," while the wind broadcasted the whisper through the night. And then that tallest tree so close to the house was himself, a black reproduction of himself with savage sap surging through its veins. It too reveled in the wildness of the night; it too exulted in being pelted by the wind-driven rain drops and in responding to the rough rhythm of the thunder-god's tom-tom.

Someone lit a light in the hall, and laid their hand on the door knob preparatory to opening the door to the Colonel's room. Then someone else across the hall called, and the first person released their grip on the knob and treaded softly away.

The Colonel fell prostrate to the floor, and attempted to burrow his head deep into the thick protective nap of the carpet. He felt an inky blackness enveloping him, his whole form seemed to be seared with some indigo stain that burned and burned like an avid acid. Then his body began to revolt against this dusky intruder, began to writhe and wriggle upon the floor, began to twitch and turn, trying to rub itself clean, trying to shed this superimposed cloak, but the blackness could not be shed—it was sprouting from the inside, and being fertilized by the night.

Time passed. Voices were heard whispering in the hall. A door closed. More whispering. Outdoors all was jubilantly mad. In the bedroom the Colonel still lay upon the floor, panting, perspiring, exhausted from his insane efforts. His reason was now completely gone. His last ounce of life was being slowly nibbled away. The blackness became more intense, and then a black crow, stranded, befuddled by the storm, sought refuge upon the window ledge, and finding none there cawed out in distress, and to the dying maniac on the floor, it seemed to caw, "nigaw, nigaw, nigaw—"

Someone opened the door, turned on the light, and screamed.

## Cordelia the Crude

*Fire!!* November 1926.

Physically, if not mentally, Cordelia was a potential prostitute, meaning that although she had not yet realized the moral import of her wanton promiscuity nor become mercenary, she had, nevertheless, become quite blasé and bountiful in the matter of bestowing sexual favors upon persuasive and likely young men. Yet, despite her seeming lack of discrimination, Cordelia was quite particular about the type of male to whom she submitted, for numbers do not necessarily denote a lack of taste, and Cordelia had discovered after several months of active observation that one could find the qualities one admires or reacts positively to in a varied hodge-podge of outwardly different individuals.

The scene of Cordelia's activities was the Roosevelt Motion Picture Theatre on Seventh Avenue near 145th Street. Thrice weekly the program changed, and thrice weekly Cordelia would plunk down the necessary twenty-five cents evening admission fee, and saunter gaily into the foul-smelling depths of her favorite cinema shrine. The Roosevelt Theatre presented all of the latest pictures, also, twice weekly, treated its audiences to a vaudeville bill, then too, one could always have the most delightful physical contacts . . . hmm . . .

Cordelia had not consciously chosen this locale nor had there been any conscious effort upon her part to take advantage of the extra opportunities afforded for physical pleasure. It had just happened that the Roosevelt Theatre was more close to her home than any other neighborhood picture palace, and it had also just happened that Cordelia had become almost immediately initiated into the ways of a Harlem theater chippie soon after her discovery of the theater itself.

It is the custom of certain men and boys who frequent these places to idle up and down the aisle until some female is seen sitting alone, to slouch down into a seat beside her, to touch her foot or else press her leg in such a way that it can be construed as accidental if necessary, and then, if the female is wise or else shows signs of willingness to become wise, to make more obvious approaches until, if successful, the approached female will soon be chatting with her baiter about the picture being shown, lolling in his arms, and helping to formulate plans for an after-theater rendezvous. Cordelia had, you see, shown a willingness to become wise upon her second visit to the Roosevelt. In a short while she had even learned how to squelch the bloated, lewd-faced Jews and eager middle-aged Negroes who might approach as well as how to inveigle the likeable little yellow or brown half-men, embryo avenue sweetbacks, with their well-modeled heads, stickily plastered hair, flaming cravats, silken or broadcloth shirts, dirty underwear, low-cut vests, form-fitting coats, bell-bottom trousers and shiny shoes with metal cornered heels clicking with a brave, brazen rhythm upon the bare concrete floor as their owners angled and searched for prey.

Cordelia, sixteen years old, matronly mature, was an undisciplined, half-literate product of rustic South Carolina, and had come to Harlem very much against her will with her parents and her six brothers and sisters. Against her will because she had not been at all anxious to leave the lackadaisical life of the little corn pone settlement where she had been born, to go trooping into the unknown vastness of New York, for she had been in love, passionately in love with one John Stokes who raised pigs, and who, like his father before him, found the raising of pigs so profitable that he could not even consider leaving Lintonville. Cordelia had blankly informed her parents that she would not go with them when they decided to be lured to New York by an older son who had remained there after the demobilization of the war time troops. She had even threatened to run away with John until they should be gone, but of course John could not leave his pigs, and John's mother was not very keen on having Cordelia for a daughter-in-law—those Joneses have bad mixed blood in 'em—so Cordelia had had to join the Gotham bound caravan and leave her lover to his succulent porkers.

However, the mere moving to Harlem had not doused the rebellious flame. Upon arriving Cordelia had not only refused to go to school and refused to hold even the most easily held job, but had also victoriously defied her harassed parents so frequently when it came to matters of discipline that she soon found herself with a mesmerizing lack of home restraint, for the stress of trying to maintain themselves and their family in the new environment was far too much of a task for Mr. and Mrs. Jones to attend to facilely and at the same time try to control a recalcitrant child. So, when Cordelia had refused either to work or to attend school, Mrs. Jones herself had gone out for day's work, leaving Cordelia at home to take care of their five-room railroad flat, the front room of which was rented out to a couple "living together," and to see that the younger children, all of whom were of school age, made their four trips daily between home and the nearby public school—as well as see that they had their greasy, if slim, food rations and an occasional change of clothing. Thus Cordelia's days were full—and so were her nights. The only difference being that the days belonged to the folks at home while the nights (since the folks were too tired or too sleepy to know or care when she came in or went out) belonged to her and to—well—whosoever will, let them come.

Cordelia had been playing this hectic, entrancing game for six months and was widely known among a certain group of young men and girls on the avenue as a fus' class chippie when she and I happened to enter the theater simultaneously. She had clumped down the aisle before me, her open galoshes swishing noisily, her two arms busy wriggling themselves free from the torn sleeve lining of a shoddy imitation fur coat that one of her mother's wash clients had sent to her. She was of medium height and build, with overly developed legs and bust, and had a clear, keen light brown complexion. Her too slick, too naturally bobbed hair, mussed by the removing of a tight, black turban was of an undecided nature, i.e., it was undecided whether to be kinky or to be kind, and her body, as she sauntered along in the partial light had such a conscious sway of invitation that unthinkingly I followed, slid into the same row of seats and sat down beside her.

Naturally she had noticed my pursuit, and thinking that I was eager to play the game, let me know immediately that she was wise, and not the least bit averse to spooning with me during the evening's performance. Interested, and, I might as well confess, intrigued physically, I too became wise, and played up to her with all the fervor, or so I thought, of an old timer, but Cordelia soon remarked that I was different from mos' of des' sheiks, and when pressed for an explanation brazenly told me in a slightly scandalized and patronizing tone that I had not even felt her legs . . . !

At one o'clock in the morning we strolled through the snowy bleakness of 144th Street between Lenox and Fifth Avenues to the walk-up tenement flat in which she lived, and after stamping the snow from our feet, pushed through the double outside doors, and followed the dismal hallway to the rear of the building where we began the tedious climbing of the crooked, creaking, inconveniently narrow stairway. Cordelia had informed me earlier in the evening that she lived on the top floor—four flights up east side rear—and on our way we rested at each floor and at each halfway landing, rested long enough to mingle the snowy dampness of our respective coats, and to hug clumsily while our lips met in an animal kiss.

Finally only another half flight remained, and instead of proceeding as was usual after our amorous demonstration I abruptly drew away from her, opened my overcoat, plunged my hand into my pants pocket, and drew out two crumpled one dollar bills which I handed to her, and then, while she stared at me foolishly, I muttered goodnight, confusedly pecked her on her cold brown cheek, and darted down into the creaking darkness.

Six months later I was taking two friends of mine, lately from the provinces, to a Saturday night house-rent party in a well-known whorehouse on 134th Street near Lenox Avenue. The place as we entered seemed to be a chaotic riot of raucous noise and clashing color all rhythmically merging in the red, smoke filled room. And there I saw Cordelia savagely careening in a drunken abortion of the Charleston and surrounded by a perspiring circle of handclapping enthusiasts. Finally fatigued, she whirled into an abrupt finish, and stopped so that she stared directly into my face, but being dizzy from the calisthenic turns and the cauterizing liquor she doubted that her eyes recognized someone out of the past, and, visibly trying to sober herself, languidly began to dance a slow drag with a lean-hipped pimply-faced yellow man who had walked between her and me. At last he released her and seeing that she was about to leave the room I rushed forward, calling "Cordelia?"—as if I was not yet sure who it was. Stopping in the doorway, she turned to see who had called, and finally recognizing me said simply, without the least trace of emotion—"'Lo kid . . ."

And without another word turned her back and walked into the hall to where she joined four girls standing there. Still eager to speak, I followed and heard one of the girls ask: "Who's the dicty kid? . . ."

And Cordelia answered: "The guy who gimme ma' firs' two bucks . . ."

Thurman is the coauthor of two extant plays. Apparently he tried his hand at a couple of others and prepared outlines for even more—including one on Mormon leader Joseph Smith. Since his two full-length plays have never been published until now, it is understandable that Thurman's place in the history of black drama has not received much critical attention. Yet Thurman was indisputably a theatrical pioneer. He brought to his dramatic ventures the same enthusiasm for discussion and debate that distinguishes his other writings. Doris Abramson notes in her *Negro Playwrights in the American Theatre* (1969) that "Wallace Thurman was the first Negro playwright who deliberately set out to write for Broadway about Negro life as only a Negro can know it" (40). In his dramatic collaborations with

William Jourdan Rapp, *Harlem: A Melodrama of Negro Life in Harlem* and *Jeremiah the Magnificent,* Thurman showcased black America for all Americans. While he was not Broadway's first black author (Garland Anderson had opened *Appearances* in 1925), Thurman brought a new realistic depiction of Negro life to the stage. Unlike plays that had preceded it, *Harlem* stood out instantly as a controversial look at black America that took ordinary life as its subject. Abramson notes that *Harlem* broke barriers in its "attempt to let sensational melodrama grow out of the *real problems* of Harlem: overcrowded apartment living, prejudices among men of color, the numbers racket, transplanted and unemployed Southern Negroes" (41). Thurman encountered both success and frustration as a playwright. Nevertheless, his translation of black life into theater represents an important legacy and a considerable achievement.

## Harlem: A Melodrama of Negro Life in Harlem

Based on Thurman's story "Cordelia the Crude," William Jourdan Rapp and Wallace Thurman's *Harlem: A Melodrama of Negro Life in Harlem* opened at the Apollo Theatre (not to be confused with the one in Harlem) on Broadway on February 20, 1929. The play ran for ninety-three performances to sold-out houses on Broadway, then went on the road and had successful runs in Detroit, Chicago, Toronto, and Los Angeles. While the play was Thurman's greatest popular success, there is considerable evidence that it did not please him aesthetically. Langston Hughes wrote to Arna Bontemps on May 18, 1962: "Wally's play in the Broadway version was more Rapp's than his (so Wally said)." Hughes and Bontemps were keen to find Thurman's own version for inclusion in a proposed anthology. In *The Big Sea,* Hughes had already stated that the play was "considerably distorted for box office purposes" (235). In addition to the distortions that came with preparing it for popular consumption, the play also went through a number of torturous rewritings to fulfill the requirements of the producers. All this is chronicled in Rapp and Thurman's "Detouring *Harlem* to Times Square," included in this section. With the many writings and rewritings this collaborative work underwent, it would be an impossible task to distinguish Thurman's specific contributions from Rapp's. In their essays about the play, Rapp and Thurman insisted on an authorial symbiosis. However, it is widely believed that Thurman provided the story, the dialogue, and the detail, while Rapp, as the more experienced playwright, structured and stitched the material together.

*Harlem,* the first part of what Rapp and Thurman had visualized as their "Color Parade" trilogy (*Harlem, Jeremiah the Magnificent,* and *Black Cinderella*), grew out their friendship. A freelance writer and magazine editor, Rapp had written a number of plays for the stage and radio, though none had been produced. It is clear that Thurman and Rapp saw their friendship as mutually beneficial, as is evident from their narrative in "The Writing of *Harlem*: The Story of a Strange Collaboration":

> Wallace Thurman, the dark side of the black and white collaboration which produced *Harlem,* was anxious to launch himself in a writing career. William Jourdan Rapp, the editor whom he sought out for advice and aid was more than anxious to help him. So the two talked over Thurman's background and decided that the richest literary field for him to exploit was Harlem, every Negro's promised land, and the actual City of Refuge for approximately a quarter million black souls. Here all sorts of strange and interesting things were happening that the world would certainly be willing to read about.
>
> Thurman wrote a number of articles on life in Harlem and Rapp sold them for him. A strong friendship developed between the two men, and soon Thurman was introducing Rapp to every aspect of Harlem existence.

Upon examining *Harlem,* it becomes clear that the basic idea of the play is essentially Thurman's. In fact, the play depends upon Thurman's cherished claim that Harlem is a diverse and dynamic community, not monotone and monolithic. The play displays the human, social, and moral variety of Harlem, as it examines the lives of blacks who had recently migrated to New York. It details the various social types of the place, including "the sweetback," "the striver," and "the chippie"—the playbill for each performance of *Harlem* included a glossary of terms that belonged to a language Rapp and Thurman called "Harlemese." The play reenacts some of Harlem's most treasured practices—the house rent party and numbers-running. *Harlem* demonstrates that there is a clash of values and viewpoints in the Harlem of the 1920s—a clash, if one judges from almost everything he wrote, that Thurman found infinitely exciting.

Originally entitled *Black Mecca* and then *City of Refuge,* the play was retitled *Black Belt* by Broadway producer Crosby Gaige, the first to show interest in the play. He put the show into rehearsal almost immediately, but this attempt at production went nowhere, as Rapp and Thurman report in "Detouring *Harlem* to Times Square": "And a month later it was taken out of rehearsal and the cast dismissed. We innocently inquired 'Why?' and Mr. Lewis [one of the producers] replied: 'You haven't a "wow" in your third act, and naturally we can't go ahead until we find one.'" The quest to get *Harlem* produced centered on the third act: producers wanted the play to end not with Cordelia leaving the Williams household but showing more explicitly her life of sin and the consequences of such a life. Perhaps taking his cue from Theodore Dreiser's *Sister Carrie,* Thurman resisted moralizing about Cordelia's choices; she is a young woman in the city tantalized by its many temptations, exploring her sexual and social possibilities. Some critics of the play—especially those in the black press—found Cordelia's lack of repentance or moral reflection a problematic way to end the play. R. Dana Skinner wrote in his 1929 review of the play: "None raised a voice to protest against the particular way which this melodrama exploits the worse features of the Negro and depends for its effect solely on the explosions of lust and sensuality. The chief desire of the authors seems to be to show as the law permits— gambling, drunkenness, sordid dancing, shooting and amours of Cordelia." Clearly, *Harlem* is a play very much in line with Thurman's well-documented opposition to propagandistic, uplifting art that ignores the realities of black life. The controversy that raged around the play highlights the fact that it was a departure from the usual treatment of blacks on the Broadway stage of the time.

Eighteen months after the original play had been written, *Harlem* was finally produced with help of director Chester Erskin. Starting with one of the original copyrighted versions of the play, Erskin interested producers in the play, which he had renamed *Harlem.* While the decade had been marked by a series of Negro musicals and revues—*Shuffle Along* (1921), *Blackbirds* (1926), *Africana* (1927),

*Hot Chocolates* (1929)—*Harlem* was a full-length drama that featured a cast of twenty-six black actors. Cordelia was played by Isabel Washington, who "wowed" the reviewers with her singing and acting. Erskin himself, so legend has it, found and trained these largely amateur actors from Harlem. As Rapp and Thurman relate in "Casting and Directing *Harlem*":

> Where did Mr. Erskin find his players? The answer is Harlem. He went around rent parties, speakeasies, cabarets, theaters, and marches up and down Lenox and Seventh Avenues looking for types. He was sure that if anyone had the outward appearance and inner quality of the role, he could get them to act it. And the performances of Isabel Washington as Cordelia Williams, Emory S. Richardson as Jasper Williams, Collington Hayes as Thaddeus Jenkins, Ernest Whitman as Kid Vamp, and of Edna Wise Barr as Arabella Williams proves that Erskin knew what he was talking about, for none of these have ever performed a dramatic role before.

Despite *Harlem*'s popular success, however, disputes between Erskin and the cast led to the abrupt closing of the New York production on May 11, 1929. Seeking higher wages, the cast threatened to strike and dared the producers to shut down the show. The producers agreed to their demands as long as the principals took a cut in pay, which almost solved the problem until Erskin showed up, yelling at everyone. The ensuing outburst of renewed dissatisfaction shut the show down.

Despite the problems with the production, the play ultimately did leave its mark on the theatrical season because of its depiction of the Harlem rent party. Like the funeral scene in DuBose Heyward's *Porgy*—a play in which Thurman, Bruce Nugent, and Dorothy West had been extras—the rent party scene in Act One of *Harlem* made it a hit. It introduced the energy of Harlem to a new audience and held a mirror up to Harlem in which Harlemites could see themselves. Some reviewers looked at the play and saw shameful excess; other reviewers saw realistic portrayals of black life. In his review of the play published in the April 1929 issue of *Opportunity,* Theophilus Lewis asserted that "*Harlem,* because it emphasizes 'I will' characters instead of the gypsy type of Negro, is a wholesome swing towards dramatic normalcy. Its characters are not abnormal people presented in an appalling light but everyday people exaggerated and pointed up for the purposes of melodrama."

*Harlem* was also the center of a frenzy of articles and essays by the two authors. Some published, some not, these articles market, explain, and justify the play. From them, one gets a picture of the debates about realism and sensationalism that surrounded the production. Despite the debate about *Harlem*'s representations, it is undoubtedly true that the play was groundbreaking as an early example of a serious play, cowritten by a black author, that brought black characters and situations to Broadway. Of course, *Harlem* was not the first black drama to make it to Broadway—Garland Anderson's *Appearances* was produced in 1925. But as Abramson notes, "Wallace Thurman was the man of his time that Garland Anderson was not" (40). *Harlem* treated the lives and concerns of ordinary black people directly and realistically. Considering its place in the history of African American theater, *Harlem* deserves greater critical attention.

### Jeremiah the Magnificent

Probably written in 1929, *Jeremiah the Magnificent*—Thurman and Rapp's second play—shows greater facility with character and plot development than *Harlem*. In particular, the portrayal of Jeremiah, the character based on Marcus Garvey, is caustic but also respectful. He is gullible, vain, and susceptible to manipulation, but he is also shown, despite his failings, to be a man of vision and great oratorical powers. Despite being a better play, however, *Jeremiah the Magnificent* was performed only once, in New York City on December 3, 1933. The second installment in the "Color Parade" trilogy, the play moves the focus away from southern migration to Harlem to larger issues of black self-esteem.

More general and more politically charged than *Harlem, Jeremiah* documents northern modes of discrimination and the complications that arise—both internally and externally—when Jeremiah tries to relocate a shipload of his followers to Africa. One surprising scene in Act Two brings Jeremiah into conversation with a "delegation of prominent Negroes." These men—J.S.B. Renard, Z. Albert Caldwell, and Major Martin—are fictionalized versions of W.E.B. Du Bois, A. Phillip Randolph, and Robert R. Moton. Thurman and Rapp use this scene to dramatize some of the differences in perspective that characterized the debate about race in the United States—and to demonstrate how little regard these middle-class race leaders had for Marcus Garvey. Garvey's appeal to the black masses frightened the middle-class leadership of the NAACP and Urban League. He had a great impact on a large number of blacks that went well beyond the scope of his plans to return to Africa. As Henry Lincoln Johnson, the attorney for Garvey's codefendants, noted at their trial, "If every Negro could have put every dime, every penny into the sea, and if he got in exchange the knowledge that he was somebody, that he meant something in the world, he would gladly do it . . . the Black Star Line was a loss in money but it was a gain in soul" (cited in Ottley, 79). *Jeremiah* successfully dramatizes the existential value of a figure like Garvey.

Beyond these debates about political vision, though, *Jeremiah the Magnificent* is also an analysis of the errors that race leaders make. Like Garvey, Jeremiah is hampered by his own failings. He abandons his talented dark-skinned wife for a younger light-skinned woman. He entrusts his safety to unscrupulous, sycophantic advisers. Nonetheless, Thurman and Rapp endowed the play with compelling dramatic moments that hint at the spiritual power a figure like Garvey had. In the long speech in Act One, for instance, the authors translate the cadences and potencies of traditional sermonizing into a high point of the drama.

> Jeremiah: My, but to have you all here gathered about me makes me feel good, as good as Christ must have felt when His disciples gathered 'round Him. It buoys me up! It makes me feel confident that "The International Fraternity of the Native Sons and Daughters of Africa" will attain magnificent heights. It is highly necessary that we all work together; that we cooperate with one another. Only thus can we establish a black brotherhood and sisterhood—a sturdy phalanx of Africa's native sons and daughters. The black man must be free.

We know from Thurman's essay on Garvey in *Aunt Hagar's Children* that the charismatic West Indian leader fascinated Thurman. He notes in his essay:

> Garvey was the first Negro leader to capture the imagination of the masses, the first whose program was designed to appeal more to the black rabble than to the intelligent Negro or the philanthropic white. Ignoring and outraging both these agencies, he went directly to the underdog. His success was phenomenal; his failure dramatic and inevitable.
>
> The scope of his original program was stupendous. Africa for the Africans. All the black men of the earth spring from common African stock. Therefore let them all join together in order once more to regain their native land. There they can establish a government of, for and by the blacks. Thus the Negro problem will be solved. Thus the Negro can assert his right to be a man among men.

The search for black self-understanding and identity that Garvey embodied invigorates the whole of Thurman's work, not just *Jeremiah*. But this play signals an evolution in Thurman's dramatic work toward a more thoughtful consideration of the yearnings and dreams that underlie the idea of black community.

## Black Cinderella

*Black Cinderella* was planned as the last part of Rapp and Thurman's "Color Parade" trilogy. In their correspondence from 1929, Rapp and Thurman make numerous references to the project. There is even the suggestion in the letters that Thurman completed a draft, which he sent to Rapp from Utah. Apparently, however, all that exists today is the synopsis:

> The third play of the COLOR PARADE trilogy is now in the process of being written. This is a short synopsis.
>
> J. Seabright Moore is a light colored Negro who has prospered as a lawyer among his people. He has a light colored wife, Anna, and two daughters. Lavinia is twenty-three and has a beautiful but dark complexion. Adelaide is twenty and fairer than either parent. In fact, Adelaide would never ordinarily be taken for a Negro. She could easily pass for an Italian or Spaniard.

The Moores have favored Adelaide, their fair daughter, in every way. Her part in life is to look beautiful and acquire learning and culture. She is relieved of all the drudging of house work. By dint of hard work and many sacrifices her parents have given her a college education. The play opens the week preceding Adelaide's graduation from Barnard College of Columbia University.

Lavinia, on the other hand, because of her dark brown skin has been made the family Cinderella. Her schooling ended with high school and she is now family cook, maid, and seamstress. Her parents and sister are a bit ashamed of her because of her dark skin and keep her in the background on all occasions. She spends most of her time in the kitchen, and Aunt Julie, the washwoman, and Buster (Robert) Brown, the elevator boy, are really closer to Lavinia than her own family. In fact, Aunt Julie is intent on making a match between Lavinia and Buster.

Buster Brown is a handsome but very black Negro. He attends trade school along with working at his job and is studying stationary engineering with the idea of eventually becoming the superintendent of an apartment building. He has character and humor. There is a bit of the poet in him and he extemporizes love songs which he sings to Lavinia in the kitchen to the accompaniment of his banjo.

The Moores hope Adelaide will marry Doctor Vincent Andrews, a brilliant young Negro surgeon who could also easily pass as a swarthy Latin. In fact, both Adelaide and her parents expect the doctor to propose that very evening of the first act.

As a foil to J. Seabright Moore who is a "striver" and a "blue blood," there is his brother Arthur Moore who is a prosperous caterer. He ridicules his brother's family's pride in their light skins and their coincident feeling of superiority over the darker members of their race. Arthur Moore especially condemns his brother's action in investing all his own savings and the intriguing of many of his clients to also invest theirs in the exploitation of a restricted Negro "blue blood" colony; i.e., a real estate development in which only "light" Negroes can own property and build homes.

The crisis of the first act develops around the discovery by Doctor Vincent Andrews of the love affair between Lavinia and Buster, and the Doctor's consequent precipitate abandonment of his courtship of Adelaide. The whole family then jumps on Lavinia and insists that she give up her dark-skinned lover. She sticks to Buster in spite of all their threats.

The second act crisis results from the financial debacle of the "blue blood" colony and the "going over" into the whites of the Doctor and Adelaide, whom he marries only on the condition that she abandon her family and her race. The Doctor has changed his name to Andree and is moving to a new city, where he will begin his practice among the whites. When the father learns of Adelaide's apostasy, he becomes violent and is stricken with an apoplectic stroke.

The third act shows J. Seabright Moore a broken man, deserted by friends and clients. He and his wife are being supported by Buster and Lavinia, now happily married. Into their tenement home comes Doctor Andree and his wife Adelaide. They have prospered greatly but their first child is brown-skinned and the Doctor forces Adelaide to give the child to Lavinia. The crisis develops around Adelaide's unwillingness to abandon the child, but the Doctor's threats conquer her mother-love, and she leaves with him, never again to see her child or her family.

From these notes, it seems clear that Thurman and Rapp intended to pursue further the themes of black individualism and identity that they had examined in *Harlem* and in *Jeremiah the Magnificent*. Presumably, they abandoned the project when there was not the same kind of enthusiasm for *Jeremiah* as there had been for *Harlem*. As was typical of any other time in his life, Thurman had many projects to pursue in 1929. In fact, he had already conceptualized and begun the writing of *Infants*, which would not be published until 1932.

The evidence suggests that Thurman wrote a number of other plays that are not extant. He mentions in a letter to Harold Jackman a play called *Goodbye New York* that he was working on in 1930, and he based his last novel, *The Interne*, on a play of the same name. Some scholars have mistakenly ascribed the plays *Singin' the Blues* (written by John McGowan) and *Savage Rhythm* (written by Harry Hamilton and Norman Foster) to him. He did not write them. On the other hand, the play *Harlem* has been overlooked by anthologists until now. Notably, the play was not included in *Lost Plays of the Harlem Renaissance: 1920–1940*, edited by James Hatch and Leo Hamalian (1996).

Thurman also wrote two screenplays for Bryan Foy Productions in Hollywood: *Tomorrow's Children* (1934) and *(Secrets of a) High School Girl* (1934). Certainly, more research is needed into the intriguing intersection of Thurman's careers as fiction writer, critic, essayist, dramatist, and screenwriter. The fact that he found success in such diverse genres testifies to his artistic ambitions and potential.

# Harlem: A Melodrama of Negro Life in Harlem

Written in 1928 with William Jourdan Rapp.

## Characters

*(in order of their appearance)*

Arabella Williams
George Williams
Maizie Williams
Mother Williams
Father Williams
Cordelia Williams
Basil Venerable
Jasper Williams
Effie
Jimmie
Jenks
Ippy Jones
Jake
Mary Lou
Roy Crowe
Kid Vamp
Jiggs
Detective Sergeant Donohue
Patrolman Sam Johnson
Patrolman Kelly
Janitress
Janitress' Daughter
White Gunmen (2)
Half a dozen dancing Negro couples for the party in the First Act

## Scenes

Act I — Living room of the Williams's tenement in Harlem. About 8 o'clock on a Saturday night in winter.

Act II — Apartment of Roy Crow in Harlem. Two hours later.

Act III — The Williams's tenement. Immediately following the end of Act II.

## Act I

**Time:** *It is a Saturday night in late November. The scene is laid in what was once the dining or living room of a five room railroad flat in Harlem in the days previous to the Negro invasion. The room is on a slant in relation to the footlights, so that the end of the rear wall on the right is nearer the front of the stage than the end of the same wall on the left. In the right side of the rear wall is a* *window, with a fire-escape, which looks out on a narrow courtyard. The window is open and through it can be heard a rumbling medley of a discordant sounds—a man swearing at a woman; a Victrola spinning forth the salacious moans of a deep toned blues singer; the screeching of a pulley line as clothes are taken off it piece by piece; and the sizzling of frying food. This exterior rumbling varies from a monotonous pianissimo to a thunderous crescendo.*

*There is a door in the middle of the rear wall leading out into the hallway landing and down the stairs into the street. Another door on the right towards the front opens into a short narrow pantry which leads into the kitchen. Practically the entire left wall is given over to a large double doorway with sliding doors which now stand wide open. This doorway leads to the three other rooms of the flat, a large part of the first being visible. It is crowded with furniture—a side board, a dining table, a brass bed, and a rocking chair can be seen. Some of these have been removed from the main room which is free of all furniture except for a player piano with piano bench against the right wall adjoining the window, and a red cretonne covered iron folding couch pushed back into the rear left hand corner. There are also three cheap wooden chairs scattered about. The rooms are feebly lit with gas lights surrounded with round blue and red gas globes.*

**At Rise:** *When the curtain rises* GEORGE, *a coffee-colored boy of twelve, and* MAIZIE, *his sister of ten, are seen dusting the rounds of the chairs. They slide along the floor from chair to chair, touching up the baseboards as they go. They are both over-developed for their age. Their clothes are clean, but cheap and shabby.*

ARABELLA, *a dark brown-skinned girl of fifteen, enters from the kitchen carrying a six-month-old black baby in her arms. She is thin and a bit awkward with the awkwardness of rapidly growing adolescence. She has on a cheap cotton dress, much too small for her, over which is a soiled gingham apron, much too big.*

*Arabella:* Is Delia still 'sleep?

*George:* Uh uh!

*Arabella:* Who she think she's—a 'sleepin' while I do all the work?

*George:* Awl, let her sleep, she gotta be up for de party.

*Arabella:* Let her sleep! She oughta come in off the streets at night an' get some sleep. Ma don't let me stay up. I hafta do all the work and go to school, while Delia—all she does is primp, look pretty, quarrel with Basil, and flout herself around with the sweetbacks on the avenue. Be up for the party! When does she ever do anything 'cept be up.

*George:* (*ignoring the tirade and beginning to sing*) My mamma doesn't two time no time, an' if she ever two times one time, I bet she don' two time me no more.

*Maizie:* (*giggling*) That's what Basil oughta sing.

*Arabella:* (*threateningly*) Yeah! Delia betta watch her step. Basil's too good for her, he is.

*George:* (*with disgust*) Aw, dat monkey chaser! What do Delia care 'bout him? She don't want him. You're the only one 'round here thinks he's som'pin'.

*Arabella:* (*passion in her voice*) Didn't Pa tell you not to call him no monkeychaser? I bet if Basil ever hears you, he'll just lam your head. An' if you says it again, I'll tell Pa. Jus' 'cause he comes from the Barbados 'stead of Carolina ain't no reason why he ain't jus' as good as us.

*George:* (*tauntingly*) Even if you're kinda sweet on him, he's a monkey chaser jes' de same.

*Maizie:* What if he is? He's betta to us than anyone else 'round here.

*George:* Whacha call being betta—owing Ma two weeks room rent; eatin' two or three meals a week widout payin' for 'em, havin' you wash his sox, and chasin' big sis when she don't wan' 'im? Stuff! (*He goes out through the door at the rear, making a face at* ARABELLA)

*Arabella:* (*angrily*) Lousy little devil! I'm goin' tell Ma! He's gettin' jes' like de rest of these fly Harlem boys.

*Maizie:* (*sitting on the couch*) I wish we was back home.

*Arabella:* (*her habitual optimism reasserting itself*) Oh, Harlem's all right if it wasn't so cold and dirty. Betta times here than home.

*Maizie:* Whadu you mean, betta times? Nobody here but Delia has betta times. Ma and Pa both workin' all de time! We nevah gets out to play! All de beds in de house rented out! And den dese Saturday night parties! Pa gets drunk; Ma cries all day Sunday; Basil and Delia fights; an' we has to go to Brother Jasper's an' sleep in the bed wid dose wettin' kids o' his.

*Arabella:* (*laughing*) They is the wettenest kids I ever saw, but you're better off sleepin' with them Saturday night than on the floor in the kitchen here with cockroaches runnin' up and down your back and folks steppin' all over your face tryin' to get into the privy.

*Maizie:* Harlem is sho' one funny place. Who eva heard of havin' a privy in de kitchen?

*George:* (*re-entering in a great hurry*) Ma's comin'. She's downstairs talkin' wid de landlord. Betta get de suppa on de table.

*Arabella:* Maizie, you go wake Delia! George, put the baby in the basket whilst I set the table!

(*They all exit.* ARABELLA *followed by* GEORGE *goes off on the right.* MAIZIE *goes off on the left, calling all the while "Delia! Delia!" The door in the rear opens and the* MOTHER *enters carrying a large bundle of clothes. She is a typical southern woman, ready to moan and pray at the slightest provocation. Harlem has intensified this tendency. She has always had to work hard, but not with so little reward as is now the case. Everything to her is a burden imposed by the Lord to make her suffer, so that she will appreciate heaven when she gets there. She is a far stronger character than her husband. He would be lost without her. All of the children save Cordelia depend upon her strength and judgment. Religion is her only way of escape from a hard reality; calling on the Lord is her only solace. Her face is heavily lined. She is thin but muscular. Her hair is a bit gray. She is dressed in a gingham wrapper over which is an old black coat and hat, a gift from one of the women for whom she does washing. She comes in slowly, closes the door and looks around as if bewildered. She is plainly tired.* MAIZIE *skips back into the room*)

*Mother:* Dem four flights sure tuckers a body out.

*Maizie:* Hello, Ma.

*Mother:* (*taking off her coat and hat and handing them to* MAIZIE) Yes, Chile—where's George? (*without waiting for an answer she calls*) George! (GEORGE *enters*) Take dese clothes in da kitchen. Tell Bella to put 'em in de tub to soak. Gotta have 'em washed by Monday mornin'. *(GEORGE exits to right with clothes)* Is de house all cleaned up for de party?

*Maizie:* Yes, Ma.

*Mother:* S'too bad the house only gets cleaned when dere's a party. I s'pose the dinner ain't ready either. It nevah is.

*(She goes off to the kitchen with MAIZIE. CORDELIA enters from the left. She is about eighteen years old and has dark brown skin and bobbed hair. She is an overmatured southern girl, selfish, lazy and sullen. She is inspired to activity or joy only when some erotic adventure confronts her or a good time is in view. She has no feeling for her parents or for her brothers and sisters. Considering herself a woman of the world, she holds their opinions and advice in contempt. She is extremely sensual and has an abundance of sex appeal. Her body is softly rounded and graceful. Her every movement and gesture is calculated to arouse a man's eroticism. She is dressed in a pink, mercerized cotton slip, black cotton stockings, and a pair of much too large felt slippers. She yawns and shakes her head in an effort fully to wake up)*

*George:* *(entering from kitchen on right)* Ma's home an' she's sore cause you didn' cook corn bread. Bella burnt it, like she always does.

*(CORDELIA shrugs her shoulders, walks over to the piano, and runs her fingers across the top, looking for dust)*

*Cordelia:* Ma's always sore 'bout somethin'. Bein' sore is good for her.

*George:* I dusted, Delia.

*Cordelia:* I thought so! I see your finger marks!

*(GEORGE opens his mouth as if to protest when the outside door in the rear opens and their FATHER enters. The FATHER is a large, surly, overworked Southerner. He likes to beget children, but does not particularly care about them after they begin to create new responsibilities. He is stern in his treatment of both them and his wife. The North has rendered him helpless. He is just a big hulk being pushed around by economic necessity. He can't realize what the hegira to Harlem has meant or what it is doing to him and his family. His blustering exterior conceals a profound sense of inadequacy in his new environment. He does all sorts of odd jobs. He wears a pair of old brown trousers, shirt and suspenders. He has been across the hall)*

*George:* Dinner's waitin' fer yuh, Pa.

*Father:* Tell yo' ma I'll be right out. *(He walks over to the radiator, rubbing his hands)* It sho' is cold. Why don' you put dis window down? *(He closes it and the medley of noises from outside suddenly ceases)* We gets little 'nuff heat as 't is, widout you tryin' to warm up de outdoors. *(CORDELIA ignores him and starts to walk off at the left. He shouts)* Delia! *(She says nothing)* Delia! *(continues to ignore him)* Don't you hear me talkin' to you? *(She stops, turns slowly, and defiantly faces him)* I'm gettin' tired of all yo' uppity ways. You got so now you don't even wanna speak to folks. Dis house is hell 'nuff as 't is widout any of yo' picayunishness.

*Cordelia:* *(calmly)* Well, I can leave.

*Father:* *(angrily)* Leave den, an' stop yo' threats. Yo won't work! Yo won't go to school! You won't mind de baby. You won' act spectable wid a man what's willin to make you a good husban'! You done turned out to be a no good chippy! Dat's what yo is, a no good chippy! Go 'head an' leave. I wish to hell ya would.

*(He stalks off to the kitchen mumbling to himself.* CORDELIA *stands still for a moment, begins to hum a tune, walks over to the window, and throws it open again. The medley of noises is heard once more)*

*Cordelia:* *(leaning out of the window and calling shrilly)* Mary Lou! Mary Lou!

*(A window is heard going opposite. A shrill voice replies)*

*Voice:* What's de matter?

*Cordelia:* Comin' over to our party tonight? It's goin' to be *too* bad, Ippy's goin' to play, an' dey'll be plenty brown-skinned papas. What? *(She pauses)*

*Voice:* Sure, I'm comin'.

*Cordelia:* Al' right, den, I'll see you. *(The window opposite closes)*

*Arabella:* *(entering from kitchen)* Delia, dinna's ready. You gonna' eat?

*Cordelia:* No, I'm not gonna eat.

*Arabella:* Well, you needn't get so hincty bout it.

*Cordelia:* I spose you don't like it.

*Arabella:* No, I don't like it. I gets tired enuff as tis doin' all the work you outer do widout you bein' so darned asterperious all the time.

*Cordelia:* Well, wha cha gonna do 'bout it. *(places hands on her hips and saunters up close to* ARABELLA, *who seems to be trying hard to think of some comeback)* Well . . .

*Arabella:* You know I ain't got no time to be argufying with you, ya better come on an' eat now if you wants it. There won't be nothin' left.

*Cordelia:* Well, one thing, there's plenty restaurants jes' round de corner.

*Arabella:* An' course they'll feed you for nothin'. Delia Williams don' need no money, she don't!

*Cordelia:* *(sneeringly)* Dat's all right what I need, hot mamma. But what Delia ain't got, someone else has.

*Arabella:* Yeah, I spose you'll make Basil spend his hard earned pennies takin' you to restaurants.

*Cordelia:* Might as well take me as someone else.

*Arabella:* You know if 'twasn't for you Basil wouldn't throw no money away.

*Cordelia:* And if it wasn't for me, Basil wouldn't be here for you to look moon-eyed at.

*Arabella:* Who looks moon-eyed at him?

*Cordelia:* *(laughing)* All right, baby girl I knows you wants a man. *(talking baby talk)* Does baby sister like Basil? Does um love him, does um?

*Arabella:* *(angrily)* Yes, I loves him, an' if you didn't done cast no spell on him—

*Cordelia:* (*interrupts*) Who done cast a spell? He likes me cause he likes me. Who'd like you, jes' look at you. Go get a mirror. Like you! Gal, you couldn't get a man if dey was givin' em away.

*Arabella:* Well, I'll show you. He will like me! Jes' wait till he finds out what you are.

*Cordelia:* He's been 'round me long enuff to know what I am. Go on! I cain't be bothered hearin' you run off at de mouth. If you wants Basil, take him. I'm sho' I don't want him. I nevah wanted him. He wanted me. And jes' to show you there's no hard feelin', I'll give you some pointers on how to handle monkey chasers. They's a hard lot.

*Arabella:* Ain't nobody hard 'round here but you. Basil's bettern you any day. Now you see some other man, you wants to throw him ovah as you did Ippy, an' he ain't got sense enuff to know you're a—

*Cordelia:* (*fiercely*) Shut up!

*Arabella:* Shut up! Huh! Cain't no one talk round here but you?

*Cordelia:* You better go on now afore you get hurt.

*Arabella:* I ain't gonna shut up an' I'm gonna tell Basil jes' whata—

(**CORDELIA** *slaps her in the face.* **ARABELLA** *staggers back.* **CORDELIA** *follows her up*)

*Cordelia:* (*hand still outspread*) Now whacha gonna tell? Huh! Whacha gonna tell?

(**ARABELLA** *begins to sob.* **CORDELIA** *laughs. She flounces off stage at the left.* **ARABELLA** *remains standing in the center of room. She gradually gets control of herself. Wipes her eyes on her apron. She slowly starts for the kitchen. The rear door opens and* **BASIL** *enters.* **BASIL VENERABLE** *is a West Indian lad of about 21 years of age. He has not been long in the U.S. He is an ambitious chap and eager to make a name for himself in the legal world. He has, however, run afoul of Cordelia and found her to be the consuming passion of his life. She has become the symbol of his success, the reward at the end of the trail. He is as much a weakling in respect to his passion as he is strong in his other ambitions. He is quick tempered and easily angered. He is extremely conscious of his being a West Indian. He is still a little bewildered by Harlem. He is dressed neatly and soberly.* **ARABELLA** *smiles bravely at* **BASIL**)

*Basil:* Hello, small sugar. Where's Delia; in the kitchen?

*Arabella:* Hello, Basil! She jest went up front, to yo' room I s'pose.

*Basil:* What's the matter, another quarrel?

*Arabella:* Yeah! You know how hincty everyone gets aroun' here whenever there is to be a party.

*Basil:* They ought not have any more.

*Arabella:* Dat's what I say, but Pa says he has to have them to pay the rent. Aren't you goin' to school tonight?

*Basil:* No! I thought I'd stay here and help with the party.

*Arabella:* But you don't wanna miss—

(**CORDELIA** *enters.* **ARABELLA** *sidles off hurriedly as* **BASIL** *leaves her to greet* **CORDELIA**)

*Basil: (smiling)* Hello, big sugar. Come, give Basil a kiss.

*Cordelia: (drawing away from him)* You always wants to kiss. Don' you nevah get tired of slobberin'? Why didn't you go to school from work like you always do?

*Basil: (still trying to be jolly)* I came to see what you were doing.

*Cordelia:* You betta get outa dat, big boy. I ain't fixin' to have no man watch me.

*Basil: (growing impatient)* Aw, forget it Delia, I was just kidding. *(pleadingly)* Why don't you be sweet like you was—well, like you was before I asked you to marry me?

*Cordelia:* Maybe I don't see anythin' to be sweet over.

*Basil: (earnestly)* But Delia, you gotta love me now, after that! Supposin' something was to go wrong—to happen to you?

*Cordelia:* Don't you worry about dat, big boy! Little Delia can take care of herself. She knows her onions, she does.

*Basil:* But Delia, I love you and I want to marry you. I feel 'shamed to face your ma and pa. I—ah—

*Cordelia:* I don't see what you gotta be shamed of. I'm not, an' I thinks I'm de one to judge de shame.

*Basil: (hopelessly)* I won't argue now. I want to change clothes and get some supper. But promise me this—will you be real sweet at the party tonight and let that sheik Roy alone.

*(She puts her arms on her hips and laughs at him)*

*Cordelia:* Think I will?

*Basil: (pleadingly)* Delia! You know that fellow Roy means you no good. He's just playing with you. Why don't you leave him be?

*Cordelia:* I don' see why I should. Seems to me, you expects too damn much.

*Basil:* What do you mean—too damn much! I don't expect any more than what is due me. Aren't we engaged? You owe me some consideration.

*Cordelia:* All right! Try and get it.

*Basil: (Infuriated, he grabs Cordelia by both arms)* Listen here, Delia, if you keeps on fooling with me, I'm going to hurt you yet.

*Cordelia: (She wrenches herself free and starts off to the left)* Lemme go. Don't you pull no rough stuff. Hurt me? Remember this! The day you hurts me is de day Marcus Garvey'll be minus a monkey hip eating countryman.

*(She turns her back on him. He grabs her by the shoulders and spins her around)*

*Basil:* Don't you make any more remarks about me or my countrymen. What if I am a West Indian? Just because I'm not a bootlicking American nigger—

*(There is a knock on the rear door, and Cordelia's elder brother walks in. JASPER is about 28 years old and is the real strong character of the family. It is he who dominates them all, and he who comes the closest to dominating CORDELIA. When all other resources have failed in the solution of any problem, it is always Jasper who is sought out. He is an idealist and believes that the Negro should take advantage of the opportunity to settle in northern industrial centers. He believes this*

*strongly, even though he and his own family have had such hard luck. He also believes—thanks to much talk heard during his sojourn in the army—that the Negro is better off in the slums of the North than on the plantations of the South. He is a faithful and steady worker, always plodding and always being frustrated. He, too, is dressed neatly and modestly. He is physically large, like his father)*

*Jasper:* Lo' folks, what's all the argument? I heard you way down the hall.

*Basil:* Nothin, Jas! It's just—

*Cordelia:* It's jes' that he wants to boss me 'round, an' I ain't goin' to let 'im.

*Jasper:* Well, I should think it's time you let someone boss you.

*Cordelia:* What business is it of yours? I'll do as I please.

*Jasper:* You've always done so.

*Cordelia:* An' I'm goin' to keep on doin' so.

*Jasper:* Aw, Delia, shut up! Say, *(turns to* BASIL*)* Basil, aren't you goin' to school? It's way after seven.

*Basil:* No, Jas, I thought I'd stay home tonight. Will you be here for the party?

*Jasper:* No, I jes' came over to take Ma and the kids over to my place for the night. I leave the parties for Pa an' you and Delia. You oughta go to school tho. This missin' classes is bad business if you wants to get on in the world.

*Basil:* Oh, I'll make it up all right.

*Jasper:* But every little bit counts. I wish I'd a' had yo' chance. Oh say, did you hear I got Jenks a job as a porter down to the loft this morning?

*Basil:* Good! Now maybe he can help me pay the room rent. *(He stands embarrassed for a moment looking at* CORDELIA. *She gives him a mean look. He stutters)* Well, I—I'm going up front to freshen up a bit. Then I'll slip on the feed bag. See you before you go, I guess. *(goes off on the left)*

*Jasper:* Where are all the folks?

*Cordelia:* In the kitchen I s'pose.

*Jasper:* Say, Delia, why don't you snap out of it? Can't you see how many folks you is makin' miserable by yo' actions?

*Cordelia:* They ought to leave me be.

*Jasper:* Who do you think you is anyway? Leave you be? I wish't I had of let you be. May the good Lord forgive me for not lettin' the whole mess of you stay back south.

*Cordelia:* None of us wanted to come up here.

*Jasper:* *(resignedly)* Well, since you're here, I guess we has to make the best of it. Only I wish you'd marry Basil. That's a smart boy an' he'd make you a good livin'. He'll be a great lawyer someday. Wouldn't you like to be known as the wife of Counsellor Venerable?

*Cordelia:* *(walking off to the kitchen)* Hell no!

*(*JASPER *looks after her a moment, shakes his head, throws his hat on the couch, and walks over to the window, which he closes. He then sits down and lights a cigar. His* MOTHER *enters from the kitchen)*

*Jasper:* Hello, Ma.

*Mother:* Hello, son, you're here mighty early, ain't you?

*Jasper:* Not so very. Thought you might like to get away before the crowd started comin' in. You looks pretty tired. Come on an' sit down. *(She sits alongside him)*

*Mother:* How's Millie an' de kids?

*Jasper:* Millie's complainin' of a misery in her back. The kids are well. Ezra's teethin'.

*Mother:* Yo' better take yo' coat off—won't feel it when we goes out, you're liable to ketch cold.

*Jasper:* Oh, all right! *(He removes his coat as* GEORGE *and* MAIZIE *enter from the kitchen)* Hello kids! Betta get ready to go with me. Party tonight, you know. *(They go off at the left whispering to each other)*

*Mother:* I'll be glad when Pa gets out of de notion of havin' dese parties. I hates to be run out of ma own house.

*Jasper:* Why don' you stay here. Jes' let me take the kids.

*Mother:* Nevah wud I stay heah. I staid for one, and dat was enuff. I tell yo, it's a shame in de sight o' God.

*Jasper:* Pa should work steady then you wouldn't have to give these parties.

*Mother:* He does deh best he can. He jes' cain't seem to find no job dat lasts. Says he might as well not work at all as work for nothin'.

*Jasper:* Maybe so. But every little bit counts. Even a few extra dollars a week would keep you from workin' so hard and always worryin'. Dey's plenty work aroun' if Pa'd get out and find it.

*Mother:* De North don' seem to agree with Pa. He jes' don't fit in.

*Jasper:* He didn't fit in down home either, did he? Wasn't he always singin' the blues about how hard it was for a nigger to get along?

*Mother:* *(hopelessly)* Oh God, if I only knew. *(starts sobbing)*

*Jasper:* *(consolingly)* Oh forget it Maw. Pa'll snap out of it soon. I guess it is hard for him to get used to things here in Harlem. *(changing subject abruptly)* Did ja rent yo' front room yet?

*Mother:* *(wiping her eyes)* Yeah, a young couple. Dey seems to be good an' steady. Dey both works.

*Jasper:* Young married couple?

*Mother:* Well, dey says dey's married. Delia says not.

*Jasper:* Oh—Delia! 'Pend on her to dish up some dirt. Whadda we goin' to do wid dat gal, anyhow?

*Mother:* *(sighing)* Wish I knew. Pa wants to put her out. Basil wants to marry her. Lawd knows I've prayed for her wicked soul ev'ry night, but somehow ma prayers in dat direction don' seem to be answered.

*Jasper:* The Lord works in mysterious ways his wonders to perform.

*Mother:* Dat's right, son. One mus' keep faith an' I s'pose things will work out; but I do wish we were all back in Carolina, out of dis crowd an' cold an' dirt an' sin.

*Jasper: (becoming emotional)* But Ma, Harlem is a wonderful place—the greatest place in the world for folks. It's the city of refuge. Why, here you—

*Mother: (interrupting)* City of Refuge! Dat's whad you wrote an' told us. Harlem is de city of refuge. Is yo' shure you don' mean City of Refuse? Dat's all dere is heah. De people! Dese dark houses made out of de devil's brick, piled up high an' crowdin' one another an' smellin' worse dan deh pig pen did back home in summer. City of Refuge! You—I—God, have mercy on our souls.

*(She starts to sob.* JASPER *attempts to comfort her)*

*Jasper:* Now, Ma . . .

*Arabella: (entering from kitchen)* Hello, Jas. What's the matta with Ma?

*Jasper:* Jes' a trifle upset. Did you finish up yo' work?

*Arabella:* You mean, did I finish up *all* the work? No one else does any aroun' here. Delia, huh, she—

*Jasper: (making a motion for her to shut up)* Get the baby and the kids ready. We mus' be goin'. Get yo' ma's things, too.

*Arabella: (exits on the left calling)* Hurry up George and Maizie. Brother Jas is waitin' for you.

*(*FATHER *enters from the kitchen carrying an old-fashioned revolver)*

*Father: (aside on seeing* MOTHER *crying)* She at it again! *(to* JASPER*)* Hello, son, how's tricks?

*Jasper:* Pretty hard scufflin', Pa! Why the pistol? Think you'll have a bad crowd at the party?

*Mother: (getting up and wiping her eyes)* Dey always has a bad crowd.

*Father:* I jes' keeps it handy case things get too rough. Last month a couple of tough babies begins scrappin' an' de neighbors called de police. I don't want dat to happen again.

*Mother:* Who' da thought I'd ever come to dis? Father in heaven! Father in heaven!

*Father: (putting the gun in his pocket)* Well, you oughtn't kick. How else would we get de rent?

*Mother:* I'd rather not get de rent, dan have to turn ma' house into a house of iniquity. 'Tain't right, I tell you. God don' love evil. He'll punish us all.

*Jasper:* If you'd keep a steady job you wouldn't need to give parties.

*Father:* Son, you ain't got no right to talk. When we was doin' well down South, you couldn't let well enuff do. You had to drag us up North here. You writes, "Come on up! Dey ain't no prejudice here, plenty jobs and the colored folks got plenty money, homes, and cars. Come on to Harlem. I comes, and what does I find. All de good jobs is held by dese West Indian fellows. Now I ain't got no prejudice 'gainst 'em. Dey's al'right with me. We're all black together, but I ain't see why dey'll take up all de jobs and work for less money den we Americans. 'Tain't fair.

*Jasper:* That's a good excuse!

*Father:* 'T'ain't no excuse! It's the truth! Deh only folks I sees up here dat has anything in deh half-white niggers who won't pay us no mind. Dis is the most unhospital place I evah saw, and I wasn't jest born yesterday. Why down home, your neighbor looks in on you, and helps you out when you needs him. I done been livin' in dis house six months an' I got my first time today to know who lives 'cross de hall. You know what's wrong wid Harlem? Dey's too many niggers! Dat's it—too many niggers!

*Jasper:* You said the same thing 'bout down home.

*Father:* (*angrily*) Dat'll be enough! I'm gettin' tired of my own flesh an' blood bawlin' me out. I ain't gonna stand no more. (*mysteriously*) Dis is de end, I tell you. Dis is de end.

*Mother:* Whacha' mean?

*Father:* I mean I'm goin' back South.

*Mother:* Pa, you know we cain't go back South; we ain't got deh money. Den de old place is gone. We ain't go not place to go to.

*Father:* I says I'm goin' back South.

*Jasper:* You know if you go back in a month or so you'll be sayin', "Wisht I was up North!" Pa, you know they ain't nothin' down South for you.

*Father:* (*sarcastically*) An' dere's so much here! Huh!—I've got a happy home! I'm rolling in wealth! I'm prosp'rous.

*Jasper:* Pa, it takes time. It takes patience. Lookit the foreigners come over here and get wealthy in a few years.

*Father:* An' dey ain't none of dem got black faces.

*Jasper:* That don't hold you back. You got to say to yourself, "I'm a man even if I am black."

*Mother:* (*excitedly*) Betta say you're a child of God.

*Father:* Yeah, a child of God livin' in hell. I'm tired of it, I tell you! Never a free minute. Nothin' but troubles and more troubles.

*Jasper:* You has to work to get free. You gotta break your chains. Why, Harlem is the greatest place in the world for Negroes. You can be a man here. You can ride in the subway and go anywhere your money an' sense can carry you.

*Father:* Yeah, an' you can slave in some dingy hell hole like dis. Dey ain't nothin' for a nigger nowhere. We's de doomed children of Ham.

(ARABELLA *enters from the left*)

*Arabella:* I cain't make George an' Maizie get ready. They wants to stay for the party.

*Mother:* (*despairingly*) Eben de little children!

*Father:* Ah, old lady, you can preach, but I don' see yo' preachin' bringin' in money fo' de rent.

*Mother:* De Lord will provide! An you jes' mark my words, nothin' ever goes under de devil's belly dat don' come up over his back. (*starts to sob again*)

*Jasper:* I'll get the kids. We'll go in a minute.

(JASPER *exits left.* ARABELLA *starts to follow*)

*Father:* Bella. *(She turns to face him)* Is Jenks home yet? I want him to get de liquor.

*Arabella:* I don' know, Pa. I'll go see. *(starts towards left)*

*Father:* Nevah mind! You go on out in de kitchen an fix dat slaw. I'll get Jenks.

*(ARABELLA goes into kitchen)*

*Mother:* Pa—you put dat gun back in de dresser where it belongs. I won't rest comf'table if I knows you're carryin' it in yo' pocket.

*Father:* Don' worry! It's jes' for emergencies and dey never comes if you're ready for 'em. *(notices paper in JASPER's coat pocket. Takes it and sits down to look it over)*

*(BASIL enters in shirt sleeves and without collar)*

*Basil:* Pardon me, Mrs. Williams, is my laundry ready?

*Mother:* Yes, son, it's on de bed in my room. Didn't have time yet to mind yo' shirts. Dat's why I didn't put it in yo' dresser drawer.

*Basil:* Thanks, I'll go get it. *(to FATHER who is reading paper)* What's the number for today?

*Father:* Four, thirty-five! Did you play?

*Basil:* No, I never play. But Jenks played. He was hoping to make a hit so he could pay his back rent and some of the money he owes me.

*Father:* What number did he play?

*Basil:* One, twenty-six.

*Father:* Huh! Dat's funny! Dat's the same number I played.

*Mother:* Is you been playin' dem numbers?

*Father:* *(surlily)* I only played a quarter!

*Mother:* So now you eben gone to gamblin'. *(FATHER petulantly folds up paper and throws it down)* Pa, didn't I ask you not to start playin' dem numbers. Didn' I hear 'bout 'em when we first come here. You might win some time, but once de fever gets you, you'll play it. All back and lots more besides.

*Father:* Folks do win. Dere's old man Jones played a dollar on six, seventy-two de odder day, and he won five hundred and forty dollars.

*Basil:* And I'll bet that he has been playing a dollar a day for the past year or more. Some people are lucky but there's really no way to beat that game. These guys that sell tips on the number are a bunch of fakers. Why it's calculated from the daily Clearing House Report and there's absolutely no way of telling ahead of time what that's going to be.

*Father:* Dat may all be so. But you can't say dat dey don't pay five hundred and forty dollars if you hits wid a dollar.

*Basil:* Sure! But you've only got one chance in a thousand to win. Like all the lotteries— the bankers clean up!

*Father:* If dat's true Basil, why does everybody play?

*Mother:* 'Cause dey's full o' de devil.

*Basil:* You aren't even sure to get paid when you do hit. Remember when seven, eleven came out? Hundreds of people had played it and should have been paid off, but were they? Huh! The bankers kept the bets and refused to pay the winners. Nothin' could be done about it!

*Mother:* Gamblin' money is bad money. Nobody ever profits through sin.

*Father:* The law ougtha tighten up on dem bankers.

*Basil:* They're outside the law. A runner is arrested once in a while. But with every cop in Harlem playing his badge number, what can you expect? Well, I've got to go finish dressing. *(He exits up front)*

*Mother:* Pa, it's bad enuff to have boarders in de house and dese parties widout you gamblin'.

*Father:* Damn it. I'm tired of yo' preachin'. Cain't a man have a minute's peace in his own house? 'Stead of preachin' to me when I'se merely tryin' to get hold of some money so we can go back South—ya betta save it fo' dat no good chippie daughter of yo's. Make somethin' o' dat no good hincty wench, den I'll think somethin' of yo' prayers an' preachin'. God damned!

*(stalks off stage calling, "Jenks—Jenks"* MOTHER *begins to moan to herself. She buries her head in her hands and sways the upper portion of her body. Finally she breaks into a prayer which is a half chant)*

*Mother:* Father in heaven! Father in heaven! Forgive dis sinful household. Lawd, fo'give dem. Save my poor wicked children. Watch over dem. Show dem de light. Guide dem, Father. Shield dem from de devil and cleanse der souls with de Holy Spirit. Amen! Father! Amen!

*(continues to moan to herself as* CORDELIA *enters all dressea up in her party finery, a made over evening dress given to her mother at one of her work places. It is quite short and molded to the body)*

*Cordelia:* You still here. Thought you'd be gone!

*Mother:* *(looking up wearily)* Yes, child! I'm still here! *(She notes the shortness of Cordelia's skirt and gazes at her horrified)* Gal, is you tryin' to get naked?

*Cordelia:* Now, don't you start preachin' 'bout my clothes.

*Mother:* I'm yo' mother, ain't I? If I cain't watch ovah you, dey ain't no one else can, is dey?

*Cordelia:* I'm old enuff to watch out fo' maself. I ain't no child. I gets tired bein' bossed 'round.

*Mother:* Now dey ain't no need in bein' hard-headed. *(softening)* Cain't you see dat I loves you gal, an' dat I only wants to see you do de right thing. Cain't you see how it wrings my heart to have all dis strife an' dissension here at home.

*Cordelia:* You wouldn't be bothered with me if you'd let me do what I wanta do.

*Mother:* But, gal, you don't always know what's best fo' you. Why won't you listen to some one older with mo' experience and mother wit.

*Cordelia:* I got mother wit. I got all I needs but freedom. Jes' cause I don' wanna tend to babies, slave, cook an' wash for Pa, or for some white woman don' mean I don't know what's bes' for me. I ain't cut out for dat. I'm cut out for something big, something more excitin' and beautiful dan bein' a washwoman or a lady's maid.

*Mother:* But chile you seems to forget—

*Cordelia:* Forget nothin'! You wants to say I cain't do nothin' else 'cause I'm black. All black women don' have to be kitchen mechanics. An' don't tell me to be no school teacher! I've had enuff of kids right here!

*Mother:* But whatcha gonna do chile? Whatcha gonna do?

*Cordelia:* *(drawing herself up)* I'm goin' on de stage. *(She begins to snap her fingers and lewdly shake her body as if going into a dance. She hums a jazzy tune)*

*Mother:* *(sharply)* Gal, you should be shamed of yoself, standin' dere in front of yo mother, wantin' to sell your soul to de devil. *(She supplicatingly raises her hands to heaven)* Oh Father, I'd rather you take my girl outta dis world!

*Cordelia:* *(disgustedly)* Some mother you are, wishin' I was dead! Well, I ain't gonna die! I'm gonna live, see! And by God, I'm gonna live high! *(She flaunts herself off towards the left)*

*Mother:* *(moaning)* Have mercy, Father, have mercy!

*(JASPER enters followed by **GEORGE** and **MAIZIE**, both dressed in street clothes. **ARABELLA** comes in carrying the baby)*

*Jasper:* *(going up to **CORDELIA** and speaking threateningly)* You at it again, Delia? Makin' Ma miserable. I've a good mind to— *(starts to strike her but she stops him with a defiant look. He turns back to his mother. **CORDELIA** goes up front)* Come on Ma, get you things on. Here's yo coat. Millie'll be lookin' fo us.

*(Helps his **MOTHER** with coat, and hands her her hat which she dons wearily. He then takes baby from **ARABELLA**, and starts for the rear door. The kids follow him sullenly. The **MOTHER** takes one last look around the room and kisses **ARABELLA**, embracing her passionately. She makes a dejected exit. **ARABELLA** goes back into the kitchen after glancing up front in the hope of seeing **BASIL**. The door in the rear opens and **EFFIE** and **JIMMIE**, the two roomers, enter. Effie and Jimmie are of a type often found in Harlem. They are a pair of youngsters who are happy-go-lucky and full of pep and joy. Tired of drifting around from room to room singly, they have decided to live together without benefit of clergy. This act is typical of their philosophical objectivity. They are not immoral, but practical. They are all for anything that can assure them a little fun. They get quite a kick out of life and out of themselves. Everything is food for their fun mill. They are cheaply but gaily dressed. As **EFFIE** comes in, we hear her exchanging greetings with **JASPER** in hall)*

*Effie:* Come on, Jimmie! You're always fo'gettin' your keys. Some day you're goin' to get out an' won't be able to get back in.

*Jimmie:* All I got to say is that—if I can't get in, then nobody else betta come out.

*Effie:* *(laughing)* Go ahead man, take these things on up to the room, while I put our groceries in the kitchen an' sees how things are progressin' fo' the party.

*Jimmie:* All right, baby! *(She hands him her coat and hat and he goes out on the left while she exits right with the groceries under her arm and humming a popular tune. At door on the left)* What do you say there, old girl? Can I get through?

*Cordelia:* *(off stage on the left)* Sho', come on!

*Jimmie:* Thought you was dressin'. Hot mama! Lawd! All perked up for de party. You oughta slay 'em tonight, Delia.

*Cordelia:* (*entering*) Watch me, big boy, watch me! (*She calls into the kitchen*) Effie! Effie!

(**JIMMIE** *disappears up front*)

*Effie:* (*offstage on the right*) Coming, kid. (**EFFIE** *enters*) Yo sho' looks good, gal.

*Cordelia:* Oh, I'm not so hot. Dis dress is too full of safety pins to be comf'table.

*Effie:* (*laughing*) S'all right. You may ketch a hot papa that'll fall fo' the dress, an' then— no mo' safety pins, eh?

*Cordelia:* Hot chance I gotta vamp, what with Basil messin' 'round an'—

*Effie:* Oh, fo'get Basil. Why don't you let him tote his little red wagon somewhere else?

*Cordelia:* You know t'aint easy to get rid of no monkey chaser.

*Effie:* Basil's all right; just a little bit crazy. He can be gotten rid of. Wisht was me. I'd—

*Father:* (*entering*) Say, Effie, is Jimmie home?

*Effie:* He's up front. Want him?

*Father:* Yeah, I want him to git de liquor. Can't wait fo' Jenks any longer.

*Effie:* Sho, he'll go.

*Cordelia:* Why doncha go yo'self?

*Father:* Now, listen heah gal—

*Cordelia:* (*interrupting testily*) Listen to nothin'. Ya talks about other folks bein' lazy, yet yo' is always aiming fo' somebody to do somethin' fo' you.

*Father:* (*controlling himself with difficulty*) Jes' three mo' words, jes' three mo' words, and I'll try mah bes' to beat de devil outta ya. I'm tired of yo' foolishness an' sass. I won't have it, d'yuh hear? I won't have it!

*Jimmie:* (*entering from left*) Hello Dad, what's de trouble?

*Effie:* (*before either* **FATHER** *or* **CORDELIA** *can speak*) Pop wants you to get him some liquor.

*Jimmie:* Whadda you want, Dad, corn, gin, or rye?

*Father:* Betta get corn. Dey seems to like dat best.

*Effie:* An' you can stretch it out farther too.

(*Someone fumbles with the knob of the door in the rear and* **JENKS** *enters. Jenks, Basil's roommate, is always out of a job and never able to find one for himself. He is obliging, but slow; willing, but without initiative; faithful and sincere. An easy-going Southerner, surprised by the rush and struggle around him. He is dressed in one of Basil's old suits and has no overcoat*)

*Jenks:* Hello, folks.

*Father:* Hi, there, workin' man! Jes' wishin' fo you.

*Jimmie:* (*as* **JENKS** *starts to drop into chair*) Don't sit down, Jenks. We've gotta go right out for liquor.

*Jenks:* I stood all the way up in de subway. Thought I never would get home. Come up on de local. The jam and push of de express was too much fo' me.

*Father:* Dat's one thing. Dey may lynch you down home, but dey shure don't squeeze you to death on no subway.

*(They all laugh save* CORDELIA, *who all during this conversation has been fingering the keys on the piano. She utters a large "Humph!" after the laughter has subsided)*

*Cordelia:* Humph! You betta be watchin' de time, an' yo' shure betta get dat liquor. Fust thing you know, somebody'll be here.

*Father:* Don' you worry 'bout what we gotta do.

*Cordelia:* Ah'm shure what you gotta do don' worry me none.

*Father: (flaring up)* Dat's de trouble wid you. You—

*Jimmie: (interrupting)* Where's the money fo' the corn? Jenks an' I'll go an' get it.

*Father:* Al'right boys. *(He pulls some bills out of his pocket and hands them to* JIMMIE*)* Git four quarts.

*Jimmie:* Wait'll I get my coat an' hat, Jenks. *(He exits on the left)*

*Father:* Ah'm goin' out to finish fixin' de pigs' feet. *(He goes off to the kitchen)*

*Jenks:* Where's Basil? At school?

*Cordelia:* That's where he's s'posed to be, ain't it?

*Jenks:* I didn' ask you where he s'posed to be. I asked you where he is. *(to* EFFIE*)* Did yo' ever see such a hincty gal in all yo' life? *(to* CORDELIA, *soothingly)* You're al'right with me, though, Delia. Guess de crowd's kicking on you.

*Cordelia:* It don' hurt 'em none to kick, an' it sho' don' cost nothin'.

*Jimmie: (entering)* Come on, Jenks. Let's go. *(They exit at rear)*

*Cordelia:* I'm gettin' so damn tired of dis house an' everything in it.

*Effie:* Calm yourself, gal. Why should you worry 'bout em? Hey, who's goin' to play the music tonight?

*Cordelia:* Ippy Jones is goin' to play de piano. He's bringin' along a drummer. Dey oughta be here most any time now.

*Effie:* Ippy? *(She laughs)* Did you think of gettin' Ippy?

*Cordelia:* Shure!

*Effie:* He's sweet on you, ain't he?

*Cordelia:* So's my old man! *(They both laugh)*

*Effie:* What's there between you and Ippy?

*Cordelia:* He had his chance and didn't take it. There's nothin' between us now.

*Effie:* But I know he's still sweet on you.

*Cordelia:* Well, I ain't sweet on him.

*Effie:* Den why you havin' him come to play de piano?

*Cordelia: (mysteriously)* Ain't you learned yet I'm a wise baby?

*Effie:* Oh, I gets you. That sheik cousin of his—dat Roy Crowe—always goes to the parties where Ippy plays. I noticed you playin' up to him at Mary Lou's party the other night. So you're sweet on him. *(She starts to sing)* "The gypsy done tol' me you had a new brown skin . . ." *(They both laugh)*

*(BASIL enters from the left)*

*Effie:* Whadda you doin' home so soon?

*Basil:* I cut class tonight so I could help finish things up for the party.

*Cordelia:* *(sulkily)* Everything's finished.

*Basil:* Effie, what's the matter with this girl? I heard her laughing way up front, but the moment I came in she sours up.

*Effie:* Yo' never can tell what love will do. *(There is a knock on the rear door)* Lawd, let me get up fron' an' put on some clothes.

*(She exits on the left. BASIL opens the rear door. The two musicians, IPPY, the piano player, and JAKE, his Trap Man, enter. Ippy is a true black playboy of Harlem. He likes women, music and whiskey. He plays at parties, not primarily to make a livelihood, but principally because he enjoys it and it offers great opportunities for every new conquest among the fair sex. Everybody likes him. He has a pleasing, punchy personality and is full of good music, good talk and good fun. Jake is a lesser edition of Ippy. Both are dressed in cheap, flashy clothes)*

*Cordelia:* Come on in, boys.

*Ippy:* Hello there, Delia! Whadda you say?

*Cordelia:* I can't say. *(to the drummer)* 'Lo, Jazz Baby.

*Jake:* Hi there, Delia. How's Good-lookin'?

*Cordelia:* Put yo' traps ovah by de piano. Basil'll show you where to put your hats and coats. *(They put their traps and music down. BASIL leads them off on the right. He is surly, CORDELIA is radiant. While they are gone she hums for a moment, then runs to the window, throws it open and yells)* Mary Lou! Mary Lou! *(A window goes up opposite. A shrill "Yeah?" is heard. CORDELIA shouts)* Come on down, kid. The music's here.

*Voice:* *(off stage)* I'm coming!

*(BASIL and the two musicians re-enter. CORDELIA pulls down the window and walks to the center of the room)*

*Cordelia:* The liquor ain't come yet, but I spec' if you boys go into de kitchen, Dad'll fix you out.

*Ippy:* Lead us to it.

*Jake:* You said it. *(Both exit on the right)*

*Basil:* I hope they stay out there. Give me a kiss, Delia. *(He walks toward her. She kisses him. He remarks surprisingly)* You can be sweet. You do love me?

*Cordelia:* Um-m-m!

*(He hugs her and kisses her again. She clings to him, enjoying the physical contact. The rear door opens and Jimmie and Jenks enter with some bundles wrapped in newspaper in their arms)*

*Jenks:* Oh me gosh! Look what's happened! *(to JIMMIE)* Didn' I tell you Delia was al'right with me.

*Cordelia:* Folks in de kitchen waitin' fo' corn.

*Jenks:* Oh, we'll let you alone if that's what you want. Come on, Jimmie. Don't intrude!
*(They exit on the right)*

*Basil:* Delia, I love you so. And you can be so mean at times.

*Cordelia:* I ain't mean, honey. I jes' get sick an' tired of Ma and Pa and I wants to get away.

*Basil:* I—I can take you away as soon as summer comes. I'm saving every penny I can, and I'll be ready to enter law school in Boston next fall. I got a job promised to me there. Just be patient until after June. It's only four more months.

*Cordelia:* *(languidly)* I don' wanna stay in this dump four more months.

*Basil:* But in June we can get married, and—

*Cordelia:* Marry? We don' hafta marry.

*Basil:* But, Delia?

*Cordelia:* But, hell. It's always marry, marry, marry! Who wants to marry? Lookit Effie and Jim. Is dey married? No! Lookit Ma and Pa. Is dey married? Yes! Who'd you rather be, Effie and Jimmie always having a good time or dem odder folks moanin' de blues? Marry? I ain't fixin' to marry nobody, nevah! Life's too short!

*Basil:* Do you mean that you would run away and live with me without getting married.

*Cordelia:* Maybe!

*Basil:* But your mother and father! What would they say?

*Cordelia:* Whadda they say now? We's livin' together right in dis house now an' I cain't see no difference. Furthermore, dey'd say dey was damned glad to get rid o' me. Maybe den de rest of de kids'd get enuff to eat.

*Basil:* But Delia . . .

*Cordelia:* Oh, but, but, but. Don' you know no other word? Cain't yo' evah say anything but "*but*"? I hates you, I does. I hates you, I does. I hates you! An' I wouldn' have you no kinda way—

*(There is a knock on the rear door. CORDELIA pushes BASIL out of her way and rushes to open the door. MARY LOU enters. She is a nice looking light-brown girl, popular and confidential with both men and women—the sort who acts as a go-between when a couple wish to get together. She is taller than Cordelia but far less attractive. She, also, wears an old, madeover evening gown)*

*Cordelia:* Hello, kid, took you a mighty long time to get dressed.

*Mary Lou:* I ain't been dressin' all this time. Jes' met a sheik downstairs waitin' to come up here.

*Cordelia:* Who was it? Where is he?

*Mary Lou:* Don' get so excited. It's jes' Ippy's cousin, Roy!

*Cordelia:* *(agitated)* Is he comin'?

*Mary Lou:* Yeah! He'll be here in a minute. Gee, Basil. Delia's such a pain when you tells her you seen someone, she don't give me a chance to talk to folks. Well, look who's here!

*(IPPY enters from the kitchen)*

*Ippy:* Hot mamma, Lawd! Gettin' sweeter ev'ry day.

*Mary Lou:* I don' miss. *(She struts around)* Say, Ippy, jes' saw your cousin.

*Ippy:* Is he coming up?

*Mary Lou:* Sure thing. He's some hot papa.

*Cordelia:* I'll say he is.

*Basil:* I don't know how hot he is, but I know he's notorious.

*Cordelia:* Course you'd find somethin' to say 'bout him— jes' 'cause he ain't one of you dicty college niggers.

*Mary Lou:* Aw, dry up, both of you.

*Cordelia:* Huh, he makes me sick.

*Basil:* Jes' because I won't get excited over one of the most notorious underworld men in Harlem—

*Ippy:* How do you get that way? 'Tain't him that's notorious. It's his pardner—Kid Vamp.

*Basil:* No matter what he is, he helps run that notorious gambling joint, and he—

*Arabella:* *(entering from kitchen and interrupting)* Basil, come on get some supper.

*Basil:* Alright baby sis! I'm coming! *(reluctantly leaves and follows* ARABELLA *into kitchen)*

*Mary Lou:* The pigs' feet ready to be sampled?

*Cordelia:* Shure! Effie an' Jimmie and the whole gang are out dere. Go on an' get a bite.

*Mary Lou:* An' I don't need no second invitation.

*(She goes into the kitchen.* IPPY *sits down at the piano.* CORDELIA *stands nearby, leaning over him)*

*Ippy:* Delia, you sure looks good tonight.

*Cordelia:* I always look good.

*Ippy:* An' don' I know it. Ya eyes ain't so bright tonight tho. Been dissipatin'?

*Cordelia:* Wouldn't you like to know?

*Ippy:* You know one thing Delia, I can't make you out. Jes' what's your game?

*Cordelia:* That's what the cops wanna know.

*Ippy:* You gotta line all right, girl, you jes' won't quit, but 'member— *(He begins playing and singing to Hesitation Blues tune)* There's a bigger fish in the ocean / Than there is in the sea / And it's a low down dog / Can't run a coon up a tree.

*Cordelia:* Your dog's gotta learn how to bite as well as bark, big boy.

*Ippy:* Somebody'll tame you down, hot mamma, an' it may be me. *(He takes her hand and tries to pull her down to him)* Oh, Delia—

*Cordelia:* Well, speak out!

*Ippy:* *(dropping her hand)* Oh, hell, what's the use?

*Cordelia:* *(She leans over him)* Why don't you try an' find out?

*Ippy:* Ain't I been tryin' for over a year?

*(He rises from the stool. His face is close to* CORDELIA*'s. He grabs her arm passionately and draws her to him. She calmly turns her head to avoid a kiss)*

*Cordelia:* Did I hear you say yo' cousin would be here tonight?

*Ippy: (drawing away)* So that's why you're so damn sweet, eh? I ain't had a chance since I let you meet him. Ev'ry time I sees you, it's—Where's your cousin? *(sits down disconsolately and begins to play with the piano keys. He breaks into a soft blues tune.* CORDELIA*'s body begins to sway.* IPPY *whips it up.* CORDELIA *starts to dance.* IPPY *watches her closely. He shouts)* Come on, kid, sing for me. *(He plays rapidly.* CORDELIA *abandons herself to the song and dance. She is radiant. Toward the end of the song* IPPY *suddenly stops. He speaks passionately)* Delia, you and I could go far together. Let's team up!

*Cordelia: (quickly putting up her defenses)* When I pulls a wagon, I pulls it alone. I don't want no excess baggage.

*(She walks away from him. The door in the rear opens and* ROY CROWE *enters. He sees* CORDELIA *and smiles broadly.* CORDELIA *rushes up to him)*

*Roy:* Hello, hot mama!

*Cordelia:* Wha' d'yer say, hot papa?

*Roy: (seeing* IPPY*)* Hello, Ippy.

*Ippy: (sullenly turning back to the piano)* Hello!

*(He plays softly for a moment, then gets up, and goes into the kitchen.* ROY CROWE *is the Man Friday of* KID VAMP*, a prominent ex-pugilist who is a leader of the Harlem underworld.* ROY *is the Harlem equivalent of the racetrack tout and drugstore cowboy rolled into one. He is rather good-looking in a slick sheikish way. He is extremely sly and slippery and is not above exploiting a woman's infatuation for financial ends)*

*Roy: (to* CORDELIA*)* Whadda you sayin' sweetie? Stood me up last night, didn't you?

*Cordelia:* Think I can come to a party every night. 'Member, I lives with mah folks.

*Roy:* Gee, I missed you.

*Cordelia:* Can that stuff. I bet you didn't even ask after me!

*Roy:* Do I has to ask after you to miss you?

*Cordelia:* Maybe not. But I knows yo' kind.

*Roy:* An' I knows yo' kind. Always standin' a fellah up. I don' like it, either.

*Cordelia:* I won' do it again, Roy. Honest, I won'.

*Roy:* Make me know it, hot mamma, make me know it. Here I had a roll that wouldn't quit— *(*ROY *exhibits a large roll of bills)* Jes' longin' for you.

*(*BASIL *enters from the kitchen. He stands in the doorway, scowling at* ROY *and* CORDELIA*)*

*Cordelia: (seeing* BASIL*, she speaks to* ROY *in a whisper)* Don't talk so loud.

*Roy:* What's the matter? This don' look like no hop joint to me. *(*CORDELIA *pinches his arm and points toward door. He sees* BASIL*, smiles sarcastically, and nods, significantly)* Oh, I see, you not only stand me up, but starts right off two-timing. Why didn't you tell me you had somebody else?

*Cordelia:* I ain't.

*Roy:* Then why is you afraid to talk up when that gink is around. I noticed him watching you over at Mary Lou's the other night. What's he to you?

*Cordelia:* He—he—

*Roy:* Out with it! If you got somebody else, make me know it. I cain't be wasting my time.

*(He starts away from* **CORDELIA,** *and comes face to face with* **BASIL** *who is advancing toward him)*

*Basil: (to* **ROY***)* Why do you come around here? You know you're not our kind!

*Roy: (belligerently)* What'er yer drivin' at?

*Basil:* This is a respectable house, and we don't want any questionable underworld characters here.

*Cordelia: (angrily)* Roy's here 'cause I want him.

*Roy: (silencing* **CORDELIA** *with a gesture)* Let me handle this guy. *(to* **BASIL***)* Now, just what're yer objections to me?

*Basil:* You help run that notorious gambling joint!

*Roy:* It don' affect you none. It's run for whites only.

*Basil:* It does affect me and every other Negro in Harlem. The sooner those kinda places go, the better off we'll all be. Why should white people use our part of town for their underworld?

*(***JIMMIE, MARY LOU, EFFIE, IPPY** *and* **JAKE,** *the drummer, pile in boisterously from the kitchen. They immediately quiet down as they see* **BASIL** *and* **ROY** *facing each other tensely)*

*Roy: (seeing the gang, and really making a speech of his reply to* **BASIL***)* Let me tell you som'pin, Mister Holier-Than-Me! I suppose you think that we in Harlem don' have no underworld, of our own? We don' have no gin mills, no buffet flats, no gambling, no whores, no numbers? Now, listen to me! Harlem niggers don' need no white folks to make an underworld. They does pretty good by themselves.

*(The crowd laughs at* **BASIL***'s expense. He, disgusted, takes a chair and moves into the hall, leaving the door open.* **ROY** *greets the rest of the company)*

*Effie:* Play away Ippy. Let's get warmed up.

*Jenks:* That's what I say. Les' get goin'.

*Effie:* Whadda you goin' to play?

*Ippy:* Anything you say, Peaches.

*Effie:* Play something low-down. Make it good to me, Ippy. Make it good to me.

*Mary Lou:* You said it, boy. Hit those keys there, Mr. Man.

*(***IPPY** *begins to play.* **JIMMIE** *dances with* **MARY LOU,** **JENKS** *with* **EFFIE,** **ROY** *with* **CORDELIA.** *A few more couples enter. They each pay* **BASIL** *a quarter, go out at left, leave their wraps and return. All begin to dance, the party is on. Every once in a while someone calls out)*

*Someone:* Oh, play it Mr. Ippy!

*(Occasionally* ROY *will yell)*

*Roy:* Shake it and break it, you dirty no-gooder.

*(The first dance is interrupted by Cordelia's* FATHER *coming in from the kitchen)*

*Father:* Jest a minute, folks. Kin I get yo' attention? *(There is much shushing and finally quiet)* Dere is eats and drinks in de rear. Don' be bashful. Plenty good pigs' feet an' juice from the corn that'll make hair grow on yo're belly an' put joy in yo' feet. Come on!

*(He exits and all the couples follow him.* CORDELIA *is left alone with* ROY *down front.* BASIL *peeps in through the door at the rear. He is watching jealously)*

*Roy:* Why didn't you explain before that you had this guy?

*Cordelia:* I tells you I ain't got him. He jes' lives here, an'—an'—

*Roy:* Jes' lives here. Sorta brother like. *(He laughs)* I know all about that kinda stuff, kid. You ain't handin' me nothin'.

*Cordelia:* I ain't tryin' to.

*Roy:* Well, if you can't do no better than that, I'm glad you ain't trying.

*Cordelia:* *(shrugging her shoulders)* Yuh don' have to b'lieve me if you don't want to. Dat much is certain.

*Roy:* Well, you see, I can't take no chances. You says you don' want this guy no more, and you says you like me. Now, make me know it.

*Cordelia:* That's all I wants to hear you say.

*Roy:* No monkey business now, understand?

*Cordelia:* Ain't I been tellin' you that all night?

*Roy:* *(with finality)* Al'right, then, its all set fo' after the party.

*Cordelia:* I'm yo' baby, big boy, and I—

*(There is a commotion at the door interrupting* CORDELIA. TOM, *one of Roy's number runners, breaks past* BASIL *into the room)*

*Tom:* *(crying)* Roy.

*Roy:* What the hell is the matter with you?

*Tom:* *(whispers)* Come here. I got to tell you somethin'. *(pulls* ROY *downstage)* I've been stuck up.

*(*ROY *indicates with head that* CORDELIA *should leave him alone. She goes into the kitchen followed by* BASIL, *who almost runs after her from the door in the rear in his anxiety to speak to her alone)*

*Roy:* *(sneeringly)* Stuck up, eh?

*Tom:* Yeah. Rafferty's gang got me again.

*Roy:* Where?

*Tom:* Right downstairs in the hallway.

*Roy:* How much did they get?

*Tom:* Forty-two bucks.

*Roy:* Go on boy—every time you has a big collection, you claims somebody sticks you up.

*Tom:* It's the God's honest truth, Roy. I swear.

*Roy:* You'd better go swear to the Kid.

*Tom:* *(frightened)* I'm scared, Roy.

*Roy:* *(significantly)* You got good reason to be.

*Tom:* What will the Kid do to me?

*Roy:* *(menacingly)* You know what he told you last time.

*Tom:* *(defiantly)* I won't tell the Kid. I'll say I didn't make any collections 'cause the bulls were trailin' me. If there's any hits, I'll just steer clear of the winners.

*Roy:* You got a small chance getting away with that. You must think the Kid is still in his diapers. Don't you know that it takes a pretty slick guy to put anything over on the Kid?

*Tom:* Jesus! What'll I do?

*(People start returning from the kitchen)*

*Roy:* Pick up some collections here. If you work fast you ought to pile up a pretty good roll. Then the Kid won't ask any questions.

*Tom:* Thanks. That's a great idea.

*Roy:* Snap into it.

*Tom:* *(turning to the crowd)* Pick your number, folks. One lucky number and you won't have to work for weeks—you can live high while you're bein' lazy.

*Jimmie:* That ought to suit you, Jenks.

*(As* **TOM** *is soliciting numbers* **ARABELLA** *circulates with a tray full of liquor which the guests pay for as they drink)*

*Male Guest:* Here y'are, Tom. A quarter on three, fourteen.

*(***TOM*** *takes the money and gives the guest a slip from a small pad on which he has recorded the number and the sum bet. These slips are made in triplicate, one for the better, one for the runner and one for the banker)*

*Tom:* There you are brother. Now who's next?

*Jimmie:* Gimme three, eleven for half a dollar.

*Tom:* That's righteous. *(He hands* **JIMMIE** *the slip and pockets the money. He spies the* **FATHER***)* Hey there, Mr. Williams. Picked your number for Monday?

*Father:* I'm through. Dere ain't no chance for a poor man to beat dat game.

*Tom:* What do you mean? It's the only way a poor man can make any money in Harlem.

*Father:* It's the easiest way for him to lose it.

*Tom:* Monday may be your lucky day. You never can tell, you know.

*Father:* Well, I'll risk two bits, an' if I don' win dis time, I'm through.

*Tom:* Now you're talking. What's the lucky number? I'm going to bet on this one myself.

*Father:* (scratches his head) Seven, seventy seven.

*Tom:* The big mystery number.

*Jenks:* That number is sure full of dynamite.

*Tom:* Here you are Mr. Williams. (He hands him the slip and pockets the quarter. He accosts JENKS. The others drift off in pairs) What d'yer say, Jenks? Try your luck.

*Jenks:* I ain't got no luck. A black cat done cross mah path.

*Tom:* Never can tell, black boy, till you give it a try.

*Jenks:* I've tried it time an' time again and it jes' don' seem to work fo' me.

*Tom:* (confidentially) How do you pick your numbers?

*Jenks:* Oh, it just sort o' comes into my head.

*Tom:* (seriously) That's bad.

*Jenks:* (interested) How should you pick it?

*Tom:* You remember the day I gave you a number?

*Jenks:* Yeah. I just missed a hit by one.

*Tom:* That shows I know what I'm talking about, don't it.

*Jenks:* Yeah, I guess it does.

*Tom:* (whispering) Well, I got another number here for you. I'm betting it myself. (looks around to see no one is listening) It's a sure hit.

*Jenks:* Yeah, I'd like to chance it, Tom, but I ain't got no money. You couldn't trust me, could you?

*Tom:* Hell, what do you think we are? The Salvation Army? (He turns away)

*Jenks:* Wait a minute. You sure dat number's gonna hit?

*Tom:* Didn't I give you one sixteen, the last time, and what won?

*Jenks:* One seventeen.

*Tom:* Can you come closer?

*Jenks:* Wait a minute. (He goes over to BASIL) Say, Basil. How much does I owe you?

*Basil:* Six seventy-five.

*Jenks:* Give me a quarter and make it seven. I'll pay it all back to you on Monday. (BASIL hands him a quarter) Thanks. (He goes back to TOM) What's dat number?

*Tom:* Nine, forty one.

*Jenks:* Put me down for a quarter. (TOM gives him a slip)

*Tom:* You'll be lousy with money, black boy.

*Jenks:* (to EFFIE) I'm not particular 'bout dat. But I need me an overcoat.

*Tom:* You're wearing it now. *(The rest of the crowd come in from the kitchen. Some more people enter from outside.* TOM *turns to them)* Number, folks. Here you are. Get your lucky number.

*Cordelia:* You folks dat want to play numbers go ahead and do it. I wanna dance with my sweet man. *(She draws* ROY *to her with a smile)* Snap into it, Ippy.

*(The music begins. They all start dancing. The party is warming up. Liquor has lit the fire and Ippy's music is fanning it to a flame. The dimly lighted room now literally surges with strange, insinuating, slow rhythms, as the dancers revel in the movements of their bodies)*

*Shouts:* Oh, shake that thing. Do it, you dirty-no-gooder. Oh, mess.

*(The last is shouted as one couple does the mess-around.* IPPY *bends his long, slick haired head until it hangs low over his chest, then he lifts his hands high in the air and as quickly lets them descend on the keyboard. There is a moment of cacophonous chaos, followed by a tantalizing melody, beginning slowly and easily. Body calls to body. They cement themselves together with limbs lewdly intertwined. A couple is kissing. Another couple is dipping to the floor and slowly shimmying belly to belly as they come back to an upright position. A slender, dark girl with wild eyes and wilder hair stands in the center of the room supported by the strong lithe arms of a long-shoreman. Her eyes are closed. Her teeth bite into her lower lip. Her trunk is bent backward until her head hangs below her waist and all the while the lower portion of her body is quivering like so much agitated Jell-O.* IPPY *begins singing)*

*Ippy:* She whips it to a jelly!

*(He bangs on the keys with renewed might. Everybody is deadly serious. Desire is sad. The music finally slows down. The dancers are hardly moving. They are just rhythmically hugging.* COR-DELIA *and* ROY *are down front. We hear them above the music)*

*Roy:* You're sure some red hot mamma!

*Cordelia:* *(shaking her body lewdly)* Like me!

*Roy:* *(hugging her with sudden frenzy)* I'm crazy 'bout yer!

*(They do an extremely passionate series of movements.* BASIL *leaves his place at the door and comes towards them. He is almost mad with jealousy and has great difficulty controlling himself)*

*Basil:* *(with suppressed passion)* Cordelia! Come on outside. I want to talk to you.

*Roy:* Can't you see she's busy. *(*CORDELIA *exaggerates the lewdness of her movements)*

*Basil:* I'm not talking to you, see!

*Roy:* *(mocking* BASIL's *West Indian accent)* Al'right, mon.

*Basil:* *(ignoring* ROY*)* Are you coming, Delia; or must I take you?

*Roy:* *(turning from* CORDELIA *and facing* BASIL *threateningly)* Cave man, eh? Listen here, buddy—

*Basil:* You keep outta this.

*(*JENKS *and* JIMMIE *have been drawing close during the argument)*

*Jimmie:* (*grabbing* BASIL *by the arm*) Don't start no fight, Basil!

(*As* BASIL *turns to* JIMMIE, ROY *blackjacks him.* BASIL *falls to the floor*)

*Jimmie:* (*starting for* ROY) You dirty rat!

(BASIL *groans. This attracts* JIMMIE *from* ROY *and he helps* BASIL *to his feet*)

*Father:* (*running from the kitchen*) What's de meanin' of all dis?

*Cordelia:* Basil started it as he always does.

*Basil:* I didn't start it!

*Cordelia:* You're a liar! You did!

*Basil:* (*starting for* ROY. JIMMIE *and* JENKS *hold him back*) I'll kill that God-damn sweet-back.

*Father:* (*to* BASIL) Keep quiet, you! (*He turns towards the door*) Shut dat door, somebody, 'fo de police comes. You— (*pointing to* ROY) —git outta here, and git quick.

*Cordelia:* I don' see why somebody who paid dere money should be put out jes' 'cause Basil wants to pick a fight wid 'em.

*Father:* Dey's a hell of a lot you don't see dat you will befo' dis night is gone.

*Cordelia:* You think so.

*Father:* I know so. (*He again addresses* ROY)

*Cordelia:* You ain't goin' to put none of mah guests out.

*Father:* None of yo' guests?

*Cordelia:* Dat's what I said.

*Father:* I'll not only put yo' guests out, but I'll put you out.

*Cordelia:* You can do it when'v'r yo' gets ready.

*Father:* An', by God, I'm ready.

*Cordelia:* Alright, den, I'll go now.

*Father:* If you starts out dat door tonight, I'll—No, I won't, you ain't worth gettin' in trouble 'bout. If you wants to go—go! Jes' don' make no more trouble 'round here. (CORDELIA *rushes up as he turns to the piano*) Start the music, Ippy. Go 'head an' dance, folks. I'm sorry dis happened.

*Basil:* (*rushing up to* FATHER *as dance begins again*) Are you going to let Delia leave here with this man?

*Father:* I don' care who she leaves with.

(CORDELIA *returns with her hat and coat. She rushes up to* ROY *who is waiting at the door*)

*Roy:* (*in a mocking tone*) Good evening, folks. (*He hugs* CORDELIA *close to him*)

*Cordelia:* (*snuggling up voluptuously to* ROY *and looking right at* BASIL) I gotta real man, now!

(BASIL *breaks away and rushes at them.* ROY *slams the door in his face. Two or three of the men grab* BASIL. *The loud, mocking laughter of* ROY *and* CORDELIA *is heard as they descend the stairs*)

*Basil:* *(yelling impotently)* You God damn yaller pimp! I'll get you! An' when I get you, I'll slit your dirty guts! I'll get you! Let me go, I tell you! Let me go! God damn it, let me go!

*(His shrieks are lost in the increasing volume of the music. The dancers are beginning to crowd the floor as the curtain falls)*

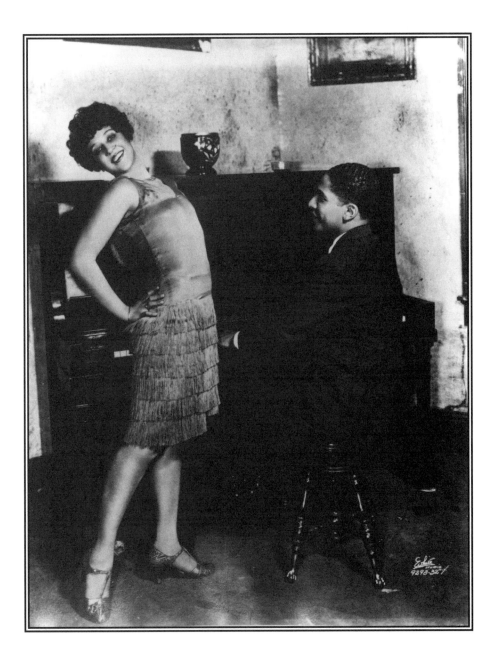

### Act II

*Scene:* *Three hours later—about midnight.*

*The scene is laid in Roy's one room kitchenette apartment. In the rear center is a curtain which is drawn aside and reveals a small alcove with a gas stove and a sink. On the right of this alcove is the entrance door from the outside. In the left wall is a double window opening on a fire-escape and in the right wall is a door leading to the bathroom. The shades of the window are down.*

*The room is furnished with a davenport, now opened out, forming a bed. It is set against the right wall front. A gate leg table is in the center of the room. Four windsor chairs, two rockers, a Victrola and two floor lamps are scattered about.*

*When the curtain rises,* CORDELIA *is busy at the sink and stove preparing coffee. She is dressed in her slip, with Roy's bathrobe thrown over it.* ROY *has on his trousers and a striped summer beachrobe. Various articles of attire are thrown on the bed.* ROY *sits at the table with a tabloid newspaper in his hand.*

*Roy:* What you think, here's Mrs. Rudolph Valentino saying as how she has talked to Rudy.

*Cordelia:* Better her 'n me.

*Roy:* Wouldn't you wanta talk to your sweet man once he was dead?

*Cordelia:* Ain't you got no sense? Who wants to talk to a dead man? When someone's dead, dey's dead. If I sees dem when I die, dat's soon enuff!—Furthermore, a sweet man ain't no sweet man no more when he's dead. Where's yo' sugar?

*Roy:* Should be right there 'side the salt and pepper.

*Cordelia:* Well, it ain't.

*Roy:* I betcha ain't looked.

*Cordelia:* Well come on, an' find it den.

*Roy:* Don't make me get up for nothin. If I comes over there an' finds it where I told you to look, it's gonna be me an' you.

*(*ROY *gets up and goes to the alcove. He moves the cans around on the shelf above the stove and finally fishes out the sugar bowl)*

*Cordelia:* Calls yo'self gittin' tight a'ready?

*Roy:* I told you you didn't look.

*Cordelia:* I s'pose now you're gonna tell me it was next to de salt shaker.

*Roy:* Gimme a kiss.

*Cordelia:* Go long, man. How you expect me to get dis coffee fixed for you?

*Roy:* (putting his arms around her) Gimme a kiss.

*Cordelia:* (protesting) Man, don' you never get tired of slobberin'?

*Roy:* Gimme a kiss.

*Cordelia:* Quit it, fool. Don' you hear dat coffee boilin'?

*Roy:* Gimme a kiss.

*(She stops her struggles and allows him to kiss her)*

*Cordelia: (breaking away)* Now, go on an' lemme be. *(She returns to her task)*

*Roy: (returning to his chair)* You know you likes it.

*Cordelia:* Think so.

*Roy:* Say you don't like it now, an' I guarantee to make you like it.

*Cordelia:* And how!

*Roy:* Must I come over and show you? *(He starts to get up)*

*Cordelia:* No, honey, I likes it. Jes' let me finish messin' wid dis coffee. *(She sets the sugar bowl and two cups on the table.* **ROY** *is continually grabbing at her as she passes. He pulls her into his lap and hugs and kisses her. She tries to look injured and sour, but manages to giggle delightedly thru it all)* Does you act like dis all de time?

*Roy:* You'll soon find out.

*Cordelia:* How do you figger—I'll soon find out?

*Roy:* Ye will if you stay here long enuff.

*Cordelia:* An' I ain't fixin' to leave soon.

*Roy:* Al' right sweet mamma, make me know it. *(As* **CORDELIA** *brings the coffee to the table, he again draws her back into his lap and kisses her. Their love making is interrupted by the ringing of the telephone which is standing on a taboret set against the short strip of the rear wall between the alcove and the entrance door on its right)* Wonder who that is?

*Cordelia:* Some of yo' other women, I guess.

*(She gets out of his lap and sits in her place at the table.* **ROY** *goes to the telephone)*

*Roy:* Hello! . . . Hello! *(He shouts)* Yes, this is Harlem 94234 . . . Who do ya want? . . . Who? Well, who said she lived here? . . . Who is this calling, anyhow? . . . No. She ain't here. What made you think so? Don't you know where she lives? . . . Now listen here, buddie, you've said enuff . . . I said she wasn't here, an' since I'm here and you ain't, I figgers I oughta know! *(He bangs down the receiver)* Somebody callin' for you! Musta been that monkey chaser.

*Cordelia:* How could it a been him? He don' know this number.

*Roy:* 'Course they don't have no telephone books. *(He turns to the table and takes his place.* **CORDELIA** *has already poured out the coffee)* What's that guy to you, anyway?

*Cordelia:* I keeps on tellin' you he ain't been nothin' to me.

*Roy:* I ain't askin' what he ain't been; I'm askin', what he is?

*Cordelia:* Roy, I keeps tellin' you dat de old lady and de old man made me go wid 'im. I don' hold no hearts for him, an' I can't help it if he holds 'em for me.

*Roy:* You prob'bly strung him along. I know your type.

*Cordelia:* Now get this straight. I ain't fixin' to keep argufying wid you 'bout dat monkey chaser. He's got mighty possessive ways, but it don't mean nothin' to me. Anyway, what if it had? Didn' I leave him to come wid you, an' ain't I gonna stay here wid you?

*Roy:* I kinda thinks you'd go off and stay wid anyone.

*Cordelia:* You ain't no slouch yo'self. I'd hate to see de woman you wouldn't chase. Ev'ry woman on de avenue knows you, but you don't hear me beefin' 'cause some chippie you chased once't in yo' life, should happen to act possessive like.

*Roy:* Al' right, we're quits; but while we're at it this time, you get me straight, too. I been in Harlem too long to be jived. I don' stan' for no two timing. You says they ain't nothin' 'tween you an' dis guy. So, get dis, as long as you ain't been nothin' to him, don' you ever be nothin' to him while you're with me. I'm broadminded. I don' mind my gal havin' a little fun. But I want you to know when to stop. I got great things planned out for us. You needs a little educatin'. Then we can work together. If you do as I say, and watch your onions, in a couple of years we'll all be set. See?

*Cordelia:* I gotcha kid; an' given a chance, I'll be a too bad mamma.

*Roy:* First we gotta get you some clothes. Any gal of mine gotta look betta than anyone else on the avenue. You just ask anyone how they all looks. You see, the Kid and I are about the biggest niggers in Harlem. He's got more runners bringing him numbers every day than any other banker up in this neck of the woods. Then, there's that house he runs for gambling. Why, it's the best in Manhattan. Nothin' goes there but the ritziest swells from Broadway. No jigs allowed. You know, they ain't no money to be made runnin' a joint for yo' folks. They nevah have no money to spend or lose. Well, I an' the Kid knows this, an' we got the best bunch of clients sewed up they ever was—folks with real money, who can lose one or two or five grand and never bat a lash. And we got a political pull that won't quit. Ya never hear of the police raiding our joint, do you?

*Cordelia: (interestedly)* I ain't never seen dis Kid. What's he like?

*Roy:* Aw, he's a big four-flusher. Wouldn't have nothin' if it wasn't for me. Used to be a prize-fighter. Got a pile of dough for laying down to some white bloke. Then never got a return bout as he was promised. The Kid just has his name an' reputation. They ain't much to him. He ain't got a connection like I got with Rafferty. Takes a nigger and a Mick to clean up in New York nowadays.

*Cordelia:* Who's Rafferty?

*Roy:* I'll introduce him to you some day. Soon as I get a stake, I'm gonna pull out an' go for myself with Rafferty.

*Cordelia:* An' when you does—?

*Roy:* I'll be the biggest nigger in the States! Gimme some more coffee.

*Cordelia:* Is you crippled?

*Roy:* Now, that's another thing you gotta learn. No man wants a girl he's gonna have to— *(The phone rings again)* God damn it! (**ROY** *gets up from the table and surlily answers the phone)* Hello. What? . . . Well, I'll be damned . . . *(to* **CORDELIA***)* It's that monkey, again.

*Cordelia:* Lemme talk to 'im. (**ROY** *starts to protest, but relinquishes the instrument to* **CORDELIA** *when she determinedly grabs it)* Well, what do you want? . . . What? . . . Is you gone plumb crazy? . . . No, I ain't fixin' to come home an' I ain't fixin' to say no more to you . . . Yeah. An' the day you does . . . Oh, go take a flying leap to hell. *(She bangs down the receiver)*

*Roy:* Now, I've had enuff of this. The next time that "mon" calls you, there's gonna be trouble, an' I don' mean maybe.

*Cordelia:* He won't call no more.

*Roy:* He betta not! *(There is a ring. Both look toward the telephone.* ROY *laughs)* That's the doorbell. Wonder who it is? Get in there.

(CORDELIA *rushes into the bathroom.* ROY *opens the door.* THE KID *slowly enters, followed by* JIGGS. THE KID *is a large, well-built, handsome brown fellow with curly black hair. He is dressed in a becoming brown suit and carries a Chesterfield overcoat on his arm.* JIGGS *is a small, snappily dressed Negro. He wears a derby and is smoking a cigar)*

*The Kid:* Oh, here you are. Where you been?

*Roy:* *(nervously)* Nowhere.

*The Kid:* Then why in the hell didn't you meet me tonight when you was supposed to, 'stead of flyin' round with my roll in your pocket. You got that money?

*Roy:* Who do you think would have it?

*The Kid:* I can't never tell where you're concerned, big boy.

*Roy:* *(pulling the roll out of his pocket)* Here. Take it. I gets damn tired of you always signifyin' that I'm light fingered. If you can't trust me, I don't see why you keeps on lettin' me hold it.

*The Kid:* Maybe I ain't gonna much longer. *(looking straight into Roy's eyes)* I had a tip that Rafferty was gonna highjack me tonight, so I could think of no one better to hold my roll than you, seein' you an' Rafferty are such good friends.

*Roy:* What are you tryin' to signify? Rafferty ain't no friend o' mine.

*The Kid:* *(seeing the table set for two)* Guess I'd better not say any more just now. You seems to have company? *(ROY moves uncomfortably)* Looks like we frightened the guest away, Jiggs. One of your boy friends in to spend the night?

*Roy:* Yeah, one of the boys—

*The Kid:* *(seeing Cordelia's dress hung across one of the rockers)* Oh, one of those kinda boys. (THE KID *and* JIGGS *laugh. Both sit down at the table)* Do you mind if we drink some of your coffee 'fore we go? I'll finish my business talk with you some time when you're alone— *(smiles)* You might have your—er—friend serve us. Let us look him over.

*Roy:* *(hesitates and then calls)* Delia. Delia. Come on out.

*Cordelia:* *(from bathroom)* Hand me my dress.

*The Kid:* Come on out, girlie, we all belong to the same family.

(CORDELIA *enters. She holds Roy's bathrobe tightly about her)*

*The Kid:* Oi! Oi! Give us a knockdown, Roy.

*Roy:* Delia, this is the Kid, and this is Jiggs.

*Cordelia:* Pleased to meetcha.

*The Kid:* An' maybe we ain't pleased to meet you. We're always glad to meet Roy's little playmates, ain't we, Jiggs?

*Jiggs:* I'll say we are, 'specially when they looks like this, an' has good coffee.

*Roy:* Give 'em some coffee, will you, Delia?

*Cordelia:* Shure!

*(She busies herself in the alcove, trying to find two more cups. THE KID watches her closely)*

*The Kid:* Where d'ja find this one, Roy?

*Jiggs:* No wonder he couldn't meet you tonight. I don't blame you, big boy.

*Roy: (proudly)* Pretty neat, no?

*The Kid:* That's one thing I have to say about you, you certainly knows how to pick 'em.

*(CORDELIA puts the cups on the table and pours the coffee)*

*Cordelia: (slowly)* May not be very hot!

*The Kid:* S'alright with me, sweetie. *(They drink their coffee. When they have finished, THE KID turns to JIGGS)* You'd better get back to the house and check up on those number slips.

*Jiggs:* Al'right, Kid. You goin' now?

*The Kid:* No, I'm gonna stick around here awhile. I think I'll have that business talk with Roy right now.

*Jiggs:* I gotcha, big boy. I gotcha. *(He takes up his hat and coat and goes to the door)* See you later, folks. *(Exits)*

*The Kid: (whispering to ROY)* Get rid of the girl. I wanna talk to you alone. *(CORDELIA begins to clear off the table. To CORDELIA)* Smart, too! Where you been all my young life?

*Roy: (interrupting)* Waitin' for me to come along and find her.

*The Kid: (turns quickly on ROY)* And that means what?

*Roy: (meaningfully)* That all others keep their hands off.

*The Kid:* You know the good book says, "In my father's house are many mansions."

*Cordelia:* An' I ain't fixin' to live in but one o' dem at a time. *(They all laugh. ROY, however, is sulky)* Is I, sugar?

*(CORDELIA stops to pat ROY on the head, smiling at THE KID as she does)*

*Roy:* Get your clothes on. I wan' cha to go to the corner to get some cigarettes.

*Cordelia:* Ain't it too late?

*Roy: (glancing at his wristwatch)* No. It's not twelve yet.

*Cordelia:* Al'right, sugar.

*(She picks up her dress from the back of the chair and goes into the bathroom)*

*The Kid:* Pretty neat, Roy, pretty neat. Where d'ja find this one?

*Roy:* She's just a broad I picked up about three weeks ago at a party.

*The Kid:* You mean to say you've had her three weeks.

*Roy:* Naw, she jest came home with me tonight.

*The Kid:* Take you three week to make her? Pretty slow. Bad technique, I would say.

*Roy:* I had to go kinda slow 'count of her family.

*The Kid:* Oh, she's a family girl. What'd she do, run off?

*Roy:* Yeah. Tonight, she told them all to go to hell, and comes home with me.

*The Kid:* Watcha gonna do with her?

*Roy:* Whadda you mean, what am I gonna do with her?

*The Kid:* Don't play dumb. I know you. *(half humorously)* Why don't you let a good guy in on this?

*Roy:* Now listen here, Kid. Watch your step. Don' try to bite me in the back. I wouldn't . . . *(hesitates)*

*The Kid:* Go ahead. You wouldn't what?

*Roy:* I wouldn't try to double cross you.

*The Kid:* I'm not so sure.

*Roy:* How come?

*The Kid:* You know damn well, how come. You know damn well that's why I'm here today, 'cause you're such a dirty double-crossin' rat.

*Roy:* Whatcha tryin' to signify now?

*The Kid:* I ain't signifyin' nothin'.

*(CORDELIA comes in dressed to go)*

*Cordelia:* What kind of cigarettes you want?

*Roy:* Chesterfields.

*Cordelia:* I smoke Murads.

*Roy:* Al'right—get both.

*Cordelia:* Gimme de money.

*Roy: (hesitating as he feels his pocket)* I ain't got no change.

*Cordelia:* But you got—

*Roy: (fiercely to CORDELIA)* Didn't you hear me say I ain't got no change? *(To THE KID nonchalantly)* Give her some cigarette money for me—will you, Kid?

*The Kid:* Cigarette money? Do I look like her papa? *(smiles broadly)* Here, kid. *(looks through his roll and finds a five. CORDELIA stares wide-eyed at the roll)* Take this, and I expect change.

*Cordelia: (laughing)* Try and get it. *(She goes out)*

*The Kid: (He towers over ROY)* Now, let's get down to business. I don't have to signify. What I says, I can back up. I says you is tryin' to double cross me. I says you been tippin' off Rafferty 'bout my number collectors—where he can find them when they's got a day's collections in their jeans. Too many of my boys is being hijacked by Rafferty's gang. He's got inside dope and you've given it to him.

*Roy:* I hopes you know what you're talkin' 'bout.

*The Kid:* You bet I know what I'm talkin' 'bout. The only runners of mine that get hijacked are those that turn into you.

*Roy:* I s'pose you calls yourself tryin' to scare me now.

*The Kid:* If you had any sense you'd already be scared.

*Roy:* Know any more good jokes?

*The Kid:* I knows that you've framed on me with Rafferty.

*Roy:* Rafferty tell you that?

*The Kid:* Never mind who told me, but I've got proof.

*Roy:* You got more'n I have.

*The Kid:* I knows that, too. An' I knows you was gettin' ready to give Rafferty a list of all my white customers from downtown so's he could invite them to his place. I knows you said: "Nigger ain't got no business makin' so good off of white folks." You lily-livered son of—

*Roy:* *(interrupting fiercely)* You've said enuff.

*The Kid:* I ain't said half, what I got to say. I could frame you, but I don' wanna do yo' no dirt. All you've got to do is disappear from Harlem. If you don', well— *(He smiles threateningly)* Must I say any more?

*Roy:* 'Pears to me you've already said too much.

*The Kid:* Now listen here, Roy. I'm tryin' to be fair. I know you're just a hard-headed, little, dumb boy, tryin' to get along in a world you don' b'long in. Harlem's no place for you. The trouble with you here is that you're out o' yo' class—a pacer in a stable of thorobred runners.

*Roy:* *(jumping up)* Ain't nobody got no class but you, has they Kid? Ain't nobody got no brains but you? *(He cries tauntingly)* Thorobred runner, huh. You mean worn-out pug! Tin horn gambler and cheap pimp! That's what you is—a cheap pimp!

*The Kid:* *(jumping up)* You double-crossin'—

*(He hits* ROY *a hard wallop on the jaw and knocks him down.* ROY *gets up slowly. He is a bit dazed)*

*Roy:* *(mumbles)* No man can hit me like that and get away with it.

*(He rushes into the alcove, grabs the butcher knife, and makes for* THE KID. THE KID *steps back and draws an automatic)*

*The Kid:* Put down that knife an' come here an' sit down. *(*ROY *stands sullenly facing* THE KID. THE KID *repeats more menacingly)* Drop that knife, an' sit down.

*(For a long moment they face each other without saying a word)*

*Roy:* *(putting the knife down)* Al'right! You and I are quits! But I'm not goin' to leave Harlem, now or any other time! Who do you think you are, setting yourself up against white folks like Rafferty? Don't you know what happens when niggers set themselves up against white folks?

*The Kid:* That's my business!

*Roy: (with complete assurance)* Well, if anything happens to me, Rafferty will make it his business!

*The Kid:* You god-damned— *(He shoots.* ROY *falls to the floor.* THE KID *rushes over to him and kneels down beside him. He pockets the revolver. The phone rings.* THE KID *is startled, thinks for a moment. Then deciding to answer, he crosses to the phone. Speaks softly)* Hello. Who do you want? Delia Williams? Naw . . . wrong number. *(He hangs up and returns to the body. The doorbell rings.* THE KID *quickly drags the body into the alcove and draws the curtains. The bell rings again. He calmly goes to the door and opens it.* CORDELIA *enters)* 'Lo, Kiddo.

*Cordelia:* Here's your change. *(holds it out to him)*

*The Kid: (waving it aside)* I ain't botherin' with that chicken feed.

*Cordelia:* Thanks. *(stockings money)* Where's Roy?

*The Kid:* Oh, he jes' stepped out. A lady 'cross the hall called him. Asked me to entertain you.

*Cordelia:* I guess you'll do dat a'right.

*The Kid:* Long as I can't help myself.

*Cordelia:* Kind hincty, eh.

*The Kid:* I'm Kid Vamp.

*Cordelia:* I'm Delia Williams.

*The Kid:* What of it?

*Cordelia:* Big boy, you know. I kinda like you.

*The Kid:* Most ladies do.

*Cordelia:* And you—

*The Kid:* I likes all the ladies.

*Cordelia:* Heavy sugar papa—eh?

*The Kid:* Have a cigarette?

*Cordelia:* No, I'll smoke one of these.

*(She takes a Murad.* THE KID *maneuvers so that he is always between* CORDELIA *and the alcove)*

*The Kid:* Come on and sit down. Talk to me awhile 'fore Roy comes back. I guess he'll be gone some little time. The lady seemed to have lots to say to him. *(They sit down)* Known Roy long?

*Cordelia:* Long enuff to be livin' with 'im.

*The Kid:* Then you plans on stayin'?

*Cordelia:* Why not?

*The Kid:* Oh, nothin'. *(He hesitates)* Only—well—it's none of my business.

*Cordelia:* Whatcha drivin' at?

*The Kid:* I'm a hard guy. It ain't often I lets my sympathies run away with me. But I hates to see a girl like you throwing herself away on a no-good like Roy.

*Cordelia:* What's de matter wid Roy?

*The Kid:* I knows that he's jivin' you along, like he does all the girls. Now, what do you think this lady 'cross the hall is to him? Didn't want you to know 'bout her. Asked me to tell you it was a man that he had business with. You see, Roy wants to be a big time sweetback. He's one of these young niggers new to city ways. Come from some little town in Virginia, got to New York, saw these sweetbacks posing 'round these corners and pool halls and decided that with his looks and his brains he could out-sweetback them all. Any time he gets a girl, he 'spects her to make money for him.

*Cordelia:* Is you tryin' to tell me Roy wants me to hustle for 'im?

*The Kid:* It'd come to that in the end. He's tried to before.

*Cordelia:* Don' see how he could, less I wanted to. De man ain't born dat can make me do somethin' I don' wanna.

*The Kid:* Aw, I've seen your kind before, get nutty 'bout some crook like Roy, and hustle right on down to the bricks for him.

*Cordelia:* Get me straight. I don' hustle for no man. I'm lookin' for somebody to put out to me. If dere's any receivin' done, little Delia's gonna do it. I ain't fixin' to keep no man. Dat's de reason dere's so many no count darkies in Harlem now, dese women takin' care of 'em, callin' 'em daddy, and payin' 'em for a little lovin'. When I wants lovin' t'ain't no business proposition, an' if they's any payin' done, it sure ain't gonna be me layin' out de jack.

*The Kid:* You got a lot of spunk, kid. I like it.

*Cordelia: (strutting her body)* Is that all you likes 'bout me?

*The Kid:* You're not a bad mamma for Roy.

*Cordelia:* Roy—he matters a damn.

*The Kid:* How come?

*Cordelia:* Roy—he—eh—

*The Kid:* Roy—what?

*Cordelia:* Well, there's other people might have a betta' go.

*The Kid:* Meanin'—

*Cordelia:* What's the use? It wouldn't do any good.

*The Kid: (taking her by the arm)* You know that?

*Cordelia:* Good as I know anything.

*(a moment of silence)*

*The Kid:* What's on your mind?

*Cordelia:* I likes de way you hold my arm.

*The Kid: (boastfully as he draws her to him)* How do you like this?

*Cordelia: (ecstatically)* Oh, daddy, daddy. *(She puts her arms around his neck, pulls herself up to his lips, and kisses him)*

*The Kid: (with a sudden and pleasing discovery)* You're not a bad kid. *(He kisses her again—this time with real passion. He now virtually explodes)* You won't stay with Roy. You're too damn good for him.

*Cordelia:* Ain't I said he don't matter? But man, you matter like hell. Kiss me again. *(This kiss leaves them exhausted. They both sigh and sit back on the couch.* CORDELIA *smiles with self-satisfaction)* Can you get me 'way from Roy?

*The Kid:* *(proudly)* You don't seem to know who I am. I'm Kid Vamp. Known from coast to coast. From now on, you travel with me.

*Cordelia:* *(playfully)* Hold on, big boy. You ain't asked me does I *want* to travel with you?

*The Kid:* I'll make you *want* to.

*Cordelia:* Say it again. I love to hear you talk dat way. *(They kiss. The door bell rings. Quickly, they draw apart)* S'pose dat's Roy come back? *(*THE KID *does not answer, but looks around excitedly. The bell rings again and again)* You'd better let 'im in.

*The Kid:* *(trying to compose himself)* I think you betta go.

*Cordelia:* How come?

*The Kid:* *(in commanding tone)* Go to that door and see who it is.

*(*CORDELIA *goes and opens the door. Her* MOTHER *and* JASPER *rush in)*

*Mother:* Delia. Delia.

*(*THE KID *stares at them. There is a tense moment of silence)*

*Jasper:* Put on yo' coat and hat. We come to take you home.

*Cordelia:* How'd you know I was here?

*Mother:* Jasper went back to deh house to find out how things wuz goin', and dey told him.

*Jasper:* *(interrupting)* Basil told me where you'd gone. Hurry up.

*Cordelia:* I ain't fixin' to go home.

*Mother:* *(throwing up her hands in supplication)* Lawd, what've I done to be punished like dis?

*Jasper:* No need bein' stubborn.

*Mother:* Put on yo' hat and coat. No one'll evah say I let ma chile go to de devil. Ain't you got no shame?

*Cordelia:* You needn't worry 'bout my shame. When I left home I said I wasn't comin' back, and I didn't mean maybe.

*Jasper:* You're goin' out of here if I has to drag you.

*The Kid:* Wait a minute there, wait a minute—who are you?

*Jasper:* I'm her brother. What's it to you?

*The Kid:* Al'right then, don't get excited, no need of an argument.

*Jasper:* You the man she left with tonight?

*The Kid:* Naw.

*Jasper:* Then what is she doin' here with you?

*The Kid:* Why, she wasn't—

*Jasper:* *(continuing belligerently)* Where's the rat who took her away?

*The Kid:* Oh, he stepped out for a minute. I was just telling her, when you folks come in, that she has made a bad bargain, and better make a change.

*Jasper:* (disgustedly) Meaning she should go with you, I s'pose?

*The Kid:* Sure.

*Mother:* Who are you? You can't be much of a man, takin' some young girl like dis 'way from home, keepin' her here in sin. You cain't be much of a man. You cain't have much respect for yo-self or for anyone else. S'pose you had a sister!

*The Kid:* I have three of 'em.

*Mother:* Would you see some man disgrace 'em?

*The Kid:* They disgraced themselves long ago, mam.

*Jasper:* Well, you're not gonna disgrace my sister. Put your coat on, Delia.

*The Kid:* I don' 'tend to disgrace nobody, bo.

*Mother:* Den let my daughter go home.

*Cordelia:* You don' see 'im holdin' me, do you? He ain't got nothin' to do wid it. I'm stayin' 'cause I wants to stay. Dat's all dey is to it.

*Jasper:* Don' you stan' dere an' sass Ma.

*The Kid:* Now why get excited 'bout dis?

*Mother:* What mother wouldn't get 'cited, seein' her daughter bindin' herself to de devil?

*The Kid:* Well, if the girl says she don' want to go home—

*Jasper:* (threateningly) You betta keep outa this.

*Cordelia:* I says I ain't goin' home, so all of you keep out.

*Jasper:* An' I says you're goin', if I has to drag you.

*Cordelia:* Well, you better begin draggin', an' you betta bring some help.

*Mother:* You mean you don' 'tend to come home no more?

*Cordelia:* I don' mean nothin' diff'rent.

*Mother:* You mean you gonna live in sin wid dis man?

*Jasper:* She'll never do it while I have a breath in my body.

*Cordelia:* You better turn up your toes, den. I ain't movin' *a'tall*.

*Mother:* (turning to THE KID in desperation) Why don't you marry my daughter? Can' you see I'm tryin' to save her from a life of sin and disgrace. Don' you think we po' colored people has trouble 'nuff widout 'curring the wrath of God? We comes up No'th to betta ourselves, and seems like we sinks further in sin. De more I pray, de more my chil'en go astray.

*The Kid:* (contritely) I'd like to marry her, mam, but I can't.

*Jasper:* I gotta picture you marryin' her, now.

*The Kid:* Sure I'd marry her, if I wasn't already married.

*Mother:* Already married. Now you *are* comin' home. No child of mine shall live in sin wid a married man. You'll come home, if I has to get de police to take you. You're not yet eighteen an' de law's on ma side, Jasper, go and get de police.

*The Kid: (nervously)* Wait a minute, mam.

*Mother:* You hear me, Jasper? Go get de police.

*The Kid: (consolingly)* Just a minute, please. Just a minute.

*Mother:* I waste no mo' minutes. Jasper, go get de police. *(JASPER starts for the door. THE KID intercepts him. He tries to quiet the MOTHER, who is now moaning over and over)* Mercy, Lawd. Mercy, Lawd. Mercy, Lawd.

*The Kid:* She'll go home. I promise you that. Don't get the police. Lemme talk to her alone for a minute. *(JASPER and MOTHER look at each other hesitatingly)* Just step outside. Please.

*Jasper:* Al'right. Just a minute. And she'd betta be ready to go when the minute's up.

*(They go out. The MOTHER is moaning)*

*The Kid:* Go on home with the folks, kid. That's the best thing to do. Then when I get things settled, I'll come after you. No need startin' a fuss. Since you're under age, you'd get us all in trouble.

*Cordelia:* I don't see no sense in goin' back dere. Dey don't want me. Dey jes' wanna keep on bossin' me. I'm as much a woman now as I'm gonna be, an' I 'tends to do as I please.

*The Kid:* Now, you listen to me. I warned you 'bout Roy. I tried to make you see how things could be 'tween you an' me. Now you go on home. I'm comin' after you before—before the day is gone. Get me. *(He pats her on the shoulders. He gets her hat and coat)*

*Cordelia:* You mean what you say?

*The Kid:* You know I does, kid. In a little while I'm comin' for you. I'll have my car. You'll be packed up. We're goin' away for a little honeymoon trip.

*Cordelia:* Hold me close, honey. Kiss me a long time. I want you, honey. I want you.

*(THE KID crushes her to him. JASPER flings open the door)*

*Jasper:* Come on.

*Cordelia: (angrily to JASPER)* I'm comin'. *(cooing to THE KID)* Don't forget, Kid. I'll be waitin'.

*(She goes out. THE KID whistles to himself, as he walks up and down nervously. He suddenly stops and goes to the telephone)*

*The Kid: (almost in a whisper)* Bradhurst 10012. *(He waits a moment)* Yes. Bradhurst 10012. *(He becomes impatient, moving his feet nervously)* Busy. Damn it. *(Bangs down the receiver. He again walks up and down. He stops and after some hesitation peeks behind the curtain. He hurriedly draws back. His face and hands are now twitching. Once more he grabs the phone. Huskily he gives the number)* Bradhurst 10012. *(With difficulty he controls himself, as he stands waiting. He again ferociously bangs down the receiver)* God damn it. Who the hell is he gassing with? *(He scratches his head, grabs his hat and overcoat, and starts for the door, just as the doorbell rings. He jumps back startled, pulls the curtain closer together, and then goes to the door. As he opens it, BASIL bursts through. THE KID grabs him by the arm)* Wait a minute there, young man. Where you goin'?

*Basil:* Where is Delia?

*The Kid:* Where's who?

*Basil:* Where's Delia? Cordelia—

*The Kid:* Oh, Cordelia, well, who's she?

*Basil:* I know you know who she is. She came here to this house last night with that little yaller sweet-back. I know she's here. I talked with her over the 'phone just a little while ago.

*The Kid:* Well, she ain't here now. Who are you, anyway?

*Basil:* I'm her fiancé, and I've come to take her home.

*The Kid:* Al'right Mr. Finacé, as you calls it. I tell you she ain't here.

*Basil:* I know she is, and if I don't find her, the law will.

*The Kid:* I told you, sonny, she ain't here. Now, will you leave, or must I throw you out?

*Basil:* I ain't going till I find her.

*(He turns and runs to the bathroom, the door of which stands wide open)*

*The Kid: (laughing)* Satisfied now? (**BASIL** *looks around wildly and sights the curtains in front of the alcove. He makes for them.* **THE KID** *bars his way)* No, no, sonny, you've looked enough. Now, beat it.

*Basil:* She's behind those curtains. That's where she is.

*The Kid:* Well, if you think so, she's there; but you won't look to find out.

*Basil:* You're hiding her behind those curtains. (**BASIL** *rushes towards the curtains.* **THE KID** *stops him with an uppercut to the jaw, knocking him down.* **BASIL** *rises slowly. For a moment he looks fiercely at* **THE KID** *as if to rush him again. Then his face lights up with an idea. He reaches into his pocket and pulls out a revolver. It is the same one that the Father had in Act I. He points it at* **THE KID***)* I said I was going to find her. I don't want to kill nobody, so you'd better get out of my way.

*The Kid: (with affected carelessness)* Al'right. I didn't know it means that much to you. Go ahead. *(He steps back against the wall alongside the telephone.* **BASIL** *starts for the curtains. As he passes* **THE KID***, the latter whips out his revolver and bangs* **BASIL** *over the head with the butt.* **BASIL** *crumples in a heap and his revolver clatters to the floor. Bending over the unconscious* **BASIL***)* You God-damned monkey-chaser. You wouldn't leave, would you? *(He kneels down alongside* **BASIL** *and listens at his heart. He looks carefully at his eyes)* I guess you're gonna sleep for a long time. I'll say you're lucky I didn't kill you. *(He picks up Basil's gun, breaks it)* Empty. Well, I'll be damned. And a thirty-eight, too. *(He thoughtfully looks first at it and then at his own, which he still holds in his other hand. He glances at the curtain of the alcove. He chuckles)* Big boy; you certainly are in luck. If there ever was a natural, this is it.

*(He takes cartridges out of his automatic and fills up the chambers of the revolver. He then searches the floor, finds the empty cartridge thrown from the automatic when he shot Roy, and pockets it along with the automatic. Thereupon, holding Basil's loaded revolver in his hand, he surveys the room. His eyes fall on the open davenport. He throws back the mattress. Carefully placing the revolver against the upholstery, he presses the mattress down on it and fires. The mattress and upholstery muffle the report to a dull thud. Now taking a silk handkerchief from*

*his breast pocket,* THE KID *carefully wipes the revolver all over, and clasps it in Basil's right hand, being careful not to touch it with his own fingers. He stands up, extremely satisfied with himself)* That ought to fix him. *(Again he picks up his coat and hat and starts for the door. Just as he is about to go out, the telephone catches his eye. He looks at it thoughtfully, regards the curtains, and then concentrates on the prostrate* BASIL. *With sudden determination, he once more sets down his hat and coat, slowly draws on his gloves, again takes out his silk handkerchief and picks up the telephone. He carefully wipes the telephone all over with the kerchief as he had previously wiped the gun. He removes the receiver with his gloved hand. Speaking into the phone)* I wanna speak to Police Headquarters . . . Al'right . . . Give me Spring 3100 . . . Is this Police Headquarters? . . . What? . . . *(He violently jiggles the receiver hook up and down with his gloved hand)* Operator. Operator . . . Yes. I want Police Headquarters . . . Yes. I did call Spring 3100 . . . *(He waits impatiently, glancing first at the curtains and then at* BASIL. *He abruptly turns to the phone)* Is this Police Headquarters? . . . I wanna report a murder . . . What's that? . . . Where? . . . Why, in 134th Street near Lenox Avenue . . . What? . . . I gotta call the precinct station? . . . What's the number? . . . Audubon 9531 . . . Al'right. *(He hangs up the receiver and sighs disgustedly. He picks up the receiver again)* Audubon 9531 . . . Yes. Hello. Is this the police station? . . . I wanna report a murder . . . *(He moves nervously and bites his lip as he waits. Suddenly he faces the phone tensely)* Where? . . . Why, in Apartment 4-D of 101 West 134th Street . . . What's my name? . . . Oh, Thompson, James Thompson . . . No. I didn't see it. I heard a shot in the apartment and then I heard a moan. I was in the hall just coming down from visitin' a friend upstairs in the same buildin' . . . What's my friend's name? Say, why do you ask me all these questions? I didn't commit the murder.

*(He bangs down the receiver, picks up his coat and hat, takes the key from the inside of the door, goes out and is heard locking the door from the outside. The telephone now begins to ring. It rings and rings. It is shrilly insistent. It penetrates the consciousness of* BASIL. *He begins to move. He moans softly. His moans become louder. He manages with difficulty to sit up. He draws his hands over his eyes and head. The revolver remains lying on the floor.* BASIL *is not even conscious of it having been clasped in his hand. The pain of touching the bruise made by the butt of* THE KID's *automatic makes him wince. He looks bewilderedly about the room. The telephone stops ringing. With the aid of a chair,* BASIL *rises to his feet. As if in a drunken stupor, he picks up his hat and puts it on his head. He arranges it carefully so it doesn't touch his bruise. He looks around the room. His face suddenly becomes animated as his eyes fall on the curtained alcove)*

Basil: Ah. *(He cries exultantly and rushes to the curtains. Throwing them aside, he looks down on the dead* ROY. *He staggers back. The curtains fall to)* Lord. Lord. *(He cries, his face, distorted with fright. He rushes to the door, turns the knob, and finds it locked. For a moment he stands stupefied, uncertain what to do. Then, frantically he pulls violently at the knob in a vain endeavor to open the door. Exhausted, he gives up and sinks into a chair)* Lord. Lord.

*(He keeps moaning, his eyes first on the curtains, and then on the door. The doorbell rings, a short ring, followed by a long one. Someone tries the handle of the door. The handle is rattled violently.* BASIL *sits watching the door as if hypnotized. There are a few hard thumps. A loud voice calls in Irish-American accents)*

Voice: *(outside)* Open, in the name of the law. *(There are a few more thumps, and then the same voice is heard)* I guess we'll have to break in. Put your shoulder to it, Sam.

*(In Negro accents comes the reply)*

*Second Voice: (Negro)* Al'right, Sergeant.

*(The door quivers as* SAM *puts his shoulder to it. This stirs* BASIL *from his stupor. He jumps up, runs to the bathroom, sees no hope of escape there, and then runs to the window on the left. The door quivers again and again as* SAM *is encouraged on)*

*Voice: (Irish)* That a boy, Sam. A couple more like that one and she's open. These houses ain't built of nothing but paper anyway.

*(*BASIL *raises the shade of the window and glances out. A look of tremendous relief passes over his face)*

*Basil: (whispering)* Thank the Lord.

*(He raises the window, slips out onto the fire-escape, closing the window after him. Just as he disappears, the door crashes in with a huge Negro patrolman hanging onto the knob. He is followed by a detective* SERGEANT *who is in plain clothes. There is also a* WHITE PATROLMAN*)*

*Sergeant:* Search the place, quick. *(The* WHITE PATROLMAN *looks in the bathroom, the colored cop under the davenport, and the* SERGEANT *in the alcove. As the latter draws the curtains apart, he cries)* Here's somebody. *(The two patrolmen come running over)* Sure looks like a stiff. *(He kneels down to examine Roy's body)* It's Murder. Shot right through the heart. *(He looks up at the patrolmen)* Either of you recognize this stiff?

*White Patrolman:* Looks like one of those poolroom cowboys. I've noticed him a couple of times since I've been on this beat.

*Sergeant:* What do you say, Sam? You know most everybody in Harlem.

*Sam: (who has been staring at the dead* ROY*)* I knows most of the crooks.

*Sergeant:* Do you know this gink? I suppose he's a crook, or he wouldn't be here in this condition.

*Sam:* If I'm not mistaken, Sergeant; this is one of Kid Vamp's gang.

*Sergeant:* Do you know his name?

*Sam:* Seems to me I pinched him once. A girl accused him of insulting her on the street. I remember it 'cause it ain't a common charge here in Harlem. Most of the girls that gets insulted, invites it, and likes it, if you're askin' me.

*Sergeant:* Do you remember his name?

*Sam:* I just don't recollect. Wait a minute. *(He takes a book out of his hip pocket)* I keeps a record of all my arrests. Sort of protection, Sergeant. We colored patrolmen, you know, are always gettin' charged with doin' things we don't do and with not doin' things we oughta do. Colored folks don't like the idea of being jacked up by colored cops. We gotta be careful, or they'll frame us. *(He keeps turning the pages of his book)* Ah. Here it is. His name is Roy Crowe. The magistrate fined him ten dollars.

*Sergeant: (getting up from his knees)* Roy Crowe, hey. Well, that helps a bit. *(He turns to the* WHITE PATROLMAN*)* Kelly, go downstairs and bring up the janitor. I want to ask him a few questions.

*Kelly:* He's probably in bed.

*Sergeant:* Roust him out.

*Kelly:* Yes, Sergeant. *(He starts out)*

*Sergeant:* And Kelly. First call up from the box on the corner and report what we found to headquarters, so they can start the medical examiner over.

*Kelly:* Al'right. *(Exits)*

*Sergeant:* Look around, Sam; and see if you can see anything of a gun. *(SAM starts searching in the bed. He lifts the pillows and starts looking under the mattress, just as the SERGEANT almost stumbles over the revolver on the floor where BASIL had left it)* Here it is. *(He picks it up and examines it. SAM comes over)* One cartridge fired. *(examining the revolver)* This is the baby that did it. Kind of careless to leave it around. I'd say we ought to stand a pretty good chance of tracing this gat. *(They are interrupted by the telephone ringing. The SERGEANT shoves the revolver in his pocket and takes up the instrument. He is about to answer, when he turns to SAM, offering him the phone)* You'd better answer, Sam. Maybe we can pick up something.

*Sam:* What'll I say?

*Sergeant:* Tell me what they say and I'll tell you what to say.

*Sam:* *(in phone)* Hello . . . The Kid . . . Just a minute.

*Sergeant:* *(clamping his hand over the transmitter and speaking in a whisper)* Is it a man or a woman?

*Sam:* A man. Wants to speak to the Kid.

*Sergeant:* *(still keeping his hand over the transmitter)* Tell him the Kid has stepped out with Roy and will be right back. Then ask him who he is?

*Sam:* *(in the phone)* The Kid has stepped out with Roy, but he said he was comin' right back. Who's this wants to speak to him?

*Sergeant:* *(again putting his hand over the transmitter)* Who is it?

*Sam:* It sounded like Prig.

*Sergeant:* What is he saying now?

*Sam:* Wants to know who I am.

*Sergeant:* Tell him you're a friend of Roy's and that the Kid left a message for him to come right over.

*Sam:* *(in phone)* I'm a friend of Roy's. The Kid asked me to tell you if you called that he wanted you to meet him here, jes' as quick as you can make it . . . No. I don't know what it's all about. But he said to make it quick. Yeah . . . Good-bye. *(He hangs up the receiver)*

*Sergeant:* *(excitedly)* What did he say?

*Sam:* He's comin' right over.

*Sergeant:* Did he repeat his name?

*Sam:* No. But it sounded to me just like Prig.

*Sergeant:* Well, we'll soon know his name and a lot more. That was a fine bit of work, Sam.

*Sam: (smiling bashfully)* Always glad to be of service, Sergeant.

*(The white patrolman, KELLY, enters with the JANITRESS, a big stout black woman in an old wrapper, and her child, a GIRL of 12)*

*Sergeant: (turning to Kelly)* What's this?

*Kelly:* The old lady wouldn't come up without the kid.

*Sergeant: (to the NEGRESS)* Are you the janitress of this building?

*Negress:* No, sir. I'm assistant superintendent.

*Sergeant: (smiling)* Who's the superintendent?

*Negress:* My old man.

*Sergeant:* Do you know the people who live in this flat?

*Negress:* No, sir. I knows nothin'.

*Sergeant:* Don't you even know the names of your tenants?

*Negress:* I knows their names, sir, but that's all I does know.

*Sergeant:* Well, what was the name of the tenant who lived here?

*Negress:* Mr. Crowe, sir. Mr. Roy Crowe.

*Sergeant:* Live here alone?

*Negress:* Yes sir, as much as any man can live alone.

*Sergeant:* Who lived with him when he didn't live alone?

*Negress:* I don' know, sir.

*Sergeant:* Did he have many visitors?

*Negress:* I don' know, sir.

*Sergeant: (a bit exasperated)* What do you know?

*Negress:* Nothin', sir.

*Sergeant:* Did anybody call to see him tonight?

*Negress:* I don'—

*(The GIRL, who has been standing alongside her mother, anxious to show her knowledge, interrupts excitedly)*

*Girl:* Jasper Williams and his mother came to see him a little while ago, and I told them what apartment Mr. Crowe—

*(The JANITRESS punches the GIRL in the ribs and she abruptly shuts up. The SERGEANT quietly takes the GIRL by the arm and leads her over to a chair where he sits her down so she faces him and can't see her mother. The GIRL is scared stiff. As the SERGEANT speaks to her soothingly, her fright quickly disappears)*

*Sergeant:* Now, little girl, nobody's going to hurt you. All I want you to do is answer my questions. *(He reaches into his pocket and brings out a roll of bills. He strips off a dollar bill and holds it up before her)* You know what this is?

*Girl:* Yes, sir. A dollar.

*Sergeant:* If you answer all my questions, I'll give it to you.

*(He turns and looks significantly at the* NEGRESS. *The* GIRL *also turns and looks inquiringly at her mother)*

*Negress:* It's al'right, dearie. Answer all the gentleman's questions.

*(The* SERGEANT *smiles and puts the roll of bills back in his pocket, holding the dollar in his hand. He stands alongside the* GIRL, *in such a position that he entirely blocks off the girl's vision of her mother even if she turns)*

*Sergeant:* Who is Jasper Williams?

*Girl: (glibly)* He's the brother of Delia Williams. Delia's sister Maizie and I go to Sunday School at the Mount Moriah mission.

*Sergeant:* Do you know where they live?

*Girl:* I knows where Delia lives.

*Sergeant:* Where's that?

*Girl:* On 132nd Street between Fifth and Lenox Avenue.

*Sergeant:* You know the number?

*Girl:* No. But I knows the house.

*Sam: (interrupting)* I think I has the number, Sergeant. *(He takes out his little book and begins going through the pages)* I remember 'bout a month ago I was called in to settle a fight at a rent party in a flat in that street. If I recollect right, the family what was givin' the party was named Williams. I took the names of everybody and I recalls there was a girl named Delia. She give me a lot of lip.

*Girl:* Dat's her. Delia's always sassin' folks.

*Sam:* Here it is, boss. Williams, 3rd floor of 30 West 132nd Street.

*Sergeant:* Seems to me, Sam—you've got everything in that book.

*Sam:* A colored patrolman sure got to be careful, Sergeant. Suppose someone wants to frame me and says on a certain night I was somewhere where a good officer of the law shouldn't be, and doin' somethin' that no honest cop should be doin', and I goes before the Judge and he says: "Sam, where was you and what was you doing there on the night of February 31st?" I couldn't answer him, boss, if I didn't have this book.

*Sergeant: (smiling)* You're a good officer, Sam.

*Sam: (modestly)* Yes sir. Thank you, Sergeant.

*Sergeant: (turns again to the* GIRL*)* You saw Jasper Williams and his mother come in?

*Girl:* Yeah. They asked me what floor Mr. Crowe lived on.

*Sergeant:* And you told them?

*Girl:* Yeah. I tol' 'em.

*Sergeant:* Did you see them come out?

*Girl:* Yeah. They came out with Delia. And she certainly looked hoppin' mad. She was bawlin' 'em out somethin' terrible.

*Sergeant:* I see. Was it long after they came in that you saw them going out?

*Girl:* No. Not so very long.

*Sergeant:* Did you see anybody else come in?

*Girl:* Lots of people was goin' in and comin' out.

*Sergeant:* Anybody for Mr. Crowe?

*Girl:* I don't know, sir. I was sitting on the stoop and lots of folks come in; but they didn't say nothin' to me.

*(The door bell rings. Everybody turns to the door. The* SERGEANT *gestures for silence and tiptoes to the door; throwing it wide open quickly.* JIGGS *is standing there)*

*Sergeant:* Well, look who's here. Come on in, Jiggsey! *(He grabs him roughly by the arm and draws him into the room. He chuckles)* So you're Mr. Prig. *(He quickly frisks him and takes an automatic from his hip pocket)* Going well-heeled these days, hey?

*Jiggs:* What's this, a frameup?

*Sergeant:* That's what we want to know from you, what is this? Sit down. I've got a few questions to ask you. *(*JIGGS *reluctantly sits down; looking furtively around the room)* No monkey business, Jiggs. You're liable to get hurt.

*Jiggs:* *(sullenly)* I know what's good for me.

*Sergeant:* *(handing the dollar bill to the* GIRL*)* You're a smart youngster. *(He turns to the* NEGRESS*)* You can go now. I'll probably want to see you again in the morning.

*Negress:* I'se always at home, sir.

*(The* WHITE PATROLMAN *opens the door. The* NEGRESS *takes the dollar from the* GIRL *and puts it inside the bosom of her wrapper as she and her youngster go out. The* WHITE COP *closes the door)*

*Sergeant:* *(examining Jiggs' revolver)* What are you carrying this for?

*Jiggs:* *(a bit defiantly)* I need it in my business. How in the hell else can we protect ourselves from Rafferty and his gang of hi-jackers? *(He goes on rapidly)* Why don't you guys get after them. They're nothin' but a bunch of hold-up men. We do a legitimate business. Can't stop coons from bettin' on the numbers. If we didn't run the bank, somebody else would.

*Sergeant:* *(slowly)* That's all very interesting, Jiggs, but I'm certainly surprised at a slick guy like you letting us find an iron on him when he's already got half a dozen convictions. Don't you know that another conviction with the Baumes Law hitting on all cylinders means a lifetime up the river?

*Jiggs:* How'd I know I was going to run into the law up here?

*Sergeant:* *(handing Jiggs' gun to the* WHITE POLICEMAN *and going up to* JIGGS*)* Now, Jiggs, you know I like to work fast. I'd do a lot for anybody who'd save me a few days' work.

*Jiggs:* What are yer drivin' at?

*Sergeant:* I'm laying all my cards on the table. If you come clean, I won't press this gun charge against you, nor will I press a more serious charge.

*Jiggs:* What are you talking about?

*Sergeant: (dramatically)* Murder.

*Jiggs: (in a frightened whisper)* Murder. How come?

*Sergeant:* Take a look behind that curtain.

(**JIGGS** *goes over to the curtains, looks behind them, and draws back horrified. He is visibly shaken. He slumps back into the chair, takes out his handkerchief, and wipes his brow)*

*Jiggs:* God. And he was alive less than an hour ago.

*Sergeant:* You saw him an hour ago?

*Jiggs: (his hands trembling)* Yeah.

*Sergeant:* Where?

*Jiggs:* Right here. *(The* **SERGEANT** *glowers at him.* **JIGGS** *looks at him, pleadingly)* Don't look at me that way, Sergeant. I had nothing to do with this. Honest to God. You don't think I'd have come here if I knew that— *(He motions towards the curtains)* My God. I've been a grifter, but I was never mixed up in murder—no, no, never, never.

*Sergeant:* You come clean and nothing will happen to you. We'll forget all about this cannon you were carrying.

*Jiggs:* I'll tell you all I know, Sergeant, but I don't know who did this.

*Sergeant:* Who was here with you an hour ago?

*Jiggs:* The Kid, and—

*Sergeant:* Kid Vamp?

*Jiggs:* Yeah.

*Sergeant:* Who else?

*Jiggs:* Some chippie—er—er *(He finds it difficult to mention* **ROY***'s name and glances towards the curtains. Suddenly he spits out the whole sentence)* Some chippie Roy had picked up.

*Sergeant:* What's her name?

*Jiggs:* I don't know. Roy called her Delia.

*Sergeant:* Did you come here alone?

*Jiggs:* No. I come with the Kid.

*Sergeant:* What was this Roy to you and the Kid?

*Jiggs:* Well, he was—er—well— *(He stutters)*

*Sergeant: (fiercely)* Come clean.

*Jiggs:* He was one of our collectors.

*Sergeant:* Your number runners turned over their collections to him?

*Jiggs:* Yeah. Some of them.

*Sergeant:* Why did you come here tonight?

*Jiggs:* To get some money Roy had for us.

*Sergeant:* Did you get it?

*Jiggs:* The Kid got it.

*Sergeant:* *(suddenly raising his voice)* Why did you call up the Kid here a little while ago?

*Jiggs:* I thought he was still here.

*Sergeant:* Then he was here when you left.

*Jiggs:* Yeah.

*Sergeant:* Why did he stay after you left?

*Jiggs:* He had some business to talk over with Roy.

*Sergeant:* What sort? *(The door bell rings. The* SERGEANT *turns quickly and, going to the door, throws it open.* ARABELLA *is standing in the hall. She has on her overcoat and hat. She appears tremendously excited. On seeing the police, she draws back. The* SERGEANT *speaks to her)* Well, miss, who are you looking for?

*Arabella:* *(hesitantly)* Guess I rang the wrong bell.

*Sergeant:* Who'd you want?

*Arabella:* I was looking for Roy Crowe's apartment.

*Sergeant:* This is it.

*Arabella:* *(a bit frightened)* Oh.

*Sergeant:* *(with affected casualness)* So you know Roy Crowe.

*Arabella:* No. But I thought maybe someone else would be here?

*Sergeant:* Well, maybe he's coming. Come on in. *(He draws her in by the arm and closes the door behind her. She is now thoroughly frightened and looks pleadingly from one face to another. The* SERGEANT *continues in his casual tone)* Who'd you expect to find here?

*Arabella:* Oh. Just a friend.

*Sergeant:* What's his name?

*Arabella:* *(stuttering)* Why, eh, eh, Basil.

*Sergeant:* *(a bit severely)* Basil who?

*Arabella:* Oh. Eh. Basil Venerable.

*Sergeant:* *(more severely)* What business'd he have here?

*Arabella:* *(now thoroughly frightened)* I thought he might have come to see Roy Crowe.

*Sergeant:* Did he tell you he was coming?

*Arabella:* No. *(She shakes her head)* No.

*Sergeant:* What made you think you'd find him here?

*Arabella:* *(flustered)* I was afraid he'd— *(She stops abruptly)*

*Sergeant:* Afraid?

*Arabella:* *(beginning to cry)* Yes. I was afraid— *(She sniffles in an effort to control herself)*

*Sergeant:* *(barking)* What are you afraid of?

*Arabella:* *(rambling)* I was afraid Basil might get hurt. *(pause)*

*Sergeant:* *(taking her by the arm and speaking significantly)* Let me show you something. *(He leads her to the curtained alcove. She instinctively shrinks back. He holds the curtains aside and points to the body)* There's Roy Crowe. He's been murdered.

*Arabella: (drawing back with a shriek, and beginning to sob hysterically)* He said he'd do it. He said he'd do it. Oh. He said he'd do it. Basil. Basil.

*(She breaks into a convulsion of sobbing as the curtain falls)*

### Act III

**Scene:** *The scene is the same as ACT I and when the curtain rises the party is at its height. The room is crowded with dancers. The music bangs away stridently. Everybody is thoroughly liquored. The dancing is lewdly abandoned and accompanied by much shouting. It is a virtual saturnalia of desire.*

*The dance goes on for four or five minutes after the rise of the curtain. It gets hotter and hotter, when suddenly the door at the rear opens and* CORDELIA *enters followed by the* MOTHER *and* JASPER.

*A few dancers cry, "Here's Delia back!" "Lo, Delia!" "Come on in, dis party's too bad!"* CORDELIA *sulks off to one side.* JASPER *stands with his back against the wall. The* MOTHER *pushes through the dancers. She is visibly agitated and alarmed at their actions.*

*Mother: (standing in the center of the room, she raises her hands above her head, and cries out passionately)* Stop! Stop dis devil's work! Stop, I say! Stop, dis instant an' get out o' ma house! *(The dancers look at her amazed.* IPPY *and the drummer stop playing. Everyone's eye in the room is turned upon her. She rushes on)* I won't have it! I won't have ma house turned into a hell of iniquity! You shan't use ma house as a door to hell! Get out! I tells you! Get out! *(The dancers begin to mutter rebelliously. On all sides are heard, "Did you ever?" "What's deh matter wid her?" "We paid our good money, didn't we?" "If I goes I want my money back!" "Ah, she's crazy!" The* MOTHER *raises her arms and eyes heavenward. She begins praying)* Oh, God, forgive me! I didn't know dey was so wicked! I didn't know—

*(She's interrupted by the crowd's taunts of "Better call on someone who can hear you!" "The Lawd don't pay niggers no mind!" "Keep prayin', Sister! Keep prayin'!" "I didn't know I was payin' good money for a prayer meetin'!" and "Aw, pull in your ears!")*

*Jasper: (defiantly facing the crowd)* Ain't you got no respect for a Christian woman? Get out of here! The lot of you! Get out!

*Mother:* Yes! Get out o' ma house befo' I calls upon de Lawd to strike you all dead!

*(The dancers now crowd up front to get their hats and coats. Their resentment has turned to amusement at the* MOTHER'*s tirade. They laugh and jeer at both her and Jasper. Such remarks fly, as: "Who-ever heard of a party ending at twelve o'clock!" "Let's go on over to Tommy's. He's ready to start a party anytime!" "Dere's a party right next door tonight!" "Let's go an' let the old lady have her prayer meetin'." They also cry out to* JASPER *as they go out, "Good night, parson!" "Hope you get a good collection!" "Better get Ippy to play a hymn!" Somebody starts to sing "Get on board little children!", and the crowd take it up. The* FATHER, *who has been in the kitchen, comes out, sensing something is wrong, by the stopping of the music, and the strange noises)*

*Father: (spying the* MOTHER*)* What you doin' here? What's goin' on!

*Mother:* You've turned ma house into a den o' sin! You sent your own daughter to deh debbil!

*Father:* (*indignantly*) Go on, old lady. You ain't got no business here. You ain't go no business mixin' in my affairs.

*Mother:* I ain't, ain't I? I ain't got no call from de Lawd to do all I can to keep ma family from goin' to hell. (*She sees that some of the guests have stopped to listen and turns violently on them*) Git out, I tells you! Git out an' don' you lemme see any o' you in here no more!

(*As the last ones leave, she rushes to the door and slams it behind them.* CORDELIA *smiles cynically at her. The* FATHER *is in such a rage that he looks around helplessly, not knowing what to do or say.* IPPY *and* JAKE *are left picking up their music and packing their belongings.* CORDELIA *starts up front. Her* FATHER *sees her*)

*Father:* Oh, I see, you found de little lamb an' brought her back here to break up de best payin' party we's had yet. I s'pose you's tryin' to save *her* from goin' to hell! Huh! De party didn' get decent till she left!

(CORDELIA *turns and glares at him*)

*Jasper:* (*soothing to* FATHER) Now Pa! (*turning to* CORDELIA) Go on up front, Delia.

*Father:* Oh, no she don'! I'm gonna show her who's boss round here. I guess you found her layin' up in some dive! Don' think I'm gonna stand for de little chippie to come back here as if she'd been to church. (*He turns on* CORDELIA *again and advances toward her*) Where you been?

(JAKE *the Drummer leaves with his trap.* IPPY *stands around uncomfortably*)

*Jasper:* (*soothingly*) Pa, let the thing drop for a while.

*Father:* You tend to your business an' I'll tend to mine. I happens to be her father, tho' the Lawd knows I wisht I wasn't. An' I also happens to pay rent here. Now! (*walking over to* CORDELIA) Where you been? (CORDELIA *ignores him. He cries louder*) Where you been?

*Mother:* I tells you I'll 'splain. Les' don' have no mo' fuss.

*Father:* I'm gonna make dat gal do as I say for oncet, if I has to kill her. Now, where you been?

*Jasper:* Cain't you see you jes' makin' lots of fuss for nothin'?

*Father:* (*turning on* JASPER) Yeah, it's for nothin'. That's jes' what she is, nothin'. She ain't nevah been nothin', an' she nevah will be nothin'.

*Cordelia:* I don' see—

*Father:* (*turning back to her*) You ain't s'pose to see nothin'. Now that you done found your tongue, tell me where you been. I wanna know.

*Cordelia:* I ain't fixin' to tell you nothin'.

(*He slaps her in the face, and draws back to hit her again, but* JASPER *grabs his arm, and the* MOTHER *rushes between him and* CORDELIA)

*Mother:* Ain't you got no pride left, hittin' her in front of folks like dat? Cain't you be reasonable an' let Jasper an' I 'tend to dis? Go to your room, Delia. *(*CORDELIA *goes out looking hatefully at her* FATHER. *The* MOTHER *turns to* IPPY, *who is at the rear door, about to leave)* I'm sorry son, dis had to happen like dis.

*Ippy:* I was just waitin' for my money.

*Father:* I'll see you in a minute. I just wanna—

*(He tries to break loose from* JASPER. *But* JASPER *holds on to him and endeavors to take him into the kitchen)*

*Jasper:* Come on out in the other room a minute, Pa. I want to talk to you.

*Father:* I don' see why we has to go in de other room. Cain't you talk in here? No need bein' 'fraid of Ippy. He knows ev'thin' already. Why, he even said he wanted to marry Delia, an' go on deh stage wid her.

*Mother:* Don' let him tease you, Ippy. Why should Ippy wanta marry Delia?

*Father:* Why should anyone want to marry her?

*Mother:* Oh, she'll settle down and marry some day.

*Father:* If she stays here, she's gonna get married right away.

*Ippy:* Who she gonna marry right away?

*Father:* She gonna marry Basil.

*Mother:* How you gonna make her marry Basil if she don' want to?

*Father:* She either marries Basil or goes into de street. She cain't stay 'round me an' be no chippie.

*Mother:* An' I ain't gonna have it said dat a daughter o' mine was chased into de streets to go to de devil.

*Father:* She done gone to de devil.

*Mother:* She don' have to stay dere. If the Lawd forgives sinners, I don' see why we can't.

*Father: (disgustedly)* Oh, you an' your Lawd!

*Mother: (regarding the room tragically)* Dis place is sure a mess. *(She looks up front)* Where's Basil? We gotta start cleanin' up. Cain't have deh house lookin' like dis for deh Lawd's day. *(She calls)* Bella. Bella.

*Father:* She ran off widout sayin' a word. She's getting' jest like deh rest of dem.

*Mother: (worriedly)* Dat ain't like Bella.

*(The front door bursts open and* BASIL *comes rushing in. Everybody jumps)*

*Father:* My God, Basil; what's de—

*Basil:* Is Delia here? Is Delia here?

*Mother:* Calm down, son; you looks as if you've seen de avengin' angel.

*Ippy:* He's pale. Yeah, Delia's here.

*(*BASIL *drops on the end of the couch. He is very weak. He holds his head between his hands, his elbows resting on his knees, so as to hide his bruised jaw)*

*Jasper:* Was somethin' chasin' you?

*Mother:* Wait'll the boy gets his breath.

*Jasper:* What chased you?

*Basil:* Nothin!

*Father:* Nothin'? Oh, you jes' takin' exercise!

*Mother: (starting for kitchen)* Let me get you a drink of water.

*Basil: (breathing hard)* I'm al'right. I just come up the stairs too fast.

**(MOTHER** *goes into the kitchen)*

*Father:* Seems like to me you always run up dem stairs, but I nevah saw you come in here lookin' like you did now, wid eyes bucked an' coat flyin'.

*Mother: (re-entering from kitchen with glass of water)* It's good and cold. *(She hands the water to* **BASIL***. He takes it and begins gulping it down quickly. His bruised jaw is revealed. The* **MOTHER** *stares at it, eyes wide open)* You been fightin'. *(She is excited)*

*Jasper:* You been fightin' with that no 'count Roy.

*Basil: (hands her back the glass)* No. No. I'm all right.

*Father: (to* **BASIL***)* Who you been scrappin' wid?

*Basil:* No one.

*Ippy:* I s'pose your jaw jes' naturally poked itself out, and got bruised on general principles.

*Jasper:* Well, if he don't want to tell us, he don't have to.

*Mother:* Take off his coat, Jasper.

**(JASPER** *and* **IPPY** *pull off Basil's overcoat)*

*Basil: (protesting)* I'm all right, I tell you. I just want to speak to Delia.

*Mother: (finally)* You can talk to Delia later. Go on out in de kitchen, and bathe your face in warm water an' put some arnica over dat bruise. Dat'll take de soreness out.

*Basil: (protesting feebly)* I tell you, I'm all right. I must speak to Delia.

*Mother:* Go 'long, take care of yourself. I tell you, you can speak to Delia afterwards.

**(BASIL** *goes into kitchen)*

*Jasper:* Can you beat it?

*Ippy:* I bet that's what happened—he went lookin' for Delia and ran into Roy.

*Father:* If I find dat little yaller bastard—

*Mother: (interrupting)* Pa.

*Father: (determinedly)* I'm gonna make Basil tell me 'bout dis. *(He rushes off into the kitchen)*

*Mother: (somberly)* Jasper, what we gonna do? What we gonna do?

*Jasper: (hopelessly)* Don' seem like they's nothin' to do, ma.

*(There is a knock on rear door)*

*Mother:* Now, who could that be? I swear if dey's any mo' 'citement— *(The door opens and the* SERGEANT *enters leading in a very scared* ARABELLA. PATROLMAN SAM *follows and closing the door, takes up a position just inside. The* MOTHER *and* JASPER *stare unbelievingly at* ARABELLA *and the police. The* MOTHER *cries)* Bella. Bella. Whacha been doin'? Whacha been doin'? Oh, God. *(She sobs)*

*Sergeant:* The girl's all right. Where's Basil Venerable? *(Nobody answers. The* SERGEANT *barks)* Is he here?

*Ippy:* He's out in the kitchen.

*(*JASPER *turns and gives* IPPY *an angry look as if to say, "Why don't you keep your mouth shut?" The* SERGEANT *strides to the kitchen door)*

*Sergeant: (shouting)* Basil Venerable. Come out here.

*Father's Voice: (angrily)* What do you want? *(He comes out and seeing the* SERGEANT, *he says meekly)* Oh, I didn't—

*Sergeant: (interrupting)* Basil Venerable.

*Father:* No. No. I'm Mister Williams.

*Sergeant: (impatiently)* Hell. *(He pushes the* FATHER *aside and strides into the kitchen)*

*Father: (looking around)* What's deh matter? *(sees Bella)* Where you been?

*(Before* ARABELLA *can answer, the* SERGEANT *strides out with* BASIL. *Everybody turns towards them)*

*Jasper: (to* SERGEANT*)* What do you want him for?

*Father:* What's he done?

*Sergeant: (brutally)* There's been a murder.

*Mother: (moaning)* Oh, Lawd. Lawd.

*(*ARABELLA *begins to cry)*

*Father, Jasper, and Ippy: (shouting almost simultaneously)* What?

*Basil: (trembling)* I didn't kill him. I didn't kill him.

*Sergeant: (taking out Basil's revolver)* Isn't this your gun?

*Basil: (stuttering)* It's Mr. Williams'!

*Father: (stepping forward)* What? *(surprised)* I left dat gun in the kitchen table drawer. It ain't loaded.

*Sergeant:* It isn't huh? *(He breaks it)* Full up and one cartridge fired.

*(The* FATHER *looks at* BASIL, *convinced of his guilt)*

*Sergeant: (to* BASIL*)* Didn't you take this gun to Roy Crowe's apartment?

*Basil:* Yes. But I didn't—

*Sergeant: (interrupting)* And this is the gun that killed Roy Crowe.

*Basil:* No. No. He was killed before I got there.

*Mother: (pleadingly)* Basil wouldn't kill nobody.

*(There is a knock on the door. Everybody turns.* SAM *opens the door)*

*The Kid: (stepping jauntily)* Howdy folks! *(He sees the* SERGEANT *and draws back)* Oh. I—

*Basil: (pointing at* THE KID *and crying hysterically)* He did it. He did it. That man there did it. *(He begins to wobble as if about to faint)*

*Sergeant:* Steady. *(He grabs* BASIL *and puts him in a chair)*

*Basil: (a bit dully)* He did it. He did it.

*The Kid:* What the hell's he talking about?

*Sergeant:* He says you murdered Roy Crowe.

*The Kid: (flaring up and starting for* BASIL*)* The God damned monkey chaser. Trying to put it off on me. I'll break his—

*Sergeant: (pushing him back)* Wait a minute.

*The Kid:* I'll tell you all about it, Sergeant. I'll tell you just how it happened.

*Cordelia: (entering)* What's all the noise? *(She sees* THE KID*)* Hello, Kid.

*The Kid: (turning)* Hello.

*Cordelia: (looking at everybody)* What's goin' on?

*The Kid:* This monkey here killed Roy and he's trying to put it off on me.

*Cordelia: (shocked)* Roy killed.

*Basil: (weakly to the* SERGEANT*)* I didn't do it. I didn't do it.

*Cordelia:* I'm sure he did it. He said right here las' night he 'tended to kill Roy.

*Mother:* Cordelia.

*Father:* Shet yo' mouth.

*Cordelia:* I won't shet my mouth. Basil did it, an' I hopes his black hide'll burn in de 'lectric chair.

*Mother: (begins moaning again)* Lawd. Lawd. What have I done? What have I done?

*Sergeant: (shouting)* For God's sake, shut up. All of you. *(They quiet down. The* SERGEANT *addresses* CORDELIA*)* Who are you?

*Cordelia: (defiantly)* Cordelia Williams

*Sergeant: (to* FATHER*)* You this girl's father?

*Father:* I don' claim her no more.

*Sergeant:* I didn't ask you that. Is she your daughter?

*Mother: (belligerently)* Shure, she's his.

*The Kid: (going up to the* SERGEANT *and speaking confidentially)* This monkey's your man, Sergeant. I saw him kill Roy.

*Sergeant: (to* THE KID*)* You *saw* this fellow kill Roy Crowe?

*The Kid:* I as good as seen him. I heard him.

*Sergeant:* Let's have the whole story.

*The Kid: (calmly)* Well, I went to see Roy tonight on some business. Cordelia, here, was with him. Roy stepped out to speak to someone in the building, and Cordelia's folks came and took her away. Then Roy came back and we were talking when this monkey chaser arrives, acting like a raving lunatic. He and Roy has one awful argument.

He wants to see Delia and Roy tells him she's gone, but he won't believe it. I tried to make peace, but the gink turned on me, callin' me every name you can imagine. I hauls off and knocks him down with a bust in the jaw. He jumps up and pulls a gun on me. Well, I got a tolerable respect for firearms, so I just backs out the door and leaves Roy and his visitor to fight it out among themselves. While I'm standin' in the hallway, I hears a shot in the apartment and then a moan. I don't want to be mixed up in no shootin', so I beats it and when I get to the corner I called up the police. That's all I know.

*Sergeant: (to* SAM*)* Well, Sam. *(nodding to* BASIL*)* This is our man al'right. Take him to the station.

*Basil: (pathetically protesting)* I didn't do it. I didn't do it.

*Sergeant:* You can tell that to the district attorney. *(*SAM *takes* BASIL *by the collar and starts him towards the door. The* SERGEANT *addresses the crowd)* All of you are liable to be subpoenaed as witnesses. We don't want to have to search for you and lock you up, so stick around.

*The Kid: (carelessly)* You always know where to find me, Sergeant.

*Sergeant: (to* SAM*)* Al'right, Sam.

*Sam: (jerking* BASIL*)* Come on.

*Basil: (straining towards* CORDELIA*)* Delia. Delia.

*(She shrugs her shoulders and turns towards* THE KID *and* SAM *pushes* BASIL *out ahead of him and the* SERGEANT *follows)*

*The Kid: (to* CORDELIA*)* All set, girlie.

*Cordelia:* Wait'll I get my hat and coat. *(*CORDELIA *starts for up front)*

*Mother:* You cain't go 'way wid dat man.

*Cordelia:* Nobody's gonna stop me dis time.

*Jasper:* Ain't you got no feelin' for your maw? Don't you see she's tryin' to keep you decent?

*Arabella: (with passion)* She can't be decent. She ain't got no feelin' for nobody. *(She goes up to* CORDELIA, *who draws back before her ferocity)* The police should've taken you 'stead of Basil. You made him kill Roy. *(She raises her clenched fists above her head as if to strike* CORDELIA*)* Oh, I hate your very guts. *(She strikes out and* JASPER *grabs her. She crumples into his arms and sobs)* Basil. Basil. My poor Basil.

*(*JASPER *takes her off up front)*

*Father: (to* CORDELIA*)* Get the hell out of here.

*Cordelia: (running up front)* An' I don't need no second invitation.

*Mother: (turning to* KID VAMP*)* Why don' you go an' tell her you'll come back when you can marry her? I don' mind her goin' if she'll be halfway 'spectable. But I cain't let her live in sin.

*The Kid:* Well, mam, it's like this—the gal seems determined to go, an' since I likes her an' she likes me, I don' see why she shouldn't leave with me as well as anyone else.

*Ippy:* I can see plenty reasons why she shouldn't leave with you.

*The Kid:* Mainly I s'pose 'cause you wants her to leave with you.

*Ippy:* You ain't got nothin' to do with that.

*The Kid:* An' you ain't got nothin' to do with me.

(**CORDELIA** *comes rushing back, carrying her hat, coat and a suit-case*)

*Cordelia:* Come on Kid—le's go!

*Ippy:* Cordelia.

*Cordelia: (antagonistically)* What do you want?

*Ippy:* You cain't go 'way with that man.

*Cordelia:* Cain't? Whachu got to do wid it? You had your chance an' didn't take it. Now, don' fret, 'cause you're a back number.

(*She again turns to* **THE KID**. **JASPER** *re-enters from up front*)

*Mother: (pleadingly to* **FATHER***)* She's gonna live in sin.

*Father:* I'd rather her live in sin dan me live in hell wid her 'roun' all de time. Let 'er go.

*Jasper: (excitedly)* She cain't go.

*Cordelia: (starting for the door)* Don' pay dem no mind, Kid. We're goin'.

*The Kid: (starting to open the door)* We sure are.

*Mother: (frantically turning to* **JASPER***)* Jasper.

*Jasper: (starting for* **THE KID***)* She can't go. I won't let her.

*The Kid:* I'd like to see any of you stop her.

*Ippy:* There's enough of us here to do it.

(**IPPY** *and* **JASPER** *start for* **THE KID**. **THE KID** *draws back and puts his hand in his pocket as if to get his gun. He realizes he hasn't got it, so he clenches his fists and prepares to meet* **IPPY** *and* **JASPER***'s attack*)

*The Kid:* I'll knock hell out of the first one of you that tries to stop us. (**IPPY** *and* **JASPER** *regard* **THE KID** *menacingly but do not attack.* **THE KID** *turns to* **CORDELIA***)* Go on, Delia.

(**CORDELIA** *goes out laughing mockingly. The Jazz can be heard starting up as the party gets going across the way*)

*Father: (derisively)* Goodbye.

*Mother: (moaning)* Oh, Lawd. Lawd. Have mercy, Lawd. Have mercy.

*Cordelia: (in hallway)* Don't shove me! *(a pause)* What are you tryin' to do?

(*She is pushed in the room by a white, tough, menacing-looking East Side* **GUNMAN***, who has his hand in his pocket, pointing an automatic at* **THE KID**. **ANOTHER GUNMAN** *appears behind him.*)

*First Gunman: (to* **KID VAMP***)* Been trailin' you for an hour and a half. Come on.

*(He nods towards the door)*

*The Kid: (eyes widening with terror)* Man. You ain't gonna take me for a ride!

*First Gunman:* Come on! Rafferty wants to see you!

*The Kid: (trembling and turning his hand so his big diamond is visible)* I'll—I'll give you that rock. I'll give you five grand—Ten grand! Don't take me. Lawd. Don't take me.

*First Gunman:* Come on or I'll give it to you now! *(He maneuvers* THE KID *out the door)* Get out!

*(*THE KID *backs out while the* FIRST GUNMAN *stands at the door covering the family. Three shots are heard in the hall above the music at the party across the way. The* SECOND GUNMAN *then reappears at the door)*

*Second Gunman: (to fellow)* Over the roof!

*First Gunman: (menacingly to the family as he backs out)* You know nothin'! See!

*(He moves his gun threateningly and then goes out, slamming the door. The two men are heard rushing upstairs to the roof.* IPPY *hurries into the hall and stumbles back)*

*Ippy:* My God! They got him!

*(*CORDELIA *also rushes out)*

*Cordelia: (drawing back into the room)* Kid! Kid! Oh! Oh! *(The* MOTHER *begins moaning hysterically.* CORDELIA *turns on the whole crowd)* You! You! It wouldn't have happened if you hadn't dragged me back here! I hate you! I hate you all! *(She turns to* IPPY*)* Come on Ippy! Take me out of here!

*Mother:* You can't go, Delia! You can't go!

*Cordelia:* Jes' try an' stop me. I'm goin', see! An' when you hears from me again, I'm gonna be livin' high, standin' in de lights above deir heads, makin' de whole world look up at me.

*(She goes out of the door with* IPPY *following and carrying her valise. The door closes behind them)*

*Mother: (moaning)* Lawd, save her soul! Save her soul! She's only a poor ign'runt sinner! *(The Jazz from the party across the way bursts out in a sudden crescendo. The* MOTHER *throws up her arms in a gesture of supplication)* Lawd! Lawd! Tell me! Tell me! Dis ain't de City of Refuge?

*(The Jazz becomes louder and louder as the curtain falls)*

# ARTICLES ABOUT *HARLEM*

## Casting and Directing *Harlem*

Unpublished; written by Rapp and Thurman.

*Harlem,* the drama of New York's Black Belt now appearing at the Apollo, excited the Broadway producers when still in manuscript almost as much as the present vivid production is exciting the public. Options on it were held by three different producers and the play was first put into rehearsal by Crosby Gaige and Al Lewis more than a year ago. The urge to produce this play stirred mightily in the breast of numerous impresarios of the theater who have won much fame and more fortune, but they all surrendered before the problem of finding the actors and training them. "Where, oh where," they asked, "can you find twenty-six Negro actors to play important roles when *Porgy* has gobbled up every Negro actor worthy of the name?" The answer was given them at the premiere of *Harlem* by twenty-four year old Chester Erskin, who cast and directed the piece.

Mr. Erskin, in addition to finding and training a cast which one critic said gives "more genuinely good acting than I have run across in one play this season," has also apparently discovered a new star in Isabel Washington, playing the tantalizing lady of the play who flaunts her erotic way through the three thrilling acts. With more experience and continued good direction, Miss Washington should achieve in Negro drama a position analogous to that held by Florence Mills, Ethel Waters, Josephine Baker, and Adelaide Hall in the Negro revues; not that Miss Washington couldn't go over and challenge these ladies in their own field for in the Rent Party scenes of *Harlem* she sings a wicked song and shakes a flippant leg to tremendous effect.

Where did Mr. Erskin find his players? The answer is Harlem. He went around rent parties, speakeasies, cabarets, theaters, and marches up and down Lenox and Seventh Avenues looking for types. He was sure that if anyone had the outward appearance and inner quality of the role, he could get them to act it. And the performances of Isabel Washington as Cordelia Williams, Emory S. Richardson as Jasper Williams, Collington Hayes as Thaddeus Jenkins, Ernest Whitman as Kid Vamp, and of Edna Wise Barr as Arabella Williams proves that Erskin knew what he was talking about, for none of these have ever performed a dramatic role before.

To select twenty-six principals of *Harlem,* Erskin read nearly a thousand individuals in the different parts. His patience was phenomenal. Many a time the producer and authors insisted that the people he had chosen could never perform satisfactorily, but he stuck to them and brought them through to a magnificently successful premiere, although he lost twenty-five pounds in the process and will need an entirely new wardrobe if he doesn't regain them soon. The producer, Edward A. Blatt, and the authors, William Jourdan Rapp and Wallace Thurman, are ready to buy him two if he wants them.

## Detouring *Harlem* to Times Square

*The New York Times* April 7, 1929; written by Rapp and Thurman.

One day, more than a year ago, two hopeful young authors turned over to a play broker six copies of a dramatization of life in Harlem entitled *Black Mecca*. Less than a week later the broker called them up and told them that Crosby Gaige and Al Lewis both wanted to buy their play, asking which one she should sell it to. The authors—that would be ourselves—answered "Both!" And so it was.

Mr. Lewis had charge of the active work of production, and within a month the play was in rehearsal under the title of *Black Belt*. And a month later it was taken out of rehearsal and the cast dismissed. We innocently inquired "Why?" and Mr. Lewis replied: "You haven't a 'wow' in your third act, and naturally we can't go ahead until we find one."

Being a bit mystified, we inquired as to just what constituted a "wow," so that we could intelligently proceed in search of one. Mr. Lewis thereupon enlightened us as follows: "It's a bit hard to explain, but I think I can best make you understand by giving you an example. Say a thief is running down the street and a cop is chasing him. The thief passes by a crowd listening to a stump speaker. He jumps up on the stump, pushes off the speaker and begins gesticulating and shouting as if he himself were running the meeting. As a result the cops run by and the thief escapes. That's a 'wow.'"

With this as an example, we tried to think of a "wow" that would fit into our third act, but somehow the muse seemed to have deserted us. But again Mr. Lewis came to the rescue. He went to see *Porgy* and returned enthusiastic about the "wow" funeral in that play, and suggested, inasmuch as *Porgy* had scooped *Black Belt* on the funeral, that the latter get its "wow" by inserting a christening in its third act. We thought this over and rejected it. Undaunted, Mr. Lewis came back with plans to "wow" the third act by putting it in a cabaret.

All this took many months and the time came around when Mr. Gaige's and Mr. Lewis's option expired. Mr. Lewis wished to renew, but feeling we never could "wow" that third act to his satisfaction, we decided to try and find another producer.

During the next weeks most of the producers on Broadway read the script and turned it down. A casting agent who had been identified with Mr. Lewis finally bought a month's option on the piece, and at the end of each succeeding week informed its palpitating authors that he was going to do the play with George Jessel, with Morley, Milliken, Throckmorton and Gribble; and with Ben Hecht and Charles MacArthur. Conferences were actually held with the Messrs. Jessel, Hecht and MacArthur.

The authors of *The Front Page* proclaimed that the first act and the beginning of the second act of *Black Belt* were swell, but that it was all wrong to bring the third act back to the Williams tenement and end the play there. Rather, the play should show Cordelia Williams going on and on along her sinful career and finally ending up disastrously, say, in Paris.

We listened attentively to the Messrs. Hecht and MacArthur, but decided we didn't want to rewrite the latter's *Lulu Belle*. So, the casting agent's option having expired, we took our script and again went forth looking for a producer.

By this time there weren't very many producers left. But we, realizing that Broadway grows producers almost as fast as the United States grows playwrights, hoped that a special one might bloom just for us. A young actor, play reader, stage manager, vaudeville sketch writer, and occasional director by the name of Chester Erskin got hold of one of the original copyrighted copies of the play. (There were more than fifty copies of the various versions of the play which had been typed at an expense of nearly $200. Most of the different versions were the result of trying to "wow" that third act to Mr. Lewis's satisfaction.) Erskin liked the play and decided he wanted to do it. He interested a theatrical publicity man by the name of C. A. Leonard. Both of them interested young Edward A. Blatt, who wanted to become a producer and bought the option sight unseen. We settled back to await its expiration. Imagine our surprise when these three young men actually produced the play under the title *Harlem*.

## *Harlem* as Educational Drama

Unpublished; written by Rapp and Thurman.

The most interesting comment on our play, *Harlem* which is now playing at the Apollo, has come from a Negro writer writing in a Negro paper. He says:

> Hitherto all American plays purporting to portray phases of Negro life, except those of O'Neill, have been written from the point of view that Aframericans seldom behave like human beings. The prevailing dramatic philosophy is that while the main motives of all other peoples are ambition, love, and the desire to get rich, Negroes are concerned solely in pleasing God and placating the conjure man. Even when it is conceded that colored folks sometimes take an interest in

worldly affairs, they are represented as having the limited intelligence of somewhat backward cretins.

The hero of *In Abraham's Bosom,* for example, rarely evinces the intellectual maturity of Caliban, while all the characters of *Porgy* are charcoal etchings of the first and second grave diggers. When sophisticated Broadway audiences observe these strange and curious creatures, their natural reaction is "What kind of people are these?" And after the play they go home with the feeling that they have spent the evening in a museum instead of a theater.

The characters of *Harlem* are people whose talk, clothes, conduct, and motives are as recognizably human as those of the characters of *Broadway* or *Four Walls.* The man in the orchestra seat may not sympathize with their motives, but he can readily understand them. And understanding these characters helps him to better comprehend the concrete Negroes he has seen in the subway or reads about in the crime columns of the newspapers. To this extent, at least, the authors of *Harlem* are educating the theater-going public.

In short, this writer maintains that most Negro dramas previous to *Harlem* dealt with what Negroes call "white folks' niggers" while *Harlem* actually presents the Negro as he is.

"White folks' niggers" consist of three distinct categories: the old servant or mammy type known derisively among the Harlemites as "Uncle Toms" and "Handkerchiefs," the lazy slow-foot type typified by such vaudevillians as Bert Williams and Miller and Lyles, and the superstitious, praying type who is always thrown into abject fear by darkness, lightning and thunder. All these types flatter the white's sense of superiority, and it pleases him to believe that all colored people are like this. The dramatist who shows them thusly is bound to be complimented for his keen understanding of the Negro.

On the other hand, the dramatist who portrays the Negro as reacting to life in the same manner as whites, except where his reactions are conditioned by his color, risks being accused of writing a white drama and having it played in black face. The fact that the black number racketeers in *Harlem* acted much like white bootleg racketeers, led many critics to question the authenticity of this part of the play. And because most of the characters are not "white folks' niggers" other critics ascribed the stark realness of the play entirely to the natural acting of the players, seemingly not realizing that those players could only have acted thus naturally because the lines they spoke and the situation they found themselves in were perfectly familiar, most of them having spoken such lines and found themselves in such situations in real life.

So perhaps *Harlem* does qualify in spite of its being as two critics stated "bang bang drama" and "as exciting as a fire," for the educational theater. At least the authors will be pleased if it helps to banish the illusion that all Negroes are "white folks' niggers."

## My Collaborator

Unpublished; written by Thurman alone.

He is the most energetic man I know being happiest when there is much work to do. He is generous of his time and money, and I have never known him to be bored or pessimistic even when *Black Belt* was in the hands of producers who, as St. John Ervine would say, ought to be in the cloak and suit business.

His enthusiasm is most infectious, affecting even blasé Broadwayites. As a suburban property entrepreneur or a public relations counsel he would be a riot. He keeps his sense of humor even when doing the hack work he must sometimes do in order to eat. He has many mysterious ways of dividing his salary check. Portions of it even being mailed to Europe, thus clinching his rights to be known as a cosmopolite.

And a cosmopolite he is, being equally at home in Harlem, on Broadway, the Lower East Side, Riverside Drive, France, The Balkans or the Near East.

Ours has been one of those collaborations you read about—no quarrels over pet ideas or sulkings because of their exclusion. Which is set down here more in admiration of Bill Rapp than in praise of myself.

I have often wished for a movietone camera during our play writing sessions. Posterity should not be deprived of the picture of Bill Rapp, excited over the possibilities or difficulties of a scene, leaping from his chair, pacing the floor, frantically gesturing the while he shouts Negro dialect with decided East Side overtones.

## Two Playwrights Look at Harlem

Unpublished; written by Rapp and Thurman.

The textbooks on playwriting say that drama arises from conflict, unrest, and change. Now, argued William Jourdan Rapp and Wallace Thurman, two playwrights in search of drama, there ought to be plenty of all these in a city of a quarter million Negroes that has sprung up in a mere decade, and is completely surrounded by cosmopolitan white New York. They, therefore, went to Harlem and took a look. Then they took many more looks. And what they found will be seen when *Harlem* opens here with a cast of sixty Negroes and a single white man.

Harlem never sleeps. That's the first fact that strikes the visitor from downtown. The restaurants, shops, and speak-easies of Seventh and Lenox Avenues are busier at four in the morning than four in the afternoon. And children of all ages play in the streets under the arc lights from sun-set to dawn.

This constant diurnal and nocturnal activity is the first sign of the over-crowding that conditions the whole of Harlem life. Pilgrims from the South are pouring into the city of refuge faster than it can extend its borders to make room for them.

The news that Harlem is the promised land has been resounding throughout the Southland, and every black soul that has chafed under Jim Crow is hurrying north to the city where Negroes can ride in the subway, and go anywhere their money and sense can carry them.

So the Black Belt of Harlem today presents a situation somewhat similar to that existing on both the East and West Sides of New York City during the height of the Irish, German, Italian, and Jewish immigrations.

A large rural population is adapting itself to urban life. And the adaptation has brought with it the characteristic conflict between the old devout generation and the young skeptics. In Harlem, the spirituals are fighting a losing battle against jazz.

Hard work, faith and family mean little to the new generation north of 125th Street. And why should they with home being an overcrowded flat full of boarders? The movies, streets, speakeasies and dance halls are more attractive than home and get Harlem's youth no matter how much the old folks may pray to the "Lawd" to keep their children from going to the devil.

And then tragedy of tragedies, the old folks are often forced to let Satan into their own homes to pay the rent.

With everybody in the family working and with the house crowded with boarders, the family still often finds it impossible to pay Harlem's high rents, so to get the necessary funds, they throw a rent party.

On a Saturday night, the house is open to outsiders for a quarter entrance fee. The youths of Harlem come and dance to the strident tones of a mechanical piano and their own shouts and moans. They eat pigs' feet and slaw. They drink corn liquor and gin. They shoot craps and play black jack. And the family profits on everything.

The rent is paid but the old folks can no longer rail against the sin from which they have profited. Economic necessity has again triumphed over morality.

Then there are the Harlem rackets which penetrate into every home no matter how lowly. The number runner will take bets as low as five cents. The "Hot Stuff Man" or seller of stolen goods finds no tenement too poor to peddle his wares. Dr. Voodoo has charms to conjure both rich and poor. The sweetback who lives off the gratuities of his many women, finds recruits for his affections not only in the dance halls and speak-easies but also at rent parties.

These things and many more the two playwrights discovered in Harlem and put them into their play. So if you want to see the throbbing and pulsating life of the Negroes' promised land, pay a visit to *Harlem* when it comes to Broadway.

# The Writing of *Harlem*:
# The Story of a Strange Collaboration

Unpublished; written by Rapp and Thurman.

About two years ago a young Negro from Salt Lake City walked into the office of a young editor from the sidewalks of New York. He bore a card of introduction from a young minister, an erstwhile missionary in Turkey, who had known the editor during his foreign correspondent days in the Near East. Out of this meeting came *Harlem,* the drama of New York's Black Belt which is now thrilling audiences at the Apollo Theater.

Wallace Thurman, the dark side of the black and white collaboration which produced *Harlem,* was anxious to launch himself in a writing career. William Jourdan Rapp, the editor whom he sought out for advice and aid was more than anxious to help him. So the two talked over Thurman's background and decided that the richest literary field for him to exploit was Harlem, every Negro's promised land, and the actual City of Refuge for approximately a quarter million black souls. Here all sorts of strange and interesting things were happening that the world would certainly be willing to read about.

Thurman wrote a number of articles on life in Harlem and Rapp sold them for him. A strong friendship developed between the two men, and soon Thurman was introducing Rapp to every aspect of Harlem existence.

Rapp's observations soon made him realize that the Negro in Harlem was going through the identical struggles of adjustment that characterized the lives of the Irish, German, and Jewish immigrants of the East Side whom he had known in his boyhood. He also learned that the Negro's psychology is similar to that of other subordinate minority groups, such as the Armenians and Greeks in Turkey, concerning whom Rapp had written a great deal.

With Rapp an inveterate playwright, having sold an option on his first play at the age of eighteen, published a dramatization of the Turkish revolution entitled *Osuman Pasha,* and had a farce *Star Struck* produced in Athens, Greece, it was only natural that the two friends should decide to do a drama of Harlem.

Now, the great time for collaborators to get together to discuss their brain child is lunch. Innumerable Broadway successes have been created during those well-fed moments when the demi-tasse and cigarettes usurp the dessert. So Rapp and Thurman decided to meet regularly for lunch. And here difficulties immediately began to arise, which proved dramatic but had nothing to do with play-writing.

Rapp would non-chalantly walk into a Broadway restaurant followed by Thurman.

And one glance by the head waiter at Thurman's dusky skin would invariably start him and his whole corps of waiters towards the entrance. Here they would surround the mixed collaborators and doing their best to hide them from the other guests maneuver them off to a remote corner where Thurman's color would be safely hidden. After a number of experiences like this, Rapp and Thurman moved their lunching ground from Broadway to Third Avenue, and a goodly part of *Harlem* was created in a Chinese restaurant under the el, for in the words of the mother in *Harlem* when she speaks of heaven, "Dere ain't no prejudice dere."

After thoroughly studying and discussing the Harlem scene, the collaborators concluded that the most novel and glamorous institution in the Black Belt was the rent party. They, therefore, decided to build their play around such a party so as to give it the "show qualities" that Broadway had indicated were needed for success. The next problem was to build into the play the outstanding conflicts in Harlem life, i.e., the conflict between the American Negro and the West Indian Negro—based largely on the fact that the West Indian Negro will work longer hours for less money than the American Negro and therefore find it easier to get jobs! The conflict between the younger generation, who are irreligious and ready to abandon themselves completely to the jazzy tempo of Harlem life, and the older generation who long for the South, and finding it impossible to adapt themselves to urban life in the North, turn to religion for solace; and the conflict within the Harlem underworld for control of the giant lottery of numbers upon which nearly one hundred thousand bets are placed daily.

That Rapp and Thurman executed as they planned is evidenced by the completed play at the Apollo, the outstanding feature of which is the rent party, but through which there runs excitingly all these various conflicts.

It took a year to write the play and it took another year to get it produced. The latter period might have been substantially cut if the authors had been willing to introduce into their script various "wow" scenes that Broadway producers told them were needed for success, but which had nothing to do with Harlem life.

If you ask Rapp and Thurman who wrote the dialogue, who developed the plot, or who decided on what atmosphere to accept or reject, they'll tell you that they don't know. They claim individual creation for but one line each in the entire play. Everything else was created together. Thurman's line is that of the indignant father when he says, "You know what's wrong wid Harlem? Dey's too many niggers!" And Rapp's line is that of the janitress who when asked by the detective if she's the janitress, says, "Aw, I'm deh assistant superintendent!" And if you analyze these lines you'll discover that the first is white in its humor and thought content, and the second is typically Negroid. So figure it out, if you can!

# Jeremiah the Magnificent

Written in 1929 with William Jourdan Rapp.

### Scenes

Act I  — Office of the "International Fraternity of the Native Sons and Daughters of Africa" in Harlem. A morning in early fall.

Act II — *Scene One* — Jeremiah Saunders' private office in Freedmen's Hall. A spring morning about six months after Act I.

*Scene Two* — Committee and Reception Room of Freedmen's Hall. That afternoon.

Act III — *Scene One* — Captain's cabin aboard the S.S. *Jeremiah,* Flag Ship of the Black Cross Line. Six weeks after Act II.

*Scene Two* — Salon of the S.S. *Jeremiah.* The following night.

*Scene Three* — Captain's cabin of the S.S. *Jeremiah.* Early the next morning.

*Note: Act II and Act III can each be done in one set. For Act II the set of Scene 2 will serve for both Scene 1 and Scene 2 if desired and for Act III the scene of Scene 2 can be used for all three scenes.*

### Characters

*(in order of their importance)*

Jeremiah Saunders
Kate Saunders
Stanley Grayson
Melissa Brown
Brandon Bell
Sarah Tyson
Jason Baker
J.S.B. Renard
Z. Albert Caldwell
Major Martin
Dora Anderson
Captain of the S.S. *Jeremiah*
Postal Inspector
Euclid Lorando
Coast Guard Captain
Knights, Ladies, Members, and Prospective Members of the "International Fraternity of the Native Sons and Daughters of Africa."

### Act I

*The setting is the office of the "International Fraternity of the Native Sons and Daughters of Africa"—a front room on the parlor floor of a "brown stone front" in Harlem. The office is spacious.*

*A big double door opens into it from a hall in the rear. When this door is open, a small table and chair are visible in the hall. Whoever is answering inquiries of callers regarding the Organization sits at this table. The entrance from the street at the top of a flight of brown-stone steps is at the right end of the hall, and at the left, behind the table, there is a flight of stairs leading to the upper floors. In addition to the double door in the rear of the office there is also a smaller door in the center of the left wall. This leads to the rear rooms of the first floor, which make up the living quarters of* JEREMIAH SAUNDERS, *founder of the "Fraternity," and his wife. A bay window, looking out on the street, is in the center of the right wall, and against the strip of wall to the right of the window is a pier mirror in a heavy mahogany frame. About the walls and facing the center of the office are rows of wooden and canvas folding chairs. More of these are stacked in the rear right corner, while in the rear left corner is a filing cabinet. There is a desk and a swivel chair near the bay window on the right, and on either side of this desk are straight back dining room chairs. On the left is a large table covered with pamphlets. A sign with big black letters reading "Africa for the Africans" hangs across the rear of the room just below the ceiling, and there is a large poster with a solid black map of Africa on a white background hung up over the files. Another poster bearing the legend "Africa is the Promised Land" hangs on the wall above the stack of folded chairs.*

*As the curtain rises,* MELISSA BROWN, *a very attractive dark brown-skin girl of about twenty-one is at the desk opening and arranging mail. She is a Harlem product, having graduated from a city high school where she took a commercial course. The streets, however, have left a more permanent imprint on her character than the classroom. She is slangy, and immoral. Her motto is "live and let live!" She is scornful of the gullibility of the uneducated Negroes, whose stupidities stir her anger, and occasionally her sympathy. Through the closed double door in the rear come voices of many people clamoring for admission.*

*Melissa:* Damn those idiots. Why don't they give us time to open the office? *(She opens a letter, glances at it, and slams it down on the desk impatiently)* Same old stuff! God, I'm tired of this!

*(She gets up from the desk and goes to the window. The double door at the rear opens slightly and* BELL *squeezes through. There is quite a hubbub heard as he enters which diminishes as he closes the door with difficulty due to the crowd pressing against it from the outside.* BELL *is a slender, dark young Negro of twenty-six or seven. He wears glasses and has the air of a scholar. He is well educated, a dreamer, and an idealist)*

*Bell:* Isn't Brother Jeremiah up yet?

*Melissa: (irritably)* Must be! *(She motions to the door on the left)* He and the old lady been argufyin' in there for half an hour.

*Bell:* I wish he'd come out and handle this crowd.

*Melissa:* You shouldn't have let them in.

*Bell:* You can't keep them out in the streets. This is their organization, you know.

*Melissa: (sarcastically)* And their funeral if they get gypped.

*Bell: (frowning and speaking a bit severely)* What do you mean?

*Melissa: (after a moment's silence in which she looks* BELL *over appraisingly)* Ask the gypsy! *(She turns back to the desk and continues opening the mail)* What about the old lady—is she or isn't she leavin' today?

*Bell:* Brother Jeremiah says she is! She says she isn't! I got reservations on the train!

*Melissa: (shrugging her shoulders)* S'too much for me.

*Bell:* She ought to go. She could do a great deal for the organization down South. That's our greatest field. And a woman can carry on down there with less danger than a man. It's really the only thing for her to do.

*Melissa:* You mean it's the only thing Brother Jeremiah sees for her to do.

*Bell:* Which is all the more reason why she should go.

*Melissa: (again facing* BELL *quizzically)* Say, boy, aren't you supposed to be a graduate of Harvard?

*Bell:* Yes.

*Melissa:* I said Harvard, not Howard!

*Bell: (mystified)* I heard you—but what—

*Melissa: (interrupting)* I just thought I'd make a mistake. Now if you were from Howard I could understand some things. You don't just expect much from those dicty nigger colleges. But you're from Harvard! *(With a shrug of her shoulders, she turns back to her letters and sighs lightly)* Oh—my, my, my!

*Bell:* What are you talking about, Melissa?

*Melissa: (carelessly)* Ask the gypsy! *(Her mood suddenly changes. She again faces him. She speaks earnestly)* Tell me! Honestly, are you really as hipped on Jeremiah and this outfit as you make out?

*Bell: (passionately)* I don't have to make out. I believe in Brother Jeremiah and his great work with my whole heart and soul. For over sixty years, my people—ever since they were freed from slavery—have been awaiting a man like Jeremiah, and a movement like this. There is no limit to what we can do for the black men and women all over the world. *(He is beginning to be carried away by his own words)* Why, for the first time in centuries, a black leader has had the courage to glorify his own color. I tell you, there's no telling—

*Melissa: (interrupting with calming sarcasm)* All right—sonnie—all right.

*Bell: (with furrowed brows)* But Melissa, you don't seem to understand.

*Melissa: (petulantly)* I understand a hell of a lot more than you give me credit for.

*Bell:* But you can't understand. If you did, you wouldn't be so skeptical. *(again beginning to preach)* Why, this movement—

*Melissa: (finishing his sentence)* Is going to lead us back to the promised land! *(She laughs at* BELL's *embarrassment)* You know I type Jeremiah's letters.

*Bell: (annoyed)* Don't be facetious!

*Melissa:* There you go usin' that Harvard lingo on me again.

*Bell: (with profound seriousness)* Melissa, let me tell you something. I know you wouldn't talk this way if you had suffered as I have just because you are a Negro.

*Melissa:* I ain't white!

*Bell:* I know. But it seems to have been my lot to face race prejudice in all its worse forms.

*Melissa:* Say, to hear you tell it, that seems to be the lot of all you educated niggers. When you're dumb like me, you go on 'bout your own business; but give a nigger a little education, and he goes peerin' through all sorts of barred windows and buttin' his head against all sorts of stone walls just a-lookin' for trouble.

*Bell: (meaningfully)* You don't have to look for trouble in America.

*Melissa:* Nor Africa either.

*Bell:* What chance has a black man in a white man's land? Think of the prejudice, the discrimination, Jim Crows, the Ku Klux, and the lynch—

*Melissa: (interrupting)* Huh! Listen, big boy! Those things may be tame compared to what's waitin' you in the jungles of Africa.

*Bell: (excitedly)* Like most Negroes, you're afraid to reach out for your rightful inheritance. You're content to let us live on unwanted in an alien land, without a flag of our own—without anything except what is allowed us by Marse George.

*Melissa:* Calm down! Calm down!

*Bell:* We must cease submitting to these gross injustices—

(**BELL** *is interrupted by a loud banging on the door*)

*Melissa:* Go tell those people something! Between you and them, and the old lady and old man quarrelling in there, I'll go nuts.

(**BELL** *goes to the double door at the rear and opens it slightly. A white man pushes by him into the room. Voices are heard crying: "We want Brother Saunders!" "We want Brother Saunders!"* **BELL** *closes the door with difficulty against the surging mob on the outside. The newcomer, a lean, Yankee featured man of forty-five, is an* **INSPECTOR** *of the United States Post Office*)

*Postal Inspector:* I want to see Jeremiah Saunders.

*Bell:* Is there anything *I* can do for you? I'm his assistant.

*Inspector:* I'd rather see Saunders personally.

*Bell:* Would you mind telling me your business?

*Postal Inspector:* I'm from the post office. There have been some complaints registered against your organization. I'm here to investigate.

*Bell: (surprised)* More investigation—why—

*Melissa: (interrupting)* We'd better let Mr. Saunders talk to this gentleman. I'll call him. *(She goes out quickly at the left)*

*Bell: (nervously)* Will you have a seat? Mr. Saunders will be right in. *(The* **INSPECTOR** *sits down, taking in the room with a glance.* **BELL** *moves about, visibly agitated)* Are these charges liable to be serious?

*Inspector:* I'm sure I can't tell you.

*Bell:* More serious than before?

*Inspector:* I don't know! *(nodding towards the rear door)* Evidently, they weren't very serious before. You still seem to be doing a thriving business. And none of you are in jail.

*Bell: (earnestly)* No, we're all right. It's just our enemies! The whites and lots of blacks, too, don't want to see an organization like this prosper.

*Inspector:* As long as you don't break the law, you've nothing to fear from Uncle Sam.

*(The door at the left opens and* **JEREMIAH** *rushes in, followed by* **KATE** *and* **MELISSA**. **JEREMIAH** **SAUNDERS** *is a tall, heavyset, powerfully built Negro, with a deep bass voice. He is a command-ing personality, possessing a high degree of personal magnetism, and a deep mystical nature. The latter is confounded with an overpowering materialistic will-to-rule. He has no especial mental powers, but possesses the clever and shrewd intuitiveness natural to some children. This shrewd-ness, however, only evinces itself in his relations with his enemies and opponents. His friends can easily lead him into openly dangerous paths so long as they flatter him. Also, like the child that he really is, he has a vivid imagination, as well as a love for ostentatious and colorful clothing. His long contact with the Bible and religious literature have given him ease and eloquence of speech.* **KATE SAUNDERS**, **JEREMIAH**'s *wife, is an honest, ambitious woman of distinctively higher men-tal caliber than her husband, and also of superior education, having graduated from Tuskegee. She, however, destroys her effectiveness in working with others by a too great suspiciousness and an accompanying desire to keep everything in her own hands even if she has to do everything herself. She is an ex-school teacher from a backward southern village where she met* **JEREMIAH**. *She is pes-simistic about her race's past and present, but extremely exalted about the future in which she sees herself as a great leader of her people. She is also deeply religious, and her sense of duty to God, her husband, and her race, are all intertwined. She is quite light-skinned but physically unattractive, being gaunt in appearance and perennially tired and overworked)*

*Kate: (pushing ahead)* What is it? What is it?

*Jeremiah: (following her up)* Just a minute, Kate! Just a minute.

*Kate: (excitedly)* Are you the postal inspector? Why, I thought this was all over. What does it mean—another investigation?

*Postal Inspector: (ignoring her and addressing* **JEREMIAH**) Are you Jeremiah Saunders?

*Jeremiah: (pompously)* I am! What can I do for you, sir?

*Inspector:* We've received several complaints concerning your organization and I've been ordered to make an investigation. You'll have to turn your books over to us.

*Kate:* But you've just finished one investigation. Mr. Grayson said everything would be all right. I don't understand.

*Inspector:* Neither do I, lady.

*Jeremiah:* The person for you to see is my lawyer.

*Inspector:* Is he here?

*Jeremiah:* No, but I'll give you his card. *(turns to* **MELISSA***)* Give this gentleman one of Mr. Grayson's cards.

*(***MELISSA** *goes to the filing cabinet, opens a small drawer at the top, and takes out a card. She hands it to the* **INSPECTOR***)*

*Kate:* Can't you tell us something definite?

*Inspector:* Lady, I can't tell you anything except to be careful of what you send through the mails. It might be a good idea to read up on postal regulations. *(He examines*

GRAYSON's *card and puts it in his card case as he takes out one of his own and hands it to* JEREMIAH) Have Mr. Grayson get in touch with me immediately. I'm at the Main Post Office. *(He starts out at the rear)* Good day!

*(He opens the door. Once more the surging crowd makes itself seen and heard. There are cries of "Where is Brother Saunders?" "We want Brother Saunders!"* BELL *pushes them back and again closes the door)*

*Bell: (to* JEREMIAH) You'll have to see these people right away, I'm afraid. They've been clamoring for an audience with you all morning. I can't keep them out much longer.

*Jeremiah:* All right, Brother Bell. I'll see them in just a moment.

*Kate: (worriedly)* Jeremiah, do you think the post office—

*Jeremiah: (interrupting and speaking confidently)* No need to worry about that. Brother Grayson attended to it before, didn't he?

*Kate: (haltingly)* I know—but it just seems—

*Jeremiah: (impatiently)* Quit your fretting, Kate. *(To* MELISSA) Is that all the mail, Melissa?

*Melissa:* Yes, sir. There's plenty of money orders and membership applications!

*Jeremiah: (stentoriously)* The Lord is good to us.

*Kate: (echoing him)* Right always prevails.

*Bell: (another echo)* There's no stopping us, I tell you.

*Jeremiah:* Kate, you and Brother Bell show those good people in while I go put on my robes. *(He goes off at the left)*

*Melissa: (To* KATE) Aren't you goin' away today, Mrs. Saunders?

*Kate: (shortly)* No! *(turning to* BELL) Brother Bell, you let them in. *(back to* MELISSA) You can take your post outside. Brother Bell will help you get their names and addresses as they leave.

*(BELL *opens the double door and steps back. The crowd surges in. There is a great deal of shoving and loud babbling. It is a motley crowd made up largely of representatives of the Negro working class—laborers, longshoremen, porters, elevator operators, washerwomen, ladies' maids, waitresses, etc. There is an occasional school-teacher and student.* KATE *stands in front of the desk attempting to quiet the crowd, while* MELISSA *and* BELL *go into the hall)*

*Kate:* Have seats, my good people, Brother Jeremiah will be right with us. *(After much confusion they are all seated.* KATE *goes to the door on the left and knocks.* JEREMIAH's *voice is heard in a thunderous "Yes!"* KATE *replies)* Brother Jeremiah, there are many good people here to see you.

*(Every eye is now turned toward the door. It opens slowly.* JEREMIAH *enters clothed in a purple robe lined with red and trimmed with imitation ermine. The entire crowd arises. Some start toward him crying greetings.* JEREMIAH *both stops and silences them with a majestic gesture)*

*Jeremiah:* Good morning, my children, good morning. Your fresh smiling faces inspire me. I'm indeed glad to see you all. It is of such stuff as you that our empire will be built. I can tell it by looking into your eyes. Among you are many of our future leaders—leaders who will show black men and women of the world the way up into the light of real freedom. *(He lays his hand on the shoulder of a small, thin middle-aged man)* My brother, what's your name?

*Man:* Grover Washington.

*Jeremiah:* And what do you do?

*Grover Washington:* I'm a Pullman porter, sir.

*Jeremiah:* Fine! You're just the sort of man we need. Indeed, yes sir, your place is right here with us. We have much work for a man of your training. In Africa we will be able to employ all your talent and experience. You and others like you will have to take charge of our railroads. You are wise to get in early. I'm indeed glad to make your acquaintance. *(He passes on to a rather tall emaciated woman before the man had quite recovered from his surprise and embarrassment)* And you, sister? What's your name? What do you do?

*First Woman: (speaking with nervous rapidity)* Molly Henderson! I'm a dressmaker. I don't have no shop. I just works at home. And—

*Jeremiah: (interrupting and drowning her out with his booming voice)* Good! You, too, will find a high place in our realm. Who knows, but you might become the maker of fashions for the whole black empire. I can see all the world envying us your creations and copying them. Black men and women have followed too long. Now we must lead. *(He turns to a rather prim, bespectacled young woman)* And you, sister?

*Second Woman: (eagerly)* I'm a school-teacher.

*Jeremiah:* Molder of the young! Trainer of minds! Inspired by our new race consciousness, you will be able to lead black youth to great mental heights. We're going to establish schools all over Africa. Yes, sir, we can't have too many school-teachers. *(He passes on to a tall, husky Negro in rough working clothes)* And you, brother?

*Second Man: (flashing a big smile which reveals an even row of large white teeth)* I'm a hod-carrier!

*Jeremiah: (a bit confidentially)* A hod-carrier! You are a most essential person. You have learned, no doubt, all the ins and outs of contracting and building. You can help us immeasurably. We have before us the great task of planning and building the capital city of the Black Empire of Africa. *(He raises his hands and looks dreamily up at the ceiling)* I can see it now rising before me, more beautiful than any other city in the whole world. *(The expression on his audience's faces is as if they, too, can see it. He now turns to a very black man in the uniform of a subway porter)* My friend, what is your vocation?

*Third Man: (flabbergasted)* Huh?

*Jeremiah:* What do you do for a living?

*Third Man:* I work in de subway!

*Jeremiah:* Well, there's a great place for you, too, in the promised land. Just keep your eyes open and watch the workings of things. It won't be long now before we'll give you a chance to use your knowledge. *(He addresses the crowd)* All of you, no matter how humble your position in the white man's world, can reach great heights in your own world, once you create it. Your period of slavery and semi-slavery here need not go for naught. You are all seeing, hearing, and learning things—valuable things. The bank messenger who keeps his ears and eyes open can get a keen insight into the inner workings of finance. A business man's valet can learn much by intently watching his employer. We will know which of the white man's tools to accept and which to reject. We can use the white man's knowledge as a foundation for a new knowledge of our own.

*(By this time he has reached the desk on the left and taken his position behind it.* KATE *sits down on one of the chairs at the side of the desk, and the crowd, taking their cue from her, resume their seats.* JEREMIAH *takes a drink from a glass of water on his desk, thus giving the crowd a chance to settle into seated quiet. As he is about to begin again, a well-dressed, handsome mulatto of about thirty-five appears in the middle of the open double door at the rear.* STANLEY GRAYSON *is a shyster lawyer. He regards himself as an intellectual and is very scornful of his own people. He regrets having to live among them and his one overwhelming desire is to get a stake, and emigrate to Europe where he can live like a gentleman and escape being forced to associate with what he terms "niggers." He is extremely handsome and might pass for an aristocratic Latin of a very swarthy skin. His dress is exceptionally smart. He is obsequious towards those whom he thinks he can get to serve his purposes in this manner. On the other hand, to dispose of those who oppose him, he is ready to stoop to most any craftiness.* JEREMIAH, *seeing* GRAYSON, *smiles a welcome, and beckons him to the empty chair at his side)*

*Jeremiah:* Good morning, Brother Grayson. Come and sit down. *(*GRAYSON *goes to the chair on the opposite side of the desk from* KATE. *The crowd follows him with their eyes.* GRAYSON *sits down with an air of importance.* JEREMIAH *helps himself to another sip of water.* KATE *looks daggers at* GRAYSON *and he returns the look with interest.* JEREMIAH *clears his throat and begins)* My, but to have you all here gathered about me makes me feel good, as good as Christ must have felt when His disciples gathered 'round Him. It buoys me up! It makes me feel confident that "The International Fraternity of the Native Sons and Daughters of Africa" will attain magnificent heights. It is highly necessary that we all work together; that we cooperate with one another. Only thus can we establish a black brotherhood and sisterhood—a sturdy phalanx of Africa's native sons and daughters. The black man must be free. No more must white men place a yoke around his neck; no more must they apply the lash, or resort to the rope, yeah—even to the flames as a means of keeping him submissive, of retaining him in slavery. The black man must be free. The British have their Isles; the Americans their United States; the Frenchmen their France; the Mexicans their Mexico; the Japanese their Japan, even the Chinese have their China; but what have we? No fatherland; no flag; no nation; no government! Is it any wonder that the white man takes advantage of us; any wonder that he makes our children eat the bitter fruits of race discrimination and race prejudice? No, I answer, No! We should hang our heads in shame! We should be ashamed to bow before our Maker in prayer! I tell you the black man must be free! It is God's Will! And since God in His mercy has led us to the gates of the royal road to freedom, I dare not—you dare not—draw back. Africa is for the Africans and the native sons and daughters of Africa shall return to their native shores. You must gird up your loins, my children; invest your souls with the Holy Spirit; pray unto your Father; and work, yes, work, as you've never worked before, work for the Black Internationale! *(He has reached a point of elegant frenzy, but now his voice suddenly becomes hushed)* Have no fear, my children, for God is with us. I have seen Him in my dreams. It is He Who is guiding us. It is He Who is showing us the way. *(He clasps his hands ecstatically over his head and prays)* Our Father, we are following your guiding star. You Who are infinitely merciful, cheer us, strengthen us, help us in our journey back to the promised land. *(There is an ecstatic chorus of such phrases as "Amen!" "Yes, Brother!")* Oh, Father, we have suffered without a complaint! We have endured great hardships and never lost faith! No, Father, we have never lost faith, not even when the lash, the rope, and the flame have been turned against us. And now we hearken to Your call—Your call to the black men and women of the world to return to Ethiopia, that it may rise again in all its glory. We are ready, Father. We are ready. Our cry henceforth is, "On to Ethiopia! On to Ethiopia!" *(There is another ecstatic chorus of the phrases, "On to Ethiopia!" "Yes, Brother!" "On to Ethiopia!" "Amen!" "Amen!"* JEREMIAH *bows his head for a moment as if in silent prayer. There is a whispering among the crowd.* JEREMIAH *raises his head. The ecstatic preacher is gone. He is once more the smiling politician. Raising his hand, he clears his throat, and again gets everybody's attention)* Those who are already members don't leave here without at least a dozen membership blanks. And you who have not already enrolled, do so at once! Brother Bell is outside to serve you.

*Kate: (jumping up and facing the crowd as it begins to rise and move towards the hall)* Brothers and sisters, it does me good, too, to see how you rally round our beloved leader. As his helpmate, his inspiration, I—

*Grayson: (hurriedly rising and interrupting)* That's right, sister, that's right. We all know how you feel. *(He turns to the crowd)* Now, my good folks, Mrs. Saunders will help Brother Bell outside and answer any questions you might wish to ask. *(*GRAYSON *skillfully leads the crowd into the hall.* KATE *glares at him)*

*Jeremiah: (to* KATE*)* Kate, be sure everybody gets a pamphlet!

*Kate: (surlily)* All right!

*(She takes a bunch of pamphlets from the table on the left and begins distributing them among the vanishing crowd. A few of the more daring of the crowd come forward and shake* JEREMIAH*'s hand)*

*Jeremiah: (as he shakes various hands)* Come again, brother, and bring all your friends. Yes, sir, it was good to see you here. We have a great work ahead of us. We need every bit of help we can get. Thank you, sister, for coming.

*(At last the whole crowd are in the hallway, and* KATE *has joined* BELL *and* MELISSA *at the table.* GRAYSON *closes the door and comes down to the desk where* JEREMIAH *is now sitting)*

*Grayson: (angrily)* Mrs. Saunders will miss her train if she doesn't hurry.

*Jeremiah:* She says she's not going.

*Grayson:* Not going! What do you mean not going? Why, all the arrangements are made. We've gotten her reservation and the reception committee in Atlanta has— *(He makes an impatient gesture)* Oh, the devil! We can't give in to a woman's whims.

*Jeremiah:* She sort'a feels as if she's being pushed aside and she kind'a resents it. You know Kate.

*Grayson:* Yes, I know Kate. But Kate— *(He throws up his arms in despair)* Haven't we fought all this out before?

*Jeremiah: (helplessly)* I thought it was settled.

*Grayson:* Settled! Nothing will ever be settled as long as we let her have the last word. Who's boss around here anyway, you or Kate?

*Jeremiah:* Oh, it isn't a matter of that.

*Grayson:* It is a matter of that! Can't you see from her every action—why, the woman's a born striver.

*Jeremiah:* You must remember, Brother Grayson, that Kate and I have been married for fifteen years. In a way, I owe everything to her. I was just a down and out country preacher when she married me.

*Grayson:* I know all that, and it has little to do with the present situation. You say she made you what you are. Could she have done it if you hadn't had the makings in you?

*Jeremiah:* *(thoughtfully)* No, I guess she couldn't.

*Grayson:* Now listen, this sort of thing happens to all great men. They marry women early in life who at the moment have advanced a trifle farther than they themselves. Such a woman often helps a man on his way up. But once he gets there, she is liable to become jealous. She's been superior to her husband so long that she doesn't ever want to admit his superiority, and she is always attempting to show her own in every possible way. Mrs. Saunders' ill-timed speech of just a minute ago is a perfect example of this sort of thing. If I hadn't interfered, she might have been talking yet. Anything to put you out of the limelight and get into it herself!

*Jeremiah:* You mean to say that Kate is jealous of me?

*Grayson:* It amounts to that. You know how she's always telling everyone that if it hadn't been for her finding that Ku Klux pamphlet pleading for a pure white America, you would never have had the idea of striving for a pure black Africa.

*Jeremiah:* *(worriedly)* Kate shouldn't be like that!

*Grayson:* *(determinedly following up this last sign of success)* No! She shouldn't be! But she is! And another thing! Your slogan is "Africa for the Blacks" and you're always preaching purity of race. What do you think your enemies and even your followers are saying about you having a mulatto wife?

*Jeremiah:* *(apprehensively)* Do you think they notice that?

*Grayson:* People notice everything, especially in a great public figure. Everyone is on the lookout for the slightest false declaration or action; your friends as well as your enemies.

*Jeremiah:* You really think there is some feeling around about—eh—about Kate's color?

*Grayson:* Of course! But that's not all. As leader of this great movement; as head of the International Fraternity; and as Emperor of the coming Black Empire of Africa, you should have a son to carry on your name and work.

*Jeremiah:* *(his imagination playing on this idea)* A son! *(He spreads his arms wide)* Jeremiah, the second!

*Grayson:* Yes, Jeremiah the second—a big, healthy black boy, bred of the purest and best in the race. Jeremiah the second—a fitting successor to his father, First Emperor of Africa.

*Jeremiah: (passing his hands before his eyes in a gesture of disillusion)* Huh! Dreams! I have no son, and Kate's already—

*Grayson: (interrupting)* Too old! You see it now. All these years and no child. She's been too ambitious for herself to let a child interfere. The future of the Saunders name meant nothing to her as long as there was hope that her own name would be before the public eye.

*Jeremiah: (thoughtfully walking up and down)* I should have a son. You're right. I should have a son. *(He stops, draws himself up to his full height, and forcefully drives his right fist into the palm of his left hand)* I *must* have a son!

*(He continues walking back and forth.* GRAYSON *smiles cynically and looks over the mail on the desk. He watches* JEREMIAH *out of the corner of his eye)*

*Grayson: (carelessly)* I see the money is still coming in!

*Jeremiah:* It's wonderful! And it's only the beginning. God has tested us for years in the fires of adversity and having found us strong and worthy is now at last smiling upon us.

*Grayson:* It's a good thing we've got Melissa here to keep books or we'd be getting into trouble again with the postal authorities. I never want to be mixed up in another mess like that last one. *(He picks up one of* MELISSA*'s account books)* She's a smart girl. Compare this book to the one Mrs. Saunders kept. Nobody, not even she, herself, could make head or tails out of it.

*Jeremiah: (fishing the postal inspector's card out of his pocket and handing it to* GRAYSON*)* Oh, I almost forgot.

*Grayson: (taking the card)* What's that?

*Jeremiah:* The postal inspector was here again.

*Grayson: (surprised)* No!

*Jeremiah:* Yes! He said they'd received some more complaints about us. He wants you to get in touch with him immediately. I hope it's nothing serious.

*Grayson: (carelessly)* Probably the old charge—using the mails to defraud!

*Jeremiah:* I don't see how anyone can say that. The Lord Himself knows this movement is dedicated to His Holy Name and obeys both God's law and man's.

*Grayson:* There's no need to worry. Just turn all the books and everything over to me. I'll straighten it all out.

*Jeremiah: (fervently)* God keep and bless you, Brother. I oft-times wonder why the Lord didn't direct me to you sooner. We could have avoided so many of our earlier mistakes. *(He clasps* GRAYSON*'s hand)*

*Grayson:* I consider it a great honor and privilege to serve you.

*Jeremiah: (with profound conviction)* Of such as you, Brother Grayson, is the Kingdom of Heaven.

*(There is a long moment of silence. Then* JEREMIAH *resumes his pacing and* GRAYSON *goes back to scanning the mail)*

*Grayson:* (*suddenly examining his watch*) Mrs. Saunders' train leaves in an hour. A delegation of sisters have arranged to see her off. It'll be a bit embarrassing if she doesn't go. There'll have to be explanations. And we'll have to wire Atlanta to call off arrangements for her reception there. (*He watches* JEREMIAH *closely as he talks*)

*Jeremiah:* (*suddenly halting in his march and speaking with great determination*) She will go! She must go!

*Grayson:* Shall I tell her that you want to see her?

*Jeremiah:* Yes, please, Brother Grayson. I'll speak to her now—right now!

(GRAYSON *goes out quickly at the rear.* JEREMIAH *stands before the pier mirror, arranging himself for the battle he senses coming*)

*Kate:* (*entering and closing the door behind her*) You wanted to see me, Jeremiah?

*Jeremiah:* (*turning towards her*) Yes, Kate. (*pauses*) It's about your going South! You've made a great mistake in deciding not to go!

*Kate:* (*firmly*) You need me here. I won't be shunted off. Just because Grayson and that little black hussy Melissa—

*Jeremiah:* (*interrupting harshly*) Kate!

*Kate:* Yes, I mean it. Everything was all right until she came.

*Jeremiah:* Oh, no it wasn't. We wouldn't have had all this trouble with the postal inspectors if she'd 'a been here in the beginning. She *can* keep books. It's not a matter of Melissa or Grayson, Kate, it's a matter of the movement.

*Kate:* (*surprised*) The movement?

*Jeremiah:* Yes, the movement, Kate. Can't you see we mustn't let our personal feelings enter into a thing like this?

*Kate:* You know it's not personal, Jeremiah. It's not personal with me. Who have I thought of first during these fifteen years, myself? No, always you! And it's you I'm thinking of now.

*Jeremiah:* I know that, Kate, and that's why I insist on your going to Atlanta. The business of our movement is bigger and more important even than me.

*Kate:* I feel that my place is here.

*Jeremiah:* Your work here, Kate, is never done; that we all know. But you can do more for us here right now by leaving than by staying.

*Kate:* Yes, I can do more for those who are trying to undermine me with you, and take the power into their own hands.

*Jeremiah:* Tut, tut—Kate! You don't know what you're saying. As a child of God, you shouldn't say such things.

*Kate:* May the Lord forgive me, Jeremiah, but—

*Jeremiah:* (*interrupting passionately*) As the Lord is my judge, these are my honest, heartfelt opinions. We're sending you to Atlanta in Georgia, the very center of the South's lynch belt. You're going into the enemies' land, where our people need our message most. It's the Southern Negroes who will be the most eager to flee these alien shores and take up their abode in their native land. Like the angels who visited Lot, you must carry to them the Heaven-sent call to flee unless destruction fall upon them.

*Kate:* I can't leave you here alone, Jeremiah! I need you and you need me!

*Jeremiah:* We cannot think of ourselves now. The movement stands at the crossroads. The South must be won. *(dramatically)* And who among our leaders can win it? *(He pauses)* You, Kate! Only you!

*Kate: (carried away by his earnest eloquence)* Jeremiah, you really think that I alone am fit for this?

*Jeremiah:* Yes! And with the Lord's aid and our prayers, your mission will be a success. You shall mobilize a black army millions strong which at a word from me will march to the sea and sail for their native land.

*Kate: (grabbing* JEREMIAH *and speaking ecstatically)* I'll go, Jeremiah, I'll go! The Lord has called! His servant must not be deaf or stubborn! My people back South—they do need us! God, how they need us! Jeremiah, forgive me for faltering. *(She raises her hands over her head and gazes upwards)* Oh, Father above, forgive me for listening to the devil's sly whisperings! Forgive me! Forgive me!

*(She rushes into* JEREMIAH's *arms. They kiss. Then, sobbing, she suddenly breaks away and goes hurriedly off at the left.* JEREMIAH *sits at the desk, wipes the perspiration off his face, leans back in his chair, meditates for a moment, and pushes the buzzer on the desk.* MELISSA *enters from the rear)*

*Melissa:* Yes, Brother Jeremiah.

*Jeremiah:* Is Brother Grayson still here?

*Melissa:* Yes, he is. Want him?

*Jeremiah:* Please send him in.

*Melissa:* Al'right!

*(She goes out at the rear and* GRAYSON *immediately reenters)*

*Grayson:* Well?

*Jeremiah:* She's leaving!

*Grayson:* Fine! Now—

*(He is interrupted by* KATE *who opens the door on the left)*

*Kate:* Jeremiah!

*Jeremiah:* Yes, Kate.

*Kate:* Come here, please. I'd like to talk to you while I pack.

*Jeremiah: (rising)* All right! I'll be right in. Pardon me, Brother Grayson, won't you?

*Grayson:* That's all right!

*(*JEREMIAH *goes out at the left.* GRAYSON *lights a cigar and struts up and down the room for a moment. He is evidently very pleased at the way things are going.* MELISSA *enters. She looks around carefully to see if* GRAYSON *is alone)*

*Melissa:* Where's the jackleg?

*Grayson:* Out back with the old lady.

*Melissa:* What's the dope?

*Grayson:* She's going.

*Melissa:* I wish 'twas me. I envy the old dear.

*Grayson:* What do you mean?

*Melissa:* I mean—well, you know what I mean. You aren't so dumb. I'm damn tired of bein' 'round here.

*Grayson:* Don't be childish.

*Melissa:* Why not? It seems the only thing to be 'round here. Next, I suppose I'll be dressing in purple robes and wearin' a crown of feathers.

*Grayson:* That'd be all right. Then we'd call you princess of the realm.

*Melissa:* It might be al'right with some folk, but it sure ain't forty with me. I'm so sick of all these dumb bozos a'gazin' upon the Reverend Jeremiah like he was Jesus Christ, Himself, arisen from the dead.

*Grayson:* That's all part of the game.

*Melissa:* And it's a damn lousy game, too.

*Grayson:* Why? Getting softhearted?

*Melissa:* Jesus, no! But it sort'a gets my goat seein' these poor old lame-back washer-women rushin' in here to give Jeremiah their pennies so's they can go back to Africa and be ladies of the court.

*Grayson:* That's still part of the game!

*Melissa:* (disgustedly) Sure! It's all right for you who only run in the office now and then, but I'm here every day all day. I didn't know there were so many suckers. All kinds of 'em! Dumb, smart, black, brown, all falling for this moonshine Brother Jeremiah gives them! I'm sick of the whole rotten business!

*Grayson:* (consolingly) I know, kid, but you've got to stick. The old lady's leaving and you know what that means.

*Melissa:* I'll say I do. It means I'll have to be patted and petted by that big black strutter until he gets up enough nerve to ask me to be the Queen of Africa.

*Grayson:* (earnestly) You've got to play him jam up, while the old lady's away. The faster you work, the sooner we both can clean out. Get me?

*Melissa:* (sarcastically) The pleasure's all mine. (She sighs) I suppose 'Lissa 'll have to do her share. (She cuddles up to him) Won't she, honey?

*Grayson:* You're my idea of a swell little partner.

*Melissa:* (kittenishly) Is that all I am?

*Grayson:* You're one grand little girl. (He kisses her) I'm crazy 'bout you.

*Melissa:* Maybe I ain't crazy about you! (She holds up her lips provokingly. He kisses her again)

*Grayson:* (looking at his watch) Getting late! I've got to go to the Post Office and get a line on these new charges.

*Melissa:* Think they're serious?

*Grayson:* Can't tell! Everything depends on the interpretation of the law, and it's never interpreted in the same way twice. If it's serious, I can probably keep things delayed until we've got ours. Everything'll be easier now with the old lady out of the way. And don't forget, the courts of justice are awfully kind to us lawyers. Bye-bye honey.

*Melissa: (invitingly)* I'll be at the house tonight.

*Grayson: (winking)* I'll be over at ten.

*Melissa:* And I'll be waiting.

*Grayson:* Ta! Ta!

*(He goes out at the rear.* MELISSA *goes to the desk and pushes the buzzer.* BELL *enters from the rear)*

*Bell:* Did you buzz?

*Melissa:* Sh—sh!

*Bell:* What's the idea?

*Melissa:* Thought you might want a rest from that crowd outside.

*Bell:* Kind of you, but as I've told you before, I enjoy talking to these folks that come in here looking for information. They're like little children, eager yet skeptical. They've all been cheated and mistreated so often that it's a wonder they've any faith at all left in them. One lady this morning told me how her brother was sent to jail and almost lynched for a crime that was committed by a white man. *(He clenches his fists and says passionately)* Oh, when I think of all the years the black man has wasted serving the white man 'stead of serving himself, my very blood boils.

*Melissa: (calmly)* Say, you ever been in love?

*Bell: (fervently)* Always!

*Melissa: (skeptically)* Oh, no!

*Bell:* Yes, in love with my people! And no other love can come into my life until I've seen them safely on the way to real freedom and real happiness.

*(Meanwhile* JEREMIAH *has entered without being seen by the other two)*

*Jeremiah: (booming and beaming)* Well spoken, my son, well spoken!

*Bell: (flustered)* Thanks.

*Melissa: (to* JEREMIAH*)* Are you going to dictate this morning?

*Jeremiah:* Not now, my child. I want to talk with Brother Bell awhile. And we'll have to get Kate off first.

*Melissa:* All right! Buzz me if you need me. *(She goes out at the rear)*

*Jeremiah: (taking his seat at the desk.)* Fine girl, fine girl! God has been very good to me. I couldn't get along without two such devoted helpmates as you and Melissa.

*Bell:* It's you who inspires us. You mustn't forget that. It's you, your personality and presence, that spurs us all on.

*Jeremiah:* I am only an humble servant of the Master trying to work out my people's salvation.

*Bell:* And you're doing it, too! Why, in just a short time you've done more than all the other so-called-Negro race leaders have ever done.

*Jeremiah:* They, my son, were selfish and self-centered. They let their passions rather than God lead them. They wasted their time tilting at windmills. They tried to fight the white man in his own land. They never thought of getting a land of their own; never considered returning to the land of their forefathers.

*Bell:* When I was in school I studied the teachings of them all: Booker T. with his back to the soil and practical education chimera; Du Bois with his social equality bug; and dozens of others. All of them gave me hope, and then let me fall. You alone have made me feel that at last victory is in sight.

*Jeremiah:* And I'm going to make all the black men in the world feel that same way! The ocean will be littered with ships carrying black men back home!

*Bell:* (*exaltedly*) And you, Jeremiah, will rule as the Nubian kings of old!

(**MELISSA** *enters from the rear*)

*Melissa:* Sister Tyson and Brother Baker are here. Do you wish to see them?

*Jeremiah:* Yes, Melissa, send them right in.(*She goes out and* **BELL** *starts to follow.* **JEREMIAH** *stops him*) You'd better stay. They've come to report about the newspaper and the hall.

*Bell:* All right.

(**SISTER TYSON** *and* **BROTHER BAKER** *enter.* **SISTER TYSON** *is a shrewish busybody, a lady of much activity and self-importance.* **BROTHER BAKER** *is a natural, bustling go-getter. Both have dark skins*)

*Jeremiah:* Good morning, my children, just wondering when you were coming in.

(*He arises. They approach the desk and shake his hand*)

*Sister Tyson:* It's my fault we're late. (*a bit sarcastically as she turns to Baker*) Guess I better tell you that 'fore Brother Baker here does.

*Brother Baker:* If I'd'a had to wait ten minutes longer, I was gonna swear off bein' on committees with women.

*Jeremiah:* You can't do without them, Brother Baker. Adam couldn't.

*Brother Baker:* (*smiling broadly*) But Ham did. The Bible says so, and aren't we black folks supposed to be the descendents of Ham? (*He laughs loudly at his own wit*)

*Jeremiah:* (*sternly*) There, there! There is mighty and pressing business at hand. What's the good report you have for me? I know it's good or else you wouldn't be so cheerful.

*Sister Tyson:* (*talking hurriedly so as to get her say in ahead of Baker*) Well, first of all, we came to an agreement on the price for the hall. We had a lot of trouble, though. It's right next to a Baptist Church, and the minister objected to its being sold to our organization.

*Jeremiah:* And they call themselves Christians!

*Bell:* They're just jealous of us, that's all.

*Brother Baker:* (*interrupting jovially*) Well, our money talked louder than that black preacher, and we can get the hall for a thousand dollars cheaper than the advertised price, too!

*Sister Tyson:* My husband will take over the contract to remodel the building. That will save us some more money. There won't be any whites to skin big profits off this job.

*Jeremiah:* (*rising and speaking fervently*) We'll call it Freedmen's Hall. It will be the black man's cradle of liberty. Merciful Father, we praise Thee!

*Sister Tyson:* (*devoutly*) Amen! Amen!

*Bell:* What about the newspaper?

*Brother Baker:* We can buy out that West Indian paper, presses and all, and cheap as dirt, too. They're sold on our organization. All they ask is that we may keep the present editor and work as hard in carrying our message to the West Indian black as to the American black man.

*Sister Tyson:* I told them our message was for black men everywhere.

*Jeremiah:* Right, sister, right. My, I'm happy this morning. God alone knows how happy. My prayers have been answered. Africa shall be free. The black man shall be free. Freedmen's Hall! Ah, inspired tongues will teach the gospel of liberty there. And in *The Black Man's International News* we will carry the burning word all over the world. Let's give thanks to the Lord. Let's all kneel and have a moment of silent communion with our Maker. (*They all kneel and raise their eyes heavenward. Their faces are rapt, as their lips move silently. After a moment,* JEREMIAH *gets to his feet. The rest follow suit*) There, my children, nothing is more refreshing than a moment of silent communion with Your Maker. Keep up the good work. I depend on you. God has his eyes on you.

*Sister Tyson:* You don't have to worry 'bout us.

*Brother Baker:* We're for you, Brother, for you and the movement.

*Jeremiah:* Now, you'd both better go over to Brother Grayson's office and give him all the details. We want to be sure that everything's legal before we go ahead.

*Brother Baker:* (*starting to go*) All right, Brother Jeremiah, just as you say.

*Sister Tyson:* (*sulkily*) I'm sure everything is all right. I know honest men when I'm dealing with them.

*Jeremiah:* But everything must be according to the letter of the law, and it takes a lawyer to know what it is.

*Brother Baker:* (*at rear door*) Come on, Sister. Good morning, Brother Jeremiah. Good morning, Brother Bell.

(SISTER TYSON *follows* BROTHER BAKER)

*Bell:* Good morning.

*Jeremiah:* Good morning, my children. (*They go out*)

*Bell:* (*exultantly*) There's no stopping us. I see my people free at last. Thanks to you, Brother Jeremiah, thanks to you!

*Jeremiah:* (*patting* BELL *on the back*) You'll hold an important position in the Black Empire. I wouldn't be surprised if some day you were my Grand Vizier.

(MELISSA *enters from the rear. There is the rumble of many voices in the hall*)

*Melissa:* There's a delegation of longshoremen here to see you. Shall I show them in?

*Jeremiah:* I must see how Kate's getting along. It's about time for her to leave. Ask them to wait awhile. Brother Bell will help take care of them.

*Melissa:* Al'right!

*(She goes out followed by* BELL. JEREMIAH *goes to the door on the left and opens it)*

*Jeremiah: (looking at watch)* Ready? Your time's about up.

*Kate: (offstage)* Coming now, dear.

*(*JEREMIAH *walks back to the center of the stage.* KATE *comes rushing in with hat and coat on and carrying two bags)*

*Kate: (setting the bags down, and looking up at him proudly)* Wasn't that getting ready to go in a hurry?

*Jeremiah:* Fine, dear. You've just time enough to make your train without rushing.

*Kate: (suddenly becoming somber)* Jeremiah, I hope you don't think it disloyal, but deep down in my heart, I really don't want to go.

*Jeremiah: (brusquely)* I thought that all settled!

*Kate: (sighing)* It is. But something keeps telling me my place is here.

*Jeremiah:* Nonsense! Your place is wherever the movement needs you most, and that place now is in the South.

*Kate:* I know you're right, Jeremiah. But it's hard for me to begin thinking that my first duty is to the movement and not to you.

*Jeremiah: (dramatically)* I am the movement!

*Kate: (She throws her arms about his neck)* My Jeremiah! Lord and Potentate over all Africa! That's my dream for you, and I'll work my nails off to make it come true. I'm ready to sacrifice everything for you, because I love you; and because through you and me, our people are going to find salvation and freedom. *(They kiss)* Good-bye, Jeremiah. Keep your faith in the One above Who has been so good to us in the past. Keep faith in me, too, Jeremiah. Remember, I'm always working, praying, working.

*Jeremiah:* My prayers follow you. *(They kiss again. As they draw apart,* JEREMIAH *goes to the desk and pushes the buzzer)* Brother Bell will see you to a taxi. *(*BELL *enters)* Is the taxi waiting?

*Bell:* Yes, Brother Jeremiah, and there's a delegation here to see Mrs. Saunders off.

*(He goes to the double door, opens it, and in walks a delegation of six women, wearing sash badges on which are printed the letters I.F.N.S.D.A., the initials of the "International Fraternity of the Native Sons and Daughters of Africa." Behind the women in the hall are seen the waiting long-shoremen)*

*Jeremiah: (standing in the center of the room with his arms around* KATE*)* My children, I thank you, for this testimonial of your respect and admiration for one with whom our organization always comes first. Sister Saunders here is going on a long and difficult journey. She will have hard work to do! She will have to face danger! But she is a soldier in the army of the Lord, and knows no fear, for He, *(*JEREMIAH *points upward)* our Heavenly Father, is always at her side. And she is also a soldier in that great army of black women who side by side with their men folk are going forth to fight—yes, to fight with the fury of a Heavenly host to reclaim their native land. Speed her on her way, Sisters. Pray for her! *(KATE has been sobbing all through this speech. JEREMIAH kisses her once more)* Good-bye, Kate.

*Kate:* Good-bye, Jeremiah.

*(Two of the delegation pick up her bags. Then at a sign from BELL, who is standing in the doorway, they all begin to sing)*

> Onward Christian Soldiers
> Marching as to War
> With the Cross of Jesus
> Going on before.

*(As they march out singing, KATE follows, head erect and step sure. BELL closes the door after them. Their voices can be heard for some time. After they have died out, JEREMIAH buzzes for MELISSA. She enters. The noise of the longshoremen's voices is again heard)*

*Melissa:* Want me?

*Jeremiah: (nodding towards the left)* I'm going inside for a moment to get a cup of coffee. It's been a strenuous morning. Tell that delegation of longshoremen I'll see them in a little while. And after they're gone, I'll want to dictate some letters.

*Melissa:* I hope you won't be long. They're a bad gang to keep quiet.

*Jeremiah:* No, I won't be long.

*(He struts proudly off at the left. Just as he exits BELL enters from the rear to the accompaniment of the buzz of husky voices)*

*Melissa:* He's gone to get some coffee.

*Bell:* He's had a pretty hard morning of it.

*Melissa:* And the worst is yet to come.

*Bell:* He doesn't mind. He can stand it. He could stand anything for the movement.

*Melissa:* And you?

*Bell:* I'd gladly give my life for it.

*Melissa:* Well, the old lady is gone to the real battlefront. If she gets out without the crackers scaring hell out of her, she'll be a lucky woman.

*Bell:* They probably won't bother a woman much. But they'd sure raise hell with any man attempting to organize their "niggers." And as for me going, I could never again stand the Southerners' attitude. Why, I'd get hotheaded and do something reckless and be lynched inside of a week.

*Melissa:* You're sure sour on white folks, aren't you?

*Bell:* Damn right! I hate their very guts! There's nothing any white man can do that I'd like!

*Melissa:* Seems to me you're as foolish and prejudiced as any Southern cracker.

*Bell: (belligerently)* I've good reason to be.

*Melissa:* Let's talk about somethin' pleasant. I've heard nothin' but movement all mornin'.

*Bell:* There's nothing else I can talk about. There's nothing else I can think about. I dream, eat, live for the movement! It's going to make my people free.

*Melissa: (looks at him out of the corner of her eye, then saunters off at the rear singing)*

> Go down, Moses,
> Way down in Egypt land.
> Tell old Pharaoh,
> To let my people go!

*(JEREMIAH enters form the left. He goes over to the mirror and preens himself)*

*Jeremiah:* You'd better send in that delegation of longshoremen. They'll be tired of waiting soon.

*Bell:* They're a little fidgety already!

*Jeremiah:* Show them in! *(JEREMIAH assumes a majestic pose at the desk. BELL throws open the rear door and in walk four big stalwart Negroes, very black, and very powerful. BELL closes the door and stands in the rear. JEREMIAH rises as they enter. The longshoremen all advance a few steps toward the center of the room and bow to him)* Greetings, gentlemen, you don't know how sorry I am to have kept you waiting.

*Spokesman: (who carries a long, paper-wrapped bundle under his arm)* We knew you was busy, sir.

*Jeremiah:* Draw chairs up near to me so we can have a nice intimate chat.

*Spokesman: (haltingly)* We will, sir, but first—first, sir, we has a little s'prise fo' you.

*Jeremiah:* Surprise for me? My, my, my! *(He beams at them)* You folks'll spoil me yet!

*Spokesman: (pulls paper off bundle revealing some red cloth)* Take dis end, Brother Spikes. *(One of the men steps forward and takes one end of the cloth)* Now walk 'cross de room.

*(As he does so, the cloth is unfurled. It is a blood-red flag with a big black cross in the center. They hold this in front of JEREMIAH, who stands rapturously gazing at it)*

*Jeremiah: (ecstatically)* Brethren!

*Spokesman:* It's our flag, sir! De banner to wave from de mast of our first ship!

*(They all smile)*

*Jeremiah:* My children! *(raises his hands and face heavenward)* My Father Who are in Heaven, hear your humble servant's thanks. *(His eyes return to the flag and the delegation)* I thank you, you who are to man our first ship. Big, stalwart black men! The pride and backbone of a rising race!

*Spokesman:* We wants you to keep dis hangin' in yo' office 'till we gets our ship.

*Jeremiah:* I shall be proud to do so. It shall hang there in that space behind me; hang there until it flies from the mast of the ship which will soon be lying in the docks waiting for the first cargo of our native sons and daughters on their way back to Africa.

*Spokesman: (laying the flag over the table on the left)* Is we gonna' have dat ship soon?

*Jeremiah:* Soon as we get the money.

*Spokesman:* We wants to know how long we gonna have to learn de ways o' ships?

*Jeremiah:* Don't worry, good brethren, God will be with us. He will pilot us. Didn't he command the seas to obey Moses? So shall he aid Jeremiah, flying with his people to the land of Ethiopia. *(He addresses* BELL*)* Brother Bell, see that this flag is hung immediately.

*Bell:* Yes, sir!

*Jeremiah: (turning back to the longshoremen)* I'll have some definite news about a ship soon. And, brethren, let me thank you once more. Let me shake every hand. Let me feel the force in your muscle; the muscle and brawn that's going to make the black man's Empire in Africa the greatest in the world. *(As he talks he shakes all their hands and skillfully maneuvers them to the rear door which* BELL *opens)* Good-day, children! Good-day!

*Chorus:* Good-day, Brother Jeremiah.

*Jeremiah:* May God bless you all. *(They are escorted out by* BELL *who closes the door after him.* JEREMIAH *starts around the room, his robes trailing behind him. He is a majestic figure, and in his present moment of exaltation, he truly looks like a monarch. He finally comes to a standstill before the mirror. He slowly begins to mumble to himself. Finally his voice gets louder and his words are more distinct. He acts as if he is addressing a great gathering, striking what he considers grand and royal poses)* Africa with its jungles, diamonds, and ivory! Africa the promised land for the black man! The Jews had their Moses, the Italians their Caesar, the French their Joan of Arc, and the Americans their George Washington. The Hindoos have their Gandhi, the Russians their Lenin, and now the black man has Jeremiah. *(He brings his hand forcibly against his chest)* God's chosen on fire with God's message!

*(*MELISSA *enters and stands listening for a moment)*

*Melissa:* Speechifyin'!

*Jeremiah: (startled)* Oh, Melissa!

*Melissa:* Ready to dictate?

*Jeremiah:* I don't feel much like it. I feel too good to work. I'm full of the spirit. Did you see the flag?

*Melissa:* Flag?

*Jeremiah:* Yes! That delegation just presented me with this flag which is to fly from the mast of the first Black Cross liner.

*Melissa: (admiring the flag)* Gorgeous! This has surely been a glorious mornin' for you, hasn't it? *(She goes over to the desk and sits down on one of the straight-backed chairs)*

*Jeremiah: (following her)* All mornings are glorious to God's children.

*Melissa:* I must be a stepchild.

*Jeremiah:* God has no stepchildren, and you least of all, you, 'Lissa.

*Melissa:* Maybe I'm wrong. *(takes up her pad and pencil)* What about the dictation? There's all those letters to be answered.

*Jeremiah:* Letters! *(moves about impatiently)* I can't sit still and dictate letters this morning. Didn't I tell you I was too full of the spirit?

*Melissa:* But we shouldn't loaf just 'cause—just 'cause—*(She hesitates)*

*Jeremiah:* Just 'cause what?

*Melissa:* Just 'cause Mrs. Saunders gone.

*Jeremiah:* That's all the more reason why we should loaf.

*Melissa:* *(surprised)* You don't mean we ought to celebrate her leavin'?

*Jeremiah:* Why not? Hasn't she fared forth on a noble mission? Hasn't she gone away clothed in glory to garner more disciples? She is going to the well of plenty, and when she comes back her jug will be full of the sparkling waters of victory. It is a time to rejoice!

*Melissa:* *(laughing)* You win! You know, with that voice of yours, you could convince anyone of anything.

*Jeremiah:* *(looking down at her in mock accusation)* What does the Scriptures say about an oily tongue?

*Melissa:* But you haven't an oily tongue! I didn't mean that! What I meant was that there's so much power and force behind what you say. You're masterful.

*Jeremiah:* Tut, tut, child. Don't flatter me!

*Melissa:* I know better than to flatter you. You're too wise!

*Jeremiah:* You know, 'Lissa, I enjoy having you around. It will be good to take you to Africa with us.

*Melissa:* Africa?

*Jeremiah:* Yes, Africa, the new Africa, the new black Africa, where our people under my guidance will erect a great empire.

*Melissa:* What could I do in Africa?

*Jeremiah:* Do? My child, there'll be plenty for everybody to do. *(He walks around the desk and sits on the corner beside her chair)* In Africa the black man and the black woman will come into their own.

*Melissa:* And me?

*Jeremiah:* You, Melissa. *(He leans close to her)* You, like my other favorite workers here in America, will have a high place in the realm.

*Melissa:* *(poutingly)* Like all the others?

*Jeremiah:* *(laughing and placing his hands on her shoulder)* No! Not like all the others! *(He beams upon her)* How'd you like to be a princess?

*Melissa:* Princess?

*Jeremiah:* Yes, a princess. Melissa, Princess of Africa—the first princess of the land under Jeremiah, Emperor of Africa.

*Melissa:* *(rising and speaking coyly)* You'd really make me a princess?

*Jeremiah: (also standing and speaking earnestly)* Why, I'd make you— *(He catches himself)* I'd make you anything you wanted to be! I'd give you anything you asked for!

*Melissa:* I don't really want much.

*Jeremiah: (majestically)* Speak, and your wish shall be granted!

*Melissa:* Well, all I want is for you to like me—a little bit!

*Jeremiah: (his hands descend to her arms and he draws her a little closer)* I like you a whole lot!

*Melissa: (snuggling close and holding up her pursed lips)* Honest?

*Jeremiah: (huskily as he struggles with his passion)* Melissa!

*(He grabs her to him and kisses her fiercely as the curtain falls)*

## Act II

### Scene One

*It is a spring morning six months later. The setting is* JEREMIAH'*s new private office in Freedmen's Hall. It is luxuriantly furnished with heavy carpet, large dark leather covered chairs and divans, and a mahogany desk, bearing two ornate French phones. The walls are literally covered with flag-draped full length pictures of* JEREMIAH *in his various costumes of State; one in an emperor's robes; another in an admiral's regalia; another in a general's outfit; and another in religious habiliment worthy of a high priest. There are also numerous trophies, such as an African death mask; a bit of primitive sculpture; and a medicine man's head-dress. A very large detail map of Africa hangs in the center of the rear wall. It is chalked up with red and blue crayon. From the appearance of the office, it is obvious that extreme good fortune has been attending the progress of the "back-to-Africa" movement. There are two doors, one at the left leading to the general outer offices, and one at the right leading to the large reception and committee room.*

*When the curtain rises,* GRAYSON *is sitting at* JEREMIAH'*s desk, leisurely puffing a cigar. The phone rings.*

*Grayson: (in phone)* Yes! . . . Bell! . . . Oh, he is! . . . All right, send him right in! *(hanging up the receiver and speaking to himself)* Now the big parade begins! *(He takes a new cigar from the desk drawer and lights it from the butt of the old. The door on the left opens and* BELL *comes in. He has on his coat and carries a traveling bag, his hat, and a briefcase.* GRAYSON *rises to meet him. They shake hands)* Welcome back, old man. Put your things down. How're you feeling anyway? You look fit as a fiddle!

*(*BELL *doesn't share* GRAYSON'*s enthusiasm. In fact, he seems somewhat depressed)*

*Bell:* I'm feeling well enough. And of course, I'm glad to be back.

*(He puts his coat, hat and briefcase on the divan, and his bag on the floor.* GRAYSON *resumes his seat at the desk, and motions* BELL *to a chair)*

*Grayson:* Sit down, old man, you act a little fagged. I bet you're bursting with news. Ever since I got your telegram, I've been impatient to see you. How was His Highness when you left him?

*Bell:* Oh, he's all right, as far as health goes.

*Grayson:* And the wife?

*Bell:* *(sharply)* I suppose she's well enough.

*Grayson:* I had a letter from Her Highness. She said you had bad weather after you left New Orleans.

*Bell:* It was pretty rough. We were a day late in getting into Norfolk.

*Grayson:* You quit the boat there?

*Bell:* Yes!

*Grayson:* What was the idea?

*Bell:* Jeremiah thought it best that I come on and see that all the arrangements for his reception home were as he wished them.

*Grayson:* We've arranged everything. I don't think you'll find anything to do. But we're glad to have you back just the same. Tell me about your trip. Was it really as successful as the reports indicate?

*Bell:* *(brightening up)* It was a triumphal tour. There has never been anything like it anywhere—any time! No conquering hero, no country's savior, no king or emperor has ever received a more rousing welcome than His Highness, Jeremiah, got at every place we visited.

*Grayson:* You certainly sent in beaucoup members!

*Bell:* The entire West Indies is ablaze with his message! *(becoming excited)* And you know what that means? It means that at last we're a truly international organization. Wherever there are black men, we now have branch offices, and alert organizers. And soon, yes, very soon, we'll be ready to follow his Royal Highness back to our native land—ten million strong!!!

*Grayson:* And they've always said "niggers" couldn't work together, couldn't organize.

*Bell:* No, they couldn't, because they've never had any goal to work towards or any leader to organize them. They couldn't any more than any other people become passionately interested in abstractions! And that's all they've been asked to give support to; a lot of phrases—social equality, economic independence, the right to be good American citizens. Bah! No wonder they never pulled together before. But now, we, Jeremiah, brings them something real, something tangible, something solid!

*Grayson:* His immortality is assured. He's the greatest leader our race has ever had.

*Bell:* *(fanatically)* He's one of the greatest men of all time. *(The phone rings)*

*Grayson:* *(answering)* Yes! . . . Who? . . . *(He is startled)* No! . . . I can't see her! . . . I don't care if she does insist! . . . I can't see her! . . . That's all there's to it! *(He jams down the phone)*

*Bell:* Was that Sister Saunders?

*Grayson:* *(hesitating)* Eh! Yes!

*Bell:* I heard at Savannah that she'd come north. You know, Grayson, I don't like this mess. It's liable to react against us. That's one reason I was glad to get here before His Highness arrives. Maybe there's something I can do to straighten matters out. You know Sister Saunders and I—

*Grayson: (interrupting and speaking with determination)* There *is* no mess! There's nothing to straighten out!

*Bell:* But there's liable to be scandal. It's just what His Highness's enemies have been looking for.

*Grayson:* There's no scandal now and there isn't going to be any. There's nothing anyone can say! His Highness has merely divorced one woman and married another.

*Bell:* Yes, but consider how it was done.

*Grayson:* In a perfectly legitimate manner!

*Bell:* Well, from your point of view, I suppose it was. You advised him to do it. But speaking frankly, Brother Grayson, I don't think it was fair to Sister Saunders.

*Grayson:* It was perfectly legal. Any number of people go to different provinces in Mexico to get divorced. Maybe you'd rather he got a divorce here in New York, where we'd have had to work up an adultery charge by having private detectives break a door down and find one or the other of them in a compromising situation. If that had happened, then there would have been some scandal, and His Highness's enemies would have had something to get hold of.

*Bell:* It's not the manner of doing it, so much as the doing of it at all. Sister Saunders is a good woman and a good worker.

*Grayson:* So she is! But don't let your sentiment get the best of you.

*Bell:* Sentiment—how?

*Grayson:* They were mismated.

*Bell: (sarcastically)* And after fifteen years—you discovered it!

*Grayson:* It was very evident that Kate Saunders was not the woman for His Highness—now!!!!

*Bell: (sneeringly)* And Melissa is?

*Grayson:* She is!

*Bell:* I'd like to know your line of reasoning!

*Grayson:* It's okay! Listen here, Bell, you know I am heart and soul behind this movement. You know I love and admire Jeremiah as I love and admire no man, living or dead. Does it seem logical that I would deliberately advise him to do something which would hurt him and in turn hurt my people all over the world?

*Bell:* No, but I just can't see Melissa as his wife.

*Grayson:* And you can see Kate?

*Bell:* Certainly!!

*Grayson:* Come now, listen to reason. Kate and Jeremiah have been married for over fifteen years. Granted she's been a helpful wife, but has she been a complete wife?

*Bell:* I don't know what you mean.

*Grayson:* Oh, yes you do! Has she borne him any children?

*Bell:* No—but—

*Grayson:* And we know that at her age there's not any likelihood of her bearing him any, is there?

*Bell: (embarrassed)* I suppose not. But I don't—

*Grayson: (hurrying on)* Now let me finish. It doesn't take any great amount of reasoning to show why it is almost a necessity for Jeremiah to have at least one child, a son, if possible, to carry on his name and his work. Why, Jeremiah is going to build a great empire. He'll go down into history as Jeremiah the Great. And do you want the same thing to happen when he dies, as happened at the death of Alexander the Great? What became of Alexander's empire? Because he had no son, it was divided among his generals, and soon sunk to decay. We don't want anything like that to happen to Jeremiah's empire—the empire to which we are all giving our lives. Do we? *(*BELL *nods and starts to speak.* GRAYSON *does not give him a chance)* All right then! Jeremiah wants a son! He must have a son! Kate can't, hasn't, wouldn't! Melissa can and will! Do you follow me?

*Bell: (sighing)* I suppose so!

*Grayson:* And doesn't that in itself justify the divorce and marriage without considering other factors such as love, compatibility, and the like?

*Bell: (reluctantly)* But it just seems—

*Grayson: (unwilling to give* BELL *the slightest opportunity to object)* Then another thing. Kate is light; Melissa is dark! And it is certainly more sensible, more in keeping with our ideals, that the first queen of Africa should be a black woman, who will have a black son.

*Bell: (thoughtfully)* I hadn't thought of that.

*Grayson: (rising and slapping him on the back)* I knew you'd see it. It's inevitable that some of our pet sentiments and prejudices be toppled over in a case like this. We must subordinate everything to the business at hand. It must be the movement first, last, and always! *(There is a knock on the door at the right.* GRAYSON *looks up with a start)* Come in!

*(*DORA ANDERSON *enters. She is* GRAYSON's *secretary and is an attractive brown-skin girl with rather a negative personality. She is naïve and trusting, and quite taken off her feet by* GRAYSON's *attentions)*

*Grayson: (curtly)* Yes!

*Dora: (reticently)* I'm sorry to interrupt. But that woman outside is making a terrible fuss. I'm afraid if you don't see her, she'll—

*Grayson: (interrupting)* I won't see her! There's no use of her hanging around.

*Dora:* I told her that, but she—

*Bell: (interrupting)* Let me see her, Brother Grayson!

*Grayson: (stalling)* No . . . eh . . . I don't think that wise! *(with sudden determination)* I'll see her! *(He addresses* DORA*)* Tell her to wait. I'll have her in shortly.

*Dora:* Yes, sir! *(She goes out)*

*Grayson: (extending his hand to* BELL*)* Well, old man, I've got lots of work to do before the boat docks this afternoon. *(They shake hands)* Don't worry about the old girl! I'll take care of her! You're going down to the dock, aren't you?

*Bell:* Yes! Of course. Er . . . but don't you think I'd better speak to Kate?

*Grayson:* After I've seen her! There'll be plenty of time!

*Bell: (picking up his stuff)* Be considerate! Remember she's—

*Grayson: (interrupting)* It's sometimes difficult, but I always act the gentleman when with a lady. *(BELL starts off at the left, but GRAYSON stops him and leads him off on the right)* It's easier to get out this way. And you'll want to see how we've decorated the throne room for His Highness's arrival.

*(He throws open the door. BELL stands on the doorsill looking off to the right with eyes and mouth wide open)*

*Bell:* It's magnificent!

*Grayson:* Thought you'd like it! You'll find we haven't forgotten anything to make His Highness's return a glorious event! *(BELL steps into the reception room)* Well, old man, I'll see you after lunch. Remember the reception committee leaves here at one o'clock!

*Bell: (offstage)* Yes! All right!

*(GRAYSON closes the door. He rubs his hands contently and strides over to the desk. Here he assumes a thoughtful pose for a moment and then presses a buzzer on his desk. DORA enters)*

*Dora:* Want to see the old lady now?

*Grayson:* Who else is out there?

*Dora:* That ship man!

*Grayson:* Hmm! How's the old lady acting?

*Dora:* She's quieted down now!

*Grayson:* Let her wait a little longer.

*Dora:* Then you want to see the ship man?

*Grayson: (grabbing DORA playfully)* No! I want to see you!

*Dora: (snuggling to him)* Did you get plenty of sleep last night?

*Grayson: (with mock severity)* You keep me out half the night, then ask me did I get plenty of sleep.

*Dora:* I kept you out.

*Grayson:* Well, if it hadn't been for you, I wouldn't have been there, would I?

*Dora: (smiling)* You don't regret it, do you?

*Grayson:* I stayed because I wanted to, because I always want to, and because— *(He hesitates)*

*Dora:* Because what?

*Grayson:* Because you're a sweet little girl and you're gonna give me a kiss!

*Dora: (drawing back coyly)* Maybe!

*Grayson:* What! I don't get any kiss! *(He draws her to him)* I'll have to take it then.

*Dora: (with a delighted giggle)* Stop! You're hurting me. *(He kisses her. She gives into his embrace)*

*Grayson:* You're sweet! *(The phone rings. GRAYSON answers)* Who? . . . Yes! . . . All right! *(He looks at DORA)* Dora will show him in! *(He hangs up the phone)* It's the postal inspector. It's very important. Have him come right in.

*Dora:* *(pouting)* Seems we can never have a moment alone.

*Grayson:* *(impatiently)* This is my busy day, what with the old man getting back and everything.

*Dora:* At last I'm going to see him.

*Grayson:* That's right! You don't know him, do you?

*Dora:* Oh, yes! I know him. *(She points at the pictures on the walls)* As general, admiral, high priest, and emperor! *(They laugh)* Has he any more costumes?

*Grayson:* Oh, yes! Lots! You'll see the whole array of royal robes before long. *(He clears his throat and arranges a few articles on his desk)* Now, you'd better show the inspector in.

*Dora:* *(unwillingly starts to go)* All right!

*(She goes out at the left. In a moment, the door opens again, and the postal* INSPECTOR *enters. He is the same inspector as in the first act)*

*Grayson:* Come right in and have a seat.

*Inspector:* I've some good news for you, Mr. Grayson. *(He sits down)*

*Grayson:* *(also sitting down)* That's what I like to hear. Have a smoke? *(He passes the* INSPECTOR *a box of cigars. He takes one.* GRAYSON *starts to light it for him when the phone rings. With a gesture of annoyance he hands the matches over to the* INSPECTOR *and answers the phone)* Hello! . . . Who? . . . Tell them His Highness will be here this afternoon. His boat is due to dock at two. He'll do his own talking to the reporters. And Miss Maxwell, tell Miss Anderson I won't see anyone but Mr. Lorando and Mrs. Saunders . . . Understand? . . . Good! *(He hangs up the phone)*

*Inspector:* Keep you pretty busy, eh?

*Grayson:* I say they do! And today promises to be the busiest day of all.

*Inspector:* Yeah!

*Grayson:* Yes! The Chief, himself, gets back today.

*Inspector:* Where's he been?

*Grayson:* On an organizing trip through the West Indies, Central America, and Mexico.

*Inspector:* You're branching out!

*Grayson:* Wherever Negroes are, that's where we are. But what about the good news?

*Inspector:* Well, everything's been squashed.

*Grayson:* *(exultantly)* We came out okay, then?

*Inspector:* You bet!

*Grayson:* *(relieved)* God, I'm glad to hear that! Have some more cigars. *(He passes over the box)*

*Inspector:* *(laughing)* Thanks! I don't mind! *(He takes a handful)* Good brand!

*Grayson:* The Chief has nothing but the best.

*Inspector:* He'll be glad to get this news.

*Grayson:* *(quietly)* I suppose he will.

*Inspector:* All your books and papers will be returned. But let me give you a bit of advice. Judging from the letters we get, you've got plenty of enemies.

*Grayson:* I'll say we have!

*Inspector:* Well, they're not even waiting for you guys to break the law to start us on your trail; so my advice is for you to bend over backwards in living up to regulations.

*Grayson:* Thanks for the tip! We'll do that!

*Inspector:* Be sure you do!

*(He rises to go. GRAYSON comes from behind the desk to accompany him to the door)*

*Grayson:* You don't how much I appreciate your advice.

*(They start to shake hands when their attention is attracted to KATE entering through the door on the right)*

*Kate: (addressing GRAYSON coldly)* Pardon my intrusion, but I could wait no longer.

*Grayson: (coldly)* I said I would see you later.

*Kate: (determinedly)* I must see you *now*.

*(She sits down on the divan. GRAYSON glares at her)*

*Inspector:* Well, I'll run along. Good-day.

*Grayson:* Good-day.

*(The INSPECTOR goes out on the left. GRAYSON walks back to the desk. KATE eyes him belligerently)*

*Grayson:* There's a man outside I must see first. After I'm through with him, I'll speak to you. There's nothing to be gained by your forcing yourself in here like this.

*Kate: (sarcastically)* There's not much to be gained around you anyway, is there? It seems to me that you're usually the sole gainer in anything you've got your hand in.

*Grayson: (coolly)* You are entitled to your opinions. I don't wish to be ungentlemanly so I politely ask you to go until I call for you. Seeing you sneaked in here through the reception hall, you might wait out there. *(sarcastically)* What do you think of the throne we have arranged for His Highness and his consort?

*(KATE gets up and walks up to GRAYSON. Only the desk is between them)*

*Kate: (shaking her finger in GRAYSON's face)* Now, Mr. Grayson, you can tell me—

*Grayson: (interrupting)* I'll tell you anything you wish to know when I'm ready.

*Kate:* You'd better get ready now.

*Grayson: (threateningly)* I tell you— *(The phone interrupts him)* Hello! Yes! *(He looks at KATE)* Send him in! *(He starts to hang up)*

*Kate: (quickly)* I'd advise you to have whoever it is wait outside because I'm going to say what I have to say and ask what I have to ask—now!!!

(**GRAYSON** *holds the phone undecided what to do. For a long tense moment they glare at each other. Finally,* **GRAYSON** *sets down the phone*)

*Grayson: (fiercely)* You can't bluff me! Now get out before I have to throw you out!!

(*The door on the left opens.* **LORANDO,** *a small, wizened Levantine enters*)

*Kate: (rushing over to* **LORANDO**) Mr. Grayson would be obliged if you'd wait outside for about five minutes. The girl must have misunderstood.

*Lorando: (apologetically)* Certainly, ma'am, certainly!

(*He goes out and* **KATE** *closes the door. She walks back to the desk and again faces* **GRAYSON,** *who sits down with affected nonchalance and lights a cigar*)

*Kate:* Mr. Grayson, perhaps you can tell me what is the exact truth about all this I read in the papers and hear about Jeremiah, myself, and this black slut, Melissa.

*Grayson: (rising and speaking with affected indignation)* Don't use such language in speaking of Her Highness. I won't permit it!

*Kate:* Then it's all true! He has married her! But it's bigamy!

*Grayson:* Not when he's divorced.

*Kate:* Divorced! It can't be legal! I don't know anything about it!

*Grayson:* I assure you it's quite legal.

*Kate:* But I wasn't notified!

*Grayson:* No notification is necessary in Mexican divorces.

*Kate: (reflectively)* You seem to have planned everything very well, Mr. Grayson! Very well!

*Grayson: (bowing in mockery)* Thank you! I usually do a thorough job!

*Kate:* But don't think you can get rid of me as easy as this. I'm going to expose you.

*Grayson: (feigning indifference)* The pleasure is yours.

*Kate: (defiantly)* You can pull the wool over Jeremiah's eyes, but I see you as you are. I see your groping fingers feeling around in here for everything you can get.

*Grayson:* Now you're talking nonsense. Everybody knows I've given up my business, everything, to aid this movement. I have sacrificed myself gladly, for my race will reap the benefits of my labor.

*Kate: (suddenly pleading)* But why do you plot and plan against me? Can't you see how much the movement means to me, too? The movement and Jeremiah are my life. *(sobbing)* Now I've lost Jeremiah, and I guess next, I'll be forced out of the movement.

*Grayson:* There's no reason why you should be. You're a good, industrious, conscientious worker.

*Kate: (again flaring up)* Good, industrious, conscientious! That's what I've been all my life! And what has it brought me? Nothing! I find myself now right where I started.

*Grayson: (soothingly)* But you're not. Think of the movement!

*Kate: (drawing herself up to her full height and facing* **GRAYSON** *with accusing eyes and tight mouth)* Mr. Grayson, I *am* thinking of the movement, and of Jeremiah, and of myself, and of you! From the very first my heart knew you for what you are—a dirty, lying thief! Now my mind knows it. And I'm going to make everybody else's heart and mind know it! I'm going to show you up if it's my last act on this earth. *(She becomes a bit hysterical)* God above will help me! He will give me back my husband! He will purge our movement of the thieves and hypocrites who now defile it. He will—

*Grayson: (fiercely)* I think you've said enough!

*Kate: (half singing in her hysteria)* Oh—ho—I haven't even begun to talk! Oh—ho—the things I will shout from the housetops! *(She waves her arms wildly)* You will have to hide your face. You will have to flee from the vengeance of my people. *(She shouts)* Vengeance! Vengeance! The vengeance of the Lord will descend upon you!

*(GRAYSON draws back before her fury. DORA bursts in through the door at the left)*

*Dora: (agitated)* What's the matter? It sounds like someone was getting killed.

*Grayson: (quickly)* This woman has lost her head! *(He goes to the door on the right and opens it)* Take her into the reception room, Miss Anderson, and calm her down!

*(The height of KATE's hysteria has passed and she is now sobbing quietly. DORA approaches her)*

*Kate: (pushing DORA away)* Away! Away! You are one of Satan's imps! *(She speaks to GRAYSON with a pitiful effort to regain her thunderous indignation)* I'll go! But I am coming back—coming back with the army of the Lord behind me. He—I—Jeremiah—we'll be revenged! We'll be revenged! *(She goes off at the right)*

*Grayson: (motioning for DORA to follow)* See that she gets out without talking to anyone. Put her in a taxi! *(DORA goes out and GRAYSON closes the door) (facing the door on the right and sneering)* Go ahead, pretty mama! Do your damnedest! I should worry! *(He stalks back to his desk. After a moment's thought, he takes up the phone)* Oh, Miss Maxwell! . . . Send Mr. Lorando in! . . . Thanks! *(He hangs up. The door on the left opens and LORANDO reenters)* Come right in. Sorry to have kept you waiting, and please excuse that little rumpus. You know how women are—sometimes!

*Lorando:* That's all right, Mr. Grayson, all right with me! I understand! *(He smiles and sits in a chair beside the desk)*

*Grayson: (sitting down and passing the box of cigars)* Have a smoke?

*Lorando:* Thank you, Mr. Grayson, thank you. *(He takes one and GRAYSON lights it for him)*

*Grayson:* How about the ship?

*Lorando:* Yes! I think I have just the thing. Would you believe it, just the thing?!

*Grayson:* That's good news. We're in a great hurry.

*Lorando:* Really! I thought you didn't expect to sail for almost a year.

*Grayson: (glancing at the door on the right through which KATE has disappeared)* We've changed our plans.

*Lorando:* Well, the boat I had in mind for you would take quite some time to condition. It's an old cabin liner—a Russian vessel. It used to run to Black Sea Ports, but has been idle since the war. Tied up by litigation, you know!

*Grayson:* How soon can it be made ready?

*Lorando:* *(hesitatingly)* Well, eh, I really ought to have four or five months!

*Grayson:* *(peremptorily)* Too long!

*Lorando:* *(temporizing)* Of course, I could get it ready sooner. But it would cost much more.

*Grayson:* *(suavely)* Now, Mr. Lorando, I think you and I understand each other. We're both, eh, business men!

*Lorando:* Of course! Of course!

*Grayson:* Well, suppose we split the profit?

*Lorando:* I don't think I quite understand—

*Grayson:* *(interrupting)* It's very simple. You and I, as partners, will sell this boat to "The International Fraternity of the Native Sons and Daughters of Africa." Naturally, I'll be a silent and also an invisible partner. Understand?

*Lorando:* *(smiling wisely)* Yes! Yes! Perfectly! Perfectly!

*Grayson:* Good! Now you get the boat as cheaply as possible, condition it as quickly as possible, and we'll buy it as— *(He completes the sentence with a gesture of his hands denoting great size and looks shrewdly at* LORANDO *with raised eyebrows and questioning smile)*

*Lorando:* Exactly! Splendid! Exactly!

*Grayson:* *(rising)* I knew we'd have no difficulty coming to terms.

*Lorando:* *(also rising)* None at all! None at all!

*Grayson:* How soon will you have something definite?

*Lorando:* Tomorrow! Is that all right?

*Grayson:* *(thoughtfully)* I'll probably be pretty busy with His Highness. You'd better give me a ring.

*Lorando:* Good! I'll do that!

*Grayson:* *(leading him to the door on the left where they shake hands)* I'll be waiting to hear from you. Good-day!

*Lorando:* Good-day! Good-day! *(He goes out)*

*(*GRAYSON *goes back to his desk, pulls up his trousers a bit, arranges his coat, and glows with self-satisfaction)*

*Grayson:* *(chuckling to himself)* Now, I guess, we're all set for the Emperor!

*(As he arranges the papers on his desk, the curtain falls)*

### Scene Two

*The afternoon of the same day in the reception and committee room of Freedmen's Hall. This is a very elaborately decorated chamber with a balcony at the rear overlooking the street. Tall French windows open on to this balcony which is surrounded by a black ornamented iron railing. The door to* JEREMIAH*'s private office—the locale of Scene One—is on the left. And on the right, there is a double door opening into a hallway at the top of the flight of stairs leading up from the street. The* *walls on all sides are adorned with various flags of the movement. They have white and red fields with the black letters I.F.N.S.D.A., a large black cross, or a black outline map of Africa superimposed. There are also a number of red shields bearing the legend, "Back to Africa!" in white letters on a black outline map of Africa. A large full-length colored portrait of* **JEREMIAH** *in his imperial robes practically covers the rear wall to the right of the French windows. A throne is set up on the left, at an angle of forty-five degrees with the footlights. It consists of a heavy, elaborately carved chair on a low rug-covered dais. A number of brilliantly colored pieces of silk are thrown over the chair. Two heavy brass and crystal chandeliers hang from the ceiling. And about the walls are a couple of gilded and gaudily tapestried divans and numerous chairs of like design. The French windows leading on to the balcony are open when the curtain goes up. A number of flagpoles bearing fluttering banners extend out from the balcony over the street, and a galaxy of pennants are strung from the balcony to the floors above. The surging noise of a great crowd collecting below is heard.* **DORA** *is standing on the balcony.* **GRAYSON** *sits carelessly on the throne, legs crossed, leisurely smoking a cigar. He is dressed in full evening dress with a wide red ribbon bearing a black cross running diagonally across his white bosom.*

*Grayson:* Any sign of His Highness?

*Dora: (looking up the street at the left)* Not yet!

*Grayson: (looking at his watch)* Wonder what's holding him up?

*Dora:* If there was a crowd like this down at the boat, I bet he don't get here for another hour. *(There is a burst of cheering from the crowd below)* It's thrilling! They act just as if they were expecting Peaches Browning. I didn't know Jeremiah had so many followers in the whole world, let alone in Harlem.

*Grayson: (leaning back in the throne chair and puffing away at his cigar)* May his tribe increase!

*Dora: (excitedly)* Here comes an auto! The mounted police are clearing a way for it.

*(There is a murmur of excitement from the crowd)*

*Grayson: (starting to get up)* Who's in it?

*Dora: (disappointedly)* It's only a taxi. *(GRAYSON falls back into his seat. DORA's voice takes on new excitement)* It's stopping here. A lady is getting out. Oh, she's all in robes!

*(GRAYSON jumps up and hurries to the window. There is cheering and clapping from the crowd)*

*Grayson:* It's Melissa!

*Dora:* Her Highness?

*Grayson:* Yes! *(He is all tense and alive. He turns to* **DORA***)* You'd better go to your office!

*Dora: (pouting)* I want to see her close up!

*Grayson:* You can see her later. I'm afraid there's something wrong—her coming on ahead like this! Hurry! I must speak to her alone.

*(DORA goes off unwillingly at the left.* **GRAYSON** *takes a last hasty glance down into the street from the balcony and then closes the French windows. The door on the right opens and* **MELISSA** *enters followed by* **BELL***.* **MELISSA** *is dressed in magnificent robes befitting a queen, while* **BELL** *is wearing a rather garish uniform, resembling somewhat the formal diplomatic dress of a Balkan court)*

*Grayson: (going up to* MELISSA) Welcome home, Your Highness!

*Melissa:* I'm ready to drop! Gimme a chair! *(*GRAYSON *swings a divan from the wall,* MELISSA *flops into it)* The crowd at the boat almost killed me.

*Bell:* They were all trying to shake His Highness's hand, and in the crush Her Highness was pretty badly knocked about. I slipped her in a taxi and we came on ahead.

*Grayson: (to* BELL) Get a glass of water! *(solicitously to* MELISSA) They didn't actually hurt you, did they?

*(*BELL *hurries out at the right)*

*Melissa: (petulantly)* It's not your fault that I ain't dead!

*Grayson:* What sort of a trip did you have?

*Melissa: (sarcastically)* Triumphal! Very much so! His Highness annexed a lot of trophies includin' *me!*

*Grayson: (soothingly)* O, you shan't regret it!

*Melissa: (fiercely)* I do!

*Grayson:* 'Lissa, please! Don't make things difficult!

*Melissa: (rapidly pouring out her woes)* Of course, nothin' is difficult for me. It's easy for me to have that big black fool pawin' and fallin' over me! Easy to kiss those pork chop lips of his! Easy to stand around dignified like and be hailed as Melissa, Queen of Africa! *(disgustedly)* Queen of Hell!!!!

*Grayson:* I know it's hard, but there's too much at stake to give up now. Sit tight, and in a few months we'll both be in Paris with enough money to live on top of the world for the rest of our lives.

*Melissa: (not to be convinced)* It listens well, but I'm God-damned tired of the whole business!

*(*BELL *reenters with the glass of water)*

*Grayson: (under his breath)* S-s-s-h!

*Bell: (approaching* MELISSA *and handing her the water)* Here you are!

*Melissa: (taking the glass)* Thanks!

*(She drinks. There is a lusty cheer from the crowd, which, with the blare of a band is heard even through the closed windows. The band is playing the "Black Internationale," the hymn of the movement)*

*Bell:* That must be His Highness!

*(*MELISSA *sighs and hands the glass back to* BELL)

*Grayson: (to* MELISSA) Go into Jeremiah's office. You can rest in there until the worst of the excitement is over.

*Melissa: (resignedly)* All right! *(*GRAYSON *takes her by the arm and they start off towards the left)*

*Grayson:* You can lie on the divan.

*(They go out)*

*(*BELL* goes to the French windows and throws them open, stepping out on the balcony. The cheers of the crowd are now hysterical. Hats and canes are seen being thrown into the air. The yells of the mob virtually drown out the band.* BELL *stands proudly looking down at the demonstration.* GRAY-SON *reenters from the left and joins* BELL. *There is a renewed and magnified burst of cheering, indicating that* JEREMIAH *is getting from his car and is entering the building.* BELL *and* GRAY-SON *come in from the balcony)*

*Bell:* (shouting so as to be heard above the din) Have you ever seen anything so glorious?!!

*(*GRAYSON *does not reply, but goes to the double door at the left and throws it open. Six very tall, husky, dark Negroes, arrayed in ornate blue uniforms with Sam Browne belts and leather leggings, and with black crosses on their sleeves, step in from the hall. At a sign from* GRAYSON *they take up their positions—three at either side of the entrance.* GRAYSON *stands waiting to the left of the rear rank of these Royal Guards. The crowd outside continues to cheer and the band is playing full blast. Suddenly* JEREMIAH *enters through the doorway at the right. His regalia is an astoundingly elaborate red and purple robe, trimmed with gold, silver, and crystal. He wears a tall, silken plumed head-dress. In his left hand, he carries a long gilded staff, fantastically carved. In spite of the outlandishness of his get-up, he looks a truly majestic figure. He enters with long strides, followed by six men and six women, Knights and Ladies of the Inner Circle, who are dressed in elaborate, many-colored costumes, which like* JEREMIAH's *and* MELISSA's, *seem half-borrowed from the court of some oriental potentate, and half-borrowed from the formal dress of a Western European count. The Knights and Ladies of the Inner Circle have all been selected from those people we saw in Act One and include* SISTER TYSON *and* BROTHER BAKER. *In* JEREMIAH's *retinue is also the leader of the longshoremen's delegation in Act One. He is dressed in Naval Captain's full-dress court uniform)*

*Grayson:* (stepping forward and bowing to JEREMIAH) Greetings, Your Highness!

*Jeremiah:* (raising his hand as if bestowing a Pontifical blessing) Ah, Brother Grayson! It's good to be back!

*(His voice is almost drowned out by another crescendo of shouting in the street. The words "Speech!" and "Jeremiah!" are occasionally distinguished above the din.* JEREMIAH *stands listening, a look of mystical ecstasy upon his face)*

*Grayson:* (when the shouting goes into pianissimo) Your Highness, I'm afraid you'll have to show yourself on the balcony.

*Bell:* (glancing out the window) They'll never move until you do!

*Jeremiah:* (advancing towards the balcony) I'll say a few words of greeting! (The moment he steps out on the balcony, there is a renewed outburst. Hats, canes and newspapers are again thrown recklessly into the air. The demonstration on the part of the crowd is one of abandoned hysteria. JEREMIAH stands rigid for a long moment, letting the crowd roar. Then he slowly raises his right arm. The gesture commands silence. He keeps the arm upraised as the uproar gradually dies down. His voice booming forth) My children! I bring you glad tidings! The black men of the whole world are with us! (There is a burst of cheers, which JEREMIAH *again quiets by raising his arm)* Today, wherever men are black, our banners are triumphantly flying, announcing a new and glorious era for our people. *(more cheers)* You have, I know, all read in the *Black International News* of my triumphal tour. But, friends, I wish you could have all been with me. It would have stirred your hearts; glorified your spirits; and forced you to your knees, as it did me, *(he strikes himself forcibly on the chest)* in humble thanks to our Heavenly Father. I was always in His hands! He watched over me! He directed my footsteps! He never abandoned me for a moment! As he guided Moses in leading the sons and daughters of Israel out of the bondage of Egypt into the Promised Land, so He is guiding me, Jeremiah, as I lead the sons and daughters of Ethiopia out of the white man's bondage into the promised land of Africa—Black Africa! *(Again the crowd cheers.* JEREMIAH *continues)* In Black Africa, where every black man and woman will have the right to life, liberty, and the pursuit of happiness; where every black man and woman can vote—and vote for black men and women; where black men and women can come into their divine inheritance—the inheritance that our Heavenly Father in His perfect wisdom has granted us! *(There are fervent shouts from the crowd of "Amen! Amen!")* Black men and women of the world unite—you have nothing to lose but your chains! We shall soon be a united people and Africa will soon be our country; the only country in the world one hundred percent black! Our cry from now on must be: "On to Ethiopia! Back to Africa!" *(*JEREMIAH *stops. The crowd roars! He raises his hand as if blessing them and steps back into the room. The band immediately blares forth.* GRAYSON *motions* JEREMIAH *to the throne on the left and he staidly mounts the dais.* GRAYSON *indicates to one of the Royal Guard to close the French windows, and the Guard does so. The crowd and band can be heard muffled through the closed window for a while, but gradually all sound from them ceases.* JEREMIAH *stands in front of the throne chair, leaning on his staff. He surveys the faces in the room)* Where is my Queen?

*Bell:* The crush of the crowd was so great at the boat, that to save Her Highness from harm, I slipped her away in a taxi.

*Grayson:* Her Highness is now resting in your private office.

*Jeremiah: (sitting down)* It's been a hard day for her!

*Grayson:* Shall I tell her to come in, Your Highness?

*Jeremiah:* No! No! Let her rest!

*Grayson:* Very good!

*Jeremiah:* Well, Brother Grayson, what's the line up?

*Grayson: (consulting his notebook)* There are a number of delegates from foreign countries who are here for our international convention. They desire an audience with you.

*Jeremiah:* I'll see them tomorrow.

*Grayson:* Very good! *(He makes a record in his notebook)* Of course, the reporters and photographers from the daily press are here. Do you wish to see them?

*Jeremiah:* They can't take any photographs! We'll give them pictures taken by our own men! They can use them! I don't want any pictures printed that I haven't seen!

*Grayson:* Very good! And the reporters?

*Jeremiah:* I'll see them in my private office.

*Grayson:* Very good! *(He again refers to his notebook.* **DORA** *hurries in from the right, and going up to* **GRAYSON**, *whispers in his ear and hands him a slip of paper.* **GRAYSON** *turns to* **JERE-MIAH**. *presenting* **DORA***)* Your Highness, this is Miss Anderson, my secretary, a new addition to our staff.

*(***DORA*** curtseys)*

*Jeremiah:* Glad to know you, my child. May you prove worthy of the work you will be entrusted with.

*Dora:* I'll do my best, Your Highness. *(She curtsies again and steps back)*

*Grayson:* Your Highness, Miss Anderson has just brought me a most startling piece of news. A delegation of prominent Negroes are here to see you. They are the men who have been fighting us all along. *(He reads from the paper)* J.S.B. Renard, Z. Albert Cald-well, and Major Martin.

*(There is an excited buzz among the members of the Inner Circle)*

*Sister Tyson:* I know they'd come around!

*Brother Baker:* There's nothing else for them to do!

*Jeremiah:* *(triumphantly)* They probably see now that nothing can stop me! Their doc-trines have proven like so many soap bubbles, while mine—mine are destined to carry my people to glory and to endure forever! *(to* **GRAYSON***)* Do you think they want to come into the fold?

*Grayson:* I hardly think that is the purpose of their visit. Renard and Caldwell have been attacking you violently in their magazines all during your trip.

*Sister Tyson:* Jealousy! That's all it is! Jealousy!

*Jeremiah:* No doubt! *(to* **GRAYSON***)* Do you think I had better see them?

*Bell:* *(interrupting)* Your Highness, I am sure these men mean neither you nor the move-ment any good. They should be sent on their way.

*Grayson:* These men have not only fought us, they have fought each other for years. Why they should be here together is a mystery that intrigues me. I'm sure we'd all like to know what has brought them.

*Jeremiah:* This is the first time they've come face to face with me. I'll see them!

*Grayson:* *(to* **DORA***)* Send them in!

*(***DORA*** hurries out. Everybody turns and faces the door on the right. They stand in attitudes of expectancy awaiting the delegation)*

*Jeremiah:* The day is not far distant when all the world will come to Jeremiah!

*(A hearty "Amen" goes up from various members of the Inner Circle. The delegation enters. The Inner Circle regards them with frank hostility. ***RENARD*** is a very dignified, middle-aged, light brown skin Negro, very much the aristocrat, with a goatee, and a cultured, well modulated voice. ***CALDWELL*** is a tall, brown skin young man, with a deep resonant voice. His speech is affected Bostonese and he carries himself with a grand manner. ***MAJOR MARTIN*** is elderly, short, heavy-set, with a dark brown skin. He is humble in speech and manner. Unlike ***RENARD*** and ***CALD-WELL***, his is not a dominating personality)*

*Jeremiah:* How do you do, gentlemen. I'm pleased to see you on this first day of my return from a most triumphant tour.

*Renard:* We thank you for granting us an audience!

*Caldwell:* We appreciate it!

*Martin:* Most assuredly!

*Jeremiah:* What is it you wish?

*Renard:* We are here, Mr. Saunders, as spokesmen for the colored people of America.

*Jeremiah: (surprised)* For the colored people of America?

*Caldwell:* Exactly, sir! We each, as you know, represent a distinct group. Together our followings comprise the very best there is within the race.

*Martin:* Quite right! Quite right!

*Jeremiah: (harshly)* What is it you want?

*Renard:* We have come, Mr. Saunders, to ask you in behalf of the colored citizens of America to cease your ridiculous campaign.

*(There is a buzz of surprise and objection from the Inner Circle)*

*Caldwell: (hurriedly)* We feel that your organization is a detriment to the whole race. It holds the Negro up to ridicule. Because of you, the world is laughing at us.

*Martin:* Quite right! Quite right!

*(Some of the members of the Inner Circle, and especially* BELL, *start indignantly for* CALDWELL*)*

*Jeremiah: (rising and waving back his cohorts)* I'll handle this! *(to the delegation)* You gentlemen come as spokesmen for the race?

*Caldwell: (defiantly)* We do!

*Jeremiah:* Well, what about my seven million followers? Are you representing them, too?

*Renard:* Speaking with absolute frankness, we don't believe you have that many members. We know that the better people of the race are against you and your organization. They can see nothing good in the doctrines you preach. Back to Africa, indeed! Whereabouts in Africa would you be welcome? The American Negroes' place is in America. This is now his native land. He has now reached the place where he can demand equality with his fellow white countryman. And equality here in America must be his goal! Not some wild plan of returning to Africa!

*Caldwell: (before* JEREMIAH *can reply)* Only the ignorant and uninformed could see a glimmer of hope in your leadership. Why don't you lend your talents to the creation of a class consciousness among Negroes? Negroes are workingmen, members of the laboring class, and they should ally themselves with other laboring men. Their salvation lies in fighting the battles of labor side by side with the white workingman!

*Martin: (a bit timidly at first but with growing warmth)* Booker T. Washington knew that the black man would eventually create a place for himself in American civilization. Booker T. told his fellow black men to prepare themselves by sticking to the soil and worming into trade and industry. Be ready for the great day, he counseled. We have been making ourselves ready. The day is almost here, and after all these years of hard work, you counsel us to run away into the black jungles of Africa; into the jungles for which we have no affinity, and to which we bear no kinship. No, sir! It is nothing but a childish nightmare.

*(All through these speeches, the Inner Circle under the leadership of* **BELL***, start to object, but* **JERE-MIAH** *keeps them quiet with constant admonitory scowls and gestures)*

*Jeremiah: (thundering)* I have listened patiently to your nonsense. Now it is my turn to speak. You gentlemen don't yet realize that the doctrines you preach are leading the Negro nowhere. You have tried and failed. And now that you see someone else, someone with divine inspiration, about to succeed, you crawl in like mice nibbling at a piece of cheese trying to eat it all away before it is placed on the table to be served the dinner guests. But, gentlemen, you are too late! And you are too unimportant for me, Jeremiah, to listen to any longer. *(He sits down)*

*Renard: (desperately)* You must listen to reason!

*Jeremiah: (sneeringly)* There is no reason in any of you!

*Martin: (almost crying)* You mean that you'll continue your ridiculous campaign; continue fooling people out of their money!

*Jeremiah:* I, and my movement, are in God's hands.

*Renard:* You're making it hard for all Negroes; making them ridiculous in the eyes of white people.

*Caldwell:* You're allied with the Ku Klux Klan! You're aiding them in their nefarious machinations to make this a pure Nordic country!

*Jeremiah:* Let them have this country! We're going to have a country of our own.

*Martin:* I tell you, you're ruining all the work we've accomplished through years of hard labor. We're losing the support of our white friends.

*Jeremiah:* The sooner you lose such support, the sooner you'll be free and independent! *(He rises again)* Now, gentlemen, I must beg you to excuse me. This is, as you know, a busy and important day for me. I've already given you too much time!

*Renard:* You must hear us out!

*Jeremiah:* I have no more time. Brother Grayson!

*Grayson:* Yes, Your Highness!

*Jeremiah:* Show these gentlemen out, and don't admit them again till they're willing to join our forces and fight for their freedom in the promised land.

*Renard: (defiantly)* We're in the promised land!

*Bell: (who all through the interview has had difficulty keeping still, now bursts forth frantically)* Promised land! Where we're lynched, spat upon, despised! It's you leeches, you traitors to the race, groveling at the white man's table for chance crumbs, who would keep us eternally enslaved! You've held sway among our people too long! But now you're through! Through! Through! Jeremiah is taking our race to the top of the ladder. He is taking us to the Promised Land, and you will be left here groveling in the dust like dogs! That's what you are—dogs! Curs!—Despicable curs!

*(RENARD, CALDWELL, and MARTIN draw back before the fury of BELL's attack)*

*Grayson: (signaling to the guards)* Escort these gentlemen out.

*(The guards surround them and maneuver them out at the right as the members of the Inner Circle talk excitedly among themselves)*

*Bell: (fiercely as the delegation disappears)* Swine!

*Jeremiah: (sitting down)* Deposed gods, Brother Bell.

*Bell:* Yes, deposed gods! That's what they are! *(then suddenly)* No, not gods, half gods! There is a poem:

> When half-gods go,
> The gods arrive!

*(He bows to JEREMIAH)*

*Jeremiah:* And after all these years of struggle, strife, and indecision, I, Jeremiah, have come to do what these half-gods have tried to do. *(THE GUARDS return and again take up their posts. JEREMIAH turns to Brother GRAYSON)* Anything else, Brother Grayson?

*Grayson:* No, Your Highness! Only the reporters! Whenever you're ready, I'll have them sent into your private office.

*Jeremiah: (addressing the Knights and Ladies of the Inner Circle)* Knights and Ladies of the Inner Circle, is there any immediate executive business to bring to my attention?

*Sister Tyson: (stepping forward)* Just one thing, Your Highness. *(She hesitates)* I wanted to know . . . Well, how are we supposed to act toward Sister Kate Saunders?

*Jeremiah: (stiffening)* Have any of you seen her?

*Grayson: (quickly before anyone else can answer)* Yes! She was in this morning and I had a talk with her.

*Jeremiah:* How—how is she?

*Grayson:* A little upset—naturally! She wanted to be here when you came. I advised her against it.

*Jeremiah:* Quite right! I'll see her in due time. Hmm! What do you feel should be done about her?

*Grayson:* I don't know just what you mean, Your Highness.

*Jeremiah:* I mean, should she be allowed to keep her place in the organization?

*Grayson:* She has often shown antagonism to me personally, but if she will stay, sir, I say keep her. After all, she is a good worker. And she feels, that in a way, it's her movement, too.

*Jeremiah: (impatiently)* That's one thing the matter with Kate! She wants all the credit! A movement can have but one leader. The rest must follow. Kate was never content to follow. She always wanted to stay two paces, in advance.

*Brother Baker: (speaks haltingly)* You see, Your Highness, Sister Kate Saunders has been to see practically all of us today. She made some—eh—well—some startling charges!

*Jeremiah:* Charges! Of what sort?

*Brother Baker: (glancing nervously at* GRAYSON*)* I don't like to speak of them. There's probably nothing in them. But out of loyalty to you and the movement, I felt—eh—well—*(He again glances apprehensively at* GRAYSON*)*

*Grayson: (confidently)* Speak out, Brother Baker. When Mrs. Saunders was in this morning, she made many wild charges. I'm sure she didn't make any to you that she hadn't already made to me.

*Jeremiah: (impatiently)* What are these charges?

*Brother Baker:* She claims, sir, that Brother Grayson is guilty of some crooked work. *(He moves nervously)*

*Jeremiah: (sharply)* Yes! What else?

*Brother Baker:* And she claims that because Brother Grayson knew that she knew he influenced you to get rid of her.

*Jeremiah: (stentoriously)* No one influences me to do anything! I act in all things through the divine inspiration of my Heavenly Father! *(He turns to* GRAYSON*)* Did Sister Saunders accuse you of this?

*Grayson: (making light of the whole matter)* She said many things. She was very angry at the time, and I didn't take her remarks seriously.

*Sister Tyson:* She was crying when she came to see me. She acted hysterical like.

*Jeremiah: (sententiously)* There is only one thing to do in a case like this. Ambition and selfishness can have no place in this movement. We must be all for one and one for all. The movement is too big for petty quarrels and petty spiteful tricks. The Lord does not love evil. Neither do we. His commandments must be obeyed. He said: "If thine eye offend thee, pluck it out!" therefore, there is only one thing for us to do. I stand ready to hear a motion to read Sister Saunders out of the organization.

*Bell: (meekly)* Don't you think we'd better give her a chance to tell her side of the story? It's best not to be too hasty!

*Grayson: (suavely)* I'm more than willing to be charitable. You've all seen that! But I hardly think we ought to dispute His Highness's judgment!

*Sister Tyson:* I make a move that Sister Kate Saunders be read out of the "International Fraternity of the Native Sons and Daughters of Africa" for disloyalty and an attempt to cause dissension in the ranks.

*Brother Baker:* And I—second the motion!

*Jeremiah:* The motion has been moved and seconded. Any question? *(pause)* Then all in favor answer "aye!"

*All: (save* BELL*)* Aye!

*Jeremiah: (rising)* I thank you, my friends, for this demonstration of loyalty and discipline. The movement can only march forward as a united body. We must never forget that in union there is strength. *(There is a murmur of approval from the Knights and Ladies of the Inner Circle. He turns to* GRAYSON*)* Is there anything further?

*Grayson:* Our International Convention opens tomorrow afternoon at two o'clock. Reservations for seats on the platform have been made for all of you. See Miss Anderson and she'll give you your tickets.

*Jeremiah:* Nothing else!

*Grayson:* No, Your Highness!

*Jeremiah: (descending from the dais)* We will now adjourn! *(The Knights and Ladies start going out on the right. The Royal Guards close the door after them, and then go out themselves.* JERE-MIAH *addresses* GRAYSON) I'll see those reporters now!

*Grayson: (to* BELL*)* Have the reporters shown into His Highness's private office.

*Bell:* Her Highness is resting in there!

*Jeremiah:* That's all right. She can come in here.

*Grayson: (motioning to the right)* You'll find reporters in the outer office!

*Bell: (going off at right)* Very well!

*Jeremiah: (patting* GRAYSON *on the shoulder)* Brother Grayson, I want to thank you for keeping things going while Brother Bell and I were away. You've proven to be a loyal and capable general.

*Grayson:* Thank you, Your Highness. I only did my duty. And now if you have a moment's time before seeing the reporters, there's several things—important things I'd like to take up with you.

*Jeremiah:* I'm listening!

*Grayson:* There is the postal matter.

*Jeremiah:* Ah, yes! You disposed of it?

*Grayson:* No! They're preparing to prosecute!

*Jeremiah: (taken aback)* Preparing to prosecute! Me? Jeremiah?

*Grayson:* I'm afraid so, Your Highness.

*Jeremiah:* But they can't! My people wouldn't stand for it!

*Grayson:* Unfortunately, we are not yet in Africa.

*Jeremiah:* But they have no charges.

*Grayson:* They can charge you with using the mails to defraud. Technically, you see, we have been selling stock in something which doesn't exist, except in our hopes and imagination.

*Jeremiah:* But it does exist! Africa is there—wild and uncultivated and waiting; pregnant with possibilities for a great empire.

*Grayson:* True enough! But according to our federal law, you can't sell stock to people via the mails in anything as uncertain as our project appears to the postal authorities. The fact that we're certain of success makes no difference.

*Jeremiah: (walking back and forth like a caged lion)* Africa belongs to the black man! It is his rightful heritage—his homeland!

*Grayson: (patiently as if talking to an irate child)* But we are in America where you have many enemies with political influence. They lie about you! And they seek out technical charges to lodge against you!

*Jeremiah:* They can gain nothing by fighting Jeremiah! I'll break them like so many clay pipe stems! *(He gestures to suit his words)* I'll destroy them!

*Grayson:* They can sow distrust among your followers and delay the carrying out of your program. Each delay will cost us both money and members!

*Jeremiah: (thundering)* If they want a battle, we'll fight! The Lord is with us!

*Grayson:* I hardly think it's wise to fight at present. There is too much at stake to risk, even the chance of defeat. Remember, although we annihilate our enemies, we can't hope to defeat the politicians in Washington who are in league with them.

*Jeremiah: (resignedly)* God's will be done! We can only do as He dictates!

*Grayson:* I have a plan. Prosecution can't begin for at least another six weeks. They will be that long collecting evidence and obtaining affidavits.

*Jeremiah: (anxiously)* And your plan?

*Grayson: (triumphantly)* By the time the authorities get ready to serve the warrant for your arrest, you will be on the high seas bound for Africa with your first contingent of colonists.

*Jeremiah: (distressed)* Is it possible to get away so soon?

*Grayson:* I think I can manage it. Of course, it won't be easy. However—

*Jeremiah: (interrupting)* But Brother Grayson, if they're preparing to arrest me, won't they surely do so when they hear I am about to leave the country?

*Grayson: (with calm assurance)* If they do, I'll get you out on bail. But they won't know that you're going. Merely a shipment of colonists! Your departure will remain a secret until the boat's well past the three mile limit and outside the jurisdiction of the United States' courts.

*Jeremiah: (almost whimpering)* How is it you didn't write me about this, Brother Grayson? It's a terrible blow—a most unexpected blow!

*Grayson:* The Postal Inspector only called this morning. I didn't know anything about it myself until then. Your Highness, I hate to say so, but I think Mrs. Saunders has been in touch with your enemies.

*Jeremiah: (with rapidly mounting anger)* She! She! *(He begins to stamp back and forth)* Kate! She, do this to me!

*Grayson:* Do not distress yourself, Your Highness. Things are not as bad as that!

*Jeremiah:* This puts us six months ahead of schedule! We have no ship! We haven't yet chosen the spot in Africa on which we were to begin building the Empire!

*Grayson:* I can arrange for the ship. And it is possible to land in Liberia. Your followers there will direct and aid you.

*Jeremiah: (eagerly)* You know of a ship?

*Grayson:* Yes, a first class liner which will become the flagship of the Black Cross Line. It's speedy, sea-worthy and cheap. I got in touch with a ship broker as soon as the postal inspector left. Fortunately, he knew of this boat which is practically ready now.

*Jeremiah: (distracted)* Haste always makes waste. I don't see how we can get away so soon. I'm afraid that—

*Grayson: (interrupting assuringly)* There'll be neither haste nor waste. You've carried your propaganda to practically every black man and woman in the western hemisphere. You have 7,000,000 adherents in the United States and corresponding numbers throughout the West Indies; you have just completed a most notable tour; you are on the eve of a most notable convention. The time to consummate your achievement is now!

*Jeremiah: (trying to reassure himself)* Yes! Perhaps you're right! No doubt this is a fitting time to send over the first colonists! But must I—eh—is it necessary that I lead them?

*Grayson:* Even disregarding the shadow of jail so unjustly hanging over you, I should say, Yes! An undertaking like this needs its leader, more at the beginning than at any other time. Here you have set up the machinery. Lesser hands can operate it. But there, the beginning is still to be made! Only an Emperor can hope to establish an Empire.

*Jeremiah: (regaining his usual self-assurance)* Yes, you're right, Brother Grayson, you're right! Moses led his children into the Promised Land! Jeremiah will do the same!

*Grayson:* I knew you would do the courageous thing!

*Jeremiah:* Yes! This is the time to strike; the time to show the doubters that Jeremiah's dreams are made of stern and durable stuff.

*Grayson: (standing at attention)* Your Imperial Majesty, Jeremiah, the Magnificent, first Emperor of Africa, I salute you!

*Jeremiah:* And you, Brother Grayson, shall keep the fires burning here while I'm winning Africa back for its rightful owners! *(MELISSA enters from the left. JEREMIAH goes over to her and takes her in his arms)* Poor child, did the crowd try to pull you to pieces?

*Melissa: (sullenly)* It mighta been a good thing if they had.

*Jeremiah: (laughing)* No! No! I need you too much to help me set up my empire.

*(BELL enters)*

*Bell:* The reporters are waiting, Your Highness.

*Jeremiah:* They can wait! I have good news for you both. We will sail for Africa within the next six weeks.

*Bell: (completely surprised)* The next six weeks!

*Melissa:* Well, I never! When did this idea hit you?

*Jeremiah:* I've been thinking about it ever since we left the West Indies. But I first wanted to see how things were here. I've found everything in ship-shape fashion. There is no need to wait longer. We can safely begin the transportation of colonists to Africa right away!

*Melissa:* And you're goin' with the first load?

*Jeremiah:* As befits an emperor—yes!

*Bell: (still surprised)* And me—I—

*Jeremiah: (to BELL)* You will accompany me. Brother Grayson will attend to things here.

*Bell: (hesitantly)* I hardly know what to say. I don't like to question your judgment. I'm sure you must be doing the right thing. Of course, whatever you decide to do, I'm with you!

*Jeremiah:* Well spoken, my son, well spoken! This is indeed a great moment in the history of our movement. Let's all unite in prayer to our Heavenly Father and thank him for His divine guidance. *(JEREMIAH raises his hands above his head and prays fer-*

*vently.* BELL *bows his head.* GRAYSON *and* MELISSA *exchange glances—questioning on her part and reassuring on his)* We thank Thee, Father, we thank Thee. Africa shall welcome her own. Ethiopia shall stretch her wings. The first shall be last and the last shall be first. We have been last, oh Lord, but no more. In your name we reclaim Africa for the Africans. We will place the black man on a pedestal so high that all must see and must respect. We go, Father! We go!

*(Slowly, during the last part of this prayer, the curtain falls)*

### Act III

### *Scene One*

*It is six weeks later. The setting is the Captain's cabin of the S.S.* Jeremiah, *flagship of the Black Cross Line. It is small, compact, and panelled with dark wood. In the back there are two small square windows looking out on the deck. The shades of these windows are down. On the right of the window is a door leading to the deck, and on the left is another door leading to another cabin, that to be occupied by* JEREMIAH *and* MELISSA. *The cabin is fitted up with a few large leather covered chairs and a leather covered couch. A flight of stairs runs up from the right to the deck above and on up to the bridge. There is also a square dark wooden table in the center of the room which is covered with papers, maps and charts. Four straight back wooden chairs with leather seats surround this table. Scattered around are many bouquets, large and elaborate and symbolic in design. They are good luck offerings from* JEREMIAH's *enthusiastic followers. The organization's flag is conspicuously displayed on the rear wall, draped about the inevitable photograph of* JEREMIAH *in his imperial robes. When the curtain goes up,* GRAYSON *and* LORANDO *are in the cabin.* LORANDO *is nervously pacing the floor.* GRAYSON *is sitting on one of the chairs beside the table.*

*Lorando:* I tell you, Grayson, I don't feel right about this thing.

*Grayson: (wearily)* You've nothing to worry about. It's not your ship any longer.

*Lorando:* No, but I—oh—*we* sold it to the organization, perfectly aware of its—well—its dangerous possibilities.

*Grayson:* You said yourself that properly manned, there'd be no danger.

*Lorando:* Oh, yes, if properly manned! But I don't believe that is the case. God, as seamen, these men are good longshoremen.

*Grayson:* They have their papers.

*Lorando: (scornfully)* And from where? Liberia!

*Grayson:* We're flying the Liberian flag.

*Lorando:* Which has nothing to do with the efficiency of your crew! Frankly, I wouldn't even go down the Hudson in a ship piloted by your captain. Huhh! Him—a captain! *(He walks back and forth a few times, then suddenly halts before* GRAYSON. *His voice is pleading)* Listen, Grayson, this thing has got on my nerves. I've been in some pretty—well, some pretty shady deals put over in my day, and I've been mixed up in a lot of them. The ship-brokering business is hardly a game for a man of many scruples. But I've never gone this far before! Why this is—it's almost murder!

*Grayson: (reassuringly)* You're just panicky from overwork. Getting the ship ready in such a quick time has worn you out. You'll feel better after you've had a week's rest.

*Lorando:* *(refusing to be side-tracked)* Let me get you a first class crew, and restore the old captain.

*Grayson:* The principles of our organization are such that this ship must be manned entirely by Negroes.

*Lorando:* But they're not capable.

*Grayson:* We believe them to be. It's our ship. You have no responsibility whatsoever.

*Lorando:* *(sighing helplessly)* Well, if a bunch of damn fools want to commit suicide, I suppose that's their business. Africa? Hell! You'll probably run aground before you pass the Statue of Liberty. *(He starts off at the rear)*

*Grayson:* *(laughing)* In which case we'll come back and start all over again. *(rising)* Don't worry, old man. These men know their business. You, like most white people, doubt the ability of Negroes. And it's the purpose of this organization to show the world that whatever white men can do, black men can do, too.

*Lorando:* *(from the cabin door)* Even unto being God-damned fools!

*(BELL enters down the stairs as LORANDO goes out. BELL is nervous and tense)*

*Bell:* Has his Highness got on board yet?

*Grayson:* Not yet.

*Bell:* Do you think that something could have gone wrong?

*Grayson:* *(calmly)* He'll be here in due time.

*Bell:* Wasn't he to slip on board before any crowd gathered?

*Grayson:* Yes!

*Bell:* Well, the pier's packed already.

*Grayson:* He's not coming on from that pier!

*Bell:* But how'll he get aboard?

*Grayson:* Now, Bell, keep your shirt on. I arranged all this. He'll be aboard in plenty of time.

*(There is a moment of silence)*

*Bell:* *(nervously clasping and unclasping his hands)* Well, I hope you're right. *(with sudden decision)* I'm going on deck.

*(He goes out at the rear. GRAYSON stands at the table looking down at the papers. MELISSA enters from the adjoining cabin on the left)*

*Melissa:* For God's sake, give me a cigarette! If I don't do somethin', I'll go nuts!

*Grayson:* *(handing her a cigarette and lighting it for her)* What're you so excited about?

*Melissa:* *(sarcastically)* What Empress about to journey to her kingdom in the African jungle wouldn't be excited?

*Grayson:* *(taking her in his arms)* Come now! You ought to be happy!

*Melissa:* Yeah! I should jump up and down and clap my hands in glee—like hell! *(breaking away from GRAYSON)* It's all right for you to talk! You don't have to sleep with that big black coon!

*Grayson: (looking around nervously)* Must you tell the whole world?

*Melissa:* I'm about ready to!

*Grayson: (once more taking her in his arms and speaking soothingly)* Don't let the excitement of being an Empress about to sail to her imperial realm make you forget more important things.

*Melissa:* Such as?

*Grayson:* Our love, darling, our love! *(He tries to kiss her)*

*Melissa: (pushing him away)* A hell of a lot you love me marrying me off to that—

*Grayson: (interrupting)* 'Lissa, you know and I know—

*Melissa: (defiantly)* What?

*Grayson:* That you're still mine!

*Melissa:* If I was sure of you as you're of me, I wouldn't be ready to rush back to the pier and wave good-bye from there instead of from the deck of this damn ship!

*Grayson:* Let's not start that all over! Didn't you promise never to mention secretaries to me again?

*Melissa:* Oh, I know that little yaller whore is after you!

*Grayson:* You shouldn't talk like that, 'Lissa. In the first place, you don't know anything about Miss Anderson; and secondly, you know I prefer my meat—dark! *(He kisses her)*

*Melissa: (with threatening innuendo)* If things go wrong, I'll—

*Grayson: (interrupting impatiently)* Yes, I know. You'll come back and kill me! You've said that before. If we don't watch our steps these last few minutes, it won't be necessary—we'll meet in hell instead of Paris.

*Melissa: (desperately)* I'd rather be in hell than bound for Africa on this ship full of lunatics.

*Grayson:* My dear, you're bound for Paris. Don't, whatever you do, forget your destination. Now, listen! *(He pulls out his watch)* The old boy is due here now.

*Melissa: (hopefully)* Maybe he's not coming! Maybe he's decided to stay behind!

*Grayson: (menacingly)* He either goes to Africa or to jail! And if I know my Jeremiah, he goes to Africa. *(He chuckles mirthlessly)* There's more chance of glory there.

*Melissa:* What makes you so sure he'll go to jail? You told me, yourself, when we came back from the West Indies, that he was okay with the government and that you were just scaring him off to Africa.

*Grayson:* He was okay then! But he isn't now! Why, if the postal authorities knew he was sailing tonight, they'd be here right this minute ready to run him in. He's been spending the organization's money like a drunken sailor during the last six weeks, and what's he's got to show for it? Nothing! I've seen to that!

*Melissa:* And what've you got to show for it?

*Grayson:* Plenty! *(They laugh)*

*Melissa:* You're some slick guy!

*Grayson:* Remember that when you start carrying out my instructions on the other side. This boat hits the Azores in about ten days. You jump the ship there and take the first boat for Lisbon. Then you go on to Paris by train.

*Melissa:* I can't see how I'm going to get off this tub without everybody knowing it.

*Grayson: (impatiently)* I've arranged all that! How many times do I have to tell you?

*Melissa:* You've arranged it, but I've got to carry it out, and it don't look so good to me.

*Grayson:* As soon as you anchor at St. Michaels, Lorando's cousin will come aboard. He'll arrange to slip you over the side at night, row you ashore, and keep you in hiding until Jeremiah and his gang are on their way to Monrovia. It's simple enough!

*Melissa:* Suppose this guy doesn't show up!

*Grayson:* He's got to show up, or Lorando doesn't get his dough, and Lorando knows it. *(MELISSA sighs)* Now are you satisfied?

*Melissa:* I suppose so! *(She throws her arms around his neck)* I hate to leave you, though!

*Grayson:* It's the only way I can get rid of Jeremiah and the rest of them. And I've got to get rid of them if I'm going to finish cleaning up quick. We'll only be separated for a couple of weeks. Why, they'll be gone before you know it. Compared to that trip through the West Indies—

*Melissa: (disgustedly)* Don't mention it!

*Grayson:* I sail tomorrow *(he laughs)* with the remains of the treasury in my pocket. Sweetie, we'll be sitting together on top of the world in Paris! *(They kiss passionately)*

*Melissa: (murmuring)* My man!

*(They cling together. Footsteps are heard on the stairs. They separate hurriedly. BELL rushes in almost before they get apart. He is so excited, however, he hardly notices them)*

*Bell: (shouting)* He's coming now!

*Grayson:* Good! *(BELL immediately rushes back up the stairs. GRAYSON turns to MELISSA and says mockingly)* Be ready, my dear, to salaam your Emperor.

*Melissa:* I *am* ready—to slam him, the black—

*(The noise of many footsteps is heard on the stairs. GRAYSON interrupts MELISSA with a gesture of silence. JEREMIAH enters. He wears a black cape, and a cap drawn down low over his eyes. He is followed by BELL, three guards in uniform, the CAPTAIN, the first mate, SISTER TYSON, and BROTHER BAKER. The CAPTAIN and mate are in typical merchant marine uniforms. The CAPTAIN is the leader of the delegation of longshoremen in Act I who appeared in the dress uniform of a captain in Act II)*

*Grayson:* Well, Your Highness, you're here right on the dot!

*Jeremiah: (taking off his cap)* Thanks to your clever plan, Brother Grayson. *(He turns to MELISSA)* My dear, you got aboard all right?

*Melissa:* Yes, Your Highness!

*Jeremiah: (to GRAYSON)* My luggage?

*Bell: (interrupting)* It's in your cabin.

*Jeremiah:* Good! *(JEREMIAH starts to remove his cape. BELL and GRAYSON rush to assist him. Underneath the cape he has on his admiral's uniform. BELL throws the cape on the lounge. JEREMIAH sits down at the table. To MELISSA)* Melissa, my darling, you'd better let Lady Tyson take you to your stateroom.

*Melissa: (wearily)* Yes, Your Highness. *(With a swift glance at* GRAYSON, *she goes off at the left followed by* LADY TYSON)

*Jeremiah: (to* CAPTAIN) Everything ready for sailing?

*Captain:* Yes, Your Highness!

*Jeremiah:* You've got your clearing papers?

*Grayson: (interrupting)* I've attended to all that, Your Highness!

*Jeremiah: (to* GRAYSON) Good! *(to* CAPTAIN) That's all, just now. Should you want me I'll be in here until sailing time. You know, of course, the importance of your task. Under the protection of you and your crew, the first independent black colonists are on their way to reclaim their homeland. It is an epic journey! You should be a proud crew! You should be a proud captain!

*Captain:* Yes, Your Highness. *(He bows and goes off up the stairs)*

*Jeremiah:* Sir Baker!

*Brother Baker:* Yes, Your Highness!

*Jeremiah:* Will you see that all is in order in my stateroom?

*Brother Baker:* Yes, Your Highness! *(He bows and goes off at the left)*

*Jeremiah: (to* BELL *and* GRAYSON) Well, my two good brothers! My staunch left and right hand *(he looks at them for a thoughtful moment)* you are indispensable to me! I can never convey to either of you the high esteem I have for you both; for your loyalty; for your devotion; for your love!

*Grayson:* Thank you, Your Highness. It's your faith in us that keeps us going.

*Bell: (passionately)* Without your genius to inspire us, our efforts would have been futile, unavailing, without—

*(The door from the back at the rear opens.* KATE *steps in.* JEREMIAH *jumps up from his chair.* BELL *and* JEREMIAH *show their surprise)*

*Jeremiah: (sternly)* Kate!

*Kate: (with a threatening calm)* I had to see you before you left.

*Jeremiah:* I refuse to grant you an audience.

*Bell: (embarrassed)* Er—Mrs. Saunders, I think it best you go.

*Kate: (firmly to* JEREMIAH) I'm not going until I've talked with you *alone*!

*Grayson:* We can have you *taken* off, Mrs. Saunders.

*Bell: (nervously)* Let's not have a scene.

*Kate: (fiercely)* Try to use force, and before this ship sails, the Federal officers will know that Jeremiah Saunders is on board, ready to take flight.

*Jeremiah: (commandingly)* Kate!

*Grayson: (advancing towards her menacingly)* You can't bluff us!

*Kate: (a note of pleading coming into her voice)* Jeremiah, I want to see you alone.

*Bell:* I think it best, Your Highness! We'll retire! *(He starts out at the rear)* Come on Brother Grayson.

*Grayson: (looking straight at* JEREMIAH) I will do as His Highness wishes!

(JEREMIAH *looks from* GRAYSON *to* BELL, *and then to* KATE)

*Jeremiah: (without looking at* GRAYSON) I'll speak to her!

(GRAYSON *hesitates a moment, frowns, then follows* BELL *out.* JEREMIAH *sits down at the table. He is plainly perturbed, not knowing what to say)*

*Kate: (looking straight at* JEREMIAH) Jeremiah, my heart is broken. *(She pauses as if expecting an answer.* JEREMIAH *remains silent)* Jeremiah, after fifteen years! *(She sobs)* I can't believe it! I can't believe it!

*Jeremiah:* We all must pay for our folly!

*Kate: (suddenly flaring up)* And mark my word, Jeremiah Saunders, you'll pay for yours.

*Jeremiah: (sententiously)* There can be no folly in the line of duty! This is a heaven-blessed movement, and nothing matters but its welfare and success!

*Kate: (sarcastically)* So I'm being sacrificed to my own movement!

*Jeremiah: (impatiently)* You're being sacrificed to your own ambition and selfishness!

*Kate: (passionately)* Me—selfish; after the fifteen years I've slaved to help put you where you are today? Me, selfish; who married a jack-leg backwoods minister, who couldn't read or write, or hold a church? Me, who sacrificed and slaved and taught preparing you for this?

*Jeremiah: (unmoved)* There is nothing I can say to you, Kate. There is no way to reason with an unreasonable woman. You should pray, Kate. God does not love evil.

*Kate:* It is your own soul which stands in danger of damnation. The Bible says: "Thou shalt not commit adultery!"

*Jeremiah:* I have not committed adultery. I was legally divorced and legally married.

*Kate:* The legal way is not always the righteous one, Jeremiah. No, God does not love evil. You should know soon! You are certain to feel his wrath!

*Jeremiah: (again taking refuge in platitudes)* Those who have no good to speak of their fellowmen should remain silent!

*Kate: (almost screaming)* I have no intention of remaining silent. I shall tell the whole world of your hypocrisy—your perfidy. You, the first Emperor of Africa, led into the paths of sin by the devil's imps.

*Jeremiah: (harshly)* Who are you talking about?

*Kate: (holding back her sobs with difficulty)* You know who I'm talking about! You know very well!

*Jeremiah: (beginning in the voice of a fatherly preacher)* Oh, Kate! *(Then he pauses and shrugs his shoulders with a gesture of hopelessness)* The boat sails in a few minutes. You'd better go ashore.

*Kate:* Don't worry! I'll get ashore! I wouldn't sail on the same ship with you and your black concubine!

*Jeremiah:* There you go, Kate, referring contemptuously to the color of someone's skin. All people can't be yellow. *(passionately)* All of us have not been bastardized by the white man! To be black is to be honorable!

*Kate:* Honor is as honor does. There are black snakes as well as yellow ones. You've gotten on top and now that your nest is now full of eggs, the snakes are climbing in. They've ousted the mother bird and are now sucking the substance from the eggs, the eggs, Jeremiah, that belong to you and me.

*Jeremiah: (sarcastically)* You speak well, Kate. The movement lost a good orator when it lost you.

*Kate:* There is no place in the movement for one who is honest and frank. There is only room for the likes of your *"friend"* Grayson and your *"wife"* Melissa. *(She pronounces the words "friend" and "wife" with biting sarcasm)*

*Jeremiah:* They have both helped mightily. We could not have progressed to where we are today without them.

*Kate: (continuing her biting sarcasm)* Is it they who've helped you progress to where you have to flee the country to keep out of jail? *(Her tone becomes a bit less caustic)* He who lays down with dogs will surely get up with fleas.

*Jeremiah:* Do you want to know whose fault it is that I stand in the shadow of prison? Yours! Yours! Tuh! With all your intelligence you couldn't keep books properly; you were ignorant of postal regulations. Who first brought the postal authorities down upon us? Why, had it been left to you, I would have been in jail long before this.

*Kate: (portentiously)* You would have saved your soul!

*Jeremiah:* My soul's in God's hands! I'm but His humble servant!

*Kate: (again turning to sarcasm)* Did God tell you to go to Mexico to get a divorce and remarry without notifying me?

*Jeremiah: (somberly)* You've never done your duty as my wife.

*Kate: (surprised)* Not done my duty? Why—

*Jeremiah: (quickly)* The Lord says: "Be ye fruitful and multiply!" And you've remained barren, knowing it's imperative that I, Jeremiah, have an heir. In marrying Melissa, I'm following the path of old Father Abraham, who conceived a child by Hagar when his wife Sara proved sterile. Abraham followed the Lord's command, and I, Jeremiah, have done likewise.

*Kate: (in bitter defense)* I'd no time for a child. You were my child. You used up all the energy and mother love I might have given to our baby. *(pleadingly)* Oh, Jeremiah, open your eyes before it is too late. Can't you see that these vultures around you are leading you to destruction? They are picking at your flesh and soon your very bones will be exposed? Oh, Jeremiah, won't you— *(A bell rings loudly out on deck)*

*Jeremiah: (relieved)* You must go!

*Kate:* No, Jeremiah, I— *(The bell rings again)*

*Jeremiah:* Go—Kate! That's the warning bell. We're about to leave!

*Kate: (fiercely)* I'm not going. I won't let them have their way! I won't let them send your body to death and your soul to hell. The devil's host shall not—

(**BELL** *and* **GRAYSON** *come rushing in followed by numerous Knights and Ladies as well as Guards)*

*Bell:* We're getting ready to pull off.

*Grayson:* *(goes up to* JEREMIAH*)* Good-bye, Your Highness. I only wish I could make the trip with you. But I'll do my share here.

*Jeremiah:* Good-bye, my son, good-bye. I fear for nothing left in your hands. May God bless you. *(They shake hands)*

*Grayson:* *(turning to* KATE*)* You'd better come, Mrs. Saunders.

*Kate:* *(determinedly)* I'm not going!

*Grayson:* *(to the Guards)* Escort Mrs. Saunders off ship!

*(Two Guards start towards Kate)*

*Kate:* *(hysterically)* Don't you dare touch me!

*(They hesitate and look at* GRAYSON*)*

*Grayson:* *(firmly)* Escort her off ship!

*(They grab* KATE *by the arms)*

*Kate:* *(fighting them and yelling)* Leave me alone! I won't go! I won't go! Jeremiah! Jeremiah!

*(They drag her off at the rear sobbing and yelling)*

*Grayson:* *(saluting the entire company)* Good-bye and God bless you! *(He goes out at the rear)*

*Jeremiah:* Brother Grayson is a good, capable, loyal helpmate! He'll see that everything goes well while we are pioneering in the Promised Land.

*(The clatter of winches and the whistling of tugs is heard)*

*Bell:* *(excitedly)* We're at last leaving the land of our persecution!

*(*MELISSA *enters from the left followed by* LADY TYSON *and* BROTHER BAKER*)*

*Jeremiah:* *(taking* MELISSA *by the hand)* Ah, my children, tears well in my eyes! A sob chokes my throat! I feel the throbbing of the engines! I hear the chugging of the tugs! I feel the beating of your hearts! We are beginning an epic journey. We are the first of a great army of blacks returning to the black man's home. God is with us lighting our path across the dark and treacherous wastes of the sea. Let us sing—the black man's song of triumph!

*(The whole assemblage bursts into the "Black Internationale" as the curtain falls)*

### Scene Two

*Twenty-four hours later in the first-class salon of the S.S.* Jeremiah *now on the high seas. The salon is a garish and elaborately decorated room, equipped with gilt furniture, bevelled mirrors, and brass ornaments. The floor is carpeted and there is a huge crystal chandelier hanging from the ceiling in the center of the room. Along the rear wall are square windows looking out onto the deck. An arched door on the left leads in from the dining salon, and a double door on the right opens into a companionway. To get the maximum dramatic effect from this scene, it is necessary that the set be built upon a rocking floor, so the rocking of the boat can be made thoroughly realistic. Supper has just ended as the curtain rises and during the first few minutes, the various diners stroll in from the dining salon on the left. As this is the first-class salon, only Knights and Ladies and other important members of the "Fraternity" frequent it. They are all dressed in elaborate evening costumes akin in design to the costumes they wore in Act II, Scene Two. The boat is rocking slightly, and the Knights and Ladies take their steps a bit gingerly. A few sit down.*

*First Knight:* Did yo' enjoy yo' meal, Lady Smith?

*First Lady:* Yes, Lord Johnson. It was mighty fine.

*Second Lady:* Jes' a bit too greasy fo' me. I like greasy pork, but when de boat is a'slippin' an' a'sliddin' 'neath you, grease don't set so well.

*Second Knight:* Oh, Lady Mordecai, you find de sea rough?

*Third Lady:* I don't miss.

*Third Knight:* You ain't done nothin' but fuss 'bout de sea since we started, and we're only one day out. What'll you do when we get in de middle of de ocean?

*First Lady:* Well, de water *is* pretty troubled.

*First Knight:* I'm gonna ask de captain if he ain't got some oil to pour upon de waves.

*Second Knight:* Dat's right. Oil does soothe de troubled waters.

*Second Lady:* Why ain't some of you all thought of dat long time ago?

*(The* CAPTAIN *enters followed by several more men and women)*

*Third Knight:* Oh, Captain!

*Captain:* Yes, milord.

*First Knight:* Will you do us the favor of comin' here for a conference?

*Captain:* Yes, milord.

*First Knight:* Ain't you got some oil?

*Captain:* Oil?

*First Lady:* Yes, oil. Coal oil.

*Captain:* I think there's some barrels on deck.

*First Knight:* Well, put some on de waves.

*Captain:* On de waves?

*First Lady:* Where else you think we'd want it?

*First Knight:* We thinks it'd help to pacify de troubled waters.

*Captain:* Oh, I see—er, er—I'd betta speak to His Highness.

*First Knight:* We're de executive committee.

*Chorus:* We sho' is.

*First Knight:* An' we expect you to obey *our* orders.

*Chorus:* Tell him, Brother.

*Captain:* Yes, milord. *(He goes out at the right)*

*Second Lady:* Who he think he is, anyway?

*Third Lady:* Don't know who he thinks he is, but he sho' got a mighty fine 'pinion o' himself.

*First Knight:* Sho' has! Eatin' at de head table wid us and His Highness.

*First Lady:* Oh, darkies like him ain't used to nobility.

*Third Lady:* Dey shure ain't. Why, dere's that Lelia Brown a'kickin' 'cause she ain't got no first-class cabin. Got real cantankerous. What was she anyway 'fore she joined de movement?

*First Knight:* It takes His Highness to get 'em tol'. I thought I'd die when he tol' 'em— everybody can't give orders, some has to take 'em.

*Second Knight:* Dem West Indian niggers is sore 'cause dey ain't among de nobility. Says dey gonna transfer to de first ship we pass.

*First Lady:* Dey can't go too soon for me. I don't see why we took those geechee niggers 'long for anyway. I can't see no reason why dey should take 'vantage o' what we American Negroes get.

*Second Lady:* Ain't it de' truth? An' down on dem islands dey's scrapin' for pot lick. Dey ain't used to livin' like us.

*Third Knight:* An' did you see dat hincty old Liberian counsel puttin' on airs? Bet he nevah wore shoes 'till he came to America an' learnt from us.

*Third Lady:* Don't he kill you, talkin' 'bout de President o' Liberia? Seems to forget dat His Highness is gin' to rule ovah all Africa.

*First Lady:* Somethin' tells me dem heathen niggers in Africa ain't goin' to be so forty.

*First Knight:* Well, they betta respect de nobility.

*Second Lady:* Don' worry! Dey knows we Americans has all de money. Dey'll be glad we came over to rule 'em.

*(There is a sudden lurch of the boat)*

*Third Lady:* (as the vessel rights itself) Wonder did dat captain pour oil on de waves? Dis ship's rockin' worse'n evah.

*First Knight:* Yes, sir! Look at dat chandelier a'swingin'!

*First Lady:* I wouldn't want it to fall on me.

*First Knight:* Mus' weigh a ton.

*Second Knight:* I wonder where we is anyhow? I don't feel any too good.

*Third Knight:* Don't do no good to ask. I done asked dat captain fifty times an' all he says is som'pin' degrees dis and som'pin' degrees dat.

*First Knight:* Dat nigger would try to be hifalutin'.

*(There is a flash of lightning followed by a distant roll of thunder)*

*First Lady:* (looking out of window) I shure don' like dis storm.

*Second Lady:* It's beginnin' to thunder and lightnin' too. God knows I ain't got no use fo' no storms at sea.

*(The ship again lurches sharply)*

*Second Knight: (putting his head on his arms which are resting on a table)* I ain't got no use fo' nothin' on nobody's sea.

*Third Knight:* Me an' you both. Here I'm jes' gettin' so I can eat, and Lawd, Lawd, Lawd, de way dis ship am a'rollin'.

*(BELL enters with MELISSA from the dining salon on the left. They all bow to MELISSA who takes a seat in the middle of the salon)*

*First Knight:* We ain't carin' much 'bout dis storm, Sir Prince.

*Bell:* Oh, it isn't much. We ran into far worse weather down in the West Indies, didn't we, Your Highness?

*Melissa:* I'll say we did! The wind just swept the deck clean, and what it didn't do to that ship.

*First Lady:* Well, when I gets to Africa, I stays dere, 'cause I'm through wid de ocean.

*Third Lady:* You tell 'em, Lady Smith.

*(There is more lightning and thunder)*

*Bell:* We'll probably run out of this storm if the captain puts on some speed.

*Second Knight: (weakly, raising his head)* He oughta put on somethin'.

*First Lady:* We tol' him to pour oil on de water.

*Bell: (mystified)* Pour oil on the water?

*Second Knight:* Yes, to pacify de troubled waves.

*Melissa:* Well, I never— *(She laughs)*

*(The group of Knights and Ladies scowl at her and draw away. The room is filling up. There are numerous little groups)*

*Bell: (to MELISSA under his breath)* You shouldn't laugh at them.

*Melissa:* Who could keep from laughing? Pour oil on the water! *(She laughs again)*

*Bell:* You must remember that they haven't had our chance. That's why we're making this journey; to give them a place where they'll have a chance to develop their minds and their talents.

*Melissa:* You can't develop what they haven't got.

*Bell:* How do we know they haven't got it? Don't forget that the first white Americans were humble folk. *(The ship gives a sudden lurch. This time the thunder follows the lightning with great rapidity and there is a deafening crash. The SECOND KNIGHT moans loudly. There's a nervous murmuring from all sides)* This storm is getting bad. I'm going to see the Captain.

*Melissa:* He can't stop the storm.

*Bell:* No, but he can steady the ship more. A little speed will help a lot. The boat doesn't seem to be moving at all.

*Melissa: (laughing)* And he might pour oil on the water.

*(BELL starts toward the right just as the CAPTAIN enters. Everyone looks at him intently. He seems harassed and excited as he scans the assembly. Seeing BELL, he runs up to him)*

*Captain:* Your Highness, Prince Bell, de engines has stopped.

*Bell:* The engines have stopped! Well, why don't you start them?

*Captain:* We can't.

*Bell:* Nonsense. Get your men to work. We can't drift about in this storm.

*Captain: (flustered)* Yes, sir! *(He hurries out)*

*First Knight: (advancing to BELL)* What's dat?

*Bell:* Just a little engine trouble!

*First Knight:* In all dis storm!

*First Lady:* Oh, Lawd!

*Bell:* Calm down. Haven't you ever been on a ship before? These things often happen.
  *(There is a renewal of excited mumbling and whispering as the FIRST KNIGHT returns to his group)*

*Melissa:* If that captain knows how to run a boat, I'm an aviator!

*Bell: (worried)* Keep quiet! We can't let these people get excited.

*Melissa:* The water'll cool them off.

*(More lurching, lightning and thunder)*

*First Knight: (again advancing to BELL)* How 'bout de life boats?

*Bell:* They're all right. But we don't need them.

*First Knight:* Where are dey?

*Bell:* On the deck.

*First Knight:* Where 'bouts on de deck?

*Bell: (impatiently)* Oh, I don't know, but they're there.

*First Knight: (turning to his group)* Ain't I tol' you? Dey ain't none.

*Melissa: (a bit worried)* You mean to tell me there're no life boats on this ship?

*Bell:* Of course there're life boats, Your Highness. They're—

*First Knight: (interrupting)* No, dey ain't none. I done looked high and low.

*Bell:* Nonsense, all ships have life boats. They won't let you sail without them.

*First Knight:* They'd let niggers sail without 'em, hopin' they'd all drown.

*(The boat now lurches more suddenly and forcefully than ever. The lightning comes fast and the thunder crashes like the rapid fire of heavy artillery. The rain beats against the windows)*

*Chorus:* Oh, Lawd! Jesus! Have mercy, Father!

*Bell: (shouting)* Now don't get alarmed. Everything will be all right in a few moments. Get the captain, somebody. *(No one moves)* Lord Zacharia, get the captain. *(The THIRD KNIGHT goes out sullenly at the right)* Now calm down, all of you. Remember the purpose of your journey. Remember—

*(He is cut short by the ship's sudden lurching. This time some are thrown to the floor and others against the wall. There is much moaning and much crying out to the Heavenly Father. The chandelier swings dangerously)*

*Melissa: (tartly)* Why isn't Jeremiah in here?

*Bell:* He went to his cabin to rest.

*Melissa:* Depend on him to rest when trouble starts.

*Bell: (pleadingly)* Your Highness—

*Melissa:* Highness be damned. Send for him to come here. He probably doesn't know there's a storm. He's probably in there gassin' to my bodyguard—that old hen, Tyson, and that big wind bag, Baker. Send for him.

*Bell:* Sir Smith, go get His Highness.

*First Knight:* I ain't particular 'bout runnin' through dis ship now.

*Bell:* We need His Highness.

*First Lady: (moaning)* We need God Almighty to save us!

*Bell: (to* FIRST KNIGHT*)* I command you to go!

*(The* FIRST KNIGHT *goes out sullenly at the right)*

*Bell: (to* MELISSA*)* Please, Your Highness, be more discreet. We are in the midst of a cyclone, and it will take— *(His words are drowned out by the thunder. The howling of the wind can now be heard between crashes. The lights go dim. The terror of the crowd increases. Supplications to Heaven are shouted on all sides)* Please, my dear people, keep calm. It is only a passing thunder shower.

*(The* CAPTAIN *enters, followed by the* THIRD KNIGHT, *and four thoroughly frightened members of the crew)*

*Bell:* What does this mean? Why aren't these men at their posts?

*Captain:* De crew is scared, sir.

*Bell:* Scared! Oh, damn these fools! *(to the men)* Get back to your posts. *(to* CAPTAIN*)* Captain, you must get the engines started.

*Captain:* We can't, sir.

*Bell: (in impotent rage)* Damn you! You! You must!

*(*JEREMIAH *enters hurriedly, followed by the* FIRST KNIGHT*)*

*Jeremiah:* Brother Bell, what's all the trouble? Children, children!

*Captain: (helplessly)* De engines stopped.

*First Knight:* An' dey ain't no life boats.

*Captain:* Yes dey is!

*Chorus: (flocking around the* CAPTAIN, JEREMIAH *and* BELL*)* Where? Where are dey? Show us. Where?

*Captain:* On de top deck.

*(There is a rush for the doors on the right. The boat lurches and men and women fall over each other. This checks the stampede)*

*Jeremiah:* Stop! Stop! I tell you. What madness is this?

*First Knight:* We ain't fixin' to stay here an' be drowned!

*Third Knight:* We sure ain't!

*Jeremiah:* Drowned! Stay yourselves! You're not going to be drowned! You're not in danger! Where is your faith in the Lord and in me?

*(As the boat rights itself, there are moans and ejaculations of "Do Jesus! Do Jesus!" and the chanting of other ejaculatory prayers, such as "Oh, Father!" and "Have mercy, Lawd!")*

*Bell:* Tell them, Your Highness. Tell them there is no danger!

*Melissa:* A lot you know about it.

*Jeremiah:* Hush, 'Lissa. *(to the* CAPTAIN*)* Get up on the bridge where you belong.

> *(The* CAPTAIN *hurries out.* JEREMIAH *raises both arms in a gesture commanding attention and speaks at the top of his voice)* Listen my children. Can't you see the Lord is testing your strength? Don't let him find you wanting! Be strong! *(His voice is almost drowned out by the noise of the storm. The ship is dipping and rolling. The rain can be heard coming down in torrents. The wind is howling, the lightning flashing, and the waves echoing the roar of the thunder. The crowd in the salon grows more and more hysterical. One group kneels and prays loudly, discordantly. Another group shouts for mercy, others weep, and others just stare, too frightened to move or speak. All are constantly thrown off their balance by the sudden incalculable movements of the ship.* MELISSA *is growing hysterical too.* BELL *is excited.* JEREMIAH *alone is calm and fighting for order)* Children, hear me! Children, hear me! Your Highness, Jeremiah, commands you to silence. Think ye the Lord would let us be destroyed; we the first pioneers; we who are going to reclaim Africa for the Africans? Have you forgotten that I, Jeremiah, God's chosen, am leading you? Think you God would suffer my loss at this time? Look at me, the black Moses, and remember that God turned back the waters of the Red Sea and let them engulf the pursuing hosts of Pharaoh so that His children could escape into the Promised Land. I tell you, God is on our side! He will do as much for us, His black children, as He did for the children of Israel!

*(Throughout this speech, the storm seems to have momentarily abated. But now the boat tosses more violently than ever. The waves beat against it. There is the sound of heavy objects falling on deck, and the tear and creaking of equipment being torn loose and swept away)*

*First Knight:* I'm goin' fo' dem life boats. *(He starts for the door on the right followed by a wild mob)*

*Bell:* *(yelling)* Stop! Stop! *(The crowd keeps going. He turns frantically to* JEREMIAH*)* For God's sake, Your Highness, stop them! They'll all be washed overboard.

*(The crowd gets wedged in the narrow doorway. They are utterly mad with fear)*

*Jeremiah:* *(thundering)* I, Jeremiah, command you to stop!

*(The crowd ignores him and all press into the companionway and disappear)*

*Melissa: (shouting at* JEREMIAH*)* A hell of an emperor you are!

*Bell:* Your Highness!

*Jeremiah: (drawing back at the hate of her attack)* Why— Oh— 'Lissa—

*Melissa:* Why don't you stop the storm? Why don't you? You, God's chosen! *(She laughs hysterically)* You God-damned black bastard! You black Jesus! Go out and command the waves to stop! *(The two men stand aghast at her as she continues wildly)* Walk on the water! Let me see you walk on the water! You! God's chosen! *(She again laughs hysterically)*

*Jeremiah:* Melissa, Melissa. *(He tries to take her in his arms)*

*Melissa: (beating him on the breast with her clenched fists)* Don't you touch me! Keep your black paws to yourself! We'll all be in hell soon and you can paw coals! *(She laughs)* Black Emperor! His Highness! You're a hell of an emperor! You don't even know what it's all about. You're blind, deaf and dumb. Goin' to Africa! Where in the hell is there an Africa for you to go? Don't you know you'll be lugged back by a British or French battleship? Liberia? They don't want you dumb American niggers there. There's only one place for the likes of you and that's down in the cottonfields of Georgia and Mississippi. Emperor! Ha! Ha! *(She laughs madly)*

*Jeremiah: (dumbfounded)* Melissa, you don't know what you're saying.

*Melissa: (tauntingly as she holds herself on her feet with difficulty)* Oh, don't I? No, I don't know anything, do I? I don't know that tomorrow when we're all in hell, Grayson sails for France with the Treasury of the International Fraternity in his pocket. I don't know that in my purse is the address in Paris where I was to meet Grayson. Where *I* was to meet him! *(She pounds her chest)* I don't know that there was no real post office charge against you until Grayson framed you to get you out of the way. You dumb son-of-a-

*Bell: (leaping at* MELISSA *and grabbing her by the throat)* You slut! *(He begins choking her. She scratches and kicks as they sway about the room)*

*Jeremiah: (crying)* No! No! Don't! Don't!

*(The storm has now reached its very peak. The salon is a wreck. The ship rocks crazily. There is moaning, chanting, shouting, and groaning from the companionway.* JEREMIAH *sways hopelessly with the motion of the ship. He is stunned.* BELL, *a veritable madman, forces* MELISSA *to the ground. The ship gives one final gigantic lurch. The lights go out. The chandelier falls with a mighty crash. There's a horrible scream)*

*Jeremiah: (his voice booming at its loudest)* Oh, God, my Father, why hast Thou forsaken me?

*(The curtain falls)*

### Scene Three

*The* CAPTAIN's *cabin the following morning at dawn. The storm has knocked the loose furniture, and the floral offerings about, giving the place the appearance of having been the scene of a brawl. As the curtain goes up,* JEREMIAH *is sitting in one of the chairs at the table. The captain of a coast guard cutter is seated opposite him. The* CAPTAIN *of the S.S.* Jeremiah *stands back of his chief.* TYSON *and* BAKER *are seated on the lounge. The other Lords and Ladies are seated about the cabin. They all look badly battered in their disheveled costumes—a vivid contrast to the spick and span* COAST GUARD CAPTAIN. *From outside on deck can be heard the low moaning of many voices—mourners for those lost in the storm.*

*Coast Guard Captain:* Well, this is sure one hell of a mess. I can't make head or tails of it. Going to Africa. Why, man, you must have been headed due South. You're just off the coast of Maryland. Lucky I ran into you, or you'd probably have drifted to hell.

*Jeremiah: (very subdued)* It is God's will.

*Sister Tyson and Brother Baker:* Amen!

*Coast Guard Captain:* We'll tow you into Newport News. The authorities there can decide what to do with you.

*Jeremiah:* It is God's will.

*Sister Tyson and Brother Baker:* Amen!

*Coast Guard Captain: (to the* CAPTAIN *of the S.S.* Jeremiah*)* Have you checked up on your crew and passengers?

*Jeremiah: (answering for the* CAPTAIN*)* Yes! *(He turns to* BAKER*)* Have you the list?

*Brother Baker:* Yes, Your Highness! *(He gives him the list)*

*Coast Guard Captain:* How many are missing?

*(*JEREMIAH *looks over the list, but* BAKER *answers before him)*

*Brother Baker:* Seventeen!

*Coast Guard Captain:* How were they lost?

*Captain:* Durin' de storm, dey lost dere head, and tried to get away in one o' de life boats. It turned over and dey all dropped in de sea.

*Coast Guard Captain:* Jesus! *(He scratches his head)* Any other casualties? *(No one answers. The* CAPTAIN, *and the Lords and Ladies all look at* JEREMIAH. *The silence is oppressive. It makes the moaning on deck suddenly seem very loud.* JEREMIAH *clears his throat ponderously. The* COAST GUARD CAPTAIN *remarks, a bit impatiently)* Well, were there any other casualties?

*Jeremiah:* My wife! Eh—she was killed in the salon. The chandelier fell on her! It also killed my chief aide, Brandon Bell. It just missed me!

*Sister Tyson and Brother Baker:* It was God's will!

*Coast Guard Captain:* Their bodies?

*Jeremiah: (pointing to the door on the left)* There!

*Coast Guard Captain:* Huh! Pretty mess! *(He addresses the* CAPTAIN*)* Why didn't you send out an S.O.S.? *(No one answers)* Isn't there a wireless aboard? Huh?

*Captain:* Yes, sir.

*Coast Guard Captain:* Then why in the hell didn't you use it!

*Captain:* Our man couldn't work it.

*Coast Guard Captain:* I'll be damned. *(to* JEREMIAH*)* You're really responsible. You're the head of this expedition. Why didn't you get men who knew their business? Wonder any of you are alive.

*Jeremiah:* God was with us.

*Coast Guard Captain:* It's a damn good thing He was too. *(He rises)* We'll tow you in. *(to* CAPTAIN*)* Get your crew together, you! Come on.

*(He exits followed by the* CAPTAIN. *There is a silence after he leaves.* BAKER *sighs, and* TYSON *moans to herself. The rest of the Lords and Ladies sniff coldly. It is easy to sense their hostility to* JEREMIAH. *This serves to stimulate him. He rises and walks back and forth)*

*Jeremiah:* *(suddenly stopping and beginning to preach)* The Lord has spoken. "Be ye not discouraged!" "Out of the ruins shall rise a new nation more strong than ever before!" We go back to begin again. We go back, but we shall return. Africa must be for the Africans. I, Jeremiah, shall fulfill the prophecy. It is so written and God's word cannot be gainsaid. *(He stops to note the effect of his words on his followers. They remain cold to his exhortings. He begins with heightened eloquence)* Come, children, let's rejoice, rejoice that the Lord has saved us for His future works. We still have our plans. You still have your Jeremiah. Rejoice that with a fresh start we shall rise to even greater heights. Jeremiah, your Emperor, God's chosen, commands you to be brave.

*First Knight:* *(jumping up)* You ain't talkin' to me! I ain't fixin' to do nothin' but get home an' stay there! Come on, Hannah!

*(The* FIRST LADY *follows him out on deck. The rest, except* TYSON *and* BAKER, *get up and follow. As they go out they remark, "You tell him, brother! Ain't it de truth! Ain't gonna miss!" etc.)*

*Jeremiah:* *(plaintively)* Children! My children! *(They ignore him. He is left alone with* TYSON *and* BAKER. TYSON *is now sobbing loudly.* JEREMIAH *raises his arms and eyes heavenward)* It is written, "Ethiopia shall stretch forth her wings." God's word cannot be gainsaid. Christ had His Judas! Christ was crucified! But His cause has triumphed. Thus shall it be with Jeremiah!

*(The moaning on deck continues as the curtain falls)*

# Part Eight

# Excerpts from the Novels

Since most of Thurman's other writings remain widely unavailable, readers know his work primarily through *The Blacker the Berry* and *Infants of the Spring*. *Blacker* has been available as a paperback since 1970, when it was reprinted by Collier-MacMillan with an introduction by Therman B. O'Daniel. *Infants* has been available as a paperback since 1992, when it was reprinted as part of the Northeastern Library of Black Literature. Both are often read in courses on the Harlem Renaissance or in survey courses in African American literature. By including in this *Reader* excerpts not only from *Blacker* and *Infants* but also from

THE BLACKER THE BERRY

WALLACE THURMAN

*The Interne*—a novel that has never been reprinted and is seldom mentioned in discussions of Thurman or the Harlem Renaissance—we hope to provide the opportunity to read these novels within the larger context of Thurman's complete oeuvre, including writings published here for the first time.

His three novels reflect Thurman's lifelong propensity for unusual subjects. *The Interne* is written in the muckraking tradition to expose the terrible conditions in public hospitals, and *Blacker* boldly explores the theme of intraracial color prejudice that most African Americans at the time would have preferred to keep a community secret. *Infants* is the only book-length satire from the period on the Harlem Renaissance: the flaws and foibles of its major participants, its unfulfilled promises, and its tragic failures.

---

*The Blacker the Berry*

---

Published in February 1929, Wallace Thurman's first novel appeared at an extremely productive time for him. Three and a half years after arriving in New York City, he had a banner year in 1929 with the publication of *The Blacker the*

*Berry* and the Broadway production of the play *Harlem: A Melodrama of Negro Life*. Like *Harlem*, *Blacker* is a testament to the energy and variety of Harlem; it has its share of sweetbacks, numbers runners, dicty Negroes, and house parties—including a literary salon that anticipates the famous gathering of writers at the Niggeratti Manor in *Infants of the Spring*. Indeed, in exploring both intraracial and interracial elements of African American identity, *Blacker* embodies one of Thurman's primary concerns in all of his novels, essays, and plays.

Thurman's contemporaries gave *Blacker* mixed reviews, complaining about Thurman's inability to dramatize his materials and his less than successful effort to give his "obsession" an "effective form." The anonymous reviewer in *The New York Times Book Review* was perhaps typical when he described the novel "as a merely competent, somewhat amorphous story. . . . [Thurman] reports where he should be dramatizing the world." Eunice Hunton Carter in *Opportunity* was critical of "the immaturity and gaucherie of the work," but Dewey R. Jones in the *Chicago Defender* declared, "Here at last is the book for which I have been waiting, and for which you have been waiting, whether you admit it or not."

The novel, whose protagonist is haunted by the darkness of her skin, has often been read in relation to Thurman's own dark skin—a reading inspired in part by Hughes's stress in *The Big Sea* on Thurman's dark skin as a source of his bitterness. In such a reading the obvious autobiographical link is overemphasized, obscuring the novel's fluent challenge to the terms and limits of the American discourses on race. More than merely a tale of the dark-skinned Emma Lou's woes within her family and community, *Blacker* comments incisively on the "tragic mulatta" tradition in African American literature. The novel's wry depictions of its heroine's racially and sexually charged encounters fashion a critique of black America's many contradictions in relation to color, class, national origin, gender, and sexuality.

In his review of *Blacker*, W.E.B. Du Bois remarked on the novel's mingling of race with "scandalous" elements of urban life. While hailing the importance of the novel's topic, Du Bois criticizes Thurman's execution. Du Bois finds the novel underdeveloped, confused between fact and fiction, hovering between low-down dirtiness and racial uplift. While astutely observing the centrality of sexuality to the novel, Du Bois does not acknowledge the significant nexus of race, gender, and sexuality in the narrative. Measured against oppositional models of identity, this racial-sexual convergence seems incomplete to Du Bois: "Indeed, there seems to be no real development in Emma's character; her sex life never becomes nasty and commercial, and yet nothing in her seems to develop beyond sex."

But this conjunction of sexuality and blackness is what makes *Blacker* a fascinating document of 1920s Afro-America. While Du Bois disapproves of the novel's licentiousness, sexual expression may in fact be the novel's essence, especially when one considers some of Thurman's other texts—for instance,

"Cordelia the Crude" and the further unfolding of that character in the play *Harlem*. The novel merges the many binaries that didactic critics such as Du Bois had created—it is both lowdown and middle class, both bohemian and proper, both factual and fictional. Rather than "uplifting the race," *Blacker* never lets the reader forget that race and sexuality are complex, slippery constructions based at least as much on perspective as on empirical realities—something Du Bois definitely knew well, having linked "double consciousness" in *Souls of Black Folk* so eloquently to "this sense of always looking at one's self through the eyes of others."

The key to reevaluating *Blacker* may be found not in Thurman's treatment of color prejudice but in his portrayals of Harlem's sexual and moral diversity, in his questioning of the parameters by which most Americans conceptualize and discuss race and its relationship to gender, class, and sexuality. In the course of her Harlem adventures, Emma Lou Morgan negotiates her way through a series of sexualized encounters that shape the whole narrative. A lesbian landlady proposes that Emma "join the fun" by moving into her house, Emma has furtive meetings in movie theaters for casual sex, and she discovers boyfriend Alva's bisexuality at the end of the book—a discovery that hastens her final change of consciousness. Like Dreiser's *Sister Carrie* and Cordelia in *Harlem,* Emma Lou, too, can be viewed as an exemplar of the young woman adrift in the city.

By placing its protagonist in conflict with an African American community that defines identity according to a hierarchy of color codes, the novel interrogates the gap between skin color and identity, questioning the nature of race and color consciousness. In a number of instances, the awareness of the discontinuity between appearance and identity is made overt. This awareness reads race as a constructed performance rather than a natural, given fact. Caught up in the hierarchy of color and performance early in the novel, Emma Lou sees her own skin as a mask that prevents her from being who she really is: "What she needed was an efficient bleaching agent, a magic cream that would remove this unwelcome black mask from her face and make her more like her fellow men." The "natural" and the "adopted" get confused in such perceptions during the course of Emma Lou's journey toward a comfortable fit between her mask and her life. In a sense, then, *The Blacker the Berry* is a study of nonessentialized, denatured constructions of the self. Thurman catalogs myriad racial and sexual performances to confound the notion that identity is monolithic, natural, essential. By painting all behaviors with the brush of performance, Thurman collapses dichotomies of identity and converges them into the tropes of mask, stage, and performance. As the novel questions the fixity of race, it situates that blackness in an environment of constructed and performative identity that allows for a diversity of experiences.

Looking at Thurman's own reactions to his novel, there is little doubt that he had expected to achieve much more than just an exploration of intraracial color prejudice. In his autobiographical "Notes on a Stepchild" (in *Aunt Hagar's Children*), he regrets that his attempt to deal with issues beyond those of color failed to rise above what he felt was propaganda. While it would be a mistake to judge the success of Thurman's published works by his own harsh statements or standards, Thurman views *Blacker* as fatally flawed:

> Thus he summed up his own people, and thus he had been most surprised to realize that after all his novel had been scorched with propaganda. True, he had made no mention of the difficulties Negroes experience in a white world. On the contrary he had concerned himself only with Negroes among their own kind, trying to interpret some of the internal phenomena of Negro life in America. His book was interesting to read only because he had lain bare conditions scarcely hinted at before, conditions to which Negroes choose to remain blind and about which white people remain in ignorance. But in doing this he realized that he had fixed the blame for these conditions on race prejudice, which manifestation of universal perversity hung like a localized cloud over his whole work.

As Theophilus Lewis puts it in an article entitled "Wallace Thurman Is a Model Harlemite"—*Blacker* is a "novel of which [Thurman] ought to proud, but isn't." It is a document of both racial and sexual self-consciousness that attempts to avoid the "universal perversity" of prejudice. Instead of avoiding this racial-sexual convergence in his next novel, *Infants of the Spring*, Thurman addresses it directly in a multiracial environment. In *Infants,* he attempts to get beyond race by staging discussions about it between his white and black characters.

---

## Infants of the Spring

---

In *Infants of the Spring* (1932), Thurman stages his analysis of the demands of art and the exigencies of race among the white and black habitués of "Niggeratti Manor"—an ironic label coined by Thurman and Hurston for Thurman's quarters at 267 West 136th Street, where younger writers and artists frequently gathered for conversation and bonhomie. Indeed, in several letters to Langston Hughes (see Part Three), Thurman expresses his desire to write about "267," complaining at one point that "my new novel is proving almost too much of a task."

It appears that Thurman had started working on this satirical account of himself and his peers as early as 1929. The book was originally accepted by another publisher in March 1930, but was not published for unknown reasons. It was finally published in February 1932 by the Macaulay Company, to mixed reviews. For instance, *The New York Times* noted condescendingly that even though Harlem novels were "somehow in a class by themselves," *Infants* was "a pretty inept book" by any standards. Martha Gruening, in *The Saturday Review of Literature,* welcomed the book, noting that Thurman was the only writer who had attempted "to debunk the Negro Renaissance." Almost a year after the book was published, Alain Locke, in his annual retrospective of new literary writing entitled "Black Truth and Black Beauty," pronounced that *Infants* had "a capital theme," making it all the more regrettable then that it lacked "fire." Thurman himself reviewed this novel in an unpublished essay (see "*Infants of the Spring*, by Wallace Thurman" in Part Four) and also commented on the circumstances of the novel in a letter dated January 30, 1932, to Granville Hicks: "I agree with you. It is a most disappointing novel. I didn't realize what a job I had undertaken until the first draft had been accepted by the publisher, and I had pocketed the advance. I tried my darnedest to back out of the deal. I wanted another year to work on the book. But I had spent the five hundred dollars. I was broke. The publisher wanted his manuscript. So there you are." From these documents, it is clear that Thurman's aspirations for the novel were not completely fulfilled.

*Infants* is a thinly veiled *roman à clef,* chronicling the parties and passions of a group of Negro artists who gather at Euphoria Blake's rooming house. These artists, with their differing temperaments, talents, and ideas embody the warring aspirations, attitudes, and expectations that swirled around Thurman and the Harlem Renaissance as he viewed it. More than *Blacker, Infants* is marked by frequent discussions between characters about art and race. In fact, one could read the novel as a series of quasi-Socratic dialogues. In particular, the discussions between Raymond, growing ever more weary of Harlem, and Stephen, the white man raised in Europe who is infatuated with Harlem, serve as focal points in the novel. In fact, the evolving but often understated relationship between the two men has been viewed by some critics as more fulfilling, more emotional, and possibly more erotic than any other in the book and is at the center of such discussions scattered throughout the novel.

As the narrative veers toward the anticipated closing of the Niggeratti Manor—where not art but booze reigns supreme—Raymond Taylor, like Thurman himself, views Harlem and its Renaissance with increasing doubt and skepticism. In Raymond's need to finally reject Harlem in order to move forward, the book reflects the love-hate relationship Thurman had with Harlem most of his adult years. The novel captures some of Thurman's initial ecstasy about the place—as expressed in his Harlem essays—as well as the later ambivalence that is reflected in his letters, his writing of the now-lost play *Goodbye New York,* and in his extended stays away from Manhattan to write. Thurman, it appears, needed the stimulation of Harlem for his creativity to flourish but also needed the more peaceful, less distracting surroundings of the suburb or the countryside to get his artistic work done. If his chronic health problems and self-destructive bohemianism had not cut his life short, Thurman might have resolved these conflicts in productive ways. But that was sadly not to be.

And yet Thurman's lifelong struggle as an artist and black man to achieve a measure of individuality against the "racial mountain" is a heroic failure that invites admiration despite its sadness. Despite their nearly opposite appraisals of the value of *Infants,* historians David Levering Lewis and Nathan Huggins have both noted the novel's focus on individuality as the price that one must pay for racialization and as a source of strength in the fight against the consequences of race. For Lewis, "the novel is so poorly done it hardly seems possible that the best-read, most brilliant, and most uncompromising of the Harlem artists could have written it," and yet he views it as "a valuable first intellectual statement of the malaise of the Renaissance"(*When Harlem Was in Vogue,* 277, 280). Lewis refers to Raymond, who declares: "Individuals will arise and escape on the ascending order of individuality. The others will remain where they are." For us, by exploring issues of individuality and race in nuanced ways, Thurman is not necessarily adopting the extreme position that "failure to become a great artist meant failure to transcend being a Negro," as Lewis would see it (278). Just as it is possible but not entirely justified to see Countee Cullen as "self-hating" on the basis of poems such as "Yet Do I Marvel," any charge of self-hatred laid at Thurman's door would be hard to support with reference to his whole oeuvre. As Raymond asserts in chapter 20 of *Infants*: "I'm sick of discussing the Negro problem, of having it thrust at me from every conversational nook and cranny. I'm sick of whites who think I can't talk about anything else, and of Negroes who think I shouldn't talk about anything else. I refuse to wail and lament. My problem is a personal one, although I most certainly do not blind myself to what it means to be a Negro."

Many readers have viewed *Infants* as an elegy to the Renaissance. Its final scene—Paul Arbian's body in the bathtub with a few of his spirit portraits hung on the wall and the pages from his manuscript strewn all over the bathroom floor, barely legible from the overflowing water—is seen as symbolic of the end of the Harlem Renaissance, mired in decadence, propaganda, and what Huggins describes as "the self-conscious promotion of art and culture typified by Alain Locke" (*Harlem Renaissance,* 241). But Huggins sees the novel as ending on a more positive note, claiming: "Thurman's message was delivered more by Raymond than by Paul." As a talented individual who struggles through his self-doubt, Raymond expects to finish his novel. When read in conjunction with Thurman's autobiographical sketches as well as his literary essays and reviews, *Infants* provides broad hints and clues on the ways in which Thurman preferred to confront the confounding challenges of racial expression and art in the U.S. context. Calling *Infants* "one of the best written and most readable novels of the period" (241), Huggins praises Raymond (Thurman) for his search for artistic integrity even as he grappled with the many issues faced by all African American artists: "the Negro as artist or advocate, the writer as individual or race man, art as self-expression or exposition of ethnic culture" (193). In *The Signifying Monkey,* Henry Louis Gates Jr. describes Thurman as a "most thoughtful literary critic" and suggests that *Infants* anticipates the critique of the 1920s that Ishmael Reed attempts in *Mumbo Jumbo* (224).

But Raymond and his relationships reflect more than the range and complexity of artistic debates in *Infants*. They also expose the sexual subtext of the novel. John A. Williams has described Ray's relationships with both Lucille and Steve as "false" and rightly observed that in this novel, "more wants to happen than the repeated topics on race relations, but it does not." Paul Arbian's openly acknowledged bisexuality, when juxtaposed against the Ray-Steve relationship, only underscores the tension between the cagey surface text and the subtext, which seeks to explore the full expression of friendship and love through sexuality. As *Blacker* corroborates, sexuality is a significant aspect of Thurman's literary imagination and his fiction anticipates in some ways the world of James Baldwin's major novels (especially beginning with *Giovanni's Room*). Like Thurman's writing, Baldwin's novels articulate themes of identity, love, and friendship to achieve, in Horace A. Porter's words, "an explicit exploration of the homosexual as the Other, to be either granted or perpetually refused the golden promise of America's democratic dreams" (150). While Cheryl Wall in her article "Paris and Harlem" has compared the Harlem of the 1920s in its liberating influence on most African American writers to Paris for the white American writers of the period, Thurman's depiction of the trial of his character Pelham Gaylord is muted. The trial and the consequent fall of the Manor provide sufficient internal evidence to appreciate the fact that Thurman could not have dealt with such explosive material more explicitly. So while issues of sexual orientation are intriguingly buried in the text of *Infants,* it would be a mistake to read the novel simply as a fictional treatise on "gay rights." And yet in *Infants* Thurman envisions a world in which the individuality and humanity of each person—regardless of color, racial background, or sexual orientation—can find full expression in friendship, in love, and even in art.

### The Interne

Thurman's last novel, *The Interne* (1932), was cowritten with Abraham L. Furman, a lawyer and brother of the publishers at the Macaulay Company. Apparently, Furman introduced Thurman to the Hollywood producer Bryan Foy. Thurman wrote two screenplays for Foy—*Tomorrow's Children* and *High School Girl*—that resemble *The Interne* in their focus on exposing the failings of social institutions.

*The Interne* is a fictional exposé of the horrible conditions at the City Hospital on Welfare Island. Thurman—who as a sickly child and youth had aspired to a medical career—often viewed medical institutions critically. The story, which has no black characters, is framed by the search for love and meaning of Carl Armstrong, a young and idealistic medical graduate about to start a career at a big public hospital. The novel documents the protagonist's "appalling situation," according to the book jacket, in the midst of "fire trap hospital buildings, in-sanitary wards, lack of men and equipment, administration riddled with politics," and much more. In his own research for this novel, Thurman was so disturbed by the conditions at the Welfare Island hospital that he vowed "never to set foot again in the place." One of the final ironies of his life was that he was forced to spend the last six months of his life in the incurable ward of the same City Hospital that he criticizes in *The Interne*.

The reviews of *The Interne* were mostly negative—typified by *The New York Times Book Review*'s description of the book as "tawdry," "squalid," and "fla-grantly exaggerated." At least one person who praised the book appears to have done so for the wrong reasons. Clifford Mitchell of *The Washington Tribune* liked the book because he found nothing "racial" about it: "[W]e grasp the fact that all characters in the book must be white, but Thurman seems quite as familiar with the motives and characteristics of the whites as his previous works indicate his familiarity with racial characters." In a letter to Granville Hicks, Thurman writes that this was his "first attempt to do something which does not concern Negroes. It may be my last."

In the same letter, Thurman mentions that he had a "tremendous idea for another Negro novel. And that is the one, I'm going to mull over and nurture, for if it is properly done it will redeem my self-esteem, and give me faith enough in myself to continue my scribblings." Sadly, this—surely a lifelong dream—was an ambition that Thurman was never to fulfill in his short life.

Like most of his plays, short fiction, and essays, Thurman's three novels repre-sent his participation in the ongoing struggle to understand and transcend the peculiar U.S. burdens of race, gender, and class that limit the aspirations, life-chances, and freedoms of individuals and their potential for personal happiness and achievement. As he writes in his "Notes on a Stepchild": "It would be his reli-gious duty to ferret deeply into himself—deeply into his race, isolating the ele-ments of universality, probing, peering, stripping all in the interests of garnering literary material to be presented truthfully, fearlessly, uncompromisingly." While each of us would come to different conclusions about how well Thurman suc-ceeded in the project he set himself, there can be no doubt that he made the attempt "truthfully, fearlessly, uncompromisingly."

# The Blacker the Berry

Published by the Macaulay Company, 1929.

## From Part 1

More acutely than ever before Emma Lou began to feel that her luscious black complexion was somewhat of a liability, and that her marked color variation from the other people in her environment was a decided curse. Not that she minded being black, being a Negro necessitated having a colored skin, but she did mind being too black. She couldn't understand why such should be the case, couldn't comprehend the cruelty of the natal attenders who had allowed her to be dipped, as it were, in indigo ink when there were so many more pleasing colors on

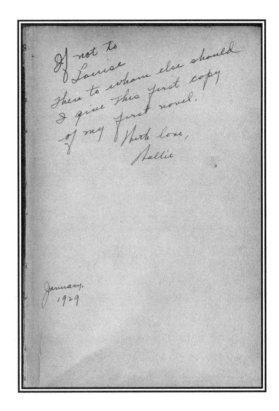

nature's palette. Biologically, it wasn't necessary, either; her mother was quite fair, so was her mother's mother, and her mother's brother, and her mother's brother's son; but none of them had had a black man for a father. Why *had* her mother married a black man? Surely there had been some eligible brown-skin men around. She didn't particularly desire to have a "high yaller" father, but for her sake certainly some more happy medium could have been found.

She wasn't the only person who regretted her darkness, either. It was an acquired family characteristic, this moaning and grieving over the color of her skin. Everything possible had been done to alleviate the unhappy condition, every suggested agent had been employed, but her skin, despite bleachings, scourgings, and powderings, had remained black—fast black—as nature had planned and effected.

She should have been a boy, then color of skin wouldn't have mattered so much, for wasn't her mother always saying that a black boy could get along, but that a black girl would never know anything but sorrow and disappointment? But she wasn't a boy; she was a girl, and color did matter, mattered so much that she would rather have missed receiving her high school diploma than have to sit as she now sat, the only odd and conspicuous figure on the auditorium platform of the Boise high school. Why had she allowed them to place her in the center of the first row, and why had they insisted upon her dressing entirely in white so that surrounded as she was by similarly attired pale-faced fellow graduates she resembled, not at all remotely, that comic picture her Uncle Joe had hung in his bedroom? The picture wherein the black, kinky head of a little red-lipped pickaninny lay like a fly in a pan of milk amid a white expanse of bedclothes.

But of course she couldn't have worn blue or black when the call was for the wearing of white, even if white was not complementary to her complexion. She would have been odd-looking anyway no matter what she wore and she would also have been conspicuous, for not only was she the only dark-skinned person on the platform, she was also the only Negro pupil in the entire school, and had been for the past four years. Well, thank goodness, the principal would soon be through with his monotonous farewell address, and she and the other members of her class would advance to the platform center as their names were called and receive the documents which would signify their unconditional release from public school.

As she thought of these things, Emma Lou glanced at those who sat to the right and left of her. She envied them their obvious elation, yet felt a strange sense of superiority because of her immunity for the moment from an ephemeral mob emotion. Get a diploma? —What did it mean to her? College? —Perhaps. A job? —Perhaps again. She was going to have a high school diploma, but it would mean nothing to her whatsoever. The tragedy of her life was that she was too black. Her face and not a slender roll of ribbon-bound parchment was to be her future identification tag in society. High school diploma indeed! What she needed was an efficient bleaching agent, a magic cream that would remove this unwelcome black mask from her face and make her more like her fellow men.

"Emma Lou Morgan."

She came to with a start. The principal had called her name and stood smiling down at her benevolently. Some one—she knew it was her Cousin Buddie, stupid imp—applauded, very faintly, very provokingly. Some one else snickered.

"Emma Lou Morgan."

The principal had called her name again, more sharply than before and his smile was less benevolent. The girl who sat to the left of her nudged her. There was nothing else for her to do but to get out of that anchoring chair and march forward to receive her diploma. But why did the people in the audience have to stare so? Didn't they all know that Emma Lou Morgan was Boise high school's only nigger student? Didn't they all know—but what was the use. She had to go get that diploma, so summoning her most insouciant manner, she advanced to the platform center, brought every muscle of her little limbs into play, haughtily extended her shiny black arm to receive the proffered diploma, bowed a chilly thanks, then holding her arms stiffly at her sides, insolently returned to her seat in that foreboding white line, insolently returned once more to splotch its pale purity and to mock it with her dark, outlandish difference.

## From Part 4

Saturday evening. Alva had urged her to hurry uptown from work. He was going to take her on a party with some friends of his. This was the first time he had ever asked her to go to any sort of social affair with him. She had never met any of his friends save Braxton, who scarcely spoke to her, and never before had Alva suggested taking her to any sort of social gathering either public or semi-public. He often took her to various motion picture theaters, both downtown and in Harlem, and at least three nights a week he would call for her at the theater and escort her to Harlem. On these occasions they often went to Chinese restaurants or to ice cream parlors before going home. But usually they would go to City College Park, find an empty bench in a dark corner where they could sit and spoon before retiring either to her room or to Alva's.

Emma Lou had, long before this, suggested going to a dance or a party, but Alva always countered that he never attended such affairs during the summer months, that he stayed away from them for precisely the same reason that he stayed away from work, namely, because it was too hot. Dancing, said he, was a matter of calisthenics, and calisthenics were work. Therefore it, like any sort of physical exercise, was taboo during hot weather.

Alva sensed that sooner or later Emma Lou would become aware of his real reason for not taking her out among his friends. He realized that one as color-conscious as she appeared to be would, at some not so distant date, jump to what for him would be uncomfortable conclusions. He did not wish to risk losing her before the end of summer, but neither could he risk taking her out among his friends, for he knew too well that he would be derided for his unseemly preference for "dark meat," and told publicly without regard for her feelings, that "black cats must go."

Furthermore he always took Geraldine to parties and dances. Geraldine with her olive colored skin and straight black hair. Geraldine, who of all people he pretended to love, really inspired him emotionally as well as physically, the one person he conquested without thought of monetary gain. Yet he had to do something with Emma Lou, and release from the quandary presented itself from most unexpected quarters.

Quite accidentally, as things of the sort happen in Harlem with its complex but interdependable social structure, he had become acquainted with a young Negro writer, who had asked him to escort a group of young artists and writers to a house-rent party. Though they had heard much of this phenomenon, none had been on the inside of one, and because of their rather polished manners and exteriors, were afraid that they might not be admitted. Proletarian Negroes are as suspicious of their more sophisticated brethren as they are of white men, and resent as keenly their intrusions into their social world. Alva had consented to act as cicerone, and, realizing that these people would be more or less free from the color prejudice exhibited by his other friends, had decided to take Emma Lou along, too. He was also aware of her intellectual pretensions, and felt that she would be especially pleased to meet recognized talents and outstanding personalities. She did not have to know that these were not his regular companions, and from then on she would have no reason to feel that he was ashamed to have her meet his friends.

Emma Lou could hardly attend to Arline's change of complexion and clothes between acts and scenes, so anxious was she to get to Alva's house and to the promised party. Her happiness was complete. She was certain now that Alva loved her, certain that he was not ashamed or even aware of her dusky complexion. She had felt from the first that he was superior to such inane truck, now she knew it. Alva loved her for herself alone, and loved her so much that he didn't mind her being a coal scuttle blond.

Sensing something unusual, Arline told Emma that she would remove her own make-up after the performance, and let her have time to get dressed for the party. This she proceeded to do all through the evening, spending much time in front of the mirror at Arline's dressing table, manicuring her nails, marcelling her hair, and applying various creams and cosmetics to her face in order to make her despised darkness less obvious. Finally, she put on one of Arline's less pretentious afternoon frocks, and set out for Alva's house.

As she approached his room door, she heard much talk and laughter, moving her to halt and speculate whether or not she should go in. Even her unusual and high-tensioned jubilance was not powerful enough to overcome immediately her shyness and fears. Suppose these friends of Alva's would not take kindly to her? Suppose they were like Braxton, who invariably curled his lip when he saw her, and seldom spoke even as much as a word of greeting? Suppose they were like the people who used to attend her mother's and grandmother's teas, club meetings, and receptions, dismissing her with — "It beats me how this child of yours looks so unlike the rest of you . . . Are you sure it isn't adopted?" Or suppose they were like the college youth she had known in Southern California? No, that couldn't be. Alva would never invite her where she would not be welcome. These were his friends. And so was Braxton, but Alva said he was peculiar. There was no danger. Alva had invited her. She was here. Anyway she wasn't so black. Hadn't she artfully lightened her skin about four or five shades until she was almost brown? Certainly it was all right. She needn't be a foolish ninny all her life. Thus, reassured, she knocked on the door, and felt herself trembling with excitement and internal uncertainty as Alva let her in, took her hat and coat, and proceeded to introduce her to the people in the room.

"Miss Morgan, meet Mr. Tony Crews. You've probably seen his book of poems. He's the little jazz boy, you know."

Emma Lou bashfully touched the extended hand of the curly-headed poet. She had not seen or read his book, but she had often noticed his name in the newspapers and magazines. He was all that she had expected him to be except that he had pimples on his face. These didn't fit in with her mental picture.

"Miss Morgan, this is Cora Thurston. Maybe I should'a introduced you ladies first."

"I'm no lady, and I hope you're not either, Miss Morgan." She smiled, shook Emma Lou's hand, then turned away to continue her interrupted conversation with Tony Crews.

"Miss Morgan, meet . . . ," he paused, and addressed a tall, dark yellow youth stretched out on the floor, "What name you going by now?"

The boy looked up and smiled.

"Why, Paul, of course."

"All right then, Miss Morgan, this is Mr. Paul, he changes his name every season."

Emma Lou sought to observe this person more closely, and was shocked to see that his shirt was open at the neck and that he was sadly in need of a haircut and a shave.

"Miss Morgan, meet Mr. Walter." A small slender dark youth with an infectious smile and small features. His face was familiar. Where had she seen him before?

"Now that you've met every one, sit down on the bed there beside Truman and have a drink. Go on with your talk folks," he urged as he went over to the dresser to fill a glass with a milk colored liquid. Cora Thurston spoke up in answer to Alva's adjuration:

"Guess there ain't much more to say. Makes me mad to discuss it anyhow."

"No need of getting mad at people like that," said Tony Crews simply and softly. "I think one should laugh at such stupidity."

"And ridicule it, too," came from the luxurious person sprawled over the floor, for he did impress Emma Lou as being luxurious, despite the fact that his suit was unpressed, and that he wore neither socks nor necktie. She noticed the many graceful gestures he made with his hands, but wondered why he kept twisting his lips to one side when he talked. Perhaps he was trying to mask the size of his mouth.

Truman was speaking now, "Ridicule will do no good, nor mere laughing at them. I admit those weapons are about the only ones an intelligent person would use, but one must also admit that they are rather futile."

"Why futile?" Paul queried indolently.

"They are futile," Truman continued, "because, well, those people cannot help being like they are—their environment has made them that way."

Miss Thurston muttered something. It sounded like "hooey," then held out an empty glass. "Give me some more firewater, Alva." Alva hastened across the room and refilled her glass. Emma Lou wondered what they were talking about. Again Cora broke the silence, "You can't tell me they can't help it. They kick about white people, then commit the same crime."

There was a knock on the door, interrupting something Tony Crews was about to say. Alva went to the door.

"Hello, Ray." A tall, blond, fair-skinned youth entered. Emma Lou gasped, and was more bewildered than ever. All of this silly talk and drinking, and now—here was a white man!

"Hy, everybody, Jusas Chraust, I hope you saved me some liquor." Tony Crews held out his empty glass and said quietly, "We've had about umpteen already, so I doubt if there's any more left."

"You can't kid me, bo. I know Alva would save me a dram or two." Having taken off his hat and coat he squatted down on the floor beside Paul.

Truman turned to Emma Lou. "Oh, Ray, meet Miss Morgan. Mr. Jorgenson, Miss Morgan."

"Glad to know you; pardon my not getting up, won't you?" Emma Lou didn't know what to say, and couldn't think of anything appropriate, but since he was smiling, she tried to smile too, and nodded her head.

"What's the big powwow?" he asked. "All of you look so serious. Haven't you had enough liquor, or are you just trying to settle the ills of the universe?"

"Neither," said Paul. "They're just damning our 'pink niggers.'"

Emma Lou was aghast. Such extraordinary people—saying "nigger" in front of a white man! Didn't they have any race pride or proper bringing up? Didn't they have any common sense?

"What've they done now?" Ray asked, reaching out to accept the glass Alva was handing him.

"No more than they've always done," Tony Crews answered. "Cora here just felt like being indignant, because she heard of a forthcoming wedding in Brooklyn to which the prospective bride and groom have announced they will *not* invite any dark people."

"Seriously now," Truman began. Ray interrupted him.

"Who in the hell wants to be serious?"

"As I was saying," Truman continued, "You can't blame light Negroes for being prejudiced against dark ones. All of you know that white is the symbol of everything pure and good, whether that everything be concrete or abstract. Ivory Soap is white. Moreover, virtue and virginity are always represented as being clothed in white garments. Then, too, the God we, or rather most Negroes worship is a patriarchal white man, seated on a white throne, in a spotless white Heaven, radiant with white streets and white-apparelled angels eating white honey and drinking white milk."

"Listen to the boy rave. Give him another drink," Ray shouted, but Truman ignored him and went on, becoming more and more animated.

"We are all living in a totally white world, where all standards are the standards of the white man, and where almost invariably what the white man does is right, and what the black man does is wrong, unless it is precedented by something a white man has done."

"Which," Cora added scornfully, "makes it all right for light Negroes to discriminate against dark ones?"

"Not at all," Truman objected, "It merely explains, not justifies, the evil—or rather, the fact of intraracial segregation. Mulattoes have always been accorded more consideration by white people than their darker brethren. They were made to feel superior even during the slave days . . . made to feel proud, as Bud Fisher would say, they were bastards. It was for the mulatto offspring of white masters and Negro slaves that the first schools for Negroes were organized, and say what you will, it is generally the Negro with a quantity of mixed blood in his veins who finds adaptation to a Nordic environment more easy than one of pure blood, which, of course, you will admit, is, to an American Negro, convenient if not virtuous."

"Does that justify their snobbishness and self-evaluated superiority?"

"No, Cora, it doesn't," returned Truman. "I'm not trying to excuse them. I'm merely trying to give what I believe to be an explanation of this thing. I have never been to Washington and only know what Paul and you have told me about conditions there, but they seem to be just about the same as conditions in Los Angeles, Omaha, Chicago, and other cities in which I have lived or visited. You see, people have to feel superior to something, and there is scant satisfaction in feeling superior to domestic animals or steel machines that one can train or utilize. It is much more pleasing to pick out some individual or group of individuals on the same plane to feel superior to. This is almost necessary when one is a member of a supposedly despised, mistreated minority group. Then consider that the mulatto is much nearer white than he is black, and is therefore more liable to act like a white man than like a black one, although I cannot say that I see a great deal of difference in any of their actions. They are human beings first and only white or black incidentally."

Ray pursed his lips and whistled.

"But you seem to forget," Tony Crews insisted, "that because a man is dark, it doesn't necessarily mean he is not of mixed blood. Now look at . . ."

"Yeah, let him look at you or at himself or at Cora," Paul interrupted. "There ain't no unmixed Negroes."

"But I haven't forgotten that," Truman said, ignoring the note of finality in Paul's voice. "I merely took it for granted that we were talking only about those Negroes who were light-skinned."

"But all light-skinned Negroes aren't color struck or color prejudiced," interjected Alva, who, up to this time, like Emma Lou, had remained silent. This was, he thought, a strategic moment for him to say something. He hoped Emma Lou would get the full significance of this statement.

"True enough," Truman began again. "But I also took it for granted that we were only talking about those who were. As I said before, Negroes are, after all, human beings, and they are subject to be influenced and controlled by the same forces and factors that influence and control other human beings. In an environment where there are so many color-prejudiced whites, there are bound to be a number of color-prejudiced blacks. Color prejudice and religion are akin in one respect. Some folks have it and some don't, and the kernel that is responsible for it is present in us all, which is to say, that potentially we are all color-prejudiced as long as we remain in this environment. For, as you know, prejudices are always caused by differences, and the majority group sets the standard. Then, too, since black is the favorite color of vaudeville comedians and jokesters, and conversely, as intimately associated with tragedy, it is no wonder that even the blackest individual will seek out some one more black than himself to laugh at."

"So saith the Lord," Tony answered soberly.

"And the Holy Ghost saith, let's have another drink."

"Happy thought, Ray," returned Cora. "Give us some more ice cream and gin, Alva."

Alva went into the alcove to prepare another concoction. Tony started the victrola. Truman turned to Emma Lou, who, all this while, had been sitting there with Alva's arm around her, every muscle in her body feeling as if it wanted to twitch, not knowing whether to be sad or to be angry. She couldn't comprehend all of this talk. She couldn't see how these people could sit down and so dispassionately discuss something that seemed particularly tragic to her. This fellow Truman, whom she was certain she knew, with all his hi-faluting talk, disgusted her immeasurably. She wasn't sure that they weren't all poking fun at her. Truman was speaking:

"Miss Morgan, didn't you attend school in Southern California?" Emma Lou at last realized where she had seen him before. So *this* was Truman Walker, the little "cock o' the walk," as they had called him on the campus. She answered him with difficulty, for there was a sob in her throat. "Yes, I did." Before Truman could say more to her, Ray called to him:

"Say, Bozo, what time are we going to the party? It's almost one o'clock now."

"Is it?" Alva seemed surprised. "But Aaron and Alta aren't here yet."

"They've been married just long enough to be late to everything."

"What do you say we go by and ring their bell?" Tony suggested, ignoring Paul's Greenwich Village wit.

"'Sall right with me." Truman lifted his glass to his lips. "Then on to the house-rent party . . . on to the bawdy bowels of Beale Street!"

They drained their glasses and prepared to leave.

"Ahhhh, sock it." . . . "Ummmm" . . . Piano playing—slow, loud, and discordant, accompanied by the rhythmic sound of shuffling feet. Down a long, dark hallway to an inside room, lit by a solitary red bulb. "Oh play it you dirty no-gooder." . . . A room full of dancing couples, scarcely moving their feet, arms completely encircling one another's bodies . . . cheeks being warmed by one another's breath . . . eyes closed . . . animal ecstasy agitating their perspiring faces. There was much panting, much hip movement, much shaking of the buttocks . . . "Do it twice in the same place." . . . "Git off that dime." Now somebody was singing, "I ask you very confidentially . . ." "Sing it man, sing it" . . . Piano treble moaning, bass rumbling like thunder. A swarm of people, motivating their bodies to express in suggestive movements the ultimate consummation of desire.

The music stopped, the room was suffocatingly hot, and Emma Lou was disturbingly dizzy. She clung fast to Alva, and let the room and its occupants whirl around her. Bodies and faces glided by. Leering faces and lewd bodies. Anxious faces and angular bodies. Sad faces and obese bodies. All mixed up together. She began to wonder how such a small room could hold so many people. "Oh, play it again . . ." She saw the pianist now, silhouetted against the dark mahogany piano, saw him bend his long, slick-haired head, until it hung low on his chest, then lift his hands high in the air, and as quickly let them descend upon the keyboard. There was one moment of cacophony, then the long, supple fingers evolved a slow, tantalizing melody out of the deafening chaos.

Every one began to dance again. Body called to body, and cemented themselves together, limbs lewdly intertwined. A couple there kissing, another couple dipping to the floor, and slowly shimming, belly to belly, as they came back to an upright position. A slender dark girl with wild eyes and wilder hair stood in the center of the room, supported by the strong, lithe arms of a longshoreman. She bent her trunk backward, until her head hung below her waistline, and all the while she kept the lower portion of her body quivering like Jell-O.

"She whips it to a jelly," the piano player was singing now, and banging on the keys with such might that an empty gin bottle on top of the piano seemed to be seized with the ague. "Oh, play it, Mr. Charlie." Emma Lou grew limp in Alva's arms.

"What's the matter, honey, drunk?" She couldn't answer. The music augmented by the general atmosphere of the room and the liquor she had drunk had presumably created another person in her stead. She felt like flying into an emotional frenzy—felt like flinging her arms and legs in insane unison. She had become very fluid, very elastic, and all the while she was giving in more and more to the music and to the liquor and to the physical madness of the moment.

When the music finally stopped, Alva led Emma Lou to a settee by the window which his crowd had appropriated. Every one was exceedingly animated, but they all talked in hushed, almost reverential tones.

"Isn't this marvelous?" Truman's eyes were ablaze with interest and excitement. Even Tony Crews seemed unusually alert.

"It's the greatest I've seen yet," he exclaimed.

Alva seemed the most unemotional one in the crowd. Paul the most detached. "Look at 'em all watching Ray."

"Remember, Bo," Truman counseled him. "Tonight you're 'passing.' Here's a new wrinkle, white man 'passes' for Negro."

"Why not? Enough of you pass for white." They all laughed, then transferred their interest back to the party. Cora was speaking:

"Didya see that little girl in pink—the one with the scar on her face—dancing with that tall, lanky, one-armed man? Wasn't she throwing it up to him?"

"Yeah," Tony admitted, "but she didn't have anything on that little Mexican-looking girl. She musta been born in Cairo."

"Saay, but isn't that one bad looking darkey over there, two chairs to the left; is he gonna smother that woman?" Truman asked, excitedly.

"I'd say she kinda liked it," Paul answered, then lit another cigarette.

"Do you know they have corn liquor in the kitchen? They serve it from a coffee pot." Aaron seemed proud of his discovery.

"Yes," said Alva, "and they got hoppin'-john out there too."

"What the hell is hoppin'-john?"

"Ray, I'm ashamed of you. Here you are passing for colored and don't know what hoppin'-john is!"

"Tell him, Cora, I don't know, either."

"Another one of those foreigners," Cora looked at Truman disdainfully. "Hoppin'-john is black-eyed peas and rice. Didn't they ever have any out in Salt Lake City?"

"Have they any chitterlings?" Alta asked eagerly.

"No, Alta," Alva replied, dryly. "This isn't Kansas. They have got pigs' feet, though."

"Lead me to 'em," Aaron and Alta shouted in unison, and led the way to the kitchen. Emma Lou clung to Alva's arm and tried to remain behind. "Alva, I'm afraid."

"Afraid of what? Come on, snap out of it! You need another drink." He pulled her up from the settee and led her through the crowded room down the long narrow dark hallway to the more crowded kitchen.

When they returned to the room, the pianist was just preparing to play again. He was tall and slender, with extra long legs and arms, giving him the appearance of a scarecrow. His pants were tight in the waist and full in the legs. He wore no coat, and a blue silk shirt hung damply to his body. He acted as if he were king of the occasion, ruling all from his piano stool throne. He talked familiarly to every one in the room, called women from other men's arms, demanded drinks from any bottle he happened to see being passed around, laughed uproariously, and made many grotesque and ofttimes obscene gestures.

There were sounds of a scuffle in an adjoining room, and an excited voice exclaimed, "You goddamn son-of-a-bitch, don't you catch my dice no more." The piano player banged on the keys and drowned out the reply, if there was one.

Emma Lou could not keep her eyes off the piano player. He was acting like a maniac, occasionally turning completely around on his stool, grimacing like a witch doctor, and letting his hands dawdle over the keyboard of the piano with an agonizing indolence, when compared to the extreme exertion to which he put the rest of his body. He was improvising. The melody of the piece he had started to play was merely a base for more bawdy variations. His left foot thumped on the floor in time with the music, while his right punished the piano's loud-pedal. Beads of perspiration gathered grease from his slicked-down hair, and rolled oleaginously down his face and neck, spotting the already damp baby-blue shirt, and streaking his already greasy black face with more shiny lanes.

A sailor lad suddenly ceased his impassioned hip movement and strode out of the room, pulling his partner behind him, pushing people out of the way as he went. The spontaneous moans and slangy ejaculations of the piano player and of the more articulate dancers became more regular, more like a chanted obligato to the music. This lasted for a couple of hours interrupted only by hectic intermissions. Then the dancers grew less violent in their movements, and though the piano player seemed never to tire there were fewer couples on the floor, and those left seemed less loathe to move their legs.

Eventually, the music stopped for a long interval, and there was a more concerted drive on the kitchen's corn liquor supply. Most of the private flasks and bottles were empty. There were more calls for food, too, and the crap game in the side room annexed more players and more kibitzers. Various men and women had disappeared altogether. Those who remained seemed worn and tired. There was much petty person to person badinage and many whispered consultations in corners. There was an argument in the hallway between the landlord and two couples, who wished to share one room without paying him more than the regulation three dollars required of one couple. Finally, Alva suggested that they leave. Emma Lou had drifted off into a state of semi-consciousness and was too near asleep or drunk to distinguish people or voices. All she knew was that she was being led out of the dreadful place, that the perturbing "pilgrimage to the proletariat's parlor social," as Truman had called it, was ended, and that she was in a taxicab, cuddled up in Alva's arms. . . .

By now Emma Lou had reached St. Nicholas Avenue and was about to cross over into the park when she heard the chimes of a clock and was reminded of the hour. It was growing late—too late for her to wander in the park alone where she knew she would be approached either by some persistent male or an insulting park policeman. Wearily she started towards home, realizing that it was necessary for her to get some rest in order to be able to be in her class room on the next morning. She mustn't jeopardize her job, for it was partially through the money she was earning from it that she would be able to find her place in life. She was tired of running up blind alleys all of which seemed to converge and lead her ultimately to the same blank wall. Her motto from now on would be "find—not seek." All things were at one's finger-tips. Life was most kind to those who were judicious in their selections, and she, weakling that she now realized she was, had not been a connoisseur.

As she drew nearer home she felt certain that should she attempt to spend another night with Alva and his child, she would surely smother to death during the night. And even though she felt this, she also knew within herself that no matter how much at the present moment she pretended to hate Alva that he had only to make the proper advances in order to win her to him again. Yet she also knew that she must leave him if she was to make her self-proposed adjustment—leave him now even if she should be weak enough to return at some not so distant date. She was determined to fight against Alva's influence over her, fight even though she lost, for she reasoned that even in losing she would win a Pyrrhic victory and thus make her life less difficult in the future, for having learned to fight future battles would be easy.

She tried to convince herself that it would not be necessary for her to have any more Jasper Cranes or Alvas in her life. To assure herself of this she intended to look John up on the morrow and if he were willing let him re-enter her life. It was clear to her now what a complete fool she had been. It was clear to her at last that she had exercised the same discrimination against her men and the people she wished for friends that they had exercised against her—and with less reason. It served her right that Jasper Crane had fooled her as he did. It served her right that Alva had used her once for the money she could give him and again as a black mammy for his child. That was the price she had had to pay for getting what she thought she wanted. But now she intended to balance things. Life after all was a give and take affair. Why should she give important things and receive nothing in return?

She was in front of the house now and looking up saw that all the lights in her room were lit. And as she climbed the stairs she could hear a drunken chorus of raucous masculine laughter. Alva had come home meanwhile, drunk of course and accompanied by the usual drunken crowd. Emma Lou started to turn back, to flee into the street—anywhere to escape being precipitated into another sordid situation, but remembering that this was to be her last night there, and that the new day would find her beginning a new life, she subdued her flight impulse and without knocking threw open the door and walked into the room. She saw the usual and expected sight: Alva, face a death mask, sitting on the bed embracing an effeminate boy whom she knew as Bobbie, and who drew hurriedly away from Alva as he saw her. There were four other boys in the room, all in varied states of drunkenness—all laughing boisterously at some obscene witticism. Emma Lou suppressed a shudder and calmly said: "Hello Alva"—The room grew silent. They all seemed shocked and surprised by her sudden appearance. Alva did not answer her greeting but instead turned to Bobbie and asked him for another drink. Bobbie fumbled nervously at his hip pocket and finally produced a flask which he handed to Alva. Emma Lou stood at the door and watched Alva drink the liquor Bobbie had given him. Every one else in the room watched her. For the moment she did not know what to say or what to do. Obviously she couldn't continue standing there by the door nor could she leave and let them feel that she had been completely put to rout.

Alva handed the flask back to Bobbie, who got up from the bed and said something about leaving. The others in the room also got up and began staggering around looking for their hats. Emma Lou thought for a moment that she was going to win without any further struggle, but she had not reckoned with Alva who, meanwhile, had sufficiently emerged from his stupor to realize that his friends were about to go.

"What the hell's the matter with you," he shouted up at Bobbie, and without waiting for an answer reached out for Bobbie's arm and jerked him back down on the bed.

"Now stay there till I tell you to get up."

The others in the room had now found their hats and started toward the door, eager to escape. Emma Lou crossed the room to where Alva was sitting and said, "You might make less noise, the baby's asleep."

The four boys had by this time opened the door and staggered out into the hallway. Bobbie edged nervously away from Alva, who leered up at Emma Lou and snarled, "If you don't like it—"

For the moment Emma Lou did not know what to do. Her first impulse was to strike him, but she was restrained because underneath the loathsome beast that he now was, she saw the Alva who had first attracted her to him, the Alva she had always loved. She suddenly felt an immense compassion for him and had difficulty in stifling an unwelcome urge to take him into her arms. Tears came into her eyes, and for a moment it seemed as if all her rationalization would go for naught. Then once more she saw Alva, not as he had been, but as he was now, a drunken, drooling libertine, struggling to keep the embarrassed Bobbie in a vile embrace. Something snapped within her. The tears in her eyes receded, her features grew set, and she felt herself hardening inside. Then, without saying a word, she resolutely turned away, went into the alcove, pulled her suitcases down from the shelf in the clothes-closet, and, to the blasphemous accompaniment of Alva berating Bobbie for wishing to leave, finished packing her clothes, not stopping even when Alva Junior's cries deafened her, and caused the people in the next room to stir uneasily.

## Infants of the Spring

Published by the Macaulay Company, 1932.

### Chapter 1

Raymond opened the door with a flourish, pushed the electric switch and preceded his two guests into the dimly illuminated room.

"Here we are, gentlemen."

"Nice diggings you have here," Stephen said.

"Damn right," Raymond agreed. "I'm nuts about 'em. Sam doesn't like my studio, though. He thinks it's decadent."

"I merely objected to some of the decorations, Ray."

"Namely, the red and black draperies, the red and black bed cover, the crimson wicker chairs, the riotous hook rugs, and Paul's erotic drawings. You see, Steve, Sam thinks it's all rather flamboyant and vulgar. He can't forget that he's a Nordic and I'm a Negro, and according to all the sociology books, my taste is naturally crass and vulgar. I must not go in for loud colors. It's a confession of my inferior race heritage. Am I right, Sam?"

"It's all Greek to me, anyhow," Stephen murmured. "I like the room . . . And these pictures are rather astonishing. Who did them?"

"The most impossible person in the world," Samuel answered.

"Wrong again," Raymond said. "Paul is one of the most delightful people in the world. I only hope he drops in before you leave, Steve. You'll enjoy meeting him."

"He certainly handles his colors well."

"But his pictures are obscene," Samuel protested. "They are nothing but highly colored phalli."

Raymond shrugged his shoulders.

"You argue with him, Steve. I'm worn to a frazzle trying to make him see the light. Everything that Sam doesn't understand, he labels depraved and degenerate. As an old friend, maybe you're willing to take up his education where I'm leaving off. Right now, I'm going to have a highball. Shall I fix three?"

Raymond retired to the alcove addition to his studio, and prepared three highballs. When he returned to the room, Stephen Jorgenson was minutely examining the various drawings which adorned the wall, while Samuel stood stiffly in front of the false fireplace, patently annoyed by his friend's interest in what he considered obscene trivia.

"I made yours weak, Sam."

"You have educated him, at that," Stephen said. "When he was at the University of Toronto, he wouldn't even take a drop of ale."

"I'm glad I didn't know him then. He's impossible enough now."

Raymond smiled maliciously at Samuel, then held his glass aloft.

"Here's to your first time in Harlem, Stephen Jorgenson, to your first day in New York, and to your first visit to these United States. Prosit."

Raymond and Stephen drained their glasses. Samuel sipped a little of the liquid, made a wry face, then placed his glass upon the ledge of the mantelpiece. "It's funny," Raymond mused, "how things happen. Three hours ago we were total strangers. Twenty-four hours ago we were not even aware of one another's existence. And now, Steve, I feel as if I have known you all my life. It's strange, too, because the acquaintanceship began under such evil auspices. First of all Samuel introduced us, and I always disliked Samuel's friends. He knows the most godawful people in the world . . . social service workers, reformed socialistic ministers, foreign missionaries, caponized radicals, lady versifiers who gush all over the place, Y.M.C.A. secretaries and others of the same dogassed ilk. They are all so saccharine and benevolent. They talk of nothing but service to mankind, not realizing that the greatest service they could render mankind would be self extermination. And you have no idea how they sympathize with me, a poor, benighted Negro.

"Consequently, when I came to dinner tonight, I was prepared to be bored and uncomfortable. Sam had told me nothing except that he had a foreign friend whom he wanted me to meet. And surprisingly enough you were foreign, foreign to everything familiar to either Samuel or myself."

"You were foreign to me, too," Stephen said.

"I know it," Raymond replied. "And you can imagine my surprise to find that you were the uncomfortable one. It rather startled me to find someone usurping my position at one of Samuel's dinners. I didn't know—I still don't know for that matter—what our host had told you about me. And of course I had no idea what you thought or felt about Negroes. I got the impression, though, that you were anticipating some sort of cannibal attack. Actually. The expression on your face as you entered the café plainly said: I hope these Negroes find their dinners ample. Otherwise they're liable to pounce upon me."

"Ray!" Samuel exclaimed. There was a note of reproof in his voice, but before he could continue, Stephen spoke:

"By golly, you're right. I was frightened. After all I had never seen a Negro before in my life, that is, not over two or three, and they were only dim, passing shadows with no immediate reality. New York itself was alarming enough, but when I emerged from the subway at 135th Street, I was actually panic stricken. It was the most eerie experience I had ever had. I felt alien, creepy, conspicuous, ashamed. I wanted to camouflage my white skin, and assume some protective coloration. Although, in reality, I suppose no one paid the slightest attention to me, I felt that everyone was sizing me up, regarding me with hostile eyes. It was ghastly. The strange dark faces, the suspicious eyes, the undercurrent of racial antagonism which I felt sweeping around me, the squalid streets, barricaded by grim tenement houses, and then that depressing public dining room in which Samuel and I were the only white persons. I was ready to bolt."

"See, Sam," Raymond said, "how unconsciously cruel you are? Of all the places to bring an innocent foreigner the moment he sets foot in America. Harlem terrifies me, and I've been here long enough to be acclimated, to say nothing of my natural affinity for the place."

"I think Steve's exaggerating."

"Exaggerating my great aunt. I'm guilty of understatement, if anything."

"Certainly," Raymond said. "There are perhaps a dozen or more things you'd like to say about your impressions, but you desist for fear of wounding me. Don't, for God's sake. I'm not the least bit self conscious about my race. And I prefer brutal frankness to genteel evasion anytime."

Stephen's keen, blue eyes once more regarded the small and slender Negro who sat opposite him, noting the smooth dark skin to which the amber colored bulbs imparted red overtones, and becoming particularly interested in the facial features. They were, Stephen thought, neither Nordic nor Negroid, but rather a happy combination of the two, retaining the slender outlines of the first, and the warm vigor of the second, thus escaping both Nordic rigidity, and African coarseness. Equally as interesting were the eyes. When in repose they seemed to be covered by some muddy mask, which rendered them dull and lifeless. But Stephen had noticed that when Raymond became animated, his eyes shed this mask and became large, brilliant, and fiery.

Samuel interrupted his reverie.

"I think it's time we were getting downtown, Steve."

"Why? Have we something else to do? This place is so restful, I don't want to move."

"Stay, then," Raymond said. "Sam's afraid I'll contaminate you if you stay around me too long."

"You're being ridiculous, Ray."

"Tell me," Stephen said. "Do you two always get along so famously?"

"Umhuh," Raymond answered. "We're really quite fond of one another, though. Otherwise I'd never countenance Samuel's Puritanism and spirit of uplift, and I'm sure he'd resent my persistent badgering. We disagree about everything. And yet there are moments when we get great pleasure out of one another's company. I need Sam's steadying influence, and he is energized by what he calls my animalism."

He smiled affectionately at the discomfited Samuel, who nervously shifted his feet, then turned to the mantelpiece, caught sight of the drink he had forgotten, and once more took the glass in his hand.

"Are you still nursing that?" Raymond asked. "I'm ready for another. How about you, Steve?"

"I can't say that I'm crazy about the taste of your gin, but I suppose the effect is desirable."

"Quite. You must get used to Harlem gin. It's a valuable and ubiquitous commodity. I couldn't do without it."

Raymond once more went into the alcove to refill the empty glasses, his mind busy contrasting the two Nordics who were his guests. Stephen was tall and fashioned like a Viking. His hair, eyes and complexion all testified to his Norse ancestry. Samuel was small, pale, anemic. His hair was blond and his eyes were blue, but neither the blondeness nor the blueness were as clearly defined or as positive as Stephen's. Samuel's ancestors had been dipped in the American melting pot, and as a result, the last of the line bore only a faint resemblance to his original progenitors.

"Tell me more about the fellow who drew these," Stephen said as Raymond returned to the room and handed him a full glass of gin and ginger ale.

"Nothing doing," Raymond replied. "Paul's a person you've got to see to appreciate. You wouldn't believe what I could tell you. It's about time he was dropping in. He knew I was going out to dinner tonight. That's why he isn't here now."

"Tell me this, then," Stephen asked, "do all these hideous Harlem houses have such nice interiors?"

"Not by a damn sight. Most of them are worse inside than out. You should see some of the holes I've had to live in. It just happens that my present landlady is a visionary as well as a business woman. She has dreams. One of them is that some day she will be a best selling author. That accounts for this house. She knew the difficulties experienced by Harlem artists and intellectuals in finding congenial living quarters, and reasoned that by turning this house over to Negroes engaged in creative work, she would make money, achieve prestige as a patron, and at the same time profit artistically from the resultant contacts."

"Is the house entirely filled with these . . . er . . . creative sprits?"

"Not yet. But we have hopes. Only the top floor remains in the hands of the Philistines. One of the ladies up there claims to be an actress, but we doubt it, and neither of her two children are precocious. The other tenant on that floor is a mysterious witchlike individual, who was living here when Euphoria leased the house, and who refused to be put out. Pelham, Eustace, Paul, and myself make up the artistic contingent. Wait til you meet the others. They're a rare collection."

Ten minutes later, Paul and Eustace entered the room.

"Oh hell," Paul said, "another Nordic. Ain't he a beauty, Eustace?"

"Cut the comedy, Paul. I want to introduce you to Stephen Jorgenson. He just arrived in America today, and this is of course his first visit to Harlem. Don't scare him to death. This is Paul, Steve. He's responsible for all these abominable drawings. And this is Eustace Savoy, actor, singer, and what have you. He runs a den of iniquity in the basement, and is also noted for his spoonerisms."

"Mad to gleet you," Eustace said, living up to his reputation.

"Have you ever been seduced?" Paul asked. "Don't blush. You just looked so pure and undefiled that I had to ask that."

Stephen looked inquiringly at Raymond.

"Don't mind Paul. He's harmless."

"I like your drawings," Stephen said.

"You should," Paul replied. "Everybody should. They're works of genius."

"You're as disgusting as ever, Paul."

"I know it, Sam, but therein lies my charm. By the way, how did you ever get to know such a gorgeous man as this . . . You know, Steve," he added abruptly, "you should take that part out of your hair and have it windblown. The hair, not the part. Plastering it down like that destroys the golden glint."

"Oh, I say . . ." Stephen began.

"That's all right. I never charge for expert advice. Where's the gin, Ray?"

"In the alcove, of course."

"But you mustn't dride the hinks," Eustace said.

"You're not at all funny," Samuel muttered.

"I'm sorry, Sam. Wait'll I have a couple of drinks. Then I'll shies and rine."

He and Paul went into the alcove.

Paul was very tall. His face was the color of a bleached saffron leaf. His hair was wiry and untrained. It was his habit not to wear a necktie because he knew that his neck was too well modeled to be hidden from public gaze. He wore no sox, either, nor underwear, and those few clothes he did deign to affect were musty and disheveled.

Eustace was a tenor. He was also a gentleman. The word elegant described him perfectly. His every movement was ornate and graceful. He had acquired his physical bearing and mannerisms from mid-Victorian matinee idols. No one knew his correct age. His face was lined and drawn. An unidentified scalp disease had rendered him bald on the right side of his head. To cover this mistake of nature, he let the hair on the left side grow long, and combed it sidewise over the top of his head. The effect was both useful and bizarre. Eustace also had a passion for cloisonné bric-a-brac, misty etchings, antique silver pieces, caviar, and rococo jewelry. And his most treasured possession was an onyx ring, the size of a robin's egg, which he wore on his right index finger.

Stephen was frankly bewildered by these two strange beings who had so unceremoniously burst into the room, and forced themselves into the spotlight. Truly, as Raymond had said, this house did harbor a rare collection of individuals.

"I hope you didn't drain the bottle," Raymond said, as Paul and Eustace pranced merrily back into the room, carefully nursing their filled glasses.

"But we thought all of that was for us," Paul said.

"Damned hogs."

"Where did you come from, Steve?" Paul asked.

"Copenhagen, Denmark."

"Oh, that's where they make snuff."

"Snuff?"

"I'm ready to go whenever you are, Steve," Samuel was restless and bored.

"But you can't take him away so soon. I haven't had a chance to talk to him yet," Paul protested. "I've got to tell him about my drawings. He looks like he might have sense enough to appreciate them."

"He's tired, Paul, and once you start to talk, we won't get home tonight."

"But I don't want to go home yet, Sam."

"See there," Paul exclaimed triumphantly, "I knew he had sense. Tell me about yourself, Steve." Paul squatted himself on the floor before Stephen's chair.

"There's nothing to tell. I was born in Canada. My father was Norwegian, my mother was a Dane. I was educated at the University of Toronto where I met Sam and identified myself as much as possible with things American. My folks moved back to Copenhagen. I spent the summer with them, and I'm here now to get a Ph.D. from Columbia."

"Why?"

"Because there's nothing else to do. If I stop going to school, I'll have to work, and the only kind of work I can do is professorial. I don't want to do that, so, as long as the old man foots the bills, I'll stay in school."

"See," Paul exclaimed. "He is one of us."

"God forbid," Samuel said, stifling a yawn.

"Now, Paul, tell me about your drawings."

"That's easy. I'm a genius. I've never had a drawing lesson in my life, and I never intend to take one. I think that Oscar Wilde is the greatest man that ever lived. Huysmans' Des Esseintes is the greatest character in literature, and Baudelaire is the greatest poet. I also like Blake, Dowson, Verlaine, Rimbaud, Poe, and Whitman. And of course Whistler, Gauguin, Picasso, and Zuloaga."

"But you're not telling me anything about your drawings."

"Unless you're dumber than I think, I've told you all you need to know."

There was a timid knock on the door.

"Come in," Raymond shouted.

Pelham sidled into the room. He was short, fat and black, and was attired in a green smock and a beret which was only two shades darker than his face.

"Hello, everybody." His voice was timid, apologetic. "I didn't know you had company."

"That's all right," Raymond reassured him. "Mr. Jorgenson, this is Pelham Gaylord. He's an artist too."

"Pleased to meet you," Stephen proffered his hand. Gingerly Pelham pressed it in his own, then quickly, like a small animal at bay, stepped back to the door, and smiled bashfully at all within the room.

"Pelham's the only decent person in the house," Samuel said.

"You mean he's the only one you can impress." It was Paul who spoke. "But I'm tired of sitting here doing nothing. There's no life to this party. We need to celebrate Steve's arrival. We need some liquor. Let's go to a speakeasy."

"Who's going to pay the bill?" Raymond asked.

"Who?" Paul repeated. "Why, Steve of course. It's his celebration, and he's bound to have some money."

"But . . ." Samuel started to protest.

"But hell . . ." Paul interrupted. "Get your hat and coat, Steve. You, too, Ray and Eustace. Let Sam stay here with Pelham. Otherwise he'll spoil the party."

"But suppose I wish to go with you?"

"And leave Pelham alone? Nothing doing, Sam. I'm sure you have lots to say to one another. And Pelham must have written some new poems today. Can't you see the light of creation in his eyes?"

All during this barrage of banter, Paul had been helping first Stephen and then Raymond into their coats. And before there could be further protest, he had ushered Stephen, Eustace and Raymond out of the room, leaving Samuel gaping sillily at the grinning Pelham.


## Chapter 3

Stephen had been in New York for a month now, and most of that month had been spent in company with Raymond. Their friendship had become something precious, inviolate and genuine. They had become as intimate in that short period as if they had known one another since childhood. In fact, there was something delightfully naïve and childlike about their frankly acknowledged affection for each other. Like children, they seemed to be totally unconscious of their racial difference. It did not matter that Stephen's ancestors were blond Norsemen, steeped in the tradition of the sagas, and that Raymond's ancestors were a motley ensemble without cultural bonds. It made no difference between them that one was black and the other white. There was something deeper than mere surface color which drew them together, something more vital and lasting than the shallow attraction of racial opposites.

Their greatest joy came when they could be alone together and talk . . . talk about any and everything. They seemed to have so much to say to one another, so much that had remained unsaid all of their respective lives because they had never met anyone else with whom they could converse unreservedly. And no matter how often these conversational communings occurred, no matter how long they lasted, there always seemed to be so much more to say.

Stephen was absorbed in learning about Harlem and about Negroes. Raymond was intrigued by the virile Icelandic sagas which Stephen read in the original and translated for his benefit. And both of them were eager to air their thoughts about literature . . .

*Ulysses* was a swamp out of which stray orchids grew. Hemingway exemplified the spirit of the twenties in America more vividly than any other contemporary American novelist. To Raymond, Thomas Mann and André Gide were the only living literary giants. André Gide was not on Stephen's list, but Sigrid Undset was. Neither liked Shaw. They agreed that Dostoyevsky was the greatest novelist of all time, but Stephen had only contempt for Marcel Proust whom Raymond swore by, but didn't read. Stendhal, Flaubert and Hardy were discussed amicably, but the sparks flew when they discussed

Tolstoy or Zola. And at the mention of Joyce's *Dubliners* and *The Portrait of the Artist as a Young Man* both grew incontinently rhapsodic. "And Hamsun: You mean you haven't read *Hunger,* Ray? God's Teeth, man, that's literature."

Their likes and dislikes in literature were sufficiently similar to give them similar philosophies about life, and sufficiently dissimilar to provide food for animated discussions. It was only when their talk veered to Harlem that they found themselves sitting at opposite poles.

Raymond prided himself on knowing all the ins and outs of Harlem. He had been resident there for over three years, during which time he had explored every nook and cranny of that phenomenal Negro settlement. It had, during this period, attained international fame, deservedly, Raymond thought. But he was disgusted with the way everyone sought to romanticize Harlem and Harlem Negroes. And it annoyed him considerably when Stephen began to do likewise. Together they had returned to the spots which Raymond had ferreted out before. They visited all of the cabarets, the speakeasies, the private clubs, the theaters, the back street rendezvous. Stephen was uncritical in his admiration for everything Negroid that he saw. All of the entertainers, musicians, singers, and actors were . . . marvelous. So were all the other Negroes who seemed to be accomplishing one thing or the other. Stephen, like all other whites who had only a book knowledge of Negroes, seemed surprised that a people who had so long been enslaved, and so recently freed, could make such progress. Raymond tried to explain to Stephen that there was nothing miraculous about the matter.

"Can't you understand, Steve," he would say, "that the Negro had to make what progress he has made or else he wouldn't have survived? He's merely tried to keep that pace set by his environment. People rave about the progress of the Negro. It is nothing near as remarkable, that is generally, as the progress made by foreign immigrants who also come to this country to find freedom from a state of serfdom and illiteracy just as stringent as that of the pre– and post–Civil War Negro. And as for Harlem. It's bound to be a startling and wondrous community. Is it not part and parcel of the greatest city in the world? Is it any more remarkable than the Ghetto or Chinatown or the Bronx? It has the same percentage of poverty, middle class endeavor, family life, and underworld, functioning under the same conditions which makes city life a nightmare to any group which is economically insecure. New York is a world within itself, and every new portion of it which gets discovered by the sophisticates and holds the spotlight, seems more unusual than that which has been discovered before."

"Jesus Christ, Ray, you don't appreciate the place. You're too much a part of it."

"Me a part of Harlem? How come?"

"You know what I mean."

"But I don't."

"Well, you live here. You've always lived here."

"Three years isn't always, Steve, even though it may seem like a lifetime."

"You're quibbling now."

"I'm not quibbling. The fault lies with you. Just because you've overcome your initial fear of the place and become fascinated by a new and bizarre environment, should you lose your reason? Harlem is New York. Please don't let the fact that it's black New York obscure your vision."

"You're both cynical and silly."

"Granted. But if you had lived in Harlem as long as I have, you would realize that Negroes are much like any other human beings. They have the same social, physical, and intellectual divisions. You're only being intrigued, as I have said before, by the newness of the thing. You should live here a while."

"That's just what I'd like to do."

"Well . . . why not?"

"By Jove, that's an idea. I'll move in with you."

"Move in with me?"

"Sure. I prefer this room to the one I have now, and certainly prefer your company to that of those nincompoops at the International House."

"White people don't live in Harlem."

"Why not?"

"It just isn't done, that's all. What would your friends say?"

"You're the only friend I give a damn about. If you want me, I don't see why I couldn't live here."

"There isn't any reason why you shouldn't, but . . ."

"But hell, I'm moving in tomorrow. O.K.?"

"I'll be damn glad to have you, Steve."

Twenty-four hours later, Stephen had moved to Harlem.

## From Chapter 5

He turned and quickly retraced his steps, thinking of what Euphoria had said about the lack of work being done in Niggeratti Manor, and also remembering a conversation which he and Stephen had had earlier in the evening. Stephen had asked:

"Just when are you going to begin work on your novel, Ray?"

"I don't know, Steve," Raymond had answered. "I can't get started. Something holds me back."

"Laziness?"

"Partially."

"Lack of material?"

"You know it isn't that. Haven't I outlined the thing for you? I know what I want to write . . . but . . ." He had shrugged his shoulders. "Something holds me back."

Stephen had shifted his gaze and lit a cigarette.

"Are you afraid," he had asked, "of exposing your own peculiar complex?"

"Complex? I have none."

"The hell you haven't," Stephen said emphatically. "Remember our talk a month ago? You pronounced yourself a Nietzschean. I pronounced you a liar. I still admit I'm at sea. I don't know whether you are or not."

"Neither do I," Raymond had admitted after a moment's pause.

"Which is just what I thought," Stephen had continued. "You'd like to be. You try very hard to be. But after all, something holds you back and that same something hinders your writing."

"Why not elucidate?"

"I can't, Ray. You baffle me. Were you like Paul or Eustace or Pelham, I could analyze you immediately. Paul has never recovered from the shock of realizing that no matter how bizarre a personality he may develop, he will still be a Negro, subject to snubs from certain ignorant people. The fact distresses him, although he should ignore both it and the people who might be guilty of such snubs. He sits around helpless, possessed of great talent, doing nothing, wishing he were white, courting the bizarre, anxious to be exploited in the public prints as a notorious character. Being a Negro, he feels that his chances for excessive notoriety à la Wilde are slim. Thus the exaggerated poses and extreme mannerisms. Since he can't be white, he will be a most unusual Negro. To say "nigger" in the presence of a white person warms the cockles of his heart. It's just a symptom of some deep set disease. You're not like that; nor are you like some of the others I've met who are so conscientiously Negroid. Like Pelham for instance, who is a natural born menial with all a menial's respect for his superiors. Or like Eustace who is ashamed of his color, and won't sing spirituals because he does not care to remind the world that he is a Negro and that his ancestors were slaves of whom he is now ashamed."

"Jesus no, Steve, you know I'm not like that. I'm just indifferent to it all. Race to me . . ."

"Yes, race to you," Stephen had interrupted impatiently, "means nothing. You stand on a peak alone, superior, nonchalant, unconcerned. I know all that. You've said it enough. Propagandists you despise. Illusions about Negroes you have none. Your only plea is that they accept themselves and be accepted by others as human beings. But what the hell does it all mean, after all? You claim to have no especial love for your race. You also claim not to despise them. The spectacle of your friends striving to be what they are not, and taking no note of their limitations, sickens you, nay revolts and angers you. Yet you, like the rest, sit about and do nothing. Are you so emancipated as you claim? Aren't you, too, hindered by some racial complex?"

"Nonsense, Steve. I know I'm a Negro and so does everyone else. I certainly cannot pass nor can I effect a change. Why worry about it? I rather love myself as I am, and am quite certain that I have as much chance to make good as anyone else, regardless of my color. In fact, I might even say that being black gives me a certain advantage which a white person of equal talent would be denied."

"Oh, I see."

"You see what?"

"Oh, nothing, let's beat it downstairs." And with this Stephen had terminated the conversation and the two of them had joined the crowd, gathered in Eustace's studio.

Raymond wondered about it now, and also about what Euphoria had said concerning the lack of creative work among Negro artists. Pelham and Eustace of course were not to be considered. They had nothing to contribute anyhow. But Paul and, he thought, himself, did have something to contribute once they made up their minds to do some actual work.

There had been throughout the nation an announcement of a Negro renaissance. The American Negro, it seemed, was entering a new phase in his development. He was about to become an important factor in the artistic life of the United States. As the middle westerner and the southerner had found indigenous expression, so was the Negro developing his own literary spokesmen.

Word had been flashed through the nation about this new phenomenon. Novels, plays, and poems by and about Negroes were being deliriously acclaimed and patronized. Blues shouters, tap dancers, high yaller chorus girls, and singers of Negro spirituals were reaping much publicity and no little money from the unexpected harvest. And yet the more discerning were becoming more and more aware that nothing, or at least very little, was being done to substantiate the current fad, to make it the foundation for something truly epochal. For the time being, the Negro was more in evidence in the high places than ever before in his American career, but unless, or so it seemed to Raymond, he, Paul, and others of the group who had climbed aboard the bandwagon actually began to do something worth while, there would be little chance of their being permanently established. He wondered what accounted for the fact that most Negroes of talent were wont to make one splurge, then sink into oblivion. Was it all the result, as Stephen had intimated, of some deep-rooted complex? Or was it merely indicative of a lack of talent?

Arriving home, Raymond decided to go to his studio before returning to the party in the basement. He could hear laughter and the clink of glasses as he climbed the stairs.

On reaching his landing, he was surprised to see rays of light gleaming through the cracks in his studio door. He was certain he had turned out the lights before he had gone out with Euphoria. Perhaps Stephen was there. He hoped so. This would be a propitious time to thrash out certain problems which were tantalizing his mind.

## Chapter 21

After Stephen's unexpected visit and their long conversation together, Raymond seemed to have developed a new store of energy. For three days and nights, he had secluded himself in his room, and devoted all his time to the continuance of his novel. For three years it had remained a project. Now he was making rapid progress. The ease with which he could work once he set himself to it amazed him, and at the same time he was suspicious of this unexpected facility. Nevertheless, his novel was progressing, and he intended to let nothing check him.

In line with this resolution, he insisted that Paul and Eustace hold their nightly gin parties without his presence, and they were also abjured to steer all company clear of his studio.

Stephen had gone upstate on a tutoring job. Lucille had not been in evidence since the donation party, and Raymond had made no attempt to get in touch with her. There was no one else in whom he had any interest. Aline and Janet he had dismissed from his mind, although Eustace and Paul had spent an entire dinner hour telling him of their latest adventures. Both had now left Aline's mother's house and were being supported by some white man, whom Aline had met at a downtown motion picture theater. They had an apartment in which they entertained groups of young colored boys on the nights their white protector was not in evidence.

Having withdrawn from every activity connected with Niggeratti Manor, Raymond had also forgotten that Dr. Parkes had promised to communicate with him, concerning some mysterious idea, and he was taken by surprise when Eustace came into the room one morning, bearing a letter from Dr. Parkes.

"Well, I'm plucked," Raymond exclaimed.

"What's the matter?" Eustace queried.

"Will you listen to this?" He read the letter aloud.

> My dear Raymond:
>
> I will be in New York on Thursday night. I want you to do me a favor. It seems to me that with the ever increasing number of younger Negro artists and intellectuals gathering in Harlem, some effort should be made to establish what well might become a distinguished salon. All of you engaged in creative work, should, I believe, welcome the chance to meet together once every fortnight, for

the purpose of exchanging ideas and expressing and criticizing individual theories. This might prove to be both stimulating and profitable. And it might also bring into active being a concerted movement which would establish the younger Negro talent once and for all as a vital artistic force. With this in mind, I would appreciate your inviting as many as your colleagues as possible to your studio on Thursday evening. I will be there to preside. I hope you are intrigued by the idea and willing to cooperate. Please wire me your answer. Collect, of course.

Very sincerely yours,
Dr. A. L. Parkes

"Are you any more good?" Raymond asked as he finished reading.

"Sounds like a great idea," Eustace replied, enthusiastically.

"It *is* great. Too great to miss," Raymond acquiesced mischievously. "Come on, let's get busy on the telephone."

Thursday night came and so did the young hopefuls. The first to arrive was Sweetie May Carr. Sweetie May was a short story writer, more noted for her ribald wit and personal effervescence than for any actual literary work. She was a great favorite among those whites who went in for Negro prodigies. Mainly because she lived up to their conception of what a typical Negro should be. It seldom occurred to any of her patrons that she did this with tongue in cheek. Given a paleface audience, Sweetie May would launch forth into a saga of the little all-colored Mississippi town where she claimed to have been born. Her repertoire of tales was earthy, vulgar and funny. Her darkies always smiled through their tears, sang spirituals on the slightest provocation, and performed buck dances when they should have been working. Sweetie May was a master of Southern dialect, and an able raconteur, but she was too indifferent to literary creation to transfer to paper that which she told so well. The intricacies of writing bored her, and her written work was for the most part turgid and unpolished. But Sweetie May knew her white folks.

"It's like this," she had told Raymond. "I have to eat. I also wish to finish my education. Being a Negro writer these days is a racket and I'm going to make the most of it while it lasts. Sure I cut the fool. But I enjoy it, too. I don't know a tinker's damn about art. I care less about it. My ultimate ambition, as you know, is to become a gynecologist. And the only way I can live easily until I have the requisite training is to pose as a writer of potential ability. *Voila!* I get my tuition paid at Columbia. I rent an apartment and have all the furniture contributed by kind-hearted o'fays. I receive bundles of groceries from various sources several times a week . . . all accomplished by dropping a discreet hint during an evening's festivities. I find queer places for whites to go in Harlem . . .

out of the way primitive churches, sidestreet speakeasies. They fall for it. About twice a year I manage to sell a story. It is acclaimed. I am a genius in the making. Thank God for this Negro literary renaissance! Long may it flourish!"

Sweetie May was accompanied by two young girls, recently emigrated from Boston. They were the latest to be hailed as incipient immortals. Their names were Doris Westmore and Hazel Jamison. Doris wrote short stories. Hazel wrote poetry. Both had become known through a literary contest fostered by one of the leading Negro magazines. Raymond liked them more than he did most of the younger recruits to the movement. For one thing, they were characterized by a freshness and naïveté which he and his cronies had lost. And, surprisingly enough for Negro prodigies, they actually gave promise of possessing literary talent. He was most pleased to see them. He was also amused by their interest and excitement. A salon! A literary gathering! It was one of the civilized institutions they had dreamed of finding in New York, one of the things they had longed and hoped for.

As time passed, others came in. Tony Crews, smiling and self-effacing, a mischievous boy, grateful for the chance to slip away from the backwoods college he attended. Raymond had never been able to analyze this young poet. His work was interesting and unusual. It was also spotty. Spasmodically he gave promise of developing into a first rate poet. Already he had published two volumes, prematurely, Raymond thought. Both had been excessively praised by whites and universally damned by Negroes. Considering the nature of his work this was to be expected. The only unknown quantity was the poet himself. Would he or would he not fulfill the promise exemplified in some of his work? Raymond had no way of knowing and even an intimate friendship with Tony himself had failed to enlighten him. For Tony was the most close-mouthed and cagey individual Raymond had ever known when it came to personal matters. He fended off every attempt to probe into his inner self and did this with such an unconscious and naïve air that the prober soon came to one of two conclusions: Either Tony had no depth whatsoever, or else he was too deep for plumbing by ordinary mortals.

DeWitt Clinton, the Negro poet laureate, was there, too, accompanied, as usual, by his *fideles achates*, David Holloway. David had been acclaimed the most handsome Negro in Harlem by a certain group of whites. He was in great demand by artists who wished to paint him. He had become a much touted romantic figure. In reality he was a fairly intelligent school teacher, quite circumspect in his habits, a rather timid beau, who imagined himself to be bored with life.

Dr. Parkes finally arrived, accompanied by Carl Denny, the artist, and Carl's wife, Annette. Next to arrive was Cedric Williams, a West Indian, whose first book, a collection of short stories with a Caribbean background, in Raymond's opinion, marked him as one of the three Negroes writing who actually had something to say, and also some concrete idea of style. Cedric was followed by Austin Brown, a portrait painter whom Raymond personally despised, a Dr. Manfred Trout, who practiced medicine and also wrote exceptionally good short stories, Glenn Madison, who was a Communist, and a long, lean professorial person, Allen Fenderson, who taught school and had ambitions to become a crusader modeled after W.E.B. Du Bois.

The roster was now complete. There was an hour of small talk and drinking of mild cocktails in order to induce ease and allow the various guests to become acquainted and voluble. Finally, Dr. Parkes ensconced himself in Raymond's favorite chair, where he could get a good view of all in the room, and clucked for order.

Raymond observed the professor closely. Paul's description never seemed more apt. He was a mother hen clucking at her chicks. Small, dapper, with sensitive features, graying hair, a dominating head, and restless hands and feet, he smiled benevolently at his brood. Then, in his best continental manner, which he had acquired during four years at European Universities, he began to speak.

"You are," he perorated, "the outstanding personalities in a new generation. On you depends the future of your race. You are not, as were your predecessors, concerned with donning armor, and clashing swords with the enemy in the public square. You are finding both an escape and a weapon in beauty, which beauty when created by you will cause the American white man to reestimate the Negro's value to his civilization, cause him to realize that the American black man is too valuable, too potential of utilitarian accomplishment, to be kept downtrodden and segregated.

"Because of your concerted storming up Parnassus, new vistas will be spread open to the entire race. The Negro in the south will no more know peonage, Jim Crowism, or loss of the ballot, and the Negro everywhere in America will know complete freedom and equality.

"But," and here his voice took on a more serious tone, "to accomplish this, your pursuit of beauty must be vital and lasting. I am somewhat fearful of the decadent strain which seems to have filtered into most of your work. Oh, yes, I know you are children of the age and all that, but you must not, like your paleface contemporaries, wallow in the mire of post-Victorian license. You have too much at stake. You must have ideals. You should become . . . well, let me suggest your going back to your racial roots, and cultivating a healthy paganism based on African traditions.

"For the moment that is all I wish to say. I now want you all to give expression to your own ideas. Perhaps we can reach a happy mean for guidance."

He cleared his throat and leaned contentedly back in his chair. No one said a word. Raymond was full of contradictions, which threatened to ooze forth despite his efforts to remain silent. But he knew that once the ooze began there would be no stopping the flood, and he was anxious to hear what some of the others might have to say.

However, a glance at the rest of the people in the room assured him that most of them had not the slightest understanding of what had been said, nor any ideas on the subject, whatsoever. Once more Dr. Parkes clucked for discussion. No one ventured a word. Raymond could see that Cedric, like himself, was full of argument, and also like him, did not wish to appear contentious at such an early stage in the discussion. Tony winked at Raymond when he caught his eye, but the expression on his face was as inscrutable as ever. Sweetie May giggled behind her handkerchief. Paul amused himself by sketching the various people in the room. The rest were blank.

"Come, come, now," Dr. Parkes urged somewhat impatiently, "I'm not to do all the talking. What have you to say, DeWitt?"

All eyes sought out the so-called Negro poet laureate. For a moment he stirred uncomfortably in his chair, then in a high pitched, nasal voice proceeded to speak.

"I think, Dr. Parkes, that you have said all there is to say. I agree with you. The young Negro artist must go back to his pagan heritage for inspiration and to the old masters for form."

Raymond could not suppress a snort. For DeWitt's few words had given him a vivid mental picture of the poet's creative hours—eyes on a page of Keats, fingers on typewriter, mind frantically conjuring African scenes. And there would of course be a bible nearby.

Paul had ceased being intent on his drawing long enough to hear "pagan heritage," and when DeWitt finished he inquired inelegantly:

"What old black pagan heritage?"

DeWitt gasped, surprised and incredulous.

"Why, from your ancestors."

"Which ones?" Paul pursued dumbly.

"Your African ones, of course." DeWitt's voice was full of disdain.

"What about the rest?"

"What rest?" He was irritated now.

"My German, English, and Indian ancestors," Paul answered willingly. "How can I go back to African ancestors when their blood is so diluted and their country and times so far away? I have no conscious affinity for them at all."

Dr. Parkes intervened: "I think you've missed the point, Paul."

"And I," Raymond was surprised at the suddenness with which he joined in the argument, "think he has hit the nail right on the head. Is there really any reason why *all* Negro artists should consciously and deliberately dig into African soil for inspiration and material unless they actually wish to do so?"

"I don't mean that. I mean you should develop your inherited spirit."

DeWitt beamed. The doctor had expressed his own hazy theory. Raymond was about to speak again, when Paul once more took the bit between his own teeth.

"I ain't got no African spirit."

Sweetie May giggled openly at this, as did Carl Denny's wife, Annette. The rest looked appropriately sober, save for Tony, whose eyes continued to telegraph mischievously to Raymond. Dr. Parkes tried to squelch Paul with a frown. He should have known better.

"I'm not an African," the culprit continued. "I'm an American and a perfect product of the melting pot."

"That's nothing to brag about," Cedric spoke for the first time.

"And I think you're all on the wrong track." All eyes were turned toward this new speaker, Allen Fenderson. "Dr. Du Bois has shown us the way. We must be militant fighters. We must not hide away in ivory towers and prate of beauty. We must fashion cudgels and bludgeons rather than sensitive plants. We must excoriate the white man, and make him grant us justice. We must fight for complete social and political and economic equality."

"What we ought to do," Glenn Madison growled intensely, "is to join hands with the workers of the world and overthrow the present capitalistic regime. We are of the proletariat and must fight our battles allied with them, rather than singly and selfishly."

"All of us?" Raymond inquired quietly.

"All of us who have a trace of manhood and are more interested in the rights of human beings than in gin parties and neurotic capitalists."

"I hope you're squelched," Paul stage whispered to Raymond.

"And how!" Raymond laughed. Several joined in. Dr. Parkes spoke quickly to Fenderson, ignoring the remarks of the Communist.

"But, Fenderson . . . this is a new generation and must make use of new weapons. Some of us will continue to fight in the old way, but there are other things to be considered, too. Remember, a beautiful sonnet can be as effectual, nay more effectual, than a rigorous hymn of hate."

"The man who would understand and be moved by a hymn of hate would not bother to read your sonnet, and, even if he did, he would not know what it was all about."

"I don't agree. Your progress must be a boring in from the top, not a battle from the bottom. Convert the higher beings and the lower orders will automatically follow."

"Spoken like a true capitalistic minion," Glenn Madison muttered angrily.

Fenderson prepared to continue his argument, but he was forestalled by Cedric.

"What does it matter," he inquired diffidently, "what any of you do so long as you remain true to yourselves? There is no necessity for this movement becoming standardized. There is ample room for everyone to follow his own individual track. Dr. Parkes wants us all to go back to Africa and resurrect our pagan heritage, become atavistic. In this he is supported by Mr. Clinton. Fenderson here wants us all to be propagandists and yell at the top of our lungs at every conceivable injustice. Madison wants us all to take a cue from Leninism and fight the capitalistic bogey. Well . . . why not let each young hopeful choose his own path? Only in that way will anything at all be achieved."

"Which is just what I say," Raymond smiled gratefully at Cedric. "One cannot make movements nor can one plot their course. When the work of a given number of individuals during a given period is looked at in retrospect, then one can identify a movement and evaluate its distinguishing characteristics. Individuality is what we should strive for. Let each seek his own salvation. To me, a wholesale flight back to Africa or a wholesale allegiance to Communism or a wholesale adherence to an antiquated and for the most part ridiculous propagandistic program are all equally futile and unintelligent."

Dr. Parkes gasped and sought for an answer. Cedric forestalled him.

"To talk of an African heritage among American Negroes *is* unintelligent. It is only in the West Indies that you can find direct descendents from African ancestors. Your primitive instincts among all but extreme proletariat have been ironed out. You're standardized Americans."

"Oh, no," Carl Denny interrupted suddenly. "You're wrong. It's in our blood. It's . . ." he fumbled for a word, "fixed. Why . . ." he stammered again, "remember Cullen's poem, 'Heritage':

>So I lie who find no peace
>Night or day, no slight release
>From the unremittent beat
>Made by cruel padded feet
>Walking through my body's street.
>Up and down they go, and back,
>Treading out a jungle track.

"We're all like that. Negroes are the only people in America not standardized. The feel of the African jungle is in their blood. Its rhythms surge through their bodies. Look how Negroes laugh and dance and sing, all spontaneous and individual."

"Exactly." Dr. Parkes and DeWitt nodded assent.

"I have yet to see an intelligent or middle class American Negro laugh and sing and dance spontaneously. That's an illusion, a pretty sentimental fiction. Moreover your songs and dances are not individual. Your spirituals are mediocre folk songs, ignorantly culled from Methodist hymn books. There are white men who can sing them just as well as Negroes, if not better, should they happen to be untrained vocalists like Robeson, rather than highly trained technicians like Hayes. And as for dancing spontaneously and feeling the rhythms of the jungle . . . humph!"

Sweetie May jumped into the breach.

"I can do the Charleston better than any white person."

"I particularly stressed . . . intelligent people. The lower orders of any race have more vim and vitality than the illuminated tenth."

Sweetie May leaped to her feet.

"Why, you West Indian . . ."

"Sweetie, Sweetie," Dr. Parkes was shocked by her polysyllabic expletive.

Pandemonium reigned. The master of ceremonies could not cope with the situation. Cedric called Sweetie an illiterate southern hussy. She called him all types of profane West Indian monkey chasers. DeWitt and David were shocked and showed it. The literary doctor, the Communist and Fenderson moved uneasily around the room. Annette and Paul giggled. The two child prodigies from Boston looked on wide-eyed, utterly bewildered and dismayed. Raymond leaned back in his chair, puffing on a cigarette, detached and amused. Austin, the portrait painter, audibly repeated over and over to himself: "Just like niggers . . . just like niggers." Carl Denny interposed himself between Cedric and Sweetie May. Dr. Parkes clucked for civilized behavior, which came only when Cedric stalked angrily out of the room.

After the alien had been routed and peace restored, Raymond passed a soothing cocktail. Meanwhile, Austin and Carl had begun arguing about painting. Carl did not possess a facile tongue. He always had difficulty formulating in words the multitude of ideas which seethed in his mind. Austin, to quote Raymond, was an illiterate cad. Having examined one of Carl's pictures on Raymond's wall, he had disparaged it. Raymond listened attentively to their argument. He despised Austin mainly because he spent most of his time imploring noted white people to give him a break by posing for a portrait. Having the gift of making himself pitiable, and having a glib tongue when it came to expatiating on the trials and tribulations of being a Negro, he found many sitters, all of whom thought they were encouraging a handicapped Negro genius. After one glimpse at the completed portrait, they inevitably changed their minds.

"I tell you," he shouted, "your pictures are distorted and grotesque. Art is art, I say. And art holds a mirror up to nature. No mirror would reflect a man composed of angles. God did not make man that way. Look at Sargent's portraits. He was an artist."

"But he wasn't," Carl expostulated. "We . . . we of this age . . . we must look at Matisse, Gauguin, Picasso and Renoir for guidance. They get the feel of the age . . . They . . ."

"Are all crazy and so are you," Austin countered before Carl could proceed.

Paul rushed to Carl's rescue. He quoted Wilde in rebuttal: Nature imitates art, then went on to blaspheme Sargent. Carl, having found some words to express a new idea fermenting in his brain, forgot the argument at hand, went off on a tangent and began telling the dazed Dr. Parkes about the Negroid quality in his drawings. DeWitt yawned and consulted his watch. Raymond mused that he probably resented having missed the prayer meeting which he attended every Thursday night. In another corner of the room the Communist and Fenderson had locked horns over the ultimate solution of the Negro problem. In loud voices each contended for his own particular solution. Karl Marx and Lenin were pitted against Du Bois and his disciples. The writing doctor, bored to death, slipped quietly from the room without announcing his departure or even saying good night. Being more intelligent than most of the others, he had wisely kept silent. Tony and Sweetie May had taken adjoining chairs, and were soon engaged in comparing their versions of original verses to the St. James infirmary, which Tony contended was soon to become as epical as the St. Louis Blues. Annette and Howard began gossiping about various outside personalities. The child prodigies looked from one to the other, silent, perplexed, uncomfortable, not knowing what to do or say. Dr. Parkes visibly recoiled from Carl's incoherent expository barrage, and wilted in his chair, willing but unable to effect a courteous exit. Raymond sauntered around the room, dispensing cocktails, chuckling to himself.

Such was the first and last salon.

## Chapter 25

It was Raymond's last night in Niggeratti Manor. Lucille had spent most of the evening with him, aiding him to pack. The studio was bare and cheerless. The walls had been stripped of the colorful original drawings contributed by Paul and Carl Denny. They were now stark and bare. The book shelves were empty, and yawned hideously in the more shaded corners. The middle of the room was filled with boxes in which his books had been packed, and in the alcove his trunk and suitcases stood at attention in military array. The rest of the house was also in a state of dishevelment. Painters and plasterers had been swarming all over the place, leaving undeniable evidence of their presence and handiwork. Niggeratti Manor was almost ready to suffer its transition from a congenial home for Negro artists to a congenial dormitory for bachelor girls.

Amid the gloom and confusion Raymond and Stephen sat, fitfully conversing between frequent drinks which had little effect. There was more bad news. Stephen had been called back to Europe. His mother was dangerously ill. There was little hope of his arriving before she died, but they insisted that he, the eldest son, start for home immediately. He was to sail the next day.

"You know, son, family is a hell of a thing. They should all be dissolved. Of course I'm perturbed at the thought of my mother's death, but I can't stop her from dying, nor can I bring her back to life should she be dead when I arrive. And yet I am dragged across an ocean, expected to display great grief and indulge in all the other tomfoolery human beings indulge themselves in when another human being dies. It's all tommyrot."

"Assuredly," Raymond agreed, "dying is an event, a perversely festive occasion, not so much for the deceased as for his so-called mourners. Let's forget it. You've got to adhere to the traditions of the clan to some degree. Let's drink to the day when a person's death will be the cue for a wild gin party rather than a signal for well meant but purely exhibitionistic grief." He held his glass aloft. "Skip ze gutter." The glasses were drained.

"And after you get in Europe?"

"I will be prevailed upon to stay at home and become a respectable schoolmaster. Now, let's finish the bottle of gin. I've got to go. It's after three and as usual we've been talking for hours and said nothing."

Raymond measured out the remaining liquor.

"O.K., Steve. Here's to the fall of Niggeratti Manor and all within."

Stephen had gone. Raymond quickly prepared himself for bed, and was almost asleep when the telephone began to ring. He cursed, decided not to get up, and turned his face toward the wall. What fool could be calling at this hour of the morning? In the old days it might have been expected, but now Niggeratti Manor was no more. There was nothing left of the old régime except reminiscences and gossip. The telephone continued to ring. Its blaring voice echoed throughout the empty house. Muttering to himself, Raymond finally left his bed, donned his bathrobe and mules, went out into the hallway, and angrily lifted the receiver:

"Hello," he grumbled.

A strange voice answered. "Hello. Is this Raymond Taylor?"

"It is."

"This is Artie Fletcher, Paul's roommate. Can you come down to my house right away? Something terrible has happened."

Raymond was now fully awake. The tone of horror in the voice at the other end of the wire both stimulated and frightened him. He had a vague, eerie premonition of impending tragedy.

"What is it? What's happened?" he inquired impatiently.

"Paul's committed suicide."

Raymond almost dropped the receiver. Mechanically he obtained the address, assured Artie Fletcher that he would rush to the scene, and within a very few moments was dressed and on his way.

The subway ride was long and tedious. Only local trains were in operation, local trains which blundered along slowly, stopping at every station, droning noisily: Paul is dead. Paul is dead.

Had Paul the debonair, Paul the poseur, Paul the irresponsible romanticist, finally faced reality and seen himself and the world as they actually were? Or was this merely another act, the final stanza in his drama of beautiful gestures? It was consonant with his character, this committing suicide. He had employed every other conceivable means to make himself stand out from the mob. Wooed the unusual, cultivated artificiality, defied all conventions of dress and conduct. Now perhaps he had decided that there was nothing left for him to do except execute self-murder in some bizarre manner. Raymond found himself not so much interested in the fact that Paul was dead as in wanting to know how death had been accomplished. The train trundled along clamoring: What did he do? What did he do? Raymond deplored the fact that he had not had sufficient money to hire a taxi.

The train reached Christopher Street. Raymond rushed out of the subway to the street above. He hesitated a moment to get his bearings, repeated the directions he had been given over the telephone, and plunged into a maze of criss-cross streets. As he neared his goal, a slender white youth fluttered toward him.

"Are you Raymond Taylor?"

"Yes."

"Come this way, please, I was watching for you."

Raymond followed his unknown companion into a malodorous, jerry-built tenement, and climbed four flights of creaky stairs to a rear room, lighted only by burning planks in the fireplace. There were several people in the room, all strangely hushed and pale. A chair was vacated for him near the fireplace. No introductions were made. Raymond lit a cigarette to hide his nervousness. His guide, whom he presumed to be Artie Fletcher, told him the details of Paul's suicide.

Earlier that evening they had gone to a party. It had been a wild revel. There had been liquor and cocaine which everyone had taken in order to experience a new thrill. There had been many people at the party and it had been difficult to keep track of any one person. When the party had come to an end, Paul was nowhere to be found, and his roommate had come home alone.

An hour or so later, he had heard a commotion in the hallway. Several people were congregated outside the bathroom door, grumbling because they had been unable to gain admittance. The bathroom, it seemed, had been occupied for almost two hours and there was no response from within. Finally someone suggested breaking down the door. This had been done. No one had been prepared for the gruesome yet fascinating spectacle which met their eyes.

Paul had evidently come home before the end of the party. On arriving, he had locked himself in the bathroom, donned a crimson mandarin robe, wrapped his head in a batik scarf of his own designing, hung a group of his spirit portraits on the dingy calcimined wall, and carpeted the floor with sheets of paper detached from the notebook in which he had been writing his novel. He had then, it seemed, placed scented joss-sticks in the four corners of the room, lit them, climbed into the bathtub, turned on the water, then slashed his wrists with a highly ornamented Chinese dirk. When they found him, the bathtub had overflowed, and Paul lay crumpled at the bottom, a colorful, inanimate corpse in a crimson streaked tub.

What delightful publicity to precede the posthumous publication of his novel, which novel, however, had been rendered illegible when the overflow of water had inundated the floor, and soaked the sheets strewn over its surface. Paul had not foreseen the possible inundation, nor had he taken into consideration the impermanency of penciled transcriptions.

Artie Fletcher had salvaged as many of the sheets as possible. He handed the sodden mass to Raymond. Ironically enough, only the title sheet and the dedication page were completely legible. The book was entitled:

**Wu Sing: The Geisha Man**

It had been dedicated:

To

Huysmans' Des Esseintes and Oscar Wilde's Oscar Wilde
Ecstatic Spirits with whom I Cohabit
And whose golden spores of decadent pollen
I shall broadcast and fertilize
It is written

Paul Arbian.

Beneath this inscription, he had drawn a distorted, inky black skyscraper, modeled after Niggeratti Manor, and on which were focused an array of blindingly white beams of light. The foundation of this building was composed of crumbling stone. At first glance it could be ascertained that the skyscraper would soon crumble and fall, leaving the dominating white lights in full possession of the sky.

## *The Interne*

Published by the Macaulay Company, 1932; written with
Abraham L. Furman.

### Chapter 1

The taxi came to a sudden halt. The driver cursed audibly. His brakes were loose and ineffectual on the slippery pavement. A long, noisy stream of traffic cluttered up the approach to the bridge, which led from the mainland to City Island.

Within the taxi sat a young man, annoyed at the delay, eager to reach his destination. Sitting on the edge of his seat, he peered anxiously through the misty windows, trying to get a glimpse of the hospital buildings on the distant island. His eyes discerned the bridge, stretching out before him, a lyric span which seemed to dissolve into nothingness. Beyond, in the center of the river, a smudge, uneven, indistinct, loomed out of the fog and rain. That must be the hospital, Carl said to himself peering ahead, and smiled hopefully as the taxi began to crawl along.

He leaned back in his seat. The taxi gathered speed. The automatic windshield wiper swished back and forth with efficient vigor. Somewhere in the foggy distance, a coal barge blew a warning whistle, which topped the prevailing sounds, and lingered like a cry of despair in Carl's ears.

Once more, he leaned forward eagerly. The taxi had gathered more speed. They were almost in the center of the bridge. City Island was now a more visible reality. And the hospital buildings emerged from the gloom as individual units.

Carl could no longer contain himself. "Dr. Carl Armstrong," he murmured aloud, "hot ziggity, boy, you're here at last."

## Chapter 14

Everyone concerned looked forward with pleasure to the internes' parties. Without exception they were always gala affairs, given once every two months and more often if the occasion presented itself, allowing the internes to forget, for the night, the unrelenting routine of the hospital, and to be their youthful, exuberant selves.

To these parties were occasionally invited privileged lay friends from the city, and such nurses as intrigued the individual young doctors.

On party nights, the recreation room was stripped of all excess furniture, only the most necessary divan and chairs being allowed to remain. And these were pushed back against the wall to allow a large, uninhabited space for dancing.

The piano from the nurses' home was borrowed on each occasion. Several times the internes had petitioned unsuccessfully for a piano of their own. The city fathers in charge of hospital appropriations refused to sanction this unmedical, esthetic request. Piqued, the boys had decided to buy a piano of their own, and had even gone so far as to create a piano sinking fund. But the money was always used for some other purpose. For the giving of a party, or the buying of liquor, or for a gambling pool to back some inspired lottery.

Tonight, Gus, as usual, was supplying the liquor and most of the food. There had been, this time, the unprecedented difficulty of delivery. Gus was going to give them a half case, and they were to buy a case themselves. All evening they had held council, trying to figure how they were going to transport the liquor from Gus' warehouse on the mainland to the hospital on the island.

Gus flatly refused to take a chance. His son, in defiance of parental warnings, had managed to smuggle in a half dozen quart bottles with the food Gus was donating. But the rest of the liquor, necessary for the party's success, remained on the other side of the bridge.

After hours of arguing, planning and meditation, Jim and Pete conceived a brilliant plan. But the party had begun before they could execute it.

Meanwhile the party progressed. There was enough liquor on hand to induce hilarity. About thirty young couples danced around the recreation room, gathered about the punch bowl and buffet in the dining room, or else migrated to dark, secluded corners on the spacious veranda.

The party was in full swing. Carl remembered he had not yet danced with Nora. He found her in the lobby. Ray was there, too, seated on a bench, a girl on each knee.

"Shall we dance?" Carl asked Nora.

"No . . . I'm going to bed."

"Why . . . what's the matter? Aren't you having a good time?"

"I am not." There was no mistaking her positiveness.

"But why?"

"You've ignored me . . . and I haven't been able to corner your pet blonde. I can hardly wait to tell her what I think of her."

"That isn't necessary."

"Maybe not necessary, but it would be damn satisfying. If I see her fixing you with that baby stare of hers again, I'll scream."

"She's my guest, Nora. I have to be nice to her."

"I'm sure she's had no cause to complain up to now."

"I can't be rude."

"So I notice."

"Now, Nora, don't be childish."

"I think you're in love with her."

"When I do fall in love . . ."

"I think you've been in love with her since that first night in my ward."

"I think you're gaga."

"I am, about you, Carl." She seemed about to soften, then remembering her role, flared up again. "But what you can see in a brainless dame like that is beyond me. I suppose you'll be marrying her soon."

"Maybe."

"Fine. Then you can hire her a nurse to see that she doesn't go out without wearing her rubber panties."

"Nora!" Ted staggered toward her. "Come on. It's time for our specialty."

"I'd like to chloroform you," she told the interloper, then allowed him to lead her back into the recreation room.

Carl stared after them, listened absent-mindedly to the music for a moment, then pulled a flask from his hip pocket, and poured a long stream of liquor down his throat. This done, he lit a cigarette, and started toward the veranda.

"Oh, Carl," it was Annabelle, petite, blonde, tiny figure encased in a billowing white evening gown. Pete was with her.

"Here she is, Carl," he said. "The best dancer in the house. Don't forget we have another soon."

"Oh, no," she gurgled. "I'll be right here . . . waiting."

"O.K." Pete left them alone.

"Gee, I'm glad I came," she said, cuddling up close to Carl.

"Having a good time?"

"I'll say I am."

"This is a jolly bunch all right."

"Couldn't be better," she agreed. "I didn't know there could be so much fun in a hospital. It must be fun to work here," she added childishly.

"These parties only happen once every two months."

"I know . . . but think of the fun you fellows must have all the time."

Carl's lips tightened into a grim, straight line as Annabelle continued to gush about the good time the hospital staff must have. Finally, he let his attention wander, and was only dimly conscious of her cooing voice and ridiculous statements. He fastened his eyes on Ray, and noted that the two girls sitting on his knees were arguing. He then remembered Ray's oft reiterated plans for the future. He was going to be a consulting specialist, dealing in neurological diseases, and interested in female clients only.

"I'll hold their hands," Ray often said, "stroke their wrists, and look soulfully into their eyes, understand all their troubles, prescribe sugar pills, and collect juicy fees. In five years, I'll be wealthy, even if I have to marry some rich heifer with chronic halitosis."

When asked why he did not forego being a doctor altogether and become a gigolo, he would answer brightly:

"Gigolos lack dignity, and they only get small fry. I'm out after big game. And Dr. Raymond Storrs can rush where gigolos fear to tread."

Carl mused that Ray certainly took advantage of every opportunity to perfect his technique. He knew how to play women one against the other, and always to his own advantage. It might behoove him, Carl, to take some lessons.

He turned his eyes and attention back to the loquacious Annabelle.

"I'd like to be a nurse," she was saying.

"I'm sure you wouldn't," Carl answered. "It's not an easy job."

Annabelle's eyes opened wide.

"With all these good-looking internes around? Why . . . I'd let the patients die while I flirted."

"Nurses don't have much chance to flirt," he told her rather curtly.

"I'm glad of that."

"Why?"

"Because I don' wanna lose my ducky-wucky to some old nurse."

"This is all rather sudden, isn't it?" He was irritated by her affected baby talk.

"No, really, I couldn't stand it."

Carl gazed at her for a moment, eyes quizzical and half-closed. Then, without saying a word, he took her by the arm and propelled her out onto the veranda.

The moon was smiling at its reflection in the slimy river. The night was hot, stuffy. The music from the main scene of the party was romantically distant. Three other couples occupied the three most secluded corners of the veranda, their shadowy figures merged into incongruous outlines.

Carl and Annabelle stood in a minute depression beside the entrance door. His hands tantalized the soft, quivering flesh beneath her thin and revealing evening gown. Their lips met. Annabelle sighed softly and let her teeth clamp down on his tongue. Abruptly, Carl shook off her embrace and rudely shoved her back into the lobby.

Annabelle affected to sulk, but before she could give voice to her surprised and petulant protestations, Jack rushed over to Carl.

"Ted's passed out," he said. "Come help me get him upstairs, will you, Carl?"

"Sure. Excuse me a moment, please, Annabelle."

They walked away. Annabelle frowned, pursed her lips, then remembering the recent scene on the veranda, decided it might be well to repair her make-up.

Carl was amazed at the noise in the recreation room. Everyone seemed to be laughing or shouting or singing drunkenly. One of the nurses, dress lifted high above the stocking line, showing her firm, naked thighs, was performing an exhibition dance, imitating Ann Pennington.

The delighted audience stamped their feet, clapped their hands and shouted encouragement and approbation in time with the music.

Ted was slumped against the wall beneath the telephone, a wan, neglected, and pitiful figure, drooling at the mouth, completely unconscious. Jack jerked him up from the floor, and taking him by the shoulders, motioned for Carl to appropriate his legs. Then, with his arms and head dangling, they carted him out of the room, up into his dormitory cubicle.

Annabelle finished repairing her make-up, and satisfied with her appearance. began to search for another admiring swain. She had not exaggerated in telling Carl of the fun she had been having. An inveterate flirt, a silly chit who thrived on the flattery and attention of predatory males, it amused her to resort to tried and true feminine tricks in order to arouse male desire. She was never subtle, and managed to be so obvious that any man with whom she came into contact immediately said to himself: This jane is on the make.

She was known among her own social group as a tease. And the truth about her accident, which landed her first in the hospital and then in Carl's hungry and inexperienced arms, had never been told. Her story to Carl was a soothing piece of fiction. The facts of the case were somewhat sordid and not at all complimentary to her. Attracted to an unknown man in a speakeasy, she had deserted her escort, and gone out with the stranger. And he, inflamed by her actions, and finding her unwilling to complete the negotiations which she herself had started, had taken matters into his own hands. Only her timely jump from the automobile had saved her from being brutally assaulted.

And now she stood in the lobby, trying to decide which of the internes she would enrapture with her charms, but before she could settle on her target, Nora, tiring of the noise and vulgarity in the circle around the untiring exhibitionist dancer, and wishing to see Carl, came out into the lobby. Sighting Annabelle, she walked over to her, and barred her from entering the recreation room.

"Hello, Miss Clarke. I've been trying to speak to you all evening."

Annabelle scrutinized her for a moment, trying hard to remember where she had seen her before.

"Oh," she said finally, "I just recognized you. You were my nurse, weren't you? You look so different now without a uniform."

"I suppose so," Nora agreed dryly. "I'm glad to see you're all right."

"Thanks." Then suddenly, "It must be nice, being a nurse."

"Would you like to be one?"

"Well . . . I didn't use to. But since meeting Carl . . ."

"Oh, I see."

"I bet I'd get more thrills."

"You certainly would." Annabelle was unconscious of the sarcasm in Nora's answer.

"Tell me about some of your experiences."

"Gladly," Nora assented.

Annabelle's face lit up. She clapped her palms together delightedly, like a small child about to receive some treasured sugar plum.

"Oh, goody," she enthused.

"Where shall I begin?"

"Any place."

"Does rape interest you?"

"Rape?"

"Well, maybe rape is a strong word. Let's say seduction."

Annabelle was justifiably mystified.

"I . . . I . . . don't understand."

"Didn't you know?" Nora asked in shocked surprise. "Why that's the first thing a nurse learns."

"What?"

"How to avoid seductions," Nora answered calmly.

"Oh," Annabelle's interest began to subside. She felt herself entertaining a growing dislike for this nurse who talked so strangely.

"Hospital beds," Nora continued inexorably, "seem always to . . . well, should I say . . . make men ambitious? Really, it's awful. You give a man a bed pan and no matter how many pains he has, when you move the bed pan, he wants you to take its place."

"Why . . . Miss . . . Miss . . ." Annabelle floundered. Her face was flushed.

"Grant's the name. Nora Grant," she supplied, then added confidentially, "You know being a doctor's wife is a little better than working in a hospital. He won't need you to help him often. You'll just have to learn how to help out a little bit. You know . . . do just simple things like giving first aid to accident victims, helping deliver children, stopping blood from flowing, and handling insane dope fiends. Now, they're interesting."

"Who?" Annabelle sought anxiously for some way to terminate this disagreeable conversation.

"Dope fiends. They're such good wrestlers. When they get a half-nelson on you the fun begins. Much more interesting than real maniacs. Why . . . I had one . . ."

"Please!"

"But I thought you wanted to hear about my experiences?"

"I . . . I think you're horrid."

"My dear, that's mild to what I think about you." Annabelle gasped as Nora continued, "And if you don't leave Carl alone, I'll . . ."

"Why . . ."

"Just watch your step, baby. Carl's my man, and if you're not careful, you're liable to get thrown out of another automobile."

And having delivered this desperate ultimatum to the amazed Annabelle, Nora turned and walked away.

## Chapter 25

It was Carl's last night in the hospital. Time had flown by quickly. And now that he had definitely decided upon the course of his future activities, he had gone about his work diligently, doing the best he could under the circumstances, and accepting that which had to be without caviling or chafing at the bit.

He and Nora had slipped away to a town seventy miles distant and been married. Nora had continued working for three more months, then quietly resigned from the hospital. No one but she and Carl was privy to their secret and no one knew the reason for her departure.

Already she and Carl had decided upon the locale where they were to make their home, and where he was to begin active professional practice. They had leased a two-family bungalow in a near-by suburban town where many manufacturing mills were located. One side of the bungalow was to be used for their home. The other would house Carl's offices. Through Dr. Mason's influence, Carl had been assured a steady income, having received an appointment to be assistant to the manufacturing combine's company physician.

He was to be in charge of the children's clinic, and to specialize in pediatrics. Nora of course would assist him once her own child was born. And since his work for the company would normally occupy only his morning hours, there would be ample time for him to build up his private practice.

Life, perhaps, would be worth living, after all. Carl had come to the conclusion that no matter what profession you entered, the odds were against you. Every living person had to fight and scuffle to exist. Despair brought you nothing but intensified misery, and impaired your usefulness. One had to set one's feet on a path, grit one's teeth, wince if forced to detour, then scramble to maintain one's footing. Life was devoid of meaning, fraught with ugliness. Humanity was ill. One could only do one's bit and hope that the little he could accomplish might minimize man's suffering.

But Carl now spent little time being philosophical. His youth asserted itself, and his love for Nora and his plans for the future infused him with hope, rendered him happy.

He gleefully entered into the fun, when a new interne was brought into the room by Cal, who with Bob and Ray, was spending six additional months in the hospital. The boy reminded Carl of himself when he had come to the hospital twelve months before. He looked both frightened and hopeful and was utterly bewildered at the barrage of talk leveled at him.

"Welcome to Internes' Hall," Ray shouted.

"Thanks."

"Can you shoot crap, fellow? Bob asked.

"Why . . . er . . ."

"How's your bridge?"

"You drink liquor, don't you?" Cal asked.

"That's why he's wearing glasses, I guess," Carl surmised. "Hooch ruins the eyes."

"Hope you haven't got double vision," Cal said. "The last fellow we had like that amputated two legs when the order was for one."

"You'll like it here, fellow," Carl said. "Where are you assigned?"

"In the morgue."

"Fine!" Carl enthused. "We've got fifteen stiffs down there now waiting for action."

"Yeah," Ray affirmed. "Some damn good cases, too."

"I'll say," Carl continued. "I sent one down this morning. Cancer. Smell it a block."

"Ever work in a hospital before?" Carl asked.

"Why . . . er . . . no."

"Isn't that too bad," Ray said.

"Oh, he'll get used to it," Carl advanced.

"Sure," Bob agreed. "He looks like he's got a strong stomach."

"Not much muscle," Ray said, feeling the bewildered boy's arm, "Some of those stiffs weigh over two hundred pounds."

"Maybe they'll put you on ward duty," Bob said.

"Sure," Carl said, "then you'll only work about fifteen hours a day. But," he warned, "there's two things you've got to watch."

"What are they?" the newcomer asked eagerly.

"Pretty patients and sex starved nurses," Carl answered. "You'll try to rape the patients, and the nurses will rape you sure as hell."

The new interne could not appreciate the outburst of laughter which followed this statement. Nor why Carl suddenly blushed, stammered, and hurriedly left the room.

# Bibliography

This bibliography includes works cited in the introductions to the eight parts of this *Reader,* general works on the Harlem Renaissance, and works specifically about Thurman.

## The Harlem Renaissance

Abramson, Doris. *Negro Playwrights in the American Theatre, 1925–1959.* New York: Columbia University Press, 1969.

Andrews, William L., ed. *Classic Fiction of the Harlem Renaissance.* New York: Oxford University Press, 1994.

Baker, Houston A. Jr. *Afro-American Poetics: Revisions of Harlem and the Black Aesthetic.* Madison: University of Wisconsin Press, 1996.

———. *Modernism and the Harlem Renaissance.* Chicago: University of Chicago Press, 1987.

Banks, William M. *Black Intellectuals: Race and Responsibility in American Life.* New York: W. W. Norton, 1996.

Bassett, John E. *Harlem in Review: Critical Reactions to Black American Writers, 1917–1939.* London: Associated University Presses, 1992.

Bernard, Emily. *Remember Me to Harlem: The Letters of Langston Hughes and Carl Van Vechten.* New York: Vintage Books, 2001.

Bone, Robert A. *Down Home: A History of Afro-American Short Fiction.* New York: Putnam, 1975.

Bontemps, Arna, ed. *The Harlem Renaissance Remembered.* New York: Dodd, Mead, 1972.

Brades, Susan Ferleger. *Rhapsodies in Black: Art of the Harlem Renaissance.* Berkeley: University of California Press, 1997.

Davis, Thadious M. *Nella Larsen: Novelist of the Harlem Renaissance.* Baton Rouge: Louisiana State University Press, 1994.

Fabre, Geneviève, and Michel Feith, eds. *Temples for Tomorrow: Looking Back at the Harlem Renaissance.* Bloomington: Indiana University Press, 2001.

Fabre, Michel. *From Harlem to Paris: Black American Writers in France, 1840–1980.* Champaign-Urbana: University of Illinois Press, 1991.

Gates, Henry Louis Jr. *The Signifying Monkey : A Theory of Afro-American Literary Criticism.* New York: Oxford University Press, 1988.

Gates, Henry Louis Jr., and Nellie Y. McKay, eds. *The Norton Anthology of African American Literature.* New York: W. W. Norton, 1997.

Gayle, Addison Jr., ed. *The Black Aesthetic.* New York: Doubleday, 1971.

———. "The Harlem Renaissance: Towards a Black Aesthetic." *MidContinent American Studies Journal* 11.2 (1970): 78–87.

Harris, Trudier, and Thadious M. Davis, eds. *Afro-American Writers from the Harlem Renaissance to 1940.* Vol. 51 of *Dictionary of Literary Biography.* Detroit: Gale, 1987.

Hatch, James V., and Leo Hamalian, eds. *Lost Plays of the Harlem Renaissance: 1920–1940.* Detroit: Wayne State University Press, 1996.

Hemenway, Robert E. *Zora Neale Hurston: A Literary Biography.* Champaign-Urbana: University of Illinois Press, 1977.

Honey, Maureen, ed. *Shadowed Dreams: Women's Poetry of the Harlem Renaissance.* New Brunswick: Rutgers University Press, 1989.

Huggins, Nathan. *Harlem Renaissance*. New York: Oxford University Press, 1971.

———, ed. *Voices from the Harlem Renaissance*. New York: Oxford University Press, 1976.

Hughes, Langston. *The Big Sea: An Autobiography*. 1940. Reprint, New York: Hill and Wang, 1963.

———. "The Negro Artist and the Racial Mountain." *Nation,* June 23, 1926, 692–694.

Hutchinson, George. *The Harlem Renaissance in Black and White*. Cambridge: Harvard University Press, 1995.

Irek, Malgorzata. *The European Roots of the Harlem Renaissance*. Berlin: John F. Kennedy-Institut für Nordamerikastudien, 1994.

Kellner, Bruce. *The Harlem Renaissance: An Historical Dictionary for the Era*. Westport, Conn.: Greenwood Press, 1984.

Knopf, Marcy, ed. *The Sleeper Wakes: Harlem Renaissance Stories by Women*. New Brunswick: Rutgers University Press, 1993.

Kornweibel, Theodore Jr. *No Crystal Stair: Black Life and "The Messenger," 1917–1928*. Westport, Conn.: Greenwood Press, 1975.

Kramer, Victor A. *The Harlem Renaissance Re-examined*. New York: AMS, 1987.

Lenz, Günter. "Symbolic Space, Communal Ritual, and the Surreality of the Urban Ghetto: Harlem in Black Literature from the 1920s to the 1960s." *Callaloo* 11 (spring 1988): 309–345.

Lewis, David Levering. *When Harlem Was in Vogue*. New York: Alfred A. Knopf, 1981.

———, ed. *The Portable Harlem Renaissance Reader*. New York: Viking, 1994.

Locke, Alain. "Black Truth and Black Beauty." *Opportunity,* January 1933, 14–18.

———. *The New Negro*. New York: Alberta and Charles Boni, 1925.

McKay, Nellie Y. *Jean Toomer, Artist : A Study of His Literary Life and Work, 1894–1936*. Chapel Hill: University of North Carolina Press, 1984.

McLendon, Jacquelyn Y. *The Politics of Color in the Fiction of Jessie Fauset and Nella Larsen*. Charlottesville: University Press of Virginia, 1995.

Mishkin, Tracy. *The Harlem and Irish Renaissances*. Gainesville: University Press of Florida, 1998.

Mitchell, Angelyn, ed. *Within the Circle: An Anthology of African American Literary Criticism from the Harlem Renaissance*. Durham: Duke University Press, 1994.

Napier, Winston, ed. *African American Literary Theory: A Reader*. New York: New York University Press, 2000.

Ottley, Roi. *"New World A-coming": Inside Black America*. Boston: Houghton Mifflin, 1943.

Patton, Venetria K., and Maureen Honey, eds. *Double-Take: A Revisionist Harlem Renaissance Anthology*. New Brunswick: Rutgers University Press, 2001.

Porter, Horace A. *Stealing the Fire: The Art and Protest of James Baldwin*. Middletown, Conn.: Wesleyan University Press, 1989.

Rampersad, Arnold. *The Life of Langston Hughes*. 2 vols. New York: Oxford University Press, 1986–1988.

Reimonenq, Alden. "Countee Cullen's Uranian 'Soul Windows.'" *Journal of Homosexuality* 26 (1993): 143–165.

Roses, Lorraine, and Ruth Randolph, eds. *Harlem's Glory: Black Women Writing, 1900–1950*. Cambridge: Harvard University Press, 1996.

Schuyler, George S. "Negro-Art Hokum." *Nation,* June 16, 1926, 662–663.

Scruggs, Charles. *The Sage in Harlem: H. L. Mencken and the Black Writers of the 1920s*. Baltimore: Johns Hopkins University Press, 1984.

Silberman, Seth Clark. "Lighting the Harlem Renaissance AFire!!: Embodying Richard Bruce

Nugent's Bohemian Politics." In *The Greatest Taboo: Homosexuality in Black Communities,* ed. Delroy Constantine Simms, 254–273. Los Angeles: Alyson Books, 2001.

Singh, Amritjit. *The Novels of the Harlem Renaissance: Twelve Black Writers, 1923–33.* University Park: Pennsylvania State University Press, 1976.

Sollors, Werner. *Interracialism: Black-White Intermarriage in American History, Literature, and Law.* Oxford: Oxford University Press, 2000.

———. *Neither Black nor White Yet Both: Thematic Explorations of Interracial Literature.* New York: Oxford University Press, 1997.

Tracy, Steven C. "'Did You Ever Dream Lucky, Wake Up Cold in Hand?' African American Vernacular Music in the Works of Langston Hughes." Paper presented as the Fourth John Roche Memorial Lecture, Rhode Island College, April 19, 2002.

———. *Langston Hughes and the Blues.* Champaign-Urbana: University of Illinois Press, 1988.

Wall, Cheryl. "Paris and Harlem: Two Culture Capitals." *Phylon* 35 (March 1974): 64–73.

———. *Women of the Harlem Renaissance.* Bloomington: Indiana University Press, 1995.

Watson, Steven. *The Harlem Renaissance: Hub of African-American Culture, 1920–1930.* New York: Pantheon Books, 1995.

### Wallace Thurman

Carter, Eunice Hunton. Review of *The Blacker the Berry. Opportunity,* May 1929.

Du Bois, W.E.B. "The Browsing Reader." Review of *The Blacker the Berry. Crisis,* July 1929, 234, 238.

Gaither, Renoir W. "The Moment of Revision: A Reappraisal of Wallace Thurman's Aesthetic in *The Blacker the Berry* and *Infants of the Spring.*" *College Language Association Journal* 37.1 (1993): 81–93.

Gruening, Martha. "Two Ways to Heaven." Review of *Infants of the Spring. Saturday Review of Literature,* March 12, 1932.

Haslam, Gerald. "Wallace Thurman: A Western Renaissance Man." *Western American Literature* 6 (spring 1971): 53–59.

Henderson, Mae G. "Portrait of Wallace Thurman." In *The Harlem Renaissance Remembered,* ed. Arna Bontemps, 146–170. New York, Dodd Mead, 1972.

Hicks, Granville. "The New Negro: An Interview with Wallace Thurman." *Churchman,* April 30, 1927.

Jones, Dewey R. "The Bookshelf." Review of *The Blacker the Berry. Chicago Defender,* March 2, 1929.

Klotman, Phyllis. "Wallace Henry Thurman." In *Afro-American Writers from the Harlem Renaissance to 1940.* Vol. 51 of *Dictionary of Literary Biography,* ed. Trudier Harris and Thadious M. Davis, 260–273. Detroit: Gale, 1987.

Lewis, Theophilus. "Harlem Sketchbook: Wallace Thurman." *New York Amsterdam News,* January 5, 1935.

———. "If This Be Puritanism." *Opportunity,* April 1929.

———. "Wallace Thurman Is a Model Harlemite." *New York Amsterdam News* (c. 1930).

McIver, Dorothy. "Stepchild in Harlem: The Literary Career of Wallace Thurman." Ph.D. diss., University of Alabama, 1983.

Mitchell, Clifford. Review of *The Interne. Washington Tribune,* August 19, 1932.

*New York Times Book Review.* "Harlem's Bohemia." Review of *Infants of the Spring.* February 28, 1932.

———. "Harlem's Negroes." Review of *The Blacker the Berry.* March 17, 1929.

———. "Hospital Staff." Review of *The Interne.* June 5, 1932.

O'Daniel, Therman B. Introduction to *The Blacker the Berry.* New York: Collier Books, 1970.

Perkins, Huel D. "Renaissance 'Renegade'? Wallace Thurman." *Black World* 25.4 (1976): 29–35.

Potter, Lawrence T. Jr. "Harlem's Forgotten Genius: The Life and Work of Wallace Henry Thurman." Ph.D. diss., University of Missouri, 1999.

Robinson, J. E. "Wallace (Henry) Thurman." In *Gay and Lesbian Literature,* ed. Sara Pendergast and Tom Pendergast, 349–350. Detroit: St. James Press, 1988.

Silberman, Seth Clark. "Looking for Richard Bruce Nugent and Wallace Henry Thurman: Reclaiming Black Male Same-Sexualities in the New Negro Movement." *In Process* 1 (1996): 53–73.

Singh, Amritjit. Foreword to *Infants of the Spring.* Boston: Northeastern University Press, 1992.

Skinner, R. Dana. "The Play." Review of *Harlem. Commonweal,* March 6, 1929.

van Notten, Eleonore. *Wallace Thurman's Harlem Renaissance.* Atlanta: Rodopi, 1994.

Walden, Daniel. "'The Canker Galls...'; Or, the Short Promising Life of Wallace Thurman." In *The Harlem Renaissance Re-Examined*, ed. Victor A. Kramer, 201–211. New York: AMS, 1987.

West, Dorothy. "Elephant's Dance: A Memoir of Wallace Thurman." *Black World* 20.1 (1970): 77–85.

Williams, John A. Afterword to *Infants of the Spring.* Carbondale: Southern Illinois University Press, 1979.

Wilson, James F. "Bulldykes, Pansies,and Chocolate Babies: Performance, Race, and Sexuality in the Harlem Renaissance." Ph.D. diss., City University of New York, 2000.

Wright, Shirley H. "A Study of the Fiction of Wallace Thurman." Ph.D. diss., East Texas State University, 1983.

Frontispiece. Wallace Thurman. Photo courtesy of Mary Louise Patterson.

Page 1. Langston Hughes and Wallace Thurman in California, 1934. Photo courtesy of the Beinecke Rare Book and Manuscript Library, Yale University, New Haven.

Page 29. Young Harlemites at the Savoy Ballroom, c. 1930. Photo courtesy of the Beinecke Rare Book and Manuscript Library, Yale University, New Haven.

Page 79. Wallace Thurman, Salt Lake City, c. 1903. Photo courtesy of the Beinecke Rare Book and Manuscript Library, Yale University, New Haven.

Page 95. Wallace Thurman, with Langston Hughes and an unidentified woman, Rockaway Beach, New York, c. 1929. Photo courtesy of the Beinecke Rare Book and Manuscript Library, Yale University, New Haven.

Pages 174–175. Facsimile of 1934 letter to Dorothy West. Photo courtesy of Dorothy West Papers, Special Collections, Boston University.

Page 177. Wallace Thurman in Hollywood, c. 1934. Photo courtesy of the Beinecke Rare Book and Manuscript Library, Yale University, New Haven.

Page 229. Silent protest parade, Fifth Avenue, New York City, July 28, 1917. Photo courtesy of the Beinecke Rare Book and Manuscript Library, Yale University, New Haven.

Page 289. Aaron Douglas, cover of *Fire!!* November 1926.

Page 305. Playbill for *Harlem,* 1929. Photo courtesy of the Beinecke Rare Book and Manuscript Library, Yale University, New Haven.

Page 339. Isabel Washington as Cordelia and Hillis Walters as Ippy in a scene from *Harlem,* 1929. Photo courtesy of the Schomburg Center for Research in Black Culture, The New York Public Library, Astor, Lenox and Tilden Foundations.

Page 379. Marcus Garvey. Photo courtesy of Delilah Jackson Papers, Special Collections and Archives, Robert W. Woodruff Library, Emory University.

Page 441. Dust jacket of the first edition of *The Blacker the Berry,* 1929, from the personal collection of Thomas H. Wirth. Photo courtesy of Thomas H. Wirth.

Page 449. Thurman's inscription to Louise Thompson in her copy of *The Blacker the Berry.* Photo courtesy of Louise Thompson Papers, Special Collections and Archives, Robert W. Woodruff Library, Emory University.

# Index of Names

Page numbers in italics refer to the introduction to the book and the part introductions.

# Index of Works

## About the Editors

**AMRITJIT SINGH**, a professor of English and African American studies at Rhode Island College, was a Fulbright Professor in 2002 at the JFK Institute of North American Studies at Freie University–Berlin. He is the author of *The Novels of the Harlem Renaissance* (1976) and coeditor of over a dozen volumes, including *The Harlem Renaissance: Revaluations* (1989), *Memory, Narrative, and Identity* (1995), and *Postcolonial Theory and the United States* (2000). He has also prepared for publication reprint editions of Thurman's *Infants of the Spring* (1992) as well as Richard Wright's *Black Power* (1995) and *The Color Curtain* (1995).

**DANIEL M. SCOTT III**, an associate professor of English and African American studies at Rhode Island College, did his undergraduate studies in Atlanta and his Ph.D. in comparative literature at the University of Illinois at Urbana–Champaign. He has written numerous essays and book reviews on African and African American literatures as well as on queer studies. His writings range from work on Charles Johnson to Wole Soyinka to Essex Hemphill.

## Joinery

In timber framing, joinery is practically the definition of the product. Without joinery, there is a structure made with posts and beams, but it is not a timber frame. It was the discovery of simple joinery that made the first frame possible, and it is intricately worked, complex wooden joinery that holds together some of the great architectural masterpieces of the world. Joinery is a complex subject, and here I'll just summarize a few basics. Look to the bibliography for more information on selecting and cutting timber-frame joints.

An important guideline in timber-frame design is that none of the joints should be more complicated than necessary. The loads and stresses of the building need to be supported by the frame and the entire frame should be securely connected through the joinery. But beyond these rules, design concentrates on using joints that can be made efficiently and that are within the skills of the joiner. Without going out of the way to pursue esoteric or convoluted joints, you will invariably find enough surprises and challenges to keep the hands and mind occupied.

Most timber joints are variations on the mortise-and-tenon—a tongue on one timber is received by a slot in the other. The simplest version of this joint is used to connect vertical timbers to the frame and for joining two beams with minimum loads; knee braces and collar ties generally use an angled variation. Mortise-and-tenon joints are locked with rounded pegs driven through holes drilled through both parts of the joint, although some sophisticated joints rely on wedges and geometry to lock. The other broad category of joints used in timber frames consists of lap joints, such as dovetails and scarfs. Variations of lap joints can be used in many parts of the frame, as we will discuss later. Some of them rely entirely on their angular shapes and the weight of the timbers to lock, others are overlapped and then locked by wedging them through mating notches.

Which joint to use where? Joints must be chosen on the basis of the tasks they are to fulfill, including locking the frame together, bearing weight and transferring forces and building loads from one timber to another. A timber transferring a compression load will need a joint with enough bearing surface to withstand crushing. A timber in tension must be firmly bound with a joint that can prevent its withdrawal. If a timber is exerting an outward thrust, as a rafter exerts on a collar tie, it needs to be restrained by buttressing within the joint.

### Considerations in choosing joints

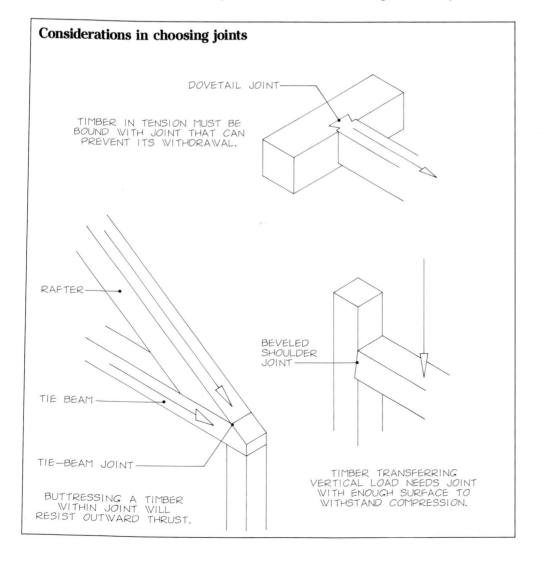

**The mortise-and-tenon**   The simple mortise-and-tenon can be very effective in resisting both tension and compression forces. We'll look at tension first.

In tension, there are three ways a joint can fail: The pegs can break, the wood between the end of the tenon and the pegs can pull out, or the fibers can fail between the edge of the mortised timber and the pegs. The perfect joint would have no one of these factors more likely to fail than the others under extreme tension—all three failures would occur simultaneously. Practically, however, we lean toward making sure there is enough wood between the pegs and the end of the tenon, because this second type of failure is abrupt and fatal; the others are gradual. For most woods, the mortise should be a minimum of 4 in. deep and the centers of 1-in.-dia. oak pegs should be placed about 1½ in. in from the edge of the mortised timber. Mortise-and-tenon widths in hardwood timbers work well at 1½ in. and should be 2 in. in softwoods.

To increase the tensile strength of the joint, increase its depth and thickness and use additional pegs when the width and length of the tenon allow. For instance, in a collar tie it might be important to increase timber depth just to increase the surface area of the tenon, allowing more pegs and leaving more wood to surround the pegs. But balance this with a consideration for other joinery and a judgment about the amount of wood removed from the receiving timber. If three beams meet at the same spot on a post, enlarging the mortises might weaken the post to an unacceptable degree. It is critical to remember that very often increasing the strength of one part of a joint decreases the strength of the other. Unless the two parts are balanced, the joint is faulty.

The ability of a joint to bear weight depends on the amount of surface area involved. Imagine for a moment the ideal situation of a load from a beam being transferred to the top of a mid-span post through a mortise-and-tenon. The joint here needs only to lock the two pieces together, and the entire remaining surface area of the connection—the shoulders of the tenoned (vertical) timber—transmits the load. The joint poses little compromise to the strength of either beam or post and, with the full width of the beam in the middle of its span bearing on the post, the beam is in little danger of shear failure.

The situation changes drastically, however, when the beam bearing the load is joined to the length of a post rather than to its top. The weight borne by the beam is transmitted through the base of the tenon, and as the load increases, it becomes necessary to find a way to increase the surface area at the base of the joint or shear failure of the tenon becomes a possibility. In a simple mortise-and-tenon, the load-bearing capacity of the joint is thus limited to the shear value of the tenon alone. (Wood species with lower allowable shear stresses would require larger timbers and also larger tenons to support the same load.) Of course there is a limit to how much a tenon can be enlarged, for the larger mortise required to house it can compromise post strength drastically.

The small bearing area of the tenon not only reduces the potential strength of the beam under load, it also gives the connection only superficial ability to resist eccentric loading, which can cause torsion. Torsion is a rotational force caused by overloading a member on one side or the other. Floor joists connected to only one side of a beam, rafters connected to a plate, and a queen post off-center on a bent girt are all examples of situations that could cause eccentric loading and resultant torsion. In fact, such conditions are much more likely to contribute to failure than is the inability of the base of the tenon to support the load.

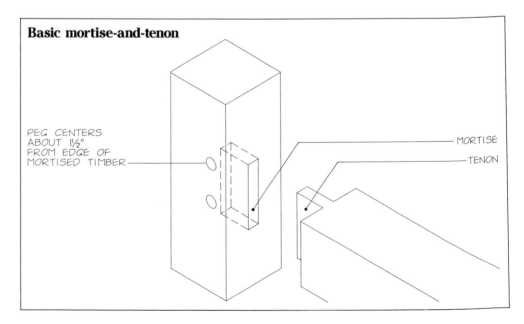

**Basic mortise-and-tenon**

PEG CENTERS
ABOUT 1½"
FROM EDGE OF
MORTISED TIMBER

MORTISE

TENON

To make the mortise-and-tenon significantly stronger in resistance to shear and torsion, we increase the surface area of the joint by letting the beam's width into the post. When the joint is angled so that only the bottom surface of the beam rests on the post, it is called a beveled shoulder joint; when the post is notched to allow the entire beam end to be received, it is called a housed joint. We generally use either a beveled shoulder or housed joint to connect all load-bearing beams, such as bent and connecting girts and summer beams, to posts. Angled variations can be used where principal rafters join to posts and for diagonal braces. We determine how much of the beam should be let into the post based on loading, torsion, other joinery in the area and wood species. We have found, for instance, that spruce tends to twist while drying, and so in frames built from this wood we house all the joints. When beams are housed or shouldered, remember that tenon depth needs to be increased or tensile strength will suffer. Remember also that while it is necessary to get sufficient bearing for the beam, it is also necessary not to weaken the post unduly. (Beam-to-beam joints are particularly problematic because very often neither can afford the strength reduction.) It's up to the designer to evaluate the problem and figure out the best compromise.

## Shouldered joints

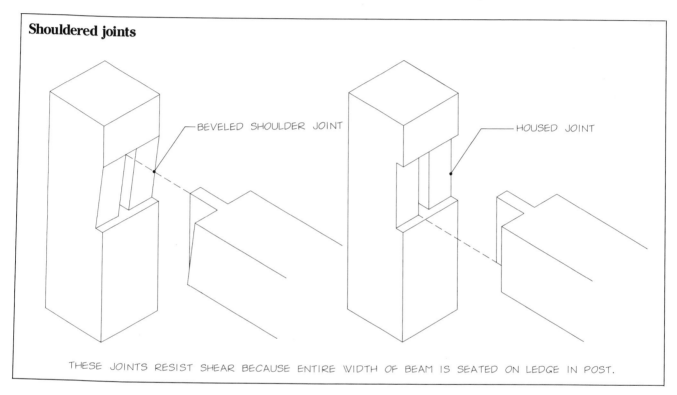

BEVELED SHOULDER JOINT

HOUSED JOINT

THESE JOINTS RESIST SHEAR BECAUSE ENTIRE WIDTH OF BEAM IS SEATED ON LEDGE IN POST.

Tusk tenons and soffit tenons are housed mortise-and-tenons used primarily for beam-to-beam connections. In both joints, mortises are cut parallel to the grain in the receiving beam to allow peg holes to be drilled across the grain. The English have a number of names for these joints, depending on the location of the tenon and the type of housing. For example, it is called a "bare-faced soffit tenon" when the tenon is cut at the bottom of the timber, a "central tenon with soffit housing" when there is a square housing at the bottom, and a "central tenon with a soffit spur" when there is a beveled housing at the bottom. In our communication with each other in the shop we don't use complicated names, so we think of the general class as either tusk or soffit tenons, with soffit tenons being those having the tenon at the beam base.

In beam-to-beam connections, where the mortise is cut mid-span on the supporting timber, it is necessary to ana-lyze loads carefully. The goal is to support the tenoned timber adequately while conserving the bearing capacity of the mortised member. Tusk tenons are useful because the mortise can be located toward the neutral axis of the timber, between the compression and tension zones (p. 27). Stronger still is a soffit tenon, because it puts the tenon and the support at the base of the beam. When the depth of the supporting beam is great enough to allow a soffit tenon to align with a mortise cut in the neutral axis, this most efficient kind of joint can be made. We have come to feel that the soffit tenon is usually the best beam-to-beam joint, and we try to use it wherever we can. The limitations on its use are based on raising technique and sequence. Obviously, it is necessary to carefully plan the raising to engage this joint on both ends. Dovetail joints, described on p. 46, are much more easily assembled than are soffit tenons.

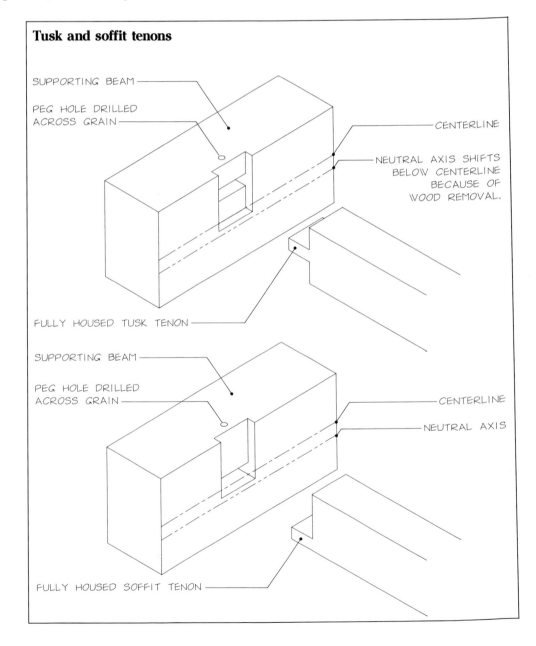

**Tusk and soffit tenons**

SUPPORTING BEAM

PEG HOLE DRILLED ACROSS GRAIN

CENTERLINE

NEUTRAL AXIS SHIFTS BELOW CENTERLINE BECAUSE OF WOOD REMOVAL.

FULLY HOUSED TUSK TENON

SUPPORTING BEAM

PEG HOLE DRILLED ACROSS GRAIN

CENTERLINE

NEUTRAL AXIS

FULLY HOUSED SOFFIT TENON

An open mortise-and-tenon (sometimes called a tongue-and-fork) is often used to connect rafters at their apex and sills at the corners. (Lap joints, alternate but inferior connectors at sill corners, are discussed on p. 46.) It's actually the mortise that is open, cut completely through the end of the timber. The tenon is cut as usual, but is larger than in most other situations. When used for rafters, the mating members simply lean into each other and are pegged—the weight of the rafters and the roofing materials help keep the joinery secure. An open mortise-and-tenon at sill corners can be exactly the same joint, but placing it there is complicated by the fact that the sill usually receives a post at that location. Load is not a problem because there is full support from the foundation. The best solution would be to use a square tenon from the post to lock the corner. Another possible variation is to make the tenon in the shape of a dovetail, a design that provides further locking, as shown below.

## Open mortise-and-tenon joints

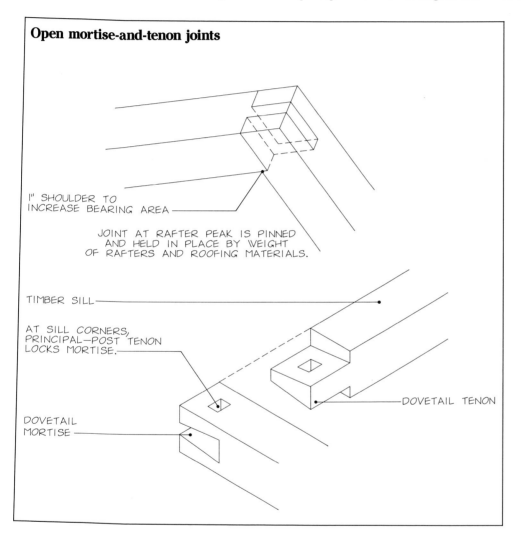

I" SHOULDER TO INCREASE BEARING AREA

JOINT AT RAFTER PEAK IS PINNED AND HELD IN PLACE BY WEIGHT OF RAFTERS AND ROOFING MATERIALS.

TIMBER SILL

AT SILL CORNERS, PRINCIPAL—POST TENON LOCKS MORTISE.

DOVETAIL TENON

DOVETAIL MORTISE

**Lap joints**    The other broad category of joints are those that lap one member over another. The receiving timber is notched to accept a tenon, or lap. Unlike the mortise-and-tenon joint, the tenon is not completely surrounded with wood when the joint is engaged; the joint is secured either with pegs or through the use of locking geometry. My own theory about lap joints is that they evolved in response to raising techniques. Early English frames, for instance, which use a great number of lap joints, were most often made by small groups of professionals and assembled one piece at a time. In this case, it would have been almost impossible to engage a mortise-and-tenon joint simultaneously at both ends, so lap joints were used instead. But as timber frames evolved in this country, most raisings involved the entire community, and raising methods and joints were developed to take advantage of this manpower. Large sections of the frame were joined together on the ground to be raised at once, using primarily mortise-and-tenons. Lap joints, which are often inferior to their mortise-and-tenon counterparts, were kept to a minimum, and usually used only for the smaller members of the floor.

Today we use lap joints primarily for beams in the floor and purlins in the roof. They can be straight laps but are more often dovetailed to resist withdrawal. These joints are frequently made with a housing or a spur to give the timber more support. The obvious problem with lap joints is that wood is removed in the compression area of the receiving beam, which can greatly reduce strength. Laps should never cut entirely across the top of the beam; even if joints are received from both sides, at least 2 in. should be left in the center. When a dovetail is used, it need not penetrate deeply into the receiving timber for good locking—perhaps no more than 2 in. If a straight lap is used, however, the tenon has to be deep enough to hold a peg, which is usually a mimimum of 3¼ in. long. The dovetail can therefore do the job with the least compromise to the supporting member. But the problem with the dovetail is that as the wood shrinks the joint becomes less effective in preventing withdrawal of the timber, which can be extreme when green timbers are used. To make sure the joint is securely locked and to reduce the effect of shrinkage, we use a tapered dovetail and drive dry wedges on both sides.

Knee braces and collar ties are also sometimes joined with straight or dovetailed laps. The best use of a diagonal dovetail lap is for the joint between a collar tie and a rafter when the tension forces will be significant. The biggest drawback to using a lap joint for knee braces or collar ties is that it has little ability to resist torsion, which could cause the joint to pull apart.

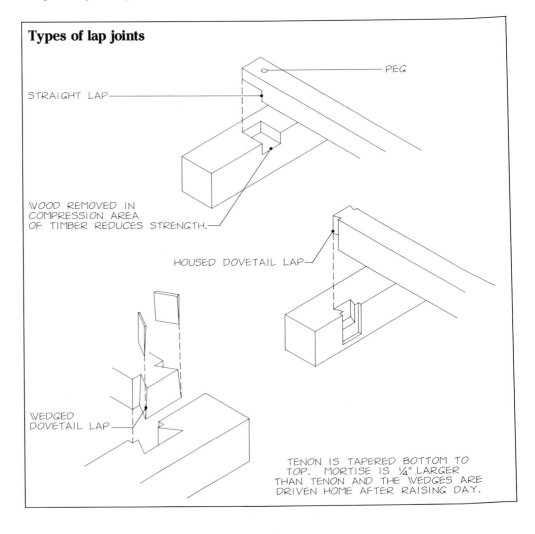

**Types of lap joints**

PEG

STRAIGHT LAP

WOOD REMOVED IN COMPRESSION AREA OF TIMBER REDUCES STRENGTH.

HOUSED DOVETAIL LAP

WEDGED DOVETAIL LAP

TENON IS TAPERED BOTTOM TO TOP. MORTISE IS ¼" LARGER THAN TENON AND THE WEDGES ARE DRIVEN HOME AFTER RAISING DAY.

Scarf joints are lap joints used to splice two or more shorter pieces into one long timber. Plates and sills most often demand scarfs because they tend to be the longest continuous members in a frame. There are hundreds of variations of scarfs, with varying degrees of complexity and effectiveness. Because so much surface area must be cut to fit, all scarfs are relatively difficult to cut, making them challenging and fun for the joiner. On the other hand, no amount of precision can fasten the halves together without compromise. A scarf-jointed timber is not as strong as a solid one and scarfs shouldn't be used to support floor or roof loads if they fall in a span between posts. The best position for a scarf is directly over a post, where it can splice effectively without compromising the load-bearing capacity of the beam. (The post tenon can be used as a key to keep the halves from slipping apart.) It is also possible to support a scarf joint with a knee

brace. With various types of keys and wedges to lock the halves together, scarfs are strongest when employed to resist tension. Given their difficulty and limitations, it is sensible to minimize the number of these joints in the frame.

One of the most fascinating joints used in timber framing is actually a combination of joints used to connect several members. The intersection of a principal post, a plate and a tie beam is known as a tying joint. Although there are many variations, almost every one includes at least two perpendicular tenons on the post, one to join to the plate and the other to join to the tie beam. The tie beam is then further connected to the plate with a dovetail lap; it is also mortised to receive rafters. This type of connection was so prevalent through much of the development of English carpentry that to not find a tying joint in a surviving timber frame is somewhat remarkable.

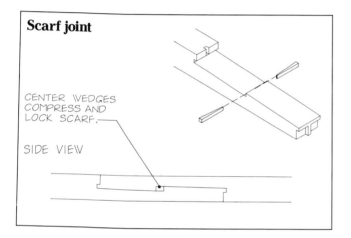

**Scarf joint**

CENTER WEDGES COMPRESS AND LOCK SCARF.

SIDE VIEW

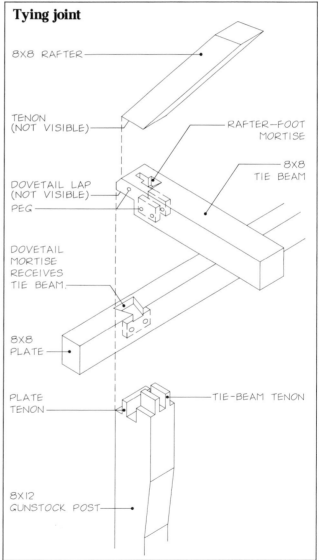

**Tying joint**

8X8 RAFTER

TENON (NOT VISIBLE)

RAFTER-FOOT MORTISE

8X8 TIE BEAM

DOVETAIL LAP (NOT VISIBLE)

PEG

DOVETAIL MORTISE RECEIVES TIE BEAM.

8X8 PLATE

PLATE TENON

TIE-BEAM TENON

8X12 GUNSTOCK POST

**This scarf joint is supported by a post. The joist is connected with a tenon that falls between the scarf halves, helping to lock the joint.**

# III. Frame Design

The timbers of the frame should be used as a design tool and meshed with the overall floor plan. The goal is to develop floor and frame plans simultaneously.

In the preceding chapter we covered a lot of theoretical ground, and now it's time to put it into practice with a series of exercises. We'll start with a somewhat traditional saltbox shape, and then sort through a range of frame options within that shape—the objective being to integrate a good frame with the desired floor plan. In reality the frame plan and the floor plan would be designed simultaneously, as discussed in the next chapter, but before you attempt that process, the following background information will be helpful. It is important to know that for any given frame there are many options available, each having benefits and trade-offs.

The frame is actually a more flexible element of the house than you might think. It has to fulfill its structural function, but it can do so in any number of configurations. It makes absolutely no sense to determine the frame according to arbitrary criteria and then squeeze the floor plan into it. To bring the magic of the frame to the living space, the timbers should be used as a design tool, their full palette of possibilities explored. And of course a great frame design would fall flat if it were not within the ability of the maker and if resources were unavailable.

As we analyze the frame options in the rest of this chapter, we'll be looking at the following issues.

**Strength**  Of course you would not deliberately design a weak frame, but it is surprising how easy it is to overlook the details. Many times a small modification yields a big improvement in terms of increasing strength. For instance, moving a summer beam a few inches might make a big difference in the amount of wood that's left between two joints. Moving a post even a foot could radically affect a beam span, eliminating a deflection problem. The frame also must be designed for a specific species of wood—what is good in oak may be a problem in fir.

**Timber length and size**  It's extremely frustrating to find that you have designed a frame around timbers that aren't available. If you are uncertain, check with the sawmills while designing. Most mills have definite limitations; small ones often can't cut anything longer than 16 ft. The trees themselves have their own natural limitations. Northeastern species, such as red oak, grow much shorter than Douglas fir and southern pine, for instance. Loggers also play a role in the availability of long timbers. If they are accustomed to cutting for veneer or furniture stock, they habitually cut long, beautiful logs into short sections. We once gave 25-ft.-long measuring tapes as Christmas presents to a few loggers, feeling certain theirs stopped at 12 ft.

**Complexity**  Do everything you can to make the frame simple and practical to cut. Justify every complication. By all means be certain that making the frame is within the capability of the joiner.

**Raising method** While designing the frame, plan the raising. Will you be lifting bents or walls? With a crane or by hand? Will there be separate roof trusses? It's necessary to know something about the site to determine what is possible. We once arrived on a much ballyhooed "beautiful" site only to discover that, yes, it was just a few hundred feet from the road, but it was practically straight up. We had a difficult enough time getting tools on the site, never mind the timbers. We seriously considered a helicopter, but finally wound up transporting the timbers one at a time with a bulldozer. The raising proceeded slowly as we hung from trees and invented lifting mechanisms.

**Raising sequence and connections** Place timbers so good joinery is possible. Be certain that each piece can be placed into the frame and pegged without undue difficulty. One of the biggest mistakes you can make is putting a timber in a position where other beams, posts or rafters make drilling or driving the pegs impossible. Don't ask how I know this.

Now we'll look at some bent options for the house shape shown below, which is a traditional saltbox except for the unevenly pitched roof. The change in pitch might have been selected purely for aesthetic reasons or because of the need for more headroom. All of the bent plans in this chapter work with this shape. Because so much of the frame is dependent on the bent timbers, we can initially study the entire frame by looking at bent plans. Major connecting timbers are indicated as darkened areas.

**Option 1** In colonial times, the saltbox was often created by adding a shed onto an existing building. This bent plan therefore represents the approach that you might see in an old building.

In this option, the A, B and D walls each have plates, so the frame would be raised as walls instead of bents. After the walls were raised and the lower bent girts and post C were in place, the tie beam would be positioned on the walls and the rafters raised.

ADVANTAGES This type of frame is easily broken into small sections for a hand-raising. Even the walls could be raised as two or three units.

The tie beam serves as the lower chord of the rafter truss, working in tension to prevent outward thrust. This is stronger tension resistance than could be achieved simply with a timber tying between rafters. The weakness in the tie beam would be the scarf joint, but the post tenon could pass through it to prevent separation.

DISADVANTAGES Unless there is a specific need to detach the shed from the frame, the frame would be stronger if the shed were integrated. On the other hand, if the shed is to be a solarium, this design would facilitate a switch to a wood that would be suited to the excess heat and moisture. (See p. 184.)

**Option 1**

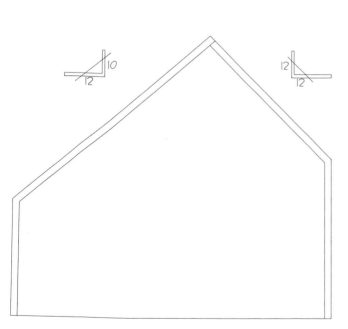

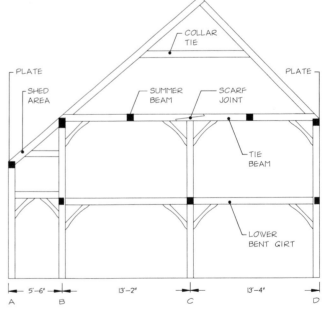

The roof is quite distinct from the rest of the frame, and unless the roof has wind braces, there will be little to prevent it from hinging.

There may be some shrinkage in total building height as the tie beam and plates shrink and the rafters settle. With the tie beam and three plates traveling the length of the building, there will be either a great many scarf joints or some extremely long timbers.

It's common for summer beams to fall mid-span between posts, but it is a condition to avoid. A mid-span point load could well require an increase in the size of the tie beam, especially when the frame is of a weaker wood.

**Diagonal wind braces, here joining from the rafter to the purlin, are sometimes needed to prevent racking perpendicular to the bent.**

**Option 2** The big difference in this design is that the rafters sit on top of posts instead of plates; connecting girts span from post to post between bents. Here the mortise-and-tenon joints between the upper bent girt and posts will need to resist tension forces from the rafters. Struts assist these joints, creating triangulation and thereby rigidity.

ADVANTAGES  Even if you could not get a timber long enough for the upper bent girt, only one scarf would be needed in each bent.

The bents could be assembled and raised as a unit, or raised lower section first, then rafters.

DISADVANTAGES  The roof and the shed are still not well integrated with the frame. By now you should understand that the word *integration* is not meant to define a socially conscious concept, but is used to describe whether the frame has sufficient transverse members at central points for connection and rigidity.

Intersecting joinery makes it difficult to put the connecting girt at the top of the wall, which means sheathing has to span between this connector and the roof sheathing.

**Option 2**

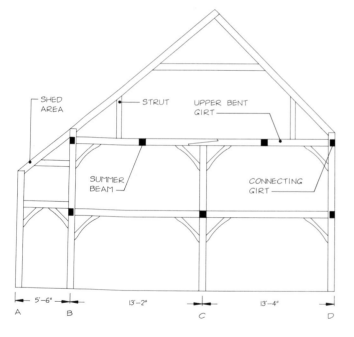

**Option 3** The difference here is that the *C* post passes through the upper bent girt and joins to the collar tie.

ADVANTAGES  Because the *C* post spans the hinging point at the top of the other posts, the bent is much more rigid here than it would be in Option 2.

The large braces joined to the collar tie support the rafters, preventing sagging and helping to limit outward thrust. I would feel safe letting these braces replace the struts in the previous bent.

Without a scarf joint on the tie beam, it is relatively easy to join the summer beam to the *C* post, which not only gives the summer beam better support, but allows the upper bent girt to be smaller, since it would not be supporting any beams mid-span. Here all major beams are supported by posts—a situation rarely achieved, but always a design goal.

DISADVANTAGES  A 22-ft.-long timber is required for the *C* post, which might have to be as large as 8x10 to receive the bent girts and summer beams. A timber this size could be hard to get, and would certainly be hard to handle.

With so much of the bent interconnected, this structure would be difficult to break into small units for a hand-raising. It begs to be completely assembled on the deck and raised with a crane.

**Option 3**

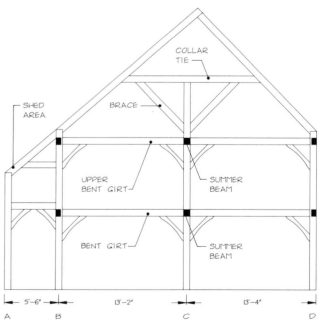

## Option 4

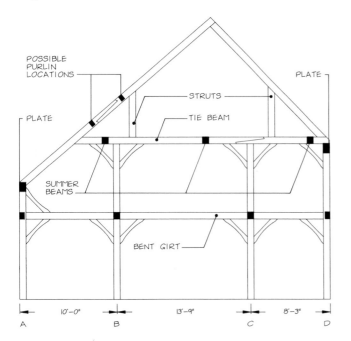

POSSIBLE
PURLIN
LOCATIONS

PLATE

PLATE

STRUTS

TIE BEAM

SUMMER
BEAMS

BENT GIRT

| 10'-0" | 13'-9" | 8'-3" |

A         B         C         D

**Option 4** Here we go back to the same general theory as Option 1, except we move the posts to accommodate some hypothetical requirements of the floor plan. To create more headroom on the upper level, we also replace the collar tie with struts. The rafter load passing through the struts would have to be carefully calculated and the tie beam sized accordingly. (A difficulty with the lexicon of timber framing is illustrated here, where the tie beam on the *D* side looks more like a collar tie on the *A* side. Since it seems to be supporting more than tying, I call it a tie beam.) The knee braces extending from the *B* and *C* posts give good support. The rafter load on the part of the tie beam cantilevered over the *B* post would also need calculating. When there is a scarf joint mid-span between supports, as there is here in the *A*-side rafter, assume that the area of the scarf can carry almost no load. This scarf is therefore in a vulnerable position. If the rafter is accepting additional loads from purlins, the purlins should be positioned away from the joint.

ADVANTAGES The shed is integrated with the rest of the frame, not only strengthening the bent, but lending more possibilities to how rooms might be arranged with posts.

DISADVANTAGES The *A*-side rafter passes through the bent levels and is theoretically a better transverse tie, but the scarf joint reduces the strength of the connection.

There is still the problem of hinging between the rafters and the tie beam, especially on the *D* side.

The tie beam is another long or scarfed timber.

The summer beams on the tie beam don't align with posts. This might happen as a result of a desire to avoid the complex joinery necessary to bring a beam into an area already consumed by a scarf. It also might fulfill a need to get load-bearing timbers close to the eaves to support the third floor. In any case, they put a strain on the tie beam, especially between posts *B* and *C,* where the span is greatest.

The summer beam between posts *C* and *D* is a big problem. Because it is directly under the rafter, the joint could not be pegged if it were a mortise-and-tenon, and there is not enough space between the rafter and the beam to position it from above, as with a dovetail. Therefore, unless it was joined prior to the placement of the rafter, the only other alternative would be to use a rather exotic blind wedge or a through tenon with wedge. This is yet another lesson I learned the hard way.

This is a complicated assembly to raise. Try to imagine the sequence: First the two outer walls are joined by the lower bent girt along with the *B* and *C* posts; then the tie beam is placed over three posts; then the *A* side rafter is installed in one or two pieces; finally the *D* side rafter. The only advantage to making a raising this hard is to satisfy sadistic impulses.

**Option 5**  First, terminology. The tie beam that looked a little like a collar tie now looks like a bent girt on one end, so I'll refer to it as an upper bent girt. There are three important changes in this option: The first is that the *B* post joins to the rafter, helping to tie the bent together and reducing the length of the bent girt. The second change is that the *C* post has been moved so that the greater span is between *C* and *D*, to complement the hypothetical floor plan. Within the limits of acceptable beam spans, moving the posts doesn't always create structural problems. The third change is that the connecting girts substitute for plates.

ADVANTAGES  With the upper bent girt reduced from 28 ft. to 22 ft., the probability of getting a full-length timber is greatly increased.

Eliminating the scarf joint means one of the summer beams can fall on a post.

The load on the *A*-side rafter is transferred immediately by the *B* post instead of going from the strut through the tie beam first, as in the previous option.

Support for the rafter scarf is improved by placing it directly over the post. (Option 4 might have been improved by placing the scarf over the strut.) In each case, the rafter would be improved if there were no scarf.

With the plates removed, this bent can be assembled on the deck and crane-raised as a unit. Besides being more efficient and ensuring better fits, a crane-raising is much safer than a hand-raising.

DISADVANTAGES  Although the scarf in the upper bent girt has been eliminated, there is still a requirement for long timbers—a 20-ft. post and a 22-ft. girt.

If a hand-raising were desired, it would be hard to separate the rafters from the rest of the bent, because you would have a scarfed 26-ft.-long rafter on one side and an 18-ft.-long rafter on the other.

Between *C* and *D* on the upper bent girt, the summer beam produces a one-point floor load almost exactly on the middle of the beam. Only one foot away is another point load from a strut, which will transmit a significant rafter load to the same beam. This would need careful analysis and it's certain that a large beam would be needed.

Because the upper bent girt lies between the rafter and the post on the *D* side, there is potentially more shrinkage here than on the *A* side where the rafter joins directly to the post. This is called differential settling, and it is a problem to be avoided in bent design.

**Option 5**

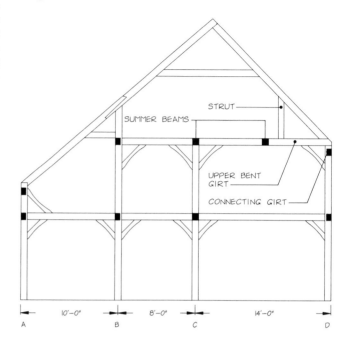

**Option 6** Here the *C* post has been moved a couple of feet toward post *B*. Perhaps the narrow aisle-like space this creates would define an area used for the chimney, stairs and halls, providing access to the large space between *C* and *D* posts. Even though the beam span between *C* and *D* posts is large, I wouldn't hesitate to use this option if it clearly benefitted the floor plan.

ADVANTAGES   Most of the major beams are still supported by posts.

By bringing the *B* post through the rafter, we've eliminated the scarf joint. This simple change in joinery means that the upper roof section is more detached and can be raised as a unit after the rest of the bent is erected.

DISADVANTAGES   The longer beam spans created by moving the *C* post require larger timbers. If the frame were made from a wood with low structural values, spans such as these might not be practical. It would depend on how the floor interacted with the bent and the resulting timber size.

Although the bird's-mouth joint used on the *B* post might be an advantage in raising, it doesn't provide the lateral strength of a long scarf joint. Once again, there is less integration of the roof with the bent.

Since the lower girt, horizontal strut and rafter all join to the *B* post, the shed is again quite distinct from the rest of the frame.

The strut and summer beam still produce a significant load in one area. The position of the summer beam is better because it is closer to the support of the knee brace but the span has also been increased, which makes the two-point load more troublesome.

Differential settling is still a concern.

**Option 7**   I would consider this the most desirable arrangement of timbers for this bent. The simple change is that both the *B* and *C* posts join to the rafters, but doing this really solves some problems.

ADVANTAGES   The vertical connection on both ends makes this bent extremely strong. There is no place where it could hinge.

The rafters join directly to posts for a strong connection and direct transfer of load. This would be a good design for some of the weaker woods because the spans are broken up and all the major beams are supported by posts.

Wood shrinks most across the grain, and because end grain joins to end grain at the rafter-post connection, shrinkage in the height of the building is reduced to an absolute minimum.

The upper bent girt is now a manageable length.

Ceiling height on the second floor is not confined to the height of the outside post. The upper bent girt can be placed anywhere along the length of the *B* and *C* posts.

Because the horizontal bracing used between the *C* and *D* posts is strong, it can be used as a design element to frame windows, divide living spaces or act as shelving.

The vaulted space between the *A* and *B* posts and the *C* and *D* posts can be used to bring a feeling of space and light, via roof windows, to the second-floor living area. This was true in other options as well, but because of the *B* and *C* posts, the outer areas are more defined.

## Option 7

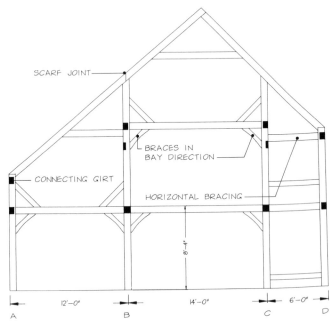

## Option 6

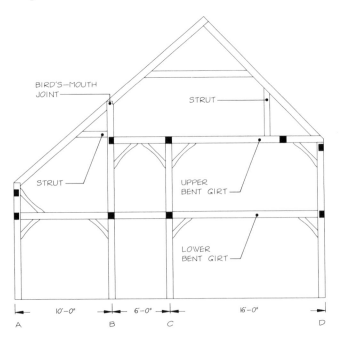

DISADVANTAGES   The major disadvantage is the need for long posts, in this case 22 ft.

This bent, almost without a doubt, would have to be crane-raised.

On raising day the connecting girts have to be joined to the upper ends of the posts. Someone has to be up on the bents to insert the tenons and one of the bents must be spread apart. It's hard work and tough on the nerves.

**Carrying the central posts to the rafters has numerous structural advantages, such as allowing loads to be transferred directly.**

**Option 8** This option has exactly the same post layout as Option 7, with most of the same structural advantages, but it is put together with plates instead of principal rafters and purlins. It could be raised either as complete horizontal walls or as bents, with the plates and rafter installed after the bents were raised.

ADVANTAGES If for some reason the purlin system was not aesthetically desirable, this would be the most obvious out. Using plates, a common rafter roof or a principal and common rafter roof are possibilities.

It would be easy to break this frame into small units for hand-raising.

When enclosing the frame from the outside, the plates give nailing surface for the sheathing, so there's no need to span between the upper connector and the top of the rafter, as there would be in Option 2. Plates also make finish work easier from the inside because they create a logical transition between the walls and roof.

This would be the system to use if you needed a framed gable-end overhang. Extend the plates beyond the posts and join common rafters to the plates and you've got it.

DISADVANTAGES Perhaps the biggest problem is that this option requires two timbers (the plates on the *B* and *C* walls) that are completely unnecessary in Option 7. In addition, the plates on the *A* and *D* walls have to be large enough to support the weight of the roof from the rafters, a structural requirement also unnecessary in Option 7. The total amounts to a great deal of extra wood.

Raising day would have all the problems of Option 7, plus there'd be additional plates to raise.

Scarf joints or timbers the length of the house are probably required for all four plates.

When joining rafters to a plate, you can't expect to develop much tying strength. Mortise-and-tenon joinery is impractical from the rafter to the plate; there is usually only a notch with a face peg. It is important with this type of framing system to make sure that the bent is strong independently of the rafters through good connections between the posts, ties and bent girts.

**Option 9** This bent demonstrates that there is flexibility in positioning posts when using the plate system. Note that with the greater span between the *C* and *D* posts, we have gone back to diagonal braces. An obvious difficulty is getting the extremely long *C* post in addition to the long plates. A frame such as this would be ideal for woods such as fir, larch or southern pine, in which long timbers are generally available. Otherwise, the advantages and disadvantages are the same as for Option 8.

### Option 9

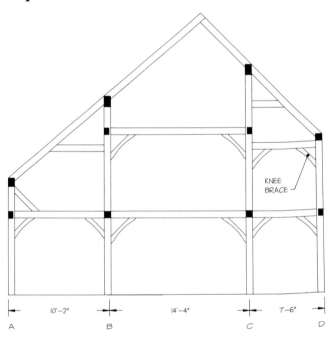

### Option 8

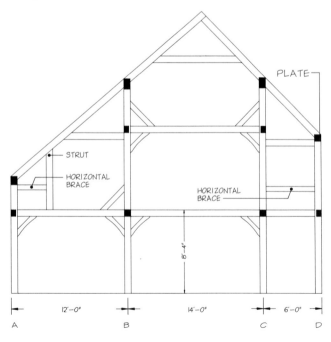

**Option 10** With the addition of a king-post truss at the roof, this is the same basic design as Option 7, and has the same advantages and disadvantages. The truss adds rigidity and strength to the rafters, although one could legitimately question whether it is necessary. But it's attractive, and might be nice if the second floor were left exposed to the roof. With braces from the king post to the rafters, the roof has good longitudinal rigidity. Placing beams on different levels avoids multiple joint intersections on the posts and opens up some interesting floor-plan possibilities.

**Option 10**

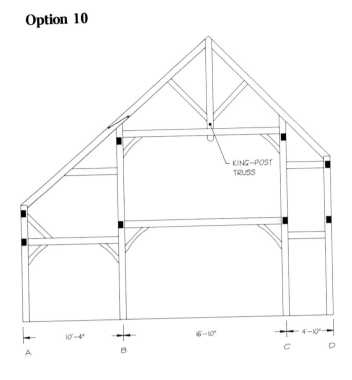

**Option 11** By passing the king post to the lower bent girt, you might be able to use that member to create a natural division between rooms on the second floor. Although a king post with braces is able to join and tie rafters, generally giving adequate resistance to sagging and spreading, here I would want to assume the worst case, which is that the lower bent girt bears some of the rafter load.

ADVANTAGES The only advantage of this layout over Option 10 has to do with relating timbers to the floor plan.

DISADVANTAGES There is a potential, depending on the quality of the joinery and loads on the upper bent girt, for the lower bent girt to need to be inordinately large to bear the one-point load for the king post.

**Option 11**

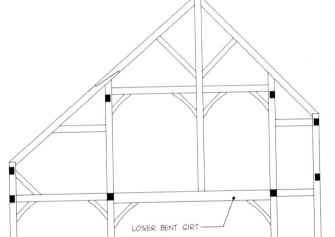

**Option 12**   This option employs a change in ceiling height without detriment to the strength or shape of the frame. Ceiling height is quite flexible with timber framing, as bent girts can be put anywhere along the length of the posts. In platform framing, people often feel compelled to stay within the normal 8-ft. stud length to keep costs down, which means ceiling height tends to fall within this range. While a great many studs will be required to frame the partitions in a timber-frame house, we try to align the partitions with major timbers; stud length is thus determined by the distance between the floor and the bottom of the timbers. Because of this, we can increase ceiling height to between 8 ft. 8 in. and 8 ft. 10 in. without going to a 10-ft.-long stud for interior partitions.

ADVANTAGES   The king post and collar tie combination strengthens the rafters and helps connect them. A tenon from the king post passes through the collar tie, keeping it from deflecting under load. The king post is also an attractive feature for a cathedral ceiling in the second floor.

Plates are used, but this bent could also be framed with a rafter-to-post connection, with a connecting girt replacing the plate. The collar tie and the diagonal strut would play a major role in limiting outward thrust at the *D* post.

DISADVANTAGES   The problem is that what is gained in ceiling height on the first floor is lost to the second.

**Option 13**   This bent forms a narrow aisle in the center of the house, which can work as an effective design element.

ADVANTAGES   The bent is strong and the parts well integrated; the *B* and *C* posts span nearly to the ridge.

The upper connectors on the *B* and *C* posts allow bracing near the top of the building in the bay direction, helping to keep the entire frame rigid.

All the major beams are joined to posts.

DISADVANTAGES   This frame requires very long timbers. The *B* and *C* posts would need to be 24 ft. The *A*-side rafter should not be scarfed until there is adequate support, so it would need to span 22 ft. from the *A* post to the scarf.

The long span between the *A* and *B* posts could also require some large timbers, especially if one of the weaker woods were used. The load from the strut is probably significant and would have bearing on the size of the beam.

## Option 13

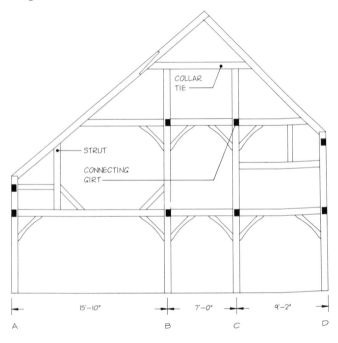

## Option 12

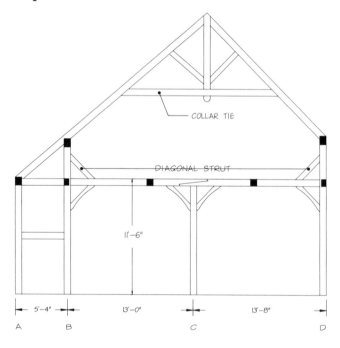

**Options 14 and 15** An asymmetrical bent plan might become necessary if the post-and-beam arrangement on the first floor needed to be different from that on the second floor. Neither of these frame plans would be my first choice, but I wouldn't hesitate to use them if it would greatly benefit the floor plan. The biggest problem is the introduction of an intermediate post on the second floor, which transfers a one-point roof and floor load to the bent girt; this timber would therefore need to be quite large. If the wood used for framing did not have good structural capacity, this option might prove to be impossible.

## Option 14

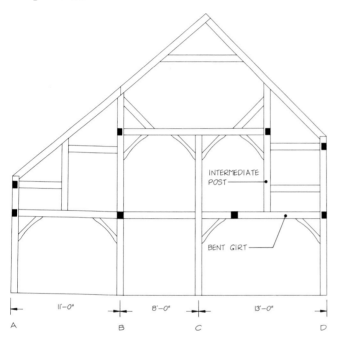

## Option 15

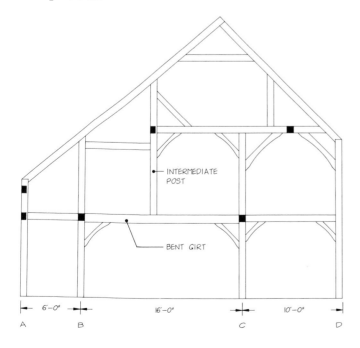

# IV. Home Design

**Getting from the frame to the floor plan is not a leap of faith, it's a heap of planning. In this interior, the hammer beams allow the living room to be spanned without posts, to create a cathedral ceiling. Timbers also frame the staircase, hallway and built-in shelves. (Photo by Bob Gere.)**

The frame is such a dominant part of a timber-frame house that it's easy to infer that a good frame automatically makes a good house. It's all too tempting to first design the frame, to make it strong and theoretically good-looking, and to superimpose the floor plan upon it later. There's a comforting linearity to this way of thinking—you take care of the "important" issues, such as timber size, joinery and beam spans first, before turning to the "niceties." And indeed, if you work this way, you will probably be rewarded with a very nice frame. You also run the risk of creating a pretty miserable home. For the "niceties" (the aesthetic and practical requirements of the occupants of a house) are more necessary to good design than might at first be apparent. To neglect these requirements is to risk making a house that sits awkwardly on its site, that has an irresolvably poor traffic pattern, or that simply fails to be attuned to people and the environment.

Good house design is hard work, made more so by the fact that the path to it is not clearly mapped. You can enter any library and find volume after volume about construction procedure, with endless detail on how to make a house strong and safe. But how to make it look and feel good? How to make it work well with the needs of the people who live in it? This is much more subjective stuff, rife with theories and opinions. Perhaps the only generalization you can make on timber-frame house design is that timbers are not necessarily a decorative benefit, and if

they enhance and simplify instead of cluttering and complicating, you can bet it was planned to be that way. It didn't just happen.

In this chapter, I'll share what I know about designing timber-frame houses, beginning with the all-important preliminary work. Although much of this material does not contain the word *timber,* I do not digress, for good timber-frame house design is based on principles and practices of design that are good for any house. To do right for the timber-frame house, it is necessary to just do right. After covering these initial design concepts, including the building program and site study, we'll look at generating floor and frame plans. Finally, I'll backtrack a little to explore some ways to use the natural spaces created by timbers, and to discuss some of the design options that various arrangements of timbers provide both for the whole house and for each room. Of course it's impossible to learn to design a house from reading one chapter in a book, but there's much to be gained from developing an understanding of each element in the design process.

## Ruling notions and the building program

It may seem obvious, but nothing is quite as important in design as understanding and addressing the reasons why a building needs to be built. Every house is born as a notion, and this original ruling notion is the seed from which the plan of the house grows and is ultimately defined. The first

Good house design balances
needs and desires with budget
and site requirements.

step is therefore to open that notion to discussion. It may be lofty or mundane; it may concern stylishness or a desire to conserve energy. It may express how a family wants to live or focus on a relationship with the outdoors, with artwork or with furniture. Most often there are several notions working in concert, but too often there is one that dominates all the others and paints its colors into every corner of the house.

A few examples illustrate how diverse and powerful ruling notions can be. One of the most common notions I see establishes a particular architectural style as a design prerequisite. In other words, the house is already a cape (or a ranch, or an underground house) before there is any understanding of the site or the living requirements of the occupants. This may seem innocent enough, but it's not. For while the strongly vertical lines of alpine architecture look great in the mountains, they look out of place on the plains. And while sprawling horizontal architecture may fit a family's image of expansive living, it might not fit their craggy hillside site. People will probably always have a preferred architectural style, but this should be considered an ambition of the design and not a ruling notion.

Another ruling notion I frequently see concerns the timber frame itself. People often approach us for a frame of a particular size and shape, but don't have any idea of how the frame will relate to the living areas—this will be worked out later, we're told. We try to point out that "a 24x36 saltbox frame" is not the right starting point for a home. Under these conditions, the features of the frame can turn into obstacles and the building process will turn into a remodeling project as the floor plan is forced to weave around existing structural ingredients.

A recent ruling notion I encountered focused on a very fine, very large Oriental rug, inherited by our clients. They had never been able to enjoy the rug because their living quarters were too small, so when they approached us about a house, they wanted to make sure there was plenty of room for the rug. A second notion was that the master bedroom should be on the first floor—no more going up and down stairs for them. Now these people were not wealthy and could not afford a large house, so you can imagine the difficulty of designing a small house with room for a 24x30 rug and a downstairs master bedroom. All early renditions of the plans had small, elongated spaces (kitchen, dining, entry, master bedroom and bath) gathered around the rug. It finally became essential to spend some time talking about these ruling notions. The rug stayed, ruling as ever, but the master bedroom went upstairs.

Good house design attempts to balance function, economy, form and energy performance into a unified package. But unless the ruling notions of a project are diagnosed and their consequences projected, the design will probably be thrown off center. Plans can still be drawn, and the house can certainly be built, but strong forces will be at work and they will elbow their way into the design time and time again. Obviously, overlooking economic realities could mean that the ceremonial first meal is cooked on a hibachi in a drafty shell. A house built to satisfy an artistic compulsion might well be better to look at than live in. Energy conservation is important, but there is a big difference between living on the inside of a solar collector and being comfortable. After defining the ruling notion, put it on the back shelf, address the many other design issues, and come back to it later. If it survives as the dominating force, it deserves to be.

The next step is to create a building program. A building program is nothing more than a list of requirements for the house, the landscape and exterior activities. It should include not only physical requirements, but also personal ideas about lifestyle. It might be important, for instance, that the living room be large enough for two settings of furniture, receive direct sunlight and have easy access to an exterior terrace. If it is important that the master bedroom be completely separate from other bedrooms and that it have enough room for morning exercise, this should be noted on the building program. After the program is completed, it's much easier to look at practical construction issues—how the house will be sited, how the rooms will be arranged—and the many decisions relating to materials.

Potential homebuilders should use the sample building program at the end of this chapter (p. 84), and think carefully about each answer. Be sure that when you've finished you have described the home you aspire to build. This does not mean that the final design will satisfy every whim. It simply means you determine the trade-offs while the plans are being drawn instead of discovering them later as faults. So be prepared for the compromises that you will inevitably have to make.

## The site study and the bubble diagram

When the building program is complete, it's time to study the site and roughly describe the house that will sit upon it. Understanding the site is an essential part of design, and having a complete house design prior to a knowledge of where it will be built is like buying the pants before you know who will be wearing them. Placement of the house should be sensitive to land contours, forestation, flora, views and sunlight. At the same time, seek summer shade, protection from the wind and isolation from noise. Economics suggest careful consideration of the location and expense of the driveway, septic system and water supply.

Laying out a building on a site is mysterious work. Whether the site brims with potential or is featureless, there is always one place for the house that is better than any other. You wander about the property like a lost divining rod in search of that precious spot. It will help, during this part of the work, to take an analytical approach. Begin by drawing a site plan that shows all the major features of the land, an example of which is shown on p. 63. The easiest way to start the site plan is to photocopy or trace the site survey map. (If a site survey does not exist, consider having one made.) At the very least, use a description from the deed and draw the site and its contours from your own measurements and observations. Show on the plan all positive and negative features, such as the trees that should be saved, lowland or swamps, creeks, views, roads, potential access directions and neighboring houses.

Of paramount importance in siting is access to the sun, so also include on the plan the direction of solar south. Achieving a good relationship with solar south is probably the simplest part of siting because it can be done scientifically and reaps measurable rewards. Using only good orientation and proper proportions of glass and insulation, passive-solar houses are no longer considered difficult or

**Listing all physical and lifestyle requirements in the building program gives the designer direction. For example, in this house sufficient mounting had to be provided for the hunting trophies, and space had to be reserved for the jukebox.**

awkward. Their design begins with the two basic siting rules that follow.

To receive the maximum amount of solar heat, the house should face as closely as possible toward solar south. Solar south is not the same as magnetic south, because of longitudinal differences and variations in the earth's magnetic field. There is a simple method for determining solar south. Go to the site at solar noon (exactly halfway between sunrise and sunset, usually listed in a local newspaper). Hang a string with a weight to make a plumb line, and mark the direction of the shadow, which will be cast toward solar north; the opposite is solar south. Determine the compass heading of solar south so you can locate it in the future. It may not be possible to orient the building directly toward solar south, but a rule of thumb is to stay within 25½ degrees, because within this range you lose less than 10% of the potential solar gain through vertical glazing.

The next rule is to try to eliminate winter shading. The angle of the sun's arc changes throughout the year, being highest in June and lowest in December. Placing yourself in the position of the house, try to follow the path of the sun from sunrise to sunset as if it were a December day. The arc you would follow depends on your latitude. (For instance, at my house the sun follows an angle of about

24½ degrees to the horizon on December 21 at solar noon.) Better yet, get out there on a December day and watch what happens. You are looking for anything that obstructs the sun, such as hills or trees. Depending on what you find, the house may have to be moved or trees may have to be thinned and pruned.

While trying to eliminate December shading, you will be grateful to have some in summer. If there are deciduous trees, the house might be positioned so it is shaded by the crown of a tree in June, but less obstructed after the leaves have dropped off in the fall. But remember that even bare branches can cause heavy shading. These may have to be pruned to allow sunlight to pass through.

A thorough understanding of the site is then blended with the building program to create the basis for the house plan. Try to fit all the important elements of the home onto the site with a bubble diagram, as shown below. In addition to living areas within the house, also locate the driveway, parking area, garage, woodshed, septic system, terraces, gardens, barn, pasture and any other required feature. The process of completing this diagram will naturally spur some decisions about the floor plan. For example, if the driveway needs to be to the east of the house to complement features of the land and reduce expense, it

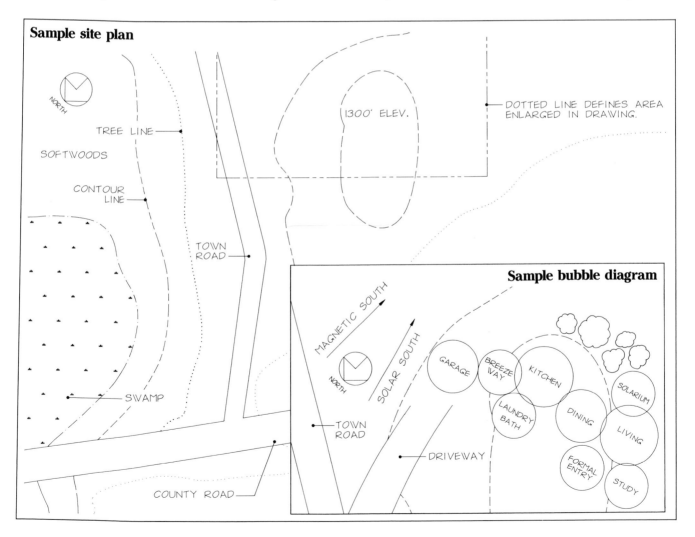

**Sample site plan**

1300' ELEV.

DOTTED LINE DEFINES AREA ENLARGED IN DRAWING.

NORTH

TREE LINE

SOFTWOODS

CONTOUR LINE

TOWN ROAD

SWAMP

TOWN ROAD

COUNTY ROAD

**Sample bubble diagram**

MAGNETIC SOUTH

SOLAR SOUTH

NORTH

GARAGE

BREEZE WAY

KITCHEN

SOLARIUM

LAUNDRY BATH

DINING

LIVING

FORMAL ENTRY

STUDY

DRIVEWAY

would not make sense to put the garage on the west. If the parking area were toward the north of the house, it would not make sense to put the main entry on the south. And if it were possible to site the house to capture a view, you would probably rather see it from the living room than the laundry room. Likewise, it is not usually prudent to waste solar-heating potential on a little-used bedroom when it could be better used in a major living area.

With the position of a few essential ingredients, many other parts of the house follow. For instance, because you would be going from the car to the house on the east side, it makes sense to put the kitchen to the east so groceries would not have to be carried across the house. For convenience, the dining area should be close to the kitchen. The kitchen is roughly located toward the southeast and the living room is toward the southwest, with the dining room in the middle. Laundry, bath and other utility areas are shifted toward the northeast, opposite the kitchen, to keep plumbing areas together and separate them from the living room. The rather extravagant formal entry falls adjacent to the living area. Eventually, a meaningful design begins to emerge, which combines the wistful desires of the building program with the attributes and limitations of the site.

Also arising out of this process should be insight into the general shape of the house. Once again, it is important to be sensitive to the land rather than to rely on preconceived notions. Obviously, the shape of the house must accommodate the living areas and their relationship to the outdoors. It should be open to the sun, buffer winds and make the views available. Perhaps most importantly, it must consider economics and energy conservation. These two concepts fit together because they have a similar goal: to pack as much as possible into a structure with a small total surface area. As a building is stretched out and convoluted, its foundation area is increased along with the surface area of the walls and roof, creating a need for more labor and materials. The additional surface area also allows more heat to move from the inside to the outside. Still, good architecture demands that we try to find a balance. The shape and form of the house should have intrinsic aesthetic value.

For timber-frame houses, intrinsic aesthetic value means that the shape of the building should be true to the language of the material. Remarkable symmetry and beauty are almost inevitable when timber framing is practiced well in its simplest form, and there can be no doubt that it begs to create rectilinear buildings that directly reflect the shape of the timbers. To ask the craft to shape a curvi-

**Classic building forms, such as this story-and-a-half colonial, are almost automatic in timber-frame construction and are seldom design failures. At right is an interior view of the same house. (Photos by Brian E. Gulick.)**

linear building, for instance, risks that the timbers will become awkward and clumsy and perhaps that the joinery will be stretched beyond its limits. Very complex buildings having numerous corners, layers, jogs, angles, turrets and dormers (as in highly decorative Victorian-style buildings) are best constructed with frames that have their fudge-factors hidden behind plaster walls. Timber-frame buildings of such complexity need difficult joinery, require the best efforts of the finest timber craftsmen and cost like the dickens.

Let the dictates of the site and the building program be the greatest influence on house shape and style. At every turn, choose simplicity over complexity—there is no shame in simple, classic building forms, which seldom disappoint and almost never fail. Most of the traditional styles that dominate architecture were born as timber-frame structures; they were refined through the centuries as the skills and aspirations of the builders evolved, but it was an evolution with no radical leaps. Instead, the master builders were confident enough in the formula, or "pattern language," that they used the same styles to construct both cottages and cathedrals. Within this ancient language lie flexibility and variations that are as suitable for 20th-century buildings as they were hundreds of years ago.

## A Standard Set of Plans

1. Site plan
2. Foundation: plan, sections, details
3. Floor deck: plan, sections, details
4. Floor plans: each level
5. Elevations: all sides
6. Building sections: where necessary
7. Interior elevations: where necessary
8. Typical wall plan
9. Kitchen plan and elevations
10. Chimney (masonry) plans and sections
11. Stair plans and sections
12. Exterior roof and wall trim details
13. Interior trim details
14. Garage plans and sections
15. Exterior deck or porch plans and sections
16. Greenhouse or solarium details
17. Electrical plans: each level
18. Plumbing plans and necessary sections
19. Bent plans: every bent
19. Timber wall plans: every wall
20. Timber floor system plans: each level
21. Panel plans (where applicable): each elevation and each roof side

## Notes on design

Generating floor and frame plans is the next step, but before embarking on this, consider these additional points.

Be prepared to do (or have done) the most meticulous, complete set of final plans possible. Houses are simply too expensive and too complicated for decisions to be left to a carpenter on a coffee break. The bitter reality of house-building is that it is like getting mugged by a thief; the house will demand the entire bankroll and frisk you again before the job is finished. The route from the bank to the completed house must thus be carefully mapped, and only with a full set of plans can you know that the house will meet your expectations and do so within budget. A full set of plans means that every item in the house has been accounted for and every detail in the assembly has been precisely explained—from the depth of the foundation hole, to the configuration of each joint in the frame, to the mounting of the last closet pole. This detailing defines the material and labor necessary before the project commences and is the only way to control costs.

At left is the standard set of plans that we use. Creating such a set of plans is likely to be quite expensive (between 5% and 10% of the total building cost), but there is no better investment. We have worked with a few homebuilders who have done a fine job with their own design and planning, but most people do not have the skills or time necessary to do a good job with blueprints. Homebuilders choosing this course should be wise enough to know what they don't know and, at the very least, retain professionals who will act as consultants.

Homebuilders choosing to work with a professional architect or designer should select this person carefully. The ability to create good architecture out of the simmering pot of information collected on the site and from the building program is a talent that does not come to every person. A good designer can stir it all together and bring to the house those tasteful lines that go down easily with the ecology of the site and the geography of the area. Enduring designs are born in humility, of the understanding that very few buildings are an aesthetic improvement over no building at all. Unfortunately, this is not a concept stressed by schools of architecture. Too often, architects emerge from their educational dens and immediately attempt to capture the world with the latest artistic snares. At the very least, make sure the designer you choose is familiar with the timber-frame system—suggest this book and others in the bibliography. Also make sure the designer has a copy of your building program and refers to it often. At interim meetings, you might even want to go over the list together as a way of analyzing decisions that have been made. And remember, working with a professional does not mean noninvolvement. A designer should not be expected to put together a house design without a great deal of input from the clients.

Of course, an alternative to designing your own home or hiring a professional designer is to use a stock design as offered by some timber-frame companies. This may seem like the ultimate design crapshoot, but if you start by going through the preliminary design steps discussed earlier and then search to find a plan that fits your requirements, the process can work quite nicely.

## Generating floor and frame plans

This stage of design is a little more complicated than the previous steps, because the designer must decide on dimensions and precise spatial relationships, and address how the house is expected to work. While I discuss the rudiments of developing frame and floor plans in this section, the details on arranging the timbers within the house are covered in the two following sections, beginning on p. 69. I chose this organization because I wanted to cover basic design methodology before going on to the more refined aspects of using timbers in a floor plan. Make sure you read those sections before charging into design work.

The birth of the floor and frame plans out of the notations in the building program, the wanderings on the site and the musings about the bubble diagram are rather magical events. Out of the chaos comes order, and sometimes no one person involved is quite sure what made it happen. But if the groundwork has been accurately laid, even the first efforts at a floor plan can be pretty accurate.

Begin with a block diagram, as shown at right. Modify that as necessary and draw a floor plan, locating major posts and beams. In these first attempts, the goal is to find an acceptable arrangement of rooms and timbers, and to see if the sizes of the spaces seem adequate. To be truly useful, however, the early floor plans should quickly evolve to include some finish and construction details, although there is no need for exhaustive detail at this stage. Stairs, chimney, partitions, closets, cabinets, plumbing fixtures, and door and window locations should be shown because the plans can't be properly analyzed otherwise. The bathroom space won't make much sense unless it is seen in relation to potential fixtures. The kitchen must be seen with cabinets, and bedrooms should have closets.

Locating the posts and beams in even the very earliest renditions of the floor plan makes sense both structurally and aesthetically. Posts can't be located without anticipating the relationship they will have with intersecting beams and studying the impact of these timbers on the overall plan. But while it's necessary to consider the floor plan, the frame plan and indeed the shape of the building as a unit, it's also important to realize that all these components should be able to stand on their own merits. This means that if it is helpful to change the building's shape to suit the floor plan, this should not be done at the risk of creating a house that is awkward or ugly. And while beams and posts can be moved to accommodate design goals, these decisions should neither cause the frame to become structurally deficient nor make the frame overly difficult to assemble. This may seem obvious, but it is the most common pitfall in timber-frame design. It's all too easy to remove a critical timber or brace, to stretch a beam beyond its normal capacity, or to contort the frame into an uncomfortable position.

To avoid these failings, it's necessary to keep redrawing floor and frame plans as you continue to refine and alter the design. Creating initial frame plans requires a concept of the shape of the house, an idea born from aesthetic judgment and an understanding of the site and the potential living spaces. The skills and intuition of a good designer are essential here. In its crudest form, the initial frame plan can be a stick representation of potential bent plans

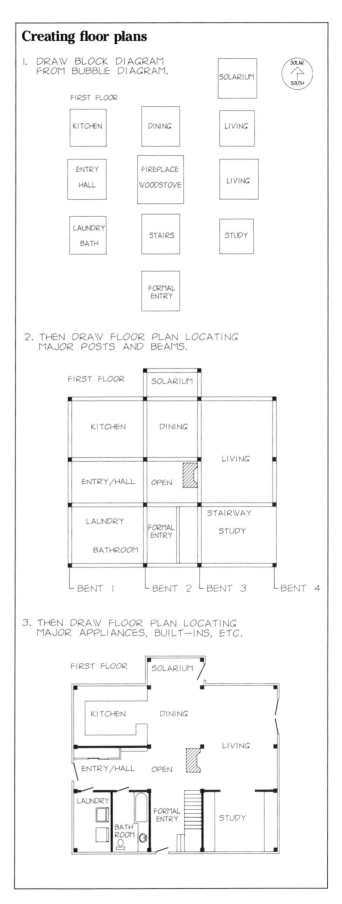

**Creating floor plans**

1. DRAW BLOCK DIAGRAM FROM BUBBLE DIAGRAM.

2. THEN DRAW FLOOR PLAN LOCATING MAJOR POSTS AND BEAMS.

3. THEN DRAW FLOOR PLAN LOCATING MAJOR APPLIANCES, BUILT-INS, ETC.

as posts and major beams are positioned on the floor plan. As design progresses, show the relationship of bent girts, summer beams and connecting girts to the floor plan with dashed lines (as shown in the drawing below). It is not necessary to go into further detail on beam arrangement until the plan is more finalized, but it is useful to shade the posts on the floor plan to show they consume space.

As illustrated in the previous chapter, the development of the bents is critical. Bents are usually the dominant structural element in the frame and they also determine the shape of the building. The way timbers are placed in the bents affects ceiling height as well as the feel of the spaces in the house. So not only should each bent be structurally sound, but as you juggle post length, beam heights and rafter intersections, you need to think about the use of space. As the floor plan becomes more defined, the frame plans should likewise develop into a complete set of plans to prevent oversights and structural inconsistencies. Eventually the bent plans must show all members, with accurate dimensions on timber locations and lengths.

As the design develops, also sketch onto the bent plans the location of windows and doors. Most people strive to center these on a wall for symmetry, using the floor plan to determine their positions. But in fact, windows and doors are best centered between braces, which aren't even visible on the floor plan. Sketching them on the bent plans makes it easier to remember that, when viewed from the inside, windows and doors are actually seen in relation to the timbers in the frame.

Designing a floor plan is filled with uncertainty because dreams finally meet hard reality. It's a time for listening to common sense and practicality. Do not design beyond the budget. Pay attention to the movement of the sun and the direction of the wind. Conserve energy. Consolidate pipes. Give people a place for privacy as well as places to congregate. Above all, even if the decisions seem complex, the floor plan should not be; you will never go wrong if you simplify. Eliminate unnecessary partitions. Straighten convoluted paths and get rid of clutter. The house can be intriguing without being mystifying.

## Preliminary floor and frame plan

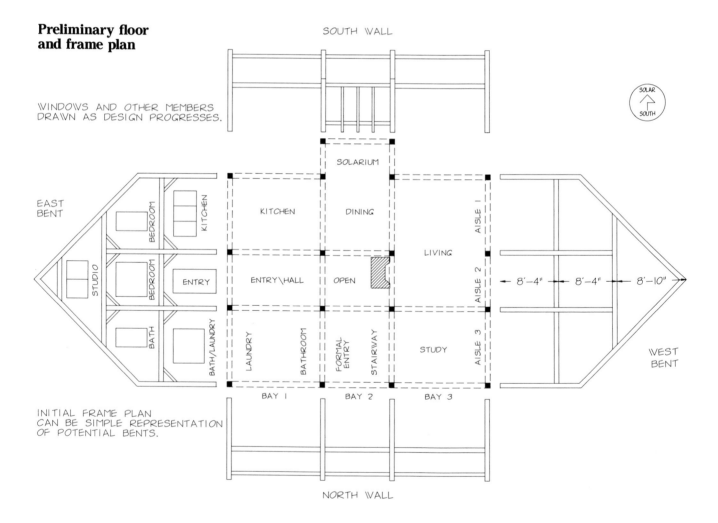

## Bays, aisles and whole-frame systems

While designing, it's important to have a good understanding of the spaces naturally created by timbers and how they can be used for maximum aesthetic and pragmatic benefit. A major objective of timber-frame design is to use the timbers for more than structure, to integrate the living areas with the timbers in a balanced and harmonious way.

The natural spaces in timber-frame houses are bays and aisles. (These spaces are shown in the drawing on the facing page.) Bays are the areas between bents, and are the dominant spaces. They run from the front to the rear of the house on each level. Most timber-frame houses have between two and four bays, the common width of which ranges from 6 ft. to 16 ft. Most typically there are three bays, with the outer two being wider and containing the bulk of the rooms. The narrower central bay contains the entrance, hall, stairway and chimney. Each of the outer bays is designed to hold compatible rooms. For instance, the more formal and private areas would be placed together in one bay, while the work and utility areas would be grouped together at the opposite end of the house. The rooms on the other levels are organized to be compatible with the rooms on the first level. Rooms with plumbing would be stacked, for instance, and second- and third-floor rooms requiring privacy would be placed over the more private rooms of the first floor. This commonsensical approach is why traditional timber-frame homes are a model for many modern designs, even for non-timber-frame houses.

A bay may contain one room, such as a living room or large bedroom, or several rooms. In the latter case, intermediate posts and summer beams or joists are used to define the areas and to provide attachment for partitions, although very often several rooms function in a single bay without partitions; here timbers suggest the transition from one space to another, as detailed in the next section. Kitchens and dining spaces work well this way, as do living

**Looking from bay to bay in this home, we see a division of spaces defined by bents.**

**Within a bay there can be one room or several. The timbers in this bay, as well as the low ceiling over the dining room, distinguish the living areas.**

rooms and study or TV areas. It's also common for areas to be open to each other among several bays—we have built a number of houses that have almost no partitions on the first floor. A single living level might contain the kitchen, dining area, living room, study, several entry halls and a bathroom. The bathroom would be enclosed, but all the other areas would be defined only by timbers. A benefit to an arrangement as open as this is that heat can move freely throughout the house, but the obvious drawback is the lack of private space.

Running perpendicular to the bays are the aisles, which, like bays, are usually not much wider than about 16 ft. If a frame is designed as a series of walls instead of bents, aisles are the more natural spaces. The posts in the walls would have to be parallel to each other, but in the bent direction interior posts would not have to align, which would offer design flexibility. If defining the spaces with aisles is more beneficial to room layout, this is the best reason of all for designing a frame with plates.

Barns are often arranged in aisles, having a central entrance for vehicles at the gable end, which allows the unloading of equipment and animal feed into the outer aisles. Most timber-frame churches and halls are also built like this, but in houses the proportion of common living area to

The living area in this house is arranged in an aisle space instead of a bay. A portion of the middle bay is left open for dramatic effect and for air circulation.

overall house area is usually more conducive to the use of bays. To illustrate this point, imagine a building about 30 ft. wide by 36 ft. long with equally spaced posts, creating three bays and three aisles. The bays would be 12 ft. on center and the aisles 10 ft. It is a lot easier to think of a potential room being half a bay (12 ft. by 15 ft.) than half an aisle (10 ft. by 18 ft.). While a living room might well consume a whole bay, for it to consume an entire aisle (10 ft. by 36 ft.) would be absurdly out of proportion.

Ease of assembly and raising also often dictate that frames be designed as bents rather than walls. Most of the rooms thus fit within bays, with one or two spaces being defined in aisles. (Since most frames have identical bents, aisle spaces are almost always available.) When the house is oriented with a south-facing eave, we often design a solarium by continuing the roofline of the main living area to create an aisle with sufficient glazing for good collection, as discussed in Chapter IX. As long as the posts continue through the frame in the same alignment, living spaces can be stacked in aisles just as they are in bays. When a room is on the outside eave wall of a building with a post length shorter than 16 ft., the space is left open to the roof for drama, but also because there would not be a great deal of added living space close to the eave anyway.

The most efficient way to use space within a house is to stack living areas from outside wall to outside wall in each bay and aisle. But what is most efficient is not always most desirable. Often it is advantageous to leave areas open through several levels, or to change floor and ceiling levels in certain areas to improve proportions or create visual divisions. We like to leave a portion of a central bay open to allow heat to travel into the other living levels, but (thinly disguised) this is also an excuse to create a view of more of the frame and to introduce a feeling of spaciousness and drama to a home. It's also possible to leave an entire bay completely open, both vertically and horizontally. This is especially effective in an otherwise small and tightly organized house. Taken to extremes, efficiently used space sometimes feels miserly and cramped, but just a small touch of playfulness can change that entirely.

Level changes in a floor plan can be used effectively to follow the slope of the site or to alter the feeling of a room. Because in timber framing we are not confined to common stud lengths, within a given timber bay or aisle it is possible to have floors (and ceilings) on several levels. This also avoids the problem of several joints intersecting at the same spot on a post, which would weaken the post. The builders of old barns almost always put floors and haylofts

**This aisle solarium was created by extending the roofline in two of three bays. The principal rafter is supported by the solarium plate.**

on several different levels, as joinery dictated, and there is no reason why this framing consideration can't be used as a design feature in houses as well. Public spaces and places for entertaining, such as living rooms, family areas and dining rooms, should have higher ceilings than places for contemplation or intimacy.

When bays and aisles are not large enough in themselves to adequately serve a design purpose, it's possible to use an entire frame to enclose an area. Used with a trussed roof system, not only is it possible to create a large, unobstructed area (usually with a cathedral ceiling), but in such a space the decorative effect of the timber frame blossoms. In most situations, the truss would have beams spanning the width of the space—a king-post truss is a good example. The drama of the open roof is best suited to living and family rooms, but we have used it for bedrooms, entries, kitchens, porches and even baths; the extra volume in the open roof might be an inefficient use of space (depending on whether there is enough area to be considered usable), but it is surprising how easy it is to trade a little economic folly for a fine feeling. Remember that just because this option calls for a whole frame,

there's no reason why the frame has to be large. The idea works just as well with small spaces.

Whole-frame areas can be either completely separate, attached to the rest of the house from just about any angle, or integrated into the other bents—same shape, different arrangement of timbers. (See Chapter IX for more details.) It's typically easier to keep the frame separate, because you can then fit it to the house without having to align it with the other bents. When all living areas will be on the same level, the entire floor plan can use whole frames. Since no living space would be needed under the roofline, the pitch of the roof could be decreased to reduce wasted space, and the trusses could be designed primarily for beauty. Living areas can be arranged any number of ways, but more variety is possible if smaller wings are appended to a larger central core. If bedrooms, baths and utility rooms can be moved into attached, appropriately sized frames, they will not have to find definition inside the main structure. Obviously, consolidation of the floor plan is indicated when the desire for heat efficiency is strong or the state of the owners' finances is weak.

## Designing with whole frames

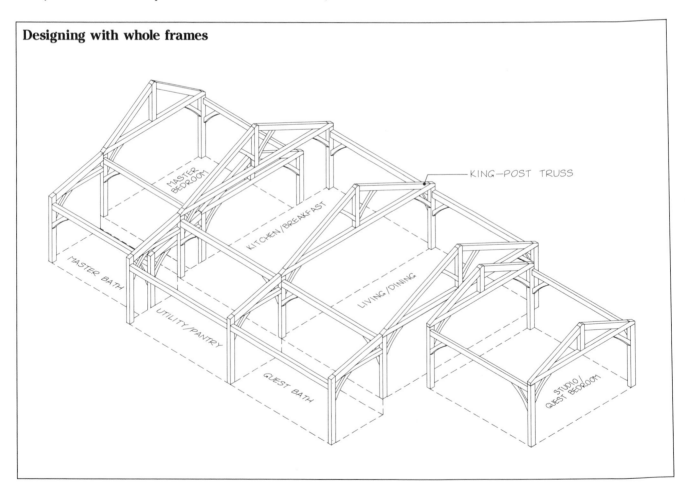

## Timbers and room design

The timbers in each wall and ceiling of a house exert enormous influence on the feel and use of the rooms. The way timbers are arranged within a room is therefore influenced as much by the occupants' planned usage and ambition for the overall mood of the house as it is by structural requirements. As we've already seen, timbers can make a space feel expansive or cozy, public or private. They can also enhance impressions of massiveness or airiness, formality or playfulness, even brightness or darkness. The wood species and general finish treatment of the timbers further heighten the effects created by their arrangement—rough-sawn pine timbers left natural will add rusticity, whereas planed oak timbers stained and oiled to a glossy finish will give the house elegance.

Some insight will inevitably be gained and ideas will no doubt be refined while you work through the building program. Narrow the choices by thinking of the rooms and their potential contents as a three-dimensional still life, with the timber frame working as a picture frame. You probably would not want to imprison a Frederic Remington in a heavily embossed gilt frame, nor would weathered barn boards suit a Renoir. If the desired ambience of the house is formal and the furniture will be ornate, give the frame that quality, too.

We have a rule that we follow to achieve the ultimate integration of a timber frame with the living area. We apply this rule to large rooms and small, important and secondary. We use it on rooms with partitions and those that will never have walls. It is a straightforward rule, stating simply that to give a space boundaries, each room must be defined with timbers. This means that rooms should always be designed with a post in each corner and beams around the perimeter, as illustrated by most of the interior photos in this book. Under the roofline, this might mean having a pair of rafters at each end of the room. As well as offering visual advantages, this rule also makes sense from the standpoint of construction. A carpenter can easily attach plates and studs for partitions to timbers in simple, rectangular wall sections. If timbers are not arranged this way, partitions often must weave around them, causing both studding and finish work to become unnecessarily complex. Of course there will be cases where it's not possible to design a post in each room corner, but these should be exceptions.

The king-post truss in this house helps create a large unobstructed space that is used here for the kitchen, dining and living areas.

Under the roof, spaces are usually defined with rafter pairs on two sides.

To further understand available options in timber arrangement, let's explore several ways to manipulate some of the framing members in the floor plan below. (This is the same house that was generated earlier by the bubble diagram.) Notice that the bents start out exactly the same, with Bents 2 and 3 having an extra post each to frame the solarium. The changes we'll be making, shown in full on p. 81, don't alter the basic floor plan, but they do affect the timbers and the use of space.

Let's look first at the passage between the kitchen and dining room on the south side of Bent 2. The posts, braces and beam form a clear and elegant division between the two spaces, a doorway without a door, a partition without a wall. Curving the braces softens the angles and makes the arch more formal. Larger braces make a stronger, bolder statement. For an even more distinct separation, we could eliminate the braces, frame the passageway with two posts and then add built-ins for china between the door frame and the outside posts. Integrating finish details in this manner helps keep the house uncluttered. We'll choose this option for our final floor plan.

When changing members of the frame, remember to pay attention to the structural requirements. In this case it's okay to remove the knee braces, because the reversed braces above the bent girt add rigidity (see the drawing on p. 79). Knee braces are sometimes used for looks where they're not needed for strength, but it is important to know for a fact when they're necessary and when they're not.

In Bent 3, we have several framing options. The original design (shown at the top of the drawing on p. 75) suggests the transition between living and dining rooms, but to further define it (and for convenience), we might add a wood-storage bin next to the chimney. For the visual benefit of the frame we would keep the bin shy of the ceiling, and if we felt that the masonry didn't integrate well with the frame, we might add a post to the bin edge, creating a narrower passageway to the living room. We'd balance this with a post on the opposite edge of the chimney. While considering these changes, we'd look at the dining area, since the posts in Bents 2 and 3 would no longer be aligned. Must they? No, for while the bents should be bal-

### First floor

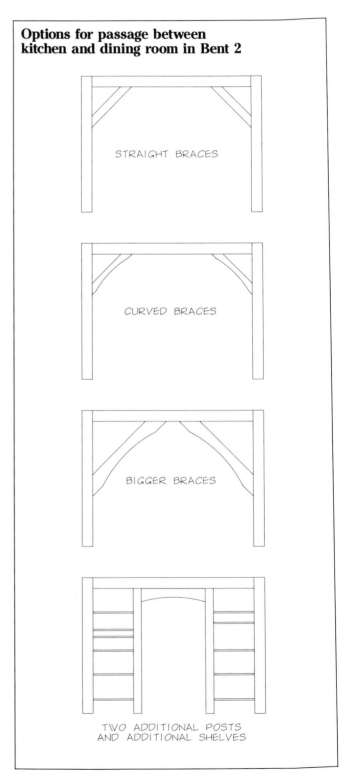

**Options for passage between kitchen and dining room in Bent 2**

STRAIGHT BRACES

CURVED BRACES

BIGGER BRACES

TWO ADDITIONAL POSTS AND ADDITIONAL SHELVES

anced, in this case the concept of integrating some of the functional necessities of the house into the frame is a balancing agent in itself. Besides, both passageways would have an archway created within the original frame. So for our final floor plan, we'll choose the last option.

The drawback to this series of changes is that the dining room now has hard boundaries, which will make it smaller.

(See the final revised floor plan, p. 81.) Without the extra posts and wood bin, the east and west sides of the dining room had more space for traffic and to accommodate an extended dining table for large groups. But given the new boundaries, the extended dining table will probably have to be arranged in the north-south orientation, with large crowds spilling over into the solarium.

**Options for passage between dining and living room in Bent 3**

## Framing a French door

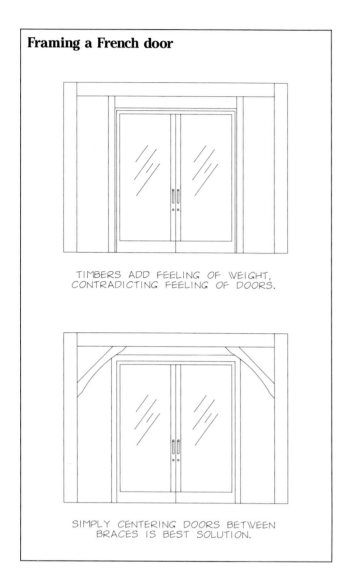

TIMBERS ADD FEELING OF WEIGHT,
CONTRADICTING FEELING OF DOORS.

SIMPLY CENTERING DOORS BETWEEN
BRACES IS BEST SOLUTION.

On the south end of Bent 4 there is a French door leading out to a deck. There are two choices here. The door can be framed with timbers or simply centered between braces. I favor the latter approach, because timbers around the door add a feeling of weight, which contradicts the intent of the door. By contrast, the arch formed by the braces suggests an opening. Simplicity wins out on this one. I would keep the door trim as soft as possible, either opting for no trim at all or painting a simple wood trim the same color as the walls.

The north end of Bent 4 is identical to the south, except there is a window instead of a door. But before making any decisions, we should consider the other ingredients in the study. Are there bookshelves and built-ins? Is there wainscoting around the perimeter of the room? Should this room have subdued light? Assuming the answer to all these questions is yes, we might consider framing this wall differently from the one having the French door. To stick with our theme of integrating finish elements within the frame, we'll cap the wainscoting with a small timber made from a compatible wood. Timbers will frame the sides of the window, and bookshelves will be built in between the window posts and the outside posts. The darkening effect of the extra timbers will also benefit the room by keeping the light from being overbearing. For symmetry, the window is centered.

On the opposite wall of the study is our first interior partition. When interior partitions align with timbers (as they should), you have to determine the side on which the knee braces will be visible. Splitting a timber with a partition requires that partition framing be individually fitted around the braces—a time-consuming task. Usually we expose the braces to the central areas to give the house more continuity. This particular wall is a toss-up because one side faces the privacy of the study and the other is somewhat hidden by the stair. What clinches the decision in favor of the stair wall is the floor-to-ceiling cabinet in the

## Framing a window

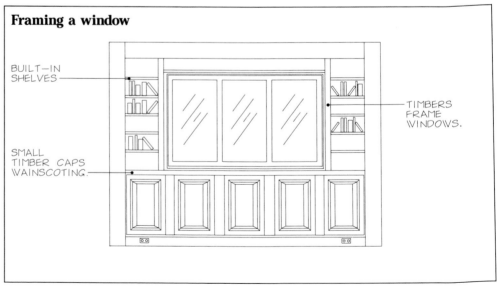

BUILT—IN
SHELVES

SMALL
TIMBER CAPS
WAINSCOTING.

TIMBERS
FRAME
WINDOWS.

study that would cover the braces anyway. So the best thing to do is to build the partition on the study-side of the braces, leaving their full thickness exposed to the stair. Still, the brace that falls on the upper part of the stair and the posts will cause some difficult finish work, because the stair parts will have to be scribed around them.

The wall of the main entry opposite the stair is a plain partition wall. For plumbing reasons (see Chapter VIII), and because so little of the frame would be visible anyway, it makes sense to move this wall toward the bath and expose the frame to the same degree as the opposite wall. When finished, this would be just a wall with timbers and braces, perhaps useful as a place for a small gallery of art.

The downstairs bathroom in this house is neither large nor luxurious, and we don't care if we lose the effect of timbers there. In the laundry room, we throw aesthetics out the window and allow the needs of the room to dictate timber and window arrangement. The only inhibiting factor is the symmetry of the exterior wall—the window should be centered between the braces.

This brings us back around to the east wall of the kitchen. Here the ingredients that need to interact with the frame are a window and the base cabinets. We want as much morning light as possible, so the window is large. To keep the arrangement simple and consistent with our other changes, we might choose to simply center the window between braces with simple trim. It would be possible to frame a timber backsplash between the posts, but it certainly would not be necessary.

## East wall of entry

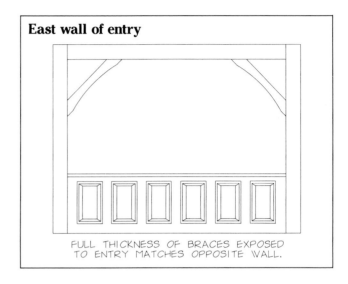

FULL THICKNESS OF BRACES EXPOSED
TO ENTRY MATCHES OPPOSITE WALL.

## Two sides of a partition wall

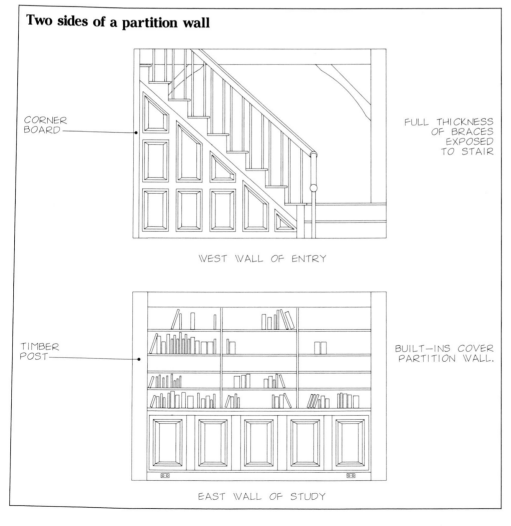

CORNER
BOARD

FULL THICKNESS
OF BRACES
EXPOSED
TO STAIR

WEST WALL OF ENTRY

TIMBER
POST

BUILT-INS COVER
PARTITION WALL.

EAST WALL OF STUDY

The arrangement and size of ceiling timbers should be designed carefully, since they have a strong effect on the feeling of a living space. Large, closely spaced timbers give this ceiling weight and symmetry.

Posts in the walls can be as much as 16 ft. apart, but ceiling joists can rarely be more than 4 ft. apart because they generally support loads from the floor above or from the roof. The impact of such closely spaced timbers (and not insignificantly, over your head) is strong; the way a room feels is greatly influenced by the ceiling beams. If they lack symmetry, the room will feel disjointed, and if the ceiling is low, the beams will make it feel lower. The structural requirements are defined by the loads that must be carried, but there is almost always a variety of ways to satisfy the requirement. In the kitchen, for instance, we could simply span between the two east-west timbers with joists. The span would be less than 10 ft., so the joists would not be large. By using a simple, repetitious pattern and increasing the span between joists as much as possible, the least expressive treatment is created.

On the other hand, if the kitchen would benefit from a focal point, particularly if there were a working island in the center, a large summer beam would be a great addition. It would further subdivide the space, allow the joists to be smaller and might even prove handy for hanging kitchen implements.

The ceiling for the living room also deserves examination, and with the first stroke of my pencil I would raise it. I want more volume for this space, partly because it ought to feel different from the other living areas and partly because I think different ceiling heights are just more interesting. Of course this could cause some finishing difficulties on the second floor, but it might be worth it. The first step in changing the height is easy—we simply move a series of bent girts and connecting girts up to another level. Bent 3 will have the greatest change because it will need to have girts on the common level for the study while also receiving the timbers for the raised living-room ceiling. But this is not difficult or complicated. The study is introspective while the living room is expansive; the change in feeling is accentuated by having these two different kinds of areas side by side.

## Raised living-room ceiling in Bent 3

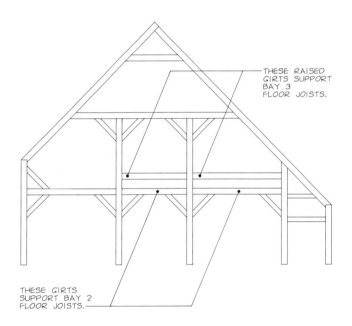

THESE RAISED GIRTS SUPPORT BAY 3 FLOOR JOISTS.

THESE GIRTS SUPPORT BAY 2 FLOOR JOISTS.

**Kitchen ceiling framing options**

FLOOR JOIST

SUMMER BEAM

With that done, let's look at four different ways of framing the living-room ceiling. The first is a series of equally spaced joists, spanning from bent to bent, a simple but dull approach. Also, because each joist would probably have to be fairly large (approximately 5x7 in oak and 6x8 in fir), this option could easily be too overbearing for some tastes.

The next arrangement uses one large summer-beam timber in the middle of the room. The ceiling is now much more interesting, and the joists can be smaller. The problem is that this solution ignores the fact that there are two other posts that could also carry beams. So I think this option would look odd, especially since the reason for the extra post is to define the fireplace—a major ingredient in the room. The next logical step would therefore be to locate a summer beam on each fireplace post. This strong, asymmetrical arrangement would influence potential furniture settings, but it could work, especially since the room seems to be aligned in this direction anyway. If this system were chosen, I'd want to keep all the joists the same size even though the shorter spans could use smaller timbers, since varying joist sizes would be visually disruptive.

The final possibility would be to add a summer beam over the wood-bin post in Bent 3, for greater symmetry. All the joists now could be quite small, which would further emphasize the summer beams. With the higher ceiling in this room, I would choose this last option, feeling the proportions of the room would benefit from the relatively close grouping of large timbers.

## Living-room ceiling framing options

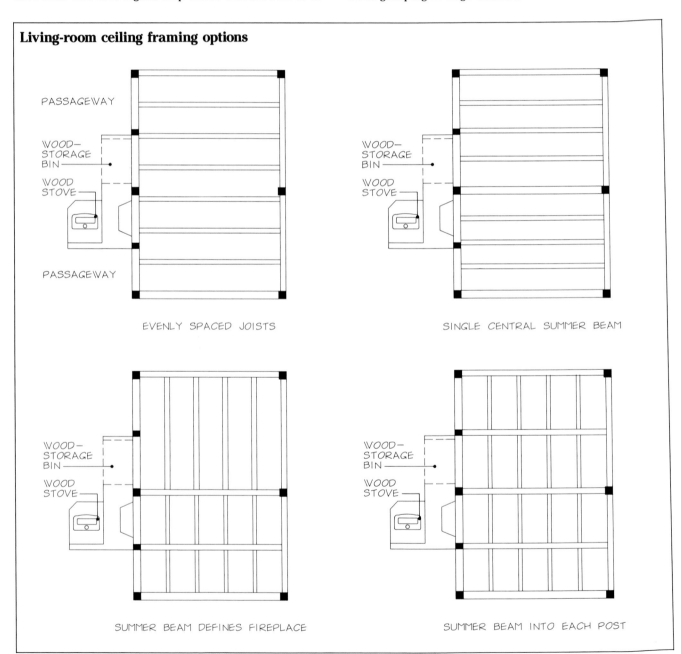

EVENLY SPACED JOISTS

SINGLE CENTRAL SUMMER BEAM

SUMMER BEAM DEFINES FIREPLACE

SUMMER BEAM INTO EACH POST

The completed floor plan is shown below. I am not endorsing it, nor do I intend to imply that good timber-frame house design requires extreme manipulation of the frame. On the contrary—very often the simplest route is best. The points I wish to illustrate are that it's worthwhile to work to make the design right, and that the construction system need not be an obstacle to that pursuit. Consider function and resources in every corner of the house, for every surface, in every room; aspire to make art. There's no shame in falling short, but it's silly not to try.

## Revised floor plan

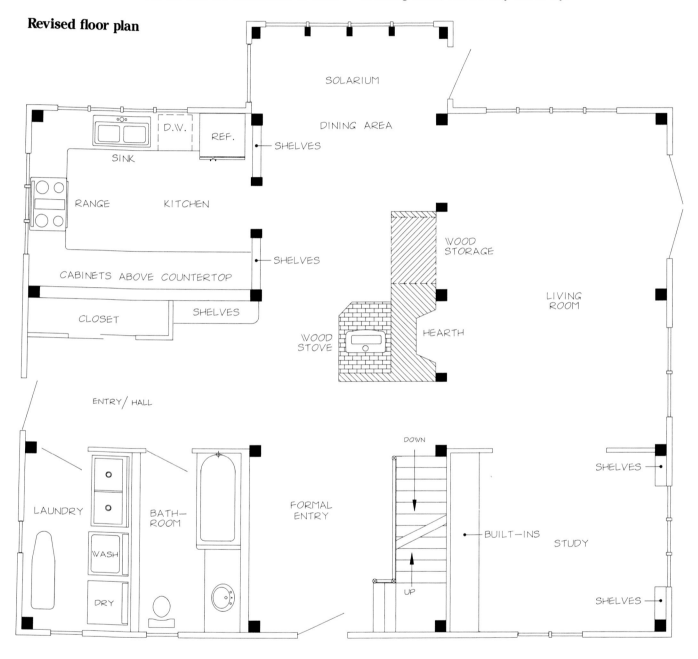

## Two more floor plans

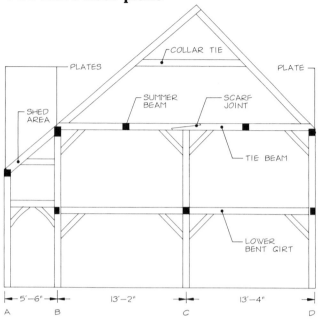

In Chapter III we analyzed 15 different bent plans of the same overall shape. Now I'd like to go back to Options 1 and 13, giving each a hypothetical floor plan to see how design requirements and floor plans might have evolved together. I'm using these only to demonstrate how floor and frame plans can work together, not to represent ideal designs. Also, these plans depict a rather large house, but this is only because the bents used in Chapter III needed to be large enough to show detail. Remember too that design begins on the site and includes many personal considerations, including budget. Different people with different sites and budgets would come up with different, but equally serviceable, plans.

The floor plan shown below follows the design logic described in this chapter, depicting a classic three-bay house oriented toward solar south. The central bay houses the stair, foyer and solarium, and gives access to the outer bays on both floors. On the first floor, only the bath and closet have partition walls; all other living areas are clearly defined by bays, timbers and other features. The south kitchen cabinets, for instance, align with bent posts and a summer beam, establishing the boundaries of the kitchen without walls. The middle bay, chimney and stairway help isolate the living room from other areas. The solarium falls between the *A* and *B* posts and could be separated from the rest of the plan if necessary. Because the shed part of the frame is distinct, it would be possible to use a different wood species for the solarium.

Since the bathroom and closet are not particularly well separated in the third bay, it would probably be helpful to add additional posts between the *D* and *C* posts to further define the partitions at the corners. I would also want to look for ways to integrate the fireplace mass and the stairway into the frame more definitively.

The second floor follows the first, with plumbing stacked and bedrooms falling over the major areas of the first floor. A rather steep stairway to the third floor could be located over the stairway to the second floor, but Bents 2 and 3 would have to be designed without collar ties to allow passage from the central bay to the outer bays between rafters. Queen posts would probably do the job, if it was certain they would not overload the tie beam.

This plan falls into place without a struggle, partly because it is large and simple, but mostly because it is based on a "pattern language" that has been used for timber-frame houses for centuries.

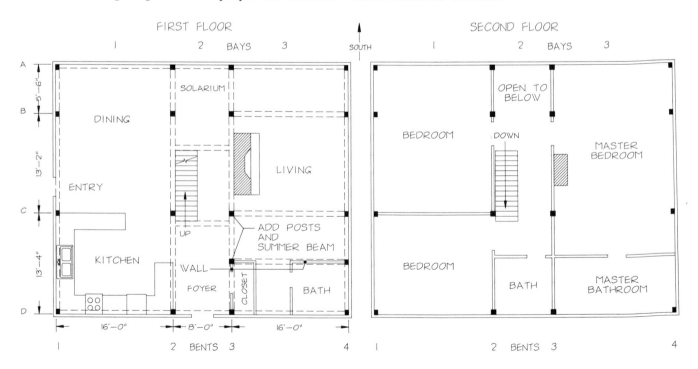

The floor plan shown below is quite different, even though the arrangement of bents and bays is basically the same, because living areas are defined primarily by aisles. The main entrance is through the gable end instead of the eave side of the building. Like the other plan, this one is organized to capture the sun in the major areas, but the southern orientation is toward a gable instead of an eave. The impetus for this arrangement would have been derived from siting considerations—in the morning the breakfast area and the kitchen would receive the sun; you would see it set in the living room. The central aisle houses the entrance and the stairway. Only the dining room and the study fall within bays. Since the stairway is already outlined with posts and beams, it offers a great opportunity to be completely integrated into the frame. Overall, the floor plan is well defined by timbers. Only the division between the study and the bath/closet area is a problem, but a large beam could be used to cap the partition.

Probably the biggest problem with the first floor is that the fireplace is poorly positioned in the living room. It would have been better to center it on the room's east side, but since the masonry can't go through the bent, it had to be placed on one side or the other. With further work, perhaps the fireplace could be pushed to the south, with the flues and mass corbeled toward the middle bay after the firebox.

On the second floor, the master bedroom falls in the aisle over the living room, but the other bedrooms and the bath fall in the outer bays. All the second-floor rooms work quite well with the frame. The biggest problem is that the bath is located over the breakfast area, creating some plumbing problems (see Chapter VIII for a complete plumbing discussion). A chase could be developed in the cabinets, but it would be easier if the bath were located over a convenient plumbing wall.

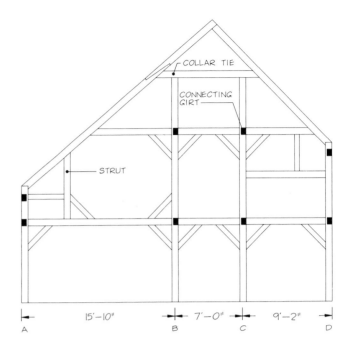

The stair is in a good location to provide access to the third floor but, as in the other plan, Bents 2 and 3 would have to be rearranged to give headroom for passage along the length of the building. An easy framing solution, given the availability of long timbers, would be to extend the *B* and *C* posts to the rafters.

Both of these plans represent good starting points for the development of frame and floor plans. They mesh well with each other, even in this rough state. As design continued, I'd expect both the floor plans and the frame plans to be altered, reflecting further integration and refinement.

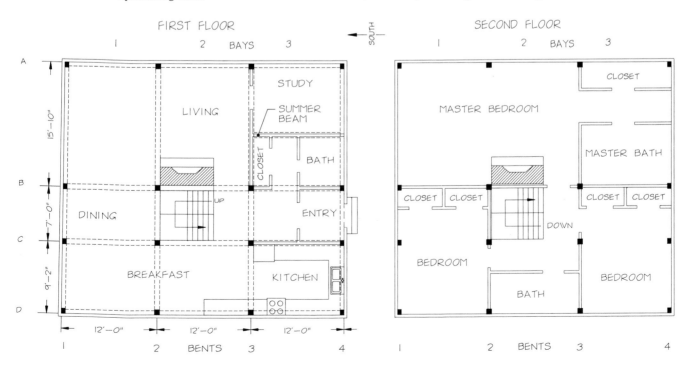

# *Sample Building Program*

## Part I

The following will help you consider the pragmatic design aspects of your home. The proposed sizes of rooms, their function, and the needs and activities of the occupants will determine the layout and dimensions of your house. Your answers to the questions below will help to determine design priorities, as well as identify any extras.

1. List projected occupants (family members, relatives, etc.).
2. List pets, livestock, etc.
3. Projected time of occupancy of each occupant (life, to 21 years, seasonal, etc.).
4. What transportation vehicles, construction equipment, farm equipment, etc., do you need to store, and what kind of shelter is required?
5. Have you determined the approximate total square footage of required living areas? If so, what is the square footage? How did you arrive at this size (existing situation, family size, number of bedrooms/baths, future needs, furniture arrangements, etc.)?
6. List any circumstances particular to you or your family that may affect design (handicapped people, medical needs, noise considerations, lighting, air quality, temperature control, etc.).
7. Should expansion flexibility be a consideration? Do you know what rooms will need to be added and how large they might be?
8. List activities and amount of time to be spent in rooms/spaces below, and approximate space required. Reference to and measurement of the spaces in which you now live may be helpful. Keep in mind furniture, equipment, storage requirements, etc.

| | |
|---|---|
| Living | Bedroom 2 |
| Family | Bedroom 3 |
| Kitchen | Bedroom 4 |
| Dining | Guest |
| Sunspace | Study/library/den |
| Baths | Project room |
| Utility/laundry | Basement |
| Pantry/storage | Garage/shop |
| Bedroom 1 | Other |

The following suggests other possible rooms and spaces to consider:

| | |
|---|---|
| Closets | Playroom |
| Greenhouse | Breakfast nook |
| Airlocks | Porch |
| Mudroom | Patio |
| Darkroom | Hot tub |
| Powder room | Pool |
| Broom closet | Sauna |
| Linen closet | Exterior wood storage |
| Dressing room | Courtyard |
| Interior wood storage | Barn |

## Part II

This group of questions is intended to help you pinpoint how you would like your life and your living space to interact. Try to consider how things feel as well as how they look, and remember the flow of people, things and food through the environment. Indicate the things that you find disturbing and the things that have a relaxing effect.

### *Bedroom*

1. Do you like to wake up in a bright and/or sunlit room?
2. Do you like to be able to look outside while lying in bed?
3. Do you prefer the bed to be a particular height off the floor?
4. Do you prefer windows, a fan or an air conditioner for ventilation?
5. What are the privacy requirements of the bedroom from the outside and from the rest of the house?
6. What are your storage requirements? Give size and general description of items.
7. Do you like a separate dressing area?
8. Does the bedroom also need to function as a study, T.V. room, sitting room, etc.?
9. What sleeping accommodations will be necessary for other family members or in the future? (Use the first 8 questions as guidelines.)
10. Would you feel comfortable if the bedroom had a cathedral area or would you prefer a flat ceiling?

### *Bathroom*

1. What are the special qualities you like in a bathroom? Does it need to be especially bright or roomy?
2. Should there be a bathroom attached to one or more bedrooms?
3. How private should the bathroom be from the rest of the house and from the outside?
4. Do you need a bathroom easily accessible from the outside?
5. Do you prefer fans or windows for ventilation?
6. Do you like to be able to see out a window while showering, washing or using the toilet?
7. Do you use the bathroom in a hurried or a leisurely manner?
8. Do you prefer showers or baths?
9. Do you bathe with others?
10. When you take a shower or bath, do you use the bathroom to dress or undress?
11. Do you have a preference for a particular brand of fixtures?
12. What requirements do you have for mirrors?
13. What are your storage requirements?
14. What sort of bathing accommodations are necessary for other family members now or in the future? (Use the first 13 questions as a guide.)

### Kitchen

1. What times of day and to what extent do you use the kitchen?
2. How many people should be able to use the kitchen at the same time?
3. Will the kitchen be used for eating as well as for food preparation?
4. What appliances are necessary?
5. Is there a particular appliance/fixture arrangement that suits you best?
6. Do you like to be able to look out the window while working in the kitchen? If so, from what area?
7. Do you like a bright or sunlit kitchen?
8. Is it important to have an exhaust fan in the kitchen?
9. How do you prefer to deal with waste and garbage?
10. Where and how would you like to store your food?
11. How much cupboard or drawer space do you need?
12. Do you like pots and pans, dishes, utensils and food to be visible or hidden?
13. Would you use timbers to hang pots and pans?
14. What non-food items do you store in the kitchen?
15. What other items (such as desk, T.V., fireplace) would you like in the kitchen?
16. Should the kitchen be separate from the rest of the house or linked to other areas?

### Dining

1. What sort of feeling should the dining area have?
2. Is the dining place a single or multipurpose area?
3. Is there a need for a guest dining area separate from the normal family dining area?
4. How many people should the dining area accommodate?
5. Is natural or artificial light, or both, appropriate?
6. What is the best way to get food and dishes to and from the dining area?

### Living

1. How do you want the living area to feel?
2. Should it be separate or integrated into other living areas?
3. What activities go on in the living area?
4. How many people should it accommodate?
5. Should it have partitions and doors?
6. Should it be bright during the day?
7. What sort of artificial light is compatible with your image of the living space?
8. Is a fireplace desirable?
9. List any equipment you think should be in the living space (stereo, books, video, etc.).
10. How do phones and T.V. figure into your life?

### Miscellaneous

1. What should the predominant feeling of the house be (rustic, informal, elegant, etc.)?
2. How would you describe the style that makes you feel most comfortable (modern, colonial, etc.)?
3. What are the attributes of the timber frame that would help to achieve the above requirements?
4. What other rooms or activity areas are necessary (study, playroom, darkroom, shop)?
5. What guest accommodations are necessary?
6. Is additional storage space necessary?
7. How much interaction between the indoors and outdoors is desirable?
8. Do you like transitional spaces (porches, decks, etc.)?
9. Do you see the exterior landscape as formal, casual or natural?
10. What requirements do you have for outdoor activities and accessory buildings (pool, tennis, gardens, horses, basketball, etc.)?
11. What is your attitude toward dirt? Where do you mind it and where is it okay? How do you clean it up and how often?
12. How many vehicles do you have and how do they fit into your life?
13. Do you have pets? How many and what kind?
14. Plants? How many? Special requirements?
15. What sort of textures do you like?
16. What colors do you like?
17. List the things you can't do in your present environment that you would like to do in your new environment.

# V. Skins and Frames

**The development of insulation systems in this century has changed our expectations of interior comfort. Higher R-values make it possible for passive solar heat to be retained for long periods. (Photo by Brian E. Gulick.)**

Through most of human history, people expected temperatures in their homes to rise and fall with the temperatures outside, to be only moderations of exterior extremes. But the development of insulation systems over the first two-thirds of this century changed that, and homeowners today expect to have full control over the weather in their rooms. Both ends of the country have been made more habitable because we have found a way to manufacture comfortable air with machines that heat, cool, humidify, dehumidify and even purify the air; it is now possible to live in both cold and hot climates without suffering. The problem, however, is that the control is illusory in most cases, achieved through the consumption of wood, coal, oil, gas and electricity. The temperature in the house is transformed temporarily, but it soon passes through the walls, roof and floors into the outside atmosphere. Our machines deliver newly tempered air before we discover the escape of the old, but while this may keep us comfortable, we all know that energy once used is forever gone.

For a few crazy years (most notably between 1925 and 1975), the actual cost and true value of fuel were far apart, and our precious energy resources went up in flames. During this time the average American home was a poorly designed energy sieve lacking only a funnel on top to make the pouring in of fuel easier. Frank Lloyd Wright himself contended that insulating the walls of a house was hardly necessary. He felt that it would be worthwhile to insulate the roof, "...whereas the insulation of the walls and the air space within the walls becomes less and less important. With modern systems of air conditioning and heating, you can manage almost any condition" *(The Natural House* [New York: Horizon Press, 1954], 155). Indeed. Why worry about conserving energy when fuel is cheap and when you can get a heating unit that "can manage almost any condition?" Unfortunately, there are homes everywhere that still use the monsters Wright was talking about.

When the energy crunch hit in the mid-1970s, it was natural to turn to the sun and wind to replace oil and gas. It wasn't immediately clear that the basic precepts of house design might need to be re-evaluated, so liquids were pumped, air was pushed and heat was stored in every conceivable medium. That's when we discovered how wasteful the average American house is. We found that solar collectors and wind generators, no matter how technologically sophisticated, are like a spit into the wind when heat is passing through the house at a full gallop.

Engineers and architects generally agree that the way to reduce energy consumption is to design houses that naturally collect heat during cold seasons and reject it during hot seasons. Then you install the best possible insulation so desired temperatures can be retained without assistance for long periods of time. It's a concept that works in both warm and cold climates, reducing heating and cooling requirements, and so theoretically simple that you really have to wonder how we managed to stray from it in the first place. But like so many other simple solutions in this era of high-tech, this one is often bypassed, for we are easily seduced by the outer limits of human understanding and tend to rely more on mysterious laws of physics and impressive mathematics than on the primal senses. My dad tells the story of the Oxford professor who was halted in his march across the campus by a couple of students who were eager to clarify some details from the classroom discussion. When the conversation was completed, the professor turned to one of the students and asked, "By the way, which direction was I coming from before you stopped me?" The student pointed down one of the paths. The professor smiled and said, "Well, good, then I've had my lunch." Lost within the labyrinth of our own single-mindedness, it's easy to forget where we are coming from and where we are going.

Perhaps the largest barrier to reducing fuel consumption in our homes is that conventional stud-framing methods are inherently contrary to good insulating practice. They were simply not invented with insulation in mind. In the mid- to late-1800s, when stud framing was being developed, people achieved thermal comfort primarily by drawing closer to the hearth: No thought was given to the system's compatibility with insulating materials because there were no insulating materials. So stud framing evolved with the single purpose of making construction easier. But because the framing, fiberglass insulation and sheathing in

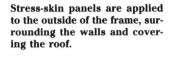

**Stress-skin panels are applied to the outside of the frame, surrounding the walls and covering the roof.**

stud construction are all installed separately, each component compromises the other. Studs suffer a reduced life because moisture from air trapped in the fiberglass causes them to rot. The fiberglass insulation would be much more effective if it weren't interrupted every 16 in. to 24 in. by a stud. The whole unit would be a better air seal if it weren't composed of hundreds of individual pieces. To prevent air infiltration through the gaps, the studs and insulation are covered with a flimsy, vulnerable sheet of plastic that is immediately punctured with thousands of screw or nail holes made in attaching the sheathing.

In addition, the fiberglass itself is not the best insulating material. It's relatively inexpensive and easy to stuff into the spaces between framing members, but unless it is more dense than the typical product, it doesn't trap air very well. Insulation value is provided by pockets of dry, still air, and the fact is that air moves through fiberglass quite easily, which is why it is used in other applications where the goal is not to trap air but to filter it. By today's energy-conserving standards, the typical batt of fiberglass amounts to being a pretty good sweater but not much of a coat.

Builders and architects are working hard to circumvent the inherent insulative deficiencies of stud framing, but most of the new solutions demonstrate that we really ought to try thinking in new ways. For example, one popular system uses two parallel stud walls with batts of insulation in both walls and more in the space between—three layers of insulation, two walls and a highway for mice. Another method is to build a conventional stud frame and then attach a light plywood truss to the outside to carry additional insulation. When completed, the composite is a fortress wall 12 in. to 16 in. thick. Stud framing is dominant throughout the world and it won't soon be replaced. But outside of the housing industry, people have developed effective insulation systems precisely because they haven't had the constraints imposed by this construction method. Walk-in freezers, coolers, thermos bottles, energy-efficient hot-water heaters and liquid gas tanks are all examples. Each one of these devices uses a continuous, uninterrupted membrane of rigid-foam insulation, a simple and sensible alternative, but one that was impractical for housing until the recent development of the new rigid foams.

In 1976, after building a few timber-frame houses and feeling limited by conventional building materials, I began to search in earnest for a better insulation system. At the time, we were using a built-up enclosure of horizontal 2x4 nailers and rigid insulation board (p. 95), and while this is still a viable technique for some projects, I was convinced

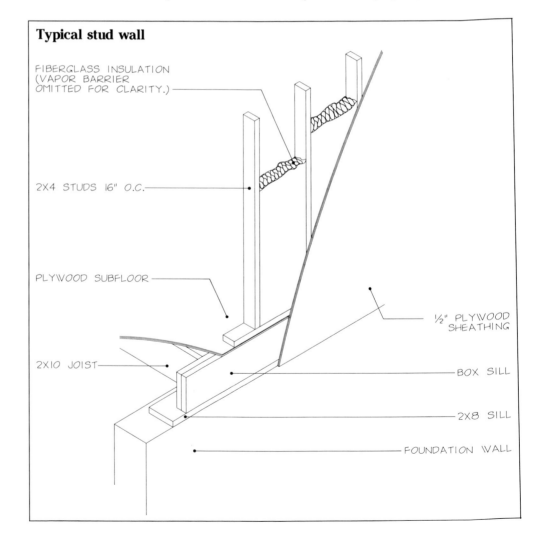

**Typical stud wall**

FIBERGLASS INSULATION
(VAPOR BARRIER
OMITTED FOR CLARITY.)

2X4 STUDS 16" O.C.

PLYWOOD SUBFLOOR

2X10 JOIST

½" PLYWOOD SHEATHING

BOX SILL

2X8 SILL

FOUNDATION WALL

that we needed something wholly different if timber-frame homes were truly to make a comeback. For our built-up method was no improvement over conventional insulation, in that it was labor-intensive and suffered from thermal breaks. In addition, it was wasteful, because a properly built timber frame can withstand all building loads without relying on sheathing for rigidity or infilling for support. In fact, what we were doing was building a frame around a frame. I knew that if we could take full advantage of the self-supporting nature of the timber frame, our buildings would be more economical to construct and the insulation values would be greatly improved.

Endless phone calls, some library research and several trips over quite a few months led me to query people who specialized in making walk-in freezers. These were people with no particular ties to any kind of product, whose business depended on using the finest insulation. Two companies were receptive to my questions; one was especially helpful and invited me to visit its research laboratory. The technicians there gave me a short course on insulation, demonstrating why interrupting it with framing members is not good practice and why fiberglass is really second-rate stuff. By using a dyed vapor and changing the atmospheric conditions on either side of a test wall, they were able to create convection currents inside the wall cavity, which caused air to filter through the fiberglass.

For its insulating walls, this company preferred a composite of materials called a stress-skin panel. The panel is nothing more than a sandwich of a thick core material bonded to two thin outer skins. According to the U.S.D.A.'s *Wood Handbook* (see bibliography, p. 229), "The facings resist nearly all the applied edgewise loads and flatwise bending moments. The thin, spaced facings provide nearly all the bending rigidity to the construction. The core spaces the facings and transmits shear between them so they are effective about a common neutral axis. The core also provides most of the shear rigidity of the sandwich construction." What this means is that the structural principle

behind a stress-skin panel is similar to that of an *I*-beam (below). The outer skins perform the same role as the beam flanges, while the foam core functions like the *I*-beam web. As in an *I*-beam, the skins resist tension and compression while the core carries the shearing forces and supports the skins against lateral wrinkling. Even if the core of the panel were not insulative, stress skins use materials economically, gaining strength without heaviness or great expense. Cardboard and hollow-core doors are examples of stress-skin panels used for light weight and efficiency of materials.

Panels used in refrigerated warehouses and freezers utilize rigid foam for the core and, usually, galvanized sheet metal for the skins. In freezers, as with timber frames, the panels support no loads. While it was exciting to see a proven product with features so perfectly suited to the needs of timber framing, a panel with metal skins is not exactly appropriate for residential construction. Our challenge was to find a way to make a panel with skins that could serve as interior finish and exterior sheathing surfaces. (For reasons of economy, we immediately dismissed the idea of screwing drywall and nail-base sheathing to the metal skins.) There were also a number of construction details to work out before stress-skin-panel technology could be successfully integrated with timber-frame homes. For instance, we would have to find a way to run wires, install windows and doors, and attach trim and finish materials. We were even uncertain about the best method for attaching panels to the frame and to each other. It was alternately exhilarating to see the potential and overwhelming to be confronted with so many obstacles.

Luckily, I wasn't alone with the thought that stress-skin panels might be suitable for residential construction. Several companies became interested in developing the panel for the burgeoning timber-frame industry. It was because of their intense efforts at the beginning of our experimentation, and continuous improvements in the performance and quality of the product, that we are now able to use stress-skin panels with confidence.

## How a stress-skin panel works

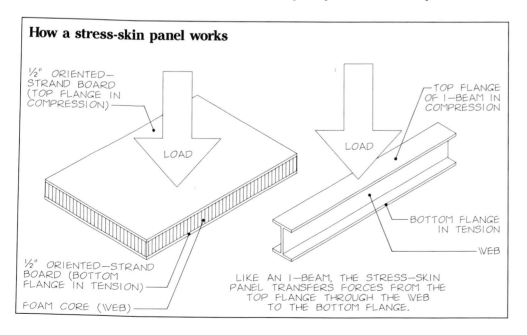

½" ORIENTED-STRAND BOARD (TOP FLANGE IN COMPRESSION)

LOAD

½" ORIENTED-STRAND BOARD (BOTTOM FLANGE IN TENSION)

FOAM CORE (WEB)

LIKE AN I—BEAM, THE STRESS—SKIN PANEL TRANSFERS FORCES FROM THE TOP FLANGE THROUGH THE WEB TO THE BOTTOM FLANGE.

TOP FLANGE OF I—BEAM IN COMPRESSION

LOAD

BOTTOM FLANGE IN TENSION

WEB

The typical panel we now use has oriented-strand board for the exterior surface and plaster-base drywall for the interior finish surface, although we occasionally use panels having oriented-strand board on both sides. (See "Selecting stress-skin panels," p. 99.) Like a human skin over a skeleton, panels are almost always applied completely to the outside of the frame, surrounding the walls and covering the roof. The frame and the insulating membrane are therefore separate, resulting in big benefits. Because the interior finish, insulation and exterior sheathing are all installed simultaneously, several construction processes are combined into one—on-site labor is saved and the building is closed in quickly. Another benefit is the absolute protection given to the frame. Since no part needs to be buried in a wall cavity or exposed to the elements, the frame is not subject to deterioration from weather or moisture. Using this system, the panels take the beating; in time, they could be replaced if necessary, but the frame should survive the ages unscathed.

Urethane and expanded polystyrene are the two foam types currently used for the panel core. Urethane cores can be laminated to the skins with an adhesive or injected in liquid form between separated skins, but expanded polystyrene cores can only be laminated to the skins. A typical core thickness for a urethane panel is 3½ in., yielding an R-value somewhere between 26 and 30, while a polystyrene core has to be 5½ in. thick to achieve a similar R-value.

Conventional wisdom suggests there should be more insulation in the roof than in the walls, so at first it might not seem to make much sense to have such high R-values in the wall panels. Theoretically, because warm air rises, air in the upper regions of the house will be substantially warmer than that at the lowest level. When heat is lost through the roof, convection currents draw up more warm air, resulting in a continual temperature differential between the upper and lower areas of the building—witness the uninsulated attic on a hot summer day. We are discovering, however, that as we slow heat loss through the roof, the temperature differential from the peak to the floor narrows to within a few degrees. What makes this especially remarkable is that most of the houses we build are open to the peak, so the distance from the top of the ceiling to the floor might be over 25 ft. In a house insulated with stress-skin panels, heat is lost only through conduction, not convection, and at a very slow rate because the rigid foam in the panels is a poor conductor. It makes sense to us to keep the entire thermal envelope at approximately the same insulative level, because if the temperature next to roof and wall surfaces tends to be about the same, R-value should be at least proportionately equal.

Out of curiosity, one of our Massachusetts clients recently monitored the winter temperature in his home for one week with a series of thermometers hung between the 25-ft.-high peak and the floor. He found that the maximum temperature difference was 5°F and the average of the period was 2°F. With the introduction of a small fan, he was able to reduce the temperature differential to one-half a degree—quite an achievement and an excellent testimony to a good insulation system.

But as effective and successful as stress-skin panels are, they are not universally correct for every circumstance. Three factors should influence the choice of enclosure system for any particular timber-frame project: structural integrity, energy performance and cost. The system needs to be strong and durable, able to resist wind loads on the walls, snow loads on the roof and the awesome load of time. Energy needs should be reduced to the lowest practical level, keeping in mind that it doesn't make much sense to spend ten dollars to save one. Relative to its value as a fuel miser and as a tough shell for the house, the system must warrant the cost of construction.

The most difficult of these criteria to determine is energy performance. Too often, there is R-value myopia, and judgments are made entirely on the basis of what the manufacturer's wrapping says you have stuffed in the cavities. Unfortunately, the wrapping goes without the warning that your actual performance may vary. A value of R-50 would be a poor joke if it were used on a north wall that was 60% glass, and all the insulation in the world won't save a house that is riddled with air leaks. I recently went to examine one of our houses that wasn't performing up to expectations, although everything seemed to be in place and done according to usual procedure. But because the windows seemed colder than normal, I decided to remove the side casing. Sure enough, not one window, door or skylight had any insulation or caulking between the jamb and rough opening. Added up, the uninsulated area amounted to a 4-ft. by 4-ft. hole in the house. Good insulation, bad workmanship; the system tilts.

There are really quite a few factors that need to be in proper balance to achieve good energy performance. Because it is so important to analyze all issues in relation to each other and because the calculations are so complex, a local engineering firm does a computerized audit of all our homes early in design. Such audits are available throughout the country (they are required by the building department in some states), through utilities, heating suppliers or private engineers. I personally like to work with a firm that expects to make money doing the audit, but not afterward.

The energy audit considers wall- and roof-surface area and orientation, window- and door-surface area and orientation, volume, available degree days, infiltration values, below-grade treatment and insulation, and heat generated by occupants and fixtures. Estimated auxiliary energy requirements are estimated on a yearly, monthly, daily or even an hourly basis. Both gains and losses are estimated. Once all data about the house is entered into the program, it is simple to manipulate values to see the effect on the bottom line. For example, we might try on paper a house with triple glazing or low-E (emissivity) glass and find that the added cost doesn't warrant the slightly improved performance. When the most desirable option is determined, the energy audit can also predict what the fuel cost will be on a yearly basis for each type of available fuel. Even if the audit does not perfectly reflect actual fuel consumption, I have found that it is usually quite accurate, if conservative. I agree with those states that have made an energy audit a requirement. If building departments are going to insist that electric outlets be no lower than 14 in. (my pet peeve), then they ought to slow the flow of fuel as it pulses through the veins of our houses.

An energy audit is also the best method to weigh the cost of the enclosure system against the cost of the heating system and the fuel to run it for the next ten or twenty years. If the long-range numbers are anywhere near close, it makes sense to choose better insulation now rather than bet on the future price of fuel. We strive to pinpoint the measures that bring us to the point of diminishing returns, the very top of the proverbial efficiency curve, so that we'll feel confident when additional insulation or better windows cannot be justified in light of the small gain.

Following the energy audit, and after a careful analysis of each component of the building's envelope and another look at the bank account, the optimum enclosure system is chosen. I discuss the selection, installation and manufacture of stress-skin panels later in this chapter, but first we'll look at a few alternate types of enclosures.

**With good insulation, rooms with high ceilings have little temperature differential between the roof and floor.**

## Alternatives to stress-skin panels

I can think of several reasons why a person might choose not to use stress-skin panels.

**1.** Small buildings under 1200 sq. ft., having little volume and surface area, are likely to reach the top of the efficiency curve with a less expensive enclosure than panels. For some buildings, we conclude that panels would be useful and cost effective on the roof where there is little waste, but that the walls should be treated differently. There are usually many penetrations in the walls for windows and doors, which generate a great deal of scrap and cause installation to be more difficult. I am not suggesting that insulation values be compromised, only that depending on the situation, it may be possible to achieve optimum performance with an alternate system.

**2.** High-quality panels are not readily available in all regions, and trucking them from one end of the country to the other isn't very practical. If the project is in an isolated backwater, it may be necessary to do some low-tech manufacturing or use an alternate system.

**3.** Panels might be too costly for the budget. We have found the installed cost of panels less expensive than the installed cost of any system with a competitive performance value. Although the cost of panels is relatively high per square foot, the installation labor is reasonable because they go on so rapidly. For most of the alternate systems the materials are less, but installation is labor-intensive. Builders who don't place a high value on their labor might therefore find other systems attractive. We've raised frames for several clients who decided to use an enclosure system they could manage with their own resources.

**4.** Some people simply have an aversion to the inorganic nature of panels. Or they can't believe panels can offer the permanence and security provided by using wooden materials for sheathing attachment. For these people, no words are persuasive. I offer only my empathy with the sentiment, and alternatives.

**5.** A building not occupied all the time, such as an office, shop or vacation home, might not justify the insulating value typical of panels. By using one of the alternatives (and lowering performance standards greatly), a lot of money can be saved. Still, it makes sense to err on the side of energy conservation.

**6.** Timber framing is an ancient method of construction and people often want to use it to create a historical reproduction. Panels are inappropriate in this situation because the added thickness outside the structure does not reproduce the look of early timber-frame homes.

With all the alternatives that follow, keep in mind that you already have a building that structurally supports itself. The insulating system needs to be stiff and offer good

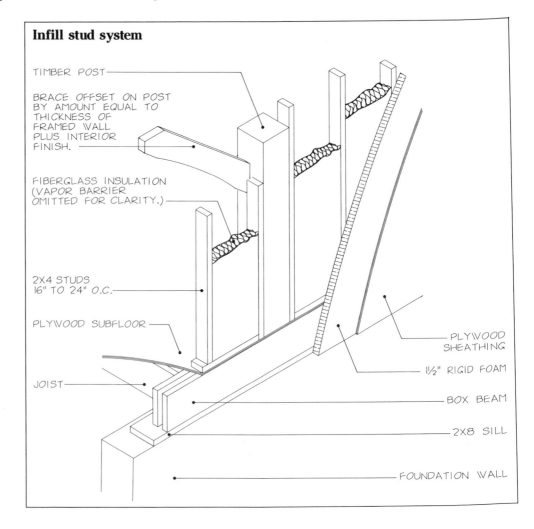

**Infill stud system**

TIMBER POST

BRACE OFFSET ON POST BY AMOUNT EQUAL TO THICKNESS OF FRAMED WALL PLUS INTERIOR FINISH.

FIBERGLASS INSULATION (VAPOR BARRIER OMITTED FOR CLARITY.)

2X4 STUDS 16" TO 24" O.C.

PLYWOOD SUBFLOOR

JOIST

PLYWOOD SHEATHING

1½" RIGID FOAM

BOX BEAM

2X8 SILL

FOUNDATION WALL

thermal value, but its structural requirements are small. Also remember that with these systems, the roof and walls will probably be treated differently. Stress-skin panels allow them to be treated identically because the same pre-manufactured components satisfy the requirements of both. But when built-up systems are used, it makes sense to customize the materials and treatment to the specific situation—timber frames typically have widely spaced timbers on the wall and only a wind-load consideration, while roof timbers tend to be more closely spaced and the loads more significant, depending on the area. Refer to the energy audit to determine if the chosen insulation values suggest using a greater R-value on the roof.

**The infill stud system for walls** This is the most difficult, least effective way to enclose a timber frame. It's also the way that most old timber frames were enclosed, so it's the method usually used when an owner wants to replicate the appearance of an old house. Basically, infilling is the process of building stud walls between the timbers and packing them with fiberglass insulation. The walls are then wrapped on the outside with a layer of 1½-in.-thick rigid foam. Because the walls are attached between timbers and not to the outside of the frame, the entire thickness of the timbers is not revealed to the living space. In addition, the

timbers wind up fitting poorly to the adjacent studs after movement and shrinkage, allowing great air infiltration. It stands to reason that there is also a potential for condensation to occur in the cavities where this air infiltration is taking place. A building publication recently reported that many of the posts and beams in a four-year-old timber-frame house enclosed with this method already had significant moisture damage. The infill stud system should only be used if the timbers are dry or if the occupants are prepared to be vigilant about caulking the gaps that develop.

Probably the one advantage of this method is that it requires only conventional materials and simple carpentry. Because the walls don't support any compression loads, single top plates are fine; headers and sills need only be a single stud. All framing can be 24 in. on center. Infilling is made much easier if the timber framer joins all minor members, such as knee braces, toward the inside of the building by the thickness of the framed wall, including the drywall. If the frame is built this way, each stud wall can then be assembled and nailed on the deck and raised as a unit. The infill stud system can be greatly improved by extending 1½ in. of each 2x4 outside the frame, allowing 1½ in. of rigid foam to be placed over the timbers and then another 1½ in. of rigid foam over the entire assembly before the exterior sheathing.

**For the infill stud system to work easily, push the braces toward the inside of the building so they won't be obstacles to studs and finishwork.**

**The exterior stud system for walls**  By simply moving the stud walls to the outside of the frame, the problems created by infilling can be avoided. The timbers can now shrink or move without causing infiltration or compromising insulation quality; finish work no longer has to fight the frame. Walls can still be conventionally framed, although as in the previous method, many structural members can be reduced or eliminated because the timber frame is self-supporting. Installation is somewhat backward, because the drywall and vapor barrier are attached to the outside of the frame first to avoid infilling between timbers. (Choosing to install the drywall after the studs means that each post will need a stud adjacent to it for sheathing attachment.) At this point the drywall will be attached only to the sills and girts of the timber frame, and will be quite flimsy. If the studs will be on 24-in. centers, choose ⅝-in.-thick drywall (instead of ½-in.) to maintain stiffness. Before the stud frame goes up, tack the vapor barrier to the drywall.

There are some inherent problems with this system. First, the drywall must be protected from the weather, which means you must work quickly and take fastidious precautions. The best protection is to have the roof on first, but if the eaves project well beyond the walls it will be hard to make a good connection to the top plate. Until the roof is on, it's a good idea to keep a tarp rolled up at the peak of the building to pull down over the drywall when necessary.

The second problem is that the stud walls can't be pushed up into position as units, because the drywall blocks the path. So whether the walls are fabricated on the deck or another level surface, they'll have to be carried to the outside of the frame and hoisted into position. It's thus best to keep the walls to a manageable size—that is, not much longer than 16 ft. or taller than 8 ft. to 10 ft. Achieve this by planning for walls to meet at the centerline of posts and girts. When the walls are positioned on the frame, toe-nail them wherever there is sufficient nailing with 20d galvanized nails. Also, spike the studs to the timbers with barn spikes at 4-ft. intervals, penetrating softwood timbers by 2 in. and hardwood timbers by 1½ in. To keep the studs from splitting, pre-drill a ¼-in. hole before spiking.

After the stud walls are in place, install the wiring and electrical boxes. Then pack in the fiberglass. For improved performance, nail sheets of rigid foam over the stud frame. If rigid foam is not used, horizontal siding can be attached directly to the studs. If the studs are 24 in. on center, the siding has to be at least ¾ in.; overlap shiplapped boards to create a conventional clapboard appearance. Vertical siding would require that horizontal furring strips be attached to the studs first.

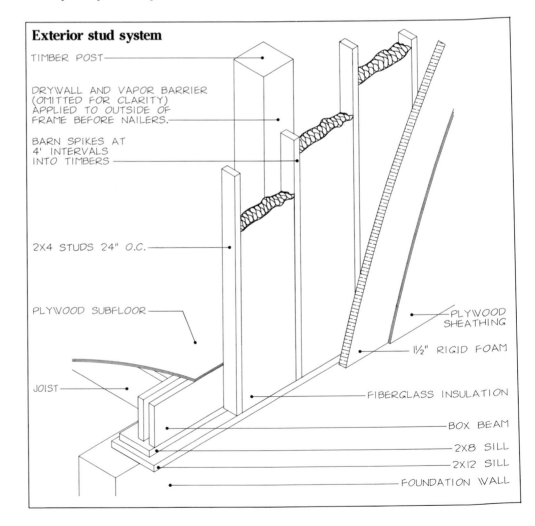

**Exterior stud system**

TIMBER POST

DRYWALL AND VAPOR BARRIER (OMITTED FOR CLARITY) APPLIED TO OUTSIDE OF FRAME BEFORE NAILERS.

BARN SPIKES AT 4' INTERVALS INTO TIMBERS

2X4 STUDS 24" O.C.

PLYWOOD SUBFLOOR

JOIST

PLYWOOD SHEATHING

1½" RIGID FOAM

FIBERGLASS INSULATION

BOX BEAM

2X8 SILL

2X12 SILL

FOUNDATION WALL

**The horizontal nailer system** This method consists of 2x4 nailers attached horizontally to the timbers with rigid-foam insulation in between. It takes full advantage of the nonstructural nature of the enclosure wall; the light-duty nailers exist only to stiffen the wall and to provide attachment for the interior and exterior sheathing. For good support, a 2x12 sill is absolutely essential. Because the sill is also the bottom nailer, it must extend beyond the frame by the same width as the other nailers.

As in the exterior stud system, the drywall and vapor barrier are applied to the outside of the frame first. Then rough window and door bucks are pre-made and toe-nailed to the 2x12 sill and timber girts on the outside with 20d galvanized nails. Rigid-foam insulation is ripped to a width that allows the nailers to be 24 in. on center—if the nailers were 1½ in. thick (typically the actual thickness of a 2x4), this dimension would be 22½ in. If you're using plywood for exterior sheathing, make sure the top edge of each 4x8 panel falls on the centerline of a nailer. For a higher R-value per inch and a better perm rate, I would choose either extruded polystyrene or urethane foam. It should be ¼ in. to ⅜ in. less in thickness than the width of the nailer to allow the exterior sheathing to breathe.

Begin application of the foam by laying a bead of caulking on the 2x12 sill. Two 1-in.-thick foam boards separated by a 1x3 spacer can be used between the sill and next nailer to create a convenient wire chase (a solid foam block would have to be routed for wires). Then lay a bead of caulking on top of each layer of foam and put on a nailer, forcing it down against the foam and toe-nailing it to the timber frame or to window and door bucks. Repeat the procedure all the way to the top of the wall. At the end of each wall, use a vertical stud to cover the ends of the spacers. Applying additional sheets of rigid insulation over the whole system, when all nailers and insulation are in place, is a good way to compensate for thermal breaks. Then screw or nail the drywall to the nailers. Wiring and electrical outlets should be run while the insulation is being installed so cuts in the foam can be made easily. Check your building code. Some will require the wiring to be in a conduit because it will be exposed to the weather for a while. Vertical siding can be nailed directly to the nailers, but horizontal siding will require vertical furring strips first.

An advantage of this method is that installation is easy and can be done one piece at a time. A crew of two can manage the entire process. In addition, layering horizontally takes advantage of gravity and makes it significantly easier to get a good seal between materials—because you have to run up and down the ladder so much, constantly trying to push the stuff together, this is difficult to do with

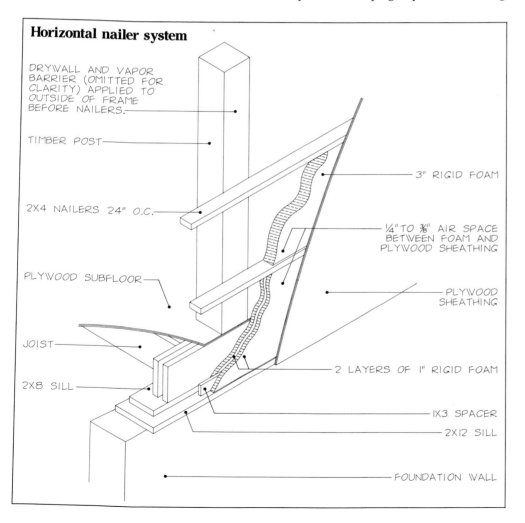

**Horizontal nailer system**

DRYWALL AND VAPOR BARRIER (OMITTED FOR CLARITY) APPLIED TO OUTSIDE OF FRAME BEFORE NAILERS.

TIMBER POST

2X4 NAILERS 24" O.C.

PLYWOOD SUBFLOOR

JOIST

2X8 SILL

3" RIGID FOAM

¼" TO ⅜" AIR SPACE BETWEEN FOAM AND PLYWOOD SHEATHING

PLYWOOD SHEATHING

2 LAYERS OF 1" RIGID FOAM

1X3 SPACER

2X12 SILL

FOUNDATION WALL

vertical studs. High-quality foam performs better than fiberglass, and caulking improves the seal. Still, there are a number of breaks because of the nailers, and the system is labor-intensive indeed.

This is a good alternate enclosure method, and you can vary the thickness of the insulation to keep costs in line. Our shop was built with a variation of this system, using furring strips for nailers with 1-in.-thick expanded styrene between them. The interior and exterior finish are vertical boards. The system gave us a decent finish, minimal insulation and minimal cost.

**The Larsen truss system for walls and roof**   The Larsen truss was developed by John Larsen specifically to address the difficulty of retrofitting older buildings with better insulation. Acknowledging that the buildings did not need additional support, just more insulation, he set out to create a rather nonstructural framework to contain insulation and hold sheathing. The solution is a light truss that uses 2x2s as flanges and plywood for the web. These trusses can be made by anyone with a tablesaw, glue, clamps and a little patience. They can be made to span the height of the walls or the length of the rafters and are lifted and managed easily by one or two people. Because the width of the plywood web is variable, you can use as much insulation

as necessary without a great deal more expense for wood. The web is usually only ⅜ in. thick and does not need to be continuous, so there isn't high conduction through the material. For a new timber-frame structure, this option would make sense only where rigid-foam plastics were either totally unavailable or totally unpalatable to the client.

Installing a Larsen truss is similar to installing exterior-stud walls except that the truss is always attached a section at a time. Drywall goes on the frame first along with the vapor barrier. A truss is then secured to the timber posts by nailing the exposed edges of the inside flanges with galvanized nails that penetrate the posts by at least 1½ in. Wiring is run and electrical boxes installed, then the fiberglass insulation is applied as tightly as possible to the plywood. It's not necessary to use additional rigid foam on the outside. Exterior sheathing is nailed directly to the outside flanges.

The principle of the Larsen truss is that it makes a cavity for a great thickness of insulation at a low price—from 12 in. to 15 in. of fiberglass is not unusual. But most of the difficulties with the system stem from this thickness, for the distance between the inside edge of the frame and the outside edge of the truss is so great that they bear almost no relationship to each other. The trusses will therefore ap-

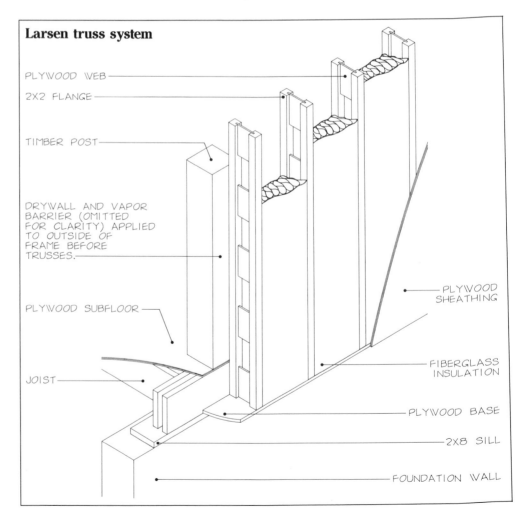

**Larsen truss system**

PLYWOOD WEB

2X2 FLANGE

TIMBER POST

DRYWALL AND VAPOR BARRIER (OMITTED FOR CLARITY) APPLIED TO OUTSIDE OF FRAME BEFORE TRUSSES.

PLYWOOD SUBFLOOR

JOIST

PLYWOOD SHEATHING

FIBERGLASS INSULATION

PLYWOOD BASE

2X8 SILL

FOUNDATION WALL

pear as unnatural appendages to the building if they over-hang the foundation, or the foundation will have to be re-considered to accommodate the trusses. I endorse the latter option. A simple solution is to design pilasters into the foundation wall to support the posts and use the wall itself as a base for the trusses. (See p. 119 for a discussion of pilasters.) Besides the aesthetic problem of the exterior walls overhanging the foundation by 12 in. or more, it's also hard to seal and protect the bottom of the insulation cavity.

Extreme wall thickness also creates some difficulties at exterior windows and doors. Jambs need to be as wide as the wall is thick, causing extra expense for materials and labor. Extreme depth at the jambs might also restrict light from spreading well as it enters the house. The condition can be eased by beveling the jamb sides back at an angle, a nice but expensive touch.

On the roof, Larsen trusses can also be used effectively. But because fiberglass roof insulation must be ventilated to avoid potential moisture buildup, the thickness of the truss and insulation must be planned to allow an air space of at least 2 in. between the insulation and exterior sheathing. The air flow must pass from the eave to the peak; holes will have to be drilled for each cavity at the eave and a ridge vent will be necessary at the peak. (Commercial materials for eave and ridge vents are available through most lumberyards.) Begin by attaching the drywall and vapor barrier to the outside of the roof timbers. To install the trusses, snap a chalkline along each eave and lay the trusses on 2-ft. centers, their bottom edges aligned with the chalkline. Trusses can be fixed to each other at the peak and the inside flanges should be attached to the timbers in the same manner as on the walls. Cut the bottoms of the trusses perpendicular to the slope of the roof. Attach planking to the bottom edges of the trusses to cap the cavity. To avoid having roof trim the same width as the trusses, attach jet blocks to the box, as shown below.

## Using Larsen trusses on the roof

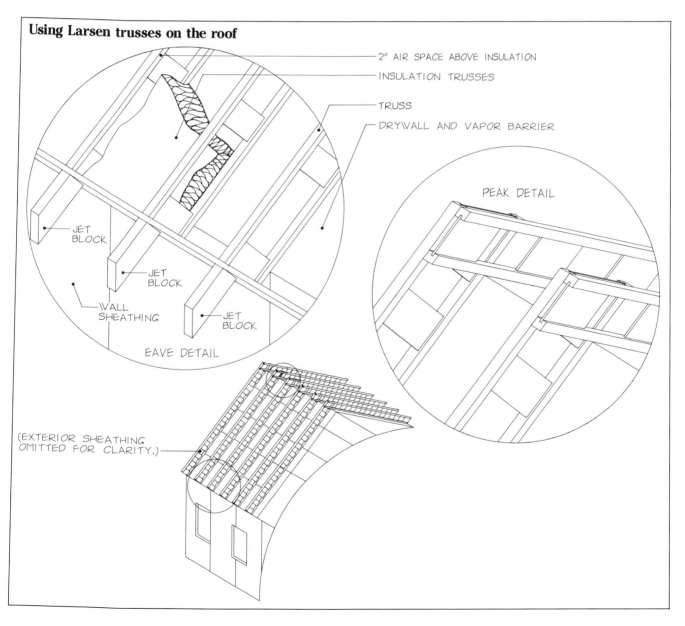

2" AIR SPACE ABOVE INSULATION

INSULATION TRUSSES

TRUSS

DRYWALL AND VAPOR BARRIER

PEAK DETAIL

JET BLOCK

JET BLOCK

WALL SHEATHING

JET BLOCK

EAVE DETAIL

(EXTERIOR SHEATHING OMITTED FOR CLARITY.)

**Foam boardstock roof insulation**  In this technique, sheets of rigid foam are laid over 2x6 planks within a box built around the roof perimeter. You use planking instead of drywall because the interior roof sheathing must be able to support its own weight between timbers without sagging.

The planks are applied directly to the roof timbers, and the box frame is built from stock that is as wide as the insulation is thick. A vapor barrier is tacked to the planks, sealing the sides of the box. All seams should be sealed or taped. The insulation is then installed in the box. Choose insulation thickness by analyzing the relative values of the foam; to prevent crushing or deflection in high-load situations, the foam should have a density of at least 2 lb. It is certainly possible to use foam not much thicker than 3 in. or 3½ in.

After installing the insulation, caulk all voids or fill them with expanding foam and tape the seams. To make an air space between the foam and the exterior sheathing, attach 2x3 or 2x4 nailers every 2 ft. on center, perpendicular to the roof timbers. Nail through the foam to the frame using the guidelines suggested for panels on p. 107. Then put on the exterior sheathing. Roof overhangs can be established using the planks and/or the nailers for support.

**Foam and nailer roof insulation**  If you want to use drywall as the interior finish instead of planking, you'll have to use a system with nailers. The basic rule here is that the nailers run perpendicular to the roof timbers (vertically over purlins and horizontally over rafters). After that, the system is similar to the horizontal nailer system for walls. Apply the drywall to the frame first, then attach the vapor barrier.

Use a nailer wide enough to allow a 2-in. air space between the foam and the exterior sheathing—if you were using 3½-in.-thick foam, for example, you would need a 5½-in.-wide nailer on edge. Place the nailers over the vapor barrier and toe-nail them to the frame with 20d galvanized nails. Then install a section of rigid foam and the next nailer. When the foam is in place and properly sealed, attach the exterior sheathing to the nailers.

While the drywall is exposed and unsupported it presents a potential safety hazard to the installer. To help prevent serious accidents and damage, work your way across or up the roof, installing all the layers at the same time. This will make it possible to stand on either the sheathing or the frame while materials are being positioned and secured.

---

**Foam boardstock roof insulation**

PLYWOOD SHEATHING

2X4 NAILERS 24" O.C.

TAPED SEAM

RIGID FOAM

2X6 PLANKS (VAPOR BARRIER OMITTED.)

2X ROOF BOX

PURLIN

---

**Foam and nailer roof insulation**

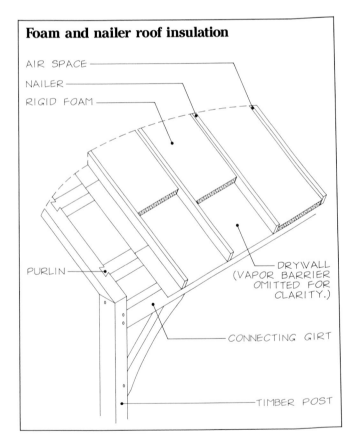

AIR SPACE

NAILER

RIGID FOAM

PURLIN

DRYWALL (VAPOR BARRIER OMITTED FOR CLARITY.)

CONNECTING GIRT

TIMBER POST

# Selecting stress-skin panels

Some of us thought we would never submit to plastic. We saw it come in our lifetime, and when we wanted to point out what was wrong with the modern world we pointed to plastic—tangible evidence of the great deterioration of all good things. Old cars were built like tanks; new cars have too much plastic. My old reliable typewriter is all metal and it still works but I can't remember where it is. There might be metal somewhere in my computer, but all that is visible is plastic. We grumble and complain, but plastic is here to stay. Perhaps a few of us would be wealthier if we had not laughed so hard at Dustin Hoffman's famous line in the film "The Graduate."

It is no coincidence that the timber-frame revival has moved just about as quickly as the development of stress-skin-panel technology. Those of us who craft timbers with mallet and chisel are often a little startled when we are reminded that the panel, a very critical component in most timber-frame building, is made mostly of plastic. The panel is only a core of foam with sheet materials bonded to either side, yet it is asked to play a major role in the construction of significant buildings. Our expectations for panels should therefore be high; they must adhere to the following tough specifications:

**1.** They have to be strong and stiff. The timber frame can have either purlins or rafters on the roof on approximately 4-ft. centers. Spans between horizontal timbers in the walls are usually no longer than 10 ft. Panels must be able to hold their own in these spaces. This can mean as much as 100 lb. per sq. ft. on the roof and a 100 mph wind bearing against the walls.

**2.** They must not delaminate. Delamination between the materials would be our worst fears come to life.

**3.** None of the materials should suffer unreasonable deterioration because of exposure to time, temperature changes or the elements. The panel should be at least as stable as wood.

**4.** Panels should have a high R-value per inch of thickness, to achieve optimum thermal performance without creating imposing fortress walls.

**5.** And they should have a low perm rate. Humidity and air pressure in tight houses are higher than normal, and it is more important to prevent the flow of humid air. If this air should penetrate the inside skin, the core material should be water-resistant enough not to soak up the excess moisture like a sponge.

**6.** They should hold up well under high heat and fire. We are all concerned about flame spread and toxicity. Panels house the foam in an oxygen-free environment, putting a flammable product in the safest possible position, but we still want to know that we would have a chance to leave the building safely should a fire start.

**7.** The core material should have a sufficient molecular bond to resist creep. Even though the panels are nailed to the frame, creep could cause a more rapid deterioration of the panel system, especially on the roof, where they are unsupported at the base and tend to carry a greater load.

**8.** The panels must allow practical, efficient installation.

**9.** Cost must be reasonable or, like many other exotic building products, stress-skin panels will be saved for commercial and public buildings and for the very rich.

Current commercially available panels are made using either urethane or expanded polystyrene as the core material. Extruded Styrofoam is too expensive, although it would make a fine panel. Phenolic foam so far is too brittle and water-absorbent, having a perm rate of 8 per inch. This is clearly unacceptable, because the panel core also has to act as a vapor barrier.

The choice between expanded polystyrene and urethane should be based on an analysis of the attributes and deficiencies of each material. Panel manufacturers tend to specialize in one foam type or the other, and they battle to claim the perfect product. In an attempt to settle the dispute, I hereby pronounce them both in need of further development—the best manufacturers are those who recognize this. Rigid foam has value only in relation to the role it plays as one of the panel components and should be judged on the basis of performance, safety, durability and cost. For more information about expanded polystyrene and urethane, see p. 110.

For the interior and exterior skins, most manufacturers offer several options, and the best option is not necessarily clear. Luckily, the problems inherent in each of the options are not necessarily severe. The most important concern should be whether or not the manufacturer has tested the specific panel to meet the buyer's local structural codes. With this aside, there are good and better choices depending on local conditions and the requirements of the building. A house on the shore of Nantucket Island should be treated differently from a house in the mountains of Colorado. Large loads and long frame spans might suggest the strongest panel; enclosing a swimming pool or a hot tub might force the decision to turn on moisture protection. Remember, too, that while it is in the best interest of the manufacturer to make and sell identical panels to keep the price down, it is often in your interest to get panels made to your specification.

In general, the exterior skin should be strong, especially under compression. It should have good nail-holding capacity for the application of siding and trim. It should be dimensionally stable, not expanding and shrinking abnormally in wet and dry cycles. Finally, it should be a capable vapor barrier, helping to prevent moisture penetration. The usual options are chipboard, oriented-strand board and plywood. Inevitably, there are trade-offs. For strength and moisture resistance, marine plywood would be wonderful, but the cost is prohibitively high. Some of the prefinished plywood sidings, such as Texture 1-11, might be durable enough, but it doesn't make sense to use a great thermal panel with a poor thermal joint—it's almost impossible to have both the expanding foam seal necessary for optimum energy performance and good looks. Regular exterior plywood is not bonded as well as it should be and tends to delaminate. Because the panel leaves no air space between the plywood and the core, moisture gets in and the plies split apart. So that leaves oriented-strand board and chipboard. Chipboard is nothing more than compressed flakes of wood held together with glue. Oriented-strand board (OSB) looks like chipboard, but the grain of the wood chips is aligned and compressed into thin sheets that are then cross-laminated in a manner similar to plywood. The advantage of OSB and chipboard over plywood is that they can't delaminate. OSB is stronger than chipboard, has better nail-holding capacity, is about the same price and has therefore been our preference.

The interior skin should have good tensile resistance to loads and serve as an air/vapor barrier to keep moisture from penetrating the core from the inside. It should also serve as a fire barrier, isolating the panel core from the house interior, and must either be the interior wall surface or be able to support it. The alternatives are usually drywall, blueboard (plaster-base drywall), greenboard or fire-resistant drywall. When we use gypsum board as the interior skin, blueboard is our first choice because it is stronger than regular drywall and holds up well when rain gets to the site before we get the roof on. Blueboard also gives our clients the option of either plastering or painting. The exterior sheathings listed previously can also be used for the interior skin, with the finish interior surface then applied as another layer—we call these panels doublechips. Although more expensive, doublechips have several advantages. First, they're stronger and give a better surface for attaching trim, cabinets and other finish elements. Second, the greater strength of the interior skin helps keep the panel from warping when the exterior skin expands in high-moisture conditions. When we use doublechips, we

specifically request the interior skin be made with little formaldehyde in the glue. Don't forget that a combination of interior panel skins might also make sense, for example, where there are high-wind conditions but little snow, drywall panels can be used for the roof and doublechips for the walls. Or doublechips can be used just in the area where cabinets need to be hung and drywall panels used for all other areas.

There are over 30 different panel manufacturers across the country, most located on the east coast. All panel manufacturers use some variation of the techniques discussed later in this chapter. Read that before deciding on the type of panel that's appropriate for your job. Aside from the differences in manufacturing technique, there are a number of manufacturer-suggested panel-to-panel connections. Several of the most common are shown at right. Panel connection is just as important a consideration as the way a panel is made, because it's the connection at the seams that is really the weak link in the system. The seams are the areas most prone to infiltration and weakness, and also where expansion and contraction are most likely to show. But there's a prevalent tendency to overlook the critical fact that insulation must be tight to be effective. Just because the panels have been pushed together does not mean a seal has been made—in fact, in almost every circumstance,

**Plaster-base drywall is generally the interior panel surface. After the panels have been installed, interior finish, insulation and exterior sheathing are completed. The foam visible between panels is in the chimney opening, which will be cut out later.**

a good seal has *not* been made. You just can't expect 90 to 200 linear inches of sheathing edge to be joined without some pretty big gaps.

Ease of installation is a big concern for panel manufacturers, many of whom use molded joints or tongue-and-groove joints for precisely that reason. We've found these systems don't give the reliable seal we're looking for, so we use the first option shown in the drawing below. The panels come to us with square edges, which we then rout out to accept two 5/8-in. by 3-in. plywood splines. Our router head is designed to take out a little extra insulation at the foam edge between splines to create a cavity into

which we can spray expandable foam. This is done after the panels have been installed by drilling a ¼-in. hole approximately every 14 in. through the splines to accept the nozzle of the foam can. We fill all roof joints as well as the wall joints. In addition, where panels join over a timber we leave a deliberate gap of at least ¼ in. during installation, to be filled with foam later. We also foam at the base between the 2x8 and 2x12 sills, at all outside corners, at the eave and rake seams between wall and roof and at the ridge. It may seem like a lot of bother, but these extra steps make the difference between good and great performance.

## Some panel-splining options

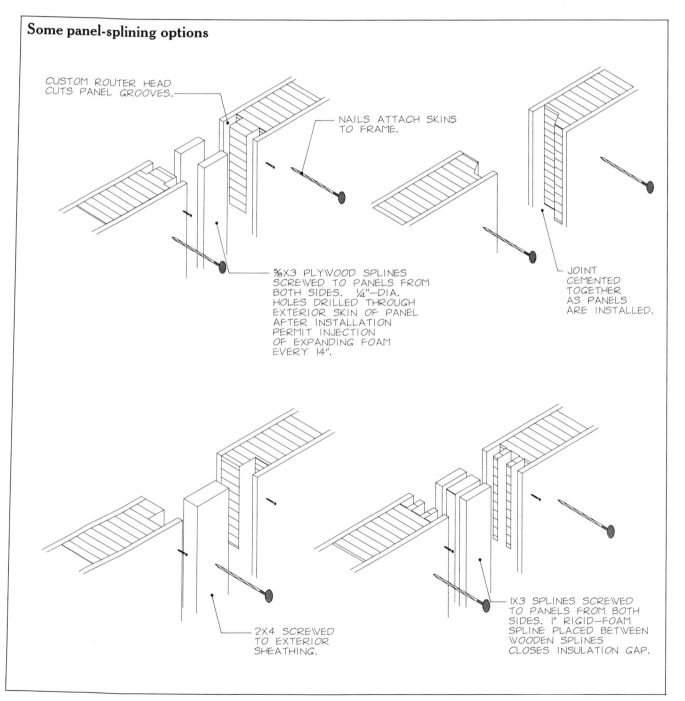

CUSTOM ROUTER HEAD CUTS PANEL GROOVES.

NAILS ATTACH SKINS TO FRAME.

⅝X3 PLYWOOD SPLINES SCREWED TO PANELS FROM BOTH SIDES. ¼"-DIA. HOLES DRILLED THROUGH EXTERIOR SKIN OF PANEL AFTER INSTALLATION PERMIT INJECTION OF EXPANDING FOAM EVERY 14".

JOINT CEMENTED TOGETHER AS PANELS ARE INSTALLED.

2X4 SCREWED TO EXTERIOR SHEATHING.

IX3 SPLINES SCREWED TO PANELS FROM BOTH SIDES. I" RIGID—FOAM SPLINE PLACED BETWEEN WOODEN SPLINES CLOSES INSULATION GAP.

# Panel installation

In theory, installing stress-skin panels is similar to installing other types of exterior sheathing except that the panels are thicker, heavier and a whole lot more expensive. To keep waste down and installation efficient, it's important to map out a strategy prior to getting started. Because panels are attached to the timber frame, the location of the timbers determines the final lengths, widths and cuts for the panels. Therefore, use a blueprint of the frame plan as the basis of the panel plan. You'll need a plan for each wall of the house and for the roof surfaces. Locate the rough openings for windows and doors using the architectural plans for reference—don't draw in any panels yet. The rough openings should be the same as those specified by the door and window manufacturers for stud-frame construction. All locations for openings should be referenced from the edges of the timbers, because the panel installers will have nothing else from which to measure.

After locating the rough openings, draw in the panels. Laying out the panels is a matter of juggling several considerations—strength, panel waste, rough openings and roof overhang. Here are a few guidelines before you begin.

First, check the requirements of the local building code and the loading characteristics of the chosen panel. Most manufactured panels are engineered to span 8 ft. to 9 ft. on the walls and 4 ft. on the roof. Be wary of a manufacturer who cannot provide documentation.

Second, you will not want to risk inordinate panel waste by forcing the spans to work out perfectly. For example, suppose the distance from the 2x12 sill to the centerline of the girt is larger than the 4x8 panels. You could certainly cut the 4x16 panels to fit, but you'd be creating a lot of debris. A better solution would be to position the 4x8 panels from where they had adequate nailing on the box beam to where they had adequate nailing on the girt, keeping in mind that 2 in. is the minimum for good nailing. To do this, you would snap a line along the face of the box beam and nail a ledger to the line to support the panels during nailing. Then you would fill in between the 2x12 sill and panels with offcuts.

The location of the timbers in the frame determines the final lengths, widths and cuts for the panels.

When the panels will not reach entirely from the sill to the first beam, nail a strap to the box beam to set a baseline. Panel offcuts are used to fill in over the box beam.

Third, panel joints that fall between the edges of rough openings should be reinforced with a 2x4 used as a spline. (I use 2x4s generically here; of course the nailers would be different dimensions if the panels were thicker or thinner.) Window openings that are wider than 3 ft. 6 in. and taller than 5 ft. and all door openings should be reinforced with a 2x4 on either side that runs from sill to girt. Smaller windows can be cut directly into the panels without a need for this reinforcement. If the window is small enough, try to locate it toward the middle of the panel, leaving 6 in. to 8 in. on either side. But if the window has to be closer to a panel edge, move the panel edge to the edge of the rough opening to put the joint in a place where it can be strengthened with a 2x4. In other words, you don't want just a little bit of insulation between the edge of the rough opening and the edge of the panel because this would create a weak area.

Fourth, non-supported overhangs on the rakes should be limited to 1½ ft. beyond the wall panel and 1 ft. should be the limit at the eave. Using doublechip panels allows an increase in overhang to 2½ ft. on the rakes and 2 ft. at the eave. (If you like, you can run the doublechips along just the roof edges and use standard panels for the rest of the roof.) Some trim details support the eave panels, making it possible to use greater overhangs with the standard panel, as discussed in Chapter X. Doublechip panels are also useful in high-load situations, because the nail-base sheathing is so much stronger in tension than drywall. When using doublechips, you can avoid infilling the interior finish sheathing by nailing it to the outside of the frame before panel installation.

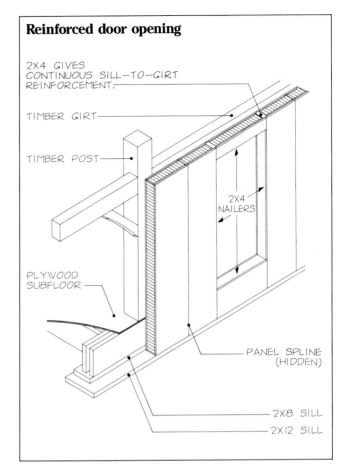

## Reinforced door opening

2X4 GIVES CONTINUOUS SILL—TO—GIRT REINFORCEMENT.

TIMBER GIRT

TIMBER POST

2X4 NAILERS

PLYWOOD SUBFLOOR

PANEL SPLINE (HIDDEN)

2X8 SILL

2X12 SILL

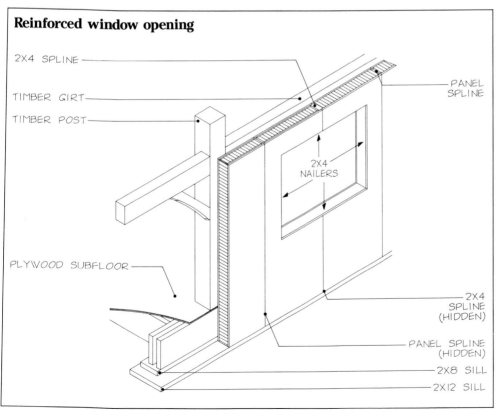

## Reinforced window opening

2X4 SPLINE

TIMBER GIRT

TIMBER POST

2X4 NAILERS

PANEL SPLINE

PLYWOOD SUBFLOOR

2X4 SPLINE (HIDDEN)

PANEL SPLINE (HIDDEN)

2X8 SILL

2X12 SILL

Here are some other things to consider when laying out the panels on the plan: Panels should run vertically on the walls, unless the spacing of the posts is closer than the spacing of the horizontal girts and plates. Next, plan to fill in above and below window and door openings with panel pieces. Work hard to have the tapered factory edges of the drywall meet at all seams, though you won't be able to achieve this everywhere. Use the same procedure on the roof, laying out full panel sections on either side of roof, window or chimney rough openings. You will have to juggle all these factors in an attempt to reduce waste. Dot in the panels over the frame plan, making sure each panel and offcut is clearly marked.

Once the plan is drawn, the panels can be cut and installed. During these stages, try to satisfy the requirements of all three panel components. Interior sheathing must be handled with proverbial kid gloves. The foam core is tougher, but it must be well sealed at all cracks and crevices before installation is finished. The exterior skin is primarily a nail base and you must make sure it provides solid, dependable nailing for all materials applied. Good installation procedure requires vigilance on all these fronts at once.

**Information and layout example from a typical panel plan**

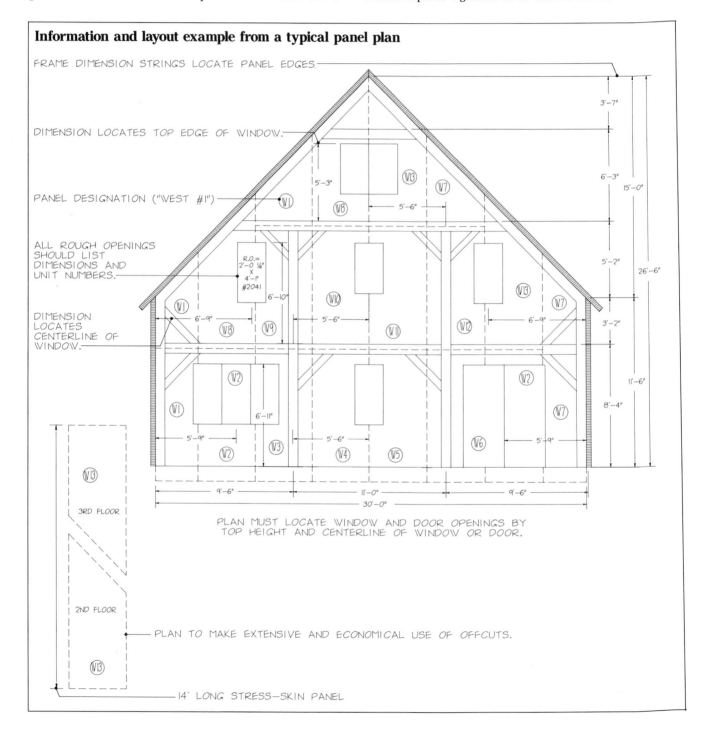

Although it's not a rule by any means, we usually attach the wall panels first. It is much easier to lift the wall panels into place without the roof panels in the way, particularly if an overhead winch or crane is used to help with lifting, as is typically the case. There are also some situations in which it would be extremely difficult to fit the wall panels under the eave panel and over the 2x12 sill simultaneously. If the eave panels extend over the upper plate or girt, nailing the top edge of the wall panel would be difficult, if not impossible. But the obvious disadvantage of installing the wall panels first is that they will be unprotected and thus vulnerable to the elements until the roof is on. You can see the need for good planning and tight scheduling.

Begin by marking panel breaks and the rough openings for windows and doors on the outside face of the sill and beams. Check that openings coincide with the frame as depicted on the plans. Then string a line along the eave to accurately represent the eave intersection. Using a chalkline or a large *T*-square, draw lines on the panels for the rough openings and all other cuts. Check all measurements before cutting. We usually lay out and cut the wall panels one at a time and precut the roof panels in preparation for crane assistance. The benefit to this system is that you can check your work on the walls as you go, whereas the possibilities of making a mistake on the roof are pretty slim, as there are usually few openings. Always use a mask and goggles when cutting. Panel dust and chips are inorganic and ugly, and should be cleaned up often. My most disturbing experience with our panel debris was when a client's beautiful cranberry bog was fouled with a frost of panel dust. The house needed insulating, not the bog.

## Tools for Installation

Panel installation requires only a couple of specialized tools. They are:

- a circular saw capable of cutting through the thickness of the panels, usually 4½ in. to 5 in.
- equipment to cut routs for the nailers and splines. Urethane foam is cut with specially made router heads that fit on portable grinders. We use two grinders so we don't have to change the heads very often. You should expect your manufacturer to help you obtain equipment compatible with the system you choose.

If the building is large and your crew is small, you may also want to use an electric winch to lift the panels to the higher areas, useful if the building is large and the crew small. If you choose to precut the panels and use a crane to lift them into place, a crane with an operator can usually be rented on a daily basis.

Other useful tools for panel installation are generally found in the typical carpenter's cache:

- a drill for wires.
- a large *T*-square.
- a power screwdriver for driving drywall screws.
- a bayonet saw for round openings and notching cuts.
- ladders long enough to reach the top of the frame.
- a good assortment of hand tools: hammer, chalkline, framing square, etc.

Wall panels are laid out and cut one at a time, after double-checking measurements and layout lines. The window opening in this panel has been routed to receive a 2x4. (Photo by Sarah Knock.)

## Panel detail at eave

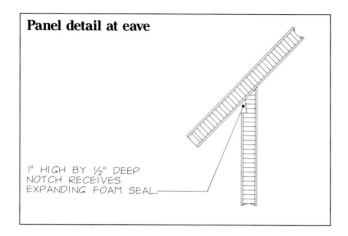

1" HIGH BY ½" DEEP
NOTCH RECEIVES
EXPANDING FOAM SEAL.

## Panel detail at corner (plan view)

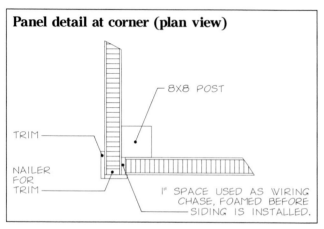

8X8 POST

TRIM

NAILER
FOR
TRIM

1" SPACE USED AS WIRING
CHASE, FOAMED BEFORE
SIDING IS INSTALLED.

The panels are heavy, so make all the cuts in a panel at the same time. Panels that will span the full height of the wall should be sawn about ⅜ in. short to make room for the foam. If your saw isn't capable of cutting through the thickness of the panel, cut in halfway from each side—any remaining foam in the middle will break off easily. Tilt the saw base to cut the roof angle. Then cut a notch in the top outside edge about 1 in. high and ½ in. deep for the expanding foam. With the grinder and spline-cutter head, rout all seams that do not fall on posts; panel edges falling on posts are sawn to allow a ¼-in. gap at the seam for the foam. The perimeters of all rough openings for windows and doors are routed to accept nailers (usually 2x4s)—do this with the nailer-cutter head in the grinder set to full depth. Also rout the outside edge on each corner for a nailer. Where the panel edges need reinforcement, as with wide windows (p. 103), the nailer head is set to one-half the width of the nailer, so the nailer can bridge the joint between panels.

After the panels have been cut and while they're still on the sawhorses, install and secure the splines to one edge of a pair of mating panels. Attach the splines first on the exterior side with 1½-in. galvanized screws and then use regular drywall screws on the other side. Screws should be on approximately 8-in. centers. The nailers on the corners of the house may be put in while the panels are on the ground, but don't put any nailers in window and door openings until the panels are installed, because you want the nailers to span the panel seams for extra reinforcement.

Before positioning the panels, start the nails through the exterior sheathing. The nails used depend on the frame

All seams that don't fall over timbers are routed using a grinder with a special spline-cutter head. Goggles and a mask should be used during cutting. (Photo by Sarah Knock.)

Before the panels are lifted into position on the frame, nails are started in the exterior sheathing. In this oak frame, spiral galvanized nails were used to avoid a tannic-acid reaction with bare metal. (Photo by Sarah Knock.)

material. For softwood frames, use an annular ring (pole barn) nail that will penetrate the timbers by at least 2 in. For hardwood frames, use spiral galvanized nails long enough to penetrate the wood by 1½ in. These should be on 8-in. to 10-in. centers around the perimeter. After starting the nails, apply a bead of construction adhesive onto the faces of the posts, beams and 2x12 sill (or face of the box sill) that the panel will be attached to. Starting with an outside corner, lift each panel into place. Slide the panels together horizontally. When installing the top panels in the wall, lift them to the eave string by prying with a flat bar. Drive in the nails.

After the first panel on each wall is in position, you will need to drive subsequent panels together to engage splines. Sometimes it is necessary to do this with a mallet, protecting the panel edge with a block of wood. If the pan-

el length is not greater than 10 ft., it is often easier to drive the splines from above rather than mating the splines edgewise, another reason to put on the wall panels first. In places where panels are stacked on top of each other, use ¼-in. plywood spacers to create a cavity for foam. Make sure to allow at least 2 in. of nailing surface in the timbers for the panel edges. Gable-end wall panels should be shy of the roof panels by the same amount for foaming.

Here the spline is being driven from above. If the panel were over 10 ft. long, this maneuver would not be possible.

These two panels were attached to the frame without pre-installed splines. With the splines in place, the panels often have to be driven together. (Photo by Sarah Knock.)

In this window opening, which has been routed to receive 2x4 nailers, the splines are seen in place. They are screwed to the interior and exterior sheathing. The gap between the splines will be filled with expanding foam later.

Before installing the panels on the roof, snap a line on the purlin or rafter that will be at the first panel seam. Following the line will help ensure that the eave or rake will be straight for the trim. Place a bead of construction adhesive on the roof timbers and the top of the beveled edge of the eave panel prior to placing the roof panels. With the nails already in position, attach the panel when it is aligned with the chalkline. Most roofs require few spline joints because the timbers are on 4-ft. centers. Space the panels ¼ in. from each other with plywood spacers, creating a cavity for foam. If you choose to ignore my instructions about foaming the wall seams, don't carry the illogi-cal thinking up to the roof, for you'd certainly be disappointed with the resulting energy performance.

When all the panels are in position and completely nailed to the frame, install the nailers in all the window and door rough openings and around the perimeter of the roof. Use long, straight material for the roof nailers and set their edges precisely to a string line. This will make installing the roof trim infinitely easier. Secure all nailers with 8d galvanized nails from the exterior skin and screw with drywall screws from the interior skin. Also make sure all the spline joints have been properly screwed and nailed.

The first roof panel is laid to a chalkline snapped on the frame. Roof-panel edges are routed for a nailer, which is installed when all the roof panels are in place. Note the notch in the wall panel to enable foam to make a good seal with the roof panel.

Nailers are fixed to the window and door openings and to the roof edges after all panels are positioned. Regular drywall screws are used for the interior sheathing and galvanized screws for the exterior. (Photo by Sarah Knock.)

Roof panels are usually lifted into place with a crane. Gaps are deliberately left between roof panels for expanding foam. (Photo by Sarah Knock.)

Now fill all the roof and wall seams with expanding foam. At each spline joint, drill a ¼-in. hole approximately every 14 in. to insert the nozzle of the foam can. Work your way up, not down, the joints. Keep the cans warm and experiment on some scraps before you start because climatic conditions affect foam expansion. We have found that a two-second count on a 14-in. spacing generally fills to the next level quite nicely, but be prepared to adjust the spacing and the count as necessary. After the foam has completely expanded and hardened, scrape the excess from the surface.

**Foam is injected into all seams. Most roof seams are open, since they usually fall on timbers. Seams that fall away from timbers must be splined, and therefore have to be drilled before foam can be injected.**

## Urethane and polystyrene panels

Urethane and polystyrene are alike in that they both are rigid-foam plastics used for insulation. Thereafter, they are mostly different from each other. Following are some of the advantages and disadvantages of each and a discussion of manufacturing techniques.

### Advantages of expanded polystyrene (beadboard)

**1.** It is easy to manufacture. Polystyrene beads are compressed together under heat and pressure to make a large block, or "bun." The bun is then slabbed with a hot knife to the required thickness.

**2.** Beadboard is light and easy to work with.

**3.** Since hot wires are used instead of a router to make cuts, there isn't a problem with dust.

**4.** Polystyrene accepts various glues and mastics, and generally bonds well to the skins.

### Disadvantages of expanded polystyrene

**1.** Because beadboard panels are so easy to make, too many people are doing it. Panels fabricated under the apple tree are sold alongside those made in well-tooled facilities with good quality control.

**2.** Polystyrene is a hazard in high heat and fire. It loses its structural integrity at about 180°F and melts at 300°F. In a fire test I witnessed, heat from the fire quickly melted the beadboard into a volatile liquid. My conclusion is that a "Type X" (half-hour fire rating) drywall product should always be used on beadboard panels.

**3.** Expanded polystyrene panels have potential for radical molecular creep. Dr. Tim Johnson, a technical researcher at Massachusetts Institute of Technology, reports that over a 16-ft. span loaded to 40 lb. per sq. ft. for 90 days, beadboard panel samples shifted 6 in. because of creep. This is a big load and little time; I worry about small loads and lots of time.

**4.** Most panel manufacturers keep costs down by using 1-lb. instead of 2-lb. density beadboard. Yet a 2-lb. density foam would give better compression strength, a higher R-value and a lower perm rate. Beadboard panels are generally the least expensive, but not because they are on sale.

**5.** Low R-value per inch means more inches of thickness are necessary for good performance. The lower price of the material is offset when the R-values are equivalent. Window and door jambs and roof edges require more lumber for framing and finish.

### Advantages of urethane

**1.** Urethane has a very high R-value. A panel made to a typical 2x4-stud thickness has an R-value of about 24 to 28.

**2.** Urethane is typically manufactured at a 2-lb. density. When properly formulated, it is stiff and resilient.

**3.** Because the panels are fairly thin, they are relatively easy to install. Nailers needed around windows and doors and at roof edges can be 2x4s instead of 2x6s or 2x8s.

**4.** With a perm rate of 1 or less per inch, the urethane itself acts as a vapor barrier.

**5.** Panels can be made either by laminating skins to slabbed bun stock or injecting liquid urethane between the skins (called pouring). Panels made from other materials can only be laminated.

**6.** Good urethane formulations will not support flame. However, they will eventually burn if the fire is fueled by another source.

### Disadvantages of urethane

**1.** Urethane foams are complex. There are hundreds of different formulas and both the manufacturing environment and technique must be precisely monitored. It is rather like Grandma's cake recipe: Leave out one little ingredient and the cake tastes terrible; open the oven for a peek at the wrong moment and it falls flat; and only Grandma knows how to make it right anyway. An increase or decrease in the blowing agent (usually freon) that causes the foam to expand can greatly affect density. If the temperature is too low during manufacture, the foam could be too brittle. There is no generic foam; physical properties change greatly from one company to another. If your supplier hasn't heard of words like isotropic, polyol and isocyanate, you are probably purchasing from the wrong company.

**2.** Although some formulations do not support flame, most urethane is flammable and the smoke is toxic.

**3.** Urethane loses some of its R-value as the gases are lost from the cellular structure. Over a period of four or five years, the panel could lose 5 points of R-value. Again, this depends on the particular formula and manufacturing system.

## Polystyrene panel construction

Polystyrene panels can be made only by laminating the skins to the slab stock. It's a simple process that helps keep the price down. This panel can literally be made by spreading glue over the surface of the foam, applying the skins, and putting weight on it while it sets.

## Laminated-urethane-panel construction

The advantage of lamination is that it isn't technologically as difficult as pouring and the slabs of foam can be pre-cured, meaning the foam should be dimensionally stable and the initial outgassing of propellant should be completed. On the other hand, pouring eliminates the potential deterioration of the adhesive and is much quicker. But because pouring technology is still new, the most important thing for consumers is to buy from reputable manufacturers.

As with polystyrene panels, a laminated urethane panel can literally be made by spreading glue or contact adhesive over the foam surfaces, applying the skins and putting weight on the assembly while it sets. Contact adhesive is used for speed; glue is used when finer technology does not exist. One company I visited used a corps of high school kids to construct giant piles of panels. The kids would mop glue on a skin, place a foam slab on the top, spread more glue and another skin, and so on, until they could no longer reach the top and were ankle deep in glue. Unused buckets of glue were then placed on top to hold down the mountain of sandwiches. The panels actually weren't bad, and though the company went out of business, it wasn't for want of glue.

I have a homespun test for panels. I put samples measuring 4 in. square on the wire cage in a canning pot with several inches of water, and bring the temperature up to 155°F. I leave it there for 12 hours—100% humidity at 155°F, and while not as harsh as what time and weather could do to a panel, it's still a pretty good test. The glue-mopped panels described previously survived the test well, but samples made with contact adhesive failed miserably.

The disadvantages of laminated panels are several. For example, pre-curing often causes tension to build up in the bun stock, resulting in warping. In addition, cutting urethane slabs from the buns leaves a film of dust on each surface because the cellular structure is somewhat brittle. Trying to bond a material to a dusty surface has predictably poor results. In one of the assembly-line gluing processes, the panels are run under a high-pressure pinch roller while contact cement is setting. The pinch roller crushes the cells on the slab surface, making even more dust. It is no wonder that this type of laminated panel almost always falls apart in my canning-pot test in less than four hours. More disconcerting than my private laboratory conclusions have been those occasions when I have found scraps of the material left on the building site that were completely delaminated after only a few weeks of exposure.

## Poured-urethane-panel construction

There are two ways to pour panels, vertically and horizontally. In vertical pouring, liquid urethane is injected between two skins set on edge in a form: Vertical simply means that the 4-ft. by 16-ft. or 4-ft. by 8-ft. sheets are standing on edge with the 4-ft. side up. This system requires simple technology and relatively inexpensive equipment, but it is necessarily slow. The skins are placed in the form manually and the foam has to rise 4 ft. before the panel is completed. Because the form has no top, vertically poured panels often show uneven cell structure and therefore would exhibit uneven strength and stability properties.

A bigger disadvantage is that the foam cells rise vertically, and tend to be elongated in the same direction. Considerable compressive strength in the flat direction is lost because of this cellular shape. Think of the cells in the foam as a cluster of eggs. We know eggs to be much stronger in length than width and if they were all lined up lengthwise, this cluster would be easy to crush. When cells are elongated like this, the foam is said to have anisotropic properties, resulting in more strength in the direction of the rise than perpendicular to it. To prevent anisotropism, the foam must have more than 2-lb. density, must be made using the proper amount of blowing agent, and both the urethane and the manufacturing environment must be at the correct temperatures.

In horizontal pouring, liquid ingredients are sprayed between two horizontal skins, one held at the bottom of a form, the other at the top. The foam is thus required to rise only the width of the panel, 3½ in. to 5 in., instead of 4 ft. Cell rise and formation occur in the same direction in which compressive strength is required. Because the foam rise is measureable and consistent, it is possible to make precise adjustments, achieving an isotropic foam and a panel with an excellent bond.

Horizontal pouring also lends itself to automation. Panels can be made on a belted assembly line, giving some assurance that the price will stay reasonable.

# VI. Getting Out of the Ground

**A timber frame built to last hundreds of years should stand on a foundation designed with the same commitment to longevity.**

Construction of a house begins with bulldozers and ends with 220-grit sandpaper, proceeding from rough to fine, from larger to smaller tolerances. It's a sensible progression. But an unfortunate corollary is the thinking that brawn is needed at the beginning of a job and brains at the end, that it's okay for dirty work to be accompanied by muddy thinking. The roar of heavy machinery, the splash of concrete and the crude piles of dirt are all uninviting, and tend to keep the project planners on the periphery until the "good" work begins. In addition, excavation and foundation work contribute little to the aesthetics of a house, so people are generally less concerned about these stages of the job. This attitude isn't surprising, but it is short-sighted, because the longevity of a building is directly related to the way it sits on the earth. A foundation is a one-shot deal: Once it is in, the die is cast. The building either has a chance to endure or it doesn't. If brains and craftsmanship are applied nowhere else in the house, they should be used to "get out of the ground."

When foundations are taken for granted, their designs are produced for the sole purpose of satisfying code requirements, and the job is usually contracted on the basis of price. On the site, critical decisions are often made by people who are not in a position to know all the criteria that should bear on design and construction. My own experience concerning faulty foundations includes a lesson or two learned the hard way. The excavation for one of our early projects in northern Massachusetts began in the late fall, well into the time when frost penetration has become a factor to consider when preparing a site. It was my job to lay out the foundation and set the benchmark (the reference point for the top of the wall). Because the site was 70 miles from my home, I left it to the homeowner and the subcontractors to see to it that the work was performed according to our plans and specifications.

On a bleak and forbidding day, a man and a bulldozer worked without enthusiasm or guidance. When they left, the area for the house was roughly excavated, but the tough digging through hard soil and cold weather resulted in an inaccurate job of grading the base of the excavation in preparation for building the footing forms. Nobody thought to protect the newly exposed earth from frost. When the crew showed up to build the footings two days later, they were faced with hard frost, hard soil and uneven terrain. What to do? I can hear them talking: "Why spend the rest of the day working like convicts with picks and shovels because of Bull Dozer's mistakes, when what we need most of all are warmer derrières and fatter wallets?" So they set the form boards without regard for anything or anybody: not for our plans, for the client's house, for the building code, for the laws of physics—not even for their own reputations. At its best, the footing was fully 12 in. deep, as called for in the plans. At its worst, the "footing" was only a 1-in.-thick concrete skim; most of it was between 2 in. and 6 in. thick. These guys had simply nailed some form boards together on top of the uneven, frozen soil and leveled out the concrete in between. To make the story even more tragic, in the next two days another crew came in and poured a wall on this footing. I discovered the crime only after the forms had been stripped. When I went into a rage about the disaster, the owner shrugged, the wall crew pointed to the footing crew, the footing crew pointed to old Bull Dozer, and he pointed back at me. Needless to say, we did not build until the problems were resolved, and I gained some of what they call experience.

In timber-frame housebuilding, the overall high standards and unique structural considerations demand that particularly close attention be paid to both foundation design and construction. Don't assume that foundations typically used with stud construction will satisfy the requirements of a timber-frame building. Unlike a stud-framed house, in which building loads are evenly transferred through hundreds of closely spaced members, a timber frame exerts point loads on a foundation through a few principal posts—between 12 and 16 in a medium to large building. The design of the foundation and the placement of the reinforcing rod within it are geared to coping with these point loads.

In this chapter I'll discuss some basic foundation considerations, as well as the modifications to conventional foundations required by timber frames. I won't go into detail on general foundation information that may be obtained elsewhere. I'll finish by discussing stick-frame and timber-frame floor-deck framing systems.

## Foundation design

Soil, climatic and geologic conditions, structural and design features of the building, and grades of the site are important factors in foundation design and are some of the reasons why foundations are not the same everywhere. Building on a delta in Louisiana suggests an entirely different approach from building on a hilltop in Vermont. If the house is to be built in California, it must be designed for earthquakes; if in Minnesota, for deep frost. As in buildings, in foundations there are no standard solutions. Instead there is a vast range of methodologies and options to be weighed against house design and site conditions.

**The soil** The real foundation for all buildings is the earth. What we put between the structure and the earth is intended to transmit the load, not support it. Support, or the lack of it, comes from somewhere in the ground beneath the building. So that the sighs and groans of the earth will not destroy the foundation, the building has to be placed on terra firma, never on loose soil or new fill because settling will cause the foundation to crack. You have to dig down to fully compacted, undisturbed earth, and once you get there you have to determine what it is composed of and the best way to build on it.

The quality of the soil is critical in foundation design. Soft soils require that the foundation have a larger bearing area (which means a broader footing) than hard soils or rock would need. The chart at right lists some types of soils and their bearing values. Solid igneous rock has a

bearing capacity of 25 to 40 tons per sq. ft. or more, while soft clay or loose sand might be able to hold only a ton or less. Because they hold water, clay and silt are the bane of good foundations and have to be treated extremely carefully. The potential for excess moisture in these soils reduces their ability to compact and allows them to expand if the moisture freezes. Loose sand, which is difficult to compact, is a problem, too. A mixture of materials is best, such as a gravel having enough clay to allow it to compact well and enough sand to allow moisture to percolate through. Solid bedrock can be great under a foundation, but it, too, has some special requirements. First, a rock surface is not generally tidy and level, so its peaks and valleys have to be carefully cleaned of dirt so that the footings won't shift or slide. Second, holes have to be drilled and rebar inserted to link the foundation to the rock. Third, if you hit rock in one corner of the foundation, you should usually dig to rock for the entire excavation, so the entire foundation will be on the same type of material. (The exception to this is when compactable soils are in a rock basin, because there the soil is contained and can't shift.)

In most cases, general local soil conditions have been well studied and standard foundation specifications for various circumstances are established as part of the code. But don't fall into the trap of allowing a standard specification to fit your non-standard situation—foundations should be specific, not generic. In an urban area, you can learn about soil conditions by looking at the foundations of adjacent buildings. In the country, use the information gathered for the percolation tests done in preparation for designing the septic system. Better yet, attend the tests to see for yourself the soil the foundation will sit on. You don't have to be an engineer to use common sense. If you can make a mud ball that holds together in your hand, the soil contains quite a bit of clay. If you can wring water out of that mud ball or if water fills the test hole on a dry day, an engineer should be consulted before the foundation is designed. Designers and architects sometimes neglect the bearing capacity of the soil; if you have one involved on

| Bearing Value for Some Soils | |
|---|---|
| **Type of soil** | **Approx. load in tons per sq. ft.** |
| Soft clay | 1 |
| Loose, medium sand | 1 |
| Compact, fine sand | 2 |
| Coarse gravel | 4 |
| Hardpan or partially cemented gravel over rock | 10 |
| Sedimentary rock, including sandstone and conglomerates | 20 |
| Solid bedrock: granite, slate, gneiss | 25 |

your project, question him or her. Frank Lloyd Wright never visited the site of his house known as Fountainhead. The building was nearly torn apart by soil expansion and needed complete restructuring in a recent renovation.

**Climate and geology**   Most of the very early timber-frame buildings in this country had foundations that were inadequate for New England winters. Having come from a much more temperate climate, the colonists simply did not know how to deal with the deep penetration of frost that moved and heaved whole buildings. Primarily for this reason, there are few timber-frame buildings left from the first years in this country, although many examples have survived from the 14th and 15th centuries and earlier throughout Europe. Putting a building on top of the ground in harsh New England weather is almost like launching it out onto the rolling sea.

The most obvious concern is freezing, because freezing soil will heave even the heaviest buildings and crack even the most massive foundations if the footing is not designed to penetrate deeper into the ground than the frost. In other words, the footing must be on soil that will never freeze. Local building codes generally have reliable information and good regulations regarding footing depth. Some extremely cold regions demand a depth of at least 6 ft., while some warm areas require only 1 ft. or so. People from the South often wonder why so many New Englanders have full basements, thinking perhaps their breeding encourages things subterranean. Actually, it's a matter of practicality. Because they have to dig 4 ft. to 6 ft. anyway, it often makes sense to go a few extra feet and gain living or storage area.

Foundations are also designed in anticipation of certain natural calamities, such as earthquakes, tornadoes or hurricanes. Local codes in areas prone to such potential disas-

ters usually give good guidance. For example, buildings in quake zones require foundations that are heavily reinforced to resist shearing forces. The idea is that the foundation walls should act as a set of rigidly interconnected beams; in the event of an earthquake, the whole building should move as a unit rather than separating or cracking. I've never experienced a significant earthquake, but people who have been through one tell me that no amount of reinforcement seems like enough in those hair-raising moments of turbulence.

In preparation for earthquakes or high winds, foundations also need a good system for keeping the house firmly attached to the foundation. This is not an overreaction to Dorothy's trip to Oz: Every year there are stories about

Lack of concern by the excavator and foundation subcontractor caused this house to be perched unnaturally high on the ground. Hundreds of yards of fill were trucked in to correct the problem.

**For the design potential of this long, low house to be fulfilled, the building needed to nestle into the ground.**

buildings that are blown off their foundations. Some attachment systems are described later in this chapter.

**Building design and site conditions** In addition to responding to the site and climate, the foundation must also meet the design and construction requirements of the building. For example, if the house is intended to conserve energy, so should the foundation. If the house has any peculiar loading situations, they need to be directed through the foundation. The foundation is the building's roots, not just a mass of masonry beneath it. Also, how the foundation is treated can drastically change the look and feel of a building. A Victorian home perched on poles would be an absurdity, but elevated on a foundation with a brick facing,

it could appear lofty. On the other hand, a traditional design perched on a tall foundation can look unconnected to its environment.

Foundations very often are stepped to follow the contour of the land. When houses are well designed, finished grades, terraces and a landscaping scheme are all addressed at the same time as the foundation. Homes built on hillside sites are commonly designed to take advantage of the terrain by gaining living area and outside access at the lowest grade level.

**Some general notes** As a preface to the discussion of the various foundation systems that follows, here are a few personal prejudices.

**With good siting and foundation design, this house fit nicely into a hillside. (Photo by Tafi Brown.)**

I favor a system with a full perimeter wall rather than one with slabs or piers. If buildings are going to be well protected from the elements and well insulated, it is practically mandatory to begin with a good seal at the base. In addition, a perimeter foundation takes advantage of the relatively constant temperature of the earth (about 55°F), making the building much easier to heat than when cold exterior air is rushing beneath it.

When using a perimeter wall, I dislike shallow crawl spaces. They are dark, dank and forbidding. People just don't like getting down on their bellies and crawling under houses, so this area usually gets little attention and maintenance. With not enough room for good air movement, mildew is encouraged, along with the subsequent deterioration of framing members. If there is going to be a space, it should be large enough for people to move around freely, and the ceiling should be high enough to allow comfortable walking. A few whacks on the head are discouraging, and as a result a perfectly good cellar seldom gets cleaned and things get dumped but never organized. I know this for a fact.

I like wood, but for first-class homes it seems a risky choice for the foundation material. There is a rising interest in pressure-treated wood foundations, based on the arguments that wood is just as strong as concrete, offers better insulation and is quick to install. But even if all of this is true, there is insufficient data about the life expectancy of pressure-treated wood. Published figures seem to place it at between 50 and 80 years. If it were advertised to last twice that long, it still wouldn't sound right to me. Especially for timber-frame buildings, foundation life spans should not be this limited.

I like poured concrete—a mixture of portland cement (lime, silica and alumina), aggregate and water. The concrete we use today is similar to the material first used by the ancient Romans. Some of their concrete buildings and roads have survived, demonstrating the durability of the material. When the Roman Empire fell, so did this wonderful recipe. It wasn't rediscovered until the late 1700s and was seldom used until the early part of this century.

Concrete is, I think, one of the most important developments in the building field. Because it is reasonable in cost, concrete makes durable buildings possible for everyone. It can be molded to fit even the most irregular contour, it hardens to a degree we can calculate and control, and it gives strength and design flexibility with few limitations. The timber-frame buildings we construct today have an opportunity to exceed the expected lifespan of those made by our predecessors in the craft, not because of any greater skill in joinery but because of better foundations. Concrete is the key.

The richer the concrete is in cement, the stronger it will be. Density is increased by adding more cement to the mix. Common concrete for houses is rated at 2,500 psi; exotic new formulations have made 15,000 to 20,000 psi possible. For our timber-frame homes, we have established 3,000-psi concrete as the minimum standard for most situations.

Properly mixed and poured, concrete is very strong in compression. A poured concrete wall does not have the shear planes inherent in walls stacked from stone, brick or concrete block (p. 119), so it has much more resistance to lateral loads. But concrete is not an elastic material and is fairly weak in shear strength. Strength can be increased many times, however, by placing metal bars (rebar) or wire mesh as reinforcement into the forms before the concrete is poured. These bars come in sizes designated by numbers 2 through 8. Each number represents one-eighth of an inch. A #2 bar is two-eighths, or ¼ in.; a #5 bar is five-eighths, and so on. For residential construction, #4 and #5 bars are the most common. Wire mesh is employed mostly for flatwork such as slabs, walks and patios.

Obviously, I am totally in favor of improving foundation standards and making the foundations for our houses better. Better usually means more difficult and more expensive, and I offer no slick techniques for circumventing this. I'd rather talk about how to save money on the roofing material or the siding or the cabinets—items that can be more easily upgraded later. To me as a building professional, the only risk in having a first-class foundation is that it may last too long, and I know of worse gambles.

## Continuous footings

Most foundations are constructed in two parts: footings first, then walls or piers. The footings allow the building load to be distributed to the earth over a larger area, as well as providing a more or less level surface for forming the rest of the foundation. A continuous footing is used for a perimeter-wall system. The footings for pier foundations are discussed with the systems they support, although many of the rules for continuous footings apply to these footings, too.

The width and thickness of the footing are determined by analyzing the weight of the building (including the concrete walls), the way loads are distributed by the structure, and the type of earth upon which the building will sit. Most codes have some general standards for footing size, concrete density and reinforcement. But these regulations should be seen for what they are—a minimum standard—and should not be applied blindly to every circumstance. Verify the situation by adding up the load that will be dis-

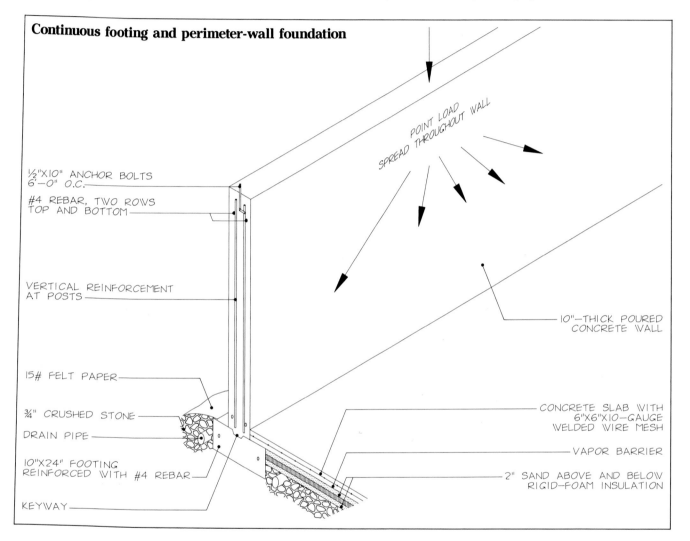

**Continuous footing and perimeter-wall foundation**

POINT LOAD SPREAD THROUGHOUT WALL

½"X10" ANCHOR BOLTS 6'—0" O.C.

#4 REBAR, TWO ROWS TOP AND BOTTOM

VERTICAL REINFORCEMENT AT POSTS

15# FELT PAPER

¾" CRUSHED STONE

DRAIN PIPE

10"X24" FOOTING REINFORCED WITH #4 REBAR

KEYWAY

10"—THICK POURED CONCRETE WALL

CONCRETE SLAB WITH 6"X6"X10—GAUGE WELDED WIRE MESH

VAPOR BARRIER

2" SAND ABOVE AND BELOW RIGID—FOAM INSULATION

tributed to each timber post. It will probably fall in the range of 3 to 5 tons at each location, including the weight of the foundation. Compare this with the bearing capacity of the soil. If the soil is soft, the footing may have to be broader and thicker to distribute the load over a larger area.

A typical continuous footing in timber-frame construction is between 8 in. and 12 in. thick and between 16 in. and 28 in. wide. Footings should have some reinforcement and a keyway (a groove along the length of the footing) to ensure a better bond with the wall. For a timber-frame building having normal loads on a site with good soil, our standard continuous footing is 10 in. thick by 24 in. wide, reinforced with two #4 bars and poured with 3,000-psi concrete. We always specify a keyway, and sometimes we even get one. Any questionable circumstances warrant a session with an engineer.

Footings must be accurately positioned and roughly level. Other than this, building the forms for them is not necessarily precise work, although it can require creativity because the rough excavation often doesn't offer an accurate base. Remember that the real foundation is the earth, and the footing must be on undisturbed ground. I once visited a project while footings were being formed and found a "professional" filling in the footing area with loose soil from the excavation to make leveling the form boards easier. His surprise and anger that I would question him about this gave me much to think about. Obviously, I was criticizing his standard practice. Such people doing some of the most important work in the house is an unnerving reality.

Footings can be poured completely in the ground, partially in the ground and partially out, or entirely on the excavation, as shown at right. If the footing is continuous, it is better to pour it on the excavation to leave room for drainage and/or insulation. The base of the footing should be fairly level across the width as well as in length—an uneven base is prone to breaking, which could reduce the bearing area of the footing to that of the wall and render the footing useless. A sloped grade could cause the footing to slip, so footings on slopes should be stepped to keep the bearing area flat. Footing forms are usually made of 2x framing stock. Wooden stakes and straps secure and stabilize the form boards.

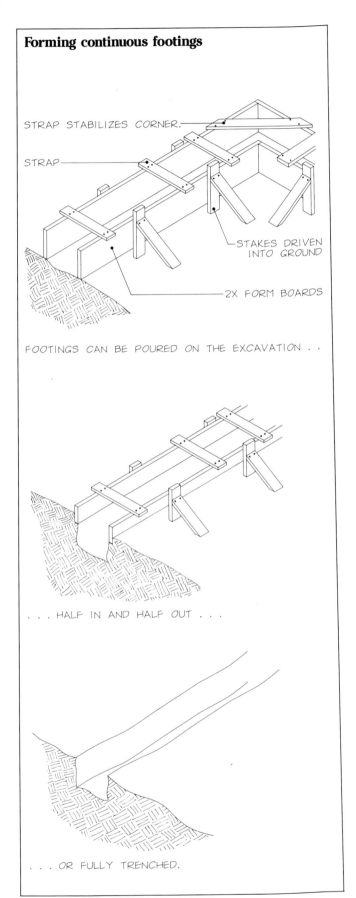

**Forming continuous footings**

STRAP STABILIZES CORNER.

STRAP

STAKES DRIVEN INTO GROUND

2X FORM BOARDS

FOOTINGS CAN BE POURED ON THE EXCAVATION . .

. . . HALF IN AND HALF OUT . . .

. . . OR FULLY TRENCHED.

## Poured concrete wall

A poured concrete perimeter wall is the foundation system we use most. In timber-frame house construction, the foundation wall must distribute the point loads from the frame to the larger area of the footing. To improve the wall's capacity to withstand the weight and distribute the load, we commonly specify two lengths of #4 rebar running horizontally at the top of the wall and two running at the bottom. These parallel bars create flanges of tensile strength, and the vertical wall section transfers the forces between. In most parts of the country, the general standard is 8-in. thick walls, poured with 2,500-psi concrete. For timber-frame buildings, the minimum standards should be increased to 10 in. and 3,000-psi concrete. Soil conditions or heavy loads often demand additional reinforcement and greater concrete density.

It's possible to make your own foundation, using site-built forms of 2x4s and plywood or boards. But because of the labor and materials savings, most concrete work is done by professionals who use specialized forms that are easily assembled. Get bids from firms recommended by other professionals and hire the one who doesn't argue about your desire to improve on the generally accepted standards.

## Concrete-block wall

A preference for using concrete block often reflects regional eccentricities or characteristics that make poured concrete difficult to use. For instance, in areas where aggregate material, which is essential to the concrete recipe, is scarce, the price of poured concrete is sometimes prohibitive. Good examples are parts of upper New York State and Florida, where clay and sand, respectively, predominate. Remote rural regions and islands might not have a concrete mixing plant because the economy cannot sup-

port it. Concrete can be mixed by hand but the work is obviously labor-intensive. In addition, the end product isn't as strong as a wall made with a continuous pour, because cold joints (incomplete bonds) are created in the wall between areas that cure at different times. And then there are areas where there is plenty of material for great concrete but the tradespeople demand what seem to be excessive prices. Accepting this and dealing with it can be just too much. In all these cases, concrete block becomes an attractive alternative.

Besides having an inherent shear problem, concrete block walls, unassisted, have about one-quarter the compressive strength of poured concrete walls. To increase wall strength for timber-frame construction, it is important to concentrate on the point loads. Long runs of block wall, subject to buckling from soil pressure, are commonly reinforced with wall extensions called pilasters at every principal post. Two purposes are thus served: The wall is strengthened against shear forces, and the bearing area of the wall is increased at the critical point-load positions. Also, because principal posts in a frame are usually in alignment with principal beams, including girders for the first floor, pilasters can also serve as a bearing surface for the girders. A common dimension for a pilaster is 16 in. by 24 in. so that typical 8x16 blocks can be woven into the wall without cutting. The pilasters are further reinforced by filling all the cores with concrete and running four pieces of #5 rebar vertically through the cores under the posts.

Because the pilasters are designed to take the full load of the building, the sections of wall between pilasters is essentially non-load-bearing, serving only to resist the lateral pressure of the soil. Eight-inch-wide blocks can be used, but the wall will still need horizontal and vertical rein-

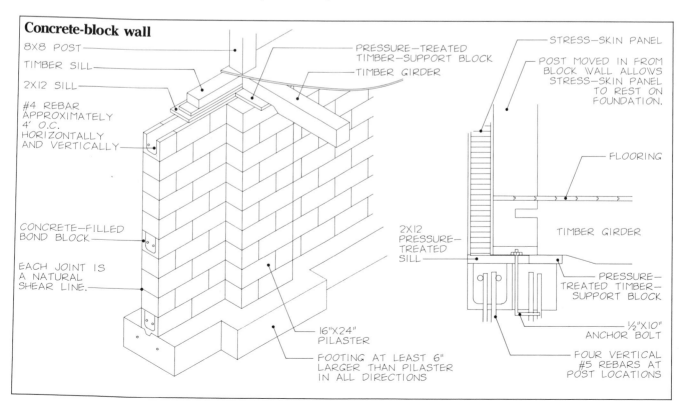

forcement to increase shear strength. For vertical reinforcement, run two lengths of #4 rebar 4 ft. on center and
fill the block cores with concrete. For horizontal reinforcement, use #4 rebar 4 ft. on center, substituting bond blocks
(special hollow blocks) for regular blocks in that course to
contain the rebar and concrete. Vertical and horizontal
bars should be tied at their intersections.

The middle of the pilaster is the strongest point, and so
it makes sense to push the timber post back from the edge
of the foundation wall. This also allows the wall to support
the stress-skin panels, rather than being overhung by
them, as is the case with poured concrete walls.

Footings for concrete block walls are poured concrete
and are identical to those for poured walls. But note that
footings are poured for pilasters at the same time as the
walls. Footings should be 6 in. larger than the proposed
pilaster in each direction in normal soil, more in soft soils.

## Perimeter wall and slab foundation

The two foundations I've just described are most often
used in buildings having full basements or crawl spaces.
These foundations can also be used for slab-on-grade construction. When there is fill on both sides of the wall, the
forces are equalized and there is less need for reinforcement against lateral pressure—pilasters on a block wall
would not be necessary, for example. Excavation for this
type of foundation proceeds quite differently, for instead of
a bulldozer to level the entire area, only a backhoe is needed to trench for the walls. As usual, footings must extend
below the frost line. Although it is expensive to put this
much effort into making a foundation that is almost entirely hidden from sight, it offers the surest strength and the
most complete protection of any slab-on-grade system.

## Grade beam and pier foundation

In this type of foundation, a reinforced concrete beam is
built over a rubble trench (a trench filled with crushed
stone) and supported by reinforced concrete piers. If a
basement or crawl space is not required and if the area to
be built upon is fairly level, this system is perfectly suitable for timber-frame structures. Although labor-intensive,
it's economical of materials and can be installed without
professional assistance. Probably the biggest drawback for
cold climates is that this foundation is hard to insulate
well. Because the beam does not extend below the frost
line, a lot of heat is lost from the house at the bottom of the
wall, where frozen soil is close to the heated house interior.

The entire building area is first excavated to remove the
topsoil and then roughly leveled to a grade beneath the
slab, allowing for insulation and drainage fill. The perimeter is then trenched another 1 ft. deep and 2 ft. wide, with
the approximate center of the beam being the center of the
trench. Holes are then drilled or dug around the perimeter
for piers—the piers should align exactly with the principal
posts in the frame and extend below the frost line. Piers
are a minimum of 10 in. in diameter. Spans greater than
12 ft. require intermediate support from additional piers in
cold climates to control potential heaving. Otherwise, the
design of the timber frame determines the number of piers
and their locations.

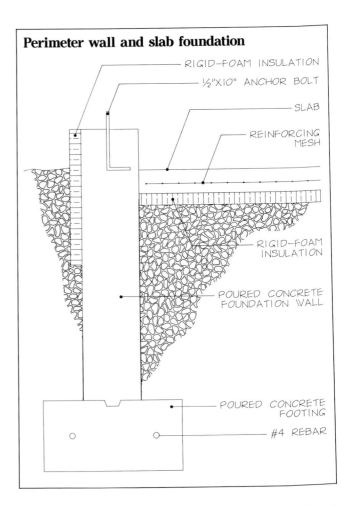

**Perimeter wall and slab foundation**

RIGID-FOAM INSULATION
½"X10" ANCHOR BOLT
SLAB
REINFORCING MESH
RIGID-FOAM INSULATION
POURED CONCRETE FOUNDATION WALL
POURED CONCRETE FOOTING
#4 REBAR

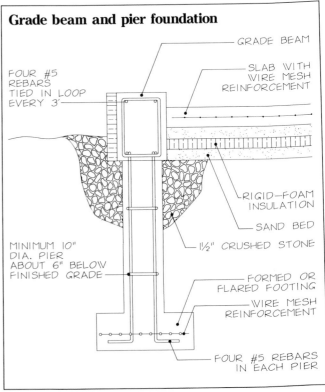

**Grade beam and pier foundation**

GRADE BEAM
SLAB WITH WIRE MESH REINFORCEMENT
FOUR #5 REBARS TIED IN LOOP EVERY 3'
RIGID-FOAM INSULATION
SAND BED
1½" CRUSHED STONE
MINIMUM 10" DIA. PIER ABOUT 6" BELOW FINISHED GRADE
FORMED OR FLARED FOOTING
WIRE MESH REINFORCEMENT
FOUR #5 REBARS IN EACH PIER

If the soil is good, it may be possible simply to flare the pier hole at the bottom for a larger bearing area. If not, it may be necessary to prepare a larger footing area and pour it separately. Plan for the top of the piers to be about 6 in. below finished grade. Most important, they must be level with each other. Forms for the piers could be round cardboard tubes made for this purpose or simple square forms made from framing stock and boards. One advantage of cardboard tubes is that since they can be backfilled before the pour, they do not require bracing. Each pier should have at least four lengths of #5 rebar that extend about 2 ft. above the top of the proposed concrete beam.

After the piers are completed, the perimeter trench is filled with 1½-in. crushed stone for drainage and then compacted. In poor soil, it may be advisable to put drain tile in the trench to carry away excess water. The bed of crushed stone should initially be leveled to the top of the piers to support the beam; more stone is added later. After these steps, the form is built for the grade beam. Typical dimensions are 12 in. wide by 18 in. high, but dimensions really depend on the specific site. Forms can be built with 2x framing stock for the sides and 1x3s for stakes and strapping. Although the building load will be concentrated on the piers, it's still important that the beam be well reinforced—usually, four #5 bars running in two parallel rows at top and bottom is adequate. Splice the perimeter rebar lengthwise and tie them in a loop (stirrup) 3 ft. on center with #2 rebar. Vertical bars from the piers are bent over at the top and tied to perimeter bars. Use 3,000-psi concrete for piers and 3,500 psi for the grade beam. A slab is poured, using conventional procedures, after footings for the interior posts are installed (see p. 122).

## Thickened-edge slab foundation

In areas where freezing is not a concern, monolithic slabs can be poured in which the outer edge is thickened to increase strength and bearing area. Thickened-edge slabs are usually used with buildings that have evenly distributed loads; the dimensions of the thick edge tend to be constant around the perimeter. For timber-frame buildings, it is necessary to modify the design slightly to create larger pads at each post location to distribute the point loads.

If the frame will be insulated with stress-skin panels, plan for the completed foundation to be 3 in. to 4 in. larger than the timber frame in each direction. Excavate the area first by scraping to a level below the slab (including insulation and drainage soil). Trenching the perimeter for the thickened edge requires some handwork, because the transition from the slab to the trench is gently sloped—slab thickness in the middle should be about 4 in., pitching to 6 in. toward the outside of the foundation. Trench dimensions are generally about 12 in. deep and 18 in. wide before tapering to the slab. At the time of excavation a pad should be prepared for each post location. Pad size depends on building loads and soil type (often pads are 24 in. square or more). When the excavation is completed, a one-sided form is built around the outside in preparation for a continuous pour of concrete for slab, trench and pads.

Four lengths of #4 rebar are run parallel around the perimeter of the trench, two on top and two on the bottom. Linear splices are overlapped and tied. Each post pad is further reinforced with a mat of #4 rebar, crisscrossed 6 in. on center. The mat should be tied to the lower perimeter rebar to help distribute the loads. The remaining slab is reinforced with 6x6, 10-gauge wire mesh, which should be tied to the rebar intersections. Concrete should be at least 3,000 psi.

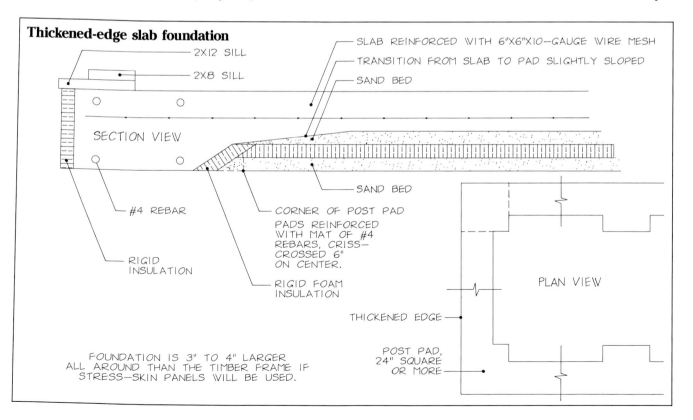

## Thickened-edge slab foundation

2X12 SILL
2X8 SILL
SLAB REINFORCED WITH 6"X6"X10—GAUGE WIRE MESH
TRANSITION FROM SLAB TO PAD SLIGHTLY SLOPED
SAND BED

SECTION VIEW

#4 REBAR

RIGID INSULATION

SAND BED

CORNER OF POST PAD
PADS REINFORCED WITH MAT OF #4 REBARS, CRISS-CROSSED 6" ON CENTER.

RIGID FOAM INSULATION

FOUNDATION IS 3" TO 4" LARGER ALL AROUND THAN THE TIMBER FRAME IF STRESS-SKIN PANELS WILL BE USED.

PLAN VIEW

THICKENED EDGE

POST PAD, 24" SQUARE OR MORE

## Interior post footings and supports

All the foundations described so far also need a series of footings to transmit the loads at interior post locations. These footings are generally between 10 in. and 12 in. thick and approximately 24 in. to 28 in. square, and are located under the slab. The forms are made using 2x framing stock and stakes. Reinforcement is usually wire mesh, but a mat of welded or tied rebar is better. These footings should also be poured on undisturbed soil with 3,000-psi concrete.

Either wood posts or concrete-filled steel piers are used to support the first floor (see p. 125 for more on interior posts). Capped steel piers can be set directly on top of the footing before the slab is poured. When the concrete sets, the pier is locked into position. When a wood post is used, a tapered, concrete plinth topped with a pressure-treated wood cap keeps the post off the concrete; otherwise, the post would draw moisture from the concrete and rot. Plinth dimensions are determined by the base of the post and the proposed thickness of the slab. The plinth top should be the same size as the post bottom, the taper should be about 20° outward to the base, and the top of the plinth should extend 3 in. above the slab. A ¾-in. metal pin (an anchor bolt without the nut will also work), set into the top of the plinth and protruding by about 4 in., keeps the post from moving.

Making the form for the plinth may seem a little tricky, but it's really just basic carpentry. Yet on our jobs, specifying this little oddity on plans for foundation contractors has resulted in almost every conceivable shape—from a round beauty created with the bottom of a tar bucket to a freeform job made by a person who probably flunked clay in kindergarten. (One contractor even refused to do the job unless we supplied the plinths already poured.) We now keep a few plinth forms on hand for such emergencies.

## Concrete pier foundation

For buildings such as vacation homes and workshops, where energy efficiency is not a priority, a concrete pier foundation can be used. Lack of integration between the piers makes this system inherently weaker than other foundations, but some people like its easy installation and economy. Follow the guidelines for constructing piers in the section on grade beams and piers, with these changes: Increase minimum diameter to 12 in., and don't extend the rebar beyond the pier top. Even with reinforcement, these piers are weak against lateral shear, so they should not extend more than 2½ ft. above finished grade.

I don't recommend this system, but if it must be used, err on the side of strength in calculations and construction. I once advised an acquaintance about the construction of a pier foundation, but when he built it the piers extended much higher above grade than I had recommended. They were also smaller in diameter and without the necessary rebar. The way I heard the story, two piers broke when he and a friend leaned on them during a visit to the site.

## A note on foundation insulation

There's no magic formula for building a house with good energy performance, and all our efforts with insulating and sealing would be misspent if we neglected the foundation level. As in the foundation itself, use the best materials. Insulation placed under a slab or outside a foundation is not likely to be replaced, so do it well the first time. For most situations, there should be 2 in. of rigid foam (R-value of 12 to 16) on the exterior of the walls and under the slab. In very cold climates, more than 2 in. may be required above grade. But wall insulation shouldn't always be installed on the outside. For instance, if the basement is unheated, I can't see how the concrete walls can be used for heat storage; worse, if the area is heated only occasionally, the cool concrete walls would have to be warmed before the room would be comfortable. Using such arguments, we often specify insulation on the inside of the foundation walls.

---

**Two ways to support interior timber posts**

8X8 TIMBER POST

PRESSURE-TREATED WOOD CAP

CONCRETE PLINTH ANCHORED BY SLAB

REINFORCING MESH

METAL PIER, 4" DIA., CONCRETE FILLED, ANCHORED BY SLAB

REINFORCED FOOTING 10" TO 12" THICK, 24" TO 28" SQUARE

⅜"-THICK METAL CAP

REINFORCING MESH

---

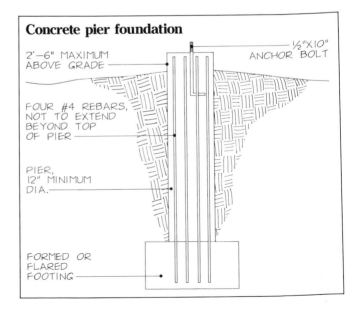

**Concrete pier foundation**

2'-6" MAXIMUM ABOVE GRADE

½"X10" ANCHOR BOLT

FOUR #4 REBARS, NOT TO EXTEND BEYOND TOP OF PIER

PIER, 12" MINIMUM DIA.

FORMED OR FLARED FOOTING

## Floor decks for timber-frame houses

To cap the foundation and form the base of the timber frame, you can use either timbers or conventional stick-framing materials. Compelling arguments can be made for the latter option. Wood placed close to the ground needs to be rot-resistant, and the species with the best natural resistance, such as white oak, chestnut, cypress and redwood, are not as available in timber dimensions as they once were. Good materials for capping the foundation, including pressure-treated lumber, are much more available in stick-framing dimensions than they are in timber dimensions. For strength, it's no longer necessary to use a system made from timbers. In the heyday of timber framing, houses were usually supported with stone foundations of dubious dependability, which racked, heaved and settled, often subjecting the floor system to incredible stresses. But with sills fully supported by the best sort of modern foundation, the only force encountered by the platform is compression. A stick floor system therefore offers little compromise in strength. Also, when the basement area is not used as a living space, the workmanship of the first-floor frame does not contribute significantly to the aesthetic quality of the home. The above considerations and the realities of building costs and budgets often lead to the use of conventional construction for the floor deck of a timber-frame house.

Having said all this, we find we are slowly going back to using timbers in the floor decks of our houses for several reasons. First, we have encountered poor workmanship when relying on local carpenters to build the decks in preparation for our arrival with the frame. Too often we truck completed frames from our shop to sites and find decks that are neither square nor level, and that have the pockets for the post tenons in the wrong places. In the time it takes to fix these mistakes, we could have cut the timber deck in our shop and installed it on the site. Second, we have never been happy with the weak link at the corners of a conventionally framed deck. The outside corners have two headers butted together and nailed, but a force parallel to the direction of the nail can easily pry the two apart. Third, using a sill made from pressure-treated lumber under the timber sill reduces the potential for rot. Fourth, it makes sense to us to continue the heirloom quality of the frame right through to the deck.

Still, if a stick-frame deck is chosen for economic reasons and the area beneath it won't be a primary living area, it's not an outrageous compromise. Following is a discussion of both stick and timber systems.

**Stick-frame floor deck**   Only a few refinements are necessary to adapt a conventional system to a timber frame. First, the arrangement of girders should coincide with the linear arrangement of posts in the frame, either parallel or perpendicular to the bents, depending on post layout. Doing this means that the piers needed to support the girders can then also be located beneath principal posts—a system necessary to transfer the point loads to the footings. The decision regarding how girders are run in relation to frame and foundation is based on giving each interior timber post direct pier support while keeping the floor-joist spans as short as possible.

**It's hard to beat the durability and the heirloom quality of a timber deck.**

## Stick-frame floor deck

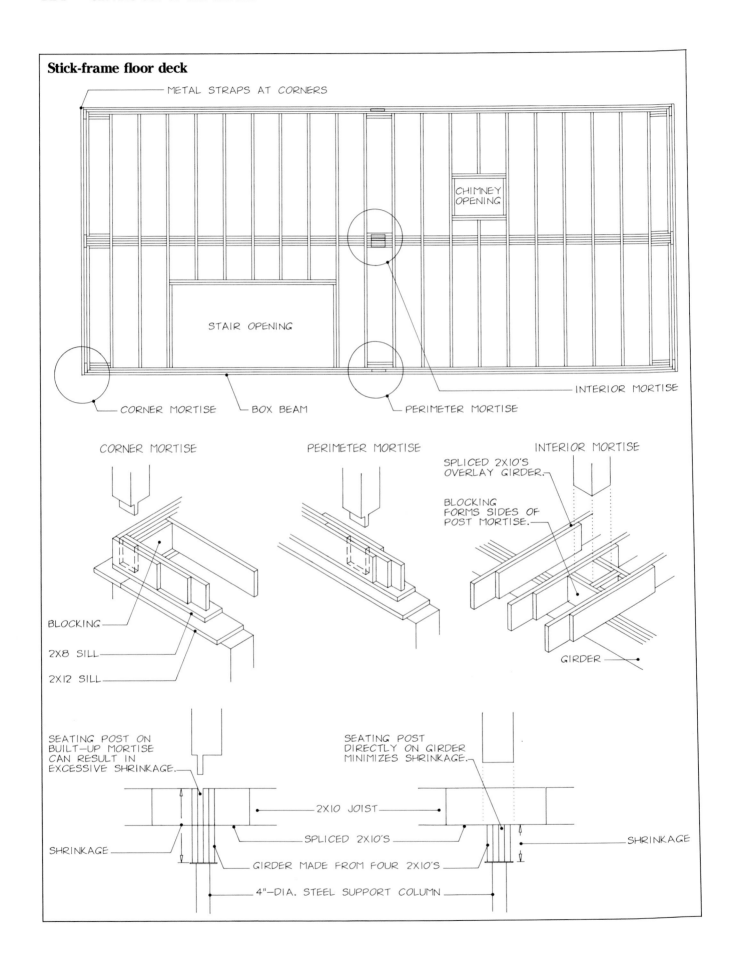

METAL STRAPS AT CORNERS

CHIMNEY OPENING

STAIR OPENING

INTERIOR MORTISE

CORNER MORTISE

BOX BEAM

PERIMETER MORTISE

CORNER MORTISE

PERIMETER MORTISE

INTERIOR MORTISE

SPLICED 2X10'S OVERLAY GIRDER.

BLOCKING FORMS SIDES OF POST MORTISE.

BLOCKING

2X8 SILL

2X12 SILL

GIRDER

SEATING POST ON BUILT-UP MORTISE CAN RESULT IN EXCESSIVE SHRINKAGE.

SEATING POST DIRECTLY ON GIRDER MINIMIZES SHRINKAGE.

2X10 JOIST

SHRINKAGE

SPLICED 2X10'S

SHRINKAGE

GIRDER MADE FROM FOUR 2X10'S

4"-DIA. STEEL SUPPORT COLUMN

Second, the bottom sill (the 2x12 sill) must extend 4½ in. to 5½ in. beyond the outer edge of the deck to cap the insulation. Use a naturally rot-resistant wood (redwood or cedar, for instance) or pressure-treated lumber. The 2x12 sills are bolted to the foundation over a termite shield and sill seal using good standard technique. A second sill, typically a 2x8, is attached to the first with its outside edge set to a chalkline representing the outside dimensions of the timber frame. Corner joints of the 2x8 sill should overlap the 2x12 sill to lock the corners. Where sills butt together end to end, stagger the joints to prevent weak areas.

Third, the perimeter rim joist must be doubled up with blocking to form a box beam. This beam helps to distribute the load to the foundation and is designed with a built-up mortise at post locations to receive post tenons. The extra blocking is extremely important because it gives the post full bearing on the foundation. Outside corners are weak and should be reinforced with metal straps.

Fourth, at interior post locations, a mortise is created in the girder by leaving a gap in one of the laminates. In this case, as well as around the perimeter, the size of the mortise-and-tenon joint is set by the thickness of the framing stock—usually 1½ in. When subfloor sheathing has good compressive strength, the post tenon can be passed through to the box beam or girder, with the post resting on top of the sheathing. When joists overlay the girder, seat the post directly on the girder rather than making a built-up mortise in the joists on top of the girder. Since wood shrinks mostly across the grain, doing this halves the potential shrinkage (bottom drawings, facing page).

Sometimes girders are to be supported with metal piers instead of wood posts. In most situations, the piers not only bear a portion of the first-floor load, they also bear a significant load from the weight of the frame, all other floor loads above the first, and loads from the roof and all the other building materials. Many tons of weight have to be supported at that one point. Do not use hollow, telescoping jack posts; these are not strong enough. Instead use heavy-duty, concrete-filled steel piers, but do not use the small, thin caps that usually come with these posts. Have top and base caps made from ⅜-in.-thick plate metal, and increase the bearing area to about 50 sq. in. I have seen a typical ³⁄₁₆-in. cap turned over like a mushroom and crushed into the girder by nearly ½ in.

When using timber posts to support the girders, use straight-grained material with few defects. In normal loading situations, posts should be at least 7 in. by 7 in. Join the posts to the girder with a mortise-and-tenon joint. The mortise is made in the same manner as for the timber-frame posts that join the girder from above.

**Here timber posts are used to support a built-up girder. The posts are joined to the girder with mortise-and-tenon joints.**

Building on pier foundations requires some modifications because the sills do not have continuous support. It is therefore necessary to build up a laminated perimeter beam to bear the floor load for the spans between piers. Use a 2x12 as the sill, extended to cap the insulation, and nail it to the perimeter beam. Other details and considerations in the contruction are similar to those described previously.

Slab-on-grade foundations, of course, have no interior floor framing, though there is still a need to receive the perimeter posts. A simple box beam, shown on the facing page, serves several purposes. The timber posts are locked into position and anchored to the foundation through other materials. The load from the posts is distributed more evenly to the foundation. The built-up beam running around the perimeter provides a good surface for attaching insulation. The 2x12 sill is bolted to the foundation as described previously. Before attaching the 2x8 box-beam base to the 2x12 sill, nail 2x4s to either side. And before nailing on the 2x8 cap, prepare a mortise at each post loca-

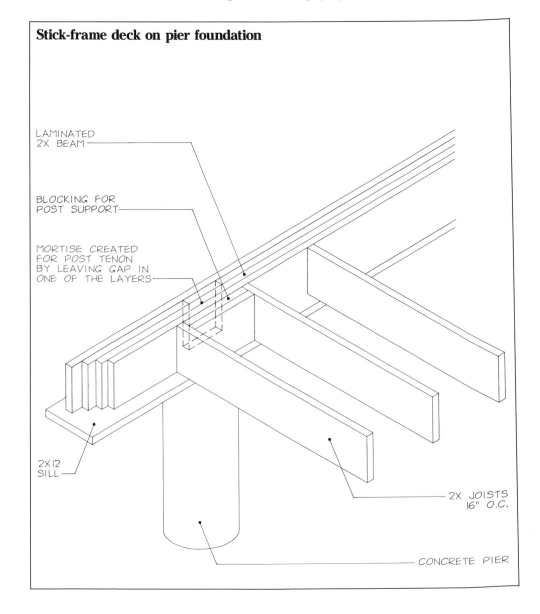

**Stick-frame deck on pier foundation**

LAMINATED 2X BEAM

BLOCKING FOR POST SUPPORT

MORTISE CREATED FOR POST TENON BY LEAVING GAP IN ONE OF THE LAYERS

2X12 SILL

2X JOISTS 16" O.C.

CONCRETE PIER

tion and add blocking for additional support. The cavity in the box beam can be used to run wiring, as discussed in Chapter VII.

If a house is built on a sloping site or uneven terrain, the top of the foundation might actually be on two or three levels to follow the contour of the land. You then need to build walls between the lower foundation levels and the first-floor level. There are two options. You can use a timber at each post location, insulating the wall the same way as the timber-frame walls above. Or you can frame conven-

tionally, but it might be necessary to go to extreme measures to keep the insulation value the same as that provided by the stress-skin panels (asssuming you use that system) on the rest of the house. And to support the girders and principal timber posts, you would also have to fabricate posts by laminating framing members together. Because there's so much complicated work in trying both to support the posts and match the insulation level of the rest of the house, there's no real time or money savings in this method—we've found it makes much more sense just to

## Box beam for slab-on-grade foundation

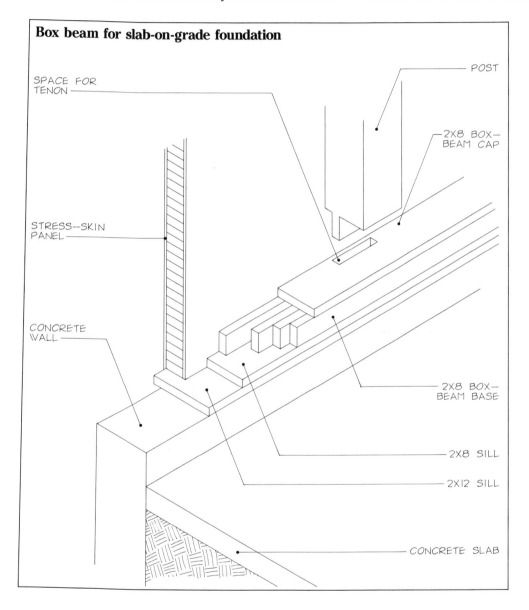

POST

SPACE FOR TENON

2X8 BOX-BEAM CAP

STRESS-SKIN PANEL

CONCRETE WALL

2X8 BOX-BEAM BASE

2X8 SILL

2X12 SILL

CONCRETE SLAB

use timber posts and continue the insulation from the frame down to the sill.

If you use posts between the lower foundation levels and the first-floor level, you'll have to make some changes in the typical sill system. Eliminate the sill overhang where the post wall on the first-floor level begins, since the insulation continues to the base sill. Remember that a box-beam sill is used only where there is full bearing on the foundation. Where the foundation drops to a different level, it is necessary to use a solid laminated beam similar to that used with a pier foundation. For the lower sill, use the box beam described for the slab-on-grade foundation. The ground-level posts are cut to length with tenons on both ends to join to mortises in the box beam on slab level and in the solid beam on floor level, as shown below. Note that knee braces can be joined from the posts to the laminated

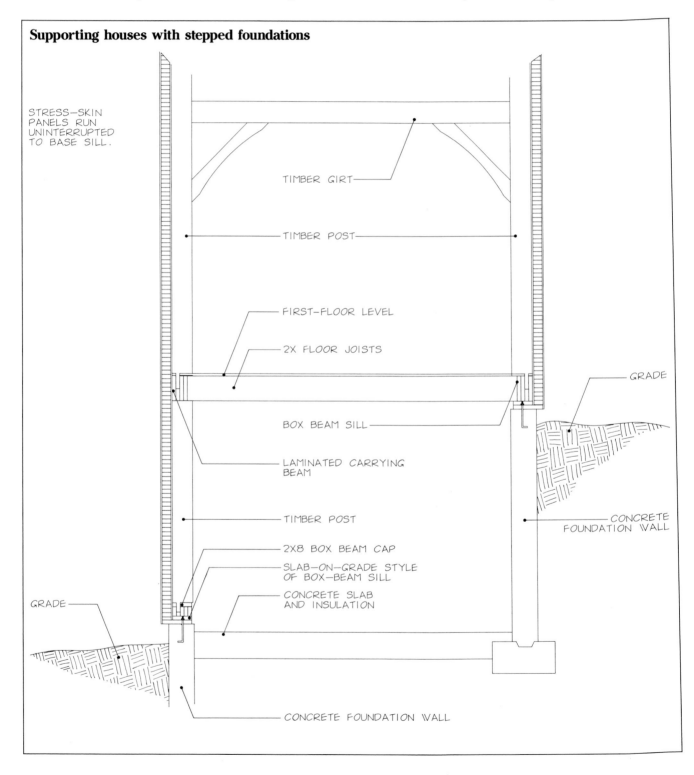

**Supporting houses with stepped foundations**

STRESS—SKIN PANELS RUN UNINTERRUPTED TO BASE SILL.

TIMBER GIRT

TIMBER POST

FIRST—FLOOR LEVEL

2X FLOOR JOISTS

GRADE

BOX BEAM SILL

LAMINATED CARRYING BEAM

CONCRETE FOUNDATION WALL

TIMBER POST

2X8 BOX BEAM CAP

SLAB—ON—GRADE STYLE OF BOX—BEAM SILL

CONCRETE SLAB AND INSULATION

GRADE

CONCRETE FOUNDATION WALL

beam using the same technique to join the tenon to the laminated beam. However, unless the post wall projects from the foundation as an ell or extension, knee braces serve only to stiffen the laminated beam; racking forces should be completely absorbed by the foundation walls.

**Timber-frame floor deck**  Using a timber floor is the surest way to maintain the integrity and strength of the rest of the frame. When the ground-floor level is used as living area (a family room, for example), the beauty of the timbers is an obvious advantage.

The system begins with a 2x12 sill and a 2x8 sill in the same manner as for a stick-frame deck. These members cap the insulation and protect the timber sill from moisture. Historically, chestnut and white oak were widely used for sills, but chestnut is now nearly extinct, and good white

Using a timber floor system gives the basement area continuity, beauty and strength. Concrete plinths protect the posts from absorbing moisture from the floor. (Photo by Gordon Cook.)

oak is hard to find. Most of the woods good for timber framing are not especially moisture resistant—redwood and cedar would be good choices for the timber sill but quality, availability and cost can be problems. Getting a suitably moisture-resistant sill in lumber dimensions is obviously much easier and less expensive. When a timber sill is used on a continuous foundation, it doesn't need to be very large because it doesn't have to carry the weight of the floor in an open span. It need only be thick enough to allow good connections to floor joists and girders, and wide enough to transfer the weight of the posts to the foundation (a timber sill for average circumstances might be about 5 in. thick by 8 in. wide). A wire chase can be created by making the sill slightly narrower than the posts.

Stick-dimension sills are anchored easily to the foundation using typical anchor bolts, but a different technique is

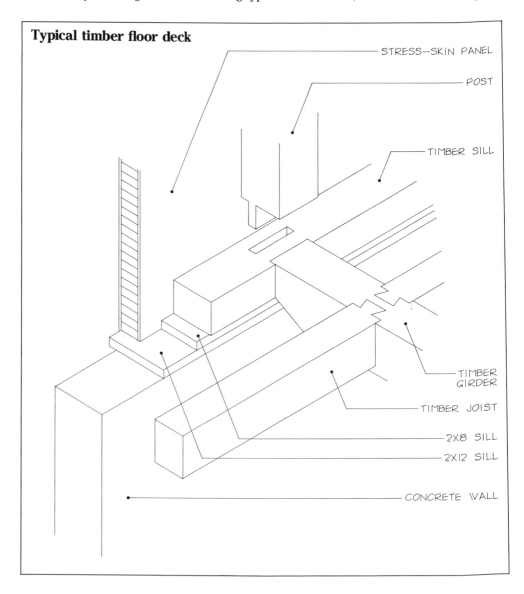

**Typical timber floor deck**

STRESS—SKIN PANEL

POST

TIMBER SILL

TIMBER GIRDER

TIMBER JOIST

2X8 SILL

2X12 SILL

CONCRETE WALL

required for anchoring timber sills. Various scarf joints are used to make linear connections in the sills, and several types of mortise-and-tenon joints are used to join sills at the corners. In every case, except for very simple (and very inferior) scarfs, these joints need to slide together horizontally, which makes it nearly impossible to set the sills over bolts. A practical solution is to nail flat metal straps to the 2x8 sill, join and position the timber sill, then bend the straps up on the outside face and onto the top of the sill, as shown below. (The timber has to be routed at each strap location to keep materials flush.) Nail the straps to the sill using 16d galvanized nails. The straps should be of at least 16-gauge material, and usually need to be located on 4-ft. centers, but check your local code.

Floor joists and girders can be joined to the timber sill with either dovetail or mortise-and-tenon joints. The timber sill is tabled (notched) at each joint area to allow the beam to rest on the 2x8 sill. When girders require more bearing than is provided by the thickness of the timber sill, the 2x8 sill can be notched to allow the girder to rest on the 2x12 sill. If a block foundation is used, the projecting pilaster can be used to support the girders, but there should still be a protective sill under the timbers. After all the floor timbers are joined, planking or a built-up floor system is installed, as described in Chapter X.

A combination system is used by a few timber framers to reduce the expense of the deck and maintain the integrity of the floor system. Timbers are used for the sills and girders, allowing for good connections and load transfer. The joists, however, are typical 2x stock, attached to the timbers with joist hangers. The conventional joists bear only the floor loads, while the timbers are used as the primary beasts of burden. The savings is in eliminating all of the joinery involved with the floor joists.

## Anchoring timber sills

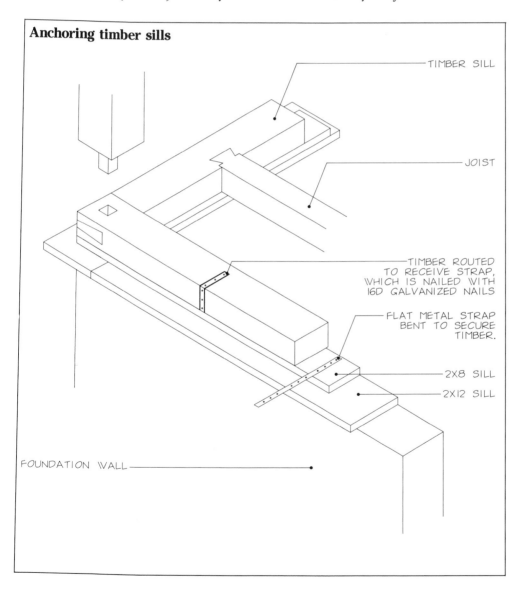

TIMBER SILL

JOIST

TIMBER ROUTED TO RECEIVE STRAP, WHICH IS NAILED WITH 16D GALVANIZED NAILS

FLAT METAL STRAP BENT TO SECURE TIMBER.

2X8 SILL

2X12 SILL

FOUNDATION WALL

# VII. Wiring and Lighting

**The warm colonial atmosphere of this area is achieved by using soft lighting and stained wood.**

Sprawled throughout our rooms like sleeping snakes, wires are an umbilical cord to the accessories of the 20th century. In my small home office alone, there are wires for two lamps; three wires for the computer; one for the telephone; two for a radio and headset; one for a drill waiting for a job in the corner; and another dangling overhead, begging for a light fixture. Thousands of linear feet of wire course through our walls, pulsing with energy people feel they can't live without. Linking us to all the things considered necessary to a civilized lifestyle, often including heating and plumbing, the electrical system is the central nervous system of the modern home. We are hooked on electricity, and although we may occasionally begrudge the price that's paid for it and perhaps even argue philosophically that life is really not finer for it, not many people would trade it in for a life unfettered.

In timber-frame homebuilding, the issue of wiring often causes puzzlement and frustration in homeowners, builders and electricians alike. "Just how do you get wires through walls built of 6x8 and 8x8 timbers enclosed with solid stress-skin panels?" many of my clients want to know. When I explain the process we use, they are usually surprised at its straightforward logic, having expected that they would have to live amid coils of exposed wires or pay for techniques of incomprehensible complexity. Still, no matter how simple wiring a timber-frame house is in theory, there is bound to be misadventure. Many of the proce-

dures for wiring timber-frame structures fall outside conventions of the trade, and a set of new rules has yet to be formalized. Timber-frame designers and builders therefore sometimes neglect to take the time to consider the electrical system, waiting until it's too late to solve problems creatively. Caught up in the excitement of building their house, homeowners often ignore the entire question of electricity, with the result that their rooms are not illuminated in the way they would have liked and outlets and switches are inconveniently located. Confused electricians, relying on intuition rather than on systematic procedures, are often driven to poke at the timbers with their drills as if the frame were a giant voodoo doll.

The problem of wiring a timber-frame house is further complicated by the fact that there is no one right way to do the job. Instead there is a wide selection of options, some more appropriate than others, depending on the particular condition. In studying electrical systems, our goal therefore is to cultivate a way of thinking that will give us the flexibility to evaluate each situation individually and to choose the very best option possible. Only with a thorough understanding of the basic principles can we hope to have the ability to make intelligent choices.

In this chapter, I will discuss the techniques of wiring a typical timber-frame house that uses stress-skin panels as its enclosure system. I focus on stress-skins because most timber-frame houses today use them, and they require the newest methodology. (When the built-up systems are used, conventional wiring practice can be employed to a greater degree.) Bear in mind, as you proceed through this chapter, that timber frames and the systems used to enclose them do not change the theories of electrical wiring or the rules governing circuit design. Nothing about the timber-frame building system alters decisions about wire size and type, switching options, safety or any other matter related to what it takes to move electrical current properly from the generating utility to the consuming appliance in the home.

## Planning the electrical system

During the design stages of the building process, it is in everyone's best interest to devote sufficient time and energy to planning the electrical system. This includes considering everything from the types of electrical devices and fixtures that will be in each room to the path of every single wire in the house. There are several strong arguments for good planning. Because running extra wires is almost impossible once the frame is enclosed, thorough planning can help guarantee that outlets and fixtures will be exactly where they're needed. Because a set of good plans makes it easier to communicate with the electrician, the job usually will run more efficiently and economically. And because good planning makes it possible for the timber framer to

make the necessary cuts for the wires while the timbers are still on sawhorses, the amount of drilling the electrician has to do will be minimal—the surest way to prevent the unnecessary marring of timbers. So planning should be the rule of thumb when dealing with an electrical system.

We once dealt with a contractor who refused to plan or to follow our plans, and the results of his bullheadedness caused serious damage to the timber frame. This fellow personified the almost universal adversarial relationship between designers and contractors—he refused to acknowledge our blueprints as anything more than a rough example of what a building of the size and shape shown might look like. And the plans for the electrical system? Well, he didn't need them, because he already knew how to wire a house. We repeatedly told him that our electrical plans were critical, given that the local building code demanded that all wires be encased in conduit, but he somehow convinced the panel crew that he had a better plan, so they enclosed the house with no consideration for where the wires would go. As it turned out, his plan was simple: Eliminate all of the inaccessible lighting locations and riddle the timber frame with notches and holes large enough for a fire hose, and wiring would be a cinch. Lest you think that such matters are always properly resolved, in this case there was no cure for the damage.

**Here, the feeling of expansiveness is reinforced through the brightness from the chandelier and floor lamps.**

Planning begins by locating lights, outlets, appliances, stereo sets, televisions, telephones and all other wire-dependent equipment on the floor plan. The designer or architect will need a great deal of input from the owner on this, and the owner will want to anticipate alternate locations for equipment that is likely to be moved occasionally, such as televisions and stereos. Running all the wires during the construction stage will provide greater flexibility later. By code, outlets will have to be about 6 ft. on center around the perimeter of the room. This is a good regulation, but it should not dictate the exact location of the outlets. While planning the electrical system, it's smart to have as many outlets and switch boxes as possible installed on the interior walls, to minimize penetrations in the insulation. In homes with high standards of energy efficiency, even small leaks can have large significance, like a pinhole in a balloon. Doing this can be difficult, however, because timber-frame houses commonly have open floor plans and few partition walls. It is inevitable that outlets will be on exterior walls, but do your best to reduce them.

Use a copy of the floor plan to create the electrical plan. (Usually a sepia is made of the floor plan and then copies of the sepia are made with a blueprint machine. Otherwise, a copy of the floor plan should be traced from the original.) As you will see later, the position of timber posts and ceiling beams are critical in determining wire paths and the lighting scheme, so these should be indicated on the electrical plan. (To keep the plan uncluttered, draw in the beams lightly or with dotted lines.) After all the fixtures are located, the designer draws the electrical plan, including wire paths. If the electrical system is simple, a designer often will leave the circuitry to the electrician after determining the paths, but in complicated situations with few or no path options, the designer will determine the circuitry as well. In the rare instances where tradesmen are contracted early in a project, we have found it valuable to involve the electrician in planning. The electrician often has helpful input and always appreciates being a part of the decision-making process and not just a victim of it.

Plans are drawn to deliver information about how an intended result will be achieved. To be effective, all the participants need to understand the role they play in implementation; the plan won't do much good as just another blueprint in the set. The joiners need to understand what the plan means to the frame. Are there timbers to notch? Holes to drill? Panel installers should be prepared for the wires that will run through panels or panel joints. Carpenters must understand the accommodations necessary for wires or fixtures. In other words, it is critical that the construction crew, the designer and the homeowner act as a team and communicate precisely about even the small details of the electrical plan. Only in this manner can a good plan be transformed into a good electrical system.

## The wiring process and prewiring

Basically, electrical wires carry currents of energy the way a pipe carries water. A single large wire brings the electricity to the house, where it is attached to a service panel, or circuit box. From the service panel, the current flows through smaller wires to various parts of the house. It does so in circuits—loops of current that flow from the service panel to the appliances and devices and back to the panel. The path for a circuit can be short and simple or long and complicated, depending on the number of devices and the distance back to the panel. Getting the main wire to the service panel, either overhead or underground, is done with standard procedure, but running the smaller wires through the house requires thought, because in timber-frame building, the wiring routes typically used in stud construction just aren't there.

The obvious difference between wiring a stud-frame house and a timber frame that will be enclosed with stress-skin panels is that in the latter there are no cavities in the exterior walls through which to run wiring. Because of this, it's necessary to do the wiring in three stages instead of in the conventional two. With a stud-frame house, the electrician comes in to rough-wire as soon as the house is enclosed by the roof, windows and exterior sheathing. Rough wiring simply means stringing the wires through the studs,

In a timber-frame house, wiring begins by stringing the wires loosely over the frame before the panels are installed.

joists and rafters to the various outlet boxes. Usually, many of the final decisions on the layout of the electrical system are left to be made during this time. The electrician then comes back to finish-wire (a matter of attaching wire ends to outlets, switches, fixtures and appliances) after the drywall is up.

But in a timber-frame house, the job starts with pre-wiring. Early in construction, just before the stress-skin panels go up, the electrician will get the wires to each general area of the house by running them loosely over the outside of the sill and frame. Where interior partitions will intersect timber posts, the electrician drills for the wires

and pulls them through. At this time the electrician also drills through rafters and purlins for track lighting, fans or chandeliers, and mounts wires and boxes for light switches.

On the first day of panel installation, the electrician will usually meet with the panel crew to review the electrical plan. It is the crew's responsibility to cut the grooves for the wires in the panels at the timber posts and to tuck the wires into place at the panel splines. When all the panels have been erected, but before the expanding foam and the 2x4 nailers have been applied, the electrician comes back to inspect the wiring and to run the wires for the switches next to exterior doors. All this is considered part of prewiring.

The panel-installation crew cuts grooves into the panels for wires and tucks the wires into spline connections. (Photo by Tafi Brown.)

After the panels have been foamed and the interior stud-wall partitions framed, the electrician will return to the house for several days to rough-wire. By this time, the carpenter will have built the extended-baseboard chases on the exterior walls, as shown on the facing page, and the electrician will be able to mount the wires and outlet boxes in them. The electrician will also mount outlet boxes on the partition-wall studs and run the wires over the floor system. After the partition walls are drywalled, finish wiring occurs in the conventional way.

It's the idea of prewiring that makes some electricians nervous about taking on a timber-frame job. But while it's true that prewiring requires a different strategy and different techniques, it's not true that different has to mean difficult or confusing. A timber-frame electrical system that is well thought out and well communicated to a responsive tradesman shouldn't be any harder to install than one for a conventional stud-frame house. In fact, electricians who have worked on several of our projects tell me that in many ways it's easier.

**After the panels are installed and interior partitions are built, the electrician can complete the rough wiring.**

## The extended-baseboard wire chase

The primary way that wires get around the perimeter of a timber-frame structure is through horizontal wire chases. Of these, the extended-baseboard chase is the most widely used. These chases are installed on the insides of exterior walls between timber posts. Wires are mounted in the cavity between furring strips, and the outlets are mounted in the front face of the baseboard. The wires pass through the timber posts on their way around the perimeter of the house through a notch cut in the back of each post—ideally, these notches were planned and cut before the frame was erected. There are no notches in the corner posts because it's too difficult to feed the wires around corners. Instead, wires are fed down through the floor and up again on the adjacent side. While commercial surface-mounted wire chases can substitute for the extended baseboard, I don't feel they are attractive enough to be used undisguised in high-quality homes.

The extended-baseboard wire chase system actually begins at the sill. If the frame is attached to a timber sill, during prewiring the electrician will mount wires in a

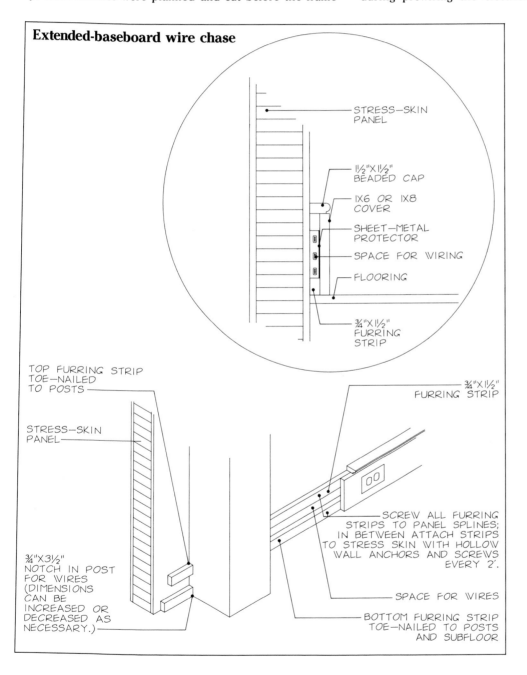

**Extended-baseboard wire chase**

STRESS—SKIN PANEL

1½"X1½" BEADED CAP

1X6 OR 1X8 COVER

SHEET—METAL PROTECTOR

SPACE FOR WIRING

FLOORING

¾"X1½" FURRING STRIP

TOP FURRING STRIP TOE—NAILED TO POSTS

STRESS—SKIN PANEL

¾"X3½" NOTCH IN POST FOR WIRES (DIMENSIONS CAN BE INCREASED OR DECREASED AS NECESSARY.)

¾"X1½" FURRING STRIP

SCREW ALL FURRING STRIPS TO PANEL SPLINES; IN BETWEEN ATTACH STRIPS TO STRESS SKIN WITH HOLLOW WALL ANCHORS AND SCREWS EVERY 2'.

SPACE FOR WIRES

BOTTOM FURRING STRIP TOE—NAILED TO POSTS AND SUBFLOOR

chase created with furring strips on the outside of the sill. This keeps the interior of the timber floor system from being ruined visually by strings of wire running in all directions. Although simple, this system consumes a great deal of wire because all wires have to pass around the perimeter. Sometimes it is acceptable to find one place for wires to pass through the house, perhaps using a dropped ceiling between joists (similar to the plumbing chase discussed on p. 159). To feed an electric device or series of outlets in the wall, the electrician pulls a tail, or wire end, into the extended-baseboard area through a gap left in the top sill furring strip. After all the wires are mounted in the sill chase, a 1/16-in.-thick sheet-metal strip is screwed or nailed over the chase for protection. This may not be required by the building inspector, but it certainly can help prevent a problem in the event of a misplaced nail.

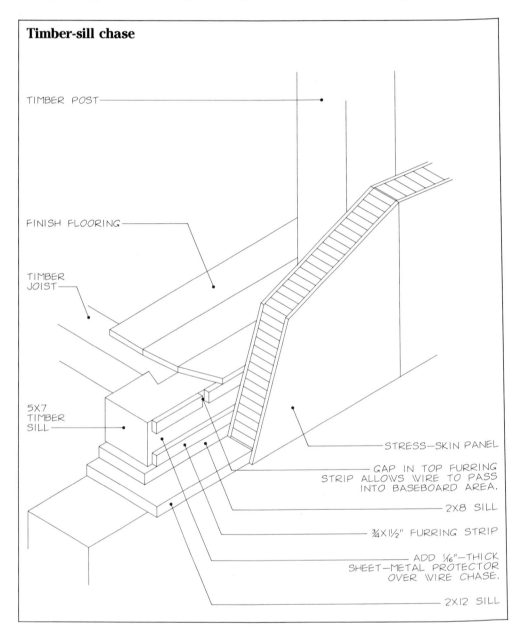

**Timber-sill chase**

TIMBER POST

FINISH FLOORING

TIMBER JOIST

5X7 TIMBER SILL

STRESS-SKIN PANEL

GAP IN TOP FURRING STRIP ALLOWS WIRE TO PASS INTO BASEBOARD AREA.

2X8 SILL

3/4 X 1 1/2" FURRING STRIP

ADD 1/16"-THICK SHEET-METAL PROTECTOR OVER WIRE CHASE.

2X12 SILL

If the timber frame is on a stick-frame floor deck, the electrician will run wires to and from the service panel in the cavities between joists during prewiring. To access the extended baseboard, the electrician drills a hole on a diagonal from the top outside corner of the subfloor toward the inside face of the perimeter box beam. One hole is drilled for each wire needed between posts. The wires are then pushed up from below at a later date, usually during rough wiring. From there the wire can be run from outlet to outlet inside the chase.

Once the frame has been prewired, it's time to assemble the extended baseboards. This is basic carpentry, although the drywall face of the typical stress-skin panel doesn't offer many opportunities for securing the furring strips. To begin, snap a line on the wall to show the top of the baseboard minus the cap. If a 1x6 or 1x8 is used for the baseboard cover, this dimension will be 5½ in. or 7½ in., respectively. Then cut the top and bottom furring strips to length, using more than one piece for each run if necessary. Apply a bead of construction adhesive to the top furring strip and fit it between posts with its top edge on the chalkline. Toe-nail it to the timber posts and screw it to the panel splines. Repeat the process with the bottom furring strip, except that you toe-nail it to the subfloor as well as to the posts. To draw the furring strips tightly to the wall, hollow wall anchors with screws are used every 2 ft. where there isn't a spline or nailer.

Using the electrical plan as a guide, next cut the holes for the outlet boxes in the baseboard cover, which is then used as a template to mark the holes on the wall. Obviously, because the space between the furring strips is narrow, duplex outlets are run horizontally. Make the cuts in the baseboard precise, because overcutting will show up as gaps around the outlet cover. Then trial-fit the baseboard between the timbers and mark the box locations on the wall between the furring strips. To keep the outlets well above the finish floor, locate the boxes directly against the bottom edge of the upper furring strip. Remove the baseboard and cut a housing in the wall for each box, keeping it as shallow as possible; you don't have to cut to the full depth of the box because of its placement in the baseboard cover. (Outlet boxes should be selected on the basis of

minimal penetration of the insulation. It's easier for the electrician if the boxes are deep, but better for the energy performance of the house if they're shallow.) Then pry the drywall and foam block loose with a flat bar or screwdriver, pulling out excess foam with a hard, sharp tool.

**If the deck is stick-framed, holes are drilled between each pair of posts from the top outside corner of the subfloor to the inside of the box beam; then the wires are run through the joists.**

When all this has been done, the electrician runs the wires through the chase and secures them with staples at the splines. If necessary, electric tape will corral several wires until the baseboard cover goes on. The carpenter then mounts the cover between timber posts, pulls the wires through the holes for the outlet boxes, and attaches the baseboard to the furring strips with 8d finish nails about 16 in. on center. The electrician mounts the boxes on the baseboard and attaches the outlets. For the benefit of safety (or the building inspector), a sheet-metal protector made to fit behind the baseboard cover will discourage future violation of the wire chase.

Some codes require electric outlets to be 14 in. off the finish floor, although the reasons for this have more to do with the convenience of the homeowner than with safety. If this issue is pressed, the extended baseboard can still be used for a wire chase while the outlet boxes are mounted higher in the wall. Set up the chase as usual, leaving the cover off until the outlets are installed. Mark the outlets on the wall no more than 16 in. off the floor, using a sample outlet box to scribe the lines. Cut the holes for the boxes as shallow as possible. Then ream a hole through the stress-skin panel from each box opening to the chase area with a ½-in.-dia. bar or pipe (rebar works well). Stay as close as possible to the interior sheathing. Intersect that hole with a hole drilled out from the chase. When you push a wire down from the outlet, it should be easy to pull it through from the chase. Pull the wire into the outlet box and mount it in the drywall with switch-box supports. Before closing up the box, fill any voids in the insulation with spray foam. Keep in mind that there is extra heat loss when the outlets are outside the wire chase, because more insulation is removed to accommodate the depth of the box.

## Other types of horizontal wire chases

Some commercial baseboard heating systems are designed with a built-in chase for one or two wires to supply outlets in a circuit. Some brands even offer inserts that include a mounting for a duplex outlet. But if the chase needs to accommodate wires for other circuits, it will probably make more sense to use the extended baseboard behind the baseboard heating unit, as shown at right. The heating unit then caps the furring strips. In this case, you can use either the outlets in the heating unit or outlets mounted in the extended baseboard between sections of heating unit.

Other interior finish elements can also be used as wire chases without substantially altering their function or design. For example, if wainscoting is intended or if a wall is to be completely paneled, the furring strips needed to attach the wainscoting or paneling to the walls create a natural wire chase. (The furring strips are installed in the same way as the extended baseboards.) In the case of paneling, where the entire wall is set up with furring strips, the vertical wires for switches and wall lights can be run as part of the same system, but make sure all the wires are in a safe position or are protected with a sheet-metal cover.

Cabinets in kitchens and baths offer another opportunity to set up a wire chase if you mount them to furring strips. Install the furring strips on the wall about 12 in. on center, locating a strip approximately 6 in. from the top and bottom of the cabinet. Because outlets for kitchen and bath cabinets tend to be located just above the cabinet back, it's necessary to leave a gap in the furring strips so a wire can be brought up from behind the cabinet. A built-up backsplash is then constructed using the method for making extended baseboards, except that ⅜-in.-thick plywood is attached to the furring strips to back the tile. Another variation is to use tile just above the counter and a more typi-

Wires are run between furring strips attached to the wall. Holes are cut in the baseboard and wall to receive the outlet boxes.

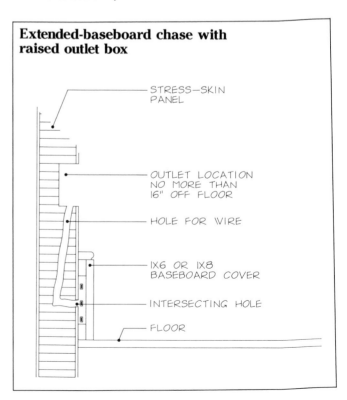

**Extended-baseboard chase with raised outlet box**

STRESS–SKIN PANEL

OUTLET LOCATION NO MORE THAN 16" OFF FLOOR

HOLE FOR WIRE

1X6 OR 1X8 BASEBOARD COVER

INTERSECTING HOLE

FLOOR

cal wire chase just above that. Using tile as the backsplash is attractive and gives better moisture protection, but to prevent accidental electric shock, always use ground-fault circuit interrupters in bathrooms and in areas close to the kitchen sink.

If it is necessary to have oulets higher than the backsplash area, use the space behind the cabinets for the chase and mount the outlet in the stress-skin panel not more than 16 in. above the cabinet, as shown on the facing page. Remember to foam all the voids around the outlets.

## Baseboard heating unit wire chase

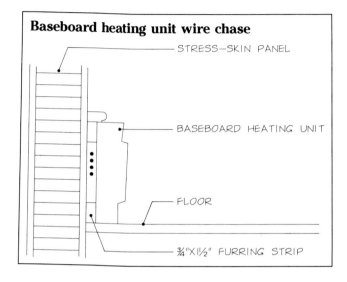

## Wainscoting wire chase

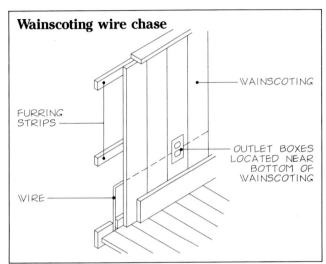

## Cabinet wire chase

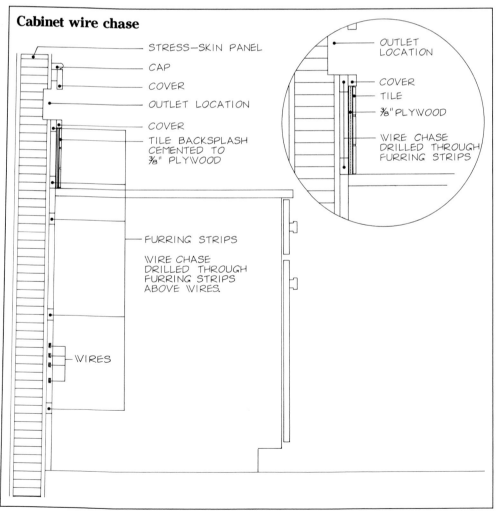

The box-beam sill used with slab-on-grade foundations, described in Chapter VI, has its own natural wire chase. Not only is there enough room for all the usual wires, there is also enough room for the outlet boxes. When the box beam is made, install 3-ft.-long pieces of 2x8 cap at the post locations. Leaving the rest of the box open until after the panels have been installed makes it possible to use it as a wire chase. At each post location, leave a 1½-in.-sq. space under the tenon for the wires to pass through. Using a scroll saw, cut rough holes within ¼ in. of the size of the outlet box on the inside face of the box beam. Run the wires through the chase, leaving adequate tails for the outlets. When all the wires are in place, finish installing the cap on the top of the box beam and add the 1x10 cover. Next, precisely cut the outlet holes in the finish baseboard, holding the baseboard over the beam to transfer the outlet locations. Feed the wires through the baseboard and nail it to the beam. For the protection of the wires, nail only to the top and bottom of the beam and not in the chase area.

In this system, exterior doorways can present a problem, because the sill is lower than the rest of the foundation wall and the box beam cannot travel around the perimeter

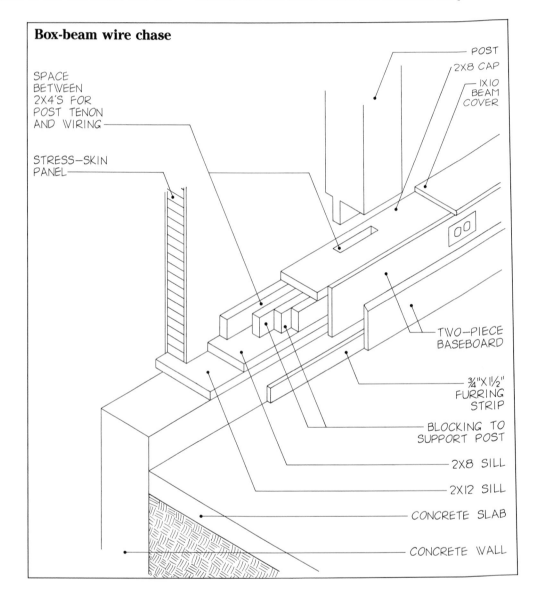

**Box-beam wire chase**

SPACE BETWEEN 2X4'S FOR POST TENON AND WIRING

STRESS—SKIN PANEL

POST

2X8 CAP

1X10 BEAM COVER

TWO—PIECE BASEBOARD

¾"X1½" FURRING STRIP

BLOCKING TO SUPPORT POST

2X8 SILL

2X12 SILL

CONCRETE SLAB

CONCRETE WALL

of the house without interruption. The most common solution is to block up the door threshold 1 in. to 1½ in. from the slab to allow electrical conduit to pass underneath. The blocking is attached to the concrete toward the inside and outside of the threshold using concrete nails or expansion bolts, and the wires are run through conduits between. (Allow a few extra inches in the width of the concrete rough opening so wires have room to drop from the foundation wall down to threshold level.) If a wooden floor is to be laid over the concrete, this technique would not alter the typical relationship between the threshold and the finish floor.

When the house is on a slab instead of on a wooden structure, it is usually necessary to have several conduits embedded in the concrete to get wires across the building. Otherwise, all the wires would have to run around the building in the chase, wasting wire and time. It would have to be clear from the plan which wires would need to be in the conduit, and the conduit would need to have a vertical riser at the ends to bring the wires to the future wire chase. This technique is familiar to almost all electricians because the situation is commonly encountered in housebuilding.

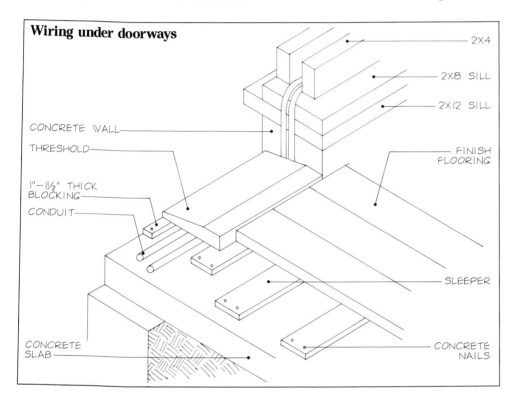

**Wiring under doorways**

## Interior wire chases

Except for the last method, which is specific to a particular foundation type, the wire chases described previously can be used on any exterior wall. They will work as conveniently on the second-floor or third-floor levels as on the first. But for interior walls, it makes sense to switch to more conventional wiring methods. Interior partition walls are framed with studs after the exterior walls are in place, and wires and boxes can be installed in them as usual.

Partition walls usually intersect exterior walls at timber posts, but if a partition and an exterior wall intersect between posts, the wires can be passed into the partition from the extended-baseboard chase as shown below. Where a partition is located at a post, the electrician must drill through the post to allow the wire to pass from the chase to the stud cavity. This happens during prewiring, before the stress-skin panels go on, and I'm always surprised at how many times these holes don't get drilled for lack of thorough planning. Without them, it can be quite difficult to get to the partition, requiring either passing the wires through the floor system or drilling completely through both the stress-skin panel and the post. The holes should be located in the posts at the level of the wire chase in the extended baseboard and drilled from the inside, which makes it easier to align them precisely with the center of the partition. The wire is fed through the post from the outside, and enough is left to wire the first outlet or switch box. Because partitions are usually not framed at this stage, it's necessary to notch the first stud around the wire and the hole drilled through the post. Subsequent studs are drilled in the conventional manner.

When it is necessary to bring a wire to an interior partition that does not intersect an exterior wall, bring the wire back to the floor system and drill through to the partition from the basement or crawl space. For timber floors, which are usually exposed to view, plan even further ahead and run this wire in a groove routed into the top of a timber.

## Vertical wire chases

In a timber-frame house, the seams between stress-skin panels make convenient chases for vertical wires. Wires can be run either in the joints between panels over timber posts, or at the panel splines. A panel joint will not always fall conveniently on a post requiring wires, but the problem is easily solved by cutting a whole panel to make a joint at the post. No structural integrity is lost and wiring is made substantially easier. At panel splines, a cavity is created that is meant to be filled with foam, but before it is filled, it's an excellent way to bring wires to the extended baseboard between posts.

**Timber-post chase** Remember that during prewiring, the wires were strung loosely in position on the timber posts, their ends directed toward horizontal wire chases or run through posts toward interior partitions. During panel installation, the crew notches the inside edges of the panels meeting over posts, to make a channel for the wires. The first panel is nailed to the frame as described in Chapter V, with the wires pulled away to avoid damage. When that panel is secured, the wires are tucked into the notch to keep them out of the way while the adjacent panel is nailed. The space between the panels and the area around the wires is then filled with foam.

This technique is also used to run wires up into the roof system. Because major rafters tend to fall on major posts, the wire can go up the outside of a post and continue along the top of a rafter without interruption. A hole can be drilled through the rafter at the proper fixture location, or the wire might be directed to a purlin to feed a light or a fan. It is generally easier to deal with wires on the roof than in the walls, because there are more timbers and therefore more open panel joints for easy access. Also, because it's usually necessary to supply electricity to only a few isolated devices rather than to the complete circuits, not a lot of wire is involved.

---

### Wiring partition walls

FURRING STRIP

SHEETROCK RUNS LONG TO COVER FURRING STRIPS AND WIRE CHASE.

STUD

DRILL THROUGH STUD.
WIRE CHASE
FURRING STRIP
STRESS-SKIN PANEL

---

### Vertical wire chase at timber post

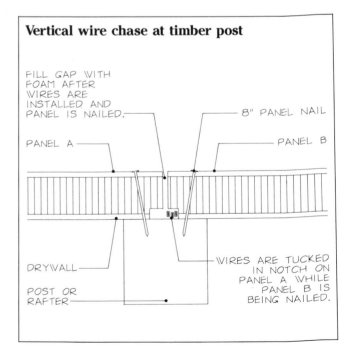

FILL GAP WITH FOAM AFTER WIRES ARE INSTALLED AND PANEL IS NAILED.

8" PANEL NAIL

PANEL A

PANEL B

DRYWALL

POST OR RAFTER

WIRES ARE TUCKED IN NOTCH ON PANEL A WHILE PANEL B IS BEING NAILED.

Single wires that need vertical access at posts can be pushed into the spaces left between panels after installation without the notching discussed previously. Wires to exterior floodlights or to feed a single circuit are often run this way. Make sure to push the wire deep into the seam and then fill the cavity with foam—if the wire isn't pushed in deep enough, it could prevent the foam from filling the joint. This technique should never be used for more than a couple of wires, as it could jeopardize both insulation quality and protection of the electrical line.

**Panel-spline chase** Wires can be run wherever two panels are splined together. After one of the pair of panels has been positioned, temporarily mount the wires on the center of its edge with a couple of light-gauge nails, as shown below. The wires must be located toward the center of the panel to protect them from possible damage by nails. From the inside and from the area of the extended-baseboard chase, drill a hole through the drywall on the installed panel and pull enough of the wire through to reach the first outlet box. Install the splines and the next panel on the frame, and then screw and nail the panels to the splines as described in Chapter V. Fill the cavity with foam, which will not only stop air infiltration but will also permanently bond the wires in place.

In a large two-story or three-story house where the path of wires from the upper areas to the service panel is unusually long and vertical access is difficult, it might make more sense to set up a sub-panel for a few of the more isolated circuits. The panel could be located in a closet or a built-in with an accessible partition. This makes it possible to run one large single line to the sub-panel and then distribute smaller wires to the circuits without having to take each line back to the main service panel.

**Wiring switch boxes next to exterior doors** These boxes tend to be quite large, because they are often control centers for interior and exterior lights. It is best to move these boxes to an interior wall where possible, since the air infiltration around them can be a large energy drain. If this is impossible, the electrician will need to mount the switch box in the stress-skin panel, and a chase must be created to get the necessary wires from the floor to the box. Because it is close to the switch location, try to create the space behind the nailer in the door's rough opening.

If the floor system is timber-framed, the wires for the switches are brought through the chase on the outside of the sill. For stick-framed floor systems, the wires are run through the joists, leaving enough wire to get to the switch box. With the rough opening for the door cut in the stress-skin panels and routed for stud nailers, create access for the wires by removing an additional depth of ¾ in. of foam behind the nailer location from the subfloor to switch-box height. The exact dimensions of the groove in the foam should be enough to accommodate all the wires with about 1 in. in width left over. This can easily be done by making a few kerfs with a power saw to the correct depth and then roughing the foam out with a wire brush mounted in a drill. If you have a router, use a straight bit and set it for about 2¼ in. deep. Cut the opening for the switch boxes from the inside using a keyhole saw started with a drill; because the switch location allows easy workability with power tools, a scroll saw could also be used to cut the switch-box opening. Make sure the cuts are precise and that a minimum amount of foam is removed. Now drill a hole from the door-jamb wire chase to the opening for the switch box.

In the back of the deepened groove at floor level, bring the wires into the chase by drilling a hole from the inside of the panel. Feed the wires into the chase and clip them to the center of the panel with light-gauge nails pressed in at an angle. Push the wires to the switch-box location, and the box is ready to mount with drywall clips. Install the 2x4 nailers around the door as described in Chapter V, then drill ¼-in.-dia. holes every 14 in. from the outside into the wire chase, being careful to stop drilling before the wires are endangered. Fill the cavity with expanding foam.

**Next to doors, wires for switches and lights can be run behind the 2x4 door nailers.**

---

**Vertical wire chase at panel spline**

FILL GAP WITH FOAM AFTER WIRES ARE INSTALLED.

WIRES ARE HELD IN PLACE BY PUSHING NAILS INTO PANEL DIAGONALLY.

SPLINE

## Wiring the floor system

One of the more notable obstacles to running wires in a timber-frame building is the lack of space in the floor system. Stud-framed structures have a natural cavity of 10 in. to 12 in. between the ceiling of one story and the subfloor of the next. But because the framing is meant to be seen in timber-frame houses, the ceiling and floor are often just opposite sides of the same material.

Good planning can keep this unique quality of a timber-frame home from becoming a liability. If there is a rule of thumb, it is to follow the path of least resistance and to avoid running wires across the floor needlessly. A wire might be fed much more easily through an interior partition than through the floor, for instance, even if the path is slightly longer. Once each wire is deemed essential, the path of the wires through the floor system can be planned.

With the electrical blueprint in hand, analyze a cross section of the floor system to determine the simplest, safest method to run the wires. (A detailed description of flooring options appears in Chapter X.) If there is only a single layer of planking over the timbers, it is probably best to groove the timbers for the wires with a router or with a power saw. But if there are other layers of flooring, there are probably simpler methods, which are discussed later in this section. If grooving is necessary, locate each one in the center of the top of the timber to give maximum area for nailing the floor on either side of the wire chase. Make the cut with a router or take several passes with a power saw and clean out the waste with a chisel. Either size the groove for a steel conduit to protect the wires or make an additional few shallow passes to allow a sheet-metal plate to cover the chase. For the sake of the timbers, remove as little wood as possible.

If the ceiling is to have a drywall finish and planking is to be used for the floor above, allow space for wires between the nailers and drywall. Before the planking is nailed on, cover the wire chase with a sheet-metal cap. Be sure the building inspector approves of this procedure first.

Here, the floor timbers were routed for wires. Before the finish flooring is installed, a metal plate should be laid over the wires.

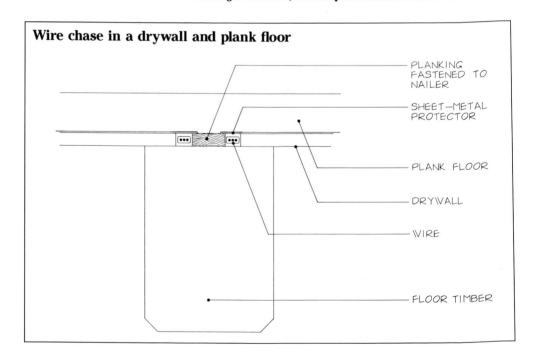

### Wire chase in a drywall and plank floor

PLANKING FASTENED TO NAILER

SHEET-METAL PROTECTOR

PLANK FLOOR

DRYWALL

WIRE

FLOOR TIMBER

When there is a layer of sound-deadening board between the floor and the ceiling, it is easy to rout a wire chase in the soft material of the board. An advantage of doing this is that the wires can be run in between timbers and presumably will be in a safer environment. Still, the wires should be protected within a metal conduit or covered with a sheet-metal cap.

One floor system, shown in the drawing below, even includes a natural cavity for running wires. Here 2x4s are used as nailers over the ceiling material and the sound-board, and the finish floor is nailed to the 2x4s. The space between the nailers is a perfect wire chase, although it is not deep enough to run wires without sheet-metal covers or conduit for protection.

Most often, wires run across floors to a location for a light fixture in the ceiling below. When the fixture is to be mounted on the timber, make sure that the timber will be large enough for the box and cover—usually, a minimum of 5 in. wide. The hole from the wire chase to the fixture should be as small as possible and drilled from the finish face of the timber for accuracy. If the fixture is to be mounted between timbers (which is not possible if there is only a single layer of planking above the timbers), the junction box will have to be mounted on the ceiling surface. If there are just one or two wires, it might be worthwhile to special-order a shallow box.

Another technique for mounting fixtures between beams is to make a fixture-mounting timber from the same wood as the frame and join it to the frame between joists or beams before the flooring is laid. This special timber can be routed for the wire and the electrical box before installation.

The small timber between the floor joists is used specifically to mount a light fixture.

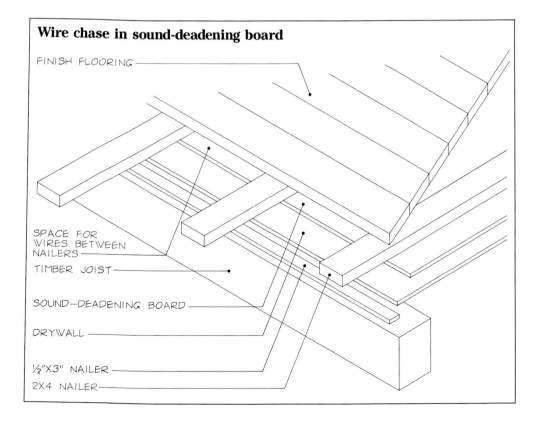

**Wire chase in sound-deadening board**

FINISH FLOORING

SPACE FOR WIRES BETWEEN NAILERS

TIMBER JOIST

SOUND-DEADENING BOARD

DRYWALL

½"X3" NAILER

2X4 NAILER

## Lighting and fixtures

Far too often, lighting is not even considered in the blueprint stage of a building project. It is left to be considered later, if at all. Yet lighting is a critical architectural element. Interesting spaces and unusual features are enhanced by proper and intelligent lighting, while rooms that are otherwise successful can fail if the lighting is poor. Much of lighting is subjective and much is a matter of common sense. I reject the notion that lighting needs to be left

**This house uses a mixture of modern track lights and a classic iron chandelier. The chandelier hangs from a purlin in the dining area, and track lights are mounted on several timbers in the kitchen.**

to the professional and can be understood only by those who have been initiated into the mysteries of the lumen. Consult books, articles and professionals, of course, but in the end permit yourself to be an expert about what is right for you. Only you know how you want a space to feel; only you know how it really will be used. If you intend to have a cozy, meditative corner, it will not need to be awash with light. On the other hand, a place where you intend to work with your hands (kitchen, sewing room or art studio) will probably require a great deal of light.

In a timber-frame home, lighting and fixtures need to co-exist with the frame, so when designing the room spaces, think about how you're going to light them, planning for the nooks and crannies as well as the soaring cathedrals. Try to avoid awkward room shapes that would be not only

hard to light, but would offer no drama or interest along with their complications. There's a difference between a dark spot in the corner of a room that creates a sense of mystery or allure, and a dark spot that is just dark. Because the flat planes of the walls and ceilings are interrupted by timbers, where you have lights you will also have shadows. Plan for them. The timber frame itself can be beautiful at night when properly illuminated, and if you think you would enjoy it, you can showcase the structure with lights.

If the floor plan includes the location of furniture and artwork, a well-conceived lighting plan will not be difficult to achieve. The first step is to determine whether it's desirable to light objects in the room, the room itself or both. Then you can consider the lighting options. For example,

**It's not always necessary to use special lighting techniques or fixtures. Switched outlets connected to lamps work adequately for many situations.**

when simple fixtures are mounted between the beams on a flat ceiling, the light will be focused onto the objects in the room. The beams will shade the light from the rest of the ceiling and cast shadows on the upper walls. Though this can be rather attractive in a dining room or kitchen, it might make a family room uninviting. To avoid the shadows, and to shift the emphasis from objects in the room to the room itself, use a shadeless fixture that hangs below the timbers; to accentuate the shadows, use a fixture having an upper shade, such as a billiard-room-style light fixture. To highlight the timbers and provide indirect lighting to the room, use uplights (having the shade on the bottom) mounted on perimeter walls and plan for the ceiling to be white between timbers to reflect the light.

**The kitchen and dining areas feature hanging lamps mounted on the ceiling between timbers.**

It is even possible to use built-in fixtures between the timber joists for subtle downlights. These lights can be either bright or muted, depending on the panel used to diffuse the light—anything from a high-intensity fiberglass diffuser to stained glass. To hold the panel, ⅝-in. rabbets are cut into the bottom of the timber headers or joists. A shallow light element (either fluorescent or tubular incandescent bulbs) is mounted to the ceiling between the joists, and then the panel is fitted into the rabbets.

Chandeliers or other fixtures hung from important beams receive added attention. But if you wish to conceal a fixture, use the timber as a valance by mounting the light on the side of a timber toward the outer perimeter of a room. Track lighting can be used creatively in this manner to wash walls with indirect light or to focus on an object.

The upper sides of roof purlins in open areas also provide good mounting for track lights. When track lights are used this way, be sure to get fixtures that can turn down past 90° or else the options for directing the light will be limited. With a full range of movement, the lights can be directed toward the ceiling to illuminate the frame or pointed down for general lighting or to highlight artwork or furniture.

Frames for large areas often have hammer beams or king-post and queen-post trusses, which can provide a mounting surface for individual fixtures or track lights. It is best in these situations if the fixtures have flexible heads, to allow the light to be focused in a number of directions. To recess tracks or boxes, dados can be cut in the timbers, but this is work best done before the frame is raised. When the tops of the timbers are not visible, it's even possible to hide a tubular light in a dado.

When houses have been well designed, the living spaces receive the natural light appropriate to their use during daylight hours. Artificial light is often intended to imitate daylight, but remember that activities and moods change during evening hours and there are new requirements involving entertainment, relaxation and romance. The easiest way to achieve the desired results while maintaining flexibility is to overlight and then use a dimmer to control the quality of light. Additional fixtures could also be installed to provide extra light when needed.

## Timber-joist ceiling fixture

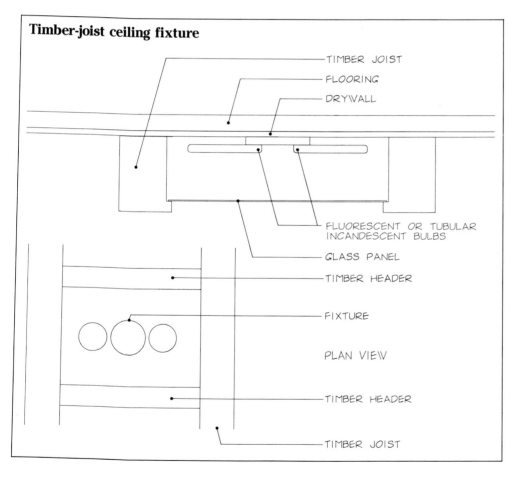

# VIII. Plumbing

**Plumbing in the timber-frame house is mostly a matter of camouflage, since the building system leaves few natural cavities for pipes—especially those that run horizontally. In this case, the horizontal pipes are hidden in the platform beneath the tub.**

I made a commitment to get more involved with the design of timber-frame houses one afternoon about ten years ago when I learned of a disaster caused by plumbing. A few months earlier we had completed a project in one of the local communities. We had helped with design, but on a very minimal level. When the frame was finished, we gave the clients a few words of advice, told them to call if necessary, and went to work on another frame. Later on I happened upon the plumber for the project, and asked him how things were going. He told me everything was fine, but, "Boy, I had trouble finding a 6-in. drill bit." I was puzzled. "What 6-in. drill bit?" I wondered. "Well, I had to go all the way to Boston and spend $90 to find a 6-in. drill bit to make a hole through your 8x8 timber for that waste line by the first-floor bath." I feel certain the color of my face and my open mouth said more than words could have. The plumber had acquired a new bit, and our frame had lost a timber. It was too late to argue or place blame, but I knew that from then on things would have to change. In a way, that plumber did me a favor. He helped me to understand that I would have to address the construction idiosyncrasies inherent in timber-frame building, and that I could not expect people in the various trades to solve problems foreign to their experience. As a result, I was forced to become more involved with the design of timber-frame houses and the way in which certain jobs, such as plumbing, were handled.

The problem with plumbing a timber-frame house is not so much how to do it, for the standards and techniques are essentially the same as those used in conventional construction, but how to hide it. In this respect, wiring and plumbing are similar. Much of the task is to create spaces in which to install the workings of the system and then to camouflage those spaces. This may sound simple in principle, but you'd be surprised at how often the principle is mishandled. By far the most common problem we see in completed plans for timber-frame houses is unresolved plumbing situations. It's as if in the heat of working out the floor plans and the aesthetics of the building, the designers (who are sometimes registered architects) simply forgot that plumbing fixtures must be attached to pipes to operate. When we ask how the problems will be solved, we're often told, "Oh, we'll just box the pipes in somehow." But while this may be appropriate in a laundry room or closet, it could ruin the looks of a living room, dining room or bedroom.

The secret to a successful plumbing job in the timber-frame home is early and careful planning. This is then followed by an intelligent selection of the best options for each particular situation. For the options to make sense, however, it's important to have at least a basic understanding of a few simple plumbing concepts, which we'll cover now. Then we'll look at planning and creating hidden spaces for pipes.

# Some basic plumbing

The fundamental aim of every plumbing system is to pump clean water in and to drain used water out. Water usually comes from either a city supply (most often a lake or reservoir) or from a well on the site. When water leaves the house, it is directed to either a city sewage-treatment center or a septic system near the building.

Water is brought to the house through a network of pipes called the supply system, and flows out after use through a network called the drainage system or sometimes DWV (for drain, waste, vent). In the supply system, water is pumped through a main trunk line to the basement or designated utility area, where one major branch pipe connects to the hot-water heater. From there smaller branch pipes carry hot and cold water to various parts of the house, feeding into even smaller pipes for individual fixtures. Most codes require supply pipes to be copper, although PVC is gaining acceptance. Because the water in the supply pipes is under pressure from the city water main or well pump, no air is required to assist the flow. This means

that the pipes can be smaller than drainage pipes (which are merely conduits for water to flow by the force of gravity)—generally no larger than 1 in. and no smaller than ⅜ in. In addition, the pipes can be run in any direction on their way from the source to the fixture.

The drainage system is more complicated than the supply system, and therefore is the focus of all our planning. When the space requirements of the drainage system have been satisfied, the requirements of the supply system will almost always automatically be satisfied as well. The drawing below shows the anatomy of a typical drainage system. Basically, it works like this: When you pull the drain plug on your bathtub, gravity pulls the water down a short vertical section of pipe to the trap, where some is captured in the loop to prevent gases in the line from rising into the living area. (Toilets don't need traps because they always contain water.) When the water passes through the trap, it drops into a pipe that is slightly pitched, called a branch drain. The branch drain directs the water to a larger verti-

## Typical drainage system

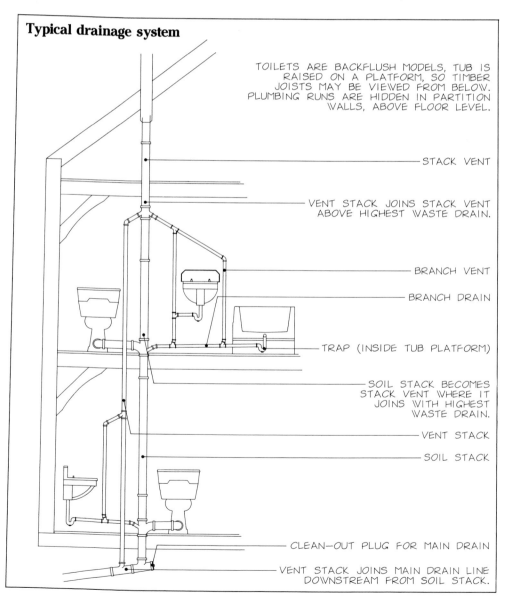

TOILETS ARE BACKFLUSH MODELS, TUB IS RAISED ON A PLATFORM, SO TIMBER JOISTS MAY BE VIEWED FROM BELOW. PLUMBING RUNS ARE HIDDEN IN PARTITION WALLS, ABOVE FLOOR LEVEL.

STACK VENT

VENT STACK JOINS STACK VENT ABOVE HIGHEST WASTE DRAIN.

BRANCH VENT

BRANCH DRAIN

TRAP (INSIDE TUB PLATFORM)

SOIL STACK BECOMES STACK VENT WHERE IT JOINS WITH HIGHEST WASTE DRAIN.

VENT STACK

SOIL STACK

CLEAN-OUT PLUG FOR MAIN DRAIN

VENT STACK JOINS MAIN DRAIN LINE DOWNSTREAM FROM SOIL STACK.

cal pipe, the soil stack. A pipe connected to the branch drain just beyond the trap allows air into the branch drain and releases the sewage gases. This branch vent rises vertically to a point above the fixtures and then pitches upward toward the upper part of the soil stack, which is called the stack vent. Essentially, drainage occurs in the lower portion of the stack while the upper part brings air to the branch vents and allows the sewage gases to escape. Meanwhile, the bath water drops through the soil stack until it reaches the main building drain, which carries all of the household waste water and sewage to the septic tank or town sewer.

Because the water in the drain pipes is not pressurized, the pipes must be oversize to allow air in the lines. The smallest drain for an individual fixture would have a 1½-in. diameter and the largest pipe, the soil stack, would have a 4½-in. diameter. (This pipe won't even fit within a typical stud wall.) Drainage pipes are commonly made of ABS plastic, although cast iron is often a better choice for the stack because it helps muffle the sound of the water as it cascades downward. Branch drains must be pitched all the way to their stacks; codes require a minimum pitch of ¼ in. per foot. Tubs, showers and toilets require the most planning because their drains are usually underneath the fixture, requiring space in the floor system. For most situations, the minimum space required beneath fixtures for drains is 6 in., but it can be much more if the run from the drain to the stack is long. For instance, if a pipe were pitched ¼ in. per foot and had to run 12 ft., an additional 3 in. would be required from the point of connection to the branch drain. Sinks are simpler to deal with because their traps and drains are above the floor. If the stack is not too far away, sink drains can be run to it without a special cavity.

## Designing the plumbing system

Plumbing design should begin at the earliest stages of house design, when there is only blank paper and a full bank account: You will want to know that everything has been done to avoid unnecessary complications. No doubt, even after careful planning there will still be situations calling for imaginative thinking, but keep these several rules of thumb in mind.

**1.** The most important rule is to consolidate plumbing areas wherever possible. Primary plumbing facilities (bath, laundry and kitchen) should be clustered in one section of the house, with plumbing areas on different floors stacked atop one another. The benefits are that pipe runs will be kept short, and there will be more opportunities for shared drains and stacks. Fewer feet of pipe minimize potential problems with the plumbing system; fewer vents passing through the roof reduce potential for heat loss and present less of a visual disturbance. Consolidation of plumbing areas also saves money. Hotel designers have this down to a science, and always place bathrooms back to back and stack them one over another.

**2.** After the plumbing areas are clustered, consolidate further by trying to get supply and drain lines to share a single chase in a common wet wall. If this isn't possible, use more than one wet wall for supply and drains, but attach the drains to only one stack. Keep working toward the simplest and most compact arrangement until the number of plumbing walls and the lengths of pipe required have been reduced to a minimum.

**3.** To reduce noise, avoid locating wet walls adjacent to major living areas. Wherever possible, run the plumbing through secondary partitions, such as those that enclose a bathtub or shower stall, or the partition between a bathroom and laundry.

Likewise, when plumbing areas can't be stacked floor to floor, avoid locating wet walls over or under open areas or primary rooms. Remember that drains go down and vents go up. When a bathroom on the second floor falls over a dining room on the first floor, the problem of hiding pipes immediately becomes difficult. Creative solutions, even design innovations, can arise from such situations, but much more often the result is disappointment. I know of a house with a hollow fake post, a camouflaged mastery that gurgles from time to time. And I've heard a story about an eccentric with an interest in the details of his household plumbing—he required Plexiglas pipes that would be revealed to the living room. When the suggestion came, my bet is that the designer wished he had located the second-floor plumbing somewhere else. I suppose exposed pipes are an option, and recently there has been a movement among some avant-garde architects to reveal all the mechanics of a building, but so far this "warehouse architecture" hasn't gained much momentum. Most of us are happiest with the more old-fashioned practices.

**4.** Partition walls that align with beams cannot be used as wet walls, because it would then be necessary to drill through the timbers. (This principle conflicts with the one established in Chapter IV, which says timbers should be used to define rooms, but supports the idea that plumbing walls should not be adjacent to primary living areas.) Instead, try to use secondary partitions, such as closet and bathroom walls, for wet walls. These secondary walls seldom have interference from timbers and are separate enough from the rest of the house so that their treatment need not match the decor of the primary living areas (see the discussion in this chapter on vertical pipe chases). Wet walls should align on each level so that pipes can run uninterrupted from the first floor to the roof.

**5.** In any plumbing area, position the fixtures as close as possible to the soil stack. When there are several fixtures, try to arrange them so that those with the most critical drainage requirements (usually the toilet first, then the tub) are closer to the stack. It would not make sense to back the sink up to the soil stack and locate the toilet in the far corner of the room, because then its drain would have to travel a long distance to the stack. Draining water from the sink is much less difficult than draining waste

from the toilet, and the branch drain from the sink can be run to the stack without interaction with the floor.

**6.** Don't plan to run plumbing through exterior walls. It would compromise insulation and allow pipes to freeze.

These guidelines may sound like a lot to keep in mind, but they actually dovetail nicely with good house design; therefore, many of them come rather naturally. Consider the following example of plumbing layout in the three-bay timber-frame house we used as an example in Chapter IV. The major living areas are gathered in one end of the house (for the sun or the view) and the secondary areas are pushed to the opposite end, toward the north. The kitchen is located near the utility entrance for convenience when bringing groceries into the house—part of the plumbing has therefore automatically been located. People doing household chores shouldn't have to run from one

## Plumbing layout in simple three-bay house

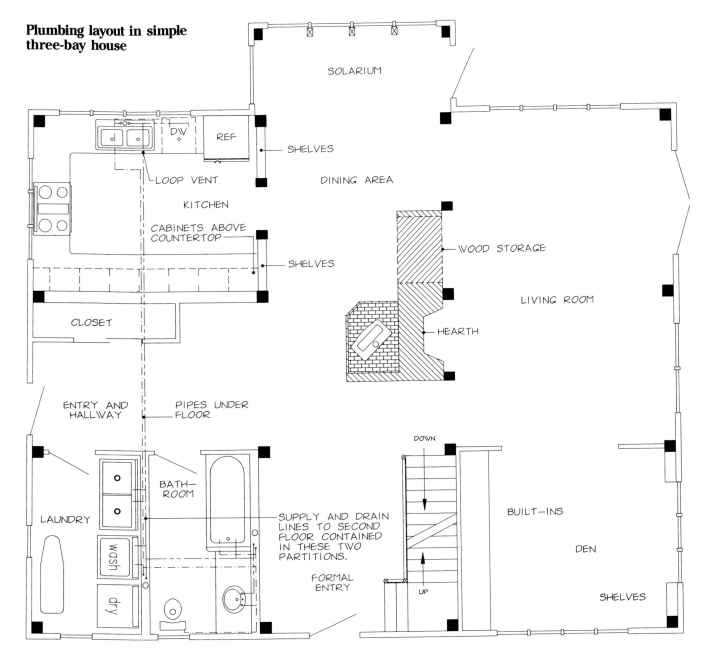

end of the house to the other, so we try to keep the laundry (if it is to be located on this level) near the kitchen. Next, we attempt to back up the bathroom to the laundry. In the best possible arrangement, the wall that contains pipes for the laundry is shared by the bath. This not only makes plumbing easier, but also removes the wet wall from the living areas. So, by following some natural design inclinations, all of the plumbing is in one bay. The rest of the house is open and unencumbered by partitions, and yet the bathroom is convenient, especially since the middle bay typically provides central access to the outer bays.

The second floor would follow the pattern of the first. Bedrooms would be located toward the view or the sun, with closets and bathrooms pushed toward the opposite end. Take the further step of stacking the second-floor bath over the first-floor plumbing area, and the difficulty of creating spaces for pipes will be minimal.

In residential construction, the set of plumbing plans itself is not usually very complicated or precise. Plan views are drawn as overlays on the floor plan. The plan views show the location of wet walls in relation to fixtures and depict pipe routes, but do not necessarily show pipe connections, elbows, etc. Typical building sections and sections specific to the plumbing area give the plumber and the carpenter an exact description of each plumbing chase. These plans and elevations should show hot and cold supply lines, drains, stacks and vents, but again it's not usually necessary to be specific about pipe connections. The plans should be drawn by the designer and reviewed carefully with both the plumber and the carpenter before work commences. This is one of the places where it's critical that the separate trades work together. The designer plans the spaces, the carpenter builds the spaces and the plumber uses the spaces. Together they can accomplish near miracles of deception. As separate forces, they can touch off a small war.

It would take a large budget, and a very skilled carpenter and plumber, to rescue a project if the design criteria I've listed are neglected. Yet even if you give planning the plumbing system your best shot, house design includes a consideration of many factors and often it will not be possible to adhere to each guideline. Final decisions on frame design and room layout will inevitably include compromises that will affect the complexity of the plumbing system. To be properly prepared for planning, it is therefore necessary to have a small arsenal of solutions for the various problems that may arise. Although vent pipes often angle diagonally upward and drain pipes always pitch toward the stack, pipe chases are always oriented either vertically or horizontally. We'll look at some typical solutions in terms of these requirements.

## Vertical pipe chases

The most obvious and common vertical chase is an interior partition. But as I've mentioned, the difficulty is that in a timber-frame house there should be a deliberate alignment of partitions and beams—a goal that conflicts with creating pipe chases. To allow the unimpeded passage of pipes, it is necessary to design wet walls to stand away from beams. This idea needs to be conceived in the planning stage; if it is, execution is easy.

When framing wet walls, remember that partitions in a timber-frame house exist only to separate rooms. They have no load-bearing function, so there need be only enough structure within the walls for stiffness and to support the finish surfaces. Therefore, try to anticipate where pipes will fall within wet walls and eliminate or reduce structure as much as possible. For instance, if the wall is being built in sections anyway, consider constructing it so there will be no obstruction to the soil stack. There certainly isn't any reason to have top and bottom plates in the wall when the hole for the stack will eliminate any structural usefulness the plates might otherwise have had.

When wet walls fall by necessity at beam locations, there are three choices, none of which calls for drilling through the timber. The first choice is to build two parallel walls between which to hide the plumbing—one wall between timbers, the other in the plumbing area. In this case, each wall supports sheathing on only one side, and the structural requirements are almost nil. To reduce the cost of lum-

This bathroom employs parallel walls to allow pipes to pass between them. Two-by-three studs for each wall would have worked as well as wider boards.

ber, frame each wall with 2x3s. Hold the inside wall away from the beam far enough to contain the pipes. The advantage of this method is that the reveal of timber to the living area is kept consistent throughout the house. The disadvantage is that the aesthetic value of the timber is lost to the room with the wet wall. In a bathroom or laundry area, this may be a worthwhile compromise.

The second choice is to build only one wall and to locate it completely on the plumbing-area side of the timbers. The timbers are then concealed from the plumbing area, and their full depth is completely exposed to the living area in the same manner as those on the exterior walls. This is most acceptable in a house having no other first-floor partitions besides those surrounding the plumbing areas (as in the floor plan on p. 155). Otherwise, the relationship of timbers to walls will be noticeably inconsistent. The walls can be created the same way as built-up exterior walls (p. 94), by applying the drywall first and the framing members afterward. Because nothing has to be fitted between timbers, this technique is easy.

The third choice is to box the pipes in a corner or in a specially designed chase, which allows the timbers to be revealed to the plumbing area as well as the living area. This works well if the plumbing can be run in the wall at the end of a shower stall or in the back of a cabinet. The partition wall is then built between timbers as usual, with a chase for drains, stacks and supply located on the inside of

This plumbing wall is built on the bathroom side of the timbers to allow pipes to be run without compromising the frame.

The partition between this sink and tub conceals a plumbing chase. It falls to the side of the timber, so that pipes can be installed easily.

the wall. Bathrooms and kitchens are sometimes especially dense with cabinets and closets, and it is occasionally possible to camouflage pipes on the inside of a room.

Despite the best intentions, it is not always possible to prevent plumbing from falling over or under living areas without convenient wet walls. The first order of business is to look for an opportunity to make an unobtrusive vertical chase from some other design element in the house, and then to use horizontal runs, described in the next section, to get the pipes to the chase. The chase can be a closet or built-in cabinet or even a chimney. (See the photo on p. 65—the built-in cabinet in the dining room conceals plumbing.) In the latter case, all that is usually necessary is to make a masonry partition within the chimney to isolate the pipes from the flues. (Check local code before proceeding.) If no enclosure is available, it may be necessary to design a built-in cabinet or shelving unit to serve the purpose. The trick, of course, is to make the built-in look natural while still providing adequate space for pipes.

To keep noise from the drain pipes from disturbing the living area, make the chase large enough to leave room for packing the pipes with fiberglass insulation. Using cast iron instead of plastic for vertical drains can also cut sound transmission.

## Horizontal pipe chases

The basic issue with horizontal pipe chases is to find a way to run pipes in the floor, given the absence of a cavity. In a timber-frame house, the floor material is usually the ceiling for the living area below, consisting either of planks over beams or a thin, built-up membrane of flooring and ceiling materials (see Chapter X). These floor systems provide aesthetic and economic benefits, but the lack of space for plumbing is a significant obstacle. With good planning, however, the problem (and the pipes) can disappear.

Assume that through careful design, the plumbing areas have been consolidated, and convenient vertical chases for supply pipes and stacks have been provided. The next step is to arrange the fixtures so they have an efficient relationship with the plumbing walls. As I mentioned earlier, by far the most difficult fixtures to accommodate are toilets, tubs and showers. The drains for these are usually in the floor and, in the case of the toilet, there is both solid waste and water. Obviously, if these fixtures are located a great distance away from the plumbing walls, the required length and slope of the pipes create difficulties. If at all possible, the toilet should be the closest fixture to the soil stack, the tub or shower next, and the sink farthest away. Furthermore, all fixtures should be as close as possible to the plumbing wall and stack. The natural extension of this rule is that vertical and horizontal chases are designed simultaneously and for the benefit of each other.

After wet walls have been determined and fixtures arranged, you must devise a way to get the horizontal pipes to connect with the vertical pipes. In most cases the horizontal pipes run in a chase built either above floor level or below it. In a few lucky situations, no chase will be required, such as when supply and drain pipes can be passed directly from the fixture to the wet wall. This is

**Raising the floor for plumbing need not inhibit the proper functioning of household systems. In this bathroom, the sink (not shown) is on the same level as the hallway, while the toilet and shower are a step up.**

quite possible with all sinks and with backflush toilets. A backflush toilet has an outlet in the rear instead of underneath. It's more expensive but worth the expense if it eliminates the carpentry involved in creating a chase. In the basement, it just might make sense to let the pipes run through the floor and under the timbers in the typical helter-skelter manner. If this area is not used for living, it is impractical to spend money and effort to hide what will seldom be seen. Anyway, it's often helpful to have the pipes exposed in the basement for maintenance, especially at the points where water enters and leaves the house.

But, more likely than not, there still will be a need for a horizontal chase. The simplest, least compromising solution is to place some of the fixtures on a raised platform while leaving most of the floor level undisturbed. Doing this separates the bathroom into distinct areas, making it a design device as well as a way to conceal pipes. If the sink is close to the entrance and the tub and toilet farthest away, the sink would remain on the common-floor level and the other fixtures would be raised. A raised tub is considered something of a luxury, and a raised toilet doesn't necessarily affect comfort or convenience. Or the sink and a backflush toilet could be hooked directly to branch drains in the wall, with just the tub elevated. Often, it will be necessary to build a chase that continues around most of the perimeter of the room, with a platform only for the fixtures. Technically, the chase for the sink plumbing has to be just large enough for its pipes, while the tub and toilet chase has to be large enough for both the fixtures and pipes. But after fulfilling this requirement, we can increase the chase area to whatever size benefits room design.

The structure of the platform is simple, since it has little load-bearing responsibility. Construct it using conventional floor-framing standards. Joists are typically 2x6s placed 16 in. on center, but if the plumbing run is long, wider joists (and therefore a deeper chase) may be required to allow the proper pitch from the fixture to the stack. The finished ceiling for the room below is installed first, whether it is planking or drywall, and then the joists are laid on top, perpendicular to the timber joists and attached by toe-nailing. Before beginning, check the manufacturer's specifications and be sure that the fixture will have adequate support.

If raising the floor is awkward or not desirable, the plumbing can be run in a cavity between timber joists. But because this system has severe limitations, we usually don't recommend it. Since plumbing can fall only between joists, the direction of travel is restricted, and the distance of travel is limited by the next intersecting beam; this means that the branch drains running between a pair of timber joists must connect with a vertical chase before they meet up with the next perpendicular beam. The distance of travel for a branch drain that runs between timber joists is also limited by the relationship between the depth of the joists and the drop of the drain as it pitches toward the stack. If it goes too low, the timber ceiling will be lost. Yet another problem is that it is unlikely that all of the fixtures will be located over the same joist bay, causing multiple branch drains where they wouldn't normally be necessary. Of course, this sort of chase has an impact on the room below. It is usually desirable to finish

the entire ceiling in the same manner as the area with the plumbing chase, in order to keep the look consistent. This does not necessarily present a problem, but it does cause extra expense.

When plumbing is consolidated, as discussed earlier, optimally there will be a bathroom or two on the second floor located directly above the plumbing on the first floor. With such dense plumbing, it is often prudent to allow these areas to have different aesthetics from the rest of the house. The most practical solution might be to drop a ceiling completely beneath the timbers, as shown in the drawing below, but this is desirable only if all the plumbing is localized and if the area is properly isolated. It is also only possible if the timber ceiling is high enough to allow a finish ceiling to be up to 6 in. below the bottom of the timber joists.

### Dropped-ceiling plumbing chase

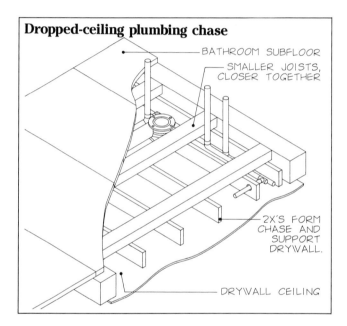

BATHROOM SUBFLOOR

SMALLER JOISTS, CLOSER TOGETHER

2X'S FORM CHASE AND SUPPORT DRYWALL.

DRYWALL CEILING

### Plumbing run between timber joists

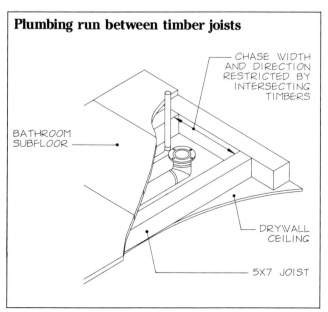

CHASE WIDTH AND DIRECTION RESTRICTED BY INTERSECTING TIMBERS

BATHROOM SUBFLOOR

DRYWALL CEILING

5X7 JOIST

When it's important to maintain the timber ceiling and the floor above can't be raised (for instance, if there are height restrictions imposed by the roofline), you can design the frame with the timber joists dropped in the area of the plumbing. The idea is to keep the timbers revealed below and the floor level consistent above by laying a built-up floor over a lowered timber ceiling—the cavity above the timber joists is the plumbing chase. This works particularly well if the area below benefits from the coziness a lowered ceiling produces: The only division between a living room and study might be the point at which the ceiling becomes lower. (The ceiling must meet code requirements for height, and usually can't be any lower than 7 ft.) To keep the perimeter timbers exposed as they would be in other parts of the frame, use deep beams (typically 12 in. to 14 in.) to receive the joists. Join the joists far enough below the tops of the timbers to leave room for the chase. A common height for the chase is 5½ in., to fit a standard 2x6 (or 2x4, if no accommodation need be made for the 4-in.-dia. waste line from a toilet). Make sure the drain pipes have an escapement from the timber-floor system at the point where they will be directed toward the vertical chase; the timber adjacent to the vertical chase should be dropped to allow the pipes to pass through. Include the finish ceiling material when considering the depth of the chase. Whether drywall or planking, it should run over the top of the joists as it does in the rest of the house.

If the frame is designed with this feature, creating the plumbing chase is easy. First install the ceiling material over the timbers as discussed in Chapter X. The built-up floor and plumbing chase is then constructed with conventional materials and techniques, with the finish floor level maintained from the timber floor to the plumbing chase area. I like to think of this system as an example of how a problem can be transformed into a design asset. It is also an example of how good planning can reward.

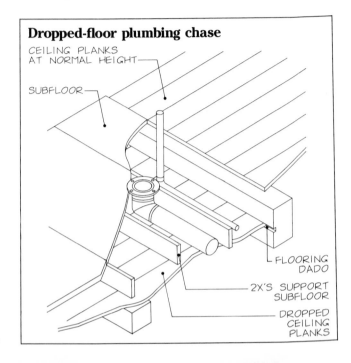

**Dropped-floor plumbing chase**

CEILING PLANKS AT NORMAL HEIGHT

SUBFLOOR

FLOORING DADO

2X'S SUPPORT SUBFLOOR

DROPPED CEILING PLANKS

**The floor joists have been lowered in this oversize bent girt to make room for a false floor above the joists. The routed groove in the girt will receive the floor planks.**

## Sink vents

Kitchen plumbing often poses a problem in timber-frame houses because the kitchen is the one space that is not easily isolated from the primary living areas. In addition, during design an attempt is usually made to make the kitchen spacious and open, which further reduces the opportunities for convenient plumbing walls.

In the kitchen, the items of concern are usually the sink and the dishwasher. Supply and drainage pipes aren't often a problem for these, because the kitchen is usually located on the first floor and plumbing can run between joists directly to the basement. Because the sink and dishwasher are almost always side by side, they can share a

**Plumbing in a kitchen is usually simple enough to make it possible to eliminate plumbing walls.**

drain and vent, which reduces the plumbing to a single hot and cold supply line and a single branch drain and vent. But when the sink is on an exterior wall or in an island, a problem arises in determining where to put the branch vent. You don't want to run a vertical pipe chase in an exterior wall because this would compromise the insulation, and you certainly don't want to run pipes directly up from an island through the room.

There are two common solutions to this problem—the loop vent and the mechanical vent—but before choosing either, check local code. With the loop vent (also called a dry vent), the pipes form a long upside-down *U* up to just under the top of the counter and then go back down to an air-intake line, as shown below. The loop vent must rise at least 6 in. above the drain to prevent siphoning waste water. The system is simple to install, uses normal plumbing fittings and usually works well, but the vent does take up space in the cabinet and some codes don't allow it.

The mechanical vent is a device that is installed into the top of a pipe that extends 6 in. above the sink drain. It works by using water pressure to open and close a valve that lets air into the line when water is draining and reseals when the drain is not operating. The disadvantage here is that the device can wear out; many codes don't allow it anyway. The advantage is that it's easy to install and easily replaced if there's a problem.

If neither of these options can be used, a vertical plumbing chase may be fabricated in a kitchen cabinet. This solution is often used but seldom successful, because the normal arrangement of kitchens and cabinets makes it difficult to hide a vertical pipe chase. The cabinet that will house the chase should reach to the ceiling and be as close as possible to the sink—examples would be a cabinet pantry or an extension of the cabinet next to the refrigerator. Modifications to the cabinet would need to be made to make room for the pipe.

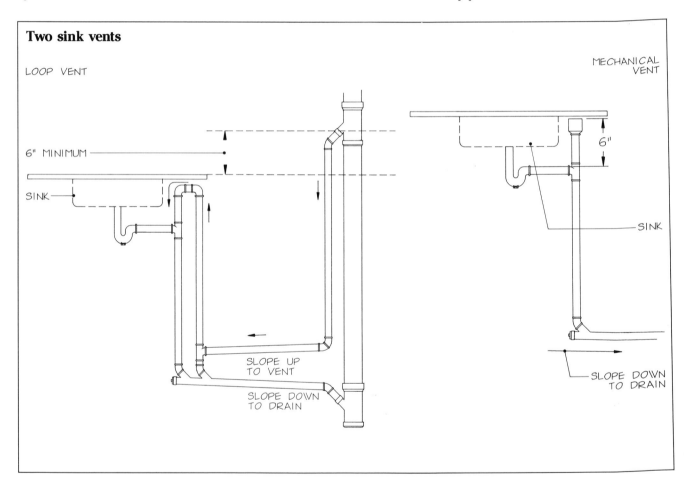

**Two sink vents**

LOOP VENT

MECHANICAL VENT

6" MINIMUM

SINK

SLOPE UP TO VENT

SLOPE DOWN TO DRAIN

6"

SINK

SLOPE DOWN TO DRAIN

# Notes on heating and cooling

Installing heating and cooling systems in timber-frame homes is somewhat troublesome, because there aren't many hidden spaces and people are usually reluctant to commit space to ductwork. In addition, cutting into a space for a heating or cooling duct spoils the definition and balance provided to the room by the timber frame. The solution in large part is to minimize equipment (and fuel consumption) by making energy efficiency a top priority in house design.

Basically, heating and cooling options will be described when the floor plan, insulation levels, and location and number of windows have been determined. The energy audit we commission on each of our houses (see p. 90), gives us a computerized series of "what ifs" that allow us to see the relative heat performance of various combinations of insulation and glazing. Before committing to any plan, we can determine with some certainty the point of diminishing returns—when money spent doesn't mean appreciably

better performance. When a strategy has been determined, the same heating engineer designs the complete system, keeping it as simple and small as possible. A problem with the conventional method of allowing the installer's supplier to lay out the system is that the result is typically three or four times larger than necessary. Let's face it, the guy pushing furnaces has little incentive to worry about future heating bills. We have found it's better to hire a person who is trained to think, not sell, and to address the problem while there is still design flexibility.

A simple, well-insulated, three-bay timber-frame house with an open area in the middle usually requires only one small, central heat source. It could be the sun, a woodstove or a small furnace. A little heat (about 30,000 to 40,000 BTUs for about 2,500 sq. ft.) emanating from the central core should keep the whole house warm. If a hot-air furnace is used, the ducts run under the first floor to strategically located registers (perhaps as few as five), which bring warm air to the first level. It is usually not necessary to run the heating system to the second or third levels, since the rooms on these floors use the open area as a plenum and receive heat through conduction and convection. The same thing can be done with hot-water or electric-baseboard heat. And that, for most houses we build, is the solution to the problem of the heating system.

Should it become necessary to run heat to various parts of the house, hot-air systems, which require extensive ductwork to move air, are not generally the first choice for timber-frame houses because of the lack of spaces. Unless a convenient cavity can be found for each duct, it is too difficult to be worthwhile. Sometimes we'll choose to use a combination of hot-air heat to the first floor with a few electric-baseboard heaters where necessary on the second floor. Considering the difference in cost between running the ducts and installing the baseboards and how rarely the electric units would actually operate, this alternative can be quite successful.

**The open area in the middle bay allows heat to rise naturally to the upper regions of the house without the use of heating ducts.**

To bring air to the furnace, a return air duct made from sheet metal is often housed in the masonry chimney. The duct must be completely isolated from flues with a solid-brick partition, and is sized according to the specifications of the heating system. The duct should be packed around the outside with fiberglass insulation to keep it from contacting masonry and to isolate it from stray smoke and flue gases. Even if the furnace does not require return air, it might make sense to put a duct into the chimney for recycling the air. The duct can be used like a fan to even out the temperature between the upper and lower levels of the house. Obviously, sheet-metal ducts, unlike flues, are capped at the peak, inside the house.

Hydronic (hot water) and electric systems are certainly easier to accommodate in the crannies of a timber-frame house than hot air. Because the pipes for hydronic systems are rarely larger than ¾-in. dia., they can be run through partitions and in some of the floor systems, using the guidelines in this chapter. For instance, avoid situations that require drilling holes through timbers. When the pipes need to pass around the outside walls, have the posts notched before the insulation is installed over the frame.

**The grill near the top of the wall covers an opening to the return air duct, which brings air down to the basement for the furnace. This duct can also be used for an air-to-air heat exchanger, or simply to circulate the house air.**

Electric heat is by far the easiest to install because all the wires can be run along with the rest of the house wiring. But unfortunately for most of the country, electric heat is also the most expensive per BTU. As I mentioned before, we often use electric heat as a backup for other systems, such as in houses having good passive-solar systems and a woodstove. A few of our clients with electric backup systems have never actually used them. But bankers who finance construction still usually like to see big, impressive heating systems, even though the most successful heating systems are quite unimpressive.

Cooling systems are hard to disguise, but the same design concepts that keep a heating system small can make a cooling system superfluous. There are many parts of the country, however, where it is almost a necessity, no matter how well the house is conceived and built. While heating systems should be placed at the bottom of the house because warm air rises, cooling systems should be placed at the top to let the cool air drop. Air conditioners require quite a bit of room, and it is almost inevitable that space for these units and their ducts will have to be designed into the plan. One of our houses uses just one supply, located close to the peak (immediately above the collar ties) in the central bay. The cool air falls through the central area and cools the outer perimeter enough to keep the house comfortable. If it is necessary to feed other areas, keep the unit high in the house and run the ducts down to the rooms. Closets, knee walls and other secondary or inaccessible areas are the best places for ducts.

The preceding paragraph probably sounds like a quick dismissal of the problem of installing cooling systems. It's not, really—there just isn't much to say. Air-conditioning ducts and equipment potentially can consume large quantities of space, so the first step is to design a well-insulated house that rejects heat. This way, only a small system will be necessary, one that does not need much space. Then, you specifically locate the places for equipment and ducts, knowing that design flexibility inevitably will be sacrificed to the necessities of the system. The theories and techniques described throughout this chapter for finding and creating cavities and avoiding timbers should apply to the cooling system, too.

**Return air duct in chimney**

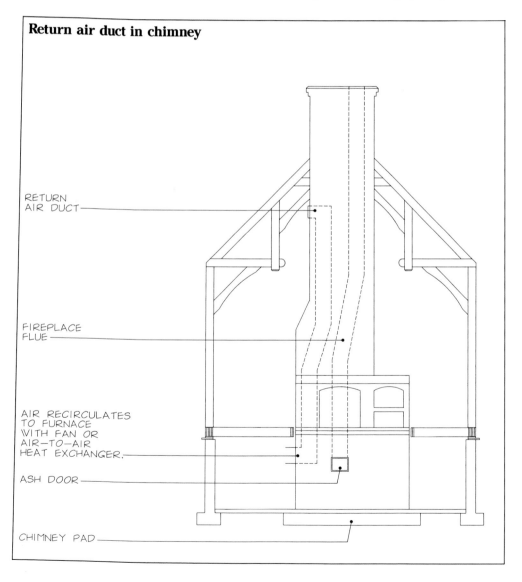

RETURN AIR DUCT

FIREPLACE FLUE

AIR RECIRCULATES TO FURNACE WITH FAN OR AIR-TO-AIR HEAT EXCHANGER

ASH DOOR

CHIMNEY PAD

# IX. Frame Details

**Pendants embellish the bottoms of the second-story posts in the framed overhang.**

In any house there are many thousands of design and construction details. Most are taken for granted; if they go unnoticed, they are successful. These are the details that keep treads and risers at a consistent dimension so people don't trip, that keep wooden posts from rotting on concrete floors and that prevent drywall from cracking at the seams. Failure to pay attention to such detail is the definition of poor construction. It causes the need for constant repair and makes the house more of an annoyance than a joy. Other kinds of details can become highlights of the decor. These are the "frivolous" details which, although technically unnecessary, are undeniably important to good house construction—the frosting on the cake. Embellishing the edges of timbers is not essential, but it can change the feeling of a living space from rustic to elegant. Designing an alcove window seat might seem to be an extra expense and bother, but could result in providing the most inviting place in the house. Inevitably, what anyone in house construction learns is this simple lesson: Everything matters. Good house construction requires attention to details both large and small.

In this chapter and the next (which is devoted to exterior and interior finish details), we will take a look at both the frosting and the cake, exploring some points that were not discussed in other chapters. Here I cover frame embellishments, roof windows, chimney and fireplace openings, exterior timbers (roof overhangs and porches), roofline changes and solariums. It is not my intention to explicate every possible circumstance; rather, I offer a smattering of ideas in a few critical categories associated with timber-frame construction. Don't look at any of these discussions as prescriptions to be used without thought of revision or modification. Instead, it is my hope that they will provide an outline for a way to think about details, and become a springboard to discovering the solutions best suited to your own home.

## Frame embellishments

Embellishments are refinements made to the edges and ends of a timber. They may be as simple as quarter-rounds or bevels to soften hard timber edges, or as complex as elaborately carved pendants at the ends of posts. Whatever the case, determining the level, type and location of embellishments is an important part of the design process that should occur during the development of the frame and floor plans. Even the decision to use no embellishment at all ought to be well considered. Timbers are milled to a severe rectilinear shape, and leaving them that way will affect frame design as surely as carving gargoyles on the rafter ends.

**Edge treatments**   For their origins, edge treatments probably owe more to practicality than decoration. Ceilings in ancient timber frames were sometimes so low that they were a problem even for our shorter ancestors. Beveling those hard timber edges was a safety precaution that grew into a design feature. In those days the work was done with a drawknife and plane, but today typical edge treatments are executed with a single router pass that is stopped before the timber end. The drawing below shows some styles, but the most popular are the quarter-round and the straight chamfer, because the end of the cut is naturally symmetrical to the timber edge. More complicated edge treatments can be done by building up a molded appearance with several passes of the router, or by using hand tools to make deeper and wider cuts than those possible by machining. Working an edge by hand also produces a less regular surface, which can be made to appear rustic or rich, depending on the desired effect.

How edges are treated has enormous influence on the look of a home. One of our recent houses was intended to have the shape and feel of a barn, and we hoped to create an atmosphere that might be described as "elegant rusticity." Because of their unrefined character, hemlock timbers were used for the frame; we planed them just enough to remove the saw marks and left the edges completely un-

**This chamfered edge treatment was done with a combination of power and hand tools, as described for the complex chamfer shown in the bottom drawing below. (Photo by Tafi Brown.)**

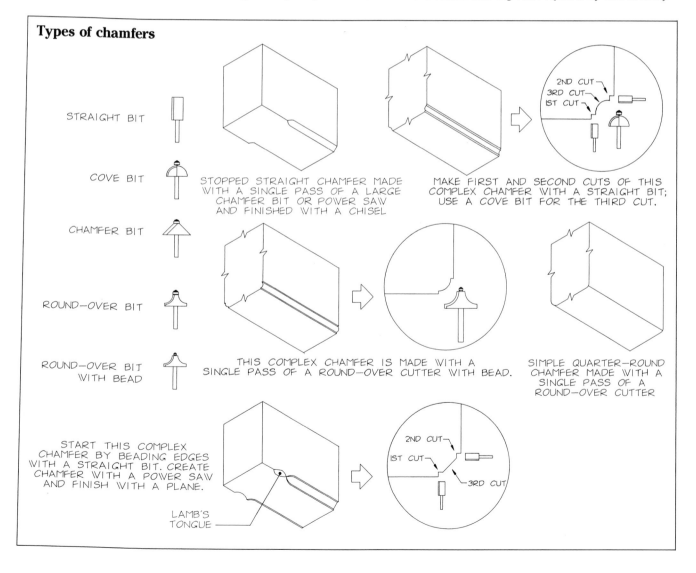

## Types of chamfers

STRAIGHT BIT

COVE BIT

CHAMFER BIT

ROUND-OVER BIT

ROUND-OVER BIT WITH BEAD

STOPPED STRAIGHT CHAMFER MADE WITH A SINGLE PASS OF A LARGE CHAMFER BIT OR POWER SAW AND FINISHED WITH A CHISEL

MAKE FIRST AND SECOND CUTS OF THIS COMPLEX CHAMFER WITH A STRAIGHT BIT; USE A COVE BIT FOR THE THIRD CUT.

2ND CUT
3RD CUT
1ST CUT

THIS COMPLEX CHAMFER IS MADE WITH A SINGLE PASS OF A ROUND-OVER CUTTER WITH BEAD.

SIMPLE QUARTER-ROUND CHAMFER MADE WITH A SINGLE PASS OF A ROUND-OVER CUTTER

START THIS COMPLEX CHAMFER BY BEADING EDGES WITH A STRAIGHT BIT. CREATE CHAMFER WITH A POWER SAW AND FINISH WITH A PLANE.

LAMB'S TONGUE

2ND CUT
1ST CUT
3RD CUT

changed. Any kind of edge embellishment would not have been true to the barn idiom, and the lack of it keeps the eye moving past the frame to the volumes and shapes of the house. By contrast, we designed another house to have some very formal rooms, which would contain chandeliers, cherry woodwork and fine furniture. To make the timbers complement this formality, we decorated them with large, beaded-edge chamfers and cut them entirely with hand tools. We also created the illusion of an arch in the summer beam by beveling the entire timber from the bottom, making the center less deep than the edges. Although these details are not overly complicated, they are sufficient to impart an elegant air. We gave the less formal rooms of this house, such as the country kitchen, a light edge treatment on the primary timbers and left the secondary timbers alone except for typical planing and oiling.

Most often we choose the middle road on edge embellishments. We frequently opt for small chamfers on all exposed edges of the secondary timbers, just enough to soften them without creating a design statement. Proportionally larger versions of the same treatment would then appear on the principal timbers. One of my favorite refinements is a simple broken edge created by whisking off the hard timber edges with a hand plane. The timbers are left looking crisp, but they feel smooth to the touch instead of splintery. (With natural edges, splinters and tears can be a real problem as the wood dries.) Another device I like for formal situations is a decorative stop called a lamb's tongue, which we usually use with quarter-rounds and chamfers (see drawing, p. 167). This is a traditional timber-framing technique but it comes to us with a history of infinite variations; modern timber framers have added their own interpretations. Lamb's tongues are always executed by hand, using a chisel, drawknife and scoop-bottom plane. We usually confine their use to hand-worked, rather than machined, edges, simply because it's difficult to rout a chamfer that is wider than 1 in., and this is insufficient room to exercise the lamb's tongue's decorative potential.

The timbers receive their edge embellishments while still on the ground, so when deciding on the subtlety or grandeur of each, it's critical to visualize the relationship of the frame to the floor plan and future partitions. A simple

**Hemlock timbers without edge embellishments were used for a house intended to have a barn-like atmosphere.**

way to generate an edge-treatment plan is to develop a color code for each type of embellishment and then indicate it in that color on copies of the frame plans (both elevations and plan views). To help in planning, we usually adhere to a few general rules. The first is that partition walls should never encroach on edge treatments. Imagine for a moment a beam divided by a partition that separates the living room and dining room. If the timber edge exposed to the living room has a large, elaborate chamfer, the partition wall will automatically be pushed toward the dining room. If the beam is a 6x8, on the edge exposed to the dining room there will be either no room for an edge treatment or room for only a very small one. Failure to consider this during design can result in a partition that overlaps the treatment on one or both edges—a mistake made obvious as the drywall climbs into the embellishment.

Another rule of good embellishment is that timbers defining the perimeter of a room should all be treated in a similar way; using a molded edge on a corner post suggests that other posts in the room should be molded, too. Large summer beams invariably dominate other timbers and therefore often receive larger, more elaborate edge treatments. The size and style of embellishments used on major beams should duplicate those used on major posts.

Edge treatments rarely continue to the timber end. Instead they stop (often with a lamb's tongue), at a prescribed distance from the intersection with other timbers in the frame. Stopping the embellishments this way is attractive and accentuates the joinery, but it also helps the joiner avoid complications. As an example, if the embellishments weren't stopped, housings to receive timbers would have to be carved to the contour of the edge treatment at the corners. Timber joinery is hard enough with square corners and right angles, so we keep edge treat-

**Edge embellishments usually stop before the intersection of other timbers.**

**Deeply embellished timbers lend a sense of formality to this house.**

The end embellishment on this hammer post was roughed out with a series of power-saw cuts and finished with a chisel and hand plane.

ments away from all timber intersections. The stops should also be noted on the edge-treatment plan.

As a last note, choose embellishments that reflect the overall ambitions of the individual house design; far too frequently, edge treatments become standardized. And make sure that they are within the abilities and time constraints of the joiner. Attempting to do too much is more likely to be in error than doing too little. One rarely goes wrong with subtlety and simplicity.

**End embellishments**   Embellishments at timber ends are actually a type of carving. They typically occur where a timber passes 6 in. to 1 ft. through a joinery intersection—leaving wood beyond the joint increases strength, but the exposed end grain is unattractive and demands further attention. When the exposed end is at the bottom of a post, it is called a pendant. The most common situations calling for a pendant are at framed overhangs, where the upper-story post passes through the second-floor girt, and where a hammer post passes by the hammer beam. When the exposed end occurs at the top of a post, especially in framing around a stairway, it is called a finial. The ends of projecting beams, such as where a plate passes by the last bent post, are also usually carved, as are rafter ends that pass through the frame and support the roof overhang. Timber framers usually cut away the bulk of the waste on end embellishments with a power saw, and then finish the work with a chisel, slick and hand plane. Work time can be reduced by simplifying the design so that it can be mostly finished with a series of power-saw cuts. Unless a portable

**Cutting a simple pendant**

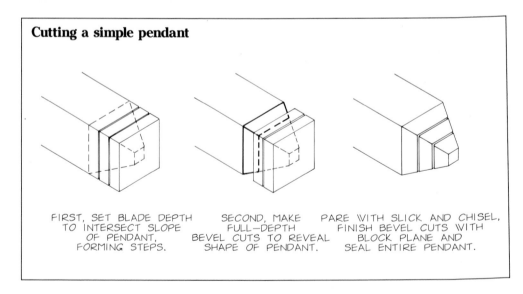

FIRST, SET BLADE DEPTH TO INTERSECT SLOPE OF PENDANT, FORMING STEPS.

SECOND, MAKE FULL-DEPTH BEVEL CUTS TO REVEAL SHAPE OF PENDANT.

PARE WITH SLICK AND CHISEL, FINISH BEVEL CUTS WITH BLOCK PLANE AND SEAL ENTIRE PENDANT.

bandsaw is available, carvings with lots of curves are time-consuming, although enjoyable, to make.

Design end embellishments to set the tone for other details of the interior finish. A highly molded design would be out of place if every other aspect of the finish were based on square edges or flat chamfers. Repeating elements of the end embellishments in the trim, stair parts or even in the furniture further integrates the frame with the rest of the house.

Traditionally, and especially during the Middle Ages, an incredible range of designs has been used for end embellishments, including meticulously detailed human and animal figures. A story I've heard a few times tells of a building in which the heads of saints is carved on the ends of the hammer beams, and each is identified with an inscription. One of the carvings did not have an inscription and could not be identified as a saint, but is now believed to be a likeness of the master carpenter on the job.

**Here the hammer-post pendant detail is repeated in the stair newel for continuity.**

## Roof windows

In the building lexicon, "roof window" seems to be replacing "skylight," probably because almost all of the new models operate like windows. Most can be outfitted with shades and screens and can be cleaned from the inside—a valued convenience. But perhaps the most important advance in roof windows is that flashing systems are now sophisticated enough that they almost never leak. In a timber-frame house, the beauty of the timberwork between the rafters encourages designers to use the space under the roofline for living rather than for storage, and roof windows are an inexpensive way to bring light and ventilation into these areas. The other alternative for bringing in light—dormer windows—adds considerably to the cost of construction.

Roof windows should be located for visibility (so you can see through them from a designated position) and for illumination of appropriate areas. In a bathroom, for example, you might want to locate a roof window over the jacuzzi, or in a bedroom over a work area. Because of their lack of horizontal members, common-rafter roofs naturally offer more choices in window placement than do rafter systems containing purlins. But most of the time, roof windows can be incorporated without changing the timber frame, since for the benefit of other building materials and roof loading, timber purlins or rafters tend to be spaced on 48-in. centers. This allows plenty of space for several standard sizes of roof windows.

If you want to trim out the roof window with wood all around, choose a size that will give you 4 in. to 6 in. between the roof-window rough opening and the timbers. Try to center the window horizontally between rafters and vertically between purlins. I prefer to choose a roof window size that will fit as tightly as possible between the timbers. The timbers then do double duty as trim, while the other two sides are trimmed with wood or, better yet, left trimless by making a drywall corner return to the roof-window sash. Remember that, like wall windows and doors, roof windows are not actually mounted in the timber frame, but instead are fastened to the stress-skin panels above the frame with nailers (see Chapter V).

Since roof windows are on a slope, it is best if the top of the opening can be trimmed level with the glass and the bottom of the opening vertical to the glass, so that maximum light can enter the living area. Purlins falling at the top or bottom of an opening have the disadvantage of slightly restricting the admission of light, but in most situations it's not worthwhile to change the framing system just to avoid this problem.

**In this bedroom, the roof window is mounted between headers that break a purlin span to make a large vertical opening in the roof. Drywall finish extends to the sash instead of to wood trim to allow the timbers to define the opening.**

Sometimes it's necessary to create an opening for a roof window by interrupting a purlin or rafter with headers, which span between two perpendicular framing members as shown below. In a purlin-roof system, headers are used to locate a roof window where there would otherwise be a purlin, or to make an opening larger than the span between purlins. Conversely, in a common-rafter roof, headers are used either to locate the roof window where there would otherwise be a rafter, or to accommodate a window that is too wide for the opening between rafters. When headers are used, they increase the roof load on the members they span, which must be considered in sizing those pieces.

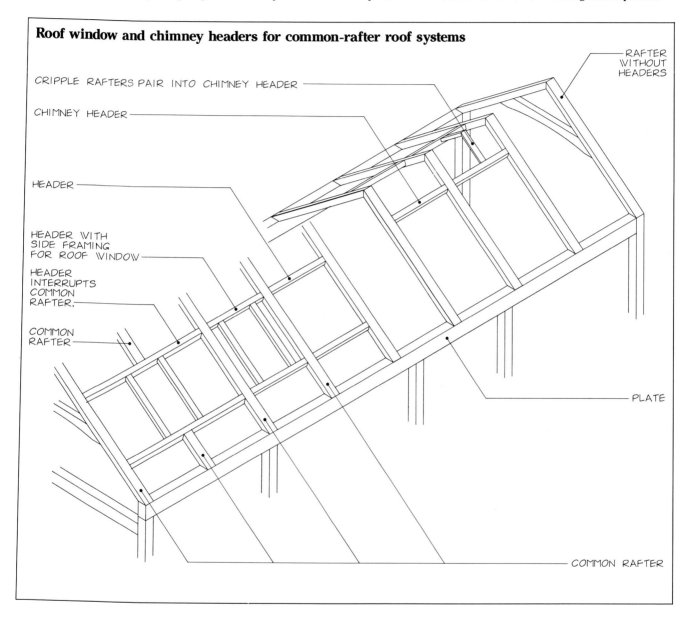

**Roof window and chimney headers for common-rafter roof systems**

RAFTER WITHOUT HEADERS

CRIPPLE RAFTERS PAIR INTO CHIMNEY HEADER

CHIMNEY HEADER

HEADER

HEADER WITH SIDE FRAMING FOR ROOF WINDOW

HEADER INTERRUPTS COMMON RAFTER.

COMMON RAFTER

PLATE

COMMON RAFTER

## Chimney and fireplace openings

Fire has always been an essential design consideration for houses. In the earliest timber-frame buildings, a hole was usually left in the roof to allow smoke to escape from the open fire that burned in the middle of the house. As I pointed out in Chapter I, usually there was no second floor in these buildings because it would have been too smoky. So the first buildings were more like elaborate tents or caves. They began to resemble our notion of a home only after chimneys were designed to exhaust the smoke.

The first thing to say about chimneys is that they must be completely independent of the timber frame; if you were to strip away the house, the chimney should stand entirely on its own. I don't know why this is, but in the preliminary plans we review, a common mistake is locating the chimney where it would pass through a bent or an exterior timber-frame wall. Wooden structures move a bit and masonry structures should not. Never use the masonry to support the frame or the frame to support the masonry. Like so many other things in a timber-frame house, it is important to plan the chimney in some detail before the frame is cut. You must know the size and location of the chimney at each floor level and where it passes through the roof. If there are fireplaces attached to the chimney, it is necessary to know how they will affect the size of the chimney opening. It's not possible to explain every detail of masonry construction here, but allow me to provide a few tips.

**1.** Know enough about typical masonry details to communicate with the professional you hire. *Audel's Guide to Masonry Construction* (see bibliography, p. 229) is a particularly good reference. Then, rely heavily on the advice of a mason you trust. If you can retain this person to help you in the design phase, you'll be a lot more confident when the timbers are being cut. After you have described all the things affecting chimney dimensions, the mason will be able to provide precise measurements based on personal construction techniques. And believe me, every mason is different. Two masons setting out to make a chimney with exactly the same specifications and using the same materials could wind up with completely different measurements, and each would tell you it was done exactly the right way. I don't pretend to understand this.

**The frame and the chimney should be designed to be completely independent of each other. Diagonal headers outline the opening on the second floor of this house, while the chimney passes behind the valley rafters.**

**2.** Despite your confusion and lack of knowledge, speak up. On one of our jobs, I argued weakly with a mason who was putting in the flues without brick partitions between them. I told him it didn't look right. He argued eloquently and persuasively that "the latest befaddle on counterflow air currents and retroactive heat flow has it that the weight of the mass is inversely proportional to the ability of the flues to do headstands," or some such gibberish. I looked at the four wobbly flues standing in his brick cavern, shook my head and just walked away. Recently, when a chimney sweep attempted to clean one of the flues, it collapsed under the mighty weight of the chimney brush. I learned that there's no sense in compounding ignorance with timidity.

**3.** Make sure every possible flue has been included in the chimney. If in doubt about a possible furnace or woodstove, put in the extra flue. Retrofitting a chimney would be a mess.

**4.** Make sure the mason is aware of the need for exterior combustion air for all fireplaces and woodstoves. Accommodation for such air passages should be considered a part of the construction of all chimneys and hearths in well-insulated houses. It is accomplished by passing a 4-in.-dia. or 5-in.-dia. galvanized pipe to the outside in the basement. The pipe brings the air to one or several openings in the hearth near the woodstove intake or just in front of the hearth for the fireplace. The air intakes for the fireplace can be hidden by dropping the base of the fire-box below finish floor level by about 5 in.; the air enters just below finish floor level in the drop.

**5.** Beware of the use of stone for interior masonry because of the possibility of bringing radon into the house.

**Framing for the chimney** For fire safety, the usual code requirement is that the chimney should be at least 2 in. away from framing materials on all sides. This means that the chimney opening in the frame will have to be a total of 4 in. larger than the proposed chimney in width and length. (This rule also ensures that the chimney will be far enough away from the frame so that it won't be damaged by wood movement.) The finish flooring should be held back by ¼ in., but the trim can be cut to fit tightly to the chimney.

Often the chimney imposes no special framing requirements on the floor system, for the spacing between joists is frequently as much as 32 in., which allows more than enough room for the width of a typical chimney. It is sometimes the case, however, that the flooring requires support around the chimney. A small timber header on either side of the chimney, perpendicular to the joists, will keep the flooring stiff. If necessary, frame between these two headers with a third header to bridge the gap between the adjacent joist and the chimney, as shown below. Try to use as few timbers as possible to keep the opening uncluttered.

Where the chimney has to interrupt the joists, use headers to create an opening in the floor. Particularly if the

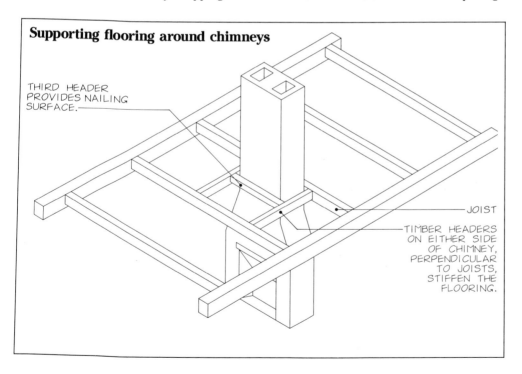

**Supporting flooring around chimneys**

THIRD HEADER PROVIDES NAILING SURFACE.

JOIST

TIMBER HEADERS ON EITHER SIDE OF CHIMNEY, PERPENDICULAR TO JOISTS, STIFFEN THE FLOORING.

opening is large, it may be smart to plan the joist layout around the chimney. If it can be done symmetrically, start by designing a joist on two sides of the chimney and then put headers between those joists on the other two sides of the chimney. Then divide the headered space equally, putting a joist in the middle. Remember that half the load borne by the headers will be transferred to the joists on either side of the chimney, so they will have to be sized to handle the increased load.

The chimney also needs to pass through the roof timbers, but this can be a little more difficult to resolve than passing it through the floor timbers. First, good geometry is critical. The opening in the roof has to fall directly above the opening in the floor and be the same size horizontally, even though it must be calculated along the roof slope. It's pretty simple math (remember Pythagoras?). I'm urging that somebody do it, with the hope that it might save wholesale embarrassment when looking up through the chimney opening on raising day. Whether the roof system is framed horizontally with purlins or vertically with rafters, there is always the possibility that no special framing will be needed. With luck, the chimney will just slide through the timbers. We always check to see if we can help luck along by changing the layout slightly. If it's close, it's worth a try. For instance, if a rafter or purlin needs to be moved only an inch or two, the change probably won't upset the bearing for materials or compromise symmetry.

When the chimney falls on one side or the other of the roof peak, the header has to span between purlins or common rafters (but never through a principal rafter, which is part of the bent) to create a larger opening. I prefer to see

the chimney pass through at the ridge because it leaves less masonry above the roofline and solves some flashing problems. (When a chimney falls below the peak, it needs a roof-shaped water diverter, called a cricket, on the upper side to prevent snow, ice and water from getting trapped behind the chimney.) When the chimney passes through the peak, the framing is a little more difficult because the header has to be a small duplicate of the top portion of the rafters (see drawing on p. 173). This small truss then frames to purlins or rafter headers on either side of the roof peak.

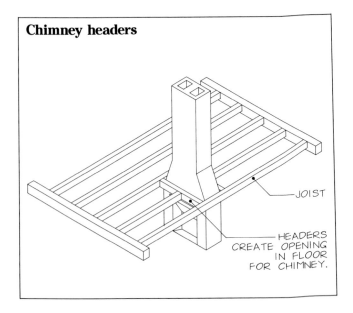

**Chimney headers**

JOIST

HEADERS CREATE OPENING IN FLOOR FOR CHIMNEY.

**An opening for the chimney is framed into the peak of this building. Two small rafters between the upper two purlins act as the headers. King-post trusses frame and decorate the area to the left, used as a music room. (Photo by Tafi Brown.)**

## Exterior timbers

In modern timber-frame construction, it's perfectly possible to protect the frame completely from the pernicious effects of weather and condensation. With the timbers snug inside the cocoon of warmth and protection offered by the exterior insulation system, the frame (and the building) should last as long as people care enough to keep the roof tight. In our designs we therefore strive to keep all the timbers inside the insulating system, but situations continually arise in which exterior porches and long roof overhangs are important, even critical, to the design of a house.

**Roof overhangs** In warm climates, long roof overhangs are used to keep the house cool; in wet climates they help protect the siding and windows from excessive moisture. A roof with a long overhang is like a hat with a decent brim. When the roof overhang is created within the enclosure system rather than by passing the timbers through it, the limit is about 2 ft., based on using structural, or double-chip, stress-skin panels (p. 103). If one of the alternate insulation systems is employed instead, especially the foam-and-nailer system (p. 98), the length of the overhang would be based on the relative strength of the framing members. For instance, 2x6s cantilevered beyond the frame could probably create an overhang of about 3 ft., depending on the roof load.

What happens when the timbers of the frame have to fall outside the protection of the insulation system? Essentially, there are three hazards. First, where timbers pass through the insulation, wood movement can cause a sizable gap to develop between the timber and the skin. Even if the gap is only ⅛ in. on both sides of a 6x8 timber, its effect on air infiltration would be comparable to bashing a 1½-in. by 1½-in. hole in the side of the house. In addition to compromising the insulation, the gaps offer a place for condensation to collect, which could cause rot in the frame. These gaps must be caulked at least once a year while the timbers are drying and stabilizing. So important is this detail that I have only half-jokingly considered carving the maintenance instructions into the timbers.

The second problem with timbers passing through the insulation is that the part of the timber that lives on the inside of the house will stabilize, while the part that lives outside will seasonally swell and shrink with changes in moisture content. Even if the joint looks tight on the inside, it should be checked on the outside—there's no way to know if there's a problem unless you climb 25 ft. up a ladder once a year.

The third problem is that the life span of the whole frame is potentially shortened by the life span of that part exposed to the elements.

Having pointed out these drawbacks, I can say that I am well aware that there is no substitute for timbers when it comes to creating a strong, beautiful roof overhang. If, after carefully weighing the options, this detail wins out, here are some things to do to minimize negative effects.

**1.** I would be much more reluctant to use a framed overhang in New Hampshire than in Arizona. Hot and dry climates are the most benign to exposed timbers, and houses in these areas suffer the least from air infiltration. Even in my native Colorado, which is sometimes cold but always

dry, timbers can be left exposed for a long time without significant damage. But here in New England, where everything is either frozen solid or supporting fungal growth, it is wise to consider every other alternative before exposing the timbers to the weather.

**2.** Our modern timber frames are more protected now than at any time in history, and because of this we can use almost any kind of wood that is structurally sound. But our forefathers in the craft, who infilled between the timbers rather than cladding from the outside (thus exposing them to the elements), chose white oak as the standard in Europe and cypress and cedar in Japan. Think like the Europeans and the Japanese if you are making a frame overhang, and use a rot-resistant wood.

**3.** Detail the roof system to protect the frame members in the overhang. Extend the roofing beyond the timber ends as much as possible to keep moisture out.

**4.** The most protected overhang is one whose timbers cantilever beyond the frame without the need for support from plates or braces. When horizontal timbers support the overhang, they have a tendency to hold moisture; diagonal braces can pull moisture back to the frame through surface tension. But if the overhang is too great, some form of bracing for the rafters will have to be used.

Principal rafter and purlin roofs are the easiest to protect at the eave for two reasons. First, only a few pieces pass through the skin. For example, in a 36-ft.-long, four-

In this frame, a plate overhangs the gable-end wall to support a truss that will be roofed, but unprotected by a wall. The problem is that the cantilevered plates, their supporting braces, the exterior truss and all the short purlins are not only exposed to the weather, but also cause numerous penetrations in the insulation. (Photo by Tafi Brown.)

bent building, there would be eight penetrations at the eaves, whereas a common-rafter system would probably require twenty. Second, at the gable end, the common rafter system requires that a plate be extended from the walls to support the last rafter pair, which causes several more penetrations and a horizontal timber. (This detail was often used on old Swiss chalets. The top of the extended plate would be protected with a little shingled roof.) In the principal rafter and purlin roof, however, the end rafters can be lowered to allow the purlins to pass over the top. We ordered extra-deep rafters for one of our projects so they could be notched across the top. No wood could be removed from the purlins because the complete cross section was needed for shear strength.

**5.** Protect all exterior-exposed wood with oil or varnish. Homeowners should remember that they will have to make a commitment to caulking the gaps between the timbers and the insulation. This should be done at least once a year for as long as the wood continues to dry and shrink.

**Porches** Porches are a different situation entirely from roofs because they must be supported by posts, and there isn't a good way to protect exposed posts to the same degree that exposed rafters or purlins can be protected. Therefore, the porch (or at least the posts that support it) simply can't last as long as the rest of the frame. Thus the best solution is to separate the porch completely from the house by isolating it on the other side of the insulation. This also makes it possible to construct the porch with a wood chosen exclusively for its weather-resistance. To make it easy to attach the roof framing to the exterior sheathing, use timbers for the front wall of the porch to support the roof, but build the roof itself with conventional framing members.

The most difficult part of the porch to protect is the base of the posts. If posts are to be attached to a concrete floor, use the plinth detail on p. 122. When the posts will rest on a wooden deck, allow water to drain by cutting any mortises to pass through the sill. (This is not a good situation for a timber sill because it is so difficult to prevent the surface from retaining moisture.) If circumstances demand the use of a timber sill, use the most water-resistant wood available and bevel the exposed top surfaces to ensure that water will run off.

**The porches in this house are supported with posts and beams, but these exterior timbers are not attached to the rest of the frame for fear they will deteriorate prematurely.**

## Changes in building height and width

For structural reasons, a change in roofline or building dimension almost always has to occur at a bent; for design purposes, the change has to be aesthetically, as well as technically, satisfactory, since the timber frame is an important part of the interior space. When one building has two or more roofline heights or any change in width, there should be a structural decision about how to support the upper and lower rafters at the break, and design decisions about which timbers will be visible and how to apply the finish and insulation.

An illustration will clarify the problem. In the floor plan below, the change in building dimension between the living room and the rest of the house results in a complete change in volume and feeling. When you take those two steps down to the living room, you move from an area with an 8-ft. ceiling to one with a ceiling that soars to about 20 ft. The three bents in the living room were designed with dramatic hammer beams, while the bents for the rest of the house had function at their core. During design, we were concerned about how the structure would look from both areas and about the transition from the grand cathedral space to the rest of the house.

As it turned out, the most practical solution was to build two independent frames—one for the living room and one for the rest of the house. If we had tried to use a single

**When you step down to the living room in this timber-frame house, you move from a low, flat ceiling to the drama of a soaring cathedral space.**

## Bent and floor plans for house designed with two widths

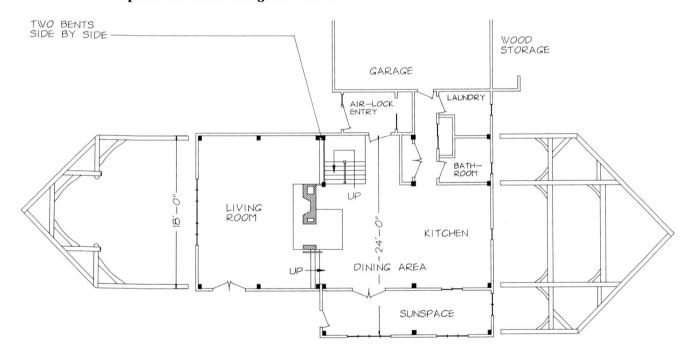

bent instead of two side-by-side, we would have needed rafters to support purlins, and posts would have been necessary for the inside corners. Each bent would have lost some distinctiveness. But by creating two separate frames, we were able to define each area completely with timbers—a design decision that was particularly important in the living room, which had a nice sense of symmetry and unity from the three hammer beams. We left a 5-in. space between the two bents at the building break for insulation and for the finish wall surface of both areas (4½ in. for the typical stress-skin panel with drywall on one side, and

½ in. for a layer of drywall to be applied to the nail-base sheathing on the other side.)

In this example, our biggest concern was to make sure the two frames were securely fastened together. While designed to be structurally independent, they also had to act as a unit. We made the tie in the insulation. The panels between the two bents are attached to the main house, and the panels on the living-room wing are routed and fitted over a nailer attached to the outside of the main-house panels. The living-room panels are screwed to the nailer and nailed to the main-house timbers.

## Attaching two separate frames

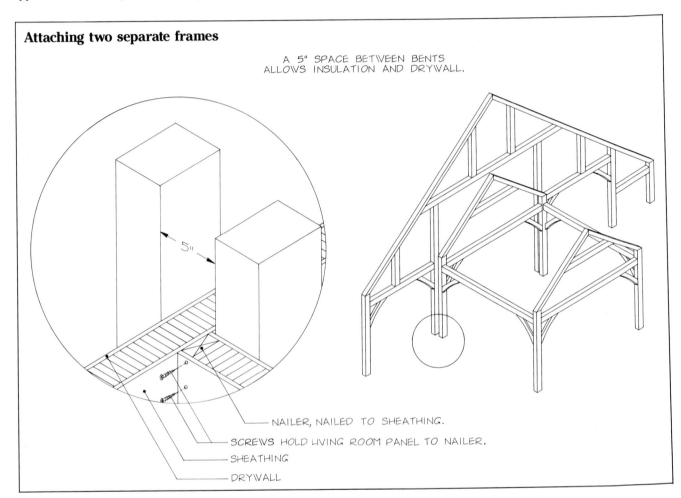

A 5" SPACE BETWEEN BENTS ALLOWS INSULATION AND DRYWALL.

5"

NAILER, NAILED TO SHEATHING.

SCREWS HOLD LIVING ROOM PANEL TO NAILER.

SHEATHING

DRYWALL

The best structural solution is to integrate elements of both bents within a single bent at the point of transition, because it ensures that everything will work as a single unit. But especially when there are internal rafters, building math and bent assembly become pretty tricky. (For that reason, these are referred to as "infernal rafters" in our shop.) There is also a finishing and insulating problem where the wall and the lower roof intersect. In the case of the house shown at right, we put the lower roof panel on first with the bottom edge held back ½ in. and routed for a nailer. After the nailer was secured, a base plate for the wall panel was precisely located and nailed to the roof panels. The wall panels were cut and routed to fit over the base plate, allowing the wall panel to be secured where there wasn't any nailing area in the frame. Finally, a drywall strip was attached from the inside to the end of the roof panel. In this frame, the internal rafter is in the same plane as the rest of the bent. It would have been better to have the internal rafter project at least a few inches toward the lower roof, to receive the panels, as shown below.

Here, an internal rafter was framed into the mid-bents to receive the purlins from the outer bents. Internal rafters solve a connection problem but involve tricky mathematical calculations.

## Detail of internal rafter and panel connection

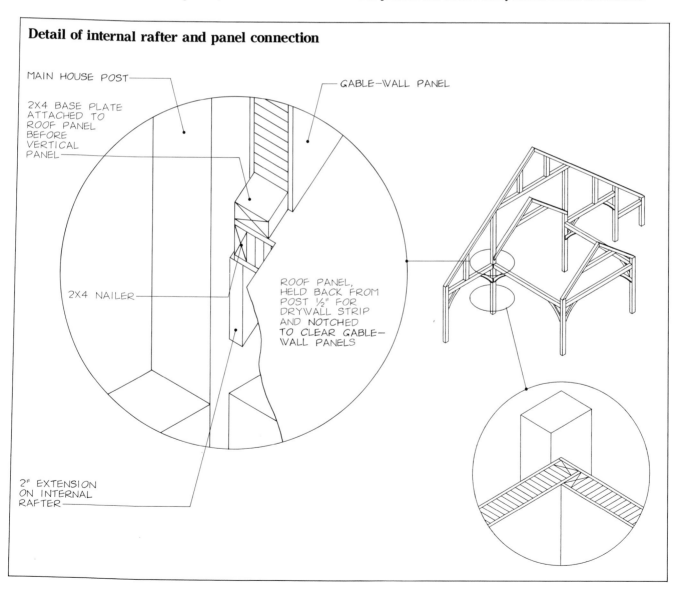

MAIN HOUSE POST

GABLE-WALL PANEL

2X4 BASE PLATE ATTACHED TO ROOF PANEL BEFORE VERTICAL PANEL

2X4 NAILER

ROOF PANEL, HELD BACK FROM POST ½" FOR DRYWALL STRIP AND NOTCHED TO CLEAR GABLE-WALL PANELS

2" EXTENSION ON INTERNAL RAFTER

Another way to solve the finishing problem at the roof break is to tie the frames together at the wall level, but hold the rafter for the lower roof away from the wall, allowing the insulation to be fitted between. This requires that the first rafter be framed to a plate, which wouldn't be a problem in a common-rafter system. In the frame shown below, the bent girt for the larger structure has to support a very large load. It carries a portion of the floor load, a point load from the upper part of the gambrel roof, and the weight of the plate and the roof system for the gable wing. Even though it is a short span, it would probably be better to have a post at the inside corner.

The bent girt of the main house not only supports a major roof load transferred through the post, but it also must support the plate and roof system for the gable wing. Luckily, the bent girt is short and well braced.

## Timbers and glass

In my years as a designer and builder of timber-frame houses, I have talked to at least a thousand people about their dream homes. Only one of them did not list the admission of sunlight as a priority. A shady spot on the north side of a hill would have suited this man just fine; the house would have let in only enough light to cause shadows. I admitted to being concerned about his dim lifestyle, and he confessed to being a Hobbit. We didn't build for this client, and I can only hope that somewhere he was able to find the dark home of his desires.

This fellow would have been perfectly content in the typical home built prior to the early-to-mid-1800s, when the use of sheet glass for windows had not yet become widespread. During that time, glass window panes were individually made by craftsmen who would blow a glass bubble and then spin it into a sheet, leaving a crown in the middle. These same glassblowers also created vases and goblets, and the window panes were likewise small and precious. While they admitted light, they did nothing for visibility, for anything viewed through the bubbles would appear distorted. Only the very wealthy could afford much glass. Most houses had a few small penetrations to allow a little light, supplemented by candles, lanterns and even the fireplace. By day and night, homes were like above-ground caves. The average modern house is a crystal palace by comparison.

There is no doubt we have come a long way, and within the last 15 years alone, significant improvements have practically revolutionized the way we think about glass. For example, by sealing air between two or three panes, windows today can have unheard-of insulating qualities; using argon gas instead of plain air in the cavity can provide an even greater R-value. Large panes can be strengthened to allow spans once requiring structural materials. Low-emissivity (low-E) coating, an infrared reflecting surface, is transparent but limits the passage of heat. Ultimate performance is achieved through a combination of technological developments, such as when argon gas is sealed between two layers of tempered, low-E glass. A recent test demonstrated that even with an exterior temperature of 0°F, the interior glass felt warm to the touch. Glazing materials and hardware have also improved over the last few years, and an array of new caulking and gaskets can give site-mounted glass reliable performance.

As a result of this technology, the creative use of glass has become a force in modern architecture. Designers have to balance the aesthetic qualities of expanses of glass against the concomitant heat loss; glass giving better energy performance means the balance shifts toward glass. By using top-shelf, low-E glass and carefully calculating optimum glazing area, significant amounts of glass can now be used without compromising energy performance. These calculations (usually computer-generated), consider solar gain, surface area of exterior walls and the roof, house volume and glass type, among many other factors. The goal is to find the point at which the heat gain from sunny days exceeds the loss from cloudy days and from the nights. Predictably, the results show that most of the glass should be concentrated on the south side of the building, with modest amounts on the east and west sides to receive light, and as little as possible on the north. Often a solarium is recommended for heat collection.

Glass and timber frames are perfect mates. The spaces between frame members can be filled in with any kind of rigid sheathing material, and glass is such a material. Because the posts carry the entire building load, windows and doors can be mounted just about anywhere in the insulating system. In this section, we'll explore a few aspects of the glass-and-timber relationship, starting with a few notes on integrating manufactured windows and doors into a timber-frame house. Then we'll cover solarium design. We'll conclude with several flashing and glazing details associated with mounting glass on timbers.

**Integrating manufactured units**  Window and door units are mounted in the insulation (as discussed in Chapter V), which is in turn secured to the frame. The only relationship between the frame and these units is therefore an aesthetic one, but it is by no means unimportant. In fact, it is critical that the frame and fenestration interact with symmetry and balance (see p. 68).

Design goals can be achieved either by altering fenestration to fit the frame or by changing the frame to suit the fenestration. In several of the houses we have designed, clients have decided that they wanted to see the frame from both the inside and the outside. Instead of mounting the windows between timbers, we located them over the timbers so visitors would meet the frame before entering the house. We have also mounted large windows between posts so that the diagonal braces could be seen from both sides. The brace pair can actually enhance the window opening, and it does not interfere with the operation of the window since it is mounted in the insulation outside the frame.

**Because the glass is mounted in the insulation outside the frame, windows can be designed to allow timber members to be visible from the outside as well as from the inside.**

Usually it is best that window and door design coincide with frame design. Fanlight windows, for instance, can be made to fit nicely between the curved arches of knee braces. Another technique is to use timbers to frame window or door openings. The unit is still mounted in the insulation, but the timbers act as the trim. For one of our projects, an architect designed almost every window and door in the house to be trimmed completely with timbers. The jambs of the units were received into rabbets cut into the perimeter of the timber opening. The timbers were dry, which allowed us to make a firm attachment between the jambs and the timbers. If they hadn't been dry, there would have been big problems.

**Solariums**   A solarium is typically thought of as a separate space used to collect heat for distribution to the house. The distribution can be accomplished through opening doors and windows, through fans and ducts, or through a masonry mass. But better glass and more intelligent sizing of glazed areas make it possible to build solariums that do not have so much heat loss that they need to be separated from other living areas. The solarium then becomes an inviting place for sitting, reading and gathering—a place of heat and light, not unlike the hearth in early homes. This is the type of solarium I am most enthusiastic about, even if it sometimes means accepting the compromise that occasionally heat loss will exceed gain.

I came to this preference the hard way. Quite a few years ago we designed a modest-sized house for a site with good solar exposure, and correctly judged that we would be able to heat with the sun on days that weren't cloudy. We added a detached solarium to the design, so that we would not lose heat through the glass when sunlight was unavailable. When the solarium was warm and the house cold, a French door and an awning window would be opened to let heat into the house. A fine idea, but we weren't prepared for the fact that since the solarium would not benefit from the house heat, it would get very cold during long cloudy periods. Also, by the time the solarium had been warmed by the sun, the house had also been warmed by solar gain through other windows and heat conducted through the solarium. Because there was no reason to open it to the house, the solarium seriously overheated in sunny periods. (The owner told me the thermometer registered over 140°F right before the mercury popped.) This solarium was either too cold or too hot for use. The solution was to reduce the amount of glass in the solarium, take out the back wall and incorporate the space into the house.

If a solarium is to be detached and useful, it must have a huge amount of mass to store the heat for night and cloudy periods. This type of solarium is usually complicated to build, always expensive and will still sometimes get too cold. Window insulation (there are many types) is a good idea, but too often it is only an idea. If people don't take the time to operate the insulation, it won't work. One of our clients figured it required five minutes in the morning and five in the evening to pull the insulation over each window. While this doesn't seem that long theoretically, it takes more time than the garbage and less than the dishes; just long enough to become another daily chore.

**In this gable-end wall, white-cedar framing was designed to coincide with custom window units. The windows will be mounted to the outside of the frame along with the panels, and the timbers will serve as interior trim.**

Solariums that are incorporated into the main living space of the house can take many forms, from a simple large window to the common three-sided addition with a large expanse of south-facing fixed glass. Operable window units are mounted on the east and west sides and often on the roof as well, making it easy to vent the solarium when there is too much heat. Since the long side of the building tends to face south anyway, the simplest way to create a solarium is to extend the roofline on the eave side. In essence, the solarium becomes an extension of a bay and is framed by passing the principal rafters over the outside posts of adjacent bays to the shorter posts at the solarium corners. From a construction standpoint, it is not important that the solarium be in the middle bay; but its placement there is usually more beneficial to heat distribution.

Using fixed glass on the south wall is also not a requirement of solariums. If you plan carefully and use stock glass sizes, it is an inexpensive way to develop a large glass area. (Stock glass is generally about one-quarter the cost per square foot of custom Thermopane glass.) We use a few different sizes that are commonly available as replacement panels for sliding glass doors. Unfortunately, the dimensions of the panels of stock glass will probably not add up to the width of the bay, so unless the bay was designed to accommodate available glass or luck prevailed, there is likely to be some insulated wall on each side of the glass.

The edges of the glass are mounted on the outside faces of the timbers, so any wood that will contact the glass must be dry, even if the rest of the frame is built from green material. Otherwise, you risk disturbing the seal between the glass and the frame or even breaking the glass as the drying wood shrinks, twists, warps or bows. Since the principal posts carry most of the building load, wood with less structural value, such as cedar and white pine, can be used to receive the glass. The solarium shown in the photo below, which has insulated walls extending past the glass to fill out to bay width, uses a few pieces of dry old-growth southern pine to accept the glass. For details on mounting glass, see p. 188.

**The solarium is often just an extension of the eave. Stock glass size helps determine wall height; dry timbers must be used where glass will be mounted. The roof window assists ventilation.**

The fixed glass area is framed below the roofline break with a plate (or connecting girt) between principal posts; the vertical members of the solarium are mortise-and-tenoned to the plate. But if fixed glass is intended for the roof slope as well as for the wall, the position of the plate becomes a problem in that it creates an awkward transition between the sloped and the vertical glazing. There are two solutions. One is to eliminate the plate entirely, connecting short rafters to the solarium posts, as shown at right. A purlin joined to the frame receives the solarium rafters at a point determined by the glass size. The roof glass meets the wall glass at a horizontal 2x2 turned round on a lathe between rafters and left square in section where it meets the rafters. When the eave is low and a plate would obstruct the view, this is a nice alternative. The other solution (if a slightly obstructed view is not a problem), is to frame common rafters from a purlin to a plate that rests on top of the solarium posts. The plate is ripped to the roof slope on the top outside edge to serve as bearing for the roof glass.

The plate system is also useful when the solarium spans more than one bay. Using a plate to connect from one corner of the solarium to the other eliminates concern about the location of the bent post, which would otherwise probably interrupt the layout of the glass panels. Solarium posts are framed into the plate at intervals determined by the glass size, and the plate also supports the bent rafters.

There are many ways to design a solarium under an eave. For example, in one contemporary-style house, we increased wall height in the solarium bay to frame a dramatic three-story window (photo facing page). Two rows of fixed glass were mounted on fir timbers and topped with a custom fanlight. Light penetrates deeply into the house, and the view is visible from a third-floor study. Roof windows operated with electric switches increase light and assist ventilation.

Where the sloped roof glass and the vertical glass come together, a special flashing detail is used instead of a timber. Here, a plant hanger is screwed to each timber and helps keep the framing rigid.

A plate sits on these solarium posts, providing a good connection between members. The plate is ripped on the top outside edge to make a bearing surface for the glass.

**Increasing the height of this wall provided room for the framing of a three-story window. Two layers of fixed glass were attached to fir timbers, and a custom fan unit, placed on top, was designed to match.**

Solariums with gable roofs can also bring sunlight deep into a building and provide a dramatic focal point. But because the compound joinery required is difficult, they are also quite expensive. Gable solariums work best when they are extensions of a single bay, if only to keep the valley framing from bisecting principal rafters. You can bet I learned about this the hard way. A building we did quite a few years ago had a gable-roofed solarium that was accessible from three levels. It began at the base with a layer of fixed glass mounted on fir timbers. Then there was a layer of operable manufactured units, reached by walkways on the second floor. Finally there was a fixed custom fanlight, through which a person on the third-floor balcony could view the outside. The only problem was that the width of the solarium extended past the middle bay into the two adjacent bays, so the valley rafters crossed two bent rafters. It is quite difficult to pre-fit this kind of joinery, so we generally make careful cuts to mathematical and geometric calculations and then just put it together. Try to imagine the situation. The valley rafter frames into the

principal rafter and then picks up on the other side, continuing to the eave where it frames into the outside wall. The math drove us silly, the cutting made us dizzy and the assembly was nearly impossible physically. But we did it like that...once.

Shed-roof solariums can also be designed at a gable end. The bent that connects to the shed has to have an additional horizontal member to receive the shed rafters, and there should be posts in the bent to align with the corner posts of the solarium. Otherwise, the framing is no different from an eave solarium. In fact, a shed solarium on the gable end is probably a little easier because the restrictions of bay width do not apply. Stock glass dimensions usually determine solarium width.

**Mounting glass on timbers**   The first thing to know is that it is not necessary to rabbet the timbers for the glass. Everyone seems to want to do this but it doesn't really help anything. Along with the insulation, it is better to keep the glass, flashing and any potential condensation

**This gable-roofed solarium is left open to the peak, but is also accessible from three levels. Walkways on either side of the second floor provide a nice overlook from the bedrooms and also allow access to the double-hung windows. A cantilevered balcony from a third-floor studio looks out through the round-top window.**

outside the frame. So rabbets are created by nailing plywood spacers to the sides, tops and bottoms of the solarium timbers just prior to installing the glass. This method is also easier than rabbeting the timbers. After all, you can't exactly pick up a timber and run it through a shaper.

Before getting into the various glazing details, let's quickly review how glass is normally installed. Basically, the glass sits in the rabbets created by the plywood spacers between two layers of glazing tape, and is capped by a metal or wood trim piece. The plywood spacers are usually built up of two layers, but if you can't find a combination that works, use a piece of plywood plus a piece of dry lumber. To calculate the correct thickness of the spacers, subtract $\frac{1}{16}$ in. from the thickness of the glass (to ensure a snug fit between the trim and the glazing tape), and add in the two pieces of $\frac{1}{8}$-in.-thick tape. In other words, if the glass is 1 in. thick, the thickness of the spacer would be $1\frac{3}{16}$ in. At least $\frac{5}{8}$ in. of glass should bear on the timber and there should be an extra $\frac{1}{4}$ in. allowed for expansion, making the depth of the rabbet $\frac{7}{8}$ in. In the plan view of a typical solarium post below, the 4x6 has nominal dimensions of $3\frac{1}{2}$ in. by $5\frac{1}{2}$ in. Because this is a middle post, receiving glass on both edges, you would subtract the $\frac{7}{8}$-in. depth of each rabbet to calculate a spacer width of $1\frac{3}{4}$ in.

After the spacers are nailed to the solarium timbers with galvanized ring-shank nails, the setting blocks, which act to level the glass and cushion it from expansion and contraction, are placed in the bottom rabbets. Each glass panel sits on two $\frac{1}{4}$-in.-thick neoprene setting blocks positioned at what are called "quarter points." To calculate these, simply divide half the width of the glass by two and measure in that far from each edge; for a 4-ft.-wide panel, the quarter points would be 1 ft. in from each edge. If further leveling is required, small strips of aluminum under one of the setting blocks make good shims.

Once the setting blocks are in, the first layer of glazing tape is carefully applied to the timbers around the perimeter of the opening. The glazing tape is $\frac{1}{8}$ in. thick by $\frac{1}{2}$ in. wide and pre-shimmed, which means there are wires embedded in it to keep it from compressing and deforming in hot weather. The tape is applied so that its edges will be flush with the edges of the $\frac{1}{2}$-in.-wide dessicant band between glass panes. It is held back $\frac{1}{8}$ in. to $\frac{3}{16}$ in. from the timber edges to receive a bead of silicone sealant. After the sealant is applied, the glass is then set in the opening, and the second layer of glazing tape is positioned so it also aligns with the edges of the dessicant band. Another bead of silicone sealant is applied. Wood or metal trim is positioned over the plywood spacers and the glass edges, applying pressure to the glazing tape and thus to the dessicant bands—it is important that no pressure be put on the glass where it is unsupported by a dessicant band. When the trim is installed, it will force excess sealant to ooze out of the glazing tape. Peel this off later.

That's the general scheme. Before proceeding, check with your supplier to ensure that the sealant is compatible with the glazing tape. There are many stories about chemical reactions causing a break in the glass seal. Now let's look at some specific details.

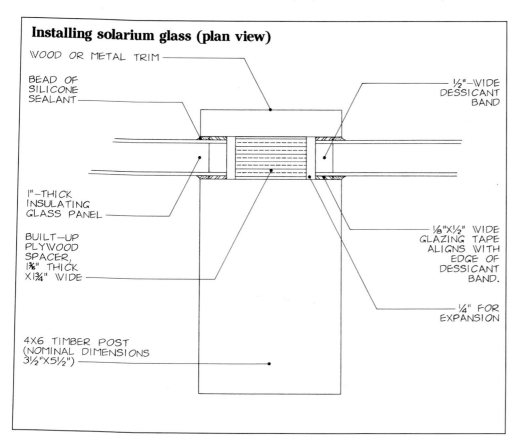

**Installing solarium glass (plan view)**

WOOD OR METAL TRIM

BEAD OF SILICONE SEALANT

$\frac{1}{2}$"-WIDE DESSICANT BAND

1"-THICK INSULATING GLASS PANEL

BUILT-UP PLYWOOD SPACER, $1\frac{3}{16}$" THICK X$1\frac{3}{4}$" WIDE

$\frac{1}{8}$"X$\frac{1}{2}$" WIDE GLAZING TAPE ALIGNS WITH EDGE OF DESSICANT BAND.

$\frac{1}{4}$" FOR EXPANSION

4X6 TIMBER POST (NOMINAL DIMENSIONS $3\frac{1}{2}$"X$5\frac{1}{2}$")

The basic theory on treating the bottoms of the glass panes comes from the people who make roof windows. They use battens on the top and sides to hold the glass in place, but generally do not cap the bottom, allowing water to run off freely. On solariums we use trim at the sides and top to hold the glass to the timbers, but at the bottom we use two different details, depending on whether or not there will be insulation beneath the glass. If there will be

insulation, it is necessary to make the transition from the thickness of the glass to the thickness of the insulation. The panel (this is assuming the use of stress-skin panels) would be held beneath the top edge of the bottom solarium beam by about 3 in. This beam is often at or near floor level. A 2-in.-wide plywood spacer is then nailed to the bottom beam, then a block of wood is ripped to slope from the top edge of the spacer to the outside edge of the panel.

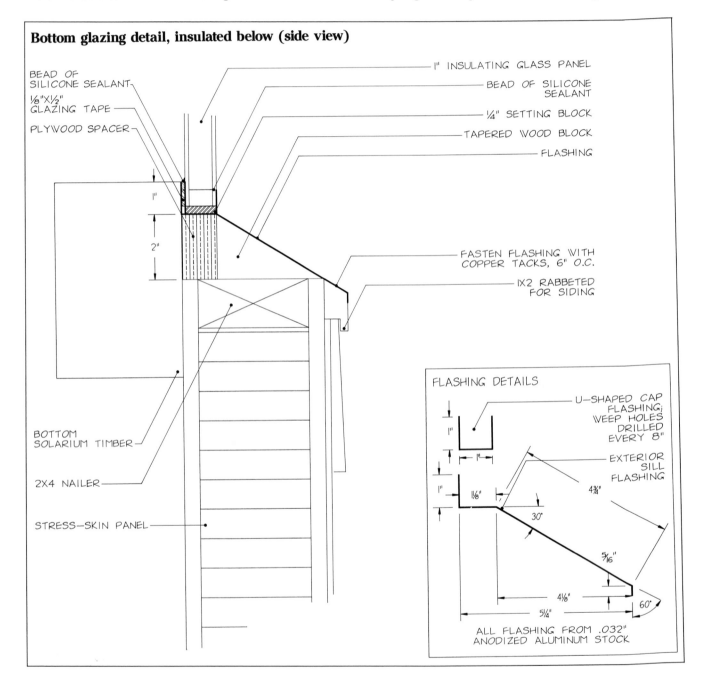

**Bottom glazing detail, insulated below (side view)**

BEAD OF SILICONE SEALANT

1/8"X1/2" GLAZING TAPE

PLYWOOD SPACER

1" INSULATING GLASS PANEL

BEAD OF SILICONE SEALANT

1/4" SETTING BLOCK

TAPERED WOOD BLOCK

FLASHING

FASTEN FLASHING WITH COPPER TACKS, 6" O.C.

1X2 RABBETED FOR SIDING

BOTTOM SOLARIUM TIMBER

2X4 NAILER

STRESS-SKIN PANEL

FLASHING DETAILS

U-SHAPED CAP FLASHING; WEEP HOLES DRILLED EVERY 8"

EXTERIOR SILL FLASHING

ALL FLASHING FROM .032" ANODIZED ALUMINUM STOCK

This block must be accurately cut and shimmed to fit precisely, because it describes the finished surface. After it is installed, some tricky flashing, generally custom-made by a sheet-metal company, goes down. The exterior sill flashing is installed first, then the *U*-shaped cap flashing accepts the neoprene setting blocks and the glass. As shown on the facing page, beads of silicone sealant prevent water from entering the assembly.

If the bottom beam of the solarium is on a slab foundation, there is usually no need for insulation below the glass. Thus the glass may rest on a trim board built out to the proper thickness for the glass plus glazing tape. A simple piece of flashing caps the top of the board and runs down about ¾ in., as shown in the drawing below. Then the *U*-shaped cap flashing is installed as specified previously.

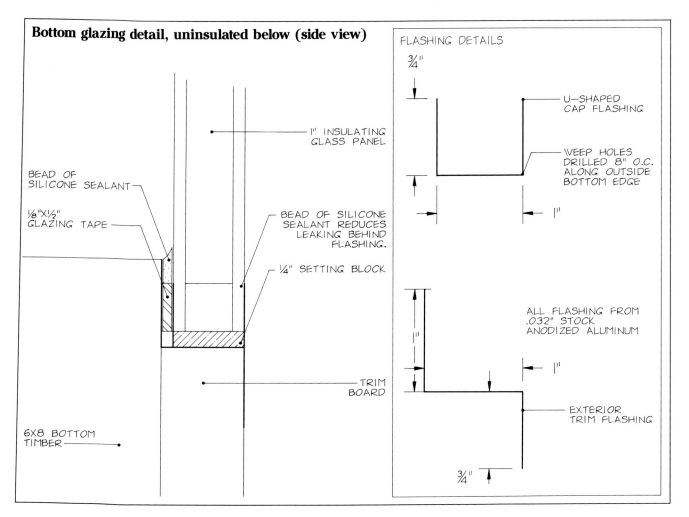

**Bottom glazing detail, uninsulated below (side view)**

1" INSULATING GLASS PANEL

BEAD OF SILICONE SEALANT

⅛"X½" GLAZING TAPE

BEAD OF SILICONE SEALANT REDUCES LEAKING BEHIND FLASHING.

¼" SETTING BLOCK

TRIM BOARD

6X8 BOTTOM TIMBER

FLASHING DETAILS

¾"

U—SHAPED CAP FLASHING

WEEP HOLES DRILLED 8" O.C. ALONG OUTSIDE BOTTOM EDGE

1"

ALL FLASHING FROM .032" STOCK ANODIZED ALUMINUM

1"

1"

EXTERIOR TRIM FLASHING

¾"

On the corners of the solarium, there are usually two detail options. In one situation (below), the insulation is mounted to the side of the post because there is no need for it on the front. In the other situation (facing page), the insulation is mounted to the front of the post because the wall surface surrounding the glass needs to be insulated. In the side detail, the trim board must cap not only the corner post but also the insulation. In the front detail, the stress-skin panel should be held back on the timber enough to allow for a plywood spacer, as shown. Three pieces of trim then cap the glass and cover the exposed panel edges. The face trim goes on first, side trim is installed second and the end trim is last. This detail might also occur at the top of the glazing when the glass does not extend all the way to the eave, as in a gable-roofed solarium. The only necessary alteration would be to add window-header flashing to the top of the end-trim board. In point of fact, a solarium is a big site-made window; when completely surrounded by the insulating system, the flashing and sealing details are the same as those on commercial windows.

As I mentioned earlier, we often create solariums by extending the roofline on the eave side. We have found that the most consistently available stock glass is 76 in. long, so we commonly design the roof extension to conclude at a height that allows for framing, flashing and trim of 76-in.-long glass. The solarium on p. 185 was designed on this basis. The roof was insulated with a typical panel overhang and trimmed as described in Chapter X. Under these circumstances, the friezeboard at the top of the wall also becomes the top batten for securing the glass, no matter whether the roof trim is based on a flat or sloped soffit. Flashing is not necessary under the eave because the roof overhang solves the moisture problem.

**Roof glass** There is justifiable argument over the wisdom of using sloped roof glass in solariums. There is more air

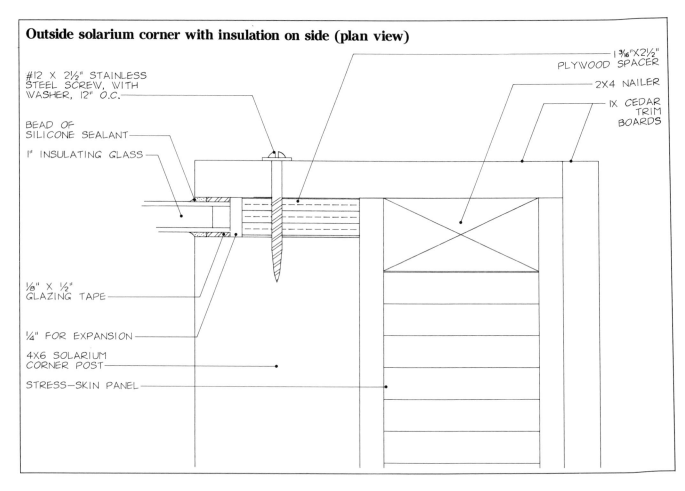

**Outside solarium corner with insulation on side (plan view)**

1 9/16"X2½" PLYWOOD SPACER

2X4 NAILER

IX CEDAR TRIM BOARDS

#12 X 2½" STAINLESS STEEL SCREW, WITH WASHER, 12" O.C.

BEAD OF SILICONE SEALANT

1" INSULATING GLASS

1/8" X 1/2" GLAZING TAPE

1/4" FOR EXPANSION

4X6 SOLARIUM CORNER POST

STRESS—SKIN PANEL

movement next to glass than next to an insulated wall, so there are likely to be more convection currents in a solarium than in the rest of the house. These currents carry lots of warm air to the solarium roof. If the roof is glass, expect the heat loss in this area to be significantly greater than if the roof were insulated. Add this to the fact that the heat gain through roof glass in summer is unnecessarily high, and it becomes hard to argue the case in practical terms. So when the design of a house suggests roof glass, I get emotional instead. I talk about a view where horizon and sky come together, a place to stargaze on a cold winter's night. I mention the effect of warm washes of sun penetrating deeply into the living area. With this out of the way, I should point out that we have used roof glass only about a dozen times on more than ten score houses. Roof glass makes the most sense when there is a relatively small solarium attached to a large living space. Roof windows are much more practical because they limit the glass

area, provide much-needed ventilation in summer and are easy to install.

If roof glass will be used, try to reduce its area by keeping the individual panes short. A strategy we commonly employ is to use pieces of full-length stock glass (usually 76 in.) for the vertical glazing, and half-sheets of the same width for the roof. Keeping width consistent is important to the design, and using a stock size for the vertical glass keeps cost down, even though the half-size glass is a custom order.

Following are several details for the application of roof glass. The first thing you might notice is that there aren't any wooden members capping the glass as there are on the wall. Using wood in an area that gets so much heat and moisture is simply an invitation to trouble. Also notice that all solarium roof glass is mounted on rafters instead of purlins. This keeps the glass size manageable and encourages water runoff.

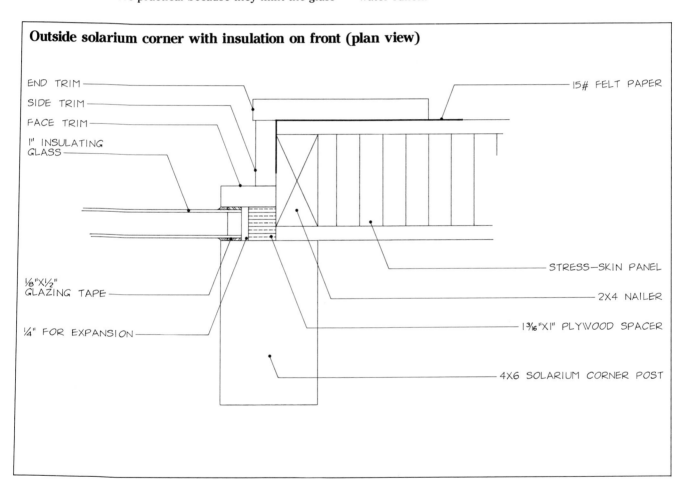

**Outside solarium corner with insulation on front (plan view)**

END TRIM
SIDE TRIM
FACE TRIM
1" INSULATING GLASS
⅛"X½" GLAZING TAPE
¼" FOR EXPANSION

15# FELT PAPER
STRESS-SKIN PANEL
2X4 NAILER
1³⁄₁₆"X1" PLYWOOD SPACER
4X6 SOLARIUM CORNER POST

Using a plate to make the transition between the wall and the roof simplifies and strengthens framing and flashing. The plate serves as a good stiffener and connects the entire assembly. It is angled to correspond with the roof slope and creates a bearing surface for both the roof glass and the vertical glass. Give the top and outside face of the plate extra moisture protection with an elastomeric membrane, positioned as shown below. These membranes have become popular for a number of different roofing applications because they remain elastic despite temperature fluctuations and tend to seal holes that might develop, even around nails and screws. Install the plywood spacers using the method to calculate thickness described on p. 189. Always use a good grade of plywood—it's important that it not absorb moisture and expand. Cut the spacer for the roof glass to fit from the outside edge of the vertical-glass spacer to about 1 in. from the top edge of the plate.

A piece of cap flashing is placed into the rabbet, with its lip bent over the top edge of the timber and securely fastened with ¾-in., #6 galvanized nails every 10 in. Flashing is a bad place to save money, so use as a minimum standard 20-oz. copper or .032 anodized aluminum. All the roof flashing should be custom-formed in a shop. The glass is laid into the channel using neoprene setting blocks, pre-

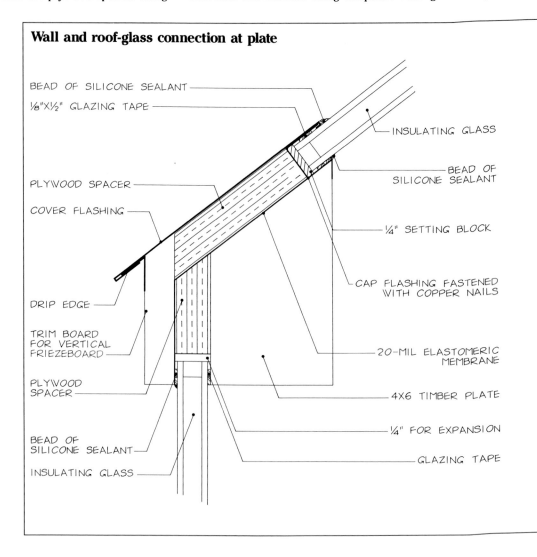

**Wall and roof-glass connection at plate**

BEAD OF SILICONE SEALANT

⅛"X½" GLAZING TAPE

PLYWOOD SPACER

COVER FLASHING

DRIP EDGE

TRIM BOARD FOR VERTICAL FRIEZEBOARD

PLYWOOD SPACER

BEAD OF SILICONE SEALANT

INSULATING GLASS

INSULATING GLASS

BEAD OF SILICONE SEALANT

¼" SETTING BLOCK

CAP FLASHING FASTENED WITH COPPER NAILS

20-MIL ELASTOMERIC MEMBRANE

4X6 TIMBER PLATE

¼" FOR EXPANSION

GLAZING TAPE

shimmed glazing tape and sealant as described on p. 189, but here the procedure differs. It's a little tricky to lift the flashing enough to allow the glass to sit on the glazing tape and sealant. The problem can be solved by sticking the glazing tape to both the top and bottom edges of the glass that will sit in the rabbet. A couple of small wedges on both sides of the glass keep the tape separate from the flashing while it enters the rabbet. After the glass is seated on the setting blocks and before the wedges are pulled out, apply sealant liberally between the flashing and the glass. Install the drip edge next, which is the small piece bent to the roof angle attached to the friezeboard at the top of the

wall. The cover flashing is bent at the ends to slip over the top of the cap flashing and over the top edge of the drip. With the drip edge, cap flashing and glass in place, fit the cover flashing over the drip edge, push it forward so that it can engage the cap flashing, then pull it back down the roof slope, locking it in place. Obviously, it is quite important that the cover flashing be tight to the glass, and there should be no ridges to trap water. Add more sealant to the drip edge and finish the sealant by beveling the excess with a putty knife to encourage runoff.

Another way to make the transition between roof and wall glass is to frame short rafters to connect with the so-

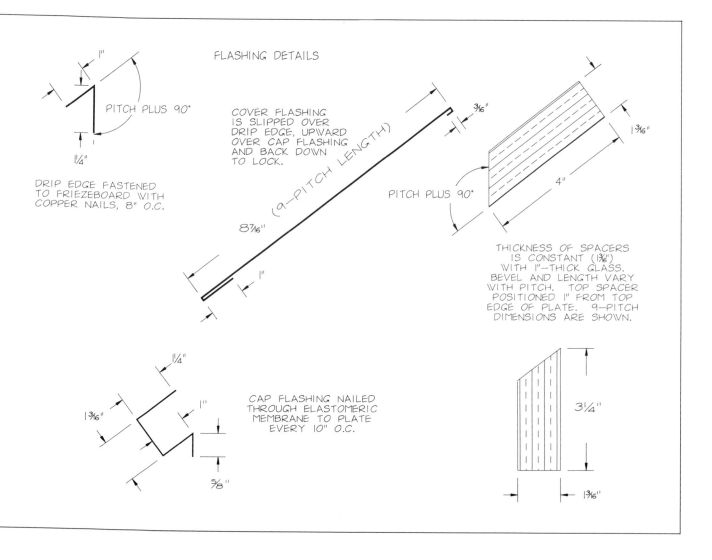

FLASHING DETAILS

PITCH PLUS 90°

1"

1/4"

DRIP EDGE FASTENED TO FRIEZEBOARD WITH COPPER NAILS, 8" O.C.

COVER FLASHING IS SLIPPED OVER DRIP EDGE, UPWARD OVER CAP FLASHING AND BACK DOWN TO LOCK.

(9-PITCH LENGTH)

3/16"

8 7/16"

1"

PITCH PLUS 90°

1 3/16"

4"

THICKNESS OF SPACERS IS CONSTANT (1 3/16") WITH 1"-THICK GLASS. BEVEL AND LENGTH VARY WITH PITCH. TOP SPACER POSITIONED 1" FROM TOP EDGE OF PLATE. 9-PITCH DIMENSIONS ARE SHOWN.

1 1/4"

1 3/16"

1"

CAP FLASHING NAILED THROUGH ELASTOMERIC MEMBRANE TO PLATE EVERY 10" O.C.

5/8"

3 1/4"

1 3/16"

larium posts, as shown in the top photo on p. 186. The roof glass meets the wall glass with no horizontal framing at the transition; flashing makes the joint leakproof. A 2x2 plant hanger mortised into each rafter serves as the horizontal connection. It is placed about 10 in. from each rafter end and 1½ in. from the inside face. The 2x2 is screwed and plugged to the rafter from the inside.

This detail is more prone to heat loss than the previous one. Because the flashing passes through to the inside, you get condensation in cold weather and ice on the inside of the glass in extreme conditions. Two pieces of flashing are needed, both of which should be made from heavy-gauge material. The vertical glass is installed to stop at a point ¼ in. below the roofline as extended from the timber

rafter, to leave room for flashing and keep the roof glass from contacting the vertical glass. A piece of *U*-shaped flashing with ⅛-in. neoprene setting blocks at the top caps the vertical glass. The setting blocks, again placed at quarter points, create a cushion should there be any pressure from the roof glass. No glazing tape is needed with this detail. The second piece of flashing is tricky and should be precisely shop-formed. The general dimensions are shown below, but the angle would change depending on the roof pitch. The weep holes are extremely important. Before the roof glass is installed, setting blocks are positioned to create space for expansion and for any water that might leak into the cavity; a bead of sealant along the top edge inhibits water entry.

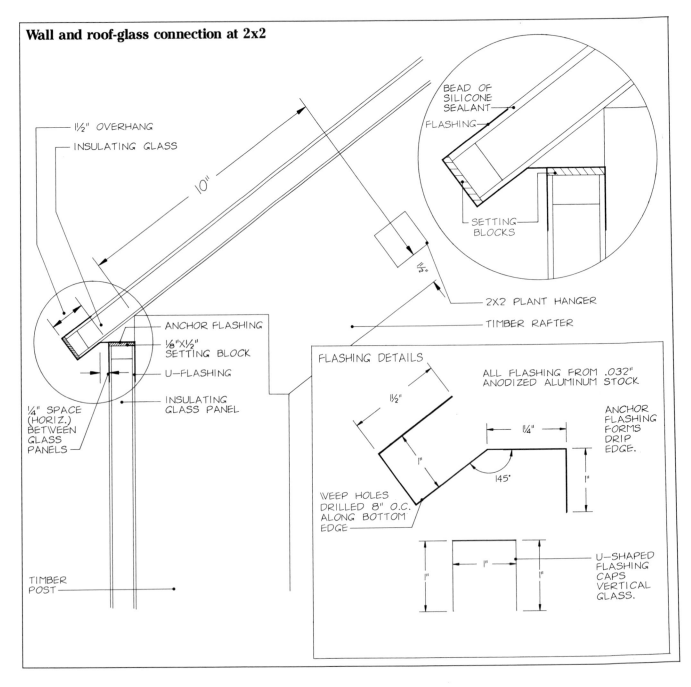

**Wall and roof-glass connection at 2x2**

1½" OVERHANG
INSULATING GLASS
10"

BEAD OF SILICONE SEALANT
FLASHING
SETTING BLOCKS

½"

2X2 PLANT HANGER
TIMBER RAFTER

ANCHOR FLASHING
⅛"X½" SETTING BLOCK
U-FLASHING
INSULATING GLASS PANEL

¼" SPACE (HORIZ.) BETWEEN GLASS PANELS

TIMBER POST

FLASHING DETAILS

ALL FLASHING FROM .032" ANODIZED ALUMINUM STOCK

1½"
1"
WEEP HOLES DRILLED 8" O.C. ALONG BOTTOM EDGE

¼"
145°
1"
ANCHOR FLASHING FORMS DRIP EDGE.

1"
1"
1"
U-SHAPED FLASHING CAPS VERTICAL GLASS.

Setting the roof glass over the rafters is identical to the procedure for vertical glazing installation, with two exceptions. First, cover the top face of the rafters with elastomeric membrane, and second, use metal for the top cap instead of wood. On the middle rafters, use ¼-in. aluminum bar stock as a stiff batten for the glass, screwing and counter-sinking it to the rafters on 8-in. centers. Take every precaution when installing the glazing tape and sealant—the potential for leakage is great. Then, for a more attractive finish, use anodized aluminum or copper to cap the bar stock. Attach the cap with #6 stainless-steel sheet-metal screws and rubber washers 2 ft. on center.

## Solarium-rafter glazing detail (exploded section perpendicular to pitch)

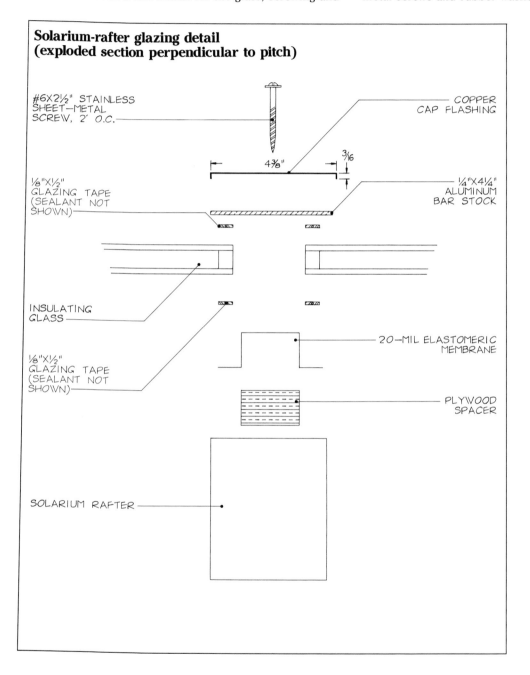

#6X2½" STAINLESS SHEET—METAL SCREW, 2' O.C.

COPPER CAP FLASHING

4⅜"

3⁄16

⅛"X½" GLAZING TAPE (SEALANT NOT SHOWN)

¼"X4¼" ALUMINUM BAR STOCK

INSULATING GLASS

20—MIL ELASTOMERIC MEMBRANE

⅛"X½" GLAZING TAPE (SEALANT NOT SHOWN)

PLYWOOD SPACER

SOLARIUM RAFTER

The detail on the outside edge of the solarium is basically the same as that described previously. It is necessary to cap both the timber and the top edge of the insulation. The plywood spacer extends to the outside edge of the exterior panel sheathing; a ¼-in.-thick plywood strip has to be added after the ¼-in.-thick aluminum batten to keep the top face flush. The flashing must span from the glass to the rake edge, and should be custom-formed as shown. Sheet-metal screws are fastened about ¾ in. from the plywood edge on the glass end. Another attachment for the flashing is made by nailing the rake edge of the flashing to the wall. Use 1-in. nails, 12 in. on center, to attach the rake flashing to the wall. Copper nails are best for copper and aluminum for aluminum.

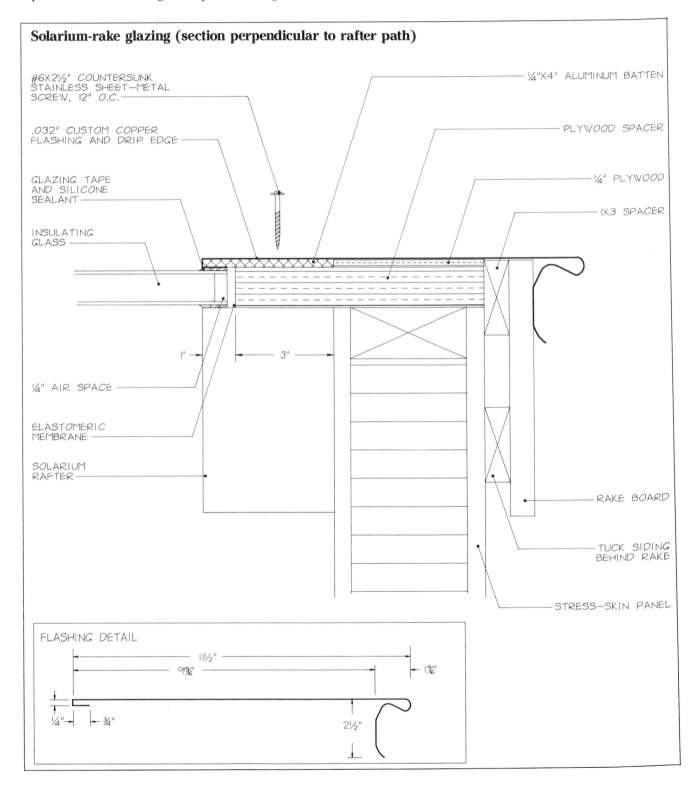

**Solarium-rake glazing (section perpendicular to rafter path)**

#6X2½" COUNTERSUNK STAINLESS SHEET-METAL SCREW, 12" O.C.

.032" CUSTOM COPPER FLASHING AND DRIP EDGE

GLAZING TAPE AND SILICONE SEALANT

INSULATING GLASS

¼" AIR SPACE

ELASTOMERIC MEMBRANE

SOLARIUM RAFTER

¼"X4" ALUMINUM BATTEN

PLYWOOD SPACER

¼" PLYWOOD

1X3 SPACER

RAKE BOARD

TUCK SIDING BEHIND RAKE

STRESS-SKIN PANEL

1"    3"

FLASHING DETAIL

11½"

9⁵⁄₁₆"

1⁹⁄₁₆"

¼"    ¾"

2½"

One of the easiest solarium details is at the top of the roof glass, where glass and insulation intersect on the purlin that receives the solarium rafters. Moisture coming off the roof falls away from the intersection, so leaks are prevented. The bottom edge of the insulation stops about 1 in. from the edge of the purlin to leave a bearing surface for the glass. A 20-mil elastomeric membrane strip is first laid into the corner, positioned between the top edge of the timber and the spacer as shown in the drawing below. The rest of the installation follows the guidelines discussed earlier; roof trim on the face of the insulation holds the glass in place. Because it is not subjected to as much weather and moisture as it might be elsewhere on the roof, wood trim works fine here.

**Purlin detail**

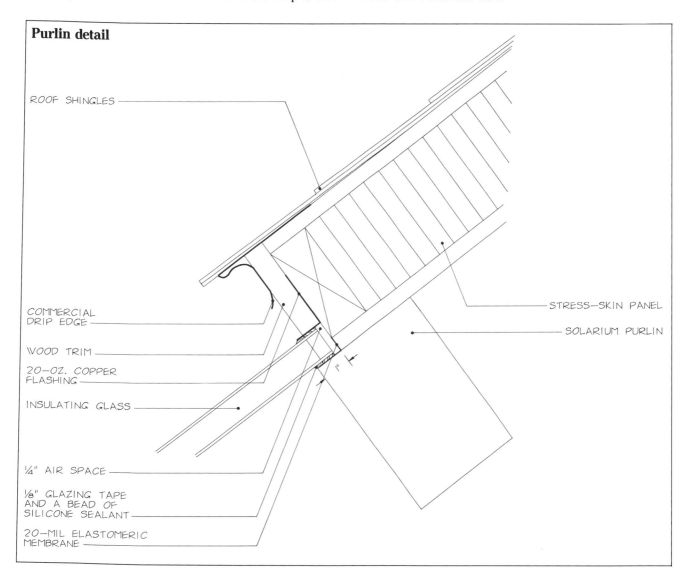

ROOF SHINGLES

COMMERCIAL DRIP EDGE

WOOD TRIM

20-OZ. COPPER FLASHING

INSULATING GLASS

¼" AIR SPACE

⅛" GLAZING TAPE AND A BEAD OF SILICONE SEALANT

20-MIL ELASTOMERIC MEMBRANE

STRESS-SKIN PANEL

SOLARIUM PURLIN

1"

# X. Finish Details

The interior detailing should match the high quality of the frame.

As we head into another chapter dense with design details, I would like to remind readers that the following information is not as rigidly prescribed as it might appear. These are not precise instructions for how to do; look at this chapter instead as a series of guidelines for how to think about how to do. Let it steer you along a relatively safe path, but do not let it stand in the way of individuality and creativity. The chapter starts with trim, both inside and out; flooring systems are next, followed by interior partitions. Finally, we'll end with a discussion of the treatment of interior surfaces.

## Window, door and baseboard trim

The process of timber framing is more likely to defy mass-production than to embrace it—the timbers are large and unwieldy, their sizes must change to suit structural conditions in the frame, and the joints are numerous and varied. In addition, timber framers themselves tend to be an independent lot. They often do what they do for reasons that are other than monetary, preferring the joy of creativity to the boredom of repetition. As a result, each timber frame takes on an individuality that becomes an important part of the character of the house. It is precisely this individuality that makes the use of stock trim materials one of my pet peeves. For if you go to just about any lumberyard in the United States for trim, you will probably be faced with many racks of trim types, but few real choices. They really

boil down to two or three types that are used all the time. I have nothing against these designs, just against their incessant and automatic use, which has brought a stifling sameness to houses all over the country. The economics of using stock trim also bothers me, because you wind up paying a lot more money for a lot less wood than if you were to buy unmachined stock. For instance, a flat 1x4 purchased from a lumberyard is typically ¾ in. by 3½ in.; when it becomes casing stock, it is further reduced to 9/16 in. by 3¼ in. The skinny little rounded-over piece of wood sold at about twice the cost of the "full-sized" material it came from is an abuse of wood and, to my mind, an insult to otherwise good construction.

Using plain, flat board stock for casing is less expensive and already more appealing than stock trim, but with surprisingly little effort, flat stock can become distinctive. Start by varying the width and thickness of the pieces. I like to make the head casing for windows and doors heavier than the side casing—the heavier board refers to a structural lintel, and the different dimensions create depth and define each piece. Play around with proportions to fit the circumstances. Consider the species, the intended color and finish of all adjacent surfaces, and the architectural and decorative style. Painted casing will look very different from wood left natural. Large, molded casing imparts a completely different feeling than small, flat casings would. Edges can be refined by molding with a router, shaper or hand plane.

My favorite trim detail is based on using 5/4 stock for the head casing and ¾-in.-thick stock for the side casing, as shown below. On windows, the stool is made from 5/4 stock and the apron from ¾-in. stock. Widths are variable. The head casing extends past the outside edge of the side casing by about ½ in. to reinforce the idea of a lintel. They meet with a simple butt joint, which can be improved with a lap on the back of the head casing. The window stool is rabbeted to fit the sash and also extends past the side casing by ½ in. Below the stool, the ¾-in. flat-stock apron is held back from the stool ends by ½ in. It's a simple scheme and easy to install. Even left unadorned, this type of trim stands apart from standard trim because it is rather raw and honest. But it can be made more elegant by adding chamfers to the inside edges of the side and head casings, to the top edge of the stool and to the bottom edge of the apron. Stop each chamfer 1½ in. before the end or, for the head casing, before it meets the side casing. When the chamfers repeat the embellishment of the timber edges, they can be especially unifying. Beads can also be routed into the face of the casing, stopped usually at the same point as the chamfers. Add a crown molding to the top edge of the header and the casing takes on more formality. The variations are endless, and that's the very point of using this type of trim. Almost anything you do will be better than the lumberyard offerings, but for less money and not a great deal of work it's possible to create a truly grand alternative.

The frame eliminates the need for crown moldings at the intersection of ceilings and walls, so in addition to window and door trim, we are primarily concerned with baseboards. These should match the material and style of the other trim, and also match each other where partitions meet ex-terior insulated walls. If the extended baseboard detail described on p. 137 and shown below is used, make sure the partition-wall baseboards are the same height as the extended baseboard plus the cap, and add a bead to match the bead on the extended-baseboard cap. Make the same kind of bead on window and door-trim edges for unity.

With little effort, flat board stock can be turned into distinctive trim. In this house, standard molding was added to side and head trim to add character and depth. The homeowner developed the idea for the curtain-rod attachment at the windows.

## Simple interior window trim

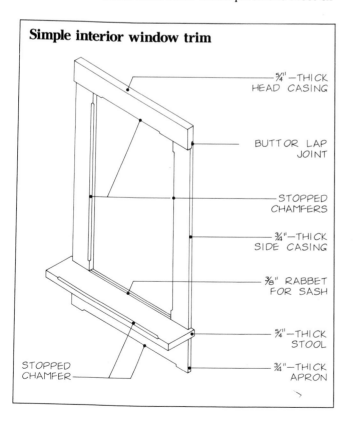

⁵⁄₄" —THICK HEAD CASING

BUTT OR LAP JOINT

STOPPED CHAMFERS

¾"—THICK SIDE CASING

⅜" RABBET FOR SASH

⁵⁄₄"—THICK STOOL

¾"—THICK APRON

STOPPED CHAMFER

## Extended-baseboard trim

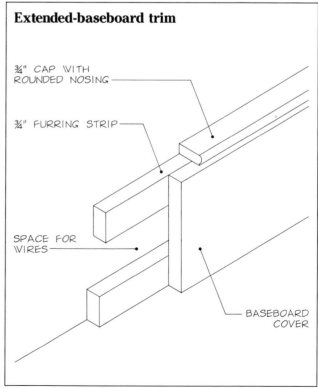

¾" CAP WITH ROUNDED NOSING

¾" FURRING STRIP

SPACE FOR WIRES

BASEBOARD COVER

## Stairs and rails

During design and construction, seize every opportunity to make the details and material of stairs and rails compatible with the timber frame. Perhaps an existing frame post could act as a newel to receive rails and stringers, or a post could be added for this purpose. At the very least, the edge and end embellishments of the frame can be repeated in the stairs.

Careful planning of stairways is absolutely necessary in timber-frame houses because their openings have to be created in the timber-floor system when the frame is being constructed. Stairs usually fall between bents, with the normal joist span broken with timber headers. They can also be designed to pass through a bent if necessary, by raising the bent girt in the area of the stair to allow headroom. The plan should detail the exact layout of risers and treads so that the frame openings will fit the stairway. Check the local building code on all stair requirements. Since stairs are a principal means of egress, the codes tend to be quite specific, although not the same from state to state.

As I have often described, in simple three-bay timber-frame homes the stair frequently falls in the middle bay. This area is commonly used for passage and transition, and so is seldom divided with partitions; sometimes it is open to the roof. To preserve this spaciousness, we often design open-riser stairs, housing the treads in the stringer and dadoing the balusters into its top edge. At the base and top of the stair, the stringer and the rail mortise into a newel post, which in turn is mortised into the stair header. The newel post has to be projected out of plane toward the stringer so that there will be enough material for a good

**In this frame, the bent girt is raised in the area of the stair to allow head room. Newel posts and stair details are designed to match frame details. Note the simple flat-stock trim around the doors.**

## Two ways to frame open-riser stairs

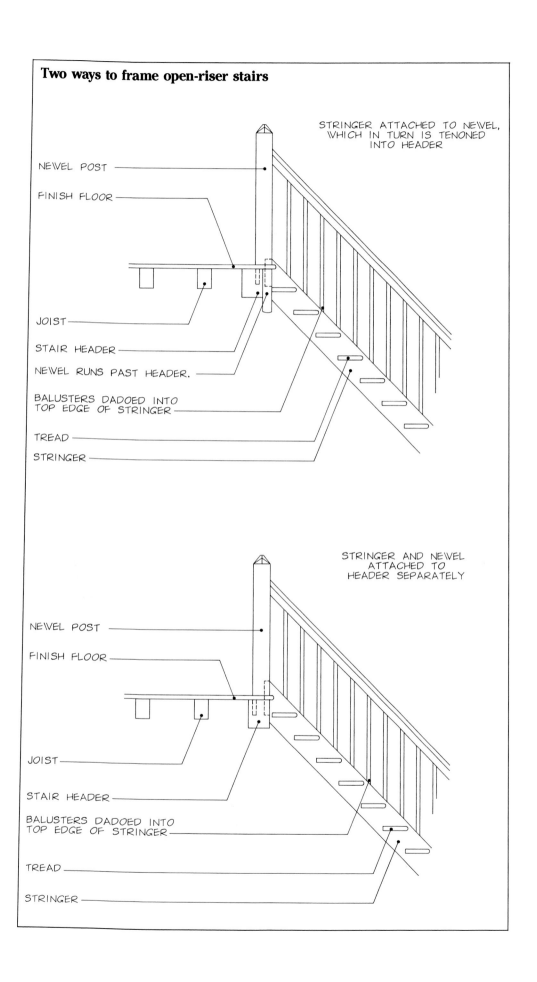

STRINGER ATTACHED TO NEWEL,
WHICH IN TURN IS TENONED
INTO HEADER

NEWEL POST

FINISH FLOOR

JOIST

STAIR HEADER

NEWEL RUNS PAST HEADER.

BALUSTERS DADOED INTO
TOP EDGE OF STRINGER

TREAD

STRINGER

STRINGER AND NEWEL
ATTACHED TO
HEADER SEPARATELY

NEWEL POST

FINISH FLOOR

JOIST

STAIR HEADER

BALUSTERS DADOED INTO
TOP EDGE OF STRINGER

TREAD

STRINGER

connection with the stringer. The projected portion of the newel extends below the stair header and is embellished much like a timber pendant. One great advantage of this system is that the stringer, newels, rails and balusters all can be shop-made and installed as a unit.

Another option is to attach the newel to the stair header separately. The newel is designed to align with the edges of the stair header and still receives the tenoned end of the stringer, but the bottom surface of the stringer bears on the face of the stair header. Do not drop the stringer so low that its end grain would be visible from below. Newel-post tenons should be at least 4 in. long to maintain rigidity.

Railings for catwalks or balconies also can be built into the frame. All that is required is that the railings align with timbers and their components (upper rail, lower rail and balusters) be designed into the frame plan. The railing can then be installed when the frame is built, making it stronger and, more important, taking full advantage of the frame as a finish element. Completely house (by at least ⅜ in.) the entire contour of rails that join to timbers, to allow for shrinkage.

**Balcony railings were built into this frame when it was raised. A housed mortise-and-tenon joint is used between rails and timbers to hide potential shrinkage.**

## Exterior trim on stress-skin panels

Exterior trim makes a statement about the significance and style of a house. It is a critical design feature, too often ignored or undervalued. If you look at old houses, you will notice that more often than not great attention was paid to the details of exterior trim, especially in the roof. Frequently, the finest woodworking was reserved for elaborate cornices and eaves with multilayered moldings. Contrast this with the modern ranch house, on which any kind of roof trim at all has been practically eliminated. I think lack of attention to this kind of detail is one reason why modern houses frequently look nondescript.

In Chapter V we discussed the stress-skin panel insulating system in detail and several alternatives in general. I didn't go into detail on the alternatives because they are based on conventional building systems and, as such, don't require much more than an explanation of procedure. The stress-skin panel system, however, is quite different from normal building practice in both theory and execution. This section will therefore touch on a few more details that relate to panels—things we didn't explore in Chapter V. Most of what follows is a direct result of the fact that typically the panels have less structure for the application of other materials than do standard insulating systems. I should note here that carpenters are by necessity creative people and often are able to deal with the following situations without advice. Indeed, most of the information in this section is comprised of things I have learned from the many carpenters who have worked on our houses over the years.

**Trim boards** Depending entirely on architectural style and siding, there potentially are quite a few different types of trim boards that might be on the walls of a house. Corner boards are the vertical trim pieces that reach from the base of the siding to the cornice work on the outside build-

ing corners. Skirt boards, used only on elaborate trim styles, run around the perimeter of the house at the base. Exterior door and window trim is just like the interior casing except that it gets much more abuse. There are usually a few other miscellaneous trim boards used for mounting such things as light fixtures and electric meters. Despite the relative lack of surfaces to nail to, mounting all these boards can become quite routine once a procedure has been established.

To secure trim boards to a panel where there is not a 2x4 edge nailer, use galvanized self-tapping screws or galvanized ring-shank nails. (Galvanized screws are used frequently for exterior decks and are available at most lumberyards.) Screws should be long enough to pass through the panel sheathing by at least ⅜ in.; nails need to pass through by about ⅝ in. Edges of boards that are 2½ in. or more away from a 2x4 nailer should be secured with screws to prevent cupping. If the board edge is less than 2½ in. away, use ring-shank nails. For instance, there is typically 1½ in. of nailing around a typical window or door opening. The trim is mounted onto the unit by at least ½ in., spans the gap between the rough opening and the nailer, and then is nailed to the nailer. Under these circumstances, unless the trim for the door or window is at least 4 in. to 4½ in. wide, no screws would be required. Near the ends of boards, always predrill for screws to prevent splitting. I see no reason why the visible head of a screw should be considered less attractive than the head of a nail, but if you disagree, take the trouble to countersink the screws and plug the holes.

**Roof trim** Of course there is no standard roof-trim detail that is appropriate for all houses. Trim is designed to fit the style, proportions and, yes, the budget. We have used numerous roof-trim details, but they fall into two basic

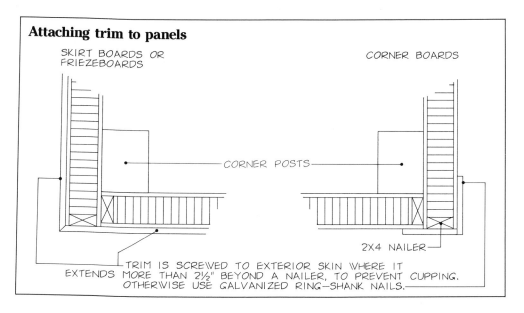

**Attaching trim to panels**

SKIRT BOARDS OR FRIEZEBOARDS

CORNER BOARDS

CORNER POSTS

2X4 NAILER

TRIM IS SCREWED TO EXTERIOR SKIN WHERE IT EXTENDS MORE THAN 2½" BEYOND A NAILER, TO PREVENT CUPPING. OTHERWISE USE GALVANIZED RING-SHANK NAILS.

categories. One has a sloped soffit at the eave, the other has a flat soffit.

Applying roof trim to stress-skin panels need not be any more difficult than trimming conventional rough framing, but it does require good planning. The easiest way to achieve the roof overhang is to extend the roof panels beyond the wall panels; the trim is attached to nailers let in to the perimeter of the roof panels. The problem is that the roof panels are then cantilevered beyond the wall panels and have limited load-bearing capacity. Generally, the overhang for the standard panel (drywall on one side) should be held to between 8 in. and 12 in. Structural panels (doublechips, p. 100), usually can extend beyond the wall panel by 16 in. to 24 in. Applying 2-in.-thick tongue-and-groove planking to the roof as the finish ceiling and extending it beyond the wall panel allows the overhang to be increased to 16 in. to 24 in. even when standard panels are used. Before making a decision about the length of the overhang, however, check the roof-load requirements in your area and ask for the panel manufacturer's recommendations about supporting the load.

**Sloped-soffit trim**  The simplest trim style uses a sloped soffit made of a solid ¾-in.-thick board or, at the very least, ⅝-in.-thick AC exterior plywood. One edge of the soffit is ripped to fit against the wall panels, and then the soffit is pushed into place—if the fit between the top of the soffit and the panels is snug, the soffit will hold itself in place temporarily while the bottom is being secured to the nail-

ers in the panel edges. (When standard panels are used, there is no way to screw or nail the top edge of the soffit to the drywall underside of the panel.) The bottom edge of the soffit must fit tightly into the fascia dado. Assuming the nailers in the panel edges are straight, simply use a combination square as a depth gauge to set dado depth; if they aren't straight, set up a string line for each run of trim and plan on shimming behind the fascia at the low points. Special nails are not needed to attach the trim to the nailer in the panel edge. Galvanized nails or galvanized ring-shank nails are usually appropriate for exterior applications.

The friezeboard, which performs the critical function of holding the top edge of the soffit in place, goes on next. The top edge of the friezeboard is ripped to the roof slope, then pushed up until it forces the soffit tightly against the panel, as shown below. Usually the bottom edge of the friezeboard is rabbeted to hold the last cut of shingles or clapboards. A built-up rabbet is sometimes used, as shown on p. 208. It will probably be necessary to use some screws (as described in the section on trim boards) to attach the friezeboard to the wall panel, since there is not likely to be good nailing between panel seams.

When installing the rake trim, the lack of an acute angle between the wall and the roof means there is nothing to hold the soffit in place before the friezeboard goes on. Solve this problem by installing the friezeboard first, using an appropriately sized spacer to set the dimension for the soffit. Remove the spacer, push the soffit into place and nail it to the nailers in the panel edges.

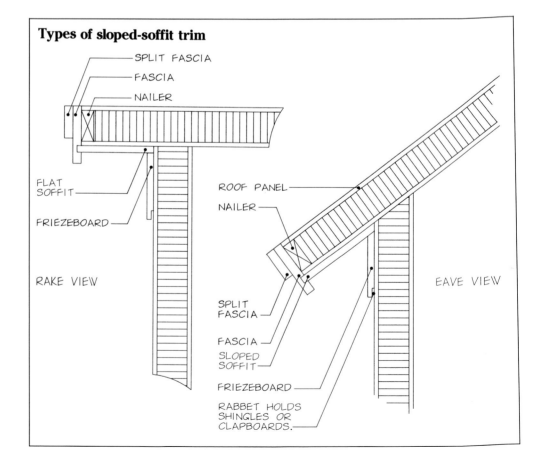

**Types of sloped-soffit trim**

SPLIT FASCIA
FASCIA
NAILER
FLAT SOFFIT
FRIEZEBOARD
RAKE VIEW

ROOF PANEL
NAILER
SPLIT FASCIA
FASCIA
SLOPED SOFFIT
FRIEZEBOARD
RABBET HOLDS SHINGLES OR CLAPBOARDS.
EAVE VIEW

The deep overhang, not the trim, is the focus of this house. Using doublechip (structural) panels made it possible to extend the overhang. Cedar 1x6s nailed to the panels form the soffit. Gable-end glass is fixed directly to the timbers of the end bent. Using dry wood is critical where glass will be attached.

When tongue-and-groove (or shiplap) planking is installed as an interior finish before the panels go on, it can be used to create the soffit. On purlin roofs, lay the planking perpendicular to the purlins, running from the eave to the peak. For expediency and accuracy, extend the planking a few inches past the end of the overhang and cut an entire roofside of planking to a chalkline. If you're going to do this, there's the potential for an awkward transition between the eave soffit and the rake soffit, because there is a lack of nailing surface on the rake side. Plan the rake soffit so that a single plank spans from the wall panel to the roof nailer, since there won't be a nailing surface between. To do this, extend the next-to-last plank beyond the last wall panel (which is the last nailing surface) far enough so that you wind up with your last plank at the outside edge of the rake. After the roof panels are installed, the last rake plank can be attached from below.

On common-rafter roofs, the planking is laid perpendicular to the rafters (parallel to the eave). In this case, extend the planking over the gable ends to form the rake soffit. Cut the plank ends all at once after they are in place, as you would at the eaves. Plan the soffit at the eave to be a single plank. Extend the first plank that is applied to the roof surface over the wall panel, as before, so that the matching can continue from the eave to the wall.

An overhang of more than 12 in. implies the use of structural panels, and since structural panels have a nail base on both sides, the problem with soffit attachment is automatically solved. Remember that codes require a fire barrier over the foam, so even if the roof will be sheathed first with planks, it will be necessary to use a material with a 15-minute fire rating (probably drywall) between the planks and the panels.

With the friezeboard and soffit in place, the fascia boards can be installed. Cut and nail the fascia boards directly to the panel nailers—the tablesawn dado in the fascia, which receives the soffit, helps with alignment. In a high-quality job, linear splices and corner joints are mitered. When the fascia boards are over 1 in. thick, we like to use finger joints or dovetails at the corners for decoration and strength. After the fascia has been attached, cut and nail the split fascia to it. The split fascia gives the trim more depth and a molded appearance.

What makes this type of sloped soffit so easy is that the trim at the eave and rake meet at right angles, resulting in relatively simple cutting and fitting. Changing species, size and treatment of the components allows the trim to be customized to suit the house. We have used everything from a single, heavy, cedar plank to small boards painted in soft colors for both the fascia and the split fascia.

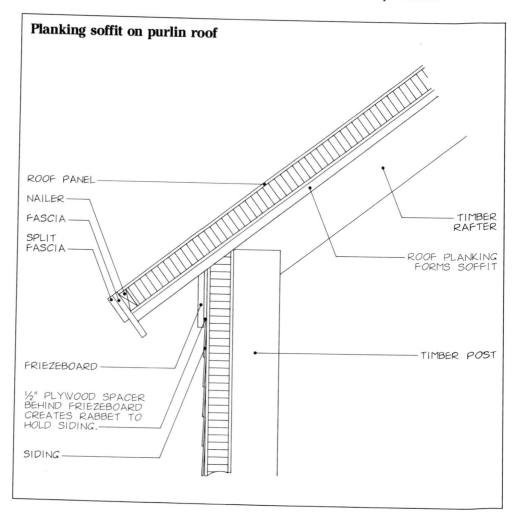

**Planking soffit on purlin roof**

ROOF PANEL

NAILER

FASCIA

SPLIT FASCIA

TIMBER RAFTER

ROOF PLANKING FORMS SOFFIT

FRIEZEBOARD

½" PLYWOOD SPACER BEHIND FRIEZEBOARD CREATES RABBET TO HOLD SIDING.

SIDING

TIMBER POST

Now we'll look at another variety of sloped-soffit trim, used when it is important to design to have a vertical eave fascia rather than one perpendicular to the roof slope—for instance, to accept gutter brackets when gutters are to be used. To give the fascia a vertical orientation, it is necessary to create an additional nailing surface after the panels have been installed. This can be accomplished either with individual wedge blocks attached to the panel nailers or a single, continuous piece ripped to the correct wedge shape. The problem with the latter method is that a fairly large piece of wood could be required, depending on the roof slope. For example, if the roof pitch were 10/12, the smallest leg of the wedge would be about 2½ in. Making individual wedge blocks takes more time, but the pieces can probably be made from scrap. When the fascia is plumb, housing the soffit in a fascia dado is much too complicated, so instead carefully rip the soffit to the plumb line and hold it about 1/16 in. beyond the wedge block to improve the odds of a good fit. Rake soffits can still be dadoed as described previously.

When a gutter is mounted on the fascia, it adds lines to the detail similar to a crown molding. To keep all of the trim consistent, a crown added to the rake trim is usually desirable. Simply add a wedge strip to the fascia (already fixed to the panel nailer) and nail on a crown.

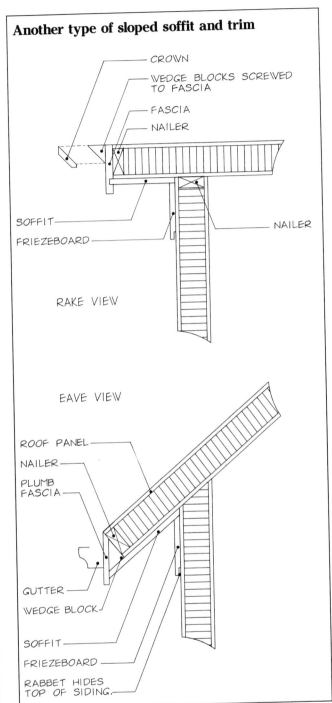

## Another type of sloped soffit and trim

**Sloped-soffit roof trim is easy and simple, but can still display nice detail. The rake boards on this house are 2-in.-wide cedar, dovetailed at all outside corners. The soffit material is 2-in.-thick tongue-and-groove planking.**

**Flat-soffit trim** In addition to having some aesthetic advantages, flat-soffit trim also provides support for the panel on the eave where the loads are likely to be the greatest. While there are too many possible variations to cover here, they all spring from a few basic procedures. The drawing on the facing page shows a built-up system that over the years has become one of our favorite details. Just as in sloped-soffit trim, the eave and rake overhangs are determined by the overhang of the roof panels. For best effect, try to design the detail to include a crown molding that follows the plane of the panel nailer and a soffit located approximately 2 in. to 4 in. below the bottom edge of the overhanging eave panel.

Start by installing the blocking, soffit and fascia. In most cases, the blocking can be cut from 2x4s; a wider fascia might require a 2x6. Each block is angled to fit against the underside of the roof slope and has a square front edge to serve as vertical nailing for the fascia. Installation is simplified by nailing a number of blocks onto a ⅝-in. plywood strip on the ground and installing the assembly as a unit. This also helps ensure a straight line at the fascia, because inconsistencies in the wall surface can be corrected by shimming between the wall panel and the plywood—it is hard to shim one block at a time. Position the plywood along a chalkline snapped on the wall panel, and screw it to the panel with galvanized self-tapping screws. Then toenail the blocking at its angled surface to the panel nailer. Shim as necessary between these two pieces to keep the front edge of the soffit level.

The next step is to provide the framing for the corner return, as shown at the left of the drawing. The corner return is optional, but I think it really enhances the richness of the roof trim by creating a visual base for the roof. When the trim is completed, the corner board should fall in the middle of the corner return, giving the whole unit the feeling of a column and capital. Because the corner return allows the eave line to continue at the gable end, the rake can finish to the top of the return instead of being mitered to the fascia—a procedure which is usually difficult, with results that are not always attractive. Set up the corner return by putting together a small assembly of plywood and blocking. It would fit against the gable end of the building, from the last eave block to the end of the return. This small assembly would be screwed to the eave block, to the panel nailer on the building corner, and to the outside sheathing of the panel beyond the corner. Then the soffit and the fascia are installed along with the eave trim. A solid 2x cap is put on the return, followed by the sloped plywood cap, which has a 15° slope to shed water. Use shims or a nailing strip on the building surface to create the slope. The 2x cap should be ripped to receive a crown at the same angle created by the roof slope on the eave side. The entire top of the corner return should be covered with flashing. It's easiest to make up a pattern and have all of the returns made by a sheet-metal shop.

For the crown to continue on the rake trim, it will be necessary to cut a beveled wedge strip to repeat the roof angle at the eave. If at all possible, this piece should be ripped from a solid piece of wood. If not, plan on installing blocks for nailing at least every 24 in.

Begin the rake trim with the soffit and freizeboard. The rake soffit would be attached using the same method described in the sloped-soffit system because of the lack of nailing on the rake overhangs. Fascias are installed after the soffits. The beveled wedge strip is next and the crown trim is last. Shim to correct alignment, especially on the fascia and crown. Good preparation when installing panels and soffit blocking saves time when the trim is attached. It is not necessary to vent the soffit because there is no air movement through the insulation.

The dimensions of this type of roof trim can be altered according to taste and circumstances. For added depth in the crown, for instance, nail a ½-in.-thick by 1½-in.-wide strip to the fascia before the crown. Generally, gutters need a flat vertical face for attachment; if they are required, make the fascia wider by using a larger block and dropping the soffit down to accommodate it. The crown usually remains on the corner return and rake when a gutter is used on the eave.

**Using only flat boards, this trim detail has a rather molded appearance in a colonial style. The corner return is centered over the corner board for balance. The extra strips on the corner board reinforce the idea of a column.**

**Elements of flat-soffit trim at corner**

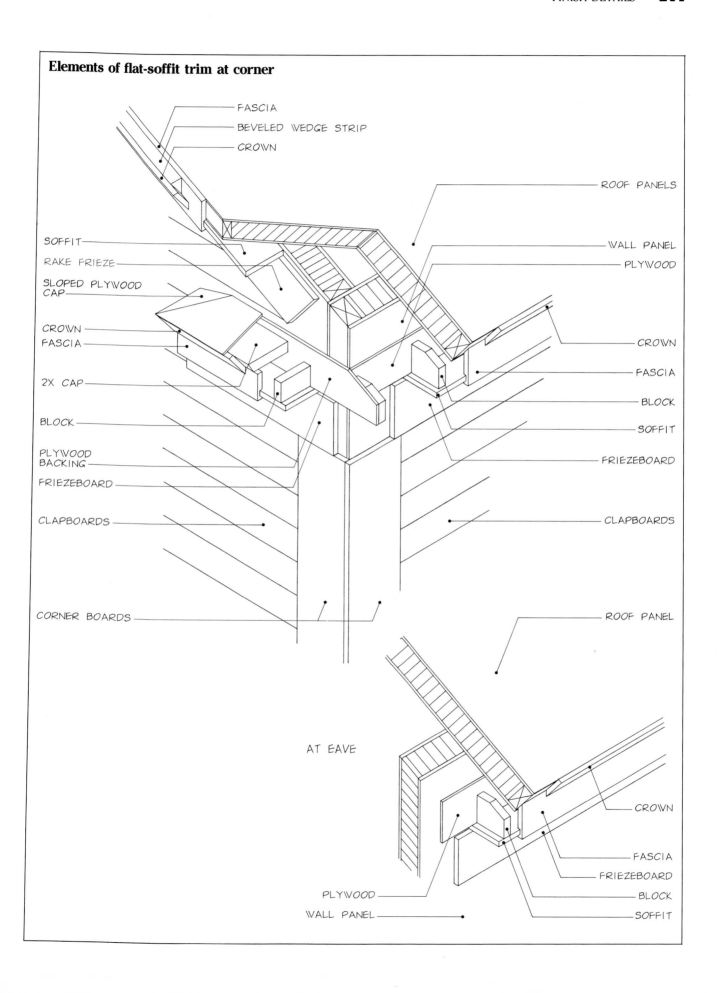

FASCIA
BEVELED WEDGE STRIP
CROWN
ROOF PANELS
WALL PANEL
PLYWOOD
SOFFIT
RAKE FRIEZE
SLOPED PLYWOOD CAP
CROWN
FASCIA
CROWN
FASCIA
2X CAP
BLOCK
BLOCK
SOFFIT
PLYWOOD BACKING
FRIEZEBOARD
FRIEZEBOARD
CLAPBOARDS
CLAPBOARDS
CORNER BOARDS
ROOF PANEL

AT EAVE

CROWN
FASCIA
FRIEZEBOARD
PLYWOOD
BLOCK
WALL PANEL
SOFFIT

**Roofing over stress-skin panels**   The typical stress-skin roof application is technically a "warm roof," meaning that one side of the panel is exposed to the heated interior and the other to the outside. When warm roofs are used in cold climates, the heat passing through the insulation can cause the snow to melt on the surface of the shingles, flow down the roof and form an ice dam at the eave; water then backs up behind the dam and leaks through the shingles. A good panel allows very little heat loss, making this problem minimal, but we still like to use 50-lb. felt or an elastomeric membrane for the first 5 ft. at the eave and for 2 ft. on either side of valleys.

Since stress-skin panels are an effective moisture barrier in both directions, it is possible for moisture to collect on the exterior surface of the panel, under the shingles. This is most likely to occur during extremes of high humidity and heat, or high humidity and cold, and while leakage is improbable, certain kinds of roofing materials might be damaged. The following types of roofing should therefore be laid over furring strips to allow air flow: wood shingles and shakes, slate, tile and standing-seam metal. (Corrugated metal can be laid directly on the panels.) The air flow keeps the roofing material away from the warmth of the panel, creating a "cold roof."

Shrinkage and movement of the timber frame cause a certain amount of movement in the panels, and even if it is slight, this movement can cause shingles to wrinkle slightly at the panel seams. For this reason I don't recommend applying lightweight asphalt or fiberglass-based shingles directly to the panels. (If you're going to use these, first nail and glue an additional sheathing membrane, such as ⅜-in.-thick plywood, to the panels to prevent movement. Of course, this extra cost would greatly exceed the cost of buying heavier shingles to begin with.) Shingles should be the heavily textured architectural type or the heaviest available weight, generally about 310 pounds per square in fiberglass and 350 pounds per square in asphalt. Asphalt shingles are becoming hard to get and are being replaced by those with a fiberglass base.

Furring strips are being attached to this roof surface in preparation for wood shingles. Since the shingles will be laid with a close exposure, the furring strips have to be quite close as well.

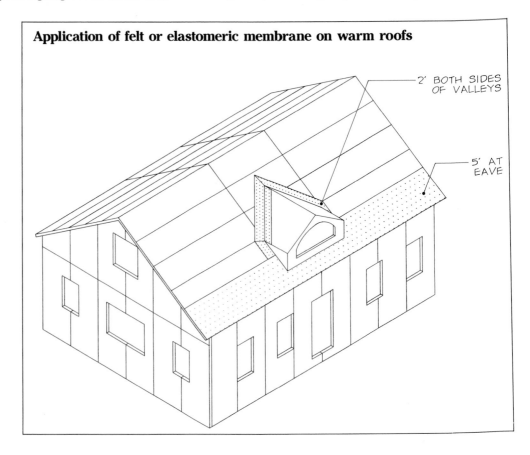

**Application of felt or elastomeric membrane on warm roofs**

2' BOTH SIDES OF VALLEYS

5' AT EAVE

# Flooring

The issue of flooring in a timber-frame house is a classic case of good news and bad news. The good news is that the floor of one story becomes the ceiling for the story below, making construction efficient and reducing costs. The bad news is that a floor with sound-deadening characteristics is not automatic, as it is when using conventional construction techniques. It is harder to deaden noise transmission through floors than walls in any event, because floors are subjected to both airborne and impact sound. The design and installation of floors and ceilings are therefore crucial not only to aesthetics, but also to comfort and privacy.

It is a very simple, but too often overlooked, observation that proximity to noise is the crux of the sound-transmission problem. In other words, if you expect to find privacy in the library, don't put it underneath your teenager's rock-and-roll headquarters. Such oversights are nearly impossible to rectify during construction. As much as possible, stack rooms that require privacy and those that are public. Let the teenager compete with the blender instead of with Mozart. It is thus sometimes possible to solve problems of sound transmission entirely during the development of the floor plans, by isolating critical areas. But such houses are necessarily large and spread out (read expensive). House plans intended to use space and volume efficiently are the most difficult to insulate against sound. I give special emphasis to this issue because, frankly, it is not one we have always dealt with well. We didn't forget about noise, we just assumed it wouldn't be a problem. Very often the house was being built for a couple not yet with children—people absolutely ignorant about the meaning of the word noise. So here is one more suggestion I offer, having learned the lesson the hard way: For the benefit of yourselves or future inhabitants, assume that someday your house will also be home to a tumbling gymnasium and a rock concert, sometimes simultaneously.

In timber-frame construction, it makes sense to install the finish floor and ceiling materials over the timber floor—applying the finish floor to one side and the ceiling to the other excludes the possibility of using the frame as finish work. The other alternative, fitting materials between timbers, turns the frame into an obstacle to efficient construction. There's a huge difference between installing flooring over timbers with 5-in. or 6-in. surfaces for fudge factors, and attempting to scribe materials to fit precisely between timbers on 30-in. to 36-in. centers. So, except to solve a few plumbing problems, we always apply flooring on top of the timber structure.

As a general guideline, it's wise to try to design timber-floor systems for nominal 2-in.-thick tongue-and-groove planks. The most commonly available planking species are pine and spruce, followed by fir and hemlock. Planking usually comes in an "adequate" grade, but hardly ever "good." It will be called something like "#2 common," which sounds pretty good by the grading book; however, every piece is likely to be as bad as is allowable and probably a little worse. Unless there is a clear demand from the client for a rustic appearance, I prefer to pay the difference for a higher grade of material, or I purchase planking from one of our small, local mills. Small mills are great. I like to sit down and have a discussion with the owner (who is

also usually the sawyer) about the qualities of good wood, and then we go outside and identify what we both think is good wood. Finally, I ask how much this good wood we've been talking about will cost. From that day forward, I can throw out the grading rules and just order "good wood."

Assuming the use of planks or a system with equivalent strength, timber joists ideally should be located about 30 in. on center to keep the flooring stiff. This is about the right spacing for pine planking (good wood); it is a little conservative for fir, hemlock or spruce. (This is not a political discussion. "Conservative" as a structural reference is a good thing for even the most liberal.) Of course, joists ought to be evenly spaced between major beams, so the trick is to find the on-center number closest to 30 in. All timbers still have to be sized to support the floor loads as suggested by the local building codes.

**Planks on timbers**  The simplest floor is a single layer of planks over timbers. Although it does nothing to deaden sound, planking is useful in such open spaces as lofts and walkways (photo below). People who wish to build their homes in affordable stages often choose to start with a layer of planks over timbers, adding other materials when they can afford them.

Tongue-and-groove planks usually come with a V-groove on the visible face, which helps to mask shrinkage or an uneven surface by creating a shadow line. For a straight match without the V-groove, just expose the opposite face. Widths of 6 in. or less are easiest to lay and cause the fewest problems.

**Because this loft is open to the rest of the house, sound-deadening insulation would not have been useful. Therefore, a single layer of planks is both the ceiling below and the flooring above.**

Wide planks (8 in. or more) are hard to straighten when they are bowed, and because there are fewer joints to absorb movement, seasonal expansion and contraction of the wood will be more noticeable. I know of a building where wide, very dry hardwood planks were laid over timbers late in the winter. When the planks took on some moisture in the summer and expanded, they pushed the walls of the building out as much as 2 in. in some places. This story gave me new respect for narrow softwood planks, and reasonable (but not extreme) seasoning. Narrow planks can also be secured by blind-nailing (angling the nail through the plank from just above the tongue), which is much easier than screwing and plugging, and much more attractive than face-nailing.

Installation begins by laying the planks perpendicular to the joists at one end of the floor system, using construction adhesive to glue the planks to the timbers. (This helps prevent squeaking and makes the floor stiffer and stronger.) Nails for blind-nailing should be 16d common; if the timbers are oak, they should be galvanized to prevent corrosion from tannic acid. When blind-nailing, use two nails directly next to each other on each timber. A good trick is to hold both nails in your hand and drive them at once; they tend to spiral around each other and provide further locking. Where posts pass through the frame, a notch made about ½-in. deep receives the plank. By simply pushing the plank into the notch, scribing is eliminated, and any gaps that might develop when the timbers shrink will be hid-

## Plank floor

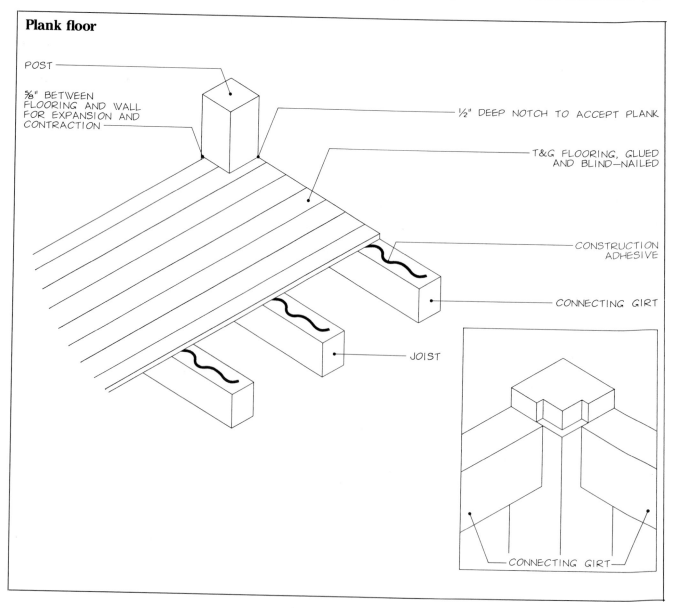

POST

⅝" BETWEEN FLOORING AND WALL FOR EXPANSION AND CONTRACTION

½" DEEP NOTCH TO ACCEPT PLANK

T&G FLOORING, GLUED AND BLIND-NAILED

CONSTRUCTION ADHESIVE

CONNECTING GIRT

JOIST

CONNECTING GIRT

den. Leave at least ⅝ in. between the planking and the wall for expansion and contraction. This space will be hidden behind the baseboard and can also be used for running electrical wire.

The ceiling side of the planks can be finished with stain, oil or both before installation. The floor side can be sanded and finished later in construction. Wall-to-wall carpeting over the planks helps absorb noise, especially that caused by impact—kids playing with blocks on a plank floor can make a deafening clatter. If you plan to use carpeting, shop as carefully for the pad as you do for the carpet, and ask for the one that would be the best sound barrier.

**Drywall over timbers** The timber frame itself brings a great amount of wood into the living environment, and adding more exposed wood on the ceiling is often detrimental to design. The alternative is usually a light-colored drywall ceiling, which makes the room brighter by reflecting light and gives the frame more distinction by avoiding wood on wood. My tendency is to use wood planks over timbers for high ceilings and cathedral ceilings, and to use drywall on low ceilings. Wood can make a large volume seem warmer and cozier. A painted ceiling in a light color makes a small volume feel larger.

To apply drywall over timbers, start by gluing and nailing a 2-in.-wide strip of plywood over the center of each

Drywall ceilings laid over the timbers help to make small rooms feel brighter and larger by reflecting light. A painted background also gives contrast to the frame.

Large, open rooms often benefit from wood ceilings. The boards absorb light and make the large space feel warmer and more comfortable.

joist, leaving approximately 1 in. of timber on each edge toward the joist bay. Use ½-in.-thick plywood for ½-in. drywall (which is what we commonly use), or ⅝-in.-thick plywood for ⅝-in. drywall. When the strips are in place, cut sheets of drywall lengthwise to fit between the plywood strips in the joist bays. Since the timbers are on approximately 30-in. centers and drywall comes in 48-in. widths, this will mean a large amount of waste. However, by using drywall that is the same length as the joist bay, seams between joists can be completely avoided, and the money saved by not having to tape and spackle seams is greater than the cost of the waste. You can eliminate yet another finishing step by prepainting the drywall before installing it over the joists.

After the drywall is down, glue and nail all the planking to the plywood strips as if they were the timbers. Finally, screw the drywall to the planking from below midway be-

tween the joists on 16-in. centers. Even if the drywall has been painted, it isn't difficult to spackle and touch up the screw heads.

It's possible to use plaster-base drywall or rock lath instead of standard drywall, with the intention of plastering between the timbers. We don't generally recommend this because it is so much more work and shrinkage eventually pulls the timbers away from the plaster edges, creating a gap. Still, there's nothing like plaster, and it's hard to argue against the warm feeling and natural texture of a good plaster job. If clients choose this option, I encourage them to wait one year before applying the plaster. The problem is that it's hard to wait if you're building on a time-limiting construction loan, or if you are simply anxious to move into a finished house. When using plaster-base drywall, don't worry about the seams, because the entire surface will be coated with plaster; try instead to reduce waste.

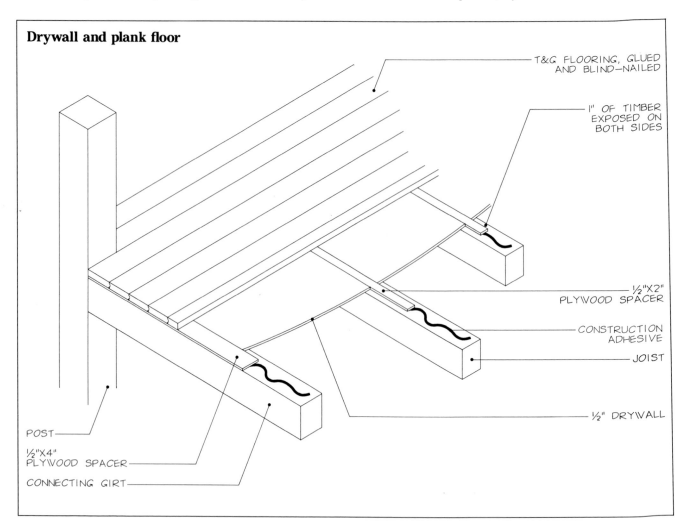

**Drywall and plank floor**

T&G FLOORING, GLUED AND BLIND-NAILED

1" OF TIMBER EXPOSED ON BOTH SIDES

½"X2" PLYWOOD SPACER

CONSTRUCTION ADHESIVE

JOIST

½" DRYWALL

POST

½"X4" PLYWOOD SPACER

CONNECTING GIRT

**Sound-deadening floors** Adding drywall to the flooring system does almost nothing for soundproofing. The interconnected posts and beams of a well-built timber frame are designed for stiffness and assembled under tension, like strings on a guitar; great strength is achieved but so is a great ability to transmit sound. To reduce sound transmission significantly, you must absorb the sound with a material of less density than the framing members.

We often use a rigid sound-deadening board such as that manufactured by the Homasote Company. Soundboard comes in 4-ft. by 8-ft. sheets and several thicknesses, of which ½ in. is the most commonly used. It has enough compressive strength to support floor loads and enough resilience to absorb some sound. To be effective, it should be installed in an unbroken layer between other materials; for instance, between the drywall and the planking in the system previously discussed. It goes over everything with no regard for joints. (It is not necessary to join the soundboard seams to timbers—just tack where possible to secure temporarily.) The soundboard is followed with blind-nailed planking, but since the nails now must pass through ½-in.-thick soundboard to get to the plywood strips, use 20d nails instead of 16d. The drywall still has to be secured to the planking through the soundboard; use 1¾-in. drywall screws. Probably the biggest problem with installation is to get the soundboard down without stepping through the ceiling. Keep yourself on a solid floor by laying the soundboard and the planking at the same time.

What's lost in this system is the structural advantage of gluing the planking to the framing. But you might want to glue the planking to the soundboard to prevent squeaking.

When soundboard is used in this way, the ceiling does not need to be drywall. Instead, the finished ceiling could be ¾-in.-thick boards (as shown in the next flooring solution, p. 218). Soundboard and planking would then be installed as described previously. In either case, the planking must span from joist to joist. Remember that it is a good idea to notch the entire thickness of the floor assembly into any posts that pass through the floor level to avoid gaps when the timbers shrink.

## Floor using soundboard

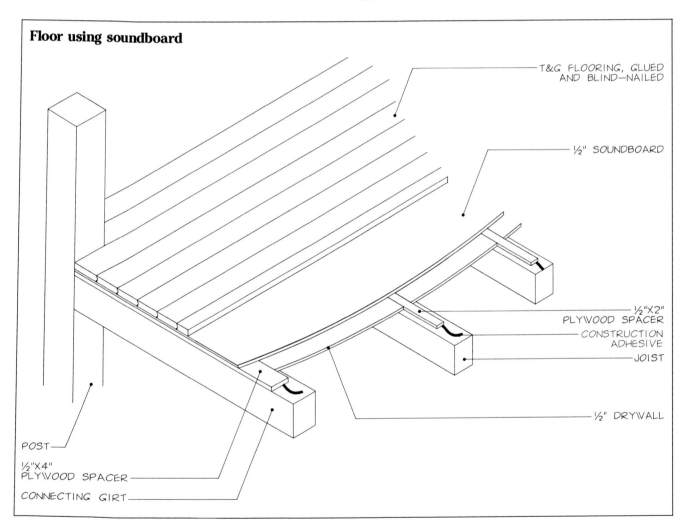

T&G FLOORING, GLUED AND BLIND-NAILED

½" SOUNDBOARD

½"X2" PLYWOOD SPACER

CONSTRUCTION ADHESIVE

JOIST

½" DRYWALL

POST

½"X4" PLYWOOD SPACER

CONNECTING GIRT

A floor system with even better sound-deadening characteristics uses the first two steps described on pages 216 and 217 (ceiling material of drywall or planking plus sound-deadening board), but then incorporates an additional step. After the ceiling material and the soundboard are in place, 2x3s are nailed or screwed 16 in. on center perpendicular to the joists. The 2x3s span the joists and are a nail base for the flooring. Both the air space between the soundboard and the finish floor and the reduced area of contact over the joists help to increase sound-deadening. If the finish floor will be either carpet or tile, a ¾-in.-thick plywood subfloor is first attached to the 2x3s. One advantage of this system is that it creates plenty of space to run wires and is even useful for some plumbing runs.

### Excellent sound-deadening floor

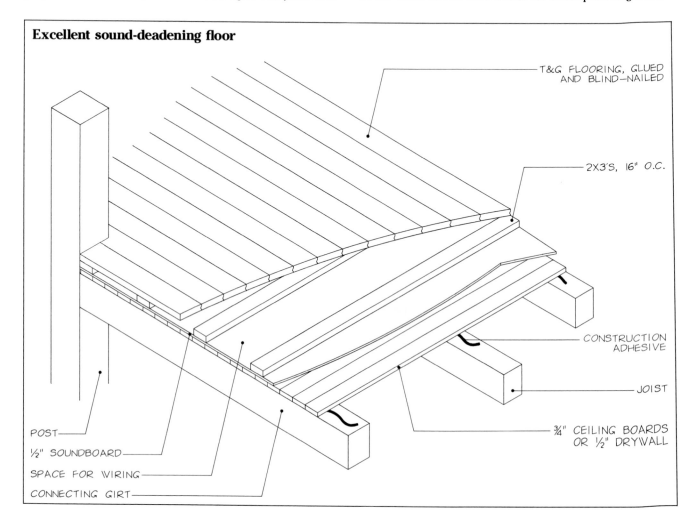

T&G FLOORING, GLUED AND BLIND—NAILED

2X3'S, 16" O.C.

CONSTRUCTION ADHESIVE

JOIST

¾" CEILING BOARDS OR ½" DRYWALL

POST

½" SOUNDBOARD

SPACE FOR WIRING

CONNECTING GIRT

# Interior partition walls

In timber-frame construction, all interior partitions are superfluous as structure. Even partitions as delicate as *shoji* (light Japanese screens made of a wooden grid overlaid with rice paper) would work. In fact *shoji* would be perfect for a timber-frame house—light, thin and movable, offering the ultimate design flexibility. But in most cases, partitions must be of sturdier construction to perform the important duties of providing privacy and blocking sound. As I pointed out in Chapter IV, the design of the timber frame should coincide with the design of the floor plan; rooms should be defined by timbers as well as walls, and therefore most walls will be aligned with timbers. In this section, I will talk a little more specifically about the relationship of walls to timbers and the various ways to install partitions.

For partitions to be integrated successfully with the frame, the designer needs to determine precisely where they will fall in relation to the timbers. Partitions usually consume a total of between 3½ in. and 4½ in., and with this in mind a decision must be made on how much of the timber will be revealed on each side of the wall. To simplify construction, try to hold partitions to one side or the other of the knee braces whenever possible, revealing the frame to the more public area and concealing it on the opposite side. The alternative—revealing the frame on both sides of a partition wall—requires that individual pieces of framing and drywall be fitted around knee braces—a laborious job. When knee braces are not an obstacle, partition walls can be constructed on the floor and installed as a unit. To the carpenter, the difference between these two methods is extreme. Note that when walls are to be constructed on one side or the other of the knee braces, the designer must specify to the framer that the knee braces be pushed toward the opposite edge of the timber to allow room for the partitions. Edge embellishments must also be considered when planning partitions (p. 169).

Timbers will shrink, which is a problem for interior partitions because it is likely that shrinkage will occur after the partitions are in place. It's not just aesthetics at stake here, for even small cracks greatly reduce the ability of the partition to deaden sound. There are only a few solutions to the shrinking-timber dilemma.

**1.** Construct the partitions and sheathe both sides, but don't finish the walls until the timbers have stabilized. If the timbers are almost dry or just in need of acclimating to the environment, this might be a reasonable wait of only a few months. However, if the frame was made of green hardwood, the wait could be up to three years.

**2.** Finish the interior walls immediately and plan on filling the spaces with paintable caulking when the cracks appear. As you can imagine, even if this does not seem like

**Partitions in a timber-frame house are not essential for the structure. Therefore, they should be built in a way that enhances the beauty of the frame, yet also provides privacy. In this house under construction, most of the partitions are built with 2x3s and will be insulated with soundboard. The framing is fitted around the knee braces, exposing the timbers to both sides of the wall.**

the best solution, it is the one most often used. The pleasure of the finished product is not deferred, and caulking the cracks is put off until the next time the walls need painting. Wallpaper should not be put on until the timbers dry because cracks around the edges are unsightly.

**3.** Make the shrinkage a design detail by installing *L*-bead where drywall meets timber. *L*-bead is basically a hard metal edge that is attached to the partition before the drywall. The drywall is cut so it roughly meets the *L*-bead and is finished after the joint is taped and spackled. When the timber shrinks, a distinct gap (usually not more than ¼ in.) is left between the edge of the *L*-bead and the timber—it shows up as a dark space. This works well in situations that will not have to be caulked, such as closet partitions. But on one of our projects an architect specified that *L*-bead be used on almost every wall and that a deliberate gap be created from the beginning, which would be filled with dark caulking.

Defining living areas with timbers instead of partitions results in an open floor plan and a feeling of spaciousness that is really a part of the definition of a timber-frame home. But this freedom of movement and of view also allows sound to travel freely. If privacy is a requisite, compensate by adding sound-deadening to all partitions in private areas.

The easiest solution is to use 2x3s to build the entire partition, and then apply sound board as the first layer of sheathing on both sides. After the sound board, attach the drywall. The only potential problem with this system is that the electrician may need to change to wide, flat switch

and outlet boxes to have enough area for wires—standard electrical boxes are too deep for a 2x3 partition.

For even better performance use a standard 2x4 for top and bottom plates, but use staggered 2x3s as studs. The 2x3s, on 8-in. centers, are alternately aligned with either plate edge, giving 16-in.-on-center nailing for each side. The advantage is that no stud touches the drywall on both sides, greatly reducing sound conduction through the wall. Half-inch soundboard is then applied to the partition on both sides before ½-in. drywall is installed. Obviously, no sound-deadening system will work effectively without a good seal between the surrounding timbers and the partition. Apply acoustical sealant to the top and bottom plates and both end studs before fastening. It may still be necessary to caulk after the timbers have stabilized.

Non-sound-deadening partitions can be made with 2x3s or 2x4s. Because standard electrical boxes are 3 in. or more in depth, 2x4s are more common.

When interior partitions don't align with timber posts, it is important that they be firmly attached to the exterior wall. If a built-up exterior insulation system is used, the problem can easily be solved by providing nailing for the stud that meets the exterior wall. But if the exterior wall is a stress-skin panel, some advance planning is necessary. Before the exterior siding goes on, screw the stud to the panel from the outside, using galvanized self-tapping screws that are long enough to penetrate the stud by at least 1 in. If this step is omitted, don't bother to tape the corner between the interior partition and the exterior wall, for it will inevitably crack.

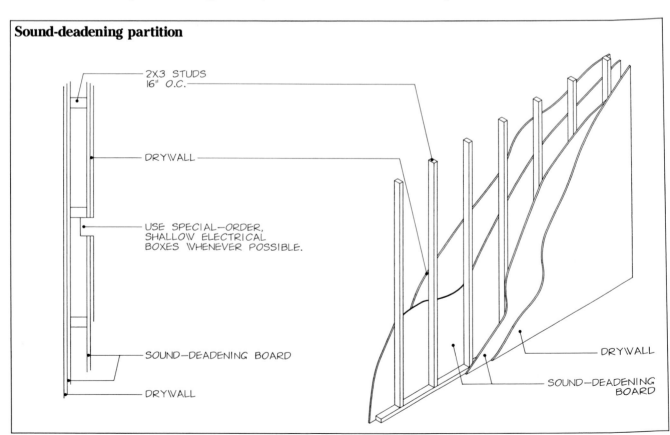

**Sound-deadening partition**

2X3 STUDS 16" O.C.

DRYWALL

USE SPECIAL-ORDER, SHALLOW ELECTRICAL BOXES WHENEVER POSSIBLE.

SOUND-DEADENING BOARD

DRYWALL

DRYWALL

SOUND-DEADENING BOARD

## Surface treatment

"There's no accounting for taste." "Beauty is in the eye of the beholder." And interior decoration has few rules. While the exterior of the house often must bend to its natural and architectural environment, the interior should be designed to satisfy the inhabitants. This is a personal and entirely subjective process. I have a friend who doesn't like things until they reach a state of nearly total decay. When plaster is cracked but still holding on, when paint is faded and beginning to peel, when floor boards are almost worn through from use, when moss is thick on the shingles, she is ecstatic; she thinks things are perfect the moment before disrepair. I know another person who requires that all things have the glitter of newness about them, that the colors be bright, that the brass and stainless steel be maintained with a showroom shine. To him, something used is old, and old things are ugly. We have designed homes to

have the look of a barn and some that were intended to have the elegance of a castle, and there's nothing wrong with either. What's important is that a timber-frame home is able to satisfy both requirements. When you hear the words "a timber-frame house," it should conjure an image of a method of construction, not a specific decorative style.

One of the important factors affecting interior style is the treatment of the interior surfaces. The timbers, walls and ceilings all should be planned carefully in advance to achieve the desired result. The exposed timber frame is an incredibly strong decorative element, and the species should be chosen mostly for its aesthetic value, since engineering should make the frame equally strong no matter what the wood. When deciding, consider that knotty wood of whatever species is more rustic than wood that is clear of knots, and straight-grained wood with tight growth rings

The timbers in the frame are usually planed and oiled, not only for appearance and ease of cleaning, but because planing the timbers to a common dimension makes joinery easier. A clear penetrating oil was used on these timbers.

is more subtle than highly figured wood with widely spaced rings. Oak and southern pine are generally coarser than pine and cedar. To make sure the wood of the frame will not clash with other design intentions, it is important to become familiar with all the options. Each person will bring to this decision personal prejudices and opinions. I once rented a basement apartment from a casket maker, where all the walls and the ceiling were paneled with reject pine from the casket business. The experience left me with very uncomfortable feelings about knotty pine. Another personal opinion I have developed is that the best timber frames are those that are the most subtle, that enhance instead of dictate.

There are two reasons why the timbers in a frame are usually planed and finished. First, it is easier to work timbers planed to specific dimensions rather than deal with varying thicknesses. Second, it is almost impossible to keep the timbers clean from the time cutting starts until the frame is raised. We usually finish-plane or sand the timbers and apply a finish to the interior surfaces just before the raising. The smooth surface also makes it easier to keep the timbers clean in the future. Although there are many possibilities, the most common surface finish is oil. If the timbers are green, it is important to apply a penetrating oil, which will allow the timbers to breathe as they dry.

Some timber framers use a mixture of linseed oil, gum turpentine and Japan drier with satisfactory results. For dry wood, a thin coat of shellac or polyurethane can be applied as a second or third coat to seal the surface and keep the mixture from bleeding. Tung oil can also be used as a base, and we have found it a little more durable while the frame is exposed to the weather. It also produces a harder finish without the addition of polyurethane or shellac. A number of commercial tung-oil finishes are available, of which our favorite for durability and appearance is Okene, made by the Pratt and Lambert Company. Your paint supplier can help you mix your own formula or choose among the different brands. One nice aspect of using an oil-based finish is that it is easy to repair marred surfaces, and new coats tend to blend in quite well. It is also possible to buy oil premixed with stain, but before applying any stain, try it on a sample of the same type of wood as the frame. "Medium walnut" (whatever that is), will look vastly different on fir than on pine, and stains are difficult to remove.

In my opinion, the first choice should be to find a wood that will naturally deliver a suitable color and texture with a clear oil finish. If that is not possible, then consider stain. A dark stain is helpful, for instance, if the frame is to blend with colonial decoration; unstained wood would probably

**Old yellow-pine timbers were re-milled to make this frame. The patina of age can't be applied; it simply takes time.**

always look too fresh. Still, the first choice might be a wood that would give an old appearance. On one of our projects, we used timbers resawn from wood salvaged from an old mill. The wood was old, and looked old.

On another of our projects, after the frame went up, one of the owners decided the timbers were too overpowering. She wanted the form and shape of the frame, but only as a background. Against my advice whitewash was applied to the timbers, allowing the grain and texture to show but masking the strong, dark wood. It was a radical solution, but the results are surprisingly good.

The wall and ceiling surfaces in a timber-frame home serve the function of a canvas backdrop, helping to highlight the artistic value of the timbers. For this reason, a successful treatment of the surfaces is very often simply a white or off-white paint, a deliberate development of the "negative space" enclosed by the "positive" surround. This is the traditional timber-frame interior style that has been used for centuries around the world. In older buildings, the walls and ceilings are usually plastered and the wood, having grown dark with age, helps to create the classic "black and white" appearance.

When a client decided the natural color of the timbers was too overpowering, she decided to whitewash them. The grain and texture of the wood still come through, but the dark color has been eliminated. (Design Associates, Architects; photo by Brian Smeltz.)

There are a couple of other specific uses for white (or very light) surfaces between timbers. They help to reflect light, making the spaces brighter and enhancing the volume. Small rooms and spaces with low ceilings therefore benefit the most from having light-colored walls and ceilings. For exactly the opposite reason, large spaces with cathedral ceilings can be made to feel more comfortable and cozy with wood ceilings to absorb the light and "bring the ceiling down." Another good reason for a white background between timbers is to create a place to hang art. In fact, it amused me to find fake timbers stuck to the wall in a large art gallery just to develop the feeling that comes naturally to a timber-frame home. The white backgrounds, framed by timbers, become individual focal areas for artwork, and the whole effect can have the atmosphere of a lively gallery. The introduction of artwork is also a great way to bring color into a house. On several occasions, clients have requested gallery space for specific paintings—a challenge we readily accept, because the timbers seem to function as an additional frame for the artwork, making it quite easy to achieve a satisfying display area.

Despite the arguments in favor of simple, light-colored surfaces between timbers, many people choose to use wallpaper with fine results. Simple print patterns seem to produce less clutter, although some of our creative clients have done striking things with rather extravagant wallpapers. As I have said, there are no rules.

All the exterior wall and ceiling surfaces are usually applied to the outside of the frame as a part of the enclosure system. In most cases, the interior surface is a drywall product. We use a plaster-base drywall, which can either receive a veneer coat of plaster or simply be taped and painted. Taping and painting the drywall is the least problematic because there is a distinct separation between the frame and the wall or ceiling surface. If the timbers shrink or twist, it isn't likely to cause any major problems. It might be necessary to touch up the paint edges as the timbers dry, but this is a much easier job than having to fill in gaps with spackling or caulking, which is almost certain to be necessary if plaster is applied to the drywall. As the timbers dry, there will be some eventual separation between the frame members and the plaster edges. Therefore, if green timbers are used for the frame, consider not using plaster at all or putting the task off until the timbers have stabilized. This same advice applies to wallpaper. It can be more than a little discouraging to fit paper precisely between the timbers only to find later that the timbers have pulled away from the edges. A better strategy is to paint the walls initially and then apply wallpaper after the frame has dried completely.

**Off-white panels between timbers create a fine background for a display of art. In timber-frame design, it is usually helpful to know something about any artwork early in the process so that space will be left in the plan.**

Ceilings, whether those of a timber floor or roof system, rarely justify plaster. The drywall is applied over the top, almost always in such a way that there are few, if any, seams. The ceiling surface easily can be finished just with paint. Because the timbers generally fall at the drywall edges, eliminating or reducing the need to tape seams, the drywall can even be painted before installation. Applying plaster on a ceiling is monumentally more difficult than paint because the timbers act as obstacles. There are also more timber edges to worry about while the frame stabi-

lizes. We often encourage clients who prefer plaster to use it on the walls but not on the ceilings. The surfaces are separate enough to have different textures, even if they are the same color. Wood ceilings are easily installed and reduce maintenance. The wood is usually oiled to bring out the grain and color. Future applications of oil will be greatly simplified if the same kind of oil is used on the ceiling boards that was used on the frame. Oil can then be applied to the entire ceiling without extra care being necessary at the edges.

**The bold print of this wallpaper contrasts with the simple lines in the frame. Because timbers shrink, it is best to apply wallpaper after the wood has dried.**

Photo: Emily Benson

# Appendix **Stress-Skin Panel Suppliers**

Stress-skin panels are a complex and expensive product. Before making a choice, get as many facts as possible. It will be worth all the time it takes to search out the best possible supplier for the project at hand. Remember as you research that materials, methods of manufacture and recommended application techniques vary widely. Remember also that a good product is backed by a dedicated support staff. The stress-skin panel industry is relatively young. Experimentation continues, and the product constantly evolves.

### Advance Energy Technologies, Inc.

| Urethane foam; foamed in place |

P.O. Box 387
Clifton Park, NY 12065
518-371-2140

### Advance Foam Plastics, Inc.

| EPS, urethane; laminated |

5250 North Sherman St.
Denver, CO 80216
303-297-3844

### Affordable Luxury Homes

| EPS; laminated |

Hwy. 224
P.O. Box 288
Markle, IN 46770
219-758-2141

### Alchem Inc.

| Polyurethane; foamed in place |

3617 Strawberry Rd.
Anchorage, AK 99502
907-243-2177

### Andrews Building Systems

| Urethane foam; foamed in place |

225 S. Price Rd.
Longmont, CO 80501
303-772-3516

### Atlas Industries

| Urethane, polystyrene; laminated |

6 Willows Rd.
Ayer, MA 01432
Outside of Massachusetts
call 1-800-343-1437
Within Massachusetts call
1-800-332-1002

### Big Sky Insulations, Inc.

| EPS; laminated |

P.O. Box 838
Belgrade, MT 59714
406-388-4146

### Branch River Foam Plastics

| Expanded polystyrene; laminated |

15 Thurber Blvd.
Smithfield, RI 02917
401-232-0270

### Century Insulation Mfg. Co.

| EPS, polyisocyanurate; laminated |

Industrial Park
P.O. Box 160
Union, MS 39365
601-774-8285

### Cheney Building Systems, Inc.

| Polyisocyanurate; polyurethane foam; foamed in place |

2755 S. 160th St.
New Berlin, WI 53151
414-784-9634

### Delta Industries

| EPS; laminated |

1951 Galaxie St.
Columbus, OH 43207
614-445-9634

### Enercept, Inc.

| EPS; laminated |

3100 9th Ave. S.E.
Watertown, SD 57201
605-882-2222

### EPS Molding, Inc.

| EPS; laminated |

2019 Brooks St.
Houston, TX 77026
713-237-9115

### Foam Plastics of New England

| EPS core manufacture only |

P.O. Box 7075
Prospect, CT 06712
203-758-6651

### Foam Products Corp.

| EPS; laminated |

2525 Adie Rd.
P.O. Box 2217
Maryland Heights, MO 63043
314-739-8100

### Futurebilt, Inc.

| EPS; laminated |

A-104 Plaza Del Sol
Wimberley, TX 78676
512-847-5721

### Homasote Co.

| Polyisocyanurate; laminated |

P.O. Box 7240
West Trenton, NJ 08628-0240
609-883-3300

### Insulated Building Systems, Inc.

| EPS; laminated |

100 Powers Court
Sterling, VA 22170
703-450-4886

### Insul-kor, Inc.

| EPS; laminated |

201 East Simonton
Elkhart, IN 46514
800-521-1402

### J-Deck Building Systems, Inc.

| EPS; laminated |

2587 Harrison Rd.
Columbus, OH 43204
614-274-7755

### Low-Temp Engineering, Inc.

| EPS; laminated |

308 East Main St.
Route 123
Norton, MA 02766
617-285-9788

**Northern Energy
Homes Inc.**

EPS; laminated

Box 463
Norwich, VT 05055
802-649-1348

**NRG Barriers, Inc.**

Polyisocyanurate; foamed in place

61 Emery St.
Sanford, ME 04073
207-324-7745

**Pacemaker Plastics Co., Inc.**

EPS; laminated

126 New Pace Rd.
Newcomerstown, OH 43832
614-498-4181

**Panel Building Systems Inc.**

EPS; laminated

431 2nd St.
Reynolds Industrial Park
Greenville, PA 16125
412-646-2400

**Pond Hill Homes Ltd.**

Polyurethane; foamed in place

Westinghouse Rd.
RD4, Box 330-1
Blairsville, PA 15717
412-459-5404

**Premier Building Systems**

EPS; laminated

19041 80th Ave. South
Kent, WA 98032
206-242-9424

**Riverbend Timber
Framing Inc.**

EPS; laminated

P.O. Box 26
Blissfield, MI 49228
517-486-4355

**T.A.S. Building Systems**

Resin-impregnated kraft-paper
honeycomb; laminated

2540 Main St. Suite G
Chula Vista, CA 92011
619-429-7151

**Thermal Foams, Inc.**

EPS; laminated

2101 Kenmore Ave.
Buffalo, NY 14207
716-874-6474

**Thermapan Industries**

EPS; laminated

Box 479
Fonthill, Ontario
Canada L0S 1E0
412-892-2675

**Therm-core Industries**

EPS; laminated

5404 Columbus Pike
Delaware, OH 43015
614-548-7990

**W. H. Porter, Inc.**

EPS, polyisocyanurate; laminated

4240 North 136th Ave.
Holland, MI 49424
616-399-1963

**Winter Panel Corp.**

Urethane; foamed in place

RR5 Box 168B
Brattleboro, VT 05301
802-254-3435

**Wisconsin EPS, Inc.**

EPS; laminated

90 Trobridge Drive
P.O. Box 669
Fond du Lac, WI 54936-0669
414-923-4146

# Bibliography

Alexander, Christopher, Sara Ishikawa and Murray Silverstein

*A Pattern Language*. New York: Oxford University Press, 1977.

*The Timeless Way of Building*. New York: Oxford University Press, 1979.

Anderson, Bruce and Malcolm Wells

*Passive Solar Energy*. Andover, Mass.: Brick House Publishing Co., 1981.

*Audel's Carpenters and Builders Guides*. 4 vols. New York: Theo Audel and Co., 1923, 1939, 1945, 1947.

Benson, Tedd and James Gruber

*Building the Timber Frame House*. New York: Charles Scribner's Sons, 1980.

Briggs, Martin S.

*A Short History of the Building Crafts*. Cambridge, England: Oxford University Press, 1925.

Brungraber, Robert (Ben) Lyman

*Traditional Timber Joinery: A Modern Analysis*. Thesis. Ann Arbor, Mich: University Microfilm Inc., 1985.

Brunskill, R.W.

*Timber Building in Britain*. London: Victor Gollancz, Ltd., 1985.

Cummings, Abbott Lowell

*The Framed Houses of Massachusetts Bay, 1625-1725*. Cambridge, Mass.: Belknap Press of Harvard University Press, 1979.

Downing, A.J.

*The Architecture of Country Houses*. New York: Dover Publications, Inc., 1969.

Fitchen, John

*The New World Dutch Barn*. Syracuse, N.Y.: Syracuse University Press, 1968.

Futagawa, Yukio

*Wooden Houses*. New York: Harry N. Abrams, Inc., 1979.

*Traditional Japanese Houses*. New York: Rizzoli International Publications, Inc., 1983.

Grosslight, Jane

*Light*. Englewood Cliffs, N.J.: Prentice-Hall, Inc., 1984.

Hansen, Hans Jurgen, ed.

*Architecture in Wood*. London: Faber and Faber, 1971.

Harris, Richard

*Discovering Timber Frame Buildings*. London: Shire Publications Ltd., 1978.

Hewett, Cecil A.

*English Historic Carpentry*. London and Chichester: Phillimore and Co., Ltd., 1980.

Isham, Norman M. and Albert F. Brown

*Early Connecticut Houses*. New York: Dover Publications, Inc., 1965

Kelly, J. Frederick

*Early Domestic Architecture of Connecticut*. New York: Dover Publications, Inc., 1952.

Kimball, Fiske

*Domestic Architecture of the American Colonies and of the Early Republic*. New York: Dover Publications, 1966.

Langdon, Philip

*American Houses*. New York: Stewart, Tabori & Chang, Inc., 1987.

Langdon, William Chauncy

*Everyday Things in American Life, 1607-1776*. New York: Charles Scribner's Sons, 1943.

Makinson, Randell L.
*Greene and Greene, Architecture as a Fine Art.* Salt Lake City, Utah: Peregrine Smith, Inc., 1977.

Mazria, Edward
*The Passive Solar Energy Book.* Emmaus, Pa.: Rodale Press, 1979.

Mullin, Ray C.
*Electrical Wiring Residential.* New York: Van Nostrand Reinhold Co., 1981.

Mullins, Lisa C., series ed.
*Architectural Treasures of Early America.* 10 vols. Harrisburg, Pa.: Historical Times Inc., 1987.

Mumford, Lewis
*Roots of Contemporary American Architecture.* New York: Dover Publications, Inc., 1972.

Packard, Robert T., A.I.A., ed.
*Architectural Graphic Standards.* New York: John Wiley and Sons, 1981.

Poor, Alfred Easton
*Colonial Architecture of Cape Cod, Nantucket and Martha's Vineyard.* New York: Dover Publications, Inc., 1932.

Price, Lorna
*The Plan of St. Gall in Brief.* Berkeley and Los Angeles, Calif.: University of California Press, 1982.

Rempel, John I.
*Building With Wood.* Toronto: University of Toronto Press, 1967.

Rybczynski, Witold
*Home: A Short History of an Idea.* New York and Canada: Viking Penguin Inc., 1986.

Schwolsky, Rick and James I. Williams
*The Builder's Guide to Solar Construction.* New York: McGraw-Hill, Inc., 1982.

Sobon, Jack and Roger Schroeder
*Timber Frame Construction.* Pownal, Vt.: Garden Way Publishing, 1984.

Stickley, Gustav
*Craftsman Homes.* New York: Dover Publications, Inc., 1979.

Tarule, Rob
"The Landscapes of England and New England in the Early Seventeeth Century." Illustrated lecture delivered at First Annual Conference of Timber Framers Guild of North America, Hancock, Mass., 1985.

West, Trudy
*The Timber Frame House in England.* New York: Architectural Book Publishing Co., 1971.

Williams, Lionel Henry and Ottalie K. Williams
*Old American Houses, 1700-1850.* New York: Crown Publishers, Inc., 1962.

*Wood Handbook: Wood as an Engineering Material.* Washington, D. C.: Superintendent of Documents, U.S. Government Printing Office, 1974.

Wood, Margaret
*The English Mediaeval House.* New York: Harper and Row, 1965.

Woodforde, John
*The Truth about Cottages.* London: The Country Book Club, 1970.

Wright, Frank Lloyd
*The Natural House.* New York: The Horizon Press, 1954.

Yoshida, Tetsuro
*The Japanese House and Garden.* New York: Fredrick A. Praeger, 1954.

# Index

| | |
|---|---|
| *Managing Editor* | Deborah Cannarella |
| *Editor* | Laura Tringali |
| *Design Director* | Roger Barnes |
| *Art Director* | Ben Kann |
| *Layout Artist* | Marianne Markey |
| *Illustration Assistant* | Deb Rives-Skiles |
| *Copy/Production Editors* | Victoria Monks, Ruth Dobsevage |
| *Director of Manufacturing* | Kathleen Davis |
| *Production Manager* | Peggy Dutton |
| *Pre-Press Manager* | Austin E. Starbird |
| *Production Coordinator* | Ellen Olmstead |
| *System Operators* | Dinah George, Nancy-Lou Knapp |
| *Production Assistants* | Lisa Carlson, Mark Coleman, Deborah Cooper |
| *Pasteup* | Marty Higham, Cynthia Lee Nyitray |
| *Indexer* | Harriet Hodges |

| | |
|---|---|
| *Typeface* | ITC Cheltenham Book, 9 point |
| *Paper* | Warrenflo, 70 lb., neutral pH |
| *Printer and Binder* | Arcata Graphics/Kingsport, Kingsport, TN |